A STUDENT'S COMP.

J. RICHARD CHRISTMAN
U.S. Coast Guard Academy
New London, CT

to accompany

VOLUMES ONE AND TWO
PHYSICS
FOURTH EDITION

DAVID HALLIDAY
Professor of Physics, Emeritus
University of Pittsburgh

ROBERT RESNICK
Professor of Physics
Rensselaer Polytechnic Institute

KENNETH S. KRANE
Professor of Physics
Oregon State University

JOHN WILEY & SONS, INC.

New York • Chichester • Brisbane • Toronto • Singapore

Copyright © 1992 by John Wiley & Sons, Inc.

All rights reserved.

Reproduction or translation of any part of
this work beyond that permitted by Sections
107 and 108 of the 1976 United States Copyright
Act without the permission of the copyright
owner is unlawful. Requests for permission
or further information should be addressed to
the Permissions Department, John Wiley & Sons.

ISBN 0-471-51873-5

Printed in the United States of America
Printed and bound by Courier Companies, Inc.
10 9 8 7 6 5 4 3 2 1

TO THE STUDENT: HOW TO USE A STUDENT'S COMPANION TO PHYSICS

A Student's Companion to Physics is designed to be used closely with the text *PHYSICS* by Halliday, Resnick, and Krane. There are 4 overview chapters, corresponding to major sections of the text: mechanics, thermodynamics, electricity and magnetism (including optics), and modern physics. Modern physics chapters are included in the extended version of the text only. Read the appropriate overview when you start to study a section and refer back to it as your study proceeds. Read it again when you finish the section. The overviews are designed to help you understand how the important topics fit together and how the text is organized.

Other chapters in the Companion correspond to chapters in the text and should be read along with the text. Most of the Companion chapters are divided into 3 sections: Basic Concepts, Problem Solving, and Notes. Many chapters contain other sections: Mathematical Skills and Computer Projects.

<u>BASIC CONCEPTS</u>. This section deals with two important types of information you should obtain from your reading of the text. The first consists of definitions of physical concepts used in the chapter; the second consists of the laws of physics, relationships between the concepts.

A firm understanding of the basic concepts is important for understanding how nature behaves, for working problems, for doing well on exams, and for understanding following chapters. Rather than just list the definitions and laws, this section asks you to do the important part of the work—write the key phrases by filling in blanks. To derive the greatest benefit, don't copy information from the text. Rather, read the text first, then try to fill in the blanks with your own words, without reference to the text. Thinking about what to write and writing it well should help you retain the important information. Comparing what you have written with what the text says will help you pinpoint any misconceptions you might have. If you have trouble expressing an idea you probably don't understand it very well. Go back and study the appropriate section of the text.

Try to write your responses carefully. The section, as completed by you, will serve later as a review. Before working a problem assignment and while studying for an exam, read over the completed section. If there are parts you don't understand you might want to write them more carefully. The better the job you do, the better this section will serve you when you review.

The concepts of physics are best learned in small doses. Try to obtain a firm understanding of each concept before moving on to the next. The Basic Concepts section will help you.

<u>PROBLEM SOLVING</u>. You cannot claim to understand a definition or law of physics unless you can apply it in a variety of different situations. The purpose of the problems is to provide you with various situations so you can test your understanding. There are three main parts to solving a physics problem. The first and probably the most difficult for many students is identifying the physical concepts or physical laws involved in the problem. Once the concepts and

laws are identified, you are ready for the second part, writing the equations associated with them. The third part of problem solving involves carrying out mathematical manipulations to obtain an answer.

This section of the Companion concentrates on helping you identify the concepts and laws involved in a problem and on helping you write the appropriate equations. The various types of problems that can be associated with the concepts of the chapter are discussed and classified, using as examples both sample problems worked in the text and additional problems given in the Companion. As you read a Problem Solving section be sure you also study the sample problems that are referenced. As you work problems, you will learn more details about the concepts and laws involved. If necessary, go back to the Basic Concepts section and revise your responses there.

MATHEMATICAL SKILLS. Here you will find a list of the mathematical skills required to solve the problems of the chapter. You will have learned most of these in algebra, trigonometry, geometry, and calculus classes. Review the list to see if you need to brush up on any of the skills. If you do, consult your math texts for complete discussions and for practice problems. Your goal should be to become facile enough with the required mathematical techniques that the math does not hinder your ability to solve problems.

COMPUTER PROJECTS. With a little programming knowledge you can use a personal computer to add a great deal to your understanding of the physical world. The Computer Projects sections of the Study Guide will help you by suggesting some things to do. A few of the projects consist simply of writing utility programs that will carry out calculations for you. Most, however, suggest ways to program a computer to investigate the behavior of an object or system of objects as various parameters that control the behavior are changed. The descriptions are written as if you were developing the programs from scratch using a high level programming language, such as BASIC or Pascal. As an alternative you might consider using a spreadsheet program, such as Lotus 123 or Quattro, or commercial problem solving software, such as Eureka, TK Solver!, MathCAD, or DERIVE. In any event, setting up or writing the program requires a clear understanding of the steps involved in the solution of the problem being considered and will probably teach you a great deal about problem solving.

Computational techniques are many and varied. Only a few can be given here. In addition, a programmer must be extremely conscious of errors that are inherent in any numerical scheme and must be well versed in techniques that reduce errors to tolerable levels. This aspect of computation is not covered in as much detail as it should be. You are strongly encouraged to consult *Computational Physics* by Steven E. Koonan (The Benjamin/Cummings Publishing Company, 1986).

NOTES. Most or all of a page is left blank for you to record any additional notes you think would be beneficial while you review. You might write some detail of a definition or law that is not covered in the Basic Concepts section and that you have trouble remembering. You might also record any details of problem solving that give you special trouble. Try to write your notes so you can understand them later when you review.

TABLE OF CONTENTS

To the Student	i
Overview I: Mechanics	1
Chapter 1: Measurement	5
Chapter 2: Motion in One Dimension	11
Chapter 3: Vectors	29
Chapter 4: Motion in Two and Three Dimensions	41
Chapter 5: Force and Newton's Laws	57
Chapter 6: Particle Dynamics	73
Chapter 7: Work and Energy	91
Chapter 8: Conservation of Energy	107
Chapter 9: Systems of Particles	123
Chapter 10: Collisions	143
Chapter 11: Rotational Kinematics	161
Chapter 12: Rotational Dynamics	171
Chapter 13: Angular Momentum	187
Chapter 14: Equilibrium of Rigid Bodies	203
Chapter 15: Oscillations	213
Chapter 16: Gravitation	231
Chapter 17: Fluid Statics	251
Chapter 18: Fluid Dynamics	265
Chapter 19: Wave Motion	277
Chapter 20: Sound Waves	297
Chapter 21: The Special Theory of Relativity	313
Overview II: Thermodynamics	333
Chapter 22: Temperature	335
Chapter 23: Kinetic Theory and the Ideal Gas	343
Chapter 24: Statistical Mechanics	357
Chapter 25: Heat and the First Law of Thermodynamics	369
Chapter 26: Entropy and the Second Law of Thermodynamics	383
Overview III: Electricity and Magnetism	399
Chapter 27: Electric Charge and Coulomb's Law	403
Chapter 28: The Electric Field	409
Chapter 29: Gauss' Law	425
Chapter 30: Electric Potential	441

Chapter 31: Capacitors and Dielectrics 455
Chapter 32: Current and Resistance 467
Chapter 33: DC Circuits 477
Chapter 34: The Magnetic Field 497
Chapter 35: Ampère's Law 511
Chapter 36: Faraday's Law of Induction 529
Chapter 37: Magnetic Properties of Matter 541
Chapter 38: Inductance 553
Chapter 39: Alternating Current Circuits 565
Chapter 40: Maxwell's Equations 579
Chapter 41: Electromagnetic Waves 587
Chapter 42: The Nature and Propagation of Light 597
Chapter 43: Reflection and Refraction at Plane Surfaces 603
Chapter 44: Spherical Mirrors and Lenses 613
Chapter 45: Interference 631
Chapter 46: Diffraction 643
Chapter 47: Gratings and Spectra 655
Chapter 48: Polarization 665

OVERVIEW I
MECHANICS

Wherever we look, from the submicroscopic world of fundamental particles to the grand scale of galaxies, we see objects in motion, influencing the motions of other objects.

Think of a gardener pushing a wheelbarrow. Each of the objects involved (the surface of the earth, the gardener, the wheelbarrow, and the air) influences the motion of the others. If the gardener stops pushing the wheelbarrow might coast for a while but it eventually stops, chiefly because of friction and air resistance. To keep himself and the wheelbarrow going the gardener must push against the ground with his feet. Both the ground and wheelbarrow push on him. For the wheels to turn the ground must push on them.

From a microscopic viewpoint, each of these objects consists of a myriad of particles, mainly electrons, protons, and neutrons, in continual motion and continually exerting an influence over the other particles.

On a grander scale the earth makes its yearly trip around the sun because the sun influences its motion. The sun itself travels around the center of our galaxy, the Milky Way, under the influence of other stars. The presence of other galaxies makes the motion of our galaxy different from what it otherwise would be.

Mechanics is the study of motion. The goal is to understand exactly what aspects of the motion of one object are changed by the presence of other objects and exactly what properties objects must have in order to influence the motion of each other. The fundamental problem of mechanics is: given the relevant properties of a group of interacting objects, what are their motions? The first 21 chapters of the text are devoted to the study of this and related problems.

Mechanics divides neatly into two parts: kinematics, a study of the description of motion, and dynamics, a study of the causes of motion.

First you will learn to describe the motion of an object and to use the ideas of displacement, velocity, and acceleration to predict changes in the position of an object. You will find that if you know the position and velocity of an object at some initial time and the acceleration of the object at all times you can predict its position and velocity at all times. In Chapter 2 you concentrate on motion in one dimension so you can master the important concepts without the geometric complications of more complex motions. In Chapter 4 the concepts are extended to describe motion in two and three dimensions.

Displacement, velocity, and acceleration in more than one dimension have direction as well as magnitude. They behave like mathematical quantities called vectors. Vectors are so important for understanding physics that the whole of Chapter 3 is devoted to explaining their properties and describing how they are manipulated mathematically.

Newton's laws of motion, introduced in Chapter 5, are at the heart of classical mechanics. Here you will learn that the interaction between two objects can be described in terms of a force and that the net force acting on an object accelerates it. That is, the net force changes the velocity of the object. The fundamental problem splits into two parts: (1) given the properties of the objects, what forces do they exert on each other and (2) given the net force on an object, what is its motion?

You will be introduced to a few simply described forces: the gravitational force of the earth on an object near its surface and the force exerted by a spring on an object in contact with one end, for example. Later on you will learn about other forces. Details of the gravitational force are described in Chapter 16, the electrical force is described in Chapter 27, and the magnetic force is described in Chapter 34.

You will also learn to calculate the acceleration of objects in a wide variety of situations. The study of Newton's laws of motion is continued in Chapter 6, where you will concentrate on frictional forces and the force required to make an object move on a circular path.

One important quantity that characterizes a system of interacting objects is its energy. Energy may take several forms. Chapter 7 introduces kinetic energy; in Chapter 8 you will learn about potential energy and internal energy. As objects move and exert forces on each other their individual energies may change and the total energy may change form, from kinetic to potential, for example. You will learn that the mechanism for these changes is the work done by the forces of interaction. You will learn how to compute the work done by a force and the changes in the energies of the objects involved.

Energy is important because, under certain conditions, the total energy of a collection (or system) of objects does not change. As the objects move the energy may change form but if certain conditions hold the total remains the same. This is one of the great conservation principles of mechanics and in Chapter 8 you will learn when it applies and when it does not. You will also learn to use it to answer questions about the motions of objects.

When the total energy of a system does change the change can be accounted for by the work done on objects in the system by objects outside the system. We may think of energy flowing into or out of the system as objects outside interact with objects inside.

Another quantity that behaves in much the same way is the momentum of a system. In Chapter 9 momentum is defined and the conditions for which it is conserved are discussed. In Chapter 10 the principle of momentum conservation is applied to collisions between objects.

Chapters 11, 12, and 13 are devoted to rotational motion. Here Newton's laws are applied to wheels spinning on fixed axes and gyroscopes, for example. The plan is like the one followed for linear motion: first kinematics, then dynamics. You will also learn about the kinetic energy of rotation and how it changes when work is done on the rotating object.

The most important concept introduced here is angular momentum, another quantity that is conserved in certain situations. You should learn to identify those situations and to use the principle of angular momentum conservation to help with your understanding of rotational motion.

In Chapter 14 Newton's laws for linear and rotational motion are used to discuss the special case of an object at rest. Here you will learn to calculate the forces that must act on an object to hold it at rest. You will also learn about the deformation of an object by the forces acting on it. These topics are enormously important for engineering applications. Bridges, buildings, and automobiles, for example, must be designed to withstand the loads to which they are subjected during use.

Motions that repeat, called oscillations, are discussed in Chapter 15 using Newton's laws and the conservation of energy principle. Oscillations are prevalent in nature and among man-made artifacts. The swaying of trees, buildings, and bridges, the motion of a clock pendulum, and the bouncing of a car as it rides over a pothole are all examples. In addition, what you learn here will be of great use when you study waves.

In Chapter 16 you will study the force of gravity, one of the fundamental forces. This discussion provides an excellent example of the dependence of a force on properties of the interacting bodies. The principles of dynamics are then applied to the motions of objects moving under the influence of gravity: planets, satellites, and spacecraft. You will bring to bear many of the concepts you learned earlier, most notably conservation of energy and angular momentum. Electrical and gravitational and forces are mathematically quite similar. Much of what you study here will be put to use when you study Chapter 27 and later chapters.

Chapters 17 and 18 deal with fluids, at rest and in motion. Density and pressure are defined first, then the principles of mechanics are used to understand the variation with depth of pressure in a body of water and in the earth's atmosphere and the variation in pressure along a pipe in which a fluid is flowing. You will also be able understand, for example, why some objects float while others sink when they are placed in a fluid and why the water speed increases when the nozzle opening of a hose is decreased.

Wave motion, in which a disturbance created at one place is propagated to another, is the basis for transmitting information (the form of the disturbance) and energy. Sound, light, radio signals, x rays, and microwaves are all examples of waves. Fundamental concepts developed in Chapter 19 are applied to sound waves in Chapter 20 and later on, in Chapter 41, the ideas are used to discuss electromagnetic radiation.

Examination of time and distance measurements reveals that if two observers are moving with different speeds, they report different values for both the temporal and spatial separation of the same two events. The discrepancy is slight and nearly undetectable for ordinary speeds but becomes acute when the relative speed of the observers is close to the speed of light. These results have profound implications for our concepts of time and distance. They also force us to redefine energy and momentum for objects moving at high speeds. The special theory of relativity deals with all these issues and is presented in the last chapter of the mechanics section, Chapter 21.

Chapter 1
MEASUREMENT

I. BASIC CONCEPTS

Physics is an experimental science and relies strongly on accurate measurements of physical quantities. All measurements are comparisons, either direct or indirect, with standards. This means that for every quantity you must not only have a qualitative understanding of what the quantity represents but also an understanding of how it is measured. A length measurement is a familiar example. You should know that the length of an object represents its extent in space and also that length might be measured by comparison with a meter stick, say, whose length is accurately known in terms of the SI standard for the meter. Make a point of understanding both aspects of each new quantity as it is introduced.

Systems of units and standards. A system of units consists of a unit for each physical quantity, organized so that all can be derived from a small number of independent base units. Standards are associated with base units but not with derived units. Ideally a standard should have the following properties: _____

The three International System <u>base</u> <u>units</u> used in mechanics are:

length: __meter__ (abbreviation: __m__)
mass: __gram__ (abbreviation: __g__)
time: __Second__ (abbreviation: __S__)

The <u>standards</u> for these units are:

length: __foot__
mass: __Pound__
time: __Second__

Notice that the SI unit for length is defined in terms of the speed of light and the time standard. The speed of light is by definition exactly $c = $ _____ m/s. Assume you have an instrument that accurately measures any time interval. Briefly explain how you can, in principle, calibrate a meter stick in terms of SI standards: _____

At the atomic level the second, non-SI, standard for mass is: _____

The atomic mass unit is related to the kilogram by $1\,u = $ _____ kg.

Another SI base unit is the mole, which is a measure of the quantity of matter in a substance. A mole of any particular type atom has as many atoms as a mole of any other type atom and, in particular, has the same number of atoms as there are in exactly _____ g of ^{12}C. This unit is used extensively in thermodynamics (see chapters 22 through 26). The number of

Chapter 1: Measurement **5**

atoms in a mole is required to convert between the standard kilogram and the atomic mass standard.

List several examples of quantities with underlined units in the International System. For each, give the units as a combination of SI base units.

QUANTITY	UNITS
_____	_____
_____	_____
_____	_____
_____	_____

To appreciate the magnitudes of quantities discussed in this course you should have an intuitive feeling for the size of a meter, kilogram, and second. Search tables 3, 4, and 5 for familiar objects and try to visualize them as you remember their sizes. Use the tables or seek elsewhere for quantities that are about 1 m long, 1 kg in mass, or 1 s in duration. List them here and remember them as examples:

objects about 1 m long: _____
objects with about 1 kg of mass: _____
intervals of about 1 s: _____

SI prefixes. SI prefixes are used extensively throughout this course. The following are used the most. For each of them write the associated power of ten and the symbol used as a prefix.

PREFIX	POWER OF TEN	SYMBOL
kilo:	10	_____
mega:	10	_____
centi:	10	_____
milli:	10	_____
micro:	10	_____
nano:	10	_____
pico:	10	_____

Memorize them. When evaluating an algebraic expression, substitute the value using the appropriate power of ten. That is, for example, if a length is given as 25 μm, substitute 25×10^{-6} m. One catch: the SI unit for mass is the kilogram. Thus a mass of 25 kg is substituted directly, while a mass of 25 g is substituted as 25×10^{-3} kg.

Dimensions. The idea behind the unit is called the dimension of a quantity. For example, the fundamental dimensions of mechanics are length (L), time (T), and mass (M). If two quantities are added or subtracted their dimensions must be the same. If they are multiplied the dimension of the product is the product of their dimensions. Thus the dimension of an area, which is the product of two lengths, is L^2. If one quantity is divided by another the dimension of the quotient is the quotient of their dimensions. The dimension of velocity, for example, is L/T. The dimensions of two quantities on opposite sides of an equal sign must be the same.

When you are trying to remember a formula or have just finished some algebra, use dimensional analysis to test the result. Reduce the dimension on each side of the equality to a product of L, T, and M, each raised to some power, then see if the dimensions match. If they do not you have done something wrong. You should also be aware that an equation that is dimensionally correct may still be wrong. A numerical factor might be incorrect or a term might be missing.

Unit conversion. Turn to Appendix G and become familiar with the conversion tables there. A good habit to cultivate is to say the words associated with a conversion. Suppose you want to convert 50 ft to meters. The length table in the appendix tells you that 1 ft is equivalent to 0.3048 m. Say "If 1 ft is equivalent to 0.3048 m, then 50 ft must be equivalent to (50 ft) × (0.3048 m/ft) = 15 m".

For practice verify the following:

2.90 in = 73.7 mm	4.50 ft = 1.37 m	2.10 mi = 3.38 km
36.0 mi/h = 57.9 km/hr	36.0 mi/h = 16.1 m/s	45.0 ft/s = 13.7 m/s
32.0 ft/s^2 = 9.75 m/s^2	100 lb = 445 N	5.10 slugs = 74.4 kg

Significant figures. Section 1-6 is an excellent discussion of significant figures. Read it carefully. Always express your answers to problems using the proper number of significant figures. Some students unthinkingly copy all 8 or 10 figures displayed by their calculator, thus demonstrating a lack of understanding. A calculated value cannot be more precise than the data that went into the calculation. Here is what you must remember about significant figures:

1. Leading zeros are not counted as significant. Thus 0.00034 has _____ significant figures.
2. Following zeros after the decimal point count. Thus 0.000340 has _____ significant figures.
3. Following zeros before the decimal point may or may not be significant. Thus 500 might contain 1, 2, or 3 significant figures. Use powers of ten notation to avoid ambiguities: 5.0×10^2, for example, unambiguously contains 2 significant figures.
4. When two numbers are added or subtracted, the number of significant figures in the result is obtained by _____.
5. When two numbers are multiplied or divided, the number of significant figures in the result is the same as _____.

II. PROBLEM SOLVING

Most of the problems at the end of this chapter are exercises in unit conversion, powers of 10 arithmetic, SI prefixes, and dimensional analysis. See the Mathematical Skills section below for a discussion of powers of ten arithmetic. Here are some examples of unit conversion.

PROBLEM 1. In Canada a liter of gasoline sells for 42 Canadian cents. If a Canadian dollar is equivalent to 82 U.S. cents what is the cost in Canada of a U.S. gallon of gasoline in U.S. dollars?

SOLUTION: You want to convert 0.42 Canadian dollars/liter to U.S dollars/gallon. Use the appendix to find the conversion from liters to gallons. Use 1 Canadian dollar = 0.82 U.S. dollar.

L = .82¢ Canadian

C1 = 82 U.S¢

[ans: $1.30]

PROBLEM 2. Express the sum $1.67\,\mathrm{m} + 50\,\mathrm{mm}$ in meters.

SOLUTION: The sum is the same as $(1.67 + 0.050)$ m.

1.67 m + 0.050 m = 1.720

[ans: 1.72 m]

Express the sum $7.50\,\mathrm{mm} + 320\,\mu\mathrm{m}$ in meters.

SOLUTION:

[ans: 7.8×10^{-3} m (7.8 mm)]

Express the sum $16\,\mathrm{km} - 5.0\,\mathrm{mi}$ in kilometers.

SOLUTION: First convert 5.0 mi to km, then subtract from 16 km.

[ans: 8.0 km]

Here are some examples of dimensional analysis. We concentrate on using dimensional analysis to find the functional relationship between given quantities. Follow the discussion of Section 1–7 to see how it is done. When you try it always reduce the dimension on each side of an equation to a product of L, T, and M, all raised to appropriate powers. All dimensions in mechanics can be written as products or quotients of these and they are independent of each other. That is, none can be written in terms of the others. Equate the exponent of L on one side of the equation to the exponent of L on the other side. Do the same for the exponents of T and M, then solve for the exponents.

PROBLEM 3. The speed of a wave on a string depends on the tension in the string and the linear mass density of the string. The tension F is a force and has SI units of kg·m/s². The linear mass density μ is the mass per unit length of the string and has SI units of kg/m. How does the wave speed depend on F and μ?

SOLUTION: Assume $v = F^n \mu^m$, where n and m are unknown exponents. The dimension of v is LT^{-1}, the dimension of F^n is $M^n L^n T^{-2n}$, and the dimension of μ^m is $M^m L^{-m}$. So $T^n \mu^m$ has dimension $M^{n+m} L^{n-m} T^{-2n}$ and this must be the same as LT^{-1}. Since M, L, and T are independent, $n + m = 0$, $n - m = 1$, and $-2n = -1$. Solve for n and m, then write the dependence of v on F and μ.

[ans: $\sqrt{F/\mu}$]

PROBLEM 4. When an object travels with uniform speed around a circular path it has an acceleration because the direction of its velocity is changing. The magnitude of the acceleration a (SI units: m/s²) depends on the speed v (SI units: m/s) and the radius r of the orbit. Use dimensional analysis to find the dependence.

SOLUTION:

[ans: v^2/r]

PROBLEM 5. When an object starts from rest and slides down a frictionless incline its speed v at the bottom depends on the acceleration g due to gravity (SI units: m/s²) and the initial height h above the bottom. Use dimensional analysis to find the dependence.

SOLUTION:

[ans: $\sqrt{gh}$]

III. MATHEMATICAL SKILLS

Although the topic is not covered in the text, you should be quite facile with powers of 10 arithmetic. If you are not, practice until you can carry out operations quickly. Later on, while trying to work a problem, you will be able to concentrate more on the problem itself and less on this detail.

When you multiply two numbers expressed as powers of ten, multiply the numbers in front of the tens, then multiply tens themselves. This last operation is carried out by adding the powers. Thus $(1.6 \times 10^3) \times (2.2 \times 10^2) = (1.6 \times 2.2) \times (10^3 \times 10^2) = 3.5 \times 10^5$ and $(1.6 \times 10^3) \times (2.2 \times 10^{-2}) = (1.6 \times 2.2) \times (10^3 \times 10^{-2}) = 3.5 \times 10 = 35$.

When you divide two numbers, divide the numbers in front of the tens, then divide the tens. The last operation is carried out by subtracting the power in the denominator from the power in the numerator. Thus $(1.6 \times 10^3)/(2.2 \times 10^2) = (1.6/2.2) \times (10^3/10^2) = .73 \times 10 = 7.3$ and $(1.6 \times 10^3)/(2.2 \times 10^{-2}) = (1.6/2.2) \times (10^3/10^{-2}) = .73 \times 10^5 = 7.3 \times 10^4$.

When you add or subtract two numbers first convert them so the powers of ten are the same, then add or subtract the numbers in front of the tens and multiply the result by 10 to the common power. Thus $1.6 \times 10^3 + 2.2 \times 10^2 = 1.6 \times 10^3 + 0.22 \times 10^3 = 1.8 \times 10^3$.

This means you must know how to write the same number with different powers of ten. Remember that multiplication by 10 is equivalent to moving the decimal point one digit to the right and division by 10 is equivalent to moving the decimal point one digit to the left. Thus $1.6 \times 10^3 = 16 \times 10^2 = 0.16 \times 10^4$. In the first case we multiplied 1.6 by 10 and divided 10^3 by 10. In the second we divided 1.6 by 10 and multiplied 10^3 by 10.

You should be able to verify the following without difficulty:

$$512 \times 10^2 = 5.12 \times 10^4$$
$$0.00512 = 5.12 \times 10^{-3}$$
$$(3.4 \times 10^2) \times (2.0 \times 10^4) = 6.8 \times 10^6$$
$$(3.4 \times 10^2)/(2.0 \times 10^4) = 1.7 \times 10^{-2}$$
$$(3.4 \times 10^4) + (2.0 \times 10^3) = (3.4 \times 10^4) + (0.20 \times 10^4) = 3.6 \times 10^4$$

IV. NOTES

Chapter 2
MOTION IN ONE DIMENSION

I. BASIC CONCEPTS

This chapter introduces you to some of the concepts used to describe motion: most importantly, position, velocity, and acceleration. Pay particular attention to their definitions and to the relationships between them.

Definitions. This chapter of the text deals with kinematics. Carefully distinguish between kinematics and dynamics: _____
_____
_____
_____

In this section of the text objects are treated as particles. Tell in your own words what a particle is. Pay particular attention to the properties of the motion. Can a particle rotate? Can a particle have parts that move relative to each other? _____
_____
_____
_____

If we treat an extended object as a particle we may pick one point on the object and follow its motion. The position of a crate, for example, means the position of the point on the crate we have chosen to follow, perhaps one of its corners.

The motion of a particle in one dimension can be described by giving its coordinate x as a function of time t. You must carefully distinguish between an <u>instant</u> of time and an <u>interval</u> of time. The symbol t represents an instant and has no extension. Thus t might be *exactly* 12 min, 2.43 s after noon on a certain day. At any other time, no matter how close, t has a different value. On the other hand, an interval extends from some initial time to some final time: *two* instants of time are required to describe it. Note that a value of the time may be positive or negative, depending on whether the instant is after or before the instant designated as $t = 0$.

Similarly, a value of x specifies a <u>point</u> on the x axis. It has no extension in space. On the other hand, two values are required to specify a <u>displacement</u>. A coordinate may be positive or negative, depending on where the point lies relative to the origin.

Suppose a particle has coordinate x_1 at time t_1 and coordinate x_2 at a later time t_2. Then its displacement Δx over the interval from t_1 to t_2 is given by

$$\Delta x =$$

The magnitude of the displacement during a time interval may be quite different from the distance traveled during the interval. Suppose a particle starts at time $t_1 = 0$ with coordinate $x_1 = 5.0$ m, arrives at $x_2 = 15.0$ m at time $t_2 = 2.0$ s, then arrives at $x_3 = 10.0$ m at time $t_3 =$

3.0 s. Note that at first it travels in the positive x direction but later it travels in the negative x direction. From t_1 to t_2 its displacement is $(15.0-5.0) = 10.0$ m; from t_2 to t_3 its displacement is $(10.0 - 15.0) = -5.0$ m; and from t_1 to t_3 its displacement is $(10.0 - 5.0) = 5.0$ m. On the other hand, the total distance traveled from t_1 to t_3 is 15 m.

The average velocity $\bar{v}$ over the interval from t_1 to t_2 is given by

$$\bar{v} =$$

Write down in words the steps you would take to find the average velocity in the interval from t_1 to t_2 if you are given the function $x(t)$: _____
_____
_____

On a graph of x vs. t the average velocity over the interval from t_1 to t_2 is related to a certain line you might draw. Describe the line and tell what property gives the average velocity: _____
_____
_____

Describe the limiting process used to obtain the instantaneous velocity at time t by applying it to a series of average velocities: _____
_____
_____

If the function $x(t)$ is known, the instantaneous velocity at any time t_1 is found by _____
_____ and evaluating the result for $t = $ _____. On a graph of x vs. t, the instantaneous velocity at any time t_1 is related to a line you might draw. Describe the line and tell what property gives the instantaneous velocity: _____
_____

The term "velocity" means instantaneous velocity. The modifier "instantaneous" is implied.

For each of the functions $x(t)$ shown graphically below, tell if the average velocity is positive or negative in the interval from t_1 to t_2. Also give the sign of the instantaneous velocity at t_1 and t_2 or state that it is zero if it is.

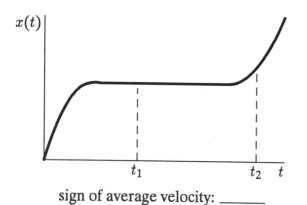

sign of average velocity: _____
sign of velocity at t_1: _____
sign of velocity at t_2: _____

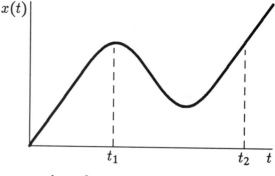

sign of average velocity: _____
sign of velocity at t_1: _____
sign of velocity at t_2: _____

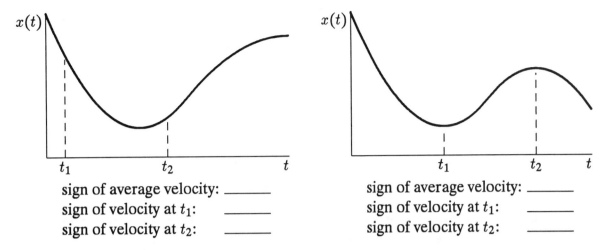

sign of average velocity: _____
sign of velocity at t_1: _____
sign of velocity at t_2: _____

sign of average velocity: _____
sign of velocity at t_1: _____
sign of velocity at t_2: _____

On each of the first three diagrams indicate a time t_3 such that the average velocity from t_1 to t_3 is zero.

On the axes below sketch graphs of the velocity $v(t)$ for the motions represented by the first two of the x vs. t graphs above. Be sure you get the sign of v correct. Also be sure your graphs indicate roughly where the speed is large and where it is small or zero.

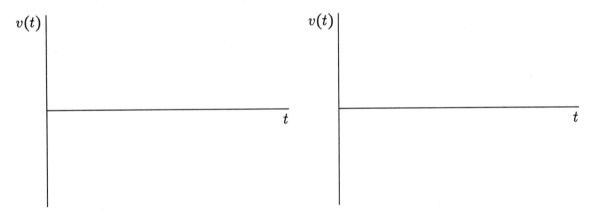

The sign of the velocity indicates the direction of travel. Describe the direction of particle motion if the velocity is positive; if the velocity is negative. Warning! Do not assume the x axis is positive to the right.

positive velocity: _____

negative velocity: _____

Define speed: _____

If the velocity of a particle changes from v_1 at time t_1 to v_2 at a later time t_2 then its average acceleration $\bar{a}$ over the interval from t_1 to t_2 is given by

$$\bar{a} =$$

Write down in words the steps you would take to find the average acceleration in the interval from t_1 to t_2, given the function $x(t)$: _____

On a graph of v vs t, the average acceleration over the interval from t_1 to t_2 is related to a line you might draw. Describe the line and tell what property gives the average acceleration: _____

Describe the limiting process used to obtain the <u>instantaneous acceleration</u> at time t by applying it to a series of average accelerations: _____

If the function $x(t)$ is known, the instantaneous acceleration at any time t_1 is found by _____ and evaluating the result for $t =$ _____. On a graph of v vs. t, the instantaneous acceleration at any time t_1 is related to a line you might draw. Describe the line and tell what property gives the instantaneous acceleration: _____

The term "acceleration" means instantaneous acceleration. The modifier "instantaneous" is implied.

Note that a positive acceleration does not necessarily mean the particle speed is increasing and a negative acceleration does not necessarily mean the particle speed is decreasing. The speed increases if the velocity and acceleration have the same sign and decreases if they have opposite signs, no matter what the signs are.

The graph to the left below shows the velocity as a function of time for an object moving along a straight line. On the t axis mark two times (t_1 and t_2, say) such that the acceleration is negative in the interval between them and label that portion of the curve "$a < 0$". Mark two times such that the acceleration is positive between them and label that portion of the curve "$a > 0$". Mark two times such that the acceleration is zero between them and label that portion of the curve "$a = 0$". There is one time for which the acceleration is zero only for that instant and is not zero for neighboring times. Label it. On the axes to the right below sketch the acceleration as a function of time.

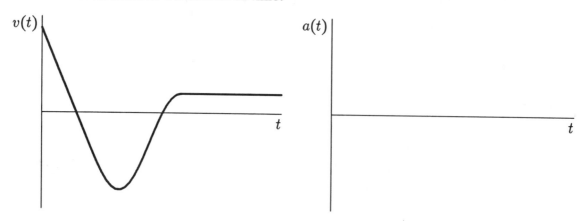

On the coordinate axes below draw possible graphs of $x(t)$ for a particle moving along the x axis with constant acceleration. Take the initial position to be $x = 0$ in all cases. For the first curve suppose the particle starts with a positive velocity and has a positive acceleration;

for the second suppose it starts with a negative velocity and has a positive acceleration; for the third suppose it starts with a positive velocity and has a negative acceleration; and for the fourth suppose it starts with a negative velocity and has a negative acceleration. Label each graph with the signs of the initial velocity and the acceleration. Also label points where the particle momentarily stops to start moving in the opposite direction.

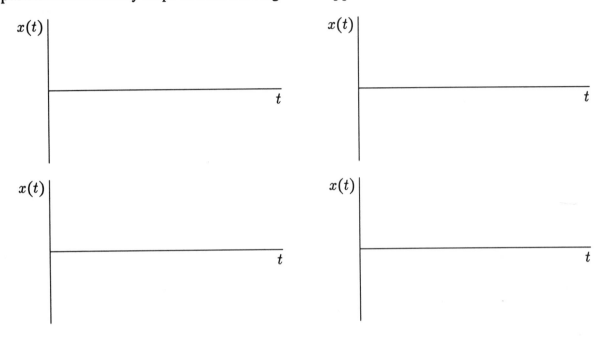

At the instant a particle is momentarily at rest its acceleration is not necessarily 0. To remind yourself that $v = 0$ does not imply $a = 0$ write down a function $x(t)$ for which $v = 0$ but $a \neq 0$ at $t = 0$:

$$x(t) =$$

A particle is initially moving in the positive x direction. Describe in words what its motion might be if at some instant it has zero velocity but non-zero acceleration: _____

For comparison, describe in words what its motion might be if at some instant it has zero velocity and thereafter it has zero acceleration: _____

If the slope of an x vs. t graph is increasing with time the acceleration is positive; if the slope is decreasing the acceleration is negative; and if the slope is constant the acceleration is zero. Go back to the four x vs. t graphs above and label the points at t_1 and t_2 with either $a > 0$, $a < 0$, or $a = 0$, as appropriate.

Motion with constant acceleration. If a particle is moving along the x axis with constant acceleration a, its coordinate is given as a function of time by

$$x(t) =$$

Chapter 2: Motion in One Dimension

and its velocity is given as a function of time by

$$v(t) =$$

The coordinate and velocity of the particle at time $t = 0$ should appear in your equations. Give the symbol used for each: coordinate at $t = 0$: _____; velocity at $t = 0$: _____.

You should memorize these equations. Since the second is the derivative with respect to time of the first and so can be derived quickly, you probably need to memorize only the first. Before using these equations always ask if the problem specifies or implies that the acceleration is constant. They are not valid if the acceleration varies with time.

Free fall. Specialize the constant acceleration kinematic equations for the case of an object in free fall, subject only to the pull of gravity. Take the y-axis to be positive in the upward direction, away from the earth, and suppose the particle moves along that axis. Then the y coordinate and the velocity of the particle, as functions of time, are given by

$$y(t) =$$
$$v(t) =$$

Here $g =$ _____ m/s^2 = _____ ft/s^2 is the magnitude of the acceleration due to gravity near the surface of the earth.

Remember that, in the absence of air resistance, *all* objects in free fall have the same acceleration, regardless of their masses. Also remember their acceleration is the same throughout their motion from the time they are released or thrown to the time they hit something. Their acceleration is g downward while they are going up, when they are at their highest points, and while they are going down.

These equations are for constant acceleration only. To emphasize this point describe a situation in which an object is thrown upward or downward near the surface of the earth and these equations are *not* valid: _____

_____

_____

II. PROBLEM SOLVING

Nearly all the problems at the end of this chapter can be solved using one or more of the following:

a. definition of average velocity
b. definition of instantaneous velocity
c. definition of average acceleration
d. definition of instantaneous acceleration
e. equations for the coordinate and velocity of a particle with constant acceleration (including the free fall equations)

Cultivate the good habit of classifying a problem according to the principle or definition that it illustrates. Look at what is given and what is asked, then see if all the ingredients are present for any given classification. Practice as you work through the following problems.

PROBLEM 1. A car goes 180 km along a straight line in 2.3 h. What is the magnitude of its average velocity?

SOLUTION: Here you use the definition of average velocity. The displacement is $\Delta x =$ _____ m; the time interval is $\Delta t =$ _____ s.

[ans: 2.17 m/s]

PROBLEM 2. A car travels at an average velocity of 90 km/h for 1.2 h, then continues at an average velocity of 60 km/h along the same line for an additional 75 km. What is the magnitude of its average velocity over the entire trip?

SOLUTION: You must compute the distance traveled and the time for each segment of the trip. Because the car continues along the same line in the same direction the magnitude of its displacement is the sum of the distances traveled in the two segments. The total time, of course, is the sum of the times for the two segments. During the first part of the trip the car takes _____ h to go _____ km. During the second part it takes _____ h to go _____ km. The total distance traveled is _____ km; the total time is _____ h, so the magnitude of the average velocity is _____ km/s. You should have obtained 74.7 km/h.

PROBLEM 3. A car travels for 0.75 h at an average velocity of 72 km/h along a straight line. It then turns around and travels halfway back in 0.50 h. What is the displacement of the final point from the initial point and what is the magnitude of the average velocity over the entire trip?

SOLUTION: The displacement during the first part of the trip is $\Delta x_1 =$ _____ km; the displacement during the second part of the trip is $\Delta x_2 =$ _____ km. The net displacement is $\Delta x = \Delta x_1 + \Delta x_2 =$ _____ km. Make sure you have the signs correct here. The displacements are in opposite directions. The total time for the trip is $\Delta t =$ _____ h. Now compute the average velocity.

[ans: 27.0 km; 21.6 km/h]

Notice that the magnitude of the net displacement is considerably less than the distance traveled. Remember that the displacement is the difference in the coordinates of the points at the end and beginning of the trip.

PROBLEM 4. A box moves across a floor. If the x axis is taken to be along its direction of motion its coordinate x in meters as a function of time t in seconds is given by $x(t) = 0.20t^2[1.0 + t]$. What is its average velocity during the time interval from $t = 1.5$ s to $t = 2.5$ s?

SOLUTION: Evaluate $x(t)$ for $t = 1.5$ s and call this x_1; evaluate $x(t)$ for $t = 2.5$ s and call this x_2. Use $\bar{v} = (x_2 - x_1)/(t_2 - t_1)$ to calculate the average velocity.

[ans: 3.25 m/s]

PROBLEM 5. The coordinate in meters of a particle is given by $x(t) = 4.0 - 6.0t + 0.75t^3$, where t is in seconds. What is its instantaneous velocity as a function of time? What is its instantaneous velocity at $t = 1.0$ s? at $t = 2.0$ s?

SOLUTION: Differentiate $x(t)$ with respect to t, then evaluate the resulting expression.

[ans: $-6.0 + 2.25t^2$; -3.75 m/s; $+3.0$ m/s]

Notice that the particle reversed its direction of motion. At $t = 1.0$ s it was going in the negative x direction, at $t = 2.0$ s it was going in the positive x direction. There must have been an instant between when $v = 0$. At what time was it stopped and where was it then?

SOLUTION: Solve $0 = -6.0 + 2.25t^2$ for t, then evaluate $x(t) = 4.0 - 6.0t + 0.75t^3$.

[ans: 1.63 s; 2.53 m]

PROBLEM 6. A particle travels along the x axis. At time $t = 2.0$ s its velocity is $+3.4$ m/s; at $t = 4.0$ s its velocity is -8.4 m/s. What is its average acceleration during this interval?

SOLUTION:

[ans: -5.9 m/s^2]

Carefully note that the particle is going faster at the end of the interval than at the beginning but the acceleration is negative.

PROBLEM 7. The coordinate of a particle is given by $x(t) = 4.0 + 2.0t^3$, where x is in meters and t is in seconds. What is its velocity as a function of time? What is its average acceleration over the interval from $t = 2.0$ s to $t = 3.0$ s?

SOLUTION: Use $v(t) = dx/dt$ to find an expression for the velocity as a function of time. Evaluate it for the beginning and end of the interval, then use $\bar{a} = \Delta v/\Delta t$ to find the average acceleration.

[ans: $6.0t^2$; 30 m/s^2]

PROBLEM 8. For time $t > 0$ the coordinate of a particle is given by $x(t) = 4.0 - 5.0t + 2.0t^3$, where x is in meters and t is in seconds. What is its acceleration as a function of time? What is its acceleration at $t = 2.0$ s? What is its acceleration when it is momentarily at rest?

SOLUTION: Differentiate $x(t)$ to find $v(t)$, then differentiate $v(t)$ to find $a(t)$. Evaluate $a(t)$ for $t = 2.0$ s. Solve $v(t) = 0$ for the time the particle is at rest, then substitute this value into the expression for $a(t)$.

[ans: $12t$; 24 m/s^2; 11.0 m/s^2]

Carefully note that an instantaneous velocity of 0 does *not* imply a vanishing acceleration. If the direction of travel changes, the velocity must vanish at some instant but the acceleration need not vanish then.

All constant acceleration problems can be solved using only the equations $x(t) = x_0 + v_0 t + \frac{1}{2}at^2$ and $v(t) = v_0 + at$. Six quantities appear in these equations: $x(t)$, $v(t)$, x_0, v_0, a, and t. In most cases all but two are given and you are asked to solve for one or both of the others. Mathematically a typical constant acceleration problem involves identifying the known and unknown quantities, then simultaneously solving the two kinematic equations for the unknowns.

Other kinematic equations are given in the text (see Table 2). They have been derived by eliminating one or another of the kinematic quantities from the expressions for $x(t)$ and $v(t)$. You may use them or solve the two fundamental equations simultaneously. The problems below show you how to use the simultaneous equation approach.

You usually have the option of selecting x_0 to be zero; this selection simply places the origin of the coordinate system at the initial position of the object. If you do this you need deal with only the other 5 kinematic quantities.

You should be aware that you may assign the value 0 to the time when the particle is at *any* point along its trajectory: when the particle starts out, when it reaches a particular point, when it has a particular velocity, or any other point. A negative value for t simply means a time before the instant you chose as $t = 0$. In the kinematic equations x_0 is the coordinate of the object and v_0 is its velocity at time $t = 0$, not necessarily when the particle starts out.

The simplest problems give the acceleration and a starting coordinate and velocity, then ask for the coordinate and velocity at some other time. If you choose $t = 0$ to correspond to the time when the object is at the starting position then x_0 is the starting coordinate, v_0 is the starting velocity and t is the later time of interest. Simply substitute the values for x_0, v_0, a, and t into $x(t) = x_0 + v_0 t + \frac{1}{2}at^2$ and $v(t) = v_0 + at$ to get the results sought.

PROBLEM 9. A particle starts at $x = 5.0$ m with a velocity of -30 m/s. If it has a constant acceleration of 4.3 m/s^2 what is its coordinate and velocity 5.0 s later?

SOLUTION:

[ans: -91 m; -8.5 m/s]

To demonstrate that the problem can be solved with another choice for $t = 0$, we choose $t = 0$ to be 5.0 s after the particle starts. This is the time for which we seek the coordinate and velocity. Then the starting time is $t = -5.0$ s, $x(-5.0\text{s}) = 5.0$ m, and $v(-5.0\text{s}) = -30$ m/s. Now the unknowns are x_0 and v_0. The velocity equation is $v(-5.0\text{s}) = v_0 + at$, so $v_0 = v(-5.0\text{s}) - at = -30 + 4.3 \times 5.0 = -8.5$ m/s. The coordinate equation is $x(-5.0\text{s}) = x_0 + v_0 t + \frac{1}{2}at^2$, so $x_0 = x(-5.0\text{s}) - v_0 t - \frac{1}{2}at^2 = 5.0 - 8.5 \times 5.0 - 0.5 \times 4.3 \times 5^2 = -91$ m.

PROBLEM 10. A car starts from rest and moves in a straight line with a constant acceleration of 0.40 m/s^2. How much time elapses before it is going 65 km/h? How far does it travel in this time?

SOLUTION: The following table should help you list the known and unknown quantities. If a quantity is known, write its value in the table; if it is unknown, write a question mark. You will probably want to place the origin of the coordinate system at the starting point of the car and take the time to be zero when the car is there. Then t is the time when the car is going 65 km/h and x is its position at that time. For convenience convert to standard SI units: enter velocities in m/s, for example.

$$x_0 = \underline{\qquad} \qquad v_0 = \underline{\qquad} \qquad t = \underline{\qquad}$$
$$x(t) = \underline{\qquad} \qquad v(t) = \underline{\qquad} \qquad a = \underline{\qquad}$$

If you followed the suggestion, you should have numerical values filled in for x_0, v_0, $v(t)$, and a. You should have question marks filled in for $x(t)$ and t.

Now solve the problem. Notice that the equation $v(t) = v_0 + at$ contains only one unknown, t. Solve it algebraically for t, then substitute numerical values. Then substitute numerical values into $x(t) - x_0 = v_0 t + \frac{1}{2}at^2$ to obtain the coordinate. Since the car does not change its direction of travel and since x is positive this is also the distance traveled.

[ans: 45.1 s; 408 m]

PROBLEM 11. A car moves in a straight line with constant acceleration. When it passes one telephone pole it is going at 35 km/h. When it passes a second pole, 200 m from the first, it is going at 65 km/h. What is the acceleration of the car? How much time does it take to go from the first pole to the second?

SOLUTION: Again we first list the known and unknown quantities. Place the origin of the coordinate system at the first pole and start the clock when the car is there. Let t represent the time the car is at the second pole. Now list the values of all kinematic quantities that are known. If a quantity in the following list is unknown, enter a question mark.

$$x_0 = \underline{\qquad} \qquad v_0 = \underline{\qquad} \qquad t = \underline{\qquad}$$
$$x(t) = \underline{\qquad} \qquad v(t) = \underline{\qquad} \qquad a = \underline{\qquad}$$

You should have written numerical values for x_0, v_0, $x(t)$, and $v(t)$. You should have entered question marks for t and a. Now solve algebraically for a. First use the equation for $v(t)$ to eliminate t from the equation for $x(t)$. $v(t) = v_0 + at$ yields the algebraic expression $t = \underline{\qquad}$. Substitute this expression into $x = v_0 t + \frac{1}{2}at^2$. After a little manipulation you should obtain $x = (v^2 - v_0^2)/2a$. Now solve for a and substitute numerical values. Finally, evaluate the expression for t.

[ans: 0.579 m/s²; 14.4 s]

PROBLEM 12. If the car of the last problem started from rest, how much time elapsed before it reached the first pole? What distance did it travel before reaching the first pole?

SOLUTION: You could place the origin at the starting point of the car and take the time to be 0 when the car is there. Instead, we shall keep the origin at the first pole and call $t = 0$ when the car is there. First find the time at which the car was at rest: solve $0 = v_0 + at$ for t.

[ans: −16.8 s]

The negative sign indicates that the car was at rest (started) 16.8 s *before* reaching the first pole.

Now substitute numerical values, including $t = -16.8$ s, into $x(t) = v_0 t + \frac{1}{2}at^2$ to find where the car was when it started.

SOLUTION:

[ans: -81.8 m]

The negative sign indicates it started 81.8 m *before* the first pole.

Here are some examples that deal with two moving objects.

PROBLEM 13. Car A is waiting at a traffic light. When it turns green A moves forward with constant acceleration. At the instant the light turns, car B, moving with constant velocity, passes car A. On the axes to the right, sketch graphs of the coordinates of the two cars as functions of time. Continue the graphs until they intersect. Label them A and B, according to the cars they represent.

On your graph use t_p to label the time at which car A passes car B and use t_v to label the time at which the cars have the same velocity.

Take the acceleration of car A to be 2.5 m/s² and the velocity of car B to be 15 m/s. Calculate the time at which they are at the same position, their velocities then, and their distance from the light.

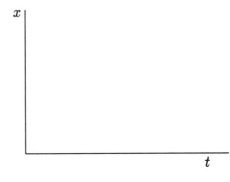

SOLUTION: Let a_A be the acceleration of A. The coordinate of A is given by $x_A(t) = \frac{1}{2}a_A t^2$. The coordinate of B is given by $x_B(t) = v_B t$, where v_B is the velocity of B. A passes B when $x_A(t) = x_B(t)$, so you should solve $\frac{1}{2}a_A t^2 = v_B t$ for the time t. Then use either $x_A(t) = \frac{1}{2}a_A t^2$ or $x_B(t) = v_B t$ to find where they are and $v_A = a_A t$ to find the velocity of A.

[ans: 12.0 s; A: 30.0 m/s, B: 15.0 m/s; 180 m]

Notice that car A is going faster than car B when A passes B. Calculate the time they have the same velocities and their distances from the light then.

SOLUTION:

[ans: 6.00 s; A: 45.0 m, B: 90.0 m]

Chapter 2: Motion in One Dimension   **21**

PROBLEM 14. Suppose a particle starts at $x = 0$ and moves in the positive x direction with a constant velocity of 4.5 m/s. At the same time a second particle starts at $x = 30$ m with an initial velocity of -1.1 m/s. It has a constant acceleration of -2.3 m/s^2. How much time elapses before they meet? Where do they meet?

SOLUTION: The coordinate of the first particle is given by $x_1(t) = v_{10}t$ and the coordinate of the second is given by $x_2(t) = x_{20} + v_{20}t + \frac{1}{2}at^2$, where a is the acceleration of the second particle. They meet when $x_1(t) = x_2(t)$, or, what is the same, $v_{10}t = x_{20} + v_{20}t + \frac{1}{2}at^2$. This is a quadratic equation for t and gives two answers, one positive and one negative. You wish the positive answer: the time they meet *after* starting. Derive the algebraic solution for t, the calculate its value. Finally substitute the value into $x_1(t) = v_{10}t$ to find the coordinate of the point where the particles meet.

[ans: 3.22 s; 14.5 m]

Free fall problems are a special class of constant acceleration problems and are solved using the same strategies. You know the acceleration even if it is not mentioned in the problem statement, so it is not an unknown. Usually the coordinate system is oriented so the y axis is up, away from the earth. The acceleration is then $-g$ and the kinematic equations become $y(t) = y_0 + v_0 t - \frac{1}{2}gt^2$ and $v(t) = v_0 - gt$.

PROBLEM 15. How fast must an object be thrown upward so its highest point is 50 m above the throwing point?

SOLUTION: Take the time to be 0 when the object is thrown and t when it reaches the highest point. Place the coordinate system with the y axis extending upward and with the origin at the throwing point. Fill in the following table with the numerical values after known quantities and question marks after unknown quantities:

$y_0 = $ _____ $v_0 = $ _____ $t = $ _____
$y(t) = $ _____ $v(t) = $ _____

You should have written values for y_0, $y(t)$, and $v(t)$; question marks for v_0 and t. The crucial step is to recognize that $v(t) = 0$ at the highest point. Here the object momentarily stops before starting to fall. Use $v = v_0 - gt$ to eliminate t from $y = v_0 t - \frac{1}{2}gt^2$, then algebraically solve for v_0. Substitute numerical values.

[ans: 31.3 m/s]

PROBLEM 16. A rock is thrown upward at 25 m/s from the edge of a cliff. How much time elapses before it hits the canyon floor 200 m below the cliff edge and what is its speed just before it hits?

SOLUTION: Take the time to be 0 when the rock is thrown and t when it hits the canyon floor. Place the origin of the coordinate system at the throwing point and take the y axis to be positive in the upward direction. Fill in the following table with the numerical values of known quantities and question marks after unknown quantities:

$y_0 = $ _____ $v_0 = $ _____ $t = $ _____
$y(t) = $ _____ $v(t) = $ _____

Chapter 2: Motion in One Dimension

Solve $y(t) = v_0 t - \frac{1}{2}gt^2$ for t. Notice the equation is quadratic in t and has two solutions. You want the positive one, corresponding to a time after the rock is thrown. Once t has been found substitute its value into $v(t) = v_0 - gt$ to find the velocity of the rock just before it hits.

[ans: 9.43 s; −67.4 m/s]

PROBLEM 17. A ball is thrown upward with a speed of 20 m/s. What is its velocity when it is halfway down, after reaching its highest point?

SOLUTION: The first job is to determine the coordinate of the highest point so you can find the coordinate of the point halfway down. Here is a table you may use to identify known and unknown quantities:

$y_0 =$ _____ $v_0 =$ _____ $t =$ _____
$y(t) =$ _____ $v(t) =$ _____

Now algebraically solve for the coordinate of the highest point, then substitute numerical values.

[ans: 20.4 m]

The coordinate of the point halfway to the highest point is $y =$ _____ m. Solve for the velocity of the ball there. Use the same initial coordinate and velocity as before and fill out the table for the new situation.

$y_0 =$ _____ $v_0 =$ _____ $t =$ _____
$y(t) =$ _____ $v(t) =$ _____

The two unknowns are v and t. Algebraically solve $v = v_0 - gt$ for t: $t =$ _____. Substitute this expression into $y = y_0 + v_0 t - \frac{1}{2}gt^2$ to obtain $y =$ _____. Finally solve for v^2. You should obtain $v^2 = v_0^2 - 2g(y - y_0)$. Note that there will be two solutions for v and they will be the negatives of each other. Since you seek the velocity of the ball on the way down you will select the negative solution. Substitute numerical values and solve for the velocity.

[ans: −14.1 m/s]

Read through all the homework problems you have been assigned for this chapter and classify each part by writing the problem number and part (e.g. 2a) in the appropriate space below.

CLASSIFICATION	PROBLEM NUMBER AND PART
definition of average velocity	_____
definition of instantaneous velocity	_____
definition of average acceleration	_____
definition of instantaneous acceleration	_____
constant acceleration kinematics equations	_____
other	_____

III. MATHEMATICAL SKILLS

The following is a listing of mathematical knowledge you will need to understand and solve problems in this chapter. Refer to a calculus or algebra text if you are rusty on any of it.

1. You should understand the notation used for a function. The symbol $f(t)$, for example, indicates that the value of f depends on the value of t. The coordinate of a moving object, for example, is different at different times so it is a function of time. Sometimes the functional dependence is given by means of an equation, such as $x(t) = v_0 t + \frac{1}{2}at^2$. Sometimes it is given by means of a graph or a table of values.

If you know the dependence of x on t you can select a value for t and find the corresponding value for x. Since you can choose the value for t, t is called the <u>independent variable</u>. Since the value for x is determined through the functional relationship by the value of t, x is called the <u>dependent variable</u>.

You must carefully distinguish between a function and its value for a particular value of the independent variable. If $x(t) = 5 - 7t$, for example, then $x(2) = 5 - 7 \times 2 = -9$. If you know the coordinate of an object as a function time and wish to find its velocity at $t = 2$ s, say, you must first find the velocity as a function of time by differentiating $x(t)$ with respect to time. Only when you have done this do you substitute numerical values into the expression. An evaluation of $x(t)$ for $t = 2$ s tells you nothing about the velocity. Similarly, an evaluation of the expression $v(t)$ for a particular instant of time tells you nothing about the acceleration.

2. You should understand the meaning of a derivative as a limit of a ratio. This is useful for a solid understanding of instantaneous velocity and acceleration. The formula $v(t) = dx/dt$ gives the velocity at any instant of time. To find the velocity at a given instant you must substitute a value for t. If the object is accelerating its velocity just before or just after the given instant is different from the velocity at the instant.

3. You should be able to write down derivatives of polynomials. For example, $d(A + Bt + Ct^2 + Dt^3)/dt = B + 2Ct + 3Dt^2$ if A, B, C, and D are constants. Your instructor may also require you to know how to evaluate the derivative with respect to t of other functions such as e^{At}, $\sin(At)$, and $\cos(At)$. The derivatives are Ae^{At}, $A\cos(At)$, and $-A\sin(At)$ respectively.

You should know the product and quotient rules for differentiation:

$$\frac{d}{dt}[f(t)g(t)] = \frac{df(t)}{dt}g(t) + f(t)\frac{dg(t)}{dt}$$

and

$$\frac{d}{dt}\left[\frac{f(t)}{g(t)}\right] = \frac{1}{g(t)}\frac{df(t)}{dt} - \frac{f(t)}{g^2(t)}\frac{dg(t)}{dt}$$

You should also know the chain rule. Suppose $u(t)$ is a function of t and $f(u)$ is a function of u. Then f depends on t through u and

$$\frac{df}{dt} = \frac{df}{du}\frac{du}{dt}$$

The chain rule was used above to find the derivatives of e^{At}, $\sin(At)$, and $\cos(At)$. In each case u was taken to be At.

4. You should be able to interpret the slope of a line tangent to a curve as a derivative. The instantaneous velocity is the slope of the coordinate as a function of time, the instantaneous acceleration is the slope of the velocity as a function of time.

5. You should be able to solve two simple simultaneous equations for 2 unknowns. For example, given any four of the algebraic symbols in $x = x_0 + v_0 t + \frac{1}{2} a t^2$ and $v = v_0 + at$, you should be able to solve for the other two.

One way is to solve one of the equations algebraically for one of the unknowns, thereby obtaining an expression for the chosen unknown in terms of the second unknown. Substitute the expression into the second equation, replacing the first unknown wherever it occurs. You now have a single equation with only one unknown. Solve in the usual way, then go back to the expression you obtained from the first equation and evaluate it for the first unknown. Several examples have been given in the Problem Solving section above.

With a little practice you will learn some of the shortcuts. If the problem asks for only one of the two unknowns, eliminate the one that is *not* requested. If one equation contains only one of the unknowns, solve it immediately and use the result in the second equation to obtain the value of the second unknown. If one equation in linear in the unknown you wish to eliminate but the other equation is quadratic, use the linear equation to eliminate the unknown from the quadratic equation, rather than vice versa.

6. You should be able to solve algebraic equations that are quadratic in the unknown. If $At^2 + Bt + C = 0$ then
$$t = \frac{-B \pm \sqrt{B^2 - 4AC}}{2A}$$

When the quantity under the radical sign does not vanish there are two solutions. Always examine both to see what physical significance they have, then decide which is required to answer the particular problem you are working. If the quantity under the radical sign is negative the solutions are complex numbers and probably have no physical significance for problems in this course. Check to be sure you have not made a mistake.

IV. COMPUTER PROJECTS

A computer can be used to calculate the position and velocity of an object when its acceleration is a known function of time. Divide the time axis into a large number of small intervals, each of duration Δt. If Δt is sufficiently small the acceleration can be approximated by a constant in each interval, perhaps with a different value in different intervals. If v_b is the velocity at the beginning of an interval then we use $v_e = v_b + a \Delta t$ to calculate the velocity at the end of the interval. If a is the average acceleration for the interval then this expression is exact. However, we do not know the average acceleration and so must approximate it. For the projects in this section we approximate a by $[a(t_b) + a(t_e)]/2$, where $a(t_b)$ is the acceleration at the beginning of the interval and $a(t_e)$ is the acceleration at the end. You will need to tell the computer how to calculate the acceleration by supplying it with the function a(t).

If x_b is the coordinate of the object at the beginning of the interval, then the coordinate at the end of the interval is given by $x_e = x_b + v \Delta t$, where v is the average velocity in the interval. Approximate v by $(v_b + v_e)/2$. Smaller values for Δt make the approximations for v_e and x_e better. We cannot, however, take Δt to be so small that significance is lost when the computer sums the terms in these equations. Note that the coordinate and velocity at the end of any interval are the coordinate and velocity at the beginning of the next interval. A skeleton program might look like this:

> input initial values: t_0, x_0, v_0
> input final time and interval width: t_f, Δt
> set $t_b = t_0$, $x_b = x_0$, and $v_b = v_0$
> calculate acceleration at beginning of first interval: $a_b = a(t_b)$
> **begin loop** over intervals
> > calculate time at end of interval: $t_e = t_b + \Delta t$
> > calculate acceleration at end of interval: $a_e = a(t_e)$
> > calculate "average" acceleration in interval: $a = (a_b + a_e)/2$
> > calculate velocity at end of interval: $v_e = v_b + a\Delta t$
> > calculate "average" velocity in interval: $v = (v_b + v_e)/2$
> > calculate coordinate at end of interval: $x_e = x_b + v\Delta t$
> > * **if** $t_e \geq t_f$ **then**
> > > print or display t_b, x_b, v_b
> > > print or display t_e, x_e, v_e
> > > exit loop
> > **end if** statement
> > set $t_b = t_e$, $x_b = x_e$, $v_b = v_e$, $a_b = a_e$ in preparation for next interval
> **end loop** over intervals
> stop

The line marked with an asterisk will be different for different applications. Sometimes you will want to display results and exit the loop when the velocity or coordinate have certain values. You will then change this line.

Because Δt is arbitrary, the end of the last interval may not correspond exactly to t_f. The first line of output corresponds to a time just before t_f and the second line corresponds to a time just after. You can force $t = t_f$ at the end of the last interval by asking the computer to recalculate Δt near the beginning of the program. First use $(t_f - t_0)/\Delta t$ to estimate the number of intervals, then round the result to the nearest integer N. Finally take $\Delta t = (t_f - t_0)/N$.

Most programming languages allow you to write a separate section of the program to define the function $a(t)$. Then the lines implementing $a_b = a(t_b)$ and $a_e = a(t_e)$ simply refer to the function definition. This is more efficient than writing the instructions for $a(t)$ twice, once for $t = t_b$ and once for $t = t_e$.

PROJECT 1. Test the program on a problem you can analyze analytically. Suppose the acceleration in m/s² is given by $12t$, for t in seconds. Take the initial coordinate (at $t = 0$) to be 5.0 m and the initial velocity to be -120 m/s. Find the coordinate and velocity at $t = 4.5$ s.

Start with $\Delta t = 0.5$ s and carry out the computation. Repeat several times, each time halving Δt, until you get the same answers to 3 significant figures on successive trials. Work the problem analytically to obtain an exact solution and carefully compare the answers with those obtained by the program. Check your program code if they differ.

PROJECT 2. Now use the program to determine when and where the particle is instantaneously stopped. You want the program to stop when the velocity changes sign in some interval. Use $v_b v_e \leq 0$ as the condition for displaying results and exiting the loop. The correct answer is between the two results displayed. If they are the same to within the number of significant figures you desire you are finished. If they are not, run the program again with a smaller value of Δt. Obtain 3 significant figure accuracy.

Now find when the particle is at $x = -100$ m and its velocity when it is there. The condition for displaying results and exiting the loop is now $(x_b + 100)(x_e + 100) \leq 0$. Obtain 3 significant figure accuracy.

PROJECT 3. Now try a more complicated example. Suppose the particle starts at $x = 0$ with a velocity of -120 m/s and has an acceleration that is given in m/s² by $a(t) = 30e^{-t/8}$, where t is the time in seconds. Find the time at which it is instantaneously at rest and its position at that time. [ans: 5.55 s; -295 m]

You can use the program to generate a list of values ready to be plotted by hand or to plot values on the monitor screen. Suppose the time interval desired between displayed points is Δt_d. Input the value of Δt_d just after the values for t_f and Δt are read, then set $t_d = t_0 + \Delta t_d$. Replace the last statements of the program, from the **if** statement, with

> **if** $t_e \geq t_d$ **then**
>   print, display, or plot x_e or v_e
>   increment t_d by Δt_d
> **end if** statement
> set $t_b = t_e$, $x_b = x_e$, $v_b = v_e$, $a_b = a_e$ in preparation for next interval
> **if** $t_e \geq t_f$ **then** exit loop
> **end loop** over intervals
> stop

Because the computer may produce values of t_e that are in error in the last place carried, the set of points you get may be somewhat different from what you anticipate. If this is intolerable, round the values of t_e so the position of the last non-zero digit is a few less than the number of digits carried by the machine.

PROJECT 4. A particle starts from rest at the origin and has an acceleration that is given in m/s² by $a(t) = 30te^{-t}$, where t is the time in seconds. Draw graphs of its coordinate and velocity as functions of time from $t = 0$ to $t = 5$ s. Plot points every 0.5 s. You should obtain 2 significant figure accuracy with $N = 100$.

The acceleration is quite small near $t = 0$ and increases as t increases. This is the effect of the factor t. It reaches a maximum at $t = 1$ s, then decreases. This is the effect of the exponential factor. Your graph of $v(t)$ should have its greatest slope in the vicinity of $t = 1$ s, then approach a line with a slope of zero. The velocity has become nearly constant. What do you think the limiting value of the velocity is?

The graph of $x(t)$ should have a slope of zero at $t = 0$ (the initial velocity is zero) but it soon curves upward and eventually approaches a straight line, the slope of which is the limiting value of the velocity.

V. NOTES

Chapter 3
VECTORS

I. BASIC CONCEPTS

You will deal with vector quantities throughout the course. In this chapter you will learn about their properties and how they can be manipulated mathematically. A solid understanding of this material will pay handsome dividends later.

Definitions. What properties distinguish a vector from a scalar? _____
_____

List below some examples of physical quantities that are scalars and some that are vectors:

SCALARS	VECTORS
_____	_____
_____	_____
_____	_____

A vector is represented graphically by an arrow in the direction of the vector, with length proportional to the magnitude of the vector (according to some scale). As an algebraic symbol, a vector is always written in boldface (**a**, for example) or with an arrow over the symbol ($\vec{a}$). The magnitude of **a** is written a, in italics (or not bold and without an arrow) or as |**a**| (or $|\vec{a}|$). Be sure you follow this convention. It helps you distinguish vectors from scalars and components of vectors. It helps you communicate properly with your instructors and exam graders. Do *not* write $a + b$ when you mean **a** + **b**, for example. They have entirely different meanings!

Displacement vectors are used as examples of vectors in this chapter. Tell in words what a displacement vector is: _____
_____

When you need an example to illustrate addition or subtraction of vectors, think of displacement vectors.

Graphical vector addition and subtraction. You will need to know how to add and subtract vectors, using both graphical and analytical techniques.

When two displacement vectors, one from point A to point B and the other from point B to point C, are added, the result is the displacement vector from _____ to _____. Except in special circumstances, the magnitude of the resultant vector is *not* the sum of the magnitudes of the vectors entering the sum and the direction of the resultant vector is *not* in the direction of any of the vectors entering the sum.

Suppose the incomplete diagram on the right is meant to demonstrate the addition of two vectors **a** and **b**. Place arrows on two sides of the triangle and label them $\vec{a}$ and $\vec{b}$. Place an arrow on the third side and label it $\vec{a} + \vec{b}$. Be sure the arrows are placed correctly so the triangle represents vector addition.

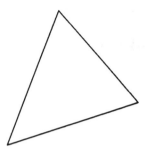

Now describe in words the steps you must take to add two vectors graphically. Be sure you mention how the vectors must be placed relative to each other and how the resultant vector is drawn: _____

_____

_____

When three or more vectors are added, the order of performing the sum is unimportant: **a** + (**b** + **c**) = (**a** + **b**) + **c**. Also remember you can reposition a vector as long as you do not change its magnitude and direction. Thus if two vectors you wish to add graphically do not happen to be placed with the tail of one at the head of the other, simply move one into the proper position.

The idea of the negative of a vector is used to define vector subtraction. How are the magnitude and direction of the negative of a vector related to the magnitude and direction of the original vector?

magnitude: _____

direction: _____

If **c** = **a** − **b**, then **c** is found by adding the negative of **b** to **a**: **c** = **a** + (−**b**). This defines vector subtraction. Write down the steps to subtract two vectors graphically (be sure to include a description of how they are positioned relative to each other): _____

_____

_____

_____

Notice that vector subtraction is defined so that if **a** + **b** = 0 then **a** = −**b** and if **c** = **a** + **b**, then **a** = **c** − **b**. Just subtract **b** from both sides of each equation. Vector subtraction is clearly useful for solving vector equations.

In the space on the right draw two vectors such that the magnitude of their sum equals the sum of their magnitudes.

In the space on the right draw two vectors such that the magnitude of their sum equals the difference of their magnitudes.

In the space to the right draw two vectors such that the magnitude of their difference equals the sum of their magnitudes.

In the space to the right draw two vectors such that the magnitude of their difference equals the difference of their magnitudes.

Analytic vector addition and subtraction. To carry out vector addition and subtraction analytically you will need to know how to find the components of a vector. For each of the vectors shown below, illustrate the x and y components by marking their lengths along the axes. Write expressions that give the components in terms of the magnitude and angle shown. Evaluate the expressions. Notice that the components of a vector can be positive or negative.

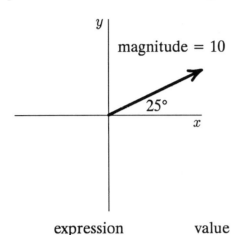

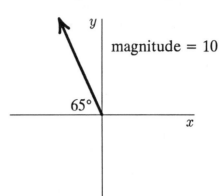

expression   value

$a_x = $ _____ = _____
$a_y = $ _____ = _____

expression   value

$a_x = $ _____ = _____
$a_y = $ _____ = _____

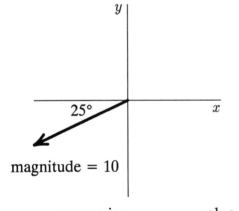

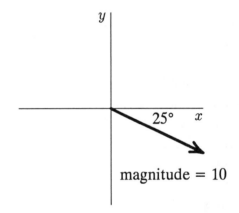

expression   value

$a_x = $ _____ = _____
$a_y = $ _____ = _____

expression   value

$a_x = $ _____ = _____
$a_y = $ _____ = _____

Describe in words the steps you can take to find the components of a vector, given its magnitude and direction and given a coordinate system: _____

_____

You must also be able to find the magnitude and orientation of a vector when you are given its components. Suppose a vector **a** lies in the xy plane and its components a_x and a_y are given. In terms of the components the magnitude of **a** is given by

$a = $

Chapter 3: Vectors   31

and the angle **a** makes with the x axis is given by

$$\phi = $$

The unit vectors **i**, **j**, and **k** are used when a vector is written in terms of its components. These vectors have magnitude 1 and are in the positive x, y, and z directions respectively. a_x **i** is a vector parallel to the x axis with x component a_x, a_y **j** is a vector parallel to the y axis with y component a_y, and a_z **k** is a vector parallel to the z axis with z component a_z. The vector **a** is given by

$$\mathbf{a} = a_x\mathbf{i} + a_y\mathbf{j} + a_z\mathbf{k},$$

where the rules of vector addition apply.
Are units associated with the unit vectors **i**, **j**, and **k**? _____
Are units associated with the components of a vector? _____
a_x **i**, a_y **j**, and a_z **k** are called the <u>vector components</u> of **a**. Do not confuse them with the components a_x, a_y, and a_z. When unit vectors are handwritten the symbols $\hat{\imath}$, $\hat{\jmath}$, and $\hat{k}$ are used.

Suppose **c** is the sum of two vectors **a** and **b** (i.e. **c** = **a** + **b**). In terms of the components of **a** and **b**: $c_x = $ _____, $c_y = $ _____, and $c_z = $ _____. Suppose **c** is the negative of **a** (i.e. **c** = **a**). In terms of the components of **a**: $c_x = $ _____, $c_y = $ _____, and $c_z = $ _____. Suppose **c** is the difference of two vectors **a** and **b** (i.e. **c** = **a** − **b**). In terms of the components of **a** and **b**: $c_x = $ _____, $c_y = $ _____, and $c_z = $ _____.

Multiplication involving vectors. Vectors can be multiplied by scalars. Let **a** be a vector and s a scalar. Then $s\mathbf{a}$ is a vector. If s is positive its direction is _____ and its magnitude is _____. If s is negative the direction of $s\mathbf{a}$ is _____ and its magnitude is _____. In terms of components $(s\mathbf{a})_x = $ _____, $(s\mathbf{a})_y = $ _____, and $(s\mathbf{a})_z = $ _____. Division of a vector by a scalar is, of course, just multiplication by the reciprocal of the scalar.

The <u>scalar product</u> (or dot product) of two vectors is defined in terms of the magnitudes of the two vectors and the angle between them when they are drawn with their tails at the same point: $\mathbf{a} \cdot \mathbf{b} = ab\cos\phi$. For each of two cases shown below write an expression for the scalar product of **a** and **b** in terms of the given quantities. To evaluate the expressions take $a = 10$ and $b = 5$.

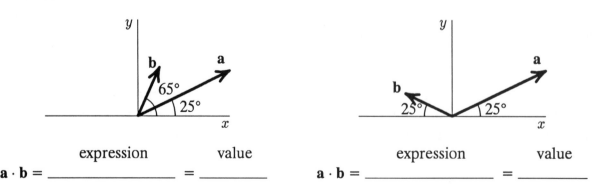

expression   value   expression   value
a · **b** = _____ = _____   **a** · **b** = _____ = _____

The scalar product can be interpreted in terms of the component of one vector along the direction of the other. For the scalar product **a** · **b**, write that interpretation in words: _____

In terms of components the scalar product is given by **a** · **b** = _____.
Notice that this expression gives **a** · **a** = $a_x^2 + a_y^2 = a^2$ for a vector in the xy plane.

Let ϕ be the angle between **a** and **b** when they are drawn with their tails at the same point. The sign of the scalar product **a** · **b** is positive if ϕ is in the range from _____ to _____ and is negative if ϕ is in the range from _____ to _____. Also **a** · **b** = 0 if ϕ is _____. Note that ϕ is always 180° or less.

Write the equation that gives the magnitude of the <u>vector</u> product **a** × **b** in terms of the magnitudes of **a** and **b**: |**a** × **b**| = _____. Carefully describe how to determine the angle that occurs in this expression: _____

For two vectors with given magnitudes the magnitude of **a** × **b** is the greatest when the angle between them is _____ and is zero when the angle between them is _____.

Describe the right hand rule used to find the direction of **a** × **b**: _____

In terms of the cartesian components of **a** and **b**, **a** × **b** = _____.

Remember that the direction of the vector product depends on the order in which the vectors appear: **a** × **b** and **b** × **a** are in opposite directions.

The magnitude of the vector product can also be interpreted in terms of the component of one vector along a certain direction perpendicular to the other vector. The direction is in the plane defined by the two vectors in the product. For the vector product **a** × **b** write the interpretation in words: _____

Remember that the scalar product of two vectors is a scalar and has no direction associated with it while the vector product is a vector and does have a direction associated with it.

II. PROBLEM SOLVING

All of the problems at the end of the chapter deal with vector manipulations: addition, subtraction, finding components, finding magnitude and direction, and the various kinds of multiplication. The vectors are given either in terms of magnitude and direction or in terms of components and answers may be requested in either of these forms. This means you may need to convert from the given form to a form suitable for the vector operation, then convert again to obtain the form required for the answer.

If a is the magnitude of a vector **a** in the xy plane and ϕ is the angle that the vector makes with the positive x axis, then the components of **a** are $a_x = a\cos\phi$ and $a_y = a\sin\phi$. If you know the components you can find the magnitude and the angle with the positive x axis. The magnitude is given by

$$a = \sqrt{a_x^2 + a_y^2}$$

and the angle is given by

$$\phi = \arctan(a_y/a_x)$$

When you have found values for a and ϕ check to be sure $a\cos\phi$ has the proper sign for the x component and $a\sin\phi$ has the proper sign for the y component. If they have the wrong signs add 180° to the value you used for the angle.

PROBLEM 1. A particle undergoes two successive displacements. It starts at the origin and goes to $x = 5.0$ m, $y = -3.0$ m. Then it goes to $x = -5.0$ m, $y = 4.0$ m. What is the magnitude and angle with the x axis of the resultant of the two displacements?

SOLUTION: Sketch the two displacements and their sum on the diagram, then fill in the table with the x and y components of the displacements and the x and y components of their sum.

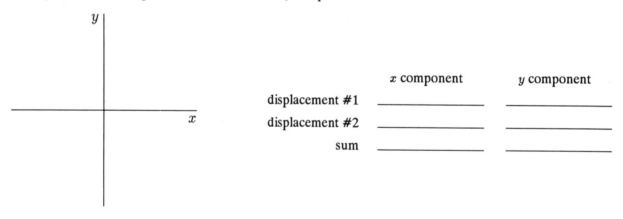

Now calculate the magnitude of the resultant and the angle it makes with the x axis.

[ans: 6.4 m; 141°]

PROBLEM 2. A car undergoes three successive displacements. First it goes 35 mi in a direction 22° north of west, then it goes 30 mi straight north, and finally it goes 15 mi east. What is the magnitude of the resultant displacement and what is the angle it makes with the x axis?

SOLUTION: On the diagram take north to be in the positive y direction and east to be in the positive x direction. Sketch the three displacements and their sum, then fill in the table with the x and y components of the displacements and the x and y components of their sum.

34 Chapter 3: Vectors

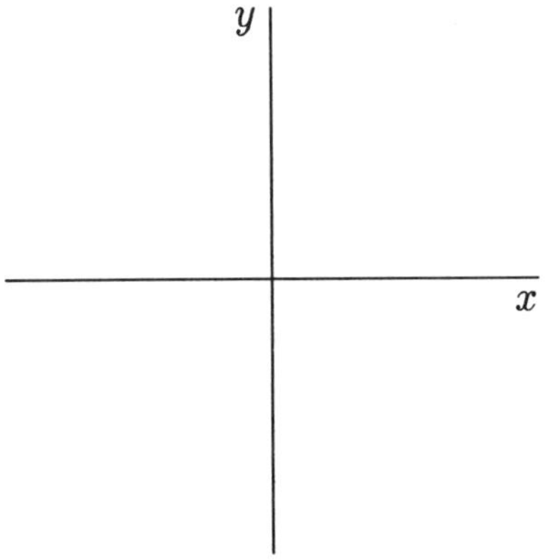

	x component	y component
displacement #1	_____	_____
displacement #2	_____	_____
displacement #3	_____	_____
sum	_____	_____

Now calculate the magnitude of the resultant and the angle it makes with the x axis.

[ans: 46.5 mi; 112°]

Vector subtraction can be used to find one of the vectors in a sum if the other vectors and their resultant are known. If $\mathbf{c} = \mathbf{a} + \mathbf{b}$, for example, then $\mathbf{a} = \mathbf{c} - \mathbf{b}$.

PROBLEM 3. A car undergoes two displacements and arrives at a place 65 mi east and 35 mi north of its starting point. The first displacement is 45 mi in a direction 67° north of east. What is the magnitude of the second displacement and what angle does it make with the positive x axis?

SOLUTION: On the diagram sketch the first displacement, the resultant displacement, and finally the second displacement. Use the table to calculate the components of the second displacement.

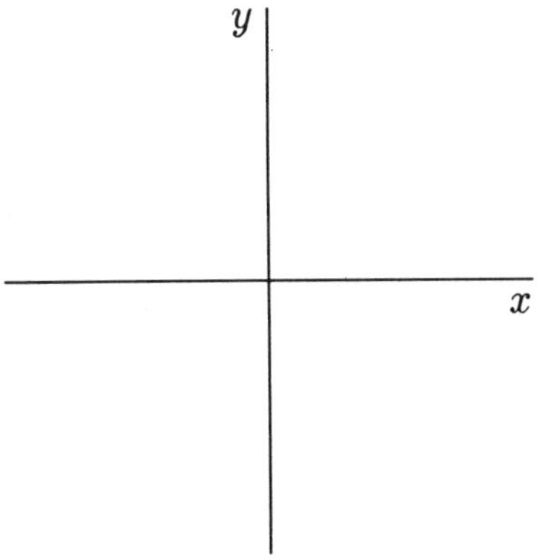

	x component	y component
displacement #1	_____	_____
resultant	_____	_____
displacement #2	_____	_____

Now calculate the magnitude of the second displacement and the angle it makes with the x axis.

[ans: 47.9 mi; −7.71°]

Chapter 3: Vectors

PROBLEM 4. Two vectors are given by $a = -6.0i + 4.0j$ and $b = 3.0i - 2.0j$. What are $a + b$, $a - b$, and $2.0a + 3.0b$?

SOLUTION:

[ans: $-3.0i + 2.0j$; $-9.0i + 6.0j$; $-3.0i + 2.0j$]

What is the magnitude of $2.0a + 3.0b$ and what angle does it make with the x axis?

SOLUTION:

[ans: 3.6; 146°]

Here's a problem you should know how to solve in preparation for the material in chapters 5 and 6.

PROBLEM 5. A crate is on a plane that makes an angle of 20° with the horizontal. Three forces act on it: the force of gravity F_g, which is 100 N downward, the normal force of the surface N, which is 94.0 N, perpendicular to the plane, and the force of friction f, which is 25 N, parallel to the plane. Forces are vectors and have SI units of newtons (abbreviated N).

Take the x axis to be horizontal and positive to the right and take the y axis to be vertical and positive upward. What are the components of the sum of the forces acting on the crate?

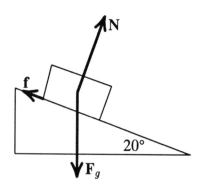

SOLUTION: You must recognize that N makes an angle of 20° with the vertical and f makes an angle of 20° with the horizontal. Use the following table to calculate the components of the total force.

	expression	value		expression	value
$F_{gx} =$	_____	= _____	$F_{gy} =$	_____	= _____
$N_x =$	_____	= _____	$N_y =$	_____	= _____
$f_x =$	_____	= _____	$f_y =$	_____	= _____
	sum = _____			sum = _____	

[ans: x: 8.7 N; y: -3.1 N]

Now take the x axis to be parallel to the plane with the positive direction down the plane and take the y axis to be perpendicular to the plane with the positive direction in the direction of N. Find the components of the sum of the forces.

SOLUTION: Use the following table.

	expression	value		expression	value
F_{gx} =	_____	= _____	F_{gy} =	_____	= _____
N_x =	_____	= _____	N_y =	_____	= _____
f_x =	_____	= _____	f_y =	_____	= _____
		sum = _____			sum = _____

[ans: x: 9.2 N; y: 0]

The scalar product of two vectors **a** and **b** can be written in two ways. The first is $\mathbf{a} \cdot \mathbf{b} = ab \cos \phi$, where ϕ is the angle between **a** and **b** when they are drawn with their tails at the same place. The second way is in terms of components: $\mathbf{a} \cdot \mathbf{b} = a_x b_x + a_y b_y + a_z b_z$. If you know the components you can find the angle between the vectors. Since $ab \cos \phi = a_x b_x + a_y b_y + a_z b_z$, $\phi = \arccos \left[(a_x b_x + a_y b_y + a_z b_z)/ab \right]$.

PROBLEM 6. Vector **a** has a magnitude of 20 units and is 37° counterclockwise from the positive x axis. Vector **b** has a magnitude of 15 units and is 132° counterclockwise from the positive x axis. What is their scalar product?

SOLUTION:

[ans: −26.1]

PROBLEM 7. Vector $\mathbf{a} = 5.5\mathbf{i} - 7.2\mathbf{j}$ and vector $\mathbf{b} = -3.5\mathbf{i} + 2.9\mathbf{j}$. What is their scalar product and what is the angle between them when they are drawn with their tails at the same point?

SOLUTION:

[ans: −40.1; 167°]

PROBLEM 8. Vector **a** has a magnitude of 20 units and is 37° counterclockwise from the positive x axis. Vector $\mathbf{b} = -3.5\mathbf{i} + 2.9\mathbf{j}$. What is their scalar product and what is the angle between them when they are drawn with their tails at the same point?

SOLUTION:

[ans: −21.0; 103°]

PROBLEM 9. Vector $\mathbf{a} = 5.5\mathbf{i} - 7.2\mathbf{j}$ and vector $\mathbf{b} = -3.5\mathbf{i} + 2.9\mathbf{j}$. What is the component of **a** along an axis that is parallel to **b**? Take the axis to be positive in the direction of **b**.

SOLUTION: Since the component of **a** along **b** is $a_b = a \cos \phi$ and $\mathbf{a} \cdot \mathbf{b} = ab \cos \phi$, the component can be written in terms of the scalar product: $a_b = \mathbf{a} \cdot \mathbf{b}/b$.

[ans: −8.83]

The magnitude of the vector product of two vectors is given by $|\mathbf{a} \times \mathbf{b}| = ab\sin\phi$, where ϕ is the angle between them when they are drawn with their tails at the same point. The direction of the vector product is given by a right hand rule. In terms of components $\mathbf{a} \times \mathbf{b} = (a_y b_z - a_z b_y)\mathbf{i} + (a_z b_x - a_x b_z)\mathbf{j} + (a_x b_y - a_y b_x)\mathbf{k}$.

PROBLEM 10. Vector **a** has a magnitude of 20 units and is 37° counterclockwise from the positive x axis. Vector **b** has a magnitude of 15 units and is 132° counterclockwise from the positive x axis. What is the magnitude and direction of the vector product $\mathbf{a} \times \mathbf{b}$?

SOLUTION:

[ans: 299; positive z direction]

What is the magnitude and direction of the vector product $\mathbf{b} \times \mathbf{a}$?

SOLUTION:

[ans: 299; negative z direction]

PROBLEM 11. Vector $\mathbf{a} = 5.5\mathbf{i} - 7.2\mathbf{j}$ and vector $\mathbf{b} = -3.5\mathbf{i} + 2.9\mathbf{j}$. What is the magnitude and direction of the vector product $\mathbf{a} \times \mathbf{b}$?

SOLUTION:

[ans: 9.25; in negative z direction]

To practice setting up problems, read the problems you have been assigned for homework, then fill in the table by naming the form in which the vectors are given (mag & dir or comp, say), the operation to be performed (addition, subtraction, multiplication by a scalar, scalar multiplication, or vector multiplication), the form you wish to write the vectors to perform the operation, and the form required for the answer. Vectors may be given in different forms in the same problem. If this is the case give both forms in the second column or just write "mixed". If the answer is a scalar just write "scalar" in the last column.

PROBLEM	GIVEN FORM	OPERATION	FORM FOR OPERATION	FORM FOR ANSWER

III. MATHEMATICAL SKILLS

Trigonometry is a large portion of the mathematics used in this chapter. Here is a listing of the important elements you should know very well.

1. The Pythagorean theorem: The square of the hypotenuse of a right triangle equals the sum of squares of the other two sides. In the diagram $C^2 = A^2 + B^2$. The theorem is true only if the triangle contains a right angle (90°). The theorem is used, for example, to calculate the magnitude of a vector in the xy plane given its x and y components.

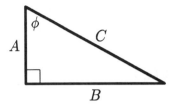

2. For the triangle shown $A = C\cos\phi$ and $B = C\sin\phi$. Remember these relations in the following form: the length of the side adjacent to an interior angle is the product of the hypotenuse and the cosine of the angle and the length of the opposite side is the product of the hypotenuse and the sine of the angle. The relations follow directly from the definition of the sine and cosine and are used to find the components of a vector, given the magnitude and the angle it makes with an axis.

Also know that for the triangle above $\tan\phi = B/A$. In a right triangle the tangent of an interior angle is the length of the opposite side divided by the adjacent side. This relationship, in the form $\phi = \arctan(a_y/a_x)$, is used to find the angle a vector makes with a coordinate axis.

WARNING! For any values of a_x and a_y the equation $\phi = \arctan(a_y/a_x)$ has two solutions for ϕ. If you use a calculator to evaluate ϕ, it will give the solution closest to 0. The other solution is the one given by the calculator plus or minus 180°. When solving for ϕ, first make a sketch of the vector pointing in roughly the right direction, so its components have the correct signs, as given, then check the answer displayed by your calculator. If necessary, add 180° to the calculator answer or subtract 180° from it to make the answer agree with the sketch. Alternatively, once ϕ has been found, calculate $a\cos\phi$ and $a\sin\phi$ to be sure they give the original components and not their negatives.

3. Memorize these special values of the trigonometric functions:

$\cos(0) = 1$ $\cos(90°) = 0$ $\cos(180°) = -1$ $\cos(270°) = 0$
$\sin(0) = 0$ $\sin(90°) = 1$ $\sin(180°) = 0$ $\sin(270°) = -1$
$\tan(0) = 0$ $\tan(90°) = \pm\infty$ $\tan(180°) = 0$ $\tan(270°) = \pm\infty$

Take special care that you don't get them confused.

4. You should also know the following trigonometric identities:

$\sin(-A) = -\sin A$
$\cos(-A) = \cos A$
$\sin(A+B) = \sin(A)\cos(B) + \cos(A)\cos(B)$
$\cos(A+B) = \cos(A)\cos(B) - \sin(A)\sin(B)$

Other trigonometric relationships are listed in an appendix of the text.

The identities given above are useful for evaluating $\sin(180° - \phi)$, $\sin(90° - \phi)$, $\sin(90° + \phi)$, $\sin(180° + \phi)$, and the corresponding cosines, for example. Using these relations you should be able to show that the components of the vector in the diagram are $a_x = a\cos(90° + \phi) = -a\sin\phi$ and $a_y = a\sin(90° + \phi) = a\cos\phi$.

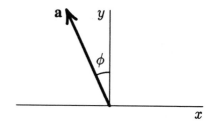

IV. NOTES

Chapter 4
MOTION IN TWO
AND THREE DIMENSIONS

I. BASIC CONCEPTS

The ideas of position, velocity, and acceleration that were introduced earlier in connection with one-dimensional motion are now extended. You should pay close attention to the definitions and relationships discussed in this chapter. There are three main topics: projectile motion, circular motion, and relative motion. First, however, some general kinematic concepts are discussed.

Definitions. Consider a particle moving in 2 or 3 dimensions. The fundamental concept used to describe its motion is its <u>position</u> <u>vector</u>. The tail of this vector is always at _____ _____ and at any instant the head is at _____. The cartesian components of the position vector are the coordinates of the particle. As the particle moves its position vector changes and so is a function of time.

A <u>displacement</u> <u>vector</u> is used to describe a change in a position vector. If the particle has position vector $\mathbf{r}_1$ at time t_1 and position vector $\mathbf{r}_2$ at a later time t_2 then the displacement vector for this interval is

$$\Delta \mathbf{r} =$$

If the particle has coordinates x_1, y_1, z_1 at time t_1 and coordinates x_2, y_2, z_2 at time t_2 then the components of the displacement vector are given by

$$\Delta r_x = \Delta x = \qquad \Delta r_y = \Delta y = \qquad \Delta r_z = \Delta z =$$

In writing these equations be sure to get the order of the subscripts right. A displacement vector is a position vector at a *later* time minus a position vector at an *earlier* time.

In terms of the displacement vector $\Delta \mathbf{r}$ the <u>average</u> <u>velocity</u> of the particle in the interval from t_1 to t_2 is

$$\mathbf{v}_{ave} =$$

The average velocity has components given by

$$v_{ave\,x} = \qquad v_{ave\,y} = \qquad v_{ave\,z} =$$

To use the definition to calculate the average velocity over the time interval from t_1 to t_2 you must know _____ for the beginning and end of the interval. Just as for one-dimensional motion, it is important to realize that the components of the position vector are coordinates and represent points on a coordinate axis. They are not necessarily related in any way to the distance traveled by the particle.

The <u>instantaneous velocity</u> **v** at any time t is the limit of the average velocity over a time interval that includes t, as the duration of the interval becomes vanishingly small. In terms of the position vector, it is given by the derivative

$$\mathbf{v} =$$

In terms of the particle coordinates its components are

$$v_x = \qquad\qquad v_y = \qquad\qquad v_z =$$

You should be aware that the instantaneous velocity, unlike the average velocity, is associated with a single instant of time. At any other instant, no matter how close, the instantaneous velocity might be different. The term "instantaneous" is usually implied: "velocity" means "instantaneous velocity".

To use the definition to calculate the instantaneous velocity you must know the position vector as a function of time. This is identical to knowing the coordinates as functions of time. The information may be given in algebraic form or as a graph. You should remember that the instantaneous velocity vector at any time is tangent to the path at the position of the particle at that time. If you are asked for the direction the particle is traveling at a certain time, you automatically calculate the components of its _____ for that time. <u>Speed</u> is the magnitude of the instantaneous velocity and, if the velocity components are given, can be calculated using $v =$ _____ .

In terms of the velocity $\mathbf{v}_1$ at time t_1 and the velocity $\mathbf{v}_2$ at a later time t_2 the <u>average acceleration</u> over the interval from t_1 to t_2 is given by

$$\mathbf{a}_{\text{ave}} =$$

In terms of velocity components the components of the average acceleration are

$$a_{\text{ave }x} = \qquad\qquad a_{\text{ave }y} = \qquad\qquad a_{\text{ave }z} =$$

To use the definition to calculate the average acceleration over the interval from t_1 to t_2 you must know _____
_____

The <u>instantaneous acceleration</u> **a** at any time t is the limit of the average acceleration over an interval that includes t, as the duration of the interval becomes vanishingly small. In terms of the velocity vector, it is given by

$$\mathbf{a} =$$

In terms of the velocity components its components are

42 *Chapter 4: Motion in Two and Three Dimensions*

$a_x =$ \hspace{2cm} $a_y =$ \hspace{2cm} $a_z =$

The terms "instantaneous acceleration" and "acceleration" mean the same thing. To use the definition to calculate the acceleration you must know the velocity vector as a function of time.

A non-zero velocity indicates that the _____ vector of the particle is changing with time. A non-zero acceleration indicates that the _____ vector of the particle is changing with time. Remember that these changes may be changes in magnitude, in direction, or both. Describe a possible motion in which the magnitude of the position vector does not change but the velocity does not vanish: _____

Describe a possible motion in which the speed does not change but the acceleration does not vanish: _____

Motion with constant acceleration. The vector equations that describe a particle moving with constant acceleration **a** are

$$\mathbf{r}(t) =$$

$$\mathbf{v}(t) =$$

In component form:

$x(t) =$ \hspace{2cm} $v_x(t) =$

$y(t) =$ \hspace{2cm} $v_y(t) =$

$z(t) =$ \hspace{2cm} $v_z(t) =$

Define the symbols you have used: _____

Projectile motion. Projectile motion with negligible air resistance is an important example of constant acceleration kinematics in 2 dimensions. The acceleration (magnitude and direction) of a particle in projectile motion with negligible air resistance is _____.

Specialize the constant acceleration equations to the case of a projectile near the surface of the earth, moving with negligible air resistance. Assume the projectile moves in the xy plane and that the y axis is vertical with the positive direction upward.

$x(t) =$ \hspace{2cm} $v_x(t) =$

$y(t) =$ \hspace{2cm} $v_y(t) =$

Chapter 4: Motion in Two and Three Dimensions

Check to be sure these equations are consistent with the magnitude and direction of the acceleration. Note that the horizontal component of the velocity does not change as the projectile moves. Also note carefully that once the projectile is fired its acceleration does not change until it hits the ground or a target.

Sometimes the initial speed v_0 and the launch angle ϕ_0 are known, rather than the x and y components of the initial velocity. Describe the launch angle: _____

_____

If the initial speed and launch angle are given, components of the initial velocity can be found using $v_{0x} =$ _____ and $v_{0y} =$ _____ .

The trajectory of a projectile is shown below. Suppose $t = 0$ when the projectile is launched. On the graph label the initial coordinates x_0 and y_0. Draw an arrow to show the initial velocity and label it $\vec{v}_0$. Label the launch angle. Draw an arrow to show the velocity of the projectile just before it hits the target and label it $\vec{v}_f$. Draw both the velocity and acceleration vectors at the point marked with a dot and at the highest point on the trajectory and label them $\vec{v}$ and $\vec{a}$, as appropriate.

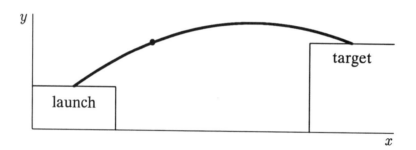

There are two special conditions you should remember. At the highest point of its trajectory the velocity of a projectile is horizontal and $v_y =$ _____ . When a projectile returns to the original launch height, $y =$ _____ . The magnitude of the displacement from the launch point to this point is called the _____ .

Uniform circular motion. Uniform circular motion, in which a particle moves around a circle with constant speed, is the second important example of motion in a plane. Remember that the velocity vector is always tangent to the path and therefore continually changes direction. This means the acceleration is *not* zero.

If the radius of the circle is r and the speed is v, the acceleration of the particle has magnitude

$$a =$$

and always points toward _____ . This means the direction of the acceleration continually changes as the particle moves around the circle. The term "centripetal acceleration" is applied to this acceleration to indicate its direction. You should not forget it is the rate of change of velocity, as are all accelerations.

Suppose a particle travels counterclockwise with constant speed around the circular path shown to the right. At each of the points A, B, and C draw a vector that gives the direction of its velocity and another that gives the direction of its acceleration. Label the vectors **v** and **a**, as appropriate.

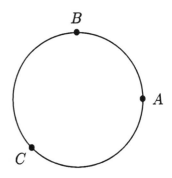

You should be aware that the magnitude of the acceleration is proportional to the *square* of the speed. If, for example, the speed is doubled without changing the orbit radius, the acceleration increases by a factor of 4.

Non-uniform circular motion. If the acceleration and velocity are always perpendicular to each other, as they are for uniform circular motion, the speed does *not* change. If the acceleration has a vector component in the direction of the velocity ($\mathbf{a} \cdot \mathbf{v} > 0$) then the speed increases and if the acceleration has a vector component in the direction opposite to the velocity ($\mathbf{a} \cdot \mathbf{v} < 0$) then the speed decreases. Illustrate these three cases below by drawing appropriately oriented acceleration vectors with tails at the particle. Label them **a**.

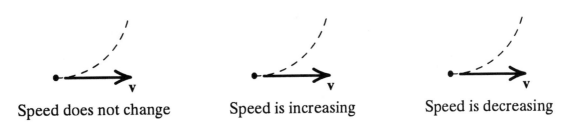

Speed does not change   Speed is increasing   Speed is decreasing

Be sure to remember when the speed changes and when it does not. The ideas will be useful when you study work and energy.

Suppose that the speed is given by some function $v(t)$. Then at time t the radial component of its acceleration is given by $a_R = $ _____ and the tangential component is given by the derivative $a_T = $ _____. You should be aware that if a_T is not zero then the speed and the magnitude of the radial component of acceleration both change. To give an example, suppose the tangential component a_T is constant in magnitude and the particle is speeding up as it moves around a circle of radius r. If it starts from rest, its speed as a function of time is given by $v(t) = $ _____ and the radial component of its acceleration is given by $a_R = $ _____. Because one component changes with time the magnitude of the acceleration is also time dependent. In fact, for this special case, the magnitude as a function of time is given by $a(t) = $ _____.

Relative motion. This topic deals with a comparison of the values obtained when the position, velocity, or acceleration of a particle is measured using two coordinate systems (or reference frames) that are moving relative to each other. Given the position, velocity, and acceleration of the particle in one frame and the relative motion of the frames, you should be able to calculate the position, velocity, and acceleration in the other frame.

According to the diagram on the right, at any instant of time the position of the particle relative to the origin of the unprimed coordinate frame is given by $\mathbf{r}_{PS}$ = _____ in terms of its position $\mathbf{r}_{PS'}$ relative to the primed origin and the position vector $\mathbf{r}_{S'S}$ of the primed origin relative to the unprimed origin.

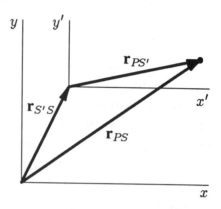

The above expression can be differentiated with respect to time to obtain $\mathbf{v}_{PS}$ = _____ for the velocity of the particle in the unprimed frame in terms of the particle velocity $\mathbf{v}_{PS'}$ in the primed frame and the velocity $\mathbf{v}_{S'S}$ of the primed frame as measured in the unprimed frame. This expression is valid even if the two frames are accelerating relative to each other.

Take special care with the subscripts. The first names an object and the second names the coordinate frame used to measure the position or velocity of the object. You should say all the words as you read the symbols. That is, when you see $\mathbf{r}_{S'S}$ you should say "the position vector of the origin of frame S' relative to the origin of frame S." You will then get acquainted with the notation fast and won't get it mixed up later.

The expression for the velocity can be differentiated with respect to time to obtain $\mathbf{a}_{PS}$ = _____ for the acceleration of the particle in the unprimed frame in terms of the particle acceleration $\mathbf{a}_{PS'}$ in the primed frame and the acceleration $\mathbf{a}_{S'S}$ of the primed frame as measured in the unprimed frame. Specialize this equation to the case when the two frames are *not* accelerating with respect to each other: $\mathbf{a}_{PS}$ = _____. Now the acceleration of the particle is the same in both frames.

If a particle has a different velocity in two reference frames, then the frames must be moving relative to each other. Similarly, if a particle has a different acceleration in two frames, then the frames must be accelerating relative to each other.

To remember what each of the symbols mean, invent a specific example that you can visualize easily. Here's one you might try. Draw an airplane on a small piece of paper and a straight line across a larger piece of paper. Lay the large paper on a table and move the airplane along the line with constant speed. The large paper represents the air and what you see is the motion of the airplane relative to the air. The speed of the airplane relative to the air can be found simply by dividing the length of the line by the time the airplane takes to fly it. Now move the large paper across the table top with constant velocity as you move the airplane along the line on the paper. You are now viewing the airplane from the ground as the wind blows. Its ground speed can be found by measuring the distance it moves on the table and dividing by the time. You might mark the starting and ending positions of the airplane with small pieces of tape on the table, then measure the distance between the pieces of tape. Convince yourself that the velocity of the airplane relative to the ground is the vector sum of its velocity relative to the air and the velocity of the air relative to the ground. Try various directions for the airplane and wind velocities.

Airplanes flying in moving air or ships sailing in moving water are often used as examples of relative motion. The airplane or ship is the particle, one coordinate frame moves with the air or water, and the other coordinate frame is fixed to the earth. The heading of the airplane or ship is in the direction of its velocity as measured relative to the air or water, *not* relative to the ground. Use your paper airplane to convince yourself of this. Its long axis is parallel to the line on the moving paper and is not necessarily parallel to the line of motion on the table.

The diagram on the right shows the velocity $\mathbf{v}_{PA}$ of an airplane relative to the air and the velocity $\mathbf{v}_{AG}$ of the air relative to the ground. Draw the vector that represents the velocity $\mathbf{v}_{PG}$ of the airplane relative to the ground. Near the midpoint of this vector draw a small airplane oriented correctly; that is, with its long axis parallel to $\mathbf{v}_{PA}$.

Relative motion at high speeds. If an object is moving at nearly the speed of light or if we compare its velocity as measured in two reference frames that are moving at nearly the speed of light relative to each other, then the results given above fail and we must use a relativistically correct equation. Consider an object P that is moving along the x axis with velocity $v_{PS'}$, as measured in reference frame S'. If S' is moving along the x axis with velocity $v_{S'S}$, as measured in another frame S, the velocity of P, as measured in S, is given by

$$v_{PS} =$$

Here c is _____ and has the value _____ m/s.

You should be able to show that if $v_{PS'} = c$, then $v_{PS} = c$. If something moves with the speed of light in one frame then it moves with the speed of light in all frames. You should also be able to show that if $v_{PS'}$ and $v_{S'S}$ are both much less than the speed of light then the correct non-relativistic result is obtained.

II. PROBLEM SOLVING

The equations $\mathbf{r}(t) = \mathbf{r}_0 + \mathbf{v}_0 t + \frac{1}{2}\mathbf{a}t^2$ and $\mathbf{v}(t) = \mathbf{v}_0 + \mathbf{a}t$, in component form, can be used to solve all constant acceleration problems, no matter what the number of dimensions. For two-dimensional problems there are 4 equations containing 11 algebraic symbols. For many problems only the displacement $\mathbf{r} - \mathbf{r}_0$ is given or asked for. The constant acceleration equations then contain 9 algebraic symbols in two dimensions. In any event, you should be able to solve the equations simultaneously for 4 quantities.

In a strictly mechanical way most kinematics problems can be solved by first identifying the quantities that are given in the problem statement, identifying the unknowns, then systematically solving the kinematic equations. The most straightforward problems give the time, initial conditions, and acceleration, then ask for the quantities on the left sides of the kinematic equations.

PROBLEM 1. Suppose a particle starts at the origin with velocity $\mathbf{v}_0 = -3.5\,\mathbf{i} + 4.7\,\mathbf{j}$, in m/s. Its acceleration is $2.1\,\mathbf{i} + 1.1\,\mathbf{j}$, in m/s². What are its coordinates and velocity 5.0 s later?

SOLUTION: Simply substitute $t = 5.0\,\text{s}$, $v_{0x} = -3.5\,\text{m/s}$, $v_{0y} = 4.7\,\text{m/s}$, $a_x = 2.1\,\text{m/s}^2$, and $a_y = 1.1\,\text{m/s}^2$ into $x(t) = v_{0x}t + \frac{1}{2}a_x t^2$, $y(t) = v_{0y}t + \frac{1}{2}a_y t^2$, $v_x(t) = v_{0x} + a_x t$, and $v_y(t) = v_{0y} + a_y t$.

[ans: $\mathbf{r} = 8.75\,\mathbf{i} + 37.25\,\mathbf{j}$ in m; $\mathbf{v} = 7.00\,\mathbf{i} + 10.2\,\mathbf{j}$ in m/s]

Once the velocity components are known you can calculate the speed and the angle between the path and the positive x axis. What are they at $t = 5.0\,\text{s}$ for this particle?

SOLUTION:

[ans: 12.4 m/s; 55.5°]

For any problem you may place the origin of the coordinate system anywhere. Convenient places are usually the initial position of the particle (its position at time $t = 0$) or its final position (where it lands, for example). For a projectile with negligible air resistance the acceleration is known, although it may not be stated explicitly. It has magnitude g and is directed toward the earth. When the highest point on the trajectory is mentioned (either as a known or unknown) you should immediately identify it as the place where the vertical component of the velocity momentarily vanishes. You should also recognize the horizontal range as the distance from the launch point to the point where the projectile returns to the launch altitude (i.e. $x = R$ when $y = y_0$ for the second time). Some problems ask for the angle between the path of the particle at some instant and the x axis, say. You should immediately recognize that you must solve for the velocity components, then use $\tan\phi = v_y/v_x$ to solve for ϕ.

Let us analyze some typical kinematics problems in detail.

PROBLEM 2. A projectile is fired with an initial velocity of 55 m/s, 25° above the horizontal. How long does it take to get to the highest point on its trajectory? How far above the launch point is the highest point? How far down range is the highest point?

SOLUTION: Kinematics problems usually describe two events. For this problem the first is the launching of the projectile. Data for the initial event consists of the time it takes place, the coordinates of the place where it takes place, and the components of the velocity. We will take the time to be zero when the projectile is fired. You will have to specify the placement of the coordinate system before you can write down the coordinates of the event. For this problem place the origin at the launch point and take the positive y axis to be upward. As for this problem you may have to calculate the velocity components from the speed and launch angle. Fill in each of the following blanks with a numerical value:

$x_0 =$ _____ m $\qquad v_{0x} =$ _____ m/s $\qquad t =$ _____ s

$y_0 =$ _____ m $\qquad v_{0y} =$ _____ m/s

The problem typically includes a second event: in this case it is the projectile reaching the highest point on its trajectory. Data for the second event consists of the time at which it took place, the coordinates of the place where it took place, and the components of the velocity. Fill in the following table for this problem. Use numerical values for known quantities and question marks for unknown values.

$x(t) =$ _____ m $\qquad v_x(t) =$ _____ m/s $\qquad t =$ _____ s

$y(t) =$ _____ m $\qquad v_y(t) =$ _____ m/s

You will also need to know the acceleration components. For this problem

$a_x =$ _____ m/s² $\qquad a_y =$ _____ m/s²

We want to solve for t, $x(t)$, and $y(t)$. First solve $v_y(t) = v_{0y} - gt$ for t, then substitute the value into $x(t) = x_0 + v_{0x}t$ and $y(t) = y_0 + v_{0y}t - \frac{1}{2}gt^2$.

[ans: 2.37 s; 118 m; 27.8 m]

Suppose that after reaching its highest point the projectile lands on a plateau 20 m above the launch point. How long is it in flight? How far down range is the landing point? What is its velocity just before it lands?

SOLUTION: We take the first event to be the same as before. The second event is the landing of the projectile. Fill in the following table:

$x(t) =$ _____ m $\qquad v_x(t) =$ _____ m/s $\qquad t =$ _____ s

$y(t) =$ _____ m $\qquad v_y(t) =$ _____ m/s

The unknowns are t, $x(t)$, $v_x(t)$, and $v_y(t)$. Solve $y(t) = v_{0x}t - \frac{1}{2}gt^2$ for t. This is a quadratic equation and has two solutions, corresponding to the two times the projectile is 20 m above the launch point, once on the way up and once on the way down. We want the greater of the two solutions. Once t is found substitute its value into $x(t) = v_{0x}t$ and $v_y(t) = v_{0y} - gt$. Note that v_x does not depend on t. It is the same as v_{0x} for all values of the time.

[ans: $t = 3.61$ s; $x = 180$ m; $v_x = 49.8$ m/s; $v_y = -12.2$ m/s]

Chapter 4: Motion in Two and Three Dimensions

Since we know the velocity components we can now calculate the speed of the projectile just before it lands and the angle its path makes with the horizontal as it lands.

SOLUTION:

[ans: 51.3 m/s; −13.8°]

Although a constant acceleration problem, once set up, can be completed with brute force algebra, a solution can usually be facilitated by giving some thought to the problem. Think of the kinematic equations for $\mathbf{r}(t)$ and $\mathbf{v}(t)$ as telling us the position and velocity of a particle at any given time. Usually one of conditions for an event is sufficient to tell us the time of the event, even if the time is not given explicitly. The problem just worked contains two examples: the highest point is reached when $v_y(t) = 0$ and the projectile reaches some given height h when $y(t) = h$. These equations, or another appropriate equation, can be solved for the time t of the event. Once t has been obtained its value can be substituted into the other kinematic equations to answer the question posed by the problem.

This is a useful way to think about the equations even if the initial conditions are among the unknowns. An algebraic expression is obtained for the time of the second event rather than a numerical value and the expression is substituted into the kinematic equations. The result is an equation that can be solved for an unknown in the problem.

PROBLEM 3. A ball is thrown at an angle of 40° above the horizontal. It passes downward through a horizontal hoop 2.0 m above the throwing point and 12 m down range. What was the initial speed of the ball?

SOLUTION: Select the throwing point to be the origin of the coordinate system and take the time to be zero when the ball is thrown. Take the x axis to be horizontal and the y axis to be vertical. Fill in the table for the first event, the throwing of the ball:

$x_0 = $ _____ m $v_0 = $ _____ m/s $t = $ _____ s

$y_0 = $ _____ m $\phi_0 = $ _____

Notice that since the problem gives the throwing angle ϕ_0 and asks for the initial speed v_0 we have decided to formulate the problem in terms of v_0 and ϕ_0, rather than v_{0x} and v_{0y}.

Fill in the table for the second event, the passing of the ball through the hoop:

$x(t) = $ _____ m $v_x(t) = $ _____ m/s $t = $ _____ s

$y(t) = $ _____ m $v_y(t) = $ _____ m/s

The unknowns are t, v_0, $v_x(t)$, and $v_y(t)$. We wish to eliminate t and solve for v_0 in terms of $x(t)$, $y(t)$, and ϕ_0. Use $x(t) = v_0 t \cos\phi_0$ to derive an expression for t in terms of v_0 and substitute the result into $y(t) = v_0 t \sin\phi_0 - \frac{1}{2}gt^2$. Find a numerical value for v_0.

[ans: 12.2 m/s]

The problem might also ask for the time of flight, the speed of the ball as it passed through the hoop, and the angle its path made with the horizontal as it passed through the hoop. Now use $x(t) = v_0 t \cos\phi_0$ to derive an expression for v_0 in terms of t, substitute the result into $y(t) = v_0 t \sin\phi_0 - \frac{1}{2}gt^2$, and solve for t. Substitute the result into $v_y(t) = v_{0y} - gt$. Also substitute $v_0 \cos\phi_0$ for v_x. Finally calculate v and ϕ.

SOLUTION:

[ans: 1.28 s; 10.5 m/s; −26.8°]

PROBLEM 4. An outfielder can throw a baseball at 35 m/s. At what angle above the horizontal should he throw it if we wants it to be caught by an infielder 95 m away? Assume the throwing and catching heights are the same.

SOLUTION: Put the origin of a coordinate system at the point where the ball is thrown and take the positive y axis to be upward. Let $t = 0$ when the ball is thrown. Fill in the following table for the first event, the throwing of the ball. Notice that, because we are asked for ϕ_0, v_0 and ϕ_0 are used rather than v_{0x} and v_{0y}.

$x_0 = $ _____ m $v_0 = $ _____ m/s $t = $ _____ s

$y_0 = $ _____ m $\phi_0 = $ _____

Fill in the following table for the second event, the catching of the ball.

$x(t) = $ _____ m $v_x(t) = $ _____ m/s $t = $ _____ s

$y(t) = $ _____ m $v_y(t) = $ _____ m/s

The unknowns are ϕ_0, t, $v_x(t)$, and $v_y(t)$. We are asked for ϕ_0. Solve $x(t) = x_0 + v_0 t \cos\phi_0$ for t and substitute the resulting expression into $y(t) = y_0 + v_0 t \sin\phi_0 - \frac{1}{2}gt^2$. After a little manipulation you should get $\sin\phi_0 \cos\phi_0 = xg/2v_0^2$.

You can solve for ϕ_0 using trial and error. Evaluate the right side of the equation. Then choose various values for ϕ_0 and calculate $\sin\phi_0 \cos\phi_0$ for each of them. Choose the values systematically. Find two values that bracket $xg/2v_0^2$, one giving a result that is too high and the other giving a result that is too low. Then pick a value between and try again.

You can use a better method if you remember the trigonometric identity $\sin(2\phi_0) = 2\sin\phi_0 \cos\phi_0$. The equation to be solved becomes $\sin(2\phi_0) = xg/v_0^2$. This can easily be solved for $2\phi_0$ and finally for ϕ_0 itself.

[ans: 24.7°]

Problems dealing with motion in a circle are relatively straightforward. Your first job is to determine if the speed is constant. Look for clues in the problem statement. If the speed is constant only three quantities come into play. They are the acceleration a, the speed v, and the circle radius r. Two must be given, then you can solve for the third, using $a = v^2/r$.

PROBLEM 5. A string tied to an object is used to whirl the object at 4.5 m/s in a horizontal circle with a 2.0-m radius. What is the magnitude of its acceleration?

SOLUTION:

[ans: 10.1 m/s^2]

PROBLEM 6. As a small airplane goes around a 500-m radius circle it experiences an inward acceleration of g. What is its speed?

SOLUTION:

[ans: 70.0 m/s]

If the motion is not uniform the tangential component of the acceleration is also involved. A second equation, $a_T = \pm dv/dt$, relates this component to rate at which the speed is changing. Be careful about signs here. Which you choose depends on whether the particle is speeding up or slowing down and on the direction you have chosen to be positive. If the particle is speeding up then the vector component $\mathbf{a}_T$ is in the same direction as the velocity; if the particle is slowing down then it is in the opposite direction. At any rate, the motion is described by 2 equations containing 4 quantities (r, v, a_R, and a_T). The equations are solved simultaneously.

PROBLEM 7. A particle goes around a 1.7-m radius circle, its speed in m/s being given by $v = 0.53t^2$, where t is the time in seconds. At $t = 2.0$ s what are the tangential and radial components of its acceleration?

SOLUTION: Use $a_T = dv/dt$ to find an expression for a_T as a function of time, then evaluate it for $t = 2.0$ s. Also evaluate $v = 0.53t^2$ to find a value for the speed at $t = 2.0$ s, then use $a_R = v^2/r$ to find a value for the radial component of the acceleration at that time.

[ans: $a_T = 2.12 \text{ m/s}^2$; $a_R = 2.64 \text{ m/s}^2$]

The radial and tangential vector components of $\mathbf{a}$ are perpendicular to each other so $a^2 = a_R^2 + a_T^2$. Calculate the magnitude of the acceleration of the particle at $t = 2.0$ s.

SOLUTION:

[ans: $3.39\,\text{m/s}^2$]

At what time will the radial and tangential components of the acceleration have equal magnitude and what is that magnitude?

SOLUTION:

[ans: $1.86\,\text{s}$; $1.97\,\text{m/s}^2$]

Relative motion problems are largely exercises in vector addition and subtraction. If $\mathbf{v}_{PS}$ is the velocity of a particle as measured in reference frame S and $\mathbf{v}_{PS'}$ is its velocity in reference frame S' then the two velocities are related by $\mathbf{v}_{PS} = \mathbf{v}_{PS'} + \mathbf{v}_{S'S}$, where $\mathbf{v}_{S'S}$ is the velocity of frame S' as measured in frame S.

In general terms most relative motion problems are solved by writing $\mathbf{v}_{PS} = \mathbf{v}_{PS'} + \mathbf{v}_{S'S}$ in component form. First pick coordinate axes. Make the x axes for the two frames parallel to each other and the y axes parallel to each other. All of the vectors are in the same plane so only two components are necessary. Most likely the vectors will be given as speeds and directions so you must convert to component form. The two component equations contain 6 quantities: 4 must be given, then values can be found for the other 2. Usually the unknowns can be found by inspection of the triangle formed by the vectors and a little trigonometry. Only rarely must you resort to brute force. At any rate, the 2 equations are solved simultaneously for the unknowns.

PROBLEM 8. Suppose a person on a train traveling forward at 35 mi/h walks toward the rear at 5 mi/h, relative to the train. What is the velocity of the person relative to the earth?

SOLUTION: A little thought reveals that it is 30 mi/h forward. Let us see how the formalism gives this answer. Let frame S be attached to the earth and frame S' be attached to the train. Both have their x axes parallel to the train's velocity, with the positive axis in the forward direction. Then the velocity of the person relative to the train is $\mathbf{v}_{PS'} = -5\,\mathbf{i}\,\text{mi/h}$ and the velocity of the train relative to the earth is $\mathbf{v}_{S'S} = +35\,\mathbf{i}\,\text{mi/h}$. Thus $\mathbf{v}_{PS} = -5\,\mathbf{i} + 35\,\mathbf{i} = +30\,\mathbf{i}\,\text{mi/h}$.

PROBLEM 9. A supertanker is heading eastward, traveling at 14 mi/h relative to the water, which is moving northward at 6.0 mi/h relative to the earth. What is the speed of the tanker relative to the earth? In what direction is the tanker traveling relative to the earth?

SOLUTION: Let frame E be attached to the earth and frame W be attached to the water. Take both positive x directions to be eastward and both positive y directions to be northward. Since the tanker is heading eastward its velocity, relative to the water is in the positive x direction, so $v_{TW} = $ _____. The velocity of the water relative to the earth is $v_{WE} = $ _____. In the space to the right draw a vector addition diagram showing v_{TE}, v_{TW}, and v_{WE}. Now find the velocity of the tanker relative to the earth. Finally calculate its speed and the angle the velocity makes with the x axis.

[ans: 15.2 mi/h; 23.2° N of E]

PROBLEM 10. An airplane can fly at 250 mi/h in still air. If the wind is blowing toward the north at 60 mi/h, in what direction should the airplane head in order to fly directly east? What is the speed of the airplane relative to the earth? Don't forget that the heading of the airplane gives the direction of its velocity relative to the air.

SOLUTION: Let v_{PE} be the velocity of the airplane relative to the earth, v_{PA} be the velocity of the airplane relative to the air, and v_{AE} be the velocity of the air relative to the earth. Take the positive x direction to be toward the east and the positive y direction to be toward the north. Draw a vector diagram showing the vector addition $v_{PE} = v_{PA} + v_{AE}$. Clearly the airplane must head slightly south of east. Let θ be the angle between the airplane's heading and the eastward direction. Then $v_{PE}\mathbf{i} = v_{PA}\mathbf{i}\cos\theta - v_{PA}\mathbf{j}\sin\theta + v_{AE}\mathbf{j}$. The two component equations are $v_{PE} = v_{PA}\cos\theta$ and $0 = -v_{PA}\sin\theta + v_{AE}$. Solve the second for θ and the first for v_{PE}.

[ans: 13.9° S of E; 243 mi/h]

III. COMPUTER PROJECTS

The program you used for Chapter 2 projects can be revised to investigate two-dimensional motion. You must now supply two components of the acceleration as functions of t and have the computer carry out calculations for each of the two components of the velocity and position vectors. The outline of a sample program might be:

input initial values: $t_0, x_0, y_0, v_{0x}, v_{0y}$
input final time and interval width: $t_f, \Delta t$
set $t_b = t_0, x_b = x_0, y_b = y_0, v_{xb} = v_{0x}, v_{yb} = v_{0y}$
calculate acceleration at beginning of first interval: $a_{xb} = a_x(t_b), a_{yb} = a_y(t_b)$
begin loop over intervals
    calculate time at end of interval: $t_e = t_b + \Delta t$
    calculate acceleration at end of interval: $a_{xe} = a_x(t_e), a_{ye} = a_y(t_e)$
    calculate "average" acceleration: $a_x = (a_{xb} + a_{xe})/2, a_y = (a_{yb} + a_{ye})/2$
    calculate velocity at end of interval: $v_{xe} = v_{xb} + a_x \Delta t, v_{ye} = v_{yb} + a_y \Delta t$
    calculate "average" velocity: $v_x = (v_{xb} + v_{xe})/2, v_y = (v_{yb} + v_{ye})/2$
    calculate coordinates at end of interval: $x_e = x_b + v_x \Delta t, y_e = y_b + v_y \Delta t$
*   **if** $t_e \geq t_f$ then
        print or display $t_b, x_b, y_b, v_{xb}, v_{yb}$
        print or display $t_e, x_e, y_e, v_{xe}, v_{ye}$
        exit loop
    end if statement
    set $t_b = t_e, x_b = x_e, y_b = y_e, v_{xb} = v_{xe}, v_{yb} = v_{ye}, a_{xb} = a_{xe}, a_{yb} = a_{ye}$
end loop over intervals
stop

As before, the line with the asterisk may be changed for other applications.

Instead of v_{0x} and v_{0y} you may wish to input the initial speed v_0 and angle ϕ_0 between the velocity and the x axis, then have the computer calculate v_{0x} and v_{0y} using $v_{0x} = v_0 \cos \phi_0$ and $v_{0y} = v_0 \sin \phi_0$. You may also wish to define $a_x(t)$ and $a_y(t)$ in a separate section of the program.

PROJECT 1. Start with a projectile motion problem that can be solved analytically. Suppose a projectile is fired over level ground at 350 m/s, at an angle of 25° above the horizontal. When does it reach the highest point? How high and how far down range is the highest point? When does it hit the ground and what is the range? Compare your answers with the analytic solutions.

If the x axis is horizontal and the y axis is positive in the upward direction then the components of the acceleration are given by $a_x = 0$ and $a_y = -9.8 \text{ m/s}^2$. To find the highest point exit the loop when $v_{ye} \leq 0$. To find the range exit the loop when $y_e \leq 0$. You must experiment a little to find an appropriate value for Δt. Start with 0.1 s and reduce it by a factor of 5 in successive calculations. Stop when you get the same results to 3 significant figures.

PROJECT 2. Now suppose the acceleration of the projectile has a horizontal component that varies with time, perhaps because a small rocket is fired horizontally. Take $a_x = 3.0t \text{ m/s}^2$ and $a_y = -9.8 \text{ m/s}^2$, where t is in seconds. Use the same initial conditions (an initial velocity of 350 m/s, 25° above the horizontal) and find the time the projectile reaches its highest point and the coordinates of the highest point, assuming it is fired from the origin. Find the range over level ground and the velocity of the object just before it lands. [ans: highest point: $t = 15.1$ s, $x = 6.51 \times 10^3$ m, $y = 1.12 \times 10^3$ m; range: $t = 30.2$ s, $x = 2.33 \times 10^4$ m, $v_x = 1.68 \times 10^3$ m/s, $v_y = -148$ m/s]

PROJECT 3. Now suppose the horizontal component of the acceleration is given in m/s² by $a_x(t) = 30te^{-t}$ while the y component is still $a_y = -9.8 \text{ m/s}^2$. The x axis is horizontal and the y axis is positive in the upward direction. The initial velocity is still 350 m/s, 25° above the horizontal. At what time does the object reach the highest point on its trajectory and what are the coordinates of that point? At what time does it return to the level of the firing point and what is the range? What are the components of its velocity just before landing? [ans:

highest point: $t = 15.1\,\text{s}$, $x = 5.18 \times 10^3\,\text{m}$, $y = 1.12 \times 10^3\,\text{m}$; range: $t = 30.2\,\text{s}$, $x = 1.04 \times 10^4\,\text{m}$; velocity: $v_x = 347\,\text{m/s}$, $v_y = -148\,\text{m/s}$]

IV. NOTES

Chapter 5
FORCE AND NEWTON'S LAWS

I. BASIC CONCEPTS

The emphasis now changes from kinematics to dynamics as you start to study how objects influence the motion of each other. This is the central chapter of the mechanics section of the text. Be sure you understand the concepts of force and mass and pay particular attention to the relationship between the net force on an object and its acceleration.

Dynamics. The fundamental problem of dynamics is to find the acceleration of an object, given the object and its environment.

The problem is split into two parts, connected by the idea of a force: the environment of an object exerts forces on the object and the net force on it causes it to accelerate. The first part of the problem is to find the net force exerted on the object, given the relevant properties of the object and its environment. An expression for a force in terms of the properties of interacting objects is called a <u>force</u> law. The second part of the problem is to find the acceleration of the object, given the net force. In this chapter we concentrate on the second part although, of necessity, some force laws are discussed. Other force laws are described in detail at appropriate points in the text.

Newton's first law. The text gives two statements of Newton's first law. Write both of them and learn them: _law of Inertia_

Statement #1: _Consider a body on which no net force acts. If the body is at rest, it will remain at rest. If the body is moving with constant velocity, it will continue to do so._

Statement #2: _If the net force acting on a body is zero, then it is possible to find a set of reference frames in which that body has no acceleration._

The first statement in the text is closer to Newton's words, the second is closer to the modern interpretation of the law.

The first law helps us define <u>inertial reference frames</u>. Suppose we have found a particle on which zero net force acts and we attach reference frame S to it. Clearly the acceleration of the particle, as measured in S, is zero. Describe a reference frame S' in which the acceleration of the particle is not zero: _S is is changing in velocity._

Which of these frames is an inertial frame? _The first statement_
Describe another inertial frame: _no net force, constant velocity, no changes in velocity either magnitude or direction_

Newton's second law. This is the central law of classical mechanics. It gives the relationship between the net force $\sum \mathbf{F}$ acting on an object and the acceleration $\mathbf{a}$ of the object:

$$\sum \mathbf{F} = m\mathbf{a},$$

where m is the mass of the object. To understand the law you must understand the definitions of force and mass.

A <u>force</u> is measured, in principle, by applying it to the standard (1 kg) mass and measuring the _acceleration_ of the standard mass. If SI units are used, the magnitudes of these quantities are numerically equal. Both are vectors in the same direction. That forces obey the laws of vector addition can be checked by simultaneously applying two forces in different directions and verifying that the result is the same as when the resultant of the forces is applied as a single force. All measurements must be made using an inertial reference frame.

The SI unit of force is _Newton_ and is abbreviated _N_. In terms of the SI base units (kg, m, s) it is $kg\,m/s^2$.

The <u>mass</u> of an object is measured, in principle, by comparing the accelerations of the object and the standard mass when the same force is applied to them. In particular, the mass of the object is given by $m =$ _1 kg_, where a is the magnitude of the acceleration of the object, a_0 is the magnitude of the acceleration of the mass standard, and m_0 is the mass of the mass standard. The accelerations must be measured using an inertial frame.

Note that small masses acquire a _greater_ acceleration than large masses when the same force is applied. Mass is said to measure <u>inertia</u> or resistance to changes in motion.

Mass is a scalar and is always positive. In the non-relativistic realm the mass of two objects in combination is the sum of the individual masses.

Newton's second law $\sum \mathbf{F} = m\mathbf{a}$ is a vector equation. It is equivalent to the three component equations

$$\sum F_x = ma_x$$

$$\sum F_y = ma_y$$

$$\sum F_z = ma_z$$

You must be aware that in these equations $\sum \mathbf{F}$ is the *total* (or *net*) force acting on the object, the vector sum of all the individual forces. This means that in any given situation you must identify all the forces acting on the object and then sum them *vectorially*.

Note that $\sum \mathbf{F} = 0$ implies $\mathbf{a} = 0$. If the resultant force vanishes then the object does not accelerate; its velocity as observed in an inertial reference frame is constant in both magnitude and direction. The resultant force may vanish because no forces act on the object or because the forces that act sum to 0. For some situations you may know that 3 forces act but are given only 2 of them. If you also know that the acceleration vanishes you can solve $\mathbf{F}_1 + \mathbf{F}_2 + \mathbf{F}_3 = 0$ for the third force.

Newton's third law. Newton's third law tells us something about forces. If object A exerts a force $\mathbf{F}_{BA}$ on object B, then according to the third law, the force exerted by object B on object A is given by $\mathbf{F}_{AB} = $ _−$\mathbf{F}_{BA}$_. Compared to the force of A on B, the force of B on A is __equal__ in magnitude and __opposite__ in direction. You should also be aware that these two forces are of the same type. That is, if object A exerts a *gravitational* force on B, then B exerts a *gravitational* force on A.

The third law is useful in solving problems involving more than one object. If two objects exert forces on each other, we immediately use the same symbol to represent their magnitudes and, in writing the second-law equations, we remember the forces are in opposite directions. In addition, we remember that the forces act on different objects. When we want to write Newton's second law for object A, one of the forces we include is the force of B on A, but emphatically *NOT* the force of A on B. The force of A on B, in addition to the other forces on B, determines the acceleration of B, not A.

Gravitational force. One force law you will use a great deal gives the force of gravity on an object. This force is called the weight and its magnitude is given by $W = $ __mg__, where m is the mass of the object and g is the magnitude of the acceleration due to gravity at the position of the object. Near the surface of the earth the direction of the weight is __toward the center downward__. Be sure you understand that mass and weight are quite different concepts. Mass is a property of an object and does not change as the object is moved from place to place or even into outer space. It is a scalar. Weight, on the other hand, is a force. It varies as the object moves from place to place and vanishes when the object is far from all other objects, as in outer space. This is because **g**, not the mass, varies from place to place.

Remember that the weight of an object is m**g** regardless of its acceleration. Weight is a force and, if appropriate, is included in the sum of all forces acting on the object. This sum equals m**a** and if other forces act then **a** is different from **g**.

The SI unit of weight is _____.

II. PROBLEM SOLVING

Some problems deal with the definitions of force and mass. If a force is applied to the standard kilogram and, as a result, the standard kilogram has an acceleration $\mathbf{a}_s$ then the magnitude of the force in newtons is numerically equal to the magnitude of the acceleration. Force is a vector and is in the direction of the acceleration. If identical forces are applied to the standard kilogram and another object and their accelerations are $\mathbf{a}_s$ and $\mathbf{a}_o$, respectively, then the mass in kg of the object is given by $m_0 = a_s/a_o$.

PROBLEM 1. The magnitude of the acceleration of the standard kilogram is 4.0 m/s² when a certain force **F** is applied to it. When the same force is applied to a certain stone the magnitude of its acceleration is 5.5 m/s². What is the magnitude of the force and what is the mass of the stone?

SOLUTION:

$(1.00)\left(\dfrac{4.0 \text{ m/s}^2}{5.5 \text{ m/s}^2}\right) = .727 \text{ kg}$

$(1.00 \text{ kg})(4.0 \text{ m/s}^2) = 4.0 \text{ kg m/s}^2 = 4.0 \text{ N}$

[ans: 4.0 N; 0.727 kg]

A definite procedure has been devised to solve dynamics problems. It ensures that you consider only one object at a time, reminds you to include all forces acting on the object you are considering, and guides you in writing Newton's second law in an appropriate form. Follow it closely. Use the list below as a check list until the procedure becomes automatic.

1. Identify the object to be considered. It is usually the object on which the given forces act or about which a question is posed.

2. Make a sketch of the object being considered. Do not include the environment of the object since this is replaced by the forces it exerts on the object.

3. On the diagram draw arrows to represent the forces exerted by the environment on the object. Try to draw them in roughly the correct directions. Label each arrow with an algebraic symbol to represent the magnitude of the force, regardless of whether or not a numerical value is given in the problem statement.

    The hard part is getting all the forces. If appropriate, don't forget to include the weight of the object, the normal force of a surface on the object, and the forces of any strings or rods attached to the object. Some notes are given below to explain how to handle these forces.

    Some students erroneously include forces that are not acting on the object. For each force you include you should be able to point to something in the environment that is exerting the force. If, for example, the object you are considering is not in contact with a surface, you should not include a normal force in your analysis.

4. Draw a coordinate system on the diagram. In principle, the placement and orientation of the coordinate system do not matter as far as obtaining the correct answer is concerned but some choices reduce the work involved. If you can guess the direction of the acceleration place one of the axes along that direction. The acceleration of an object sliding on a surface such as a table top or inclined plane, for example, is parallel to the surface. Once the coordinate system is drawn, label the angle each force makes with a coordinate axis. This will be helpful in writing down the components of the forces later.

    The diagram, with all forces shown but without the coordinate system, is called a <u>free-body diagram</u> (or a <u>force diagram</u>). We add the coordinate system to help us carry out the next step in the solution of the problem.

5. Write Newton's second law in component form: $\sum F_x = ma_x$, $\sum F_y = ma_y$, and if necessary $\sum F_z = ma_z$. The left sides of these equations should contain the appropriate components of the forces you drew on your diagram. You should be able to write the equations by inspection of your diagram. Use algebraic symbols to write them, not numbers; most problems give or ask for force magnitudes so you should write each force component as the product of a magnitude and the sine or cosine of an appropriate angle.

6. Identify the known quantities and solve for the unknowns. Sometimes one or more of the acceleration components are given or implied. Constant velocity, of course, means all acceleration components vanish. If you picked the x axis along the direction of the acceleration then $a_y = 0$ and $a_z = 0$.

Here is an example for you to try.

PROBLEM 2. A painter working on the outside of the second story uses a rope to lower some equipment with a downward acceleration of 1.5 m/s². If the equipment has a mass of 7.3 kg what is the force of gravity on it? What is the force of the rope on it?

SOLUTION: The force of gravity is mg, downward. You know that the rope is pulling up on the equipment since the acceleration is less than g in magnitude. In the space to the right draw a free-body diagram showing the two forces acting on the equipment. Label the force of gravity with its magnitude mg and the force of the rope with its magnitude T. The first is downward; the second is upward. Take the positive axis of the coordinate system to be downward. Then Newton's second law is $mg - T = ma$. Solve for T.

[ans: 71.5 N, down; 60.6 N, up]

Carefully go over the sample problems of Section 5–10 in the text. They have a great deal to teach you about the forces acting on objects.

The idea of a <u>normal force</u> is introduced in Sample Problem 5. Remember that "normal" here means perpendicular. When an object is in contact with a rigid surface, the surface exerts a force on the object that prevents the object from penetrating the surface and the force must be normal to the surface. If a book rests on a horizontal table top, for example, the force of gravity pulls down on the book and the normal force of the table pushes up on it. The magnitude of the normal force adjusts so it is exactly right to prevent the book from moving through the table. Unless the object adheres to the surface the normal force must be directed *away* from the surface.

Clearly the direction of a normal force depends on the orientation of the surface. If the table top makes the angle θ with the horizontal then the normal force makes the same angle θ with the vertical. If you hold a book against a vertical wall by pushing on it, the normal force of the wall on the book is horizontal.

Don't assume you know the magnitude of the normal force until you have solved the second-law equations. As Sample Problem 5 of the text makes clear, the magnitude of a normal force depends on the magnitudes and directions of other forces acting on the object. Notice that when the direction of the applied force changes the magnitude of the normal force changes.

Usually a normal force is not mentioned in problem statements. You must be aware that it exists if the object of interest is in contact with a surface; then include it in your free-body diagram and your second-law equations.

PROBLEM 3. A 25-kg crate is being pushed along a horizontal floor with a force of 60 N, at an angle of 25° above the horizontal. The floor exerts a 15-N force of friction, parallel to the floor and opposite to the direction of the crate's motion. What is the acceleration of the crate and what is the normal force of the floor on the crate?

SOLUTION: In the space to the right draw a free-body diagram for the crate. The forces acting on it are the force of the person pushing, the normal force of the floor, the force of friction, and the force of gravity. The force of gravity has magnitude mg and is directed downward. Use F to label the force of the person, f to label the force of friction, N to label the normal force, and mg to label the force of gravity. These symbols represent magnitudes.

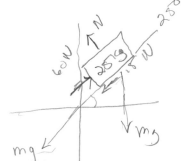

Add a coordinate system to the diagram. Take the x axis to be parallel to the floor, positive in the direction of the crate's motion. Take the y axis to be positive in the upward direction. Use θ to label the angle between the force of the person and the x axis.

Fill in the following table giving the x and y components of the various forces. Use algebraic symbols, not numbers.

FORCE	X COMPONENT	Y COMPONENT
force of person	60 N	
force of friction	−15 N	
force of gravity		9.8
normal force		

The sum of the x components is _____75 N_____. Set this equal to ma_x. The sum of the y components is _____. Set this equal to ma_y. Numerical values of known quantities are $F =$ __45__, $\theta =$ __25__, $f =$ __15__, $m =$ __2.5g__, $g =$ __9.8__, and $a_y =$ _____. The unknown quantities are N and a_x. Solve $F\cos\theta - f = ma_x$ for a_x and $F\sin\theta - mg + N = 0$ for N.

$75 N = ma_x$ $F\cos(25°) - 15 = ma_x$
$N(9.8) = ma_y$ $F\sin(25) - (2.5 kg)(9.8 m/s^2) + 75 = N$

[ans: 1.58 m/s²; 220 N]

You can use the known direction of a normal force to tell when an object leaves a surface. Suppose a book rests on a table top and you lift up on it with a force of magnitude F that is less than its weight. The book remains in contact with the table and, according to Newton's second law, $F + N - mg = 0$. As you increase F the normal force of the table on the book decreases in magnitude and when $F = mg$ it vanishes. With a slight additional increase in F the book leaves the table and its acceleration is no longer zero. That is, after the book leaves the table $N = 0$ and $F - mg = ma$. The minimum force F that will pick the book off the table can be found by placing $N = 0$ and $a = 0$ in Newton's second law.

PROBLEM 4. A 2.5-kg box is being pushed along a horizontal floor with a force of 40 N, at an angle θ above the horizontal. The floor exerts a 15-N force of friction, parallel to the floor and opposite to the direction of the crate's motion. What is the greatest value of θ for which the box does not lift off the floor? What is the acceleration of the box if θ has this value?

SOLUTION: The free-body diagram is the same as before. Use the same coordinate system. Fill in the following table giving the x and y components of the various forces. Use algebraic symbols, not numbers.

FORCE	X COMPONENT	Y COMPONENT
force of person		
force of friction		
force of gravity		
normal force		

62   Chapter 5: Force and Newton's Laws

The sum of the x components is _____. Set this equal to ma_x. The sum of the y components is _____. Set this equal to ma_y.

We suppose θ is adjusted so the box is about to lift off the floor. Then the numerical values of the known quantities are $F =$ _____, $f =$ _____, $m =$ _____, $g =$ _____, $N =$ _____, and $a_y =$ _____. The unknown quantities are θ and a_x. Solve $F\sin\theta - mg + N = 0$ for $\sin\theta$, then find the value of θ. Solve $F\cos\theta - f = ma_x$ for a_x.

[ans: 54.8°; 3.23 m/s²]

Sample Problem 7 of the text deals with a passenger standing on a scale in an elevator. The normal force is provided by the scale and the scale reading gives its value. The normal force here does the same job as in other situations: it prevents an object (the passenger) from moving through the surface (the scale platform). Note that the condition entered into the second law is that the acceleration of the passenger is the same as the acceleration of the elevator. When the acceleration of the elevator is upward, the force of the scale must be greater than the weight of the passenger; when its acceleration is downward, the force must be less than the weight of the passenger; and when the elevator is moving with constant velocity, the force equals the weight. These statements are true regardless of whether the elevator is moving up or down.

PROBLEM 5. A boy responding to a dare uses a large bathroom scale as a sled. He stands on it and slides down an icy hill that makes an angle of 10° with the horizontal. As he is going down he notices that the scale reads 360 N. What is his weight and acceleration?

SOLUTION: Use the space to the right to draw the free-body diagram for the boy. Also draw the coordinate system, with one axis parallel to the incline. Use θ to designate the angle of the incline, N to designate the magnitude of the normal force, and W to designate the magnitude of the boy's weight. Since the hill is icy, the force of friction is negligible.

Fill in the following table using algebraic symbols, not numbers.

FORCE	X COMPONENT	Y COMPONENT
force of gravity	_____	_____
force of scale	_____	_____

Use these symbols to write Newton's second law in component form:

x component:

y component:

Enter values for known quantities: $\theta =$ _____, $g =$ _____, $m =$ _____, and $N =$ _____. Solve for the weight and acceleration.

[ans: 366 N; 1.70 m/s²]

The idea of <u>tension</u> in a string is introduced in Sample Problem 4 of the text. This is the force exerted by the string on the object to which it is attached. The force is always parallel to the string and the string always pulls on the object. It cannot push. In some problems a string is attached to an object and a person pulls on the string. If the mass of the string is negligible the string simply transmits the force. That is, the force of the string on the object is the same as the force of the person on the string. You may, if you like, think of the person as applying the force directly to the object. The string, of course, serves to define the direction of the force.

PROBLEM 6. A 2.6-kg box is pulled up a frictionless incline by means of a rope that is parallel to the incline. The incline makes an angle of 30° with the horizontal and the box speeds up at a rate of 1.2 m/s². What is the tension in the rope and what is the normal force of the incline on the box?

SOLUTION: In the space to the right draw a free-body diagram for the box. Label the forces acting on it: T for the tension in the rope, N for the normal force of the incline, and mg for the force of gravity. These symbols represent magnitudes. Orient the coordinate system so the positive x axis is up the incline, in the direction of the acceleration, and the positive y axis is normal to the incline and points away from it. Use θ to label the angle the incline makes with the horizontal. Notice that the normal force is in the positive y direction, the tension in the rope is in the positive x direction, and the force of gravity makes the angle θ with the vertical.

Fill in the following table giving the x and y components of the various forces. Use algebraic symbols, not numbers.

FORCE	X COMPONENT	Y COMPONENT
tension in rope	_____	_____
force of gravity	_____	_____
normal force	_____	_____

The sum of the x components is _____. Set this equal to ma_x. The sum of the y components is _____. Set this equal to ma_y. Numerical values of known quantities are $\theta =$ _____, $m =$ _____, $g =$ _____, $a_x =$ _____, and $a_y =$ _____. The unknown quantities are N and T. Solve $T - mg\sin\theta = ma_x$ for a_x and $N - mg\cos\theta = 0$ for N.

[ans: 15.9 N; 22.1 N]

Take care about the direction of the acceleration. It need not be the same as the direction of motion.

PROBLEM 7. Work the same problem but assume the box is slowing down at the rate 1.2 m/s².

SOLUTION: Use the same free-body diagram and coordinate system. Now, however, $a_x = $ _____. Solve for T. The normal force is the same as before.

[ans: 9.62 N; 22.1 N]

We now apply Newton's second law to situations in which two or more objects influence each other's motions. The technique is straightforward. Carry out the instructions listed at the beginning of this section for each object separately. Don't forget the force exerted by each object on the others or the forces of any strings or rods that might connect the objects. Also don't forget that the forces exerted by the objects on each other are related by Newton's third law. You will obtain one set of second-law equations, in component form, for each object. You must also be aware that the accelerations of the objects might be related to each other. This occurs, for example, if the objects are connected by a rod or string. In your list of equations to be solved you must then include an equation that describes the relationship. The second-law equations, one for each object, and the equation relating the accelerations are then solved simultaneously.

Study the sample problems in Section 5–11 of the text and the problems given below. Each has something to teach you about setting up and solving problems.

Sample Problem 8 deals with a block of mass m_1 on a frictionless horizontal surface. One end of a light string is attached to the block, then runs horizontally to a pulley at the edge of the surface, goes over the pulley, and hangs vertically. A second block, with mass m_2, is attached to the hanging end. See Fig. 18 of the text.

In the space to the right draw the free-body diagram for block 1. Label the force of the string T_1, the normal force N, and the force of gravity $m_1 g$. These symbols represent magnitudes. Orient the coordinate system so the positive x axis is to the right, parallel to the string, and the positive y axis is upward. Fill in the following table in terms of T_1, N, and $m_1 g$:

FORCE	X COMPONENT	Y COMPONENT
force of string	_____	_____
normal force	_____	_____
force of gravity	_____	_____
sum	_____	_____

The force of the string should be in the positive x direction, the normal force should be in the positive y direction, and the force of gravity should be in the negative y direction. The acceleration of block 1 is along the x axis. Equate the sum of the x components to $m_1 a_1$ and

Chapter 5: Force and Newton's Laws

the sum of the y components to 0:

In the space to the right draw the free-body diagram for block 2. Label the force of the string T_2, and the force of gravity m_2g. Orient the coordinate system so the positive y axis is downward. Fill in the following table in terms of T_2, N, and m_2g:

FORCE	X COMPONENT	Y COMPONENT
force of string		
force of gravity		
sum		

The force of the string is in the negative y direction and the force of gravity is in the positive y direction. The acceleration of block 2 is along the y axis. Equate the sum of the x components to 0 and the sum of the y components to m_2a_2:

Notice that different coordinate systems were used for the two objects. This is always permissible and the algebra is usually simplified if the coordinate system for each object is oriented so one axis is parallel to the acceleration of the object.

The accelerations of the two objects are related and you must include the relationship in your specification of the problem. Since the string is inextensible the blocks must move together and the magnitudes of their accelerations must be the same. You need the relationship between their components, however. This is particularly easy to describe if the coordinate system for each object is chosen so one axis is parallel to the acceleration of the object, as we have done above. Then the acceleration component for one is either equal to the acceleration component of the other or to its negative. Imagine the objects in motion to decide which. Suppose block 2 moves downward, in the positive y direction (using the coordinate system we associated with block 2). Then block 1 moves to the right, in the positive x direction (using the coordinate system we associated with block 1). Thus $a_1 = a_2$. Had we chosen the positive y axis for block 2 to be upward, then the relationship would have been $a_1 = -a_2$.

For this problem we must also consider the role of the pulley. Since it is essentially massless and frictionless, its sole influence is to change the direction of the string. We have already taken this into account in two ways: when we related the acceleration components of the two blocks and when we entered the components of the forces of the string on the blocks into the second-law equations. When you study rotational dynamics you will learn how to deal with pulleys that have mass.

Finally you must recognize that, since the string is essentially massless the tension is the same everywhere in the string. This means the magnitude of the force exerted by the string on block 1 is the same as the magnitude of the force exerted by the string on block 2: $T_1 = T_2$. The forces are in different directions, of course.

To understand this conclusion let us ignore the pulley for a moment and analyze the motion of a string connecting two blocks that move along the same line. Let T_1 be the magnitude of the force exerted by the string on block 1 and T_2 be the magnitude of the force exerted by the string on block 2. Then, by Newton's third law, T_1 is the magnitude of the force of block 1 on the string and T_2 is the magnitude of the force of block 2 on the string. These are in opposite directions so the net force on the string is $T_1 - T_2$ and, according to the second law, this must equal the product of the mass and acceleration of the string. Since the mass is negligible the two forces very nearly cancel and we may take $T_1 = T_2$. The magnitude of each of these is denoted by T and called the tension in the string. A massless string pulls on each of the objects at its ends with the same force T, in different directions of course.

Now return to the problem and rewrite the second-law equations for the two blocks. You may omit subscripts on the string forces and accelerations. That is, let $T_1 = T$, $T_2 = T$, $a_1 = a$, and $a_2 = a$. The second law equations become

x component for block 1:

y component for block 1:

y component for block 2:

You are now ready to solve some problems. Notice that you start with the same set of equations for every problem dealing with this situation.

PROBLEM 8. Suppose $m_1 = 5.0$ kg and $m_2 = 2.0$ kg. What is the acceleration of each block, what is the tension in the string, and what is the normal force of the surface on block 1?

SOLUTION: The three equations must be solved simultaneously. First use $T = m_1 a$ to eliminate T from $m_2 g - T = m_2 a$, then solve algebraically for a and evaluate the resulting expression.

[ans: $2.80 \, \text{m/s}^2$]

Once a is known, substitute its value into $T = m_1 a$ to find T.

[ans: $14.0 \, \text{N}$]

Lastly, evaluate $N - m_1 g = 0$ to find N.

[ans: $49.0 \, \text{N}$]

If the acceleration and one of the masses are known the other mass can be found. Sometimes the kinematic equations, rather than the second-law equations, must be used to find the acceleration. The appearance of coordinates, velocities, or time in a problem is a strong clue that the kinematic equations will be used for some purpose.

PROBLEM 9. Suppose mass $m_2 = 4.0$ kg is hung on the vertical portion of the string. It starts from rest and falls 2.0 m in 1.3 s. What is m_1?

SOLUTION: First use the kinematic equation $y = \frac{1}{2}at^2$ to find the acceleration of the blocks, then solve the Newton's second law equations for m_1.

[ans: $a = 2.37$ m/s^2; $m_1 = 12.6$ kg]

Sample Problem 10 of the text is very much the same, except that block 1 is placed on a frictionless inclined plane. See Fig. 20 of the text.

In the space to the right draw the free-body diagram for block 1. Label the force of the string T, the normal force N, and the force of gravity m_1g. These symbols represent magnitudes. Orient the coordinate system so the positive x axis is up the plane and the positive y axis is normal to the plane. Designate the angle of the plane by θ. Fill in the following table in terms of T, N, and m_1g:

FORCE	X COMPONENT	Y COMPONENT
force of string	_____	_____
force of gravity	_____	_____
normal force	_____	_____
sum	_____	_____

The acceleration of block 1 is along the x axis. Equate the sum of the x components to m_1a and the sum of the y components to 0.

In the space to the right draw the free-body diagram for block 2. Label the force of the string T (the same as for block 1) and the force of gravity m_2g. Orient the coordinate system so the positive y axis is downward. Fill in the following table in terms of T, N, and m_2g:

FORCE	X COMPONENT	Y COMPONENT
force of string	_____	_____
force of gravity	_____	_____
sum	_____	_____

The force of the string is in the negative y direction and the force of gravity is in the positive y direction. The acceleration of block 2 is along the y axis. Equate the sum of the y components to m_2a. Notice that the notation you have used takes into account the proper relationship between the accelerations of the blocks and the transmission of the force by the string. Also note that you do not need to know which way the blocks actually move. If a is positive block 1 moves up the plane and block 1 falls. If a is negative block 1 moves down the plane and block 2 rises.

Write the second law equations:

x component for block 1:

y component for block 1:

y component for block 2:

You are now ready to solve some problems. All deal with the same physical situation but they differ in the quantities that are given.

PROBLEM 10. Suppose $m_1 = 5.0$ kg, $m_2 = 3.0$ kg, and $\theta = 25°$. What is the acceleration of the blocks, what is the tension in the string, and what is the normal force of the incline on block 1?

SOLUTION:

[ans: 1.09 m/s²; 26.1 N; 44.4 N]

PROBLEM 11. Suppose m_1 is the same but m_2 is changed to 2.0 kg. What then is the acceleration of the blocks?

SOLUTION:

[ans: −0.158 m/s²]

PROBLEM 12. Notice that when m_2 is 3.0 kg the acceleration of block 1 is up the plane but when m_2 is 2.0 kg it is down the plane. For what value of m_2 does the acceleration vanish?

SOLUTION:

[ans: 2.11 kg]

PROBLEM 13. Block 1 has a mass of 5.0 kg and block 2 has a mass of 1.5 kg. They start with identical initial velocities, with block 1 going up the plane, and come to rest 1.7 s later. What was the initial velocity? The angle of the incline remains 25°.

SOLUTION: Use the second-law equations to find the value of the acceleration, then substitute the result into the appropriate kinematic equation and solve for the initial velocity.

[ans: 1.57 m/s]

PROBLEM 14. Block 1 has a mass of 5.0 kg and block 2 has a mass of 1.5 kg. What should the angle of the incline be so the acceleration of the blocks vanishes? What then is the tension in the string?

SOLUTION:

[ans: 17.5°; 14.7 N]

Chapter 5: Force and Newton's Laws

If two objects exert forces on each other, Newton's third law must be invoked. Suppose the objects are labelled A and B. The force of B on A is included in the sum of forces acting on A and the force of A on B is included in the sum of forces acting on B. Use the same symbol for the magnitudes of the forces and be sure your diagram indicates that the forces are in opposite directions. When you write Newton's second law in component form the sign you write in front of the magnitude symbol is determined by the direction of the force.

PROBLEM 15. A student holds a 2.0-kg book in his hand and while holding it lets his hand and book fall with an acceleration of 3.6 m/s², downward. What is the force of the earth on the book?

SOLUTION: This is the force of gravity $m\mathbf{g}$.

[ans: 19.6 N, down]

What is the force of the book on the earth?

SOLUTION: This is a gravitational force on the earth and is the reaction force to the gravitational force of the earth on the book.

[ans: 19.6 N, up]

What is force of the hand on the book?

SOLUTION: Let F be the force of the hand on the book and take the positive axis to be upward. Write Newton's second law in the form $F - mg = ma$ and solve for F.

[ans: 12.4 N, up]

What is the force of the book on the hand?

SOLUTION:

[ans: 12.4 N, down.]

Note especially that neither the force of the hand on the book or the force of the book on the hand is equal in magnitude to the force of gravity on the book. In fact, they cannot be if the book is to have an acceleration that is less than the acceleration due to gravity alone.

Suppose now that the student holds the book at rest in his hand. What then is the force of the earth on the book and the force of the hand on the book?

SOLUTION:

[ans: 19.4 N, up; 19.4 N, down]

These forces are equal in magnitude and opposite in direction but they are emphatically *not* an action-reaction pair. The most obvious indication that they cannot be is that they act on the same body, the book. In addition, the only reason they are equal and opposite is that they are the only forces acting on the book and the book has zero acceleration. If a third force acts vertically or if the book accelerates then these forces cannot be equal in magnitude. On the other hand, the gravitational forces of the earth on the book and the book on the earth are always equal in magnitude and opposite in direction, no matter what other forces act and no matter what the acceleration of the book.

PROBLEM 16. Suppose a 60-kg man pulls a 20-kg wheelbarrow so they both have an acceleration of 1.5 m/s² in the forward direction. In addition to a normal force, the ground exerts a horizontal (frictional) force of 10 N in the backward direction on the wheelbarrow. What is the force of the man on the wheelbarrow?

SOLUTION: Draw the free-body diagram for the wheelbarrow. Take the positive x direction to be in the direction of motion. Let m_w be the mass of the wheelbarrow, F be the magnitude of the force exerted by the man on it, and f be the magnitude of the horizontal force exerted by the ground on it. The man pulls forward and the ground pulls backward, so $F - f = m_w a$. Solve for F.

[ans: 40 N, in the forward direction]

What is the force of the wheelbarrow on the man?

SOLUTION: Here's where the third law enters. The force of the wheelbarrow on the man is equal in magnitude and opposite in direction to the force of the man on the wheelbarrow. The answer is therefore 40 N, in the rearward direction.

What is the horizontal force of the ground on the man?

SOLUTION: Let m_M be the mass of the man, F the force exerted on him by the wheelbarrow, and f' the force exerted on him by the ground. Then $f' - F = m_M a$. Solve for f'.

[ans: 130 N, in the forward direction]

Note that we took the x component of the force to be f' and entered it into Newton's second law with a plus sign. The solution turned out to be positive, indicating that the force is indeed in the positive x direction. To generate this force the man pushes back on the ground with a force of 130 N.

PROBLEM 17. A weightlifter lifts a 125-kg barbell over his head, then lets it down to the floor. During its descent its acceleration is $5.6 \, \text{m/s}^2$, downward. What is the force of the barbell on the weightlifter's hands?

SOLUTION: We cannot profitably apply Newton's second law to the weightlifter's hands since we don't know either their mass or the force of the arms on them. We can, however, apply the law to the barbell and use it to find the force of the hands on barbell. The force of the barbell on the hands is equal in magnitude and opposite in direction.

[ans: 525 N, down]

PROBLEM 18. Two gliders are in contact with each other on an essentially frictionless horizontal air track. The one to the left has mass m_1 and the one to the right has mass m_2. A person pushes to the right on the left glider with a horizontal force F. Derive expressions for the acceleration of the gliders and for the magnitude of the force one glider exerts on the other.

SOLUTION: In the space to the right draw a free-body diagram for each glider. Use f to label the force exerted by glider 1 on glider 2. Assume it is a push rather than a pull. The force of glider 2 on glider 1 has the same magnitude, so use the same label. The two forces are oppositely directed, of course. Take the positive x axis to be horizontal and to the right.

The gliders stay in contact as they move, so their accelerations are the same. Use a for the common acceleration and write Newton's second law for each glider.

glider 1:

glider 2:

Use one of these equations to eliminate f from the other, then solve for a. Substitute the expression you find for a into either of the second-law equations and solve for f.

[ans: $a = F/(m_1 + m_2)$; $f = m_2 F/(m_1 + m_2)$]

We might consider the two gliders as a single object, with mass $m_1 + m_2$ and acted on by the force F. This way of looking at the gliders would allow us to calculate the acceleration easily but it would not allow us to calculate the contact force. To do that we must consider the gliders separately.

PROBLEM 19. For the gliders of the previous problem suppose $m_1 = 3.0$ kg and $m_2 = 5.0$ kg. If $F = 50$ N what are the acceleration and contact force?

SOLUTION:

[ans: 6.25 m/s^2; 31.25 N]

Suppose the gliders are interchanged so $m_1 = 5.0$ kg and $m_2 = 3.0$ kg. If F is again 50 N what are the acceleration and contact force?

SOLUTION:

[ans: 6.25 m/s^2; 18.75 N]

Notice that in each case the contact force is just right to give block 2 the same acceleration as block 1. Since m_2 is less in the second example, the contact force is also less.

III. NOTES

Chapter 6
PARTICLE DYNAMICS

I. BASIC CONCEPTS

This chapter contains a great many applications of Newton's laws, with special emphasis frictional and centripetal forces. You will also learn how to solve problems involving variable forces.

Force laws. A force law is a rule that tells us how to calculate the force exerted by one object on another if we know certain properties of the two objects and their relative positions. The properties required are different for different type forces. Examples are: gravitational mass gives rise to gravitational forces, electric charge gives rise to electric and magnetic forces. Other force laws (not as fundamental) allow us to compute the force of friction when one object slides or rolls on another and to calculate the force of a spring on an object.

Friction. Two macroscopic objects in contact may exert frictional forces on each other. Friction is unavoidable when the objects are sliding on each other, although lubricants and air films may make it small. Even when the objects are stationary with respect to each other, they exert frictional forces if other forces present would otherwise cause them to slide. Explain in words the mechanism that gives rise to a frictional force: _____
_____

Although all frictional forces arise from the same fundamental phenomenon two types are discussed in the text: static and kinetic. Tell how you can identify the situation in which each is acting.

static: _____
kinetic: _____

When the two objects are not moving relative to each other the force of static friction is determined by the condition that their accelerations be equal. Usually, but not always, an object rests on a surface (a table top or an inclined plane, for example) that is as rest. Then the force of static friction on the object is just sufficient to hold it at rest. Mathematically the frictional force is determined, via Newton's second law, by the condition that the component of acceleration parallel to the surface is zero. This condition is analogous to the condition used to determine the normal force. The difference is that the normal force of one object on another is perpendicular to the surface of contact while the frictional force is parallel to it.

The magnitude f of the force of static friction exerted by one surface on another must be less than a certain value, determined by the nature of the surfaces and by the magnitude of the normal force one surface exerts on the other. In particular, $f \leq$ _____, where μ_s is the coefficient of static friction and N is the magnitude of the normal force. If the force of friction required to hold the surfaces at rest with respect to each other is greater than the maximum allowed, then the surfaces slide over each other. Once this happens the magnitude of the force of friction is given by $f = \mu_k N$, where μ_k is called _____.

The normal force that appears in the expressions for the force of kinetic friction and the maximum force of static friction must be computed for each situation using Newton's second law. As you know by now the magnitude of the normal force depends on the directions and magnitudes of other forces acting.

Uniform circular motion. An object in uniform circular motion has a non-zero acceleration because the direction of its velocity changes with time. A force must be applied to the object in order to produce its acceleration. If m is the mass of the object, then the applied force must have magnitude $F =$ _____, where r is the radius of the orbit and v is the speed of the object. The force is directed _____ and, because of its direction, is called a _____ force. Acquire the habit of pointing out to yourself the object that exerts the force. It might be a string, for example.

Variable forces. Sections 6–4 and 6–5 of the text outline the usual problem in dynamics, valid for any force, even non-constant forces that depend on the time or on the velocity of the object. Position dependent forces are covered later when energy techniques are discussed.

Suppose the total force acting on an object of mass m depends on the time and is given by the function $F(t)$. Then _____ gives the acceleration as a function of time. Over the interval from time t_0 to time t_1 the change in the velocity is given by the integral expression $v(t) - v_0 =$ _____. Once the velocity is known as a function of time the coordinate of the object can be calculated using the integral $x(t) - x_0 =$ _____. You should be able to find expressions for $v(t)$ and $x(t)$ if the force is any polynomial of the time. The steps are carried out in Sample Problem 4 of the text.

When the force is a function of the velocity or is a complicated function of time, numerical methods must often be used to integrate Newton's second law, but the fundamental relationships given above are valid.

Projectile motion with air resistance. Newton's second law can be integrated for a projectile subject to a force of air resistance (drag force) that is proportional to the velocity of the projectile: $\mathbf{D} = -b\mathbf{v}$, where b is the drag coefficient. The drag coefficient is determined by the interaction between the air and the object and does not depend on the velocity of the object.

First consider one-dimensional motion (a projectile fired straight up or down). If the positive y axis is taken to be downward the sum of the forces on the projectile is given by $\sum F =$ _____. Since, by Newton's second law, this must equal ma, the acceleration of the projectile is given by $a =$ _____. The general idea is to rearrange $\sum F = m\, dv/dt$ so the velocity dependent functions are all on one side of the equation, then integrate. The result, in integral form, is

$$t =$$

Don't forget to include limits on the integral. Sample Problem 5 of the text gives the details for an object dropped from rest, subject to a drag force that is proportional to v. The velocity as a function of time is given by

$$v(t) = \underline{}$$

You should note that the velocity of a falling object subject to drag approaches a limit, called the terminal velocity. An expression for the terminal velocity can be found by allowing t to become large without limit in the expression for $v(t)$. If the magnitude of the drag force is given by bv, the terminal velocity is given by $v_T = \underline{}$. For a falling object the terminal velocity is always downward.

You should also be able to find the expression for v_T by examining Newton's second law. When the object is traveling at terminal velocity the force of gravity is balanced by the \underline{} force and the acceleration vanishes. Use $D - mg = 0$ to find that the magnitude of the terminal velocity is given by $v_T = \underline{}$ if the magnitude of the drag force is bv and is given by $v_T = \underline{}$ if the magnitude of the drag force is given by bv^2. Note that an object can be started with an initial velocity that is greater than, less than, or the same as the terminal velocity. In any case, the velocity of the object approaches the terminal velocity as it falls.

Expressions for the velocity and position of a projectile moving in two dimensions can also be found if the drag force is proportional to the velocity. Take the positive y axis to be upward and the x axis to be horizontal in the plane of motion. Then, in terms of the drag force $-b\mathbf{v}$ and the force of gravity $m\mathbf{g}$, Newton's second law becomes:

x component: y component:

The velocity components as functions of time, found by integrating the second law, are

$$v_x(t) = v_{0x} e^{-bt/m}$$

$$v_y(t) = (mg/b + v_{0y})e^{-bt/m} - mg/b.$$

Differentiate these expressions with respect to time to verify that they satisfy the second law equations;

Also verify that the expressions predict $v_x(0) = v_{0x}$ and $v_y(0) = v_{0y}$. Carefully note that the horizontal component of the velocity is not constant as it would be in the absence of air resistance. It decreases exponentially with time.

In the limit of large t, these expressions predict that $v_x = \underline{}$ and $v_y = \underline{}$. The projectile is now falling straight down with a speed equal to the terminal speed.

Validity of Newton's laws. Newton's laws of motion are valid for a wide range of physical phenomena, but they do have limits and the laws must be replaced or modified in two domains. Relativity theory must be used if \underline{} and quantum theory must be used if _____

II. PROBLEM SOLVING

Many problems of this chapter deal with frictional forces. Proceed as before: draw the free-body diagram and write down Newton's second law in component form, just as for any other problem. Use an algebraic symbol, f say, for the frictional force. You must now decide if the frictional force is static or kinetic. If static friction is involved, f is probably an unknown but the acceleration is known or is related to another known quantity in the problem. If the object is at rest on a stationary surface its acceleration is zero. If it is at rest relative to an accelerating surface its acceleration is the same as that of the surface. Kinetic friction is involved if one surface is sliding on the other. Then the magnitude of the frictional force is given by $\mu_k N$.

If you do not know that the object is at rest relative to the surface, assume it is and use Newton's second law, with the acceleration of the object equal to the acceleration of the surface, to calculate both the force of static friction f_{rest} that will hold it at rest and the normal force. Compare f_{rest} with $\mu_s N$. If $f_{\text{rest}} < \mu_s N$ the object remains at rest relative to the surface and the force of friction has the value you computed. That is, $f = f_{\text{rest}}$. If $f_{\text{rest}} > \mu_s N$ then the object moves relative to the surface. Go back to the second law equations and set $f = \mu_k N$, then solve for the acceleration.

PROBLEM 1. A chair of mass m rests on a horizontal floor. When a person pushes to the right with a force of magnitude F, at an angle θ above the horizontal, the chair does not move. What is the force of friction exerted by the floor on the chair?

SOLUTION: In the space to the right draw a free-body diagram for the chair. Designate the force of friction by f. It is parallel to the floor and acts to the left. Don't forget the weight of the chair and the normal force of the floor on the chair. Take the x axis to be positive to the right and the y axis to be positive upward. The component form of Newton's second law, with explicit expressions for the forces, is:

x component: y component:

Solve for f.

[ans: $F\cos\theta$]

Notice that f depends on θ. The force of friction exactly balances the horizontal component of the applied force. Suppose the chair has a mass of 10 kg and the coefficient of static friction is 0.25. Does the chair move if the person pushes with a force of 27 N, directed 55° above the horizontal?

SOLUTION: The force of friction that is needed to hold the chair at rest is $f_{rest} = F\cos\theta =$ _____ N. The normal force of the floor on the chair is $N = mg - F\sin\theta =$ _____ N. The maximum frictional force that the floor can generate is $\mu_s N =$ _____ N. Since $f_{rest} < \mu_s N$ the chair does not move.

Does the chair move if the person pushes horizontally with the same magnitude force?

SOLUTION:

[ans: yes]

If the coefficient of kinetic friction between the chair and the floor is 0.20 what is the acceleration of the chair when the applied force is horizontal?

SOLUTION: The horizontal component of Newton's second law is $F - f = ma$ and the vertical component is $N - mg = 0$. Since the chair is moving $f = \mu_k N$. Solve these equations for a.

[ans: 0.250 m/s²]

PROBLEM 2. A 5.0-kg block is released from rest on a 25° inclined plane. The coefficients of friction are $\mu_s = 0.25$ and $\mu_k = 0.20$. Does the block start sliding? If it does, what is its acceleration?

SOLUTION: In the space to the right draw the free-body diagram for the block. Designate the force of friction by f, the mass of the block by m, the normal force by N, the force of gravity by mg, and the angle of the plane by θ. The force of friction is up the plane. Orient the coordinate system so the positive x axis is down the plane. Let a be the acceleration of the block and write the components of Newton's second law in terms of m, g, f, N, a, and θ:

x component: y component:

Set $a = 0$ and $f = f_{rest}$, then solve for f_{rest} and N. Finally calculate $\mu_s N$.

[ans: 20.7 N; 44.4 N; 11.1 N]

Since 20.7 N is greater than 11.1 N the block slides. When f is replaced by $\mu_k N$ in the second-law equations they become:

x component: y component:

Solve for a.

[ans: 2.37 m/s²]

PROBLEM 3. A block of mass $m_1 = 3.0$ kg is on a horizontal surface. A second block, of mass $m_2 = 2.0$ kg, is on top of the first, as shown. The coefficient of static friction between the blocks is 0.40. What is the maximum acceleration m_1 can have without sliding from under m_2?

SOLUTION: Assume m_2 moves with m_1 and let their common acceleration be a. Use Newton's second law to find an expression in terms of a for the force of friction f exerted by m_1 on m_2. For m_2 to move with m_1, f must be less than $\mu_s N$, where N is the normal force of m_1 on m_2. The maximum acceleration has the value for which $f = \mu_s N$.

In the space to the right draw a free-body diagram for m_2. Include the force of friction, the force of gravity, and the normal force. Write the vertical and horizontal components of Newton's second law:

horizontal component: vertical component:

Solve for f and N in terms of a:

Now substitute these expressions into $f = \mu_s N$ and solve for a.

[ans: 3.92 m/s²]

Notice that the value of m_2 is immaterial.

Assume the surface under m_1 is frictionless and solve for the magnitude of the maximum horizontal force with which a person can push on m_1 without m_1 sliding from under m_2.

SOLUTION: In the space to the right draw a free-body diagram for m_1. Don't forget that m_2 exerts a backward frictional force on m_1, equal in magnitude to the frictional force m_1 exerts on m_2. It also exerts a downward normal force on m_1, equal in magnitude to the upward normal force m_1 exerts on m_2. Now write the second-law equations and solve for the applied force.

[ans: 19.6 N]

Sometimes the direction of the static force of friction is unknown. Consider an object on an inclined plane that is tilted so the object will slide down if you do not exert a force on it. Suppose, however, you pull on it with a force F that is parallel to the plane and directed up the plane. You will find that you can apply a fairly wide range of forces without having the object move. If F is small the force of friction is up the plane; if F is large the force of friction is down the plane. The static frictional force can have any value from $\mu_s N$ down the plane to $\mu_s N$ up the plane (including 0), depending on the value of F.

If you are given F and asked to find the force of friction, the solution is straightforward. Simply place the x axis along the plane, enter f into the x component of the second law equation just as if the force were in the positive x direction, then solve for f. If you obtain a negative number the force is actually in the negative x direction.

The problem is a little trickier if you are asked to find the maximum or minimum applied force for which the object remains at rest. When the applied force is a minimum the force of friction is in the same direction as the applied force and has its maximum magnitude, $\mu_s N$. If the positive x axis is in the direction of the applied force substitute $+\mu_s N$ for the force of friction in the second law equation. When the applied force is a maximum the force of friction is opposite the applied force and has its maximum magnitude. Now substitute $-\mu_s N$ for the force of friction. In each case solve for F.

PROBLEM 4. A person applies a 15-N force to a block initially at rest on a 25° incline. The force is directed up the plane. If the coefficient of static friction is 0.25 what is the force of friction?

SOLUTION: In the space to the right draw the free-body diagram for the block. Designate the force of friction by f, the mass of the block by m, the normal force by N, the force of gravity by mg, and the angle of the plane by θ. Assume the force of friction is up the plane and orient the coordinate system so the positive x axis is in that direction. Take the acceleration of the block to be zero and write the x and y components of Newton's second law in terms of m, g, f, N, and θ:

x component: y component:

Solve for f and N, then calculate $\mu_s N$.

[ans: 5.71 N; 44.4 N; 11.1 N]

Since 5.71 N is less than 11.1 N the block does not slide and the force of friction is 5.71 N, up the plane. If the plane were frictionless a 15-N force would not be enough to keep the block from sliding down. Friction is required to help hold the block.

PROBLEM 5. For the block of the previous problem what minimum applied force F, up the plane, keeps the block from sliding down?

SOLUTION: The force of friction must now be directed up the plane and have its maximum magnitude. Replace f with $\mu_s N$ in the second-law equations. They become:

x component: *y* component:

Solve for F.

[ans: 9.60 N]

PROBLEM 6. For the block of the previous problem what maximum force F, up the plane, can be applied without moving the block?

SOLUTION: The force of friction must now be directed down the plane and have its maximum magnitude. Replace f by $-\mu_s N$ in the second-law equations. They become:

x component: *y* component:

Solve for F.

[ans: 31.8 N]

For any applied force between 9.60 N and 31.8 N the block does not slide.

Frictional forces are parallel to the surface where they are exerted. If the surface is vertical, for example, the force of friction is vertical. Here's an example.

PROBLEM 7. A teacher holds a 150-g chalkboard eraser against a vertical chalkboard. The coefficient of static friction between the eraser and board is 0.55. If the teacher exerts a force of 2.0 N perpendicularly to the board, does the eraser fall?

SOLUTION: In the space to the right draw a free-body diagram for the eraser. The normal force of the chalkboard on the eraser is horizontal and the force of friction is upward. Write the components of Newton's second law and solve for the force of friction f_{rest} that will hold the eraser and for the normal force N. Compare f_{rest} and $\mu_s N$.

[ans: the eraser falls]

What is the minimum perpendicular force the teacher must exert to keep the eraser from falling?

SOLUTION:

[ans: 2.67 N]

80   *Chapter 6: Particle Dynamics*

Uniform circular motion problems are solved in much the same way as any other second law problem. Carry out the set of instructions given in the previous chapter of this Student's Companion. Place the coordinate system so one of the axes is in the direction of the acceleration, pointing from the object toward the center of its orbit. For most problems you will want to substitute v^2/r for the magnitude of the acceleration. Here v is the speed of the object and r is the radius of its orbit.

To see how it is done you should carefully study the three examples discussed in the text: a conical pendulum, an amusement park rotor, and a banked roadway. For a conical pendulum the horizontal component of the tension in the string provides the centripetal force required to keep the object in its circular orbit. For a rotor it is the normal force of the wall and for a banked curve it is the horizontal component of the normal force.

PROBLEM 8. A 2.5-kg object is attached to one end of a light rod with a length of 0.75 m. The other end of the rod is pivoted so the object swings in a horizontal circle with radius equal to the length of the rod. If the speed of the object is 3.6 m/s what is the force of the rod on it? What is the force of the object on the rod?

SOLUTION:

[ans: 43.2 N, inward; 43.2 N, outward]

The rod may not be responsible for the only force on the object. Suppose the circular path is vertical. Then at the top of the swing the force of gravity is toward the center of the circle and aids the rod in getting the object around while at the bottom of the swing it is away from the circle center and hinders the rod. As a result, the force of the rod must be greater at the bottom than at the top if the motion is to be uniform.

PROBLEM 9. The length of the rod and the speed of the object are the same as before but now the circular path is vertical. Calculate the force of the rod on the object at the bottom of the swing and at the top.

SOLUTION: In the space to the right draw a free-body diagram for the object at the bottom of the swing. Label the force of gravity mg and the force of the rod F. The first is directed downward while the second is directed upward. Take the positive axis to be upward and write Newton's second law for the object. Substitute v^2/r for the acceleration and solve for F.

[ans: 67.7 N, up]

Now draw a free-body diagram for the object at the top of its swing. Take the force of the rod on the object to be downward. Take the positive axis to be downward and write Newton's second law for the object. Substitute v^2/r for the acceleration and solve for F.

[ans: 18.7 N, down]

PROBLEM 10. If the object moves sufficiently slowly the rod may push outward at the top of the swing. This occurs if the force of gravity alone is greater than the force required to get the object around the top of the circle at the given speed. Consider the object and rod of the last problem and calculate the range of speeds for which the force of the rod is outward at the top of the swing.

SOLUTION: If the positive axis is taken to be downward this is the same as asking for the range of speeds for which the force of the rod F is negative. Look at the second-law equation you wrote for the previous problem, corresponding to the top of the swing. Note that when $v = 0$ then F is clearly negative. In fact, $F = -mg$. Also note that as v increases the magnitude of F decreases until it vanishes. If v increases still more F becomes positive and its magnitude increases. The upper limit of the range of speeds for which F is negative is determined by the condition $F = 0$. Set $F = 0$ and solve for v.

[ans: $0 < v < 2.71$ m/s]

If a string is used instead of a rod the object will not go around the circle if its speed is in this range. The string can only pull inward, not push outward. As a too-slow object nears the top of the swing the string goes slack.

Some problems deal with the banking of roads. Notice that as the angle of banking is made greater the horizontal component of the normal force increases. The trick is to set the angle just right so the normal force provides the correct centripetal force to get cars around the curve at the chosen speed. Then a frictional force is not required.

PROBLEM 11. At what angle should a road be banked so cars traveling at 30 km/h can round a 200-m radius curve without relying on friction?

SOLUTION: In the space to the right draw a free-body diagram for a car. It will look exactly like the diagram for an object on an inclined plane. Take the incline to be at an angle θ above the horizontal. Include the force of gravity and the normal force of the road on the car. Take the x axis to be horizontal and positive toward the center of the circular path. This is the direction of the acceleration if the car does not slide. Take the y axis to be vertical. The second-law equations are:

x component: y component:

Substitute v^2/r for the acceleration a. You should obtain $N\sin\theta = mv^2/r$ and $N\cos\theta - mg = 0$. Eliminate N, solve for $\tan\theta$ ($= \sin\theta/\cos\theta$), and calculate the value of θ.

[ans: 19.2°]

In some problems the force is given as a function of time. The acceleration is then also a function of time: for one-dimensional motion $a(t) = F(t)/m$. The acceleration can be integrated to find an expression for the velocity, again as a function of time, and the velocity can be integrated to find an expression for the coordinate as a function of time. Initial conditions are used to evaluate the constants of integration. Then the usual kinematic questions can be answered.

PROBLEM 12. An object of mass m moves along the x axis, subject to a force given by $F = At^3$, where A is a constant. At time $t = 0$ it is at the origin and its velocity is v_0. Derive expressions for its velocity and coordinate as functions of time.

SOLUTION: The acceleration of the object is given by $a(t) = (A/m)t^3$. Its velocity is given by $v(t) = \int a(t)\,dt = \int (A/m)t^3\,dt = (A/4m)t^4 + C_1$, where C_1 is a constant of integration, determined by the requirement that $v(0)$ be v_0. Since $v(0) = C_1$, this means $C_1 = v_0$. Thus $v(t) = v_0 + (A/4m)t^4$.

The coordinate is given by $x(t) = \int v(t)\,dt = \int v_o\,dt + \int (A/4m)t^4\,dt$. Evaluate the integral and use the condition that $x(0) = 0$ to find a value for the constant of integration.

[ans: $v(t) = v_0 + (A/4m)t^4$; $x(t) = v_0 t + (A/20m)t^5$]

The force is applied to slow a 5.0-kg object initially moving at 17 m/s in the positive x direction. What value should the force constant A have so the object momentarily comes to rest in 12 s? How far does the object travel in this time?

SOLUTION: Solve $v_0 + (A/4m)t^4 = 0$ for A, then evaluate $x(t) = v_0 t + (A/20m)t^5$.

[ans: -0.0164 N/s^3; 163 m]

PROBLEM 13. An object of mass m moves along the x axis, subject to a force given by $F(t) = Ae^{-\alpha t}$, where A and α are constants. At time $t = 0$ its velocity is v_0 and its coordinate is x_0. Find expressions for its velocity and coordinate as functions of time.

SOLUTION: The acceleration of the object is $a(t) = (A/m)e^{-\alpha t}$ and the velocity is $v(t) = \int a(t)\,dt = \int (A/m)e^{-\alpha t}\,dt$. Evaluate the integration constant by requiring that $v(0) = v_0$. The coordinate is given by $x(t) = \int v(t)\,dt$. Evaluate the constant of integration by requiring that $x(0) = x_0$.

[ans: $v(t) = v_0 + (A/m\alpha)(1 - e^{-\alpha t})$; $x(t) = x_0 + [v_0 + (A/m\alpha)]t - (A/m\alpha^2)(1 - e^{-\alpha t})$]

A 50-kg crate is sliding across essentially frictionless ice at 12 m/s. A person attempts to stop the crate by pushing on it with a force given by $50e^{-0.055t}$, where the force is in newtons and the time t is in seconds. The initial force is 50 N but the person rapidly tires and the force decreases exponentially. How much time is required to stop the crate and how far does the crate go in that time?

SOLUTION: First solve $v_0 + (A/m\alpha)(1 - e^{-\alpha t}) = 0$ for t. You should obtain $t = -(1/\alpha)\ln[1 + (v_0 m\alpha/A)]$. Now substitute $m = 50$ kg, $v_0 = 12$ m/s, $A = -50$ N, and $\alpha = 0.055\,\text{s}^{-1}$ and obtain a numerical result for t. Note that the force opposes the motion so A is negative if v_0 is positive. To find the distance traveled take $x_0 = 0$ and substitute values into $x(t) = [v_0 + (A/m\alpha)]t - (A/m\alpha^2)(1 - e^{-\alpha t})$.

[ans: 19.6 s; 96.9 m]

Some problems deal with projectiles experiencing a drag force. The text handles the special case for which the drag force is proportional to the velocity: $\mathbf{D} = -b\mathbf{v}$, where b is a constant. If an object of mass m is dropped from rest then its velocity is given by

$$v(t) = \frac{mg}{b}\left[1 - e^{-bt/m}\right],$$

where the downward direction was taken to be positive. You should be able to integrate this expression to show that the y coordinate is given by

$$y(t) = y_0 + \frac{mg}{b}t - \frac{m^2 g}{b^2}\left[1 - e^{-bt/m}\right]$$

where $y = 0$ is the initial position. First evaluate the indefinite integral $\int v(t)\,dt$, then evaluate the constant of integration by requiring that $y(0) = y_0$.

PROBLEM 14. A 0.50-kg ball is dropped from rest in air. Take the force of the air to be proportional to the velocity with a drag coefficient of $b = 0.12$ kg/s. Compare its speed after 5.0 s with the speed of a similar ball dropped in a vacuum. Also compare the distances traveled during this time by the two balls.

SOLUTION:

[ans: $v_{\text{air}} = 28.5$ m/s, $v_{\text{vac}} = 49.0$ m/s; $y_{\text{air}} = 85.3$ m, $y_{\text{vac}} = 122.5$ m]

You should know that the terminal speed is given by $v_T = mg/b$.

PROBLEM 15. What is the terminal speed of the ball of the previous problem? If dropped from rest how long does the ball take to reach 80% of its terminal speed? How far does it fall in that time?

SOLUTION: Calculate the terminal speed by evaluating mg/b. Now derive an expression for the time to reach any speed v. You should obtain $t = -(m/b)\ln(1 - bv/mg)$. Evaluate this expression for $v = 0.8mg/b$. Finally calculate y.

[ans: 40.8 m/s; 6.71 s; 138 m]

Here is a projectile problem that takes air resistance into account.

PROBLEM 16. A 1.7-kg projectile is fired with a velocity of 250 m/s, at an angle 35° above the horizontal. The force of air resistance is given by $-bv$, with $b = 0.10$. Take the origin of the coordinate system to be at the firing point, with the positive y axis upward and the x axis horizontal and in the plane of motion. What are the velocity components of the projectile 20 s after firing?

SOLUTION: Use the expressions you derived in the Basic Concepts section: $v_x(t) = v_{0x}e^{-bt/m}$ and $v_y(t) = (mg/b + v_{0y})e^{-bt/m} - mg/b$.

[ans: $v_x = 63.1$ m/s; $v_y = -71.0$ m/s]

For contrast, what would these quantities be if air resistance were absent?

SOLUTION:

[ans: $v_x = 205$ m/s; $v_y = -52.6$ m/s]

When subjected to air resistance what is the ratio of its speed to terminal velocity after 20 s?

SOLUTION:

[ans: 0.570]

Chapter 6: Particle Dynamics

III. MATHEMATICAL SKILLS

The mathematical skills you will need for most parts of this chapter are old standbys. Much depends on your ability to resolve vectors into components and to solve simultaneous algebraic equations.

1. To fully understand Section 6–5 of the text you will also need to know how to evaluate definite integrals of polynomials. In particular you should know that the definite integral with respect to t of t^n is given by

$$\int t^n \, dt = \frac{t^{n+1}}{n+1} + C$$

if n is not -1. Here C is a constant of integration. If $n = -1$ then

$$\int \frac{dt}{t} = \ln t + C.$$

These integrals are used to evaluate

$$v_2 - v_1 = \int_{t_1}^{t_2} a(t) \, dt$$

for the change in velocity of a particle with an acceleration that is a polynomial function of time.

2. To understand Section 6–7 of the text you should be familiar with integrals of the form

$$\int \frac{dv}{g - bv/m} = -\frac{m}{b} \ln(g - bv/m) + C.$$

The integral is evaluated by letting $u = g - bv/m$. Then $dv = -(m/b) \, du$ and the indefinite integral becomes

$$-\frac{m}{b} \int \frac{du}{u} = -\frac{m}{b} \ln u.$$

The result is obtained when $g - bv/m$ is substituted for u.

This integral is used to evaluate

$$t = \int \frac{dv}{a(v)}$$

when the acceleration has the form $a = mg - bv$, appropriate for a projectile acted on by a force of air resistance proportional to the velocity.

Evaluating the constant of integration and solving for v may be a little tricky in this case. If $v = v_0$ at $t = 0$ then $C = (m/b) \ln(g - bv_0/m)$, so the equation becomes $t = -(m/b) \ln(g - bv/m) + (m/b) \ln(g - bv_0/m) = -(m/b) \ln \left[\frac{g - bv/m}{g - bv_0/m} \right] = -(m/b) \ln \left[\frac{mg - bv}{mg - bv_0} \right]$, where we have used the identity $\ln A - \ln B = \ln(A/B)$. Multiply both sides by $-m/b$ to obtain $-mt/b = \ln \left[\frac{mg - bv}{mg - bv_0} \right]$. Now use each side of this equation as the exponent of e, the base of the natural logarithms, and recognize the $e^{\ln \alpha} = \alpha$. You should obtain $e^{-bt/m} = (mg - bv)/(mg - bv_0)$. Finally, solve for v: $v(t) = (mg/b) + (v_0 - mg/b)e^{-bt/m}$.

IV. COMPUTER PROJECTS

The computer program you wrote for the projects of Chapter 2 must be revised if the acceleration is a function of position or velocity. The average acceleration in an interval cannot be approximated by $a = (a_b + a_e)/2$ because a_e cannot be computed until v_e or x_e are known. You can, however, use the acceleration at the beginning of the interval and write $v_e = v_b + a_b \Delta t$. This is usually a poor approximation to the average acceleration, so compared to the program of Chapter 2, much smaller intervals must be used to obtain the same accuracy.

Errors arise when a large number of intervals are used because the computer normally carries only a small number of significant figures (eight or ten) and the last is often in error. These so-called round-off errors accumulate. You can decrease the effect significantly by carrying out the calculation in double precision.

The outline of a possible program for one-dimensional motion is:

> input initial values: t_0, x_0, v_0
> input final time and interval width: t_f, Δt
> set $t_b = t_0$, $x_b = x_0$, $v_b = v_0$
> **begin loop** over intervals
>   calculate acceleration at beginning of interval: $a_b = a(t_b)$
>   calculate velocity at end of interval: $v_e = v_b + a_b \Delta t$
>   calculate "average" velocity: $v = (v_b + v_e)/2$
>   calculate coordinate at end of interval: $x_e = x_b + v\Delta t$
>   calculate time at end of interval: $t_e = t_b + \Delta t$
> * **if** $t_e \geq t_f$ **then**
>   print or display t_b, x_b, v_b
>   print or display t_e, x_e, v_e
>   exit loop
> **end if** statement
> set $t_b = t_e$, $x_b = x_e$, $v_b = v_e$
> **end loop** over intervals
> stop

The program lines following the asterisk may be different for different applications. For some projects you will want to display results for more than one time. Replace the exit loop statement with a statement that increments t_f. For some projects you may want to plot points on the monitor screen instead of displaying or printing values. For either of these modifications see the programs of the Chapter 6 Computer Projects section.

Sometimes air resistance must be taken into account when an object moves in the air. For many objects the acceleration is then given by $a = -g - (b/m)v$, where the positive direction is upward, m is the mass of the object, and b is a constant that depends on the interaction of the object with the air.

PROJECT 1. Suppose an object with $(b/m) = 0.100\,\text{s}^{-1}$ is fired upward from ground level with an initial speed of 100 m/s. On separate graphs plot the coordinate and velocity every 0.5 s from $t = 0$ to $t = 17\,\text{s}$. First test the program to see what interval width it needs to obtain 2 significant figure accuracy over that range of time.

How much time does the projectile take to get to the highest point on its trajectory and how high is that point? How long does it take to get back to the ground and what is its velocity just before it reaches the ground? You might use your graph to obtain approximate answers, then use the program to refine them. [ans: highest point: 7.03 s, 311 m; ground: 16.2 s, 58.9 m/s]

Notice that the projectile takes longer to fall from the highest point than it does to reach that point and that it returns to ground level with a speed that is less than the firing speed. Compare your answers with those you would obtain if the object were in free fall. The time to reach the highest point is _____ and the highest point is _____ with air resistance than without. The total flight time is _____ and the speed on impact is _____ with air resistance than without.

PROJECT 2. Here's a problem for which the acceleration depends on position. Starting from rest at $x = 0$, a 3.5 kg box is dragged along the ground in the positive x direction by a constant force of 12 N acting horizontally. The ground is rougher toward larger x and the coefficient of kinetic friction increases according to $\mu_k = 0.070\sqrt{x}$. Use the program to find its position and velocity every second from $t = 0$ to the time it comes to rest again. At what time does the box come to rest? How far has it been dragged? [ans: 11.4 s; 56.2 m]

What is the maximum speed of the box? When does it have this speed? Where is it when it has this speed? [ans: 7.56 m/s; 4.84 s; 25.0 m from the starting point]

For the special case of an acceleration that is linear in the velocity $a = (a_b + a_e)/2$ *can be used* to calculate the velocity at the end of the interval. Here's how to do it. The acceleration at the beginning of the interval is given by $a_b = -g - (b/m)v_b$ and the acceleration at the end of the interval is given by $a_e = -g - (b/m)v_e$, so $a = (a_b + a_e)/2 = -g - (b/2m)v_b - (b/2m)v_e$. The velocity at the end of the interval is given by $v_e = v_b + a\Delta t = v_b - g\Delta t - (b\Delta t/2m)v_b - (b\Delta t/2m)v_e$. Solve this expression for v_e. The result is $v_e = [v_b(1 - b\Delta t/2m) - g\Delta t]/(1 + b\Delta t/2m)$. If $h = b\Delta t/2m$ this becomes $v_e = [v_b(1 - h) - g\Delta t]/(1 + h)$. An outline of a program is:

> input initial values: t_0, x_0, v_0
> input final time and interval width: t_f, Δt
> calculate parameter: $h = b\Delta t/2m$
> set $t_b = t_0$, $x_b = x_0$, $v_b = v_0$
> **begin loop** over intervals
>   calculate velocity at end of interval: $v_e = [v_b(1 - h) - g\Delta t]/(1 + h)$
>   calculate "average" velocity: $v = (v_b + v_e)/2$
>   calculate coordinate at end of interval: $x_e = x_b + v\Delta t$
>   calculate time at end of interval: $t_e = t_b + \Delta t$
> *   **if** $t_e \geq t_f$ **then**
>   print or display t_b, x_b, v_b
>   print or display t_e, x_e, v_e
>   exit loop
>   **end if** statement
>   set $t_b = t_e$, $x_b = x_e$, $v_b = v_e$
> **end loop** over intervals
> stop

PROJECT 3. Use this program to find the position and velocity of the projectile of the first project at the end of 17 s. Notice that far fewer intervals are required to obtain 2 significant figure accuracy.

Now use the program to investigate the influence of the air on a projectile fired straight upward. Make a table of the time to reach the highest point, the coordinate of the highest point, the time to reach the ground, and the velocity just before reaching ground, all as functions of the resistance parameter b/m. Try $b/m = 1.00\,\text{s}^{-1}$, $0.100\,\text{s}^{-1}$, and $0.0100\,\text{s}^{-1}$.

This exercise is designed to give you some idea of the effect of air resistance. A javelin has a small value of b/m, a basketball has a larger value, and a ping-pong ball has a still larger value. The larger b/m the greater the effects.

You might try to find a value for the resistance coefficient b/m of a ping-pong ball or other object. Shoot it into the air with a spring gun, measure the initial speed with a photogate timer, and time its return to the firing level. Then try various values of b in your computer program until you find the value that reproduces the experimentally determined time of flight.

PROJECT 4. You can use the program to investigate the approach to terminal velocity. Suppose an object with $b/m = 0.100\,\text{s}^{-1}$ is dropped from a high cliff. Use the program to plot its velocity every 2 s over the first minute of its fall. The speed should approach the value $mg/b = 98\,\text{m/s}$. Repeat for objects with $b/m = 0.0500\,\text{s}^{-1}$ and $0.500\,\text{s}^{-1}$.

The program can be modified to deal with a projectile moving in two dimensions, and subjected to a drag force that is proportional to its velocity. The acceleration components are now given by $a_x = -(b/m)v_x$ and $a_y = -g - (b/m)v_y$. The outline of a program is:

> input initial values: $t_0, x_0, y_0, v_{0x}, v_{0y}$
> input final time and interval width: $t_f, \Delta t$
> calculate parameter: $h = b\Delta t/2m$
> set $t_b = t_0, x_b = x_0, y_b = y_0, v_{xb} = v_{0x}, v_{yb} = v_{0y}$
> **begin loop** over intervals
> > calculate velocity at end of interval:
> > $$v_{xe} = v_{xb}(1-h)/(1+h)$$
> > $$v_{ye} = [v_{yb}(1-h) - g\Delta t]/(1+h)$$
> > calculate "average" velocity: $v_x = (v_{xb} + v_{xe})/2$, $v_y = (v_{yb} + v_{ye})/2$
> > calculate coordinates at end of interval: $x_e = x_b + v_x \Delta t$, $y_e = y_b + v_y \Delta t$
> > calculate time at end of interval: $t_e = t_b + \Delta t$
> > * **if** $t_e \geq t_f$ **then**
> > > print or display $t_b, x_b, y_b, v_{xb}, v_{yb}$
> > > print or display $t_e, x_e, y_e, v_{xe}, v_{ye}$
> > > exit loop
> > **end if** statement
> > set $t_b = t_e, x_b = x_e, y_b = y_e, v_{xb} = v_{xe}, v_{yb} = v_{ye}$
> **end loop** over intervals
> stop

PROJECT 5. A projectile with $b/m = 0.100\,\text{s}^{-1}$ is fired over level ground with an initial velocity of 100 m/s, 45° above the horizontal. Find its coordinates every 0.5 s from $t = 0$ (the time of firing) to $t = 12.5\,\text{s}$. Plot its trajectory. First find the value of Δt required to obtain 3 significant figure accuracy over the entire time interval.

Find the time the projectile reaches the highest point and the coordinates of the highest point. Take the x axis to be horizontal and the y axis to be positive in the upward direction. [ans: $t = 5.43\,\text{s}$, $x = 296\,\text{m}$, $y = 175\,\text{m}$]

Find the time of flight and range. What are its velocity components just before it lands? [ans: $t = 12.1\,\text{s}$, $x = 495\,\text{m}$, $v_x = 21.2\,\text{m/s}$, $v_y = -47.5\,\text{m/s}$]

PROJECT 6. If air resistance is significant, maximum range is obtained for a firing angle that is different from $45°$. Consider a projectile with $b/m = 0.100\,\text{s}^{-1}$, fired with an initial speed of $100\,\text{m/s}$ over level ground. Find the range to 3 significant figures for firing angles of $30°$, $35°$, and $40°$. You already know the range for $45°$. The firing angle for maximum range is between _____ and _____. You may wish to refine the interval to obtain the firing angle to 2 significant figures. [ans: $510\,\text{m}$; $519\,\text{m}$; $514\,\text{m}$]

PROJECT 7. Suppose the projectile of the last project ($b/m = 0.100$) is fired from the edge of a high cliff instead of over level ground. Assume the same initial conditions (fired from the origin with a speed of $100\,\text{m/s}$, $45°$ above the horizontal) and plot its trajectory for the first minute of its flight.

Notice that the trajectory is not symmetric about the highest point but is blunted in the forward direction. Near the end of the time interval the projectile is falling nearly straight down. Both the horizontal and vertical components of the velocity approach limiting values. The limiting value of the horizontal component is _____ and the limiting value of the vertical component is _____.

V. NOTES

Chapter 7
WORK AND ENERGY

I. BASIC CONCEPTS

The central concept of this chapter is the idea of <u>work</u>. You should learn and understand the definition of work, you should learn how to calculate the work done by forces in various situations, and you should learn how work is related to the change in the kinetic energy of a particle (the work-energy theorem).

Definition of work. Several definitions of work are given, for situations of increasing complexity. You should remember them and the situations to which they apply.

1. The particle moves through a displacement **s**. The force **F** being considered is constant and makes the angle ϕ with the displacement when **F** and **s** are drawn with their tails at the same point. Then the work done by **F** is given by $W =$ _____. Alternatively, if the particle has displacement Δx, along the x axis, and the force has x component F_x then the expression for the work can be written $W =$ _____.

2. The particle moves along a straight line (the x axis). The force **F** is parallel to the x axis and is not constant. Then the work it does as the particle moves from x_1 to x_2 is given by the integral $W =$ _____. Be sure your definition allows for a force in the same direction as the displacement and for one in the opposite direction. To evaluate the integral, the x component of the force must be known as a function of _____. The work done by the force is the area under the graph of _____ vs. _____.

3. A particle moves in a plane, subjected to a variable force **F**. Write the integral definition of the work here: $W =$ _____. Explain how the integral can be evaluated, in principle, by dividing the path into a large number of segments. Don't forget to give the quantity to be evaluated for each segment. _____

This expression, generalized slightly to three dimensions, is the general definition of work. The expressions you wrote above in 1 and 2 for special situations can be derived from it.

If several forces act on the particle you may calculate the work done by each separately. The total work done by all forces is the sum of the individual works and is the same as the work done by the resultant force.

A person carrying a heavy box horizontally with constant velocity does no work on the box because _____.

The normal force of a surface on a sliding crate does no work, no matter what the orientation of the surface, because _____.

A string used to whirl an object around a circle with constant speed does no work because _____.

Work can be positive or negative. Each of the four diagrams below show a block moving on a table top. On each of the upper two show how you would apply a force so it does positive work. On each of the lower two show how you would apply a force so it does negative work. In each case direct the force so it is not parallel to the velocity and assume the block continues to move in the same direction, at least for a short while after the force is applied.

Force **F** does
positive work

Force **F** does
negative work

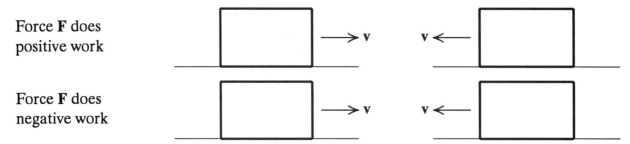

Work is a scalar. It does not have a direction associated with it. When several forces act on an object and you want to find the total work done, you simply add the works done by the individual forces. You must, of course, include the appropriate sign for each work. The direction of each force and the direction of the displacement are important for calculating the work done by a force but neither these directions or any others must be taken into account when the individual works are summed.

The SI unit of work is _____. In terms of SI base units it is _____. To discuss atomic phenomena physicists often use a unit of work called an electron volt. In terms of the SI unit for work 1 eV is _____ J.

Work done by gravity. You should know how to calculate the work done by the force of gravity and the work done by the force of an ideal spring. In addition to being excellent examples of a constant and a variable force, respectively, they enter many problems.

When an object of mass m falls from height y_i to height y_f near the earth's surface gravity does work $W = $ _____. When the mass is raised from height y_i to height y_f gravity does work $W = $ _____. The two expressions you wrote should be identical. On the way down the sign of the work done by gravity is _____ while on the way up it is _____. If a ball is thrown into the air, falls, and is caught at the height from which it was thrown, the total work done by gravity on the round trip is _____.

You should know that the work done by gravity is the same no matter what path is taken between the initial and final points. In addition, all that counts is the initial and final altitudes. The two positions need not have the same horizontal coordinate.

Remember that the expression you wrote for the work done by gravity is valid even if the object experiences air resistance. If air resistance is present this expression does not, of course, give the total work done by all forces.

Work done by an ideal spring. The force exerted by an ideal spring on an object is a variable force. Its direction depends on whether the spring is extended or compressed and its magnitude depends on the amount of extension or compression.

Assume the spring is along the x axis with one end fixed and the other attached to the object, as shown. When the object is at $x = 0$ the spring is neither extended or compressed. This is the equilibrium configuration. When the object is at any coordinate x, the force exerted by the spring is given by $F_s = $ _____. Here k is the <u>force constant</u> of the spring. The _____ the spring, the larger the force constant.

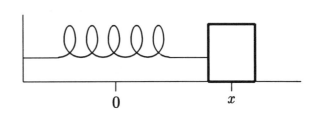

When the spring shown above is extended the sign of the force it exerts is _____ and the force tends to pull the object toward _____. When the spring is compressed the sign of the force is _____ and the force tends to push the object toward _____. This behavior is summarized by calling the force a <u>restoring</u> force.

Sample Problem 3 of the text shows how the force constant of an ideal spring can be measured by hanging a mass m from its end and measuring the elongation s of the spring with the mass at rest. In terms of m, s, and k the net force on the mass is given by _____ and since the mass is in equilibrium this must be zero. Thus $k = $ _____.

The SI units of a force constant are _____.

As the object moves from some initial coordinate x_i to some final coordinate x_f the work done by the spring is

$$W_s = \int_{x_i}^{x_f} -kx \, dx = $$

Give an example in which the spring does positive work: _____

Give an example in which the spring does negative work: _____

In general, the spring does positive work whenever the object attached to it is moving toward the equilibrium point and does negative work whenever the object is moving away from the equilibrium point.

Suppose a mass is attached to a horizontal spring and is free to move on a frictionless table top, as in the diagram above. Now you pull on the mass with a force given by $F_{\text{ext}} = +kx$. As the mass goes from $x = x_i$ to $x = x_f$ the work you do is given by $W_{\text{ext}} = $ _____. Of course, you may pull on the mass with any force you like, not necessarily $+kx$. The external force $+kx$, however, has special significance because the net force on the mass is then _____ and the mass does not accelerate during the pulling. When this external force is applied, the total work done by the spring and your force, taken together, is _____.

Work-energy theorem. The significance of work is found in the work-energy theorem, which shows us that work tends to change the kinetic energy of a particle. First, the kinetic energy of a particle with mass m and speed v is defined by

$$K = $$

Kinetic energy is a scalar and so does not have a direction associated with it. Sometimes you will be given the velocity components v_x and v_y for an object moving in the xy plane and asked to compute the kinetic energy. In terms of the velocity components the kinetic energy is given by $K =$ _____. You cannot interpret the two terms in this expression as components of a vector. They are not.

The work-energy theorem is: during any portion of a particle's motion the *total* work done by *all* forces acting on the particle equals the change in the particle's kinetic energy. Let W_total be the total work done on a particle during some portion of its motion. If K_i is the initial kinetic energy and K_f is the final kinetic energy for that portion, then $W_\text{total} =$ _____. Be sure you perform the subtraction in the correct order.

If the total work is negative, then the kinetic energy and speed of the particle both _____; if the total work is positive, then the kinetic energy and speed _____; and if the total work is zero then the kinetic energy and speed _____. Go back and review the exercise in Chapter 5 of this Student's Companion showing that a particle speeds up when the acceleration has a vector component in the direction of the velocity and slows down when the acceleration has a vector component opposite the velocity. In the first case the net force is doing positive work and in the second it is doing negative work.

You should be able to carry out the proof of the work-energy theorem for a particle moving in one dimension. Start with $F = ma$ and consider the velocity v to be a function of coordinate x, which in turn is a function of time t. Use the chain rule of the calculus to show that $F = mv\,dv/dx$:

Then substitute into $W = \int_i^f F\,dx$ and show that $W = \int_i^f mv\,dv$:

Finally evaluate the integral to show that $W = \tfrac{1}{2}mv_f^2 - \tfrac{1}{2}mv_i^2$:

This derivation shows that the work-energy theorem is a direct consequence of Newton's second law. It also explains why the *total* work (done by *all* forces acting on the particle) enters the theorem.

If two observers use different inertial frames (moving relative to each other) to measure the change in kinetic energy of an object over some interval, they will obtain different values.

They will also obtain different values for the total work done on the object, but the work-energy theorem will still be valid for both frames. Each will conclude that the change in kinetic energy equals the total work done.

You should understand that the work-energy theorem is valid only for a single particle or an object that can be treated as a particle. If the object becomes distorted or the positions of individual particles within the object change relative to each other then the theorem cannot be applied directly to the object as a whole.

Give an example of a situation in which no work is done by external forces on an object but the object comes to rest: _____
_____
_____

In this example the object cannot be treated as a single particle and the work-energy theorem is not valid for the object as a whole.

Frictional forces arise when microscopic welds between two surfaces in contact are made, distorted, or broken. If $\mathbf{F}$ is the force at a weld and $\Delta \mathbf{s}$ is the displacement of the weld then the work done by the force is $\mathbf{F} \cdot \Delta \mathbf{s}$, but $\Delta \mathbf{s}$ is not the same as the displacement of the object as a whole. Thus the work-energy theorem in the form appropriate for a particle cannot be applied to an object on which frictional forces act.

Power. In words, power is _____.
Suppose the function $W(t)$ represents the work done by a force from time 0 to time t. Then the instantaneous power delivered by the force is given by $P =$ _____. The SI unit of power is _____. In terms of SI base units the unit for power is _____.

Consider a particle that at some instant of time is moving with velocity $\mathbf{v}$ and is acted on by a force $\mathbf{F}$. The power delivered to the particle by the force is given by $P =$ _____. Be sure your equation is valid even if the force is not parallel to the velocity.

Relativistic kinetic energy. When an object is moving with a speed near the speed of light c another definition of kinetic energy must be used. If m is the mass and v is the speed then the kinetic energy is defined by

$$K =$$

II. PROBLEM SOLVING

In general, the work done by a force $\mathbf{F}$ on an object is defined by the integral $W = \int_i^f \mathbf{F} \cdot d\mathbf{s}$, taken along the path of the object from some initial point to some final point. You should know how to evaluate this expression for several special cases.

If the force is constant along the path, the integral can evaluated immediately, with the result $W = \mathbf{F} \cdot \mathbf{s}$, where $\mathbf{s}$ is the displacement during the interval, a vector from the initial to the final point. Recall that the scalar product is given by $\mathbf{F} \cdot \mathbf{s} = Fs \cos \phi$, where ϕ is the angle between $\mathbf{F}$ and $\mathbf{s}$ when they are drawn with their tails at the same point.

PROBLEM 1. A 50-N force is used to pull crate a distance of 15 m in a straight line across a horizontal floor. Find the work done by the force for two orientations: (a) horizontal and (b) directed 25° above the horizontal.

SOLUTION: In both (a) and (b) $F = 50$ N and $s = 15$ m. In (a) $\phi = 0$ while in (b) $\phi = 25°$. Since the force is constant use $W = Fs \cos\phi$ to calculate the work.

[ans: 750 J; 680 J]

The force being considered may be the only force acting, one of several forces acting, or the resultant force.

PROBLEM 2. Person A pulls to the right on the crate of the previous problem with a horizontal force of 50 N while person B pulls to the left with a horizontal force of 40 N. As the crate moves 15 m to the right, what is the work done by A, what is the work done by B, and what is the total work done on the crate by A and B together?

SOLUTION:

[ans: 750 J; −600 J; 150 J]

The negative sign appears because the force of B is opposite the displacement. The angle between them is 180° and $\cos(180°) = -1$.

PROBLEM 3. A 5.0-kg knapsack is carried up one side of a 100-m high plateau, across the horizontal top, and halfway down the other side. Calculate the work done by gravity during each segment of the trip and the total work done by gravity over the whole trip. Compare the total work to the work that would have been done if the knapsack had been carried halfway up directly.

SOLUTION:

[ans: −9800 J; 0; 4900 J; −4900 J; they are the same]

PROBLEM 4. An inclined plane rises from left to right at 25° above the horizontal. A 5.0-kg crate is pushed 5.0 m up the incline. What is the magnitude of the force of gravity? What is the angle between the force of gravity and the displacement when they are drawn with their tails at the same point? What is the work done by gravity? What is the change in altitude of the crate? What would the work done by gravity be if the crate were lifted straight up through the same change in altitude?

SOLUTION:

[ans: 49.0 N; 115°; −104 J; 2.11 m; −104 J]

Close to the surface of the earth, where the force of gravity may be taken to be uniform, the magnitude of the work done by gravity is always mgh, where h is the magnitude of the change in altitude of the object. The work is positive if the altitude decreases and negative if the altitude increases.

To solve some problems you must use Newton's second law to calculate the force. Here is an example in which the second law is used to calculate the force applied to a box on an incline.

PROBLEM 5. A force **F** is applied to a 7.0-kg crate as it slides 5.0 m down a frictionless incline that makes an angle of 20° with the horizontal. If the acceleration of the box is 2.5 m/s², down the incline, what is the component of the applied force along the incline? What work is done by the applied force and by gravity?

SOLUTION: Draw a free-body diagram, choose a coordinate system, write Newton's second law in component form, and solve for the component of **F** along the incline. Only this component contributes to the acceleration and to the work done. The normal component determines the normal force of the incline on the box but we are not interested in that here. Once the relevant force component has been calculated, compute the work done by the applied force. Finally, calculate the change in altitude of the box and use this to compute the work done by gravity.

[ans: 5.96 N, up the plane; −29.8 J; 117 J]

PROBLEM 6. Work the same problem for an acceleration of 4.0 m/s², down the incline.

SOLUTION:

[ans: 4.54 N, down the plane; 22.7 J; 117 J]

If the force changes in either magnitude or direction (or both) along the path you must evaluate the integral $\int_i^f \mathbf{F} \cdot d\mathbf{s}$. We start with the work done by a variable force when the motion is along a line.

PROBLEM 7. A crate is pushed in the positive x direction along a frictionless horizontal floor with a horizontal force. The magnitude of the applied force is given by $F = 0.50x$, where F is in newtons and x is in meters. What work does the force do as the crate moves from $x = 0$ to $x = 2.5$ m? What work does it do as the crate moves from $x = 2.5$ m to $x = 5.0$ m?

SOLUTION: First write $F = Ax$ and carry out the integration

$$W = \int_{x_1}^{x_2} F\, ds = A \int_{x_1}^{x_2} x\, dx =$$

Then evaluate the result for the two cases.

[ans: 1.56 J; 4.69 J]

Chapter 7: Work and Energy

Notice that the distance traveled is the same but the work is greater in the second case. This is because the crate is further along the x axis, where the force is greater.

A spring provides a good example of a variable force. If one end is fixed and the other end is moved so the spring is either extended or compressed from its equilibrium length, then the force exerted by the spring is given by $F = -kx$, where k is the force constant. The coordinate x of the spring end is measured with the origin at the position of the movable end when the spring has its equilibrium length. If the end of the spring is moved from x_1 to x_2 the work done by the spring is $-\frac{1}{2}(x_2^2 - x_1^2)$.

Some problems deal with finding the force constant.

PROBLEM 8. A spring is hung vertically alongside a meter stick. When 5.0 kg is placed on the spring its equilibrium position is at the 20 cm mark on the meter stick. When 15 kg is hung on the spring its equilibrium point is at the 50 cm mark. What is the force constant and where is the end of the spring when no mass is placed on it?

SOLUTION: Place the origin of the coordinate system at the 0 mark on the meter stick and take the y axis to be positive downward. The equilibrium position of the spring end when no mass is attached to it is not necessarily $y = 0$. If we let y_0 be this equilibrium position then the force of the spring on the mass is given by $-k(y - y_0)$ and equilibrium with mass m attached occurs when $k(y - y_0) = mg$. Write this equation twice, once for $m = 5.0$ kg and once for $m = 15$ kg. Then solve the two equations simultaneously for k and y_0.

SOLUTION:

[ans: 327 N/m; at the 5.0 cm mark]

PROBLEM 9. When a 4.0-kg mass is hung on the end of vertical spring and gently lowered to its new equilibrium point, the spring extends 35 cm. What is the force constant? What work was done by the spring as it was extended? What work was done by gravity? If a person pulls the spring downward an additional 35 cm what additional work is done by the spring?

SOLUTION: Take the origin to be at the lower end of the spring before the mass is hung on it. At the new equilibrium position the force of the spring balances the force of gravity, so $mg = kx$, where x is the extension of the spring. Use this to find a value for k. Once k is known the work done by the spring can be computed easily.

[ans: 112 N/m; −6.86 J; 13.7 J; −46.5 J]

Now consider the work done by a force when the motion is in a plane.

PROBLEM 10. A force of constant magnitude $F = 20$ N is applied to a particle that moves from one end to the other of a 1.5-m radius semicircle. First suppose the force is constant in direction, always parallel to the diameter and in the direction of the displacement. What work does it do? Then suppose the force changes direction so it is always tangent to the path, in the direction of the velocity. What work does it do now? (Clearly this force is not the only force acting on the particle. At the least a radial force must be applied to keep the particle on the path.)

SOLUTION: In the first case the force is constant in magnitude and direction, so you can use $W = \mathbf{F} \cdot \mathbf{s}$. The magnitude of $\mathbf{s}$ is the diameter of the semicircle. In the second case you must evaluate the integral $\int_i^f F \cos\phi \, ds$, where ϕ is the angle between $\mathbf{F}$ and the infinitesimal displacement ds. This angle is always 0 so $W = F \int_i^f ds = Fs$, where s is the path length traveled (πr, where r is the radius).

[ans: 60 J; 94.2 J]

PROBLEM 11. A 0.75-kg ball hangs by a 1.7-m string from the ceiling. A horizontal force is applied to push it until the string makes the angle $\theta_f = 25°$ with the vertical. If the magnitude of the applied force is continually adjusted so the ball moves with constant speed what work does it do?

SOLUTION: The first job is to find the magnitude of the force. Draw a free-body diagram for the ball when the string makes an arbitrary angle θ with the vertical. Take one coordinate axis to be tangent to the circular path of the ball and the other to be along the string. Write Newton's second law in component form and show that, since the tangential acceleration of the ball is zero, the magnitude of the applied force is given by $mg \tan\theta$.

Now compute the work. The tangential component of the applied force is $F \cos\theta$, so $W = \int_i^f F \cos\theta \, ds$, where ds is an element of path length along the circle. If θ is measured in radians then $ds = r \, d\theta$. Substitute $r \, d\theta$ for ds and carry out the integration. You should obtain $W = mgr(1 - \cos\theta_f)$. Evaluate this result.

[ans: 1.17 J]

You can also solve this problem using the work-energy theorem. Since the ball moves with constant speed, the change in kinetic energy is zero and the total work done on it is also zero. Only the applied force and the force of gravity do work, so $W_F + W_g = 0$. A little geometry shows the ball rises through the distance $r(1 - \cos\theta_f)$, so the work done by gravity is $W_g = -mgr(1 - \cos\theta_f)$. Thus the work done by the applied force is $+mgr(1 - \cos\theta_f)$, in agreement with the result given above.

The work-energy theorem tells us that the total work W done on a particle is equal to the change in the kinetic energy of the particle. That is, $W = \Delta K$, or since the kinetic energy is given by $\frac{1}{2}mv^2$, $W = \frac{1}{2}m(v_f^2 - v_i^2)$. Here m is the mass of the particle, v_i is its speed at the beginning of the interval, and v_f is its speed at the end of the interval.

PROBLEM 12. A 7.0-kg crate starts from rest and slides 1.7 m down a frictionless incline that makes an angle of 25° with the horizontal. What is its speed when it reaches the bottom?

SOLUTION: Here the only forces acting on the crate are the force of gravity and the normal force of the incline. Only the force of gravity does work. How much? $W_g = $ _____. The initial kinetic energy is 0 since the crate is at rest. The final kinetic energy is given by $\frac{1}{2}mv_f^2$. Equate W_g and $\frac{1}{2}v_f^2$, then solve for v_f.

[ans: 3.75 m/s]

PROBLEM 13. An engine on a train is used to pull a 1200-kg freight car around 90° of a 300-m radius curve. As it goes around it speeds up uniformly from 40 km/h to 60 km/h Use the work-energy theorem to calculate the tangential component of the force on it, assumed to be constant in magnitude.

SOLUTION: Since we know the initial and final speeds we can use $W = \Delta K$ to calculate the work done by the force. On the other hand the force is tangent to the path and has constant magnitude so the work it does is given by $W = F_T s$, where s is the distance the car travels. In terms of the circle radius r, $s = \pi r/2$. Substitute this expression for s into the expression for the work done and equate the result to the value you found for W. Solve for F.

[ans: 196 N]

PROBLEM 14. A 0.15-kg ball is dropped from a height of 2.0 m, hits the floor, and rebounds to a height of 1.2 m. Find the minimum work done by the floor on the ball.

SOLUTION: You need to know how fast the ball was going just before it hit the floor and just after it left the floor on the way up. As the ball drops gravity does work _____ J. Equate this to $\frac{1}{2}mv_b^2$, where v_b is the speed of the ball just before it hits the floor. Solve for v_b. As the ball rises gravity does work _____ J. Equate this to $-\frac{1}{2}mv_a^2$, where v_a is the speed of the ball just after it leaves the floor. Solve for v_a. Now the speeds are known and you can use the work-energy theorem to find the minimum work done by the floor.

[ans: −0.138 J]

Any additional work done by the floor on the ball goes to deform the ball and floor or to change the kinetic energies of the particles in the ball and floor without changing the kinetic energy associated with the motion of the ball as a whole. If the floor had not done any work then the ball would have rebounded to the height from which it was originally dropped.

PROBLEM 15. A extremely stiff spring (force constant = 7.0×10^4 N/m), at its equilibrium length, rests vertically with one end on the ground. A 0.75-kg stone is dropped onto the spring and becomes attached to its end. As a result the spring is compressed 5.0 cm before the stone comes momentarily to rest. From what height above the spring was the stone dropped?

SOLUTION: Two forces do work: gravity and the spring. The stone starts with a speed of 0 and ends with a speed of 0 so the change in kinetic energy is 0. This means the total work done is 0: $W_g + W_s = 0$. If the stone is dropped from a height h and the spring is compressed a distance x then the work done by gravity is $mg(h+x)$ and work done by the spring is $-\frac{1}{2}kx^2$. Solve $mg(h+x) - \frac{1}{2}kx^2 = 0$ for h.

[ans: 11.9 m]

PROBLEM 16. A stone is tied to a string and whirled in a vertical circle with a radius of 1.8 m. The string remains radial and no tangential force is applied but, of course, the force of gravity has a tangential component over most of the path. What minimum speed must the stone have at the bottom of its swing to stay on the circular path at the top?

SOLUTION: We first find the minimum speed at the top that will keep the stone on its circular path. At the top both the force of the string and the force of gravity are downward, toward the center of the circle. So $T + mg = mv_t^2/r$, where T is the tension in the string and v_t is the speed of the stone at the top. The minimum allowed speed occurs when $T = 0$. So, in terms of the circle radius r, $v_t = $ _____. Now use the work-energy theorem to find the speed at the bottom. When the stone is at the top its displacement from the bottom is $2r$ and is vertical. On the way up gravity does work _____, where m is the mass of the stone. Equate this to $\frac{1}{2}m(v_t^2 - v_b^2)$ and solve for v_b. Notice that you do not need to know the mass of the stone. Now substitute numerical values.

[ans: 9.39 m/s]

III. MATHEMATICAL SKILLS

1. Scalar products are used extensively in this chapter. Be sure you know how to evaluate them. Review the appropriate sections of Chapter 3 of the text and this Student Companion.

2. You must also know how to evaluate definite integrals of simple functions, chiefly polynomials. In particular know that

$$\int_{x_1}^{x_2} x^n \, dx = \frac{1}{n+1} x^{n+1} \Big|_{x_1}^{x_2} = \frac{1}{n+1} \left[x_2^{n+1} - x_1^{n+1} \right]$$

for $n \ne -1$ and

$$\int_{x_1}^{x_2} \frac{dx}{x} = \ln \frac{x_2}{x_1}.$$

These are essentially the same integrals that were given in the last chapter. The variable of integration is now x instead of t. Integrals such as these might be needed to find the work done by a variable force.

3. The integrals used to evaluate the work done by a force are a special type of integral called a path integral and are usually evaluated by converting them to ordinary integrals. For motion in a plane you may write

$$W = \int_a^b \mathbf{F} \cdot d\mathbf{s} = \int_a^b \left[F_x(x,y) \, dx + F_y(x,y) \, dy \right]$$

Here x and y are related to each other by the equation of the path. The path might be a straight line, in which case $y = A + Sx$, where A is the intercept and S is the slope. Thus $dy = S \, dx$ and we may write

$$W = \int_{x_a}^{x_b} \left[F_x(x, A + Sx) + S F_y(x, A + Sx) \right] dx$$

Notice that the integrand is now a function of a single variable.

In some cases the path is a portion of a circle. Place the origin at the center and let θ be the angle between the positive x axis and the line from the origin to the particle. At each point on the path, the important component of the force is the component along a line tangent to the circle. You must know this component as a function of θ. The length of a path segment is $R\,d\theta$, where R is the radius of the circle and $d\theta$ is the angle subtended by the segment, measured in radians. The integral for the work becomes

$$W = R \int_{\theta_a}^{\theta_b} F_T(\theta)\,d\theta ,$$

where F_T is the tangential component of the force.

4. You will also need to know the binomial theorem to derive the non-relativistic expression for the kinetic energy from the relativistic expression. The first few terms of $(A + B)^n$ are

$$(A+B)^n = A^n + nA^{n-1}B + \frac{n(n-1)}{2}A^{n-2}B^2 + \frac{n(n-1)(n-2)}{2\cdot 3}A^{n-3}B^3 + \ldots$$

and, in general, term r in the sum is

$$\frac{n!}{r!(n-r)!}A^{n-r}B^r$$

where $r!$ is the factorial of r: $r! = 2 \cdot 3 \cdot 4 \ldots r$.

The exponent n may be any number, positive or negative. To estimate the square root of $A + B$ take $n = 1/2$ and to estimate the reciprocal of the square root take $n = -1/2$.

The series has a finite number of terms if n is a positive integer, the last term being B^n. Otherwise the series never ends. Notice that in successive terms of the series the exponent of A decreases while the exponent of B increases. Since you will want a sum in which successive terms are smaller than previous terms, take A to be the larger of the two quantities and B the smaller.

IV. COMPUTER PROJECTS

A force with x component $F(x)$ does work given by $W = \int_{x_i}^{x_f} F(x)\,dx$ as the object on which it acts moves from x_i to x_f along the x axis. A computer can be used to evaluate integrals of this form. One of the simplest techniques to use is Simpson's rule. An interval of the x axis, from x_b to x_e, is divided into two equal parts of width Δx. Let F_b be the value of the force at the beginning of the interval ($x = x_b$), F_m be the value at the middle ($x = x_b + \Delta x$), and F_e be the value of the end ($x = x_e$). Then according to Simpson's rule the integral over the interval can be approximated by

$$\int_{x_b}^{x_e} F(x)\,dx = \frac{\Delta x}{3}(F_b + 4F_m + F_e)$$

This expression can be derived easily by fitting the function $F(x)$ to a quadratic of the form $F(x) = A_0 + A_1(x - x_b) + A_2(x - x_b)^2$. The coefficients are chosen so the quadratic yields the correct values of the function at $x = x_b$, $x = x_m$, and $x = x_e$. They are $A_0 = F_b$, $A_1 = -(F_e - 4F_m + 3F_b)/2\Delta x$, and $A_2 = (F_e - 2F_m + F_b)/2(\Delta x)^2$. The quadratic can be integrated easily. The approximation becomes better as the interval becomes narrower.

To evaluate an integral for an extended portion of the x axis divide the region from x_i to x_f into N intervals, where N is an even number. The interval width is given by $\Delta x = (x_f - x_i)/N$. Label the points $x_0 (= x_i)$, x_1, x_2, ..., $x_N (= x_f)$ and the corresponding values of the force F_0, F_1, F_2, ..., F_N. Apply the formula given above to each pair of intervals; for example, x_0 is the first point, x_1 is the second point, and x_2 is the third point in the first application; x_2 is the first point, x_3 is the second point, and x_4 is the third point in the next application. Except for x_0 and x_N, each point with an even label enters the final formula twice: once as the first point in an integration and once as a third point. Each point with an odd label enters as a midpoint of an integration. Thus

$$\int_{x_i}^{x_f} F(x)\,dx = \frac{\Delta x}{3}\left[(F_N - F_0) + 2(F_0 + F_2 + F_4 + \ldots + F_{N-2}) + 4(F_1 + F_3 + \ldots + F_{N-1})\right]$$

F_0 is subtracted at the beginning of the equation because it is also included in the sum over values with even labels. There it is multiplied by 2 but it should be included only once. F_N is not included in any of the sums but it must be included once in the final equation, so it is added at the beginning.

Write a computer program to evaluate a work integral for one-dimensional motion. Input the lower and upper limits of the integral and the number of intervals. You can force N to be an even integer by dividing the value read by 2, rounding the result to the nearest integer, then multiplying by 2. Calculate Δx. Now write a loop to sum all the force values with even labels and sum all the force values with odd labels. Instructions in the loop will be executed $N/2$ times. An outline might be:

> input limits of integral: x_i, x_f
> input number of intervals: N
> replace N with nearest even integer
> calculate interval width: $\Delta x = (x_f - x_i)/N$
> initialize quantity to hold sum of values with even labels: $S_e = 0$
> initialize quantity to hold sum of values with odd labels: $S_o = 0$
> set $x = x_i$
> **begin loop** over intervals: counter runs from 1 to $N/2$
> > calculate $F(x)$ and add it to the sum of values with even labels:
> > > replace S_e with $S_e + F(x)$
> >
> > increment x by Δx
> > calculate $F(x)$ and add it to the sum of values with odd labels:
> > > replace S_o with $S_o + F(x)$
> >
> > increment x by Δx
>
> **end loop**
> calculate force for upper and lower limits: $F_0 = F(x_i)$, $F_N = F(x_f)$
> evaluate integral: $W = (\Delta x/3)(F_N - F_0 + 2S_e + 4S_o)$
> display result and stop

You may want to define the force function $F(x)$ in a separate section of the program.

PROJECT 1. This problem can be solved analytically. Use it to check your program. A 2.00 kg block moves along the x axis, subjected to the force $\mathbf{F} = (6 - 4x)\mathbf{i}$, where $\mathbf{F}$ is in newtons and x is in meters. What work is done by this force as the block moves from $x = 0$ to $x = 1.00$ m? from $x = 0$ to $x = 5.00$ m? Try $N = 2$ and double its value on successive runs until you get the same first three significant figures. Check the answers by working the problem analytically. [ans: 4.00 J; −20.0 J]

If $\mathbf{F}$ is the total force acting on the block then the work that it does is equal to the change in the kinetic energy of the block. Suppose the block has a speed of 6.00 m/s when it is at $x = 0$. What is its speed when it is at $x = 1.00$ m? at 5.00 m? [ans: 6.32 m/s; 4.00 m/s]

PROJECT 2. A certain non-ideal spring exerts a force that is given by $F(x) = -250x - 125xe^{-x}$, where x is its extension (if positive) or its compression (if negative). A 2.3 kg block is attached to one end and the other end is held fixed. The block is pulled out so the spring is extended by 25 cm, then it is released from rest. How much work does the spring do on the block as the block moves from its initial position to the position for which the spring is neither extended or compressed? If the force of the spring is the only force acting on the block what is its speed as it passes this point? [ans: 11.1 J; 3.11 m/s]

The block continues to move, compressing the spring. By how much is it compressed when the block comes to rest instantaneously? Since the block is at rest both initially and finally the total work done by the spring over the entire trip is 0. Use trial and error to find the final value of x. Alternatively you might imbed the program in a loop in which x_f is incremented each time around and search for the values of x_f that straddle $W = 0$. [ans: −0.237 m]

If the force is given as a function of time and you are asked for the work it does over a given time interval you must change your strategy somewhat. If the force is the total force you can use one of the programs of Chapters 2 or 4 to find the velocity. The total work done, of course, is just the change in kinetic energy. If the force is not the total force you can use the power equation $dW/dt = Fv$ to write $W = \int Fv\, dt$. The program is the same as before except the integrand is now Fv. You must know the velocity as a function of time to use this method.

PROJECT 3. The position as a function of time for a 25-kg crate sliding across the floor (along the x axis) is given by $x(t) = 3te^{-0.60t}$, where x is in meters and t is in seconds. Numerically evaluate the integral for the work done by the total force during the first 2.0 s. You must find $v(t)$ and $F(t)$ by differentiating $x(t)$ and using $F = ma$. Select an integration interval for 3 significant figure accuracy. [ans: −112 J]

Use the expression you found for $v(t)$ to calculate the change in the kinetic energy during the first 2.0 s. Your result should agree with the result of the numerical integration.

PROJECT 4. A 2.0-kg block starts from rest at the origin and moves along the x axis. The total force acting on it is given by $F(t) = (8.0 + 12t^2)$, where F is in newtons and t is in seconds. Use analytical means to show that $v(t) = 4.0t + 2.0t^3$ and $x(t) = 2.0t^2 + 0.50t^4$.

One of the forces acting on the block is given by $F_1(t) = 3.0t^2$. Find the work done by F_1 during the first 3.0 s of the motion. Do this by analytical evaluation of $\int Fv\, dt$ and by numerical integration. [ans: 972 J]

PROJECT 5. You learned that the velocity of a body with mass m dropped from rest near the surface of the earth is given by $v(t) = (mg/b)(1 - e^{-bt/m})$, where b is the drag coefficient and down was taken to be the positive direction. Suppose an object with $m = 5.0$ kg and $b = 75$ kg/s is dropped from the edge of a high cliff. Use numerical integration to find the work done by gravity during the first 5.0 s. [ans: 5280 J]

The work done by gravity, of course, is given by mgs, where s is the distance fallen. Use $s(t) = (mg/b)[t - (m/b) + (m/b)e^{-bt/m}]$ to calculate the distance fallen in 5.0 s, then use $W = mgs$ to calculate the work done by gravity. You should obtain agreement with your numerical calculation.

If the object moves in two dimensions the integral for the work becomes $W = \int \mathbf{F} \cdot d\mathbf{s} = \int (F_x\, dx + F_y\, dy)$, where $d\mathbf{s} = \mathbf{i}\, dx + \mathbf{j}\, dy$ is an infinitesimal segment of the path. It is a vector tangent to the path. Because the object is on a specified path in the xy plane the infinitesimals dx and dy are related to each other. You must know the path to determine the relationship.

There are several ways of specifying a path. One way is to give the functional relationship between the coordinates of points on the path: $y = g(x)$, where $g(x)$ is a specified function. Then $dy = (dg/dx)\, dx$ and the work integral becomes $W = \int [F_x + (dg/dx)F_y]\, dx$. For example, a straight line with slope S and intercept A is given by $y = A + Sx$ and the work is given by $W = \int (F_x + SF_y)\, dx$. The limits of integration are the x coordinates of the points at the beginning and end of the path.

You must use the functional relationship between x and y in another way. The components of the force will be given as functions of both x and y but there can be only a single variable of integration. You must substitute the function $g(x)$ wherever y occurs in the expressions for the force components. The following is an example.

PROJECT 6. A particle moves along a straight line from $x = 1.5$ m, $y = 2.0$ m to $x = 3.0$ m, $y = 5.0$ m. One of the forces acting on it is given by $\mathbf{F} = 3x^2y^2\mathbf{i} + 2x^3y\mathbf{j}$. What work does this force do?

The path is given by $y = -1.0 + 2.0x$, so $dy = 2.0\, dx$. The force components are given by $F_x = 3x^2y^2 = 3x^2(-1 + 2x)^2$ and $F_y = 2x^3y = 2x^3(-1 + 2x)$. The work integral is $\int_{1.5}^{3.0}[3x^2(-1 + 2x)^2 + 4x^3(-1 + 2x)]\, dx$. The integration program can now be used to calculate the work done. [ans: 662 J]

The method does not work if the path is parallel to the y axis and it requires a large number of intervals if the path is nearly parallel to the y axis. Then you should use y as the variable of integration. The expression for work becomes $\int [(dx/dy)F_x + F_y]\, dy$.

A path can also be specified by giving both x and y as functions of some parameter. The parameter need not have any physical significance. For the straight line path of the previous project you might write $x = 1.5 + 1.5s$ and $y = 2.0 + 3.0s$. When the parameter $s = 0$ then $x = 1.5$ and $y = 2.0$. When $s = 1$ then $x = 3.0$ and $y = 5.0$. Now $dx = 1.5\, ds$ and $dy = 3.0\, ds$. The integral for the work becomes $W = \int_0^1 [(dx/ds)F_x + (dy/ds)F_y]\, ds$. You must now substitute for both x and y in terms of s in the expressions for F_x and F_y.

If the path is an arc of a circle you might use the angle between the x axis and the radial line to the point as a parameter. Thus $x = R\cos\theta$ and $y = R\sin\theta$, where R is radius of the path. The integral for the work becomes $W = \int [-F_x \sin\theta + F_y \cos\theta]R\, d\theta$. Notice that $-F_x \sin\theta + F_y \cos\theta$ is the component of the force along a line that is tangent to the path and $R\, d\theta$ is the length of a path segment that subtends the angle $d\theta$ at the center of the circle.

PROJECT 7. A particle travels counterclockwise around a circular path with a radius of 2.0 m, centered at the origin. What work is done by the force $\mathbf{F} = 3x^2y^2\mathbf{i} + 2x^3y\mathbf{j}$ as the particle goes from $x = 2.0$ m, $y = 0$ to $x = -\sqrt{2.0}$ m, $y = \sqrt{2.0}$ m? This is an arc of $3\pi/4$ radians or $135°$. [ans: -5.66 J]

V. NOTES

Chapter 8
CONSERVATION OF ENERGY

I. BASIC CONCEPTS

The closely related concepts of a conservative force and a potential energy are central to this chapter. Pay close attention to their definitions. If *all* forces exerted by objects in a system on each other are conservative and no net work is done on them by an outside agent then the mechanical energy of the system (the sum of the kinetic and potential energies) does not change. When non-conservative forces act another energy, called the internal energy of the system, must be included in the sum for the principle of energy conservation to hold.

Definitions Two ways are given in the text to test a force to see if it is <u>conservative</u>. One of them is: _____
_____
_____

The other is: _____
_____

The two tests are equivalent to each other. If a force meets either of them then it automatically meets the other. If it fails either of them then it automatically fails the other.
  Give some examples of conservative forces: _____
_____
Give at least one example of a non-conservative force: _____
_____

  Consider a block attached to a horizontal spring and on a rough horizontal surface. Suppose the system starts with the spring neither extended nor compressed. An external force is applied so the block moves from its initial position to a position for which the spring extension is d, then back to its initial position. You should be able to show that the spring does zero work during this motion and that the force of friction does non-zero work.

work done by spring on outward trip: $W_s =$ _____
work done by spring on inward trip: $W_s =$ _____
sign of the work done by friction on outward trip: _____
sign of the work done by friction on inward trip: _____

  Because a force of friction is actually the sum of a large number of forces, acting at the welds that form between two surfaces, and the welds move through displacements that are different from the displacement of the object, the work done by friction is not given by $\int \mathbf{f} \cdot d\mathbf{s}$, where d**s** is an infinitesimal displacement of the object. You cannot calculate the work done by friction without a detailed model of the surfaces but you should be able to argue that the work done by friction cannot be zero over a round trip. You should also be able to argue that the work done by friction on an object sliding on a stationary surface is negative.

A potential energy function is associated with a conservative force. Here we consider forces that are given as functions of the positions of the interacting objects and not, for example, as functions of their velocities or the time. For a given force we consider the system consisting of the two objects that are interacting via the force (the earth and a projectile, a spring and a mass, for example). Some configuration of the system, called the reference configuration, is selected and the potential energy is arbitrarily assigned the value 0 for this configuration. The potential energy for any other configuration is taken to be the negative of the work done by the force as the system moves from the reference configuration to the configuration being considered. For the force of gravity a configuration is specified by giving the positions of the objects relative to each other. For the force of a spring a configuration is specified by giving the extension or compression of the spring or, alternatively, the position of the object attached to the spring. Later on, when you study rotational motion, a specification of configuration may also include the orientations of the objects.

The work done by a conservative force as the system goes from any initial to any final configuration is the negative of the change in the potential energy. That is, $W =$ _____, where U_i is the potential energy associated with the initial configuration and U_f is the potential energy associated with the final configuration. The path taken is immaterial since the work done by a conservative force is independent of the path. Every time the system reaches the same configuration it has the same potential energy. This property is at the very heart of the potential energy concept. Carefully note that a potential energy is associated with at least two objects, not with a single object.

To test your understanding of potential energy, explain in words why a potential energy cannot be associated with a frictional force. After all, the force does work. Why not just define the potential energy to be the negative of the work? _____
_____

Potential energy is a scalar. If two or more conservative forces act, the total potential energy is simply the sum of the individual potential energies.

The SI unit for potential energy is _____ and is abbreviated _____. In terms of SI base units this unit is _____.

Only changes in potential energy have physical meaning. If the same arbitrary value is added to the potential energy for every configuration of the system, the motions of the objects in the system do not change. This is why almost any configuration can be selected as the reference configuration. Selection of a new reference configuration simply adds the same value to the potential energy for every configuration.

Gravitational potential energy. The gravitational potential energy of a system consisting of the earth and a mass m, near its surface, is given by $U =$ _____, where y is the vertical coordinate of the object, measured relative to the earth, and the reference configuration is
_____.

Gravitational potential energy is often ascribed to m alone, but it is actually a property of the earth-mass system.

Remember that the gravitational potential energy depends only on the altitude of the object above the surface of the earth, even if the object moves horizontally as well as vertically. When the object is raised a distance Δy and simultaneously moved horizontally by Δx, the potential energy increases by $mg\Delta y$. You should be able to show this, starting with $\Delta U = -\mathbf{F} \cdot \mathbf{s}$, where $\mathbf{s}$ is the displacement $\Delta x\,\mathbf{i} + \Delta y\,\mathbf{j}$. Here the positive y direction was taken to be upward.

If the altitude of an object above the earth is increased, the sign of the work done by gravity is _____ and the sign of the potential energy change is _____. If the altitude is decreased, the sign of the work done by gravity is _____ and the sign of the potential energy change is _____. See Sample Problem 1 of the text for a calculation of gravitational potential energy and carefully note the sign of the potential energy change as the elevator cab goes up.

Spring potential energy. The potential energy of a system consisting of a mass attached to an ideal spring with force constant k is given by $U =$ _____, where the coordinate x of the mass is measured relative to an origin at _____. The reference configuration is _____.

When a spring is elongated by pulling the mass outward from its equilibrium position the sign of the work done by the spring is _____ and the sign of the change in the spring potential energy is _____. When the spring is compressed by pushing the mass inward from its equilibrium position the sign of the work done by the spring is _____ and the sign of the change in the spring potential energy is _____. Suppose the mass starts with coordinate x_1, measured from the equilibrium position, and is moved to $-x_1$. The work done by the spring is _____ and the change in the spring potential energy is _____.

Calculation of force from potential energy. You have learned how to compute the potential energy associated with a given conservative force. The reverse calculation can also be carried out. If the potential energy function is known, the force can be computed by evaluating its derivatives with respect to coordinates. Consider an object moving along the x axis and acted on by a conservative force F. If $U(x)$ is the potential energy as a function of the object's coordinate, then $F =$ _____.

For an object moving in three dimensions, the potential energy is a function of three coordinates. If $U(x, y, z)$ is the potential energy, then the components of the force are given by

$F_x =$ $F_y =$ $F_z =$

Conservation of mechanical energy. We consider a system of objects that interact with each other via conservative forces and suppose no net work is done on objects of the system by outside agents. The change in the total potential energy as the system changes configuration is the negative of the total work done by all forces. According to the work-energy theorem the total work is also equal to the change in _____. Thus if all forces are conservative the sum of the _____ and _____ energies does not change as the system changes configuration. In symbols, $\Delta K + \Delta U = 0$.

Chapter 8: Conservation of Energy

The mechanical energy of a system is defined by $E = \underline{} + \underline{}$ and if all forces are conservative then $\Delta E = 0$. You should recognize that, in general, K is the sum of the kinetic energies of all objects in the system. When, for example, an object moves in the earth's gravitational field, K is strictly the sum of the kinetic energies of the object and the earth. If, however, the object is must less massive than the earth then the kinetic energy of the earth does not change significantly and can be omitted from the energy equation. Similarly, an ideal spring is considered to be massless and so does not contribute to the kinetic energy of any system of which it is a part.

Conservation of energy can be used to solve some problems. If you know the potential and kinetic energies for one configuration of the system, then you can calculate the total mechanical energy for that configuration. If mechanical energy is conserved then it has the same value for all configurations. Now suppose you are given or can calculate the potential energy for a second configuration. Then the conservation of energy principle allows you to compute the kinetic energy for the second configuration. Likewise, if you know the kinetic energy for the second configuration then the conservation of energy principle allows you to compute the potential energy.

For example, suppose a mass m is attached to a spring with force constant k. Initially it is pulled out so the spring is elongated by x_0 and it is given a speed v_0 to start its motion. In terms of these quantities the mechanical energy of the spring-mass system is $E = \underline{}$. Suppose the speed of the mass is v_1 when it is at x_1. In terms of these quantities the mechanical energy is $E = \underline{}$. Equate the two expressions for the mechanical energy to obtain $\frac{1}{2}mv_0^2 + \frac{1}{2}kx_0^2 = \frac{1}{2}mv_1^2 + \frac{1}{2}kx_1^2$. This equation can be solved for any one of the quantities that appear in it.

Suppose a ball of mass m is thrown into the air with an initial speed of v_0. Take the zero of gravitational potential energy to be at the release point and neglect the motion of the earth. The mechanical energy of the earth-ball system is then $E = \underline{}$. What is the speed of the ball when it a distance y above the release point? The potential energy is then $U(y) = \underline{}$ and if we let v be the speed of the ball its kinetic energy is $K = \underline{}$. Equate the two expressions for the mechanical energy to obtain $\frac{1}{2}mv_0^2 = \frac{1}{2}mv^2 + mgy$. This can be solved for v. The speed is the same whether it is going up or coming down. You can also find how high the ball goes before falling back again. Put $v = 0$ and solve for y.

Potential energy curves. You should know how to obtain information about the motion of an object from a potential energy curve, the potential energy plotted as a function of its coordinate. In thinking about these curves you may assume that the agent exerting the force is stationary. Only the object whose motion is being considered has kinetic energy.

Consider the potential energy function shown below for an object that moves along the x axis. Suppose the object has mechanical energy E, represented by a dotted line, and suppose further that mechanical energy is conserved. Indicate on the graph the potential and kinetic energies of the object when it is at x_1. Use x_{min} to label the minimum coordinate of the object in its motion and use x_{max} to label the maximum coordinate of the object in its motion. Look at the slope of the potential energy function and mark with an arrow the directions of the force when the object is at x_{min} and at x_{max}. x_{min} and x_{max} are called the $\underline{}$ of the motion.

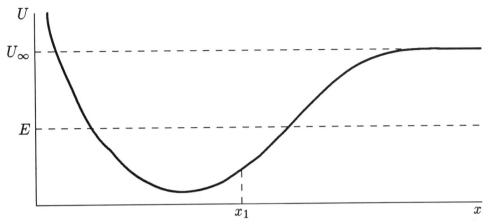

Suppose the object starts at x_1 and travels in the negative x direction. Qualitatively describe its subsequent motion, telling where it is speeding up, where it is slowing down, where it momentarily stops, and what it does after it stops: _____

_____

_____

If the energy of the object is increased slightly, what happens to the minimum and maximum coordinates? The minimum coordinate _____ and the maximum coordinate _____.

Suppose now that the object is far to the right and is traveling in the negative x direction with a mechanical energy that is slightly greater than U_∞ on the graph. Describe its subsequent motion: _____

_____

Equilibrium points are coordinates for which the force vanishes. The slope of the potential energy curve is _____ at these points. When an object is released near a point of stable equilibrium its subsequent motion is _____. When an object is released near a point of unstable equilibrium its subsequent motion is _____. When an object is released near a point of neutral equilibrium its subsequent motion is

_____.

Given a potential energy function $U(x)$ for one dimensional motion we can find the equilibrium points by solving _____ for x. We can check to see if the equilibrium is stable or unstable by evaluating _____ at the equilibrium point. The value is _____ for a stable equilibrium point, _____ for an unstable equilibrium point, and _____ for a neutral equilibrium point.

The graph above shows an equilibrium point just to the left of x_1. Is it a point of stable, unstable, or neutral equilibrium? _____

Energy conservation can be used to solve for the position of an object as a function of time. Consider a particle of mass m moving along the x axis and suppose the potential energy U is known as a function of x. Since $E = K + U$, the speed of the particle when it is at x is given by $v = $ _____. Since $v = dx/dt$, $dt = dx/\sqrt{2[E - U(x)]/m}$. Suppose the

Chapter 8: Conservation of Energy   111

particle has coordinate x_0 at time t_0 and coordinate x at time t. Integrate the left side from t_0 to t and right side from x_0 to x to obtain

$$t - t_0 = \int_{x_0}^{x}$$

In principle the integral on the right side can be evaluated, if not in closed form then numerically by computer. In either case the results can be used to generate a table of the particle coordinate x as a function of time t. If the integral can be evaluated in closed form, an algebraic expression can sometimes be found for $x(t)$.

Systems of particles. Consider a system of particles that may interact with each other and with other particles outside the system. The energy equation can be written $\Delta E = W$, where E is the total energy of the system and W is the total work done *on* particles of the system by outside agents. You should realize that for a given physical situation you might choose the system in any of several different ways. If a ball is thrown upward in the air the system might consist of the ball alone, the ball and earth, or the ball, earth, and air.

The energy E of a system can be written as the sum of several terms. Suppose the system consists of macroscopic objects like a ball and the earth. Each has a kinetic energy and if they interact via a conservative force a potential energy U is associated with their interaction. In addition, the system has an <u>internal energy</u>, ultimately associated with the motions of the particles in the ball and earth. Thus the energy equation becomes

$$\Delta K + \Delta U + \Delta E_{int} = W$$

Here K is the sum of the kinetic energies of objects in the system, U is the sum of the potential energies associated with interactions between objects in the system, E_{int} is the total internal energy of the system, and W is the total work done on particles of the system by outside agents.

The internal energy is a new term that must be included if either internal or external forces change the motions of particles within any object. When a gas is compressed, for example, its internal energy changes. Non-conservative forces, such as friction, always change the internal energy of the object on which they act because the welds between the surfaces deform and the microscopic surface structure changes as time goes on. Thus both the macroscopic kinetic energies and the internal energy of the system change. Similarly, air resistance results from collisions of many air molecule with particles in an object. These change both the macroscopic kinetic energy of the object and its internal energy.

Carefully note that if no net work is done on objects of the system by outside agents, then $\Delta K + \Delta U + \Delta E_{int} = 0$. Energy, in the form $K + U + E_{int}$, is conserved. Forces between particles of the system may change the kinetic energies of the objects, the potential energies of their interactions, or the internal energy, but the sum retains the same value. Energy may change form but the total does not change.

You should be able to interpret the various terms in the energy equation. The text uses the example of a block attached to a spring and moving on a horizontal table top. For each of the following systems tell what bodies do work on the system and what bodies contribute to the macroscopic kinetic energy, to the macroscopic potential energy, and to the total internal energy:

System: block alone
W: _____
K: _____
U: _____
E_{int}: _____

System: block and spring
W: _____
K: _____
U: _____
E_{int}: _____

System: block, spring, and table top
W: _____
K: _____
U: _____
E_{int}: _____

An increase in the internal energy of an object results in an increase in the temperature of that object. When the temperature of an object is different from the temperature of its surroundings, energy is transferred as heat and another term must be included in the energy equation given above. You will learn more about this when you study Chapter 25.

Whenever the total mechanical energy is not conserved physicists have been able to restore the principle of energy conservation by defining another form of energy and including it in the energy balance, just as internal energy restores the energy conservation principle when an object is deformed or frictional forces act. The energy associated with electromagnetic radiation (visible light, for example) and the energy associated with the rest mass of a particle are other examples of energies that sometimes must be taken into account. The change in energy associated with a mass change Δm is given by $\Delta E =$ _____.

Energy quantization. A particle is said to be bound if $E \geq U$ only in a finite region of space. Quantum mechanically the energy of a bound particle is quantized: it may have only certain discrete values; other values are not allowed. For a spring-mass system the separation of allowed energy values is given by $\Delta E =$ _____, in terms of the force constant k, the mass m, and the Planck constant h.

For macroscopic spring-mass systems the separation between allowed energy values is too small to be detected, but for atomic masses it has profound effects.

The energy of an atomic system can change from one allowed value to another with the emission or absorption of a quantum of electromagnetic radiation, called a photon. The relation between the energy of the photon and the wavelength λ of the wave associated with it is given by $E =$ _____.

II. PROBLEM SOLVING

Some problems of this chapter involve the calculation of a potential energy change, given the force and the end points of the path. Since the change in potential energy is just the negative of the work done by the force, the calculation proceeds like the calculations of work in the last chapter. If you are asked for the potential energy itself, the reference configuration (for which the potential energy is zero) must be identified, either in the problem statement or by you. You then calculate the negative of the work done on the system as the system changes from the reference configuration to the configuration of interest.

PROBLEM 1. A particle moving along the x axis is subjected to a force given by $F = kx^3$, where k is a constant. Find an expression for the potential energy as a function of the coordinate x, with the zero of potential energy at $x = 0$. Check your answer using $F = -dU/dx$.

SOLUTION: The potential energy is given by the integral $U(x) = U(0) - \int_0^x F(x)\,dx$. Select $U(0) = 0$ and evaluate the integral for the given force.

[ans: $-(k/4)x^4$]

If $k = 1200\,\text{N/m}^3$ what is the change in potential energy as the particle moves from the origin to $x = 0.35\,\text{m}$?

SOLUTION:

[ans: $-4.50\,\text{J}$]

The negative sign indicates that the potential energy decreases. Note that the force does positive work.

PROBLEM 2. A particle moving along the x axis is subjected to a force given by $F = k/x^2$, where k is a constant. Find an expression for the potential energy as a function of the coordinate x. Take the potential energy to be 0 in the limit as x becomes very large. Check your answer using $F = -dU/dx$.

SOLUTION: The reference point is now at $x = \infty$ and $U(x)$ is given by $U(x) = -\int_\infty^x (k/x^2)\,dx$.

[ans: k/x]

Some problems involve the computation of potential energy for a particle moving in two dimensions. Use $U = -\int (F_x\,dx + F_y\,dy)$, where the path of integration is from the reference point (where $U = 0$) to the point of interest. The path may be split into segments for convenience. For any segment x and y are related to each other by the path: if the path is along a line with slope s then $y = y_0 + s(x - x_0)$ and $dy = s\,dx$; if the path is parallel to the x axis then y is constant and $dy = 0$; if the path is parallel to the y axis then x is constant and $dx = 0$. Use these expressions to convert the path integral into an integral in one variable.

PROBLEM 3. As a particle moves in the xy plane it is subjected to a conservative force given by $\mathbf{F} = 3x^2y\,\mathbf{i} + x^3\,\mathbf{j}$. Take the potential energy to be zero at the origin and derive an expression for the potential energy at the general point x, y.

SOLUTION: Use a two segment path with one segment along the x axis from the origin to $x, 0$ and the second segment along a line parallel to the y axis from $x, 0$ to x, y. The change in potential energy is zero for the first segment since the x component of the force vanishes along the x axis. For the second segment $\Delta U = -\int_0^y x^3\,dy$.

[ans: $-x^3y$]

Suppose the particle has a mass of 0.12 kg and starts at the origin with a speed of 3.5 m/s. Later it passes through the point $x = 0.75$ m, $y = 1.2$ m. What is its speed then?

SOLUTION: Use $K_1 + U_1 = K_2 + U_2$, with $K_1 = \frac{1}{2}mv_1^2$, $U_1 = 0$, $K_2 = \frac{1}{2}mv_2^2$, and $U_2 = -x^3y$. Solve for v_2.

[ans: 4.55 m/s]

Problems involving the conservation of mechanical energy are relatively straightforward. You must first select a system of objects. Normally you will pick the system so that no outside agents do work on objects in the system. You must then decide if all the forces of interaction between objects of the system are conservative. The force of gravity and the force of an ideal spring are; the force of friction and drag forces are not. In one dimension forces that are uniform over all possible paths are conservative. For other forces you may have to apply one of the tests for a conservative force.

If all forces that do work are conservative, then conservation of energy yields $K_1 + U_1 = K_2 + U_2$, where K_1 and U_1 are the kinetic and potential energies for one configuration and K_2 and U_2 are the kinetic and potential energies for another. You must identify the two configurations from the problem statement. You will be given enough information to calculate three of the four energies that appear in the energy equation and can use the conservation of energy to compute the fourth.

In many problems you are asked to find some parameter that occurs in the expression for the kinetic or potential energy at the second configuration. For example, you may be asked for the compression of a spring, a force constant, the speed of an object, or its mass. You first find the relevant energy (kinetic or potential), then solve for the parameter. Study Sample Problem 2 of the text for an example of this type. Here are some other examples.

PROBLEM 4. A 0.55-kg particle moves along the x axis. When its coordinate is x the potential energy is $U(x) = 300x^4$, where U is in joules and x is in meters. No non-conservative forces act. If it starts at the origin with a velocity of 7.9 m/s, in the positive x direction, where does it first stop, even if only momentarily? What is its speed when it is halfway to its stopping point? Assume the total kinetic energy of the system can be ascribed to the particle.

SOLUTION: Write $K_1 + U_1 = K_2 + U_2$. In configuration 1 the particle is at the origin and has a speed of 7.9 m/s; in configuration 2 it has an unknown coordinate x_2 and its speed is zero. Evaluate K_1, U_1, and K_2. Write

$U_2 = 300x_2^4$ and solve the conservation of energy equation for x_2. To answer the second question take the third configuration to be when the particle has coordinate $x_3 = x_2/2$ and speed v_3. Solve $K_1 + U_1 = K_3 + U_3$ for v_3.

[ans: 0.489 m; 7.65 m/s]

What is the force on the particle when it stops? What does it do immediately after it stops?

SOLUTION: Use $F = -dU/dx$ to find the force when $x = x_2$.

[ans: −140 N; it starts moving in the negative x direction]

PROBLEM 5. A 0.55-kg mass moves on a horizontal frictionless table top. It is attached to one end of a spring with force constant $k = 400$ N/m, the other end of the spring being fixed. The mass is pulled 3.0 cm from its equilibrium position and released with a velocity of 1.5 m/s, back toward the equilibrium position. What is its speed as it passes the equilibrium position? What is the maximum compression of the spring? What is the acceleration of the mass when the spring has maximum compression?

SOLUTION: The only force doing work on the mass, the force of spring, is conservative, so the mechanical energy of the spring-mass system is conserved. The potential energy is given by $U = \frac{1}{2}kx^2$, where x is the elongation or compression of the spring. In configuration 1 the spring is elongated by 3.0 cm and the mass has a speed of 1.5 m/s. In configuration 2 the spring is neither compressed or elongated and the speed of the mass is unknown. Use conservation of energy to solve for the speed. In configuration 3 the spring is compressed the maximum amount, so the mass has zero speed. Use conservation of energy to solve for the compression.

[ans: 1.70 m/s; 6.32 cm]

The mass now starts moving the other way and the spring elongates. What is the maximum elongation of the spring?

SOLUTION:

[ans: 6.32 cm]

PROBLEM 6. A spring gun is constructed by placing a spring inside a firmly anchored tube, closed at one end. A pellet is placed on the spring and the spring is compressed. When released the spring pushes the pellet out the open end of the tube. Suppose the pellet has a mass of 0.050 kg, the spring has a force constant of 800 kg/m, and it is initially compressed 25 cm. Neglect the force of friction between the inner surface of the tube and the pellet and suppose the gun is fired horizontally. What is the muzzle speed of the pellet?

SOLUTION: Consider the system consisting of the spring and the pellet. Its mechanical energy $K + U$ is conserved. Let x be the extension or compression of the spring, positive if an extension and negative if a compression, as usual. Let v be the speed of the pellet. In configuration 1 the spring is compressed and the pellet is at rest. The total mechanical energy for this configuration is $\frac{1}{2}kx^2$, where $x = $ _____ m. The spring and the pellet part ways when $x = 0$. At that point the end of the spring starts slowing down whereas the pellet continues forward with undiminished speed. The total mechanical energy for configuration 2 is $\frac{1}{2}mv^2$. Equate the expression for the energies of the two configurations and solve for v.

[ans: 31.6 m/s]

If the gun is fired vertically is the muzzle speed reduced significantly by the force of gravity? Use the initial conditions given in the previous problem and calculate the speed of the pellet as it leaves the spring.

SOLUTION: Now you must include gravitational potential energy. Write $E = K + U_s + U_g$, where $K = \frac{1}{2}mv^2$, $U_s = \frac{1}{2}ky^2$, and $U_g = mgy$. For the initial configuration $v = 0$ and $y = -0.25$ m; for the final configuration $y = 0$.

[ans: 31.5 m/s]

The effect of gravity on the muzzle speed is very small.

PROBLEM 7. A frictionless roller coaster is going 3.3 m/s at the top of a hill 20 m above ground. How fast does it go over the next hill, 15 m above the ground? How fast is it going when it reaches ground level again?

SOLUTION: The system is the roller coaster, the track, and the earth. The track does no work and the force of gravity is conservative so mechanical energy K + U is conserved. The mass of the roller coaster is not given so you cannot calculate either the kinetic or potential energy. However, when you solve the problem algebraically you will find the mass cancels out and you can obtain a numerical answer. For configuration 1 the mechanical energy is given by $\frac{1}{2}mv_1^2 + mgh_1$, where v_1 is the initial speed of the roller coaster and h_1 is its initial altitude above ground level. For configuration 2 the mechanical energy is $\frac{1}{2}mv_2^2 + mgh_2$. Equate the two expressions and solve for v_2. Carry out a similar calculation for v_3, the speed when it returns to ground level. Then the mechanical energy is $\frac{1}{2}mv_s^2$.

[ans: 10.4 m/s; 20.1 m/s]

PROBLEM 8. A woman athlete can throw a 0.25-kg ball 125 m straight up. How high can she throw a 2.0-kg ball? Assume she imparts the same kinetic energy to each ball and neglect air resistance.

SOLUTION: Use kinematics or the conservation of energy to show that the height reached by the ball is $v_0^2/2g$, where v_0 is the initial speed. Thus the ratio of the heights for the two balls is $h_1/h_2 = v_{10}^2/v_{20}^2$. Since the initial kinetic energies are the same h_1/h_2 can be expressed as a ratio of the masses. Do this and solve for the height of the heavier ball.

[ans: 15.6 m]

PROBLEM 9. A projectile is launched at 150 m/s. If air resistance is negligible what is its speed when it is 400 m above the launching point?

SOLUTION: The system consists of the projectile and the earth. Take the zero of gravitational potential energy to be at the launching point. Then the initial potential energy is 0 and the initial kinetic energy is $\frac{1}{2}mv_0^2$. The potential energy at altitude h is mgh and the kinetic energy is $\frac{1}{2}mv^2$, where v is the speed at that altitude. Equate the initial and final energies, notice that the mass cancels out, and solve for v.

[ans: 121 m/s]

Sometimes the second configuration is described, not by giving a coordinate or speed, but by giving some other condition that can be used to compute a coordinate or speed, usually by means of Newton's second law.

PROBLEM 10. A small paper clip is placed on top of an upside down hemispherical bowl and slides down. At any instant its position is given by the angle θ, as shown. At what value of θ does the clip lose contact with the bowl?

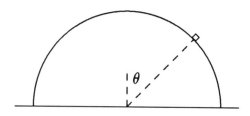

SOLUTION: As the clip slides down the curved surface, the proper centripetal force is provided by the radial component of the force of gravity $mg\cos\theta$, reduced by the normal force of the bowl, pushing outward on it.

Remember the condition for losing contact: the normal force of the bowl on the clip vanishes. Since we are dealing with forces here we must use Newton's second law, not conservation of energy. Write $mg\cos\theta - N = mv^2/R$, where R is the radius of the bowl. At low speeds $mg\cos\theta$ is greater than mv^2/R and the bowl pushes out on the clip. As the speed increases N decreases and the clip loses contact when $N = 0$ or, what is the same, when $v^2 = gR\cos\theta$.

Now use the conservation of energy to find its speed for any angle θ. Take the zero of potential energy to be at the ground. At the top of the bowl the potential energy is mgR and the kinetic energy is 0. When the radial line makes the angle θ with the vertical the potential energy is $mgR\cos\theta$ and the kinetic energy is $\frac{1}{2}mv^2$. Thus $mgR = mgR\cos\theta + \frac{1}{2}mv^2$. Solve $v^2 = gR\cos\theta$ and $mgR = mgR\cos\theta + \frac{1}{2}mv^2$ simultaneously for θ.

[ans: 48.2°]

If the bowl has a radius of 8.0 cm how fast is the clip going when it loses contact?

SOLUTION:

[ans: 0.723 m/s]

If an external force acts on the system then you must use $\Delta K + \Delta U = W$, where W is the work done by the external force.

PROBLEM 11. Two gliders, each having a mass of 300 g, are coupled by a spring with a force constant of 30.0 N/m. They are placed at rest on a horizontal frictionless air track, with the spring at its equilibrium length. The first glider is at $x = 0$ and the second is at $x = 0.200$ m. A student pushes on the first glider with a constant force of 4.00 N. At the end of 0.75 s the first glider is at $x = 1.920$ m and the second is at $x = 2.030$ m. What work was done by the student?

SOLUTION: Use $W = F\Delta x$, the appropriate expression for a constant force. Δx is the displacement of the first glider.

[ans: 7.68 J]

What was the change in the spring potential energy during the 0.75 s interval?

SOLUTION: Calculate the elongation or compression of the spring for each configuration.

[ans: 0.12 J]

What was the change in the total kinetic energy of the system consisting of the two gliders and the spring?

SOLUTION: Solve $W = \Delta K + \Delta U$ for ΔK.

[ans: 7.56 J]

Some problems ask you to find equilibrium points and classify them as stable, unstable, or neutral. Use $dU(x)/dx = 0$ to find the points and evaluate $d^2U(x)/dx^2$ at the point to classify them. Turning points can be found by solving $U(x) = E$ for x.

PROBLEM 12. As a particle moves along the x axis the potential energy is given by $U(x) = x^3 - 5.85x^2 + 9.72x$, where U is in joules and x is in meters. Find the coordinates of the equilibrium points.

SOLUTION:

[ans: 1.2 m, 2.7 m]

Are these points of stable, unstable, or neutral equilibrium?

SOLUTION:

[ans: unstable at $x = 1.2$ m, stable at $x = 2.7$ m]

What are the turning points for a mechanical energy of 4.0 J?

SOLUTION: Use trial and error to solve $U(x) = 4.0$ J for x. There are three turning points. A crude sketch of the function might help.

SOLUTION:

[ans: 0.616 m, 2.02 m, 3.21 m]

What is the force on the particle when it is at each of the turning points?

SOLUTION: Evaluate $F = -dU/dx$.

[ans: −3.65 N, +1.67 N, −3.08 N]

If the particle is started with $x < 0.616$ m, traveling in the positive x direction, it turns around at $x = 0.616$ m and never returns. If it is started between $x = 2.02$ m and 3.21 m it oscillates between these two points.

When non-conservative forces act the mechanical energy changes. You can calculate the change if you know the mechanical energy for two configurations, usually by computing the kinetic and potential energies for the configurations.

PROBLEM 13. A 1.0-kg projectile is launched at 150 m/s. When it is 200 m above the launching point its speed is 105 m/s. How much mechanical energy is lost through air resistance?

SOLUTION: Here you have enough information to compute the mechanical energy $K + U$ for two configurations. Take the difference to find the mechanical energy lost.

[ans: 3780 J]

What is the change in the internal energy of the system consisting of the projectile, the earth, and the air?

SOLUTION: Include the air in the system and assume it is closed, so $\Delta K + \Delta U + \Delta E_{int} = 0$. You have already computed $\Delta K + \Delta U$.

[ans: 3780 J]

PROBLEM 14. A 15-kg crate, starting with a speed of 5.5 m/s, slides 2.4 m up a 20° incline before it comes to rest. What mechanical energy was lost by the earth-crate system?

SOLUTION:

[ans: 106 J]

If the work done by friction on the crate was −62 J, what was the change in the internal energy of the crate?

SOLUTION: Use $\Delta E + \Delta E_{int} = W_f$, where W_f is the work done by friction.

[ans: 44 J]

The work done by friction is not given by $-fs$, where f is the force of friction and s is the distance traveled. It is somewhat less in magnitude than fs and the difference between the actual value and $-fs$ appears as an increase internal energy.

PROBLEM 15. The magnitude of the force of air resistance acting on a certain 0.15-kg ball is given by bv, where $b = 0.020$ N·s/m and v is the speed of the ball. The ball is dropped from rest at the edge of a cliff and falls for 4.1 s before hitting the ground below. How much mechanical energy is lost through air resistance?

SOLUTION: See Chapter 6 for the equations that give the coordinate and velocity as functions of time. Calculate the changes in the gravitational potential and kinetic energies, then use $\Delta E = \Delta K + \Delta U$ to calculate the mechanical energy lost.

[ans: 29.9 J lost]

III. COMPUTER PROJECTS

The force $\mathbf{F} = 3x^2y^2\mathbf{i} + 2x^3y\mathbf{j}$ is a conservative force. You can see what this means by computing the work it does as the particle on which it acts goes via different paths from the same initial point to the same final point. The Simpson's rule integration program outlined in the Computer Projects section of Chapter 7 can be used without modification.

PROJECT 1. Use the program to calculate the work done by the force $\mathbf{F} = 3x^2y^2\mathbf{i} + 2x^3y\mathbf{j}$ as the particle goes from $x = 2.0$ m, $y = 0$ to $x = 4.0$ m, $y = 2.0$ m along each of the following paths.
  a. Along the x axis from $x = 2.0$ m to $x = 4.0$ m, then along the line $x = 4.0$ m from $y = 0$ to $y = 2.0$ m. Carry out the integration in two parts, one for each segment.
  b. Along the line $x = 2.0$ m from $y = 0$ to $y = 2.0$ m, then along the line $y = 2.0$ m from $x = 2.0$ m to $x = 4.0$ m.
  c. Along the straight line that joins the two points. You might take $x = 2.0 + 2.0s$ and $y = 2.0s$, where s is a parameter that varies between 0 and 1.
  d. Along the perimeter of a 2.0-m radius circle centered at $x = 4.0$ m, $y = 0$. For this path $x = 4.0 - 2.0\cos\theta$ and $y = 2.0\sin\theta$, where θ varies from 0 to $\pi/2$ radians (90°).

Notice that in every case the work done is 256 J.

PROJECT 2. Since the work done by the force of the previous project is independent of the path a potential energy function is associated with it. The value of this function at any point is the negative of work done by the force as the particle moves from some reference point to the point in question. The table below is actually a grid of points in the xy plane. The x coordinate is given at the top and the y coordinate is given at the left side. Take the origin to be the reference point ($U = 0$ there) and calculate the potential energy associated with each of the other points. You may want to automate the process by reading in the coordinates x_f and y_f of the point. Use the path given by $x = x_f s$ and $y = y_f s$, where s is a parameter than varies from 0 to 1. Since $dx/ds = x_f$ and $dy/ds = y_f$ the potential energy is given by $U(x_f, y_f) = -\int_0^1 (F_x x_f + F_y y_f)\,ds$. You must write the force components in terms of s. Fill in the table.

$y(m)$ \ $x(m)$	0	1	2	3	4
4					
3					
2					
1					
0	0				

The analytic form for the potential energy function is $U(x, y) = -x^3y^2$. You can easily check this since the derivative of U with respect to x must give the negative of the x component of the force and the derivative of U with respect to y must give the negative of the y component. Use the exact analytic expression to check the results of your numerical calculations. You might construct a program consisting of two loops, one over values of the x coordinate and the other over values of the y coordinate, to compute and display values of U.

Use the table to compute the following quantities:

a. The change in the potential energy when the particle moves from $x = 2$ m, $y = 1$ m to $x = 3$ m, $y = 3$ m. [ans: -235 J]
b. The work done by the force when the particle moves from $x = 3$ m, $y = 4$ m to $x = 1$ m, $y = 1$ m. [ans: -431 J]
c. The change in the kinetic energy of the particle when it moves from $x = 3$ m, $y = 1$ m to $x = 2$ m, $y = 4$ m. Assume only one force acts on it. [ans: $+101$ J]

PROJECT 3. By way of contrast, the force $\mathbf{F} = (3x^2 - 6)y^2 \mathbf{i} + 2x^3 y \mathbf{j}$ is not conservative. Use the program to calculate the work done by the force as the particle goes from $x = 2.0$ m, $y = 0$ to $x = 4.0$ m, $y = 2.0$ m along each of the following paths.

a. Along the x axis from $x = 2.0$ m to $x = 4.0$ m, then along the line $x = 4.0$ m from $y = 0$ to $y = 2.0$ m.
b. Along the line $x = 2.0$ m from $y = 0$ to $y = 2.0$ m, then along the line $y = 2.0$ m from $x = 2.0$ m to $x = 4.0$ m.
c. Along the straight line that joins the to points.
d. Along the perimeter of a 2.0-m radius circle centered at $x = 4.0$ m, $y = 0$.

Notice that the work done is different for different paths. You can easily see that it is impossible to assign a potential energy to each point in space so that the work done by the force equals the difference in the values assigned to the end points of the path. [ans: 256 J; 208 J; 240 J; 224 J]

IV. NOTES

Chapter 9
SYSTEMS OF PARTICLES

I. BASIC CONCEPTS

In this chapter you consider systems of more than one particle. The center of mass is important for describing the motion of the system as a whole and its velocity is closely related to the total momentum of the system. External forces acting on the system accelerate the center of mass and when the net external force vanishes the velocity of the center of mass is constant. The total momentum of the system is then conserved.

Coordinates of the center of mass. You should know how to calculate the coordinates of the center of mass of a collection of particles. If particle i has mass m_i and coordinates x_i, y_i, and z_i then the coordinates of the center of mass are given by

$$x_{\text{cm}} =$$

$$y_{\text{cm}} =$$

$$z_{\text{cm}} =$$

where M is the total mass of the system. In vector notation the position vector of the center of mass is given by

$$\mathbf{r}_{\text{cm}} =$$

where $\mathbf{r}_i$ is the position vector of particle i. The center of mass is not necessarily at the position of any particle in the system.

For a *continuous* distribution of mass the coordinates of the center of mass are given by the integrals

$$x_{\text{cm}} =$$

$$y_{\text{cm}} =$$

$$z_{\text{cm}} =$$

If the mass is distributed symmetrically about some point or line then the center of mass is at that point or on that line. For example, the center of mass of a uniform spherical shell is at its _____, the center of mass of a uniform sphere is at its _____, the center of mass of a uniform cylinder is at the midpoint of its _____, the center of mass of a uniform rectangular plate is at its _____, and the center of mass of a uniform triangular plate is at the intersection of _____ and halfway through its thickness.

Sometimes an object with a complicated shape can be thought of as a group of parts such that the center of mass of each part can be found easily. Then each part is replaced by a particle with mass equal to the mass of the part, positioned at the center of mass of the part. The center of mass of the whole object is at the position of the center of mass of these particles. This idea can also be used to find the center of mass of an object with a hole. See Sample Problem 3 of the text.

Velocity and acceleration of the center of mass. As the particles that comprise the system move, the coordinates of the center of mass might change. Consider a system of discrete particles and differentiate the expression for the center of mass position vector to find the velocity of the center of mass in terms of the individual particle velocities:

$$\mathbf{v}_{cm} =$$

where $\mathbf{v}_i$ is the velocity of particle i.

Consider two objects moving along the x axis, one with mass m_A and coordinate given by $x_A(t) = x_{A0} + v_A t$ and the other with mass m_B and coordinate given by $x_B(t) = x_{B0} + v_B t$, where v_A and v_B are constant velocities. Use the definition of the center of mass to find an expression for the coordinate of the center of mass as a function of time: $x_{cm}(t) = $ _____ _____. Now differentiate this expression to find an expression for the velocity of the center of mass: $v_{cm} = $ _____. How should the velocities of the objects be related for the center of mass to be at rest? _____

As the particles accelerate the center of mass might accelerate. Differentiate the expression for the velocity of the center of mass to find the acceleration of the center of mass in terms of the accelerations of the individual particles:

$$\mathbf{a}_{cm} =$$

Newton's second law for the center of mass. Since each individual particle obeys Newton's second law, $\sum \mathbf{F}_i = M\mathbf{a}_{cm}$, where $\mathbf{F}_i$ is the vector sum of all forces on particle i, some of which might be due to other particles in the system and some of which might be due to the environment of the system. The sum $\sum \mathbf{F}_i$ is the vector sum of all forces acting on all particles of the system but it reduces to $\sum \mathbf{F}_{ext}$, the sum over all *external* forces acting on particles of the system, i.e. those due to the environment of the system. Explain in words why these two sums are equal: _____

_____

Thus the center of mass obeys a Newton's second law: $\sum \mathbf{F}_{ext} = M\mathbf{a}_{cm}$. The mass that appears in the law is the total mass of the system (the sum of the masses of the individual particles) and the force that appears is the vector sum of all external forces acting on all particles of the system.

If the total external force acting on a system is zero, then the acceleration of the center of mass is _____ and the velocity of the center of mass is _____. If the center of mass of the system is initially at rest and the total external force acting on the system is zero, then the velocity of the center of mass is always _____ and the center of mass remains at its initial position.

These statements are true no matter what forces the particles of the system exert on each other. The particles might, for example, be fragments that are blown apart in an explosion or they might be objects that collide violently. On the other hand, they might interact with each other from afar via gravitational or electrical forces and never touch.

Momentum. The momentum of a particle of mass m, moving with a velocity $\mathbf{v}$ (assumed to be much less than the speed of light) is given by $\mathbf{p} =$ _____. Momentum is a vector. Its SI units are _____.

In terms of the magnitude of its momentum the kinetic energy of a particle is given by $K =$ _____. This is identical to $K = \frac{1}{2}mv^2$. In terms of momentum, Newton's second law for a particle is $\sum \mathbf{F} =$ _____, where $\sum \mathbf{F}$ is _____. You should be able to prove this formulation is exactly the same as $\sum \mathbf{F} = m\mathbf{a}$.

If the particle speed is close to the speed of light, relativistically correct expressions must be used. The momentum of a particle of mass m moving with velocity $\mathbf{v}$ is then given by $\mathbf{p} =$ _____ and the kinetic energy is then given by $K =$ _____.

The total momentum $\mathbf{P}$ of a system of particles is the vector sum of the individual momenta. Since, for non-relativistic particles, $\mathbf{P} = m_1\mathbf{v}_1 + m_2\mathbf{v}_2 + \ldots$, the total momentum is given by $\mathbf{P} = M\mathbf{v}_{cm}$, where M is the total mass of the system. That is, the total momentum of the system is identical to the momentum of a single particle with mass equal to the total mass of the system, moving with a velocity equal to the velocity of the center of mass. Furthermore the net external force changes the total momentum of the system according to

$$\sum \mathbf{F}_{ext} =$$

Conservation of momentum. State in words the principle of conservation of total momentum for a system of particles. Your statement should have the form "If ... , then ... ." _____

_____

Since momentum is a vector this principle is equivalent to the three equations: $P_x =$ constant if $\sum F_{ext\,x} = 0$, $P_y =$ constant if $\sum F_{ext\,y} = 0$, and $P_z =$ constant if $\sum F_{ext\,z} = 0$. One component of momentum may be conserved even if the others are not. See, for example, Sample Problem 7 of the text.

If the momentum of a system is conserved then the acceleration of the center of mass of the system is _____ and the velocity of the center of mass is _____.

Variable mass systems. $\sum \mathbf{F}_{ext} = M\mathbf{a}_{cm}$ and $\sum \mathbf{F}_{ext} = d\mathbf{P}/dt$ hold only for systems of constant mass. Recall that when they were derived, the number of particles in the system was held constant. In Section 9–8 the derivation of an expression for $d\mathbf{P}/dt$ is carefully carried out for a system that loses mass. As another application of the ideas presented in this section let us carry out the derivation for a system that gains mass.

Suppose that at time t a solid object with mass M is moving with velocity $\mathbf{v}$. Nearby is some mass ΔM that will join the object in time Δt. Initially ΔM is traveling with velocity

u but when it joins, it and the object will have the same velocity, different from the original velocity of the object. The object exerts a force on the mass to bring it to the appropriate velocity and the mass exerts a force of equal magnitude but opposite direction on object.

We consider the object *and* the nearby mass ΔM as a constant mass system. Its total momentum at initial time t is $\mathbf{P}_i =$ _____. Now suppose the mass joins the object: at time $t + \Delta t$ the combination has mass $M + \Delta M$ and velocity $\mathbf{v} + \Delta \mathbf{v}$. The total momentum is now $\mathbf{P}_f =$ _____.

Use the expressions you gave above to write $(\mathbf{P}_f - \mathbf{P}_i)/\Delta t$ in terms of M, $\mathbf{v}$, $\mathbf{u}$, $\Delta \mathbf{v}$, and ΔM, then evaluate the expression in the limit as $\Delta t \to 0$. The term $\Delta M \Delta \mathbf{v}/\Delta t$ vanishes in this limit since it is proportional to an infinitesimal. Because Newton's second law is valid, the result equals the net external force on the object. In the space below carry out the steps to show that $\sum \mathbf{F}_{\text{ext}} = M d\mathbf{v}/dt - (\mathbf{u} - \mathbf{v})dM/dt$:

One question that might be asked is: what external force must act on the object if it is to move with constant velocity as mass is added? If this question is asked, the following quantities must be given: _____.

Another possible question is: if no external forces act, what is the acceleration of the object as mass is added? If this question is asked the following quantities must be given: _____.

Translational and internal energy. The total kinetic energy of a system of particles can be divided into two parts. One is associated with motion of the center of mass and is called the <u>kinetic energy of the system as a whole</u>. The other is associated with the motions of the various particles relative to the center of mass. It, along with the potential energy associated with mutual interactions of particles in the system, make up the <u>internal energy</u>.

For a system of particles with total mass M and center of mass velocity $\mathbf{v}_{\text{cm}}$, the translational kinetic energy of the system as a whole is given by $K_{\text{cm}} =$ _____. Changes in this kinetic energy are related to the net external force acting on the system and the displacement of the center of mass. The system is replaced by a single particle of mass M, located at the position of the center of mass and acted on by the net external force $\sum \mathbf{F}_{\text{ext}}$. The change in the translational kinetic energy then equals the work W_{cm} done by this force.

Suppose the center of mass of a system moves from x_i to x_f along the x axis, during which motion the net external force is F_{ext}, a constant. The work done by this force on the "particle" that replaces the system is given by $W_{\text{cm}} =$ _____ and the change in the translational kinetic energy of the system is given by $\Delta K_{\text{cm}} =$ _____.

Be sure you realize that W_{cm} is not necessarily the actual work done by the resultant external force. In some cases the actual work done by $\sum \mathbf{F}_{\text{ext}}$ may change the internal energy of the system as well as change the translational kinetic energy. In other cases $\sum \mathbf{F}_{\text{ext}}$ may not actually do any work but W_{cm} is not zero and the kinetic energy changes. This occurs, for example when the point of application of an external force does not move but the center of mass does. Then the actual work that changes the kinetic energy is done by internal forces. If you are asked for a change in energy use the energy equation $\Delta K_{\text{cm}} + \Delta U + \Delta E_{\text{int}} = W$, where W is the actual work done by external forces. If you are asked about a change in the kinetic energy use $W_{\text{cm}} = \Delta K_{\text{cm}}$.

Be sure you understand the motion of a skater pushing off from a railing, as discussed in the text. Assume the skater's hands do not leave the railing. When filling in the blanks below consider what happens from the point of view of an observer at rest with respect to the railing.

The force that accelerates the skater is: _____

Does this force do any work? _____

The forces that do work are: _____

Does the kinetic energy of the skater change? _____

II. PROBLEM SOLVING

When asked to calculate the coordinates of the center of mass, first decide if the system consists of discrete particles, with the masses and coordinates of each given. If it does, the calculation proceeds by straightforward evaluation of $\mathbf{r}_{\text{cm}} = (1/M) \sum m_i \mathbf{r}_i$.

PROBLEM 1. A system consists of 4 particles, all in the xy plane. The first has a mass of 3.91 kg and coordinates $x = 0.38$ m, $y = 1.53$ m; the second has a mass of 2.13 kg and coordinates $x = 1.28$ m, y = 0.45 m; the third has a mass of 1.29 kg and coordinates $x = -2.45$ m, y = 1.22 m; and the fourth has a mass of 1.80 kg and coordinates $x = 3.12$ m, $y = -1.77$ m. What are the coordinates of the center of mass?

SOLUTION:

[ans: 0.730 m, 0.584 m]

Notice that no particle is located at the center of mass of the 4-particle system.

If, on the other hand, the system can be considered to be a continuous distribution of mass, next decide if the mass is uniformly distributed. Look for symmetry if it is. The object may have a plane of symmetry, such as the plane that bisects a plate halfway through its thickness or the plane that is the perpendicular bisector of a cylinder axis. The center of mass must be on such a plane. If there is more than one, the center of mass is somewhere on their intersection.

Also look for axes of symmetry. Every diameter of a uniform sphere or uniform spherical shell is an axis of symmetry as is the axis of a uniform cylinder or cylindrical shell. The center of mass must lie on these. If two axes of symmetry intersect or if one intersects a plane of symmetry the center of mass is at the intersection.

PROBLEM 2. A uniform steel plate is in the shape of an equilateral triangle with each side having a length of 1.2 m. It is 1.20 cm thick. If a coordinate system is placed so the origin is at one vertex, then a second vertex is on the x axis at $x = 1.20$ m and the third vertex is at $x = 0.60 = 0.60$ m, $y = 1.04$ m. What are the coordinates of the center of mass?

SOLUTION: In the space to the right draw the triangle and coordinate system. Now draw a line from each vertex to the center of the opposite side. The x and y coordinates of the center of mass are the same as the x and y coordinates of the intersection of these lines. Clearly the x coordinate is $x = L/2$, where L is the length of one side of the triangle. If d is the distance from the origin to the intersection then the coordinates of the center of mass are $x = d\cos 30°$ and $y = d\sin 30°$. The center of mass is halfway through the thickness of the plate.

[ans: $x = 0.60$ m, $y = 0.35$ m, $z = 0.60$ cm]

Perhaps you can partition the object into several smaller objects, each with an easily identifiable center of mass. Replace each of these with a single particle, with mass equal to the mass of the object it replaces and located at the center of mass of the object. Then find the center of mass of the particles.

PROBLEM 3. A U-shaped object is constructed of three uniform rectangular plates of different materials, with the dimensions shown and all with the same thickness, 0.020 cm. Plate A has a mass of 0.167 kg, plate B has a mass of 0.214 kg, and plate C has a mass of 0.315 kg. Where is the center of mass?

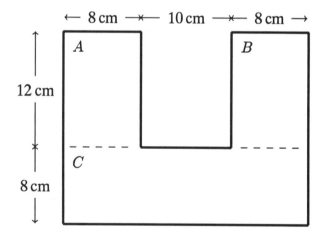

SOLUTION: Put the origin of a coordinate system at the lower left corner of the object, with the x axis extending to the right and the y axis extending upward. Assume the lower face of the object is in the xy plane and the z axis points out of the page. The center of mass of each plate is at the geometric center of the plate. For example, the center of mass of plate A alone has coordinates $x_A = 4.0$ cm, $y_A = 14$ cm. Treat each of the plates as if they were particles located at the center of mass of the plate. The x coordinate of the center of mass of the entire object, for example, is given by $x_{cm} = (m_A x_A + m_B x_B + m_C x_C)/(m_A + m_B + m_C)$.

[ans: $x_{cm} = 9.84$ cm, $y_{cm} = 11.3$ cm, $z_{cm} = 0.010$ cm]

Some objects have holes in them. Sample Problem 3 of the text gives an example and its solution provides an excellent model for solving many problems of this type. Fill the hole with material of the same density and find the center of mass of the composite. Also find the center of mass of the material filling the hole. Replace the material filling the hole and the original object with particles. You must use algebraic symbols for the coordinates of the center of mass of the original object since they are unknown. Equate each coordinate of the center of mass of the composite to the corresponding coordinate of the center of mass of the two particles and solve for the coordinate of the original object.

PROBLEM 4. A 0.800-kg plate is in the shape of a rectangle 40.0 cm long, 30.0 cm wide, and 0.0300 cm thick. A square hole with an edge of 10.0 cm is cut through it 15.0 cm from each of the two sides and 5.00 cm from the lower edge, as shown. Where is the center of mass?

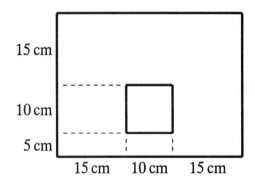

SOLUTION: The center of mass clearly lies on the plane that bisects the length of the rectangle. If the origin is at the lower left corner, the x axis is horizontal, and the y axis is vertical, then $x_{cm} = 20.0$ cm. Suppose the hole is filled with material of the same mass density as the rectangle. Let m_h be the mass that fills the hole and let y_h be the coordinate of the center of mass of this material. It is $y_h = 5.00 + 5.00 = 10.0$ cm. Let y_f be the coordinate of the center of mass of the filled rectangle. It is $y_f = 15.0$ cm. Finally, let y_r be the center of mass of the rectangle with the hole. This is the unknown. From the definition of the center of mass $y_f = (m_r y_r + m_h y_h)/(m_r + m_h)$.

The mass that fills the hole can be calculated by taking a ratio of areas. Let a and b be the sides of the rectangle and let c be the side of the square hole. Then $m_h = c^2 m_r/(ab - c^2)$. The denominator, of course, is the area of the rectangle with the hole.

Now all the masses and all the center of mass coordinates except y_r are known. Solve for y_r.

[ans: 15.5 cm]

When all else fails resort to Eq. 18 in the text. If the object has an axis of symmetry place one of the coordinate axes along it. Divide the object into strips perpendicular to the axis of symmetry, each with infinitesimal width. Suppose the y axis is along a line of symmetry and dy is the width of a strip. Find the mass in each strip as a function of its position y along the axis. It will be proportional to dy. Then integrate over y.

PROBLEM 5. A plate of mass M has the shape of a semicircle with radius R. Its thickness is T. Find the center of mass.

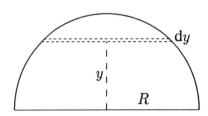

SOLUTION: Place the origin of a coordinate system at the center of the circle, with the x axis along the straight edge of the plate. By symmetry the x coordinate of the center of mass is $x_{cm} = 0$. Divide the semicircle into infinitesimal strips, each of width dy. One is shown on the diagram. The area of a strip is $2\sqrt{R^2 - y^2}\,dy$ and the mass it contains is $dm = 4M\sqrt{R^2 - y^2}\,dy/\pi R^2$ so the y coordinate of the center of mass is given by $y_{cm} = (4/\pi R^2)\int_0^R y\sqrt{R^2 - y^2}\,dy$. Evaluate the integral.

$$\left[\text{ans: } x_{cm} = 0, y_{cm} = 4R/3\pi, z_{cm} = T/2\right]$$

Suppose $R = 1.2$ m and $T = 0.20$ cm. What are the coordinates of the center of mass?

SOLUTION:

$$\left[\text{ans: } x_{cm} = 0, y_{cm} = 38.2 \text{ cm}, z_{cm} = 0.10 \text{ cm}\right]$$

Some problems deal with the motion of the center of mass. The simplest gives the external forces acting on each particle of a system and asks for the acceleration of the center of mass. Use $\sum \mathbf{F}_{ext} = M\mathbf{a}_{cm}$, where M is the total mass of the system. Simply add (vectorially) all the external forces and divide the resultant by the total mass of the system. Sample Problem 1 of the text is an example. Here are some others.

PROBLEM 6. Two gliders are initially at rest on a horizontal air track, one at each end. The one at the left end has a mass of 0.150 kg and the one at the right end has a mass of 0.220 kg. Students push the gliders toward each other by applying a horizontal force of 0.0230 N to the glider at the left and a horizontal force of 0.0300 N to the glider at the right. What is the acceleration of the center of mass of the two glider system? What is its speed 2.5 s after the students start pushing? How far has it traveled in that time?

SOLUTION: The net external force is $\sum F_{ext} = 0.300 - 0.0230 = 0.0070$ N, to the left. The total mass is $0.150 + 0.220 = 0.370$ kg.

$$\left[\text{ans: } 0.0189 \text{ m/s}^2; 0.0473 \text{ m/s}; 59.1 \text{ cm}\right]$$

The answers would be the same if the gliders interact with each other. Suppose they are connected by a spring and the same external forces are applied. The acceleration of the center of mass would still be 0.0189 m/s^2. The motions of the individual gliders, however, are different in the two situations.

PROBLEM 7. A ball is thrown straight up with an initial speed of 25 m/s. At the same instant an identical ball is dropped from a point 100 m above the first ball. At that instant where is the center of mass and what is its velocity?

SOLUTION: Use $y_{cm} = (m_1 y_1 + m_2 y_2)/(m_1 + m_2)$ to find the coordinate and $v_{cm} = (m_1 v_1 + m_2 v_2)/(m_1 + m_2)$ to find the velocity.

[ans: 50.0 m above the lower ball; 12.5 m/s]

Where is the center of mass and what is its velocity 1.0 s later?

SOLUTION: Since the total external force on the two-ball system is $(m_1 + m_2)g$ the acceleration of the center of mass is g, down. Use the kinematic equations $y_{cm} = y_{cm\,0} + v_{cm\,0} t - \frac{1}{2} g t^2$ and $v_{cm} = v_{cm\,0} - gt$, where the positive y direction was chosen to be upward.

SOLUTION:

[ans: 57.6 m above the initial position of the lower ball; 2.70 m/s, upward]

What is the highest position reached by the center of mass and what are the positions of the balls when the center of mass is there?

SOLUTION:

[ans: 58.0 m; 23.9 m, 92.0 m]

In a variant of this type problem, the system consists of several objects with a common acceleration or a common magnitude of acceleration, which is the unknown. This occurs when the objects slide together on a surface or are tied together by a string. The acceleration of the center of mass can be calculated if the net external force is known. Once you have done this, substitute the value into the relationship between the acceleration of the center of mass and the unknown acceleration and solve for the unknown acceleration.

Sample Problem 2 of the text is an example. Carefully note that the net external force is the vector sum of the force exerted by the table on the pulley and the force of gravity on block 2. The force of gravity and the normal force of the table on block 1 are also external forces but they cancel. Since the pulley does not accelerate, the force of the table on it is balanced by the force of the cord on it. This condition is used to write the external force of the table in terms of the tension in the cord.

PROBLEM 8.   An Atwood machine is constructed by placing a light string over a light frictionless pulley. Mass $m_1 = 0.35$ kg is attached to one end and mass $m_2 = 0.50$ kg is attached to the other. Find the acceleration of the center of mass of the two-mass system and the accelerations of the individual masses.

SOLUTION: Take the positive y axis to be upward, in the direction of the force of the string on each mass. Then the total force on the system is $F = 2T - (m_1 + m_2)g$, where T is the tension in the string. The acceleration of the center of mass is given by $a_{cm} = F/(m_1 + m_2) = 2T/(m_1 + m_2) - g$. The relationship between the acceleration of the center of mass and the acceleration of m_1 is provided by $a_{cm} = (m_1 a_1 + m_2 a_2)/(m_1 + m_2)$. Take $a_1 = a$ and $a_2 = -a$. Then $a_{cm} = a(m_1 - m_2)/(m_1 + m_2)$ and $a(m_1 - m_2)/(m_1 + m_2) = 2T/(m_1 + m_2) - g$. This equation contains 2 unknowns, T and a. Newton's second law for m_1 gives $T - m_1 g = m_1 a$ so $T = m_1(g + a)$. Now you can solve for a.

[ans: $a_{cm} = -0.305$ m/s^2; $a = 1.73$ m/s^2]

Some problems deal with two-body systems for which the center of mass is initially at rest and the net external force vanishes. One of the bodies changes position and the problem asks for the change in position of the other. Since the net external force is zero the velocity of the center of mass is a constant and since it is initially zero it remains zero. The center of mass does not move. Write an expression for the center of mass in terms of the initial coordinates of the bodies, write a second expression in terms of the final coordinates, and equate the two expressions. Then solve for the final coordinates of the second body.

PROBLEM 9. A 75-kg man sits in the stern of a 95-kg row boat that is motionless is still water. He then moves to the bow, 3.1 m away, and sits there. How far and in what direction does the boat move? Neglect the force of the water on the boat.

SOLUTION: Let x_{bi} be the initial coordinate of the boat, x_{mi} the initial coordinate of the man, x_{bf} the final coordinate of the boat, and x_{mf} the final coordinate of the man. Since the center of mass of the man-boat system does not move, $m_b x_{bi} + m_m x_{mi} = m_b x_{bf} + m_m x_{mf}$ or $m_b \Delta x_b = -m_m \Delta x_m$. You must realize that these coordinates are relative to a fixed coordinate system. The problem gives the distance the man moves relative to the boat, not relative to the fixed reference frame. If d is this distance then $\Delta x_m = \Delta x_b + d$. Substitute $\Delta x_b + d$ for Δx_m in $m_b \Delta x_b = -m_m \Delta x_m$, solve for Δx_b, and evaluate the expression you obtain.

[ans: 1.37 m, opposite the direction the man moves]

Another short calculation shows the man moves 1.73 m relative to the fixed coordinate system.

In a variant of this type problem the initial and final coordinates of both objects are given, along with the mass of one of them, and you are asked for the mass of the other.

The equation $\sum \mathbf{F}_{\text{ext}} = M\mathbf{a}_{\text{cm}}$ is a vector equation. In some cases the x component of the external force vanishes while the other components do not. The x component of the velocity of the center of mass is then constant while the other components are not. If, in addition, the center of mass started at rest the x coordinate of the center of mass remains at the same place. For example, in Sample Problem 5 of the text the external force is in the vertical direction so only the horizontal coordinate of the center of mass remains at rest as the ball rolls inside the shell.

PROBLEM 10. A large wedge with a mass of 15 kg rests on a frictionless horizontal surface. The inclined surface is 20° above the horizontal, as shown. A 5.0-kg crate is placed at the top and allowed to slide down. After the crate has gone 2.5 m down the incline, how far has the wedge moved?

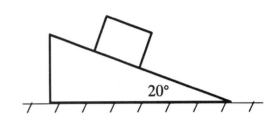

SOLUTION: Take the system to consist of both the crate and wedge. The only external forces acting are the forces of gravity on the crate and wedge and the normal force of the horizontal surface on the wedge. These are all vertical, so the horizontal component of the acceleration of the center of mass vanishes. Since the objects started at rest the horizontal coordinate of the center of mass does not move as the crate slides down and the wedge moves left. Place the origin of a coordinate system at the initial position of the center of mass of the wedge alone. Take the x axis to the right and the y axis upward. Let m_c be the mass of the crate, m_w the mass of the wedge, and x_{ci} the initial x coordinate of the crate. The x coordinate of the center of mass of the crate-wedge system is then given by $x_{\text{cm}} = m_c x_{ci}/(m_c + m_w)$. Let x_{cf} be the x coordinate of the crate and x_{wf} be the x coordinate of the wedge after the crate has gone a distance d down the incline. Then $x_{\text{cm}} = (m_c x_{cf} + m_w x_{wf})/(m_c + m_w)$. The two expressions must be equal, so $m_c x_{ci}/(m_c + m_w) = (m_c x_{cf} + m_w x_{wf})/(m_c + m_w)$ or, after the denominators are canceled, $m_c x_{ci} = m_c x_{cf} + m_w x_{wf}$.

A little geometry shows that if the crate slides a distance d down the incline while the wedge moves to x_{wf} then relative to the fixed coordinate system $x_{cf} - x_{ci} = x_{wf} + d\cos\theta$, where θ is the angle of the incline. Solve this equation simultaneously with $m_c x_{ci} = m_c x_{cf} + m_w x_{wf}$ for x_{wf}. You should get $x_{wf} = -[m_c/(m_c + m_w)]d\cos\theta$. Now substitute numerical values.

[ans: 0.587 m]

Carefully note that the answer does not depend on whether or not the inclined surface of the wedge is frictionless. Friction between the crate and the wedge is an internal force and does not change the motion of the center of mass of crate-wedge system. The answer, however, is valid only if the horizontal surface is frictionless. Friction between the horizontal surface and the wedge is external to the crate-wedge system and does change the motion of its center of mass.

Many problems of this chapter can be worked using the principle of momentum conservation. If you suspect the principle can be used, first check for external forces: if there are none or if they add to zero, total momentum is conserved. For some problems only one component of momentum is conserved. If the problem asks about motion in that direction, then the principle can be used.

To use it you must identify the system you will consider. Clearly the system must not be acted on by a net external force. If a net force acts on object A, then the momentum of A is *not* conserved. Perhaps object B provides the only force on A and perhaps the sum of the forces on A and B vanishes. Then the total momentum of A and B is conserved and you will take the system to consist of *both* A and B.

As a cannonball is fired horizontally from a cannon on a frictionless horizontal surface, for example, the horizontal component of the ball's momentum is not conserved. The ball is accelerated by the gunpowder explosion. However, the horizontal component of the total momentum of the ball and cannon together is conserved, since the net external force on the ball-cannon system has no horizontal component. See Sample Problem 7 of the text.

Conservation of momentum problems take several forms. In the simplest, an event occurs in which two objects with known masses exert equal and opposite forces on each other. The velocities of the objects before the event and the velocity of one object after the event are given. You are asked for the velocity of the second object after the event. Write the momentum of each object, both before and after the event, as the product of a mass and velocity. Equate the total momentum before the event to the total momentum after and solve for the unknown velocity.

PROBLEM 11. A 75-kg ice boat is stationary near the shore of a frozen lake. A 40-kg boy running at 6.2 m/s jumps into the boat and holds onto its rigging. If the ice is essentially frictionless what is the speed of the boat and boy?

SOLUTION: The horizontal component of the total momentum of the boy and boat is conserved. It is given by $P = mv_i$, where m is the mass of the boy and v_i is his initial velocity. It is also given by $P = mv_f + Mv_f$, where M is the mass of the boat and v_f is the final velocity of both the boy and boat. Solve $mv_i = (m + M)v_f$ for v_f.

[ans: 2.16 m/s]

PROBLEM 12. A 25-kg cannon, initially at rest on a fixed horizontal frictionless surface, fires a 2.0-kg cannonball horizontally with a muzzle velocity of 150 m/s. What is the recoil velocity of the cannon? What is the speed of the ball relative to the surface?

SOLUTION: Here the horizontal component of the total momentum of the system consisting of the cannon and cannonball is conserved. It is zero before, during, and after the firing. Let M be the mass and V the velocity of the cannon after firing. Let m be the mass and v be the velocity of the ball, relative to the fixed surface. Then conservation of momentum yields $MV + mv = 0$. Be careful about the term "muzzle velocity". It means the velocity of the cannonball *relative to the cannon*. In writing the conservation of momentum equation, velocities must be expressed relative to an inertial frame. Thus the velocity of the cannonball relative to the fixed surface is $v = v_m + V$, where v_m is the muzzle velocity. Solve the two equations simultaneously for V and v.

[ans: -11.1 m/s; 139 m/s]

Sample Problem 6 of the text is a variant. Here we know the horizontal component of the momentum of each bullet before it is imbedded in the block so we know the total momentum of the system consisting of the block and bullets. After the bullets are imbedded, they and the block all have the same velocity and we write the total momentum in terms of the common velocity. Finally we use conservation of momentum to solve for the final velocity.

In another case, a bullet may not become imbedded. Here's an example.

PROBLEM 13. A 2.0-kg wooden block rests on a horizontal frictionless surface. A 3.8-g bullet is fired horizontally into the block at 1100 m/s. It passes through and afterwards the block is found to have a speed of 0.80 m/s. What is the speed of the bullet after it emerges?

SOLUTION:

[ans: 679 m/s]

Some problems ask for the final kinetic energy of an object, rather than for its velocity. Just because kinetic energy is mentioned in the problem, do not assume energy conservation can be invoked to solve it. Check the external forces to see if momentum is conserved and if it is, use momentum conservation to find the velocity, then calculate the kinetic energy. Sample Problem 8 of the text is an example. Here's another.

PROBLEM 14. How much kinetic energy is lost by the bullet of the previous problem? How much kinetic energy is gained by the block?

SOLUTION:

[ans: 1420 J; 0.640 J]

The difference goes chiefly to increase internal energy.

To deal with a variable mass system, extend it so the new system has fixed mass, then apply $\sum \mathbf{F}_{ext} = d\mathbf{P}/dt$. This usually means including the mass that will be added or taken away from the original system over some time interval, perhaps infinitesimal.

Chapter 9: Systems of Particles   135

PROBLEM 15. A supermarket conveyor track consists of a series of small light frictionless wheels on which baskets move, as shown. A 0.65-kg basket is moving on the track at 0.35 m/s when a 1.3-kg bag of groceries is dropped into it. What then is the speed of the basket?

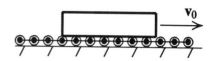

SOLUTION: We take the system to be the basket and groceries. The net external force, composed of the downward force of gravity and the upward force of the track on the basket, has a zero horizontal component. Hence the horizontal component of the total momentum is conserved. Let M be the mass of the basket, m the mass of the groceries, v_0 the initial velocity of the basket, and v_f the final velocity of the basket. Momentum conservation leads to $Mv_0 = (M+m)v_f$. Solve for v_f. Notice that before the bag of groceries enters the basket it does not contribute to the horizontal component of the total momentum because its velocity is downward.

[ans: 0.117 m/s]

PROBLEM 16. As an industrial conveyor belt moves horizontally at 0.27 m/s, sand is dropped onto it at the rate 1.7 kg/s. What power must the belt motor supply to keep the belt moving at a constant velocity?

SOLUTION: Suppose mass Δm is added to the belt in time Δt. Initially the horizontal component of its velocity is 0; after it is on the belt the horizontal component is the same as the speed v of the belt. The change in the horizontal component of its momentum is $v\Delta m$. The momentum of the belt itself does not change, so the change in momentum of the system is $\Delta P = v\Delta m$ and the rate of change of the total momentum is $\Delta P/\Delta t = v\Delta m/\Delta t$. This must equal the external force, which is the force of the motor on the belt. Thus $F = v\Delta m/\Delta t$. The power that must be supplied is $Fv = v^2\Delta m/\Delta t$. Evaluate this expression.

SOLUTION:

[ans: 0.124 W]

Not all of the work goes to increase the kinetic energy of the sand. Calculate the rate of increase of kinetic energy as the sand hits the belt and speeds up.

SOLUTION:

[ans: 0.0620 W]

Some problems deal with the energy of a system of particles. You must be able to identify the kinetic, potential, and internal energies and understand how to calculate changes in them.

PROBLEM 17. A 1200-kg automobile on a level road accelerates uniformly from rest to 80 km/h in 30 s. What is the horizontal component of the force of the road on the tires?

SOLUTION: If air resistance can be neglected the force of the road provides the only horizontal external force on the car and is the force that accelerates the car. Use kinematics to find the acceleration, then use $\mathbf{F}_{\text{ext}} = M\mathbf{a}_{\text{cm}}$ to find the force.

[ans: 889 N]

What is the change in the kinetic energy of the car?

SOLUTION: Use $\Delta K = \frac{1}{2}Mv_f^2 - \frac{1}{2}Mv_i^2$.

[ans: 2.96×10^5 J]

If the tires do not slip on the road the point of application of the force is instantaneously at rest when the force is applied so the road does 0 work on the car. The entire increase of kinetic energy comes from the internal energy of the car.

Calculate W_{cm} and compare the result with the change in kinetic energy.

SOLUTION: Since the force of the road is constant use $W_{\text{cm}} = F\Delta x$, where Δx is the distance traveled. Use $\Delta x = \frac{1}{2}at^2$ to compute Δx.

[ans: 2.96×10^5 J]

PROBLEM 18. Now suppose the car of the last problem accelerates from rest to 80 km/h in 30 s while doing down a 2.0° incline. What is the decrease in the internal energy of the car?

SOLUTION: Use $\Delta K + \Delta E_{\text{int}} = W$, where W is the work done by gravity. Use kinematics to find the change h in altitude of the car, then $W = mgh$.

[ans: 1.59×10^5 J]

In the following problem a model is used to show you how the work done by an external force might change the kinetic energy and internal energy of a system. Two gliders are used to represent atoms in an object and a spring coupling them is used to represent an interatomic interaction. A student pushes on one of the gliders and, as a result, the center of mass of the system moves and the system as a whole acquires translational kinetic energy. But the work done by the student is greater than the change in kinetic energy. The extra work brings about a further increase in the total kinetic energy that cannot be associated with the motion of the system as a whole but must be associated with the motions of the parts (the gliders) relative to the center of mass. It also changes the spring length, so the potential energy increases. Both the additional kinetic energy and the spring potential energy are part of the internal energy of the system.

PROBLEM 19. Two gliders, each having a mass of 300 g, are coupled by a spring with a force constant of 30.0 N/m. They are placed at rest on a horizontal frictionless air track, with the spring at its equilibrium length. The first glider is at $x = 0$ and the second is at $x = 0.200$ m. A student pushes on the first glider with a constant force of 4.00 N. At the end of 0.75 s the first glider is at $x = 1.920$ m and the second is at $x = 2.030$ m. During the 0.75 s interval what was the change in the translational kinetic energy of the two-glider system as a whole? The numbers given here are the same as those for a problem in the last chapter.

SOLUTION: Use $\Delta K = F \Delta x_{cm}$, where Δx_{cm} is the displacement of the center of mass of the two-glider system: $x_{cm} = (x_1 + x_2)/2$ since the gliders have the same mass.

[ans: 7.50 J]

Notice that this is not the same as the sum of the changes in the kinetic energies of the gliders, which you calculated to be 7.56 J.

What was the change in the internal energy of the two-glider system?

SOLUTION: Use $\Delta K + \Delta E_{int} = W$, where W is the actual work done by the student. You calculated this to be 7.68 J.

[ans: 0.18 J]

Of this increase, 7.56 − 7.50 = 0.06 J can be characterized as kinetic and 0.12 J can be characterized as potential.

III. COMPUTER PROJECTS

You can use a computer to follow individual objects in a system as they interact with each other and to follow the center of mass of the system. As an example consider two carts connected by an ideal spring on a horizontal frictionless air track. If the spring has a natural length ℓ_0 (when it is neither compressed or extended), then when its length is ℓ it exerts a force of magnitude $k|\ell - \ell_0|$ on each of the carts. Here k is the force constant. If the spring is extended ($\ell > \ell_0$) then both forces are toward the center of the spring. If the spring is compressed ($\ell < \ell_0$) then both forces are away from the center of the spring.

Let x_1 be the coordinate of one cart and x_2 be the coordinate of the other and take $x_2 > x_1$. Then $\ell = x_2 - x_1$. The force on cart 1 is given by $F_1 = k(x_2 - x_1 - \ell_0)$ and the force on cart 2 is given by $F_2 = -k(x_2 - x_1 - \ell_0)$.

A program that will calculate the coordinate and velocity of each cart as functions of time can be modeled after the first program of Chapter 6. In addition, the program calculates the coordinate and velocity of the center of mass. Here's an outline.

> input initial values: t_0, x_{10}, v_{10}, x_{20}, v_{20}
> input final time and interval width: t_f, Δt
> input display interval: Δt_d
> set $t_b = t_0$, $t_d = t_0 + \Delta t_d$, $x_{1b} = x_{10}$, $v_{1b} = v_{10}$, $x_{2b} = x_{20}$, $v_{2b} = v_{20}$
> calculate coordinate and velocity of center of mass:
> $\quad x_{cm} = (m_1 x_{1b} + m_2 x_{2b})/(m_1 + m_2)$
> $\quad v_{cm} = (m_1 v_{1b} + m_2 v_{2b})/(m_1 + m_2)$
> display x_{1b}, x_{2b}, x_{cm}
> **begin loop** over intervals
> $\quad$ calculate force on 1 at beginning of interval: $F_1 = k(x_{2b} - x_{1b} - \ell_0)$
> $\quad$ calculate accelerations at beginning of interval: $a_1 = F_1/m_1$, $a_2 = -F_1/m_2$
> $\quad$ calculate velocities at end of interval: $v_{1e} = v_{1b} + a_1 \Delta t$, $v_{2e} = v_{2b} + a_2 \Delta t$
> $\quad$ calculate "average" velocities: $v_1 = (v_{1b} + v_{1e})/2$, $v_2 = (v_{2b} + v_{2e})/2$
> $\quad$ calculate coordinates at end of interval: $x_{1e} = x_{1b} + v_1 \Delta t$, $x_{2e} = x_{2b} + v_2 \Delta t$
> $\quad$ calculate time at end of interval: $t_e = t_b + \Delta t$
> $\quad$ **if** $t_e \geq t_d$ **then**
> $\quad\quad$ calculate coordinate and velocity of center of mass:
> $\quad\quad\quad x_{cm} = (m_1 x_{1e} + m_2 x_{2e})/(m_1 + m_2)$
> $\quad\quad\quad v_{cm} = (m_1 v_{1e} + m_2 v_{2e})/(m_1 + m_2)$
> $\quad\quad$ * display x_{1e}, x_{2e}, x_{cm}
> $\quad\quad$ increment t_d by Δt_d
> $\quad$ **end if** statement
> $\quad$ **if** $t_e \geq t_f$ **then** exit loop
> $\quad$ set $t_b = t_f$, $x_{1b} = x_{1e}$, $v_{1b} = v_{1e}$, $x_{2b} = x_{2e}$, $v_{2b} = v_{2e}$
> **end loop** over intervals
> stop

For some applications the line marked with an asterisk is replaced by: display v_{1e}, v_{2e}, v_{cm}. You will need to use a large number of extremely small intervals, so truncation errors may be significant. If possible, use double precision variables.

PROJECT 1. Take $m_1 = 250$ g, $m_2 = 600$ g, $\ell_0 = 0.50$ m, and $k = 5.00$ N/m. Initially ($t = 0$) cart 1 is at rest at the origin and cart 2 is at rest at $x_2 = 0.70$ m. Plot the coordinates of the carts and the coordinate of the center of mass every 0.10 s from $t = 0$ to $t = 2.0$ s. You will need to take $\Delta t = 0.02$ s to obtain 2 significant figure accuracy.

Notice that the center of mass is initially at rest and remains at rest even though the carts oscillate back and forth. Now repeat the calculation with $v_{20} = 0.30$ m/s. The carts again oscillate but as they do the center of mass moves with constant velocity in the positive x direction. Check that the calculation gives the same result as $v_{cm} = (m_1 v_{10} + m_2 v_{20})/(m_1 + m_2)$ for the velocity of the center of mass.

For both of the situations considered the total momentum of the system, given by $P = (m_1 + m_2)v_{cm}$, is conserved. Now suppose an external force of 0.50 N is applied to in the positive x direction the first cart. The acceleration of that cart is computed using $a_1 = (F_1 + 0.50)/m_1$, where F_1 is still the force of the spring on cart 1. The acceleration of cart 2 is $a_2 = -F_1/m_2$, as before. Now plot the coordinates of the carts and the coordinate of the center of mass every 0.10 s from $t = 0$ to $t = 2.0$ s. The curve representing the coordinate of the center of mass should be parabolic. In fact, $x_{cm} = \frac{1}{2}a_{cm}t^2$, where $a_{cm} = F_e/(m_1 + m_2)$. Here F_e is the external force, 0.50 N. Plot v_{cm} as a function of time and verify that it is a straight line with the correct slope.

A computer program can also be used to investigate the motion of a rocket. If the rocket and fuel have mass $M(t)$ and the fuel is ejected with velocity $\mathbf{u}$, measured relative to the rocket, then the acceleration of the rocket is given by

$$M\frac{d\mathbf{v}}{dt} = \mathbf{F} + \frac{dM}{dt}\mathbf{u},$$

where $\mathbf{F}$ is the external force acting on the rocket (the force of gravity, for example).

First consider a rocket that is fired straight upward and assume the acceleration due to gravity is uniform. If the positive y axis is upward then $\mathbf{u} = -u\mathbf{j}$, $\mathbf{v} = v\mathbf{j}$, and $\mathbf{F} = -Mg\mathbf{j}$. If the fuel is expended uniformly you may write $M(t) = M_0 - Kt$, where K is a constant. The differential equation becomes

$$(M_0 - Kt)\frac{dv}{dt} = uK - (M_0 - Kt)g.$$

This expression gives the acceleration as a function of time and can be used in conjunction with the computer program of Chapter 2 to calculate the position and velocity of the rocket at any time until the rocket runs out of fuel.

The differential equation above can also be solved analytically to obtain

$$v(t) = v_0 - gt + u \ln \frac{M_0}{M_0 - Kt},$$

where v_0 is the velocity at $t = 0$. You will use this expression in the next computer project to check your program.

PROJECT 2. Consider a rocket that carries 80% of its original mass as fuel and ejects it uniformly for 5.0 s, at which time the fuel is gone and the engine shuts off. The rocket starts from rest and the speed of the fuel relative to the rocket is $u = 5000$ m/s. Use the program of Chapter 2 to find the speed and altitude of the rocket at burnout. First you need to find a value for K. At the end of 5.0 s M is the mass of the rocket alone and this is $0.20M_0$. Thus $M_0 - 5.0K = 0.20M_0$, so $K = 0.16M_0$. While fuel is being expended the acceleration of the rocket is given by $a(t) = -g + 0.16u/(1 - 0.16t)$. Use this expression in the program. Check your answer by using the analytic expression above to calculate the speed. [ans: 8.00×10^3 m/s; 1.48×10^4 m]

The next project deals with a rocket that moves in two dimensions. If the acceleration due to gravity is uniform then

$$\frac{d\mathbf{v}}{dt} = -\frac{K\mathbf{u}}{M_0 - Kt} - g\mathbf{j},$$

where the y axis is positive in the upward direction. Since **u** is always opposite to **v** we may write $\mathbf{u} = -u\mathbf{v}/v$, where the unit vector $\mathbf{v}/v$ is in the direction of **v**. Thus

$$\frac{d\mathbf{v}}{dt} = \left[\frac{Ku}{M_0 - Kt}\right]\frac{\mathbf{v}}{v} - g\mathbf{j}.$$

This equation holds while fuel is being ejected. After burnout the rocket becomes an ordinary projectile and $d\mathbf{v}/dt = -g\mathbf{j}$.

You will write a program to investigate the motion of a rocket fired from rest over level ground. Since each component of the acceleration depends on both components of the velocity you cannot use $\mathbf{s} = (\mathbf{a}_b + \mathbf{a}_e)/2$ to approximate the average acceleration in an interval. Use instead the acceleration at the beginning of the interval. For the first interval the program will have trouble with the unit vector since it involves division by 0. For that interval replace v_x/v by $\cos\phi_0$ and v_y/v by $\sin\phi_0$, where ϕ_0 is the firing angle. At the end of the interval, after the velocity components have been computed, calculate $v = \sqrt{v_{xe}^2 + v_{ye}^2}$ and the ratios v_{xe}/v and v_{ye}/v, in preparation for the next interval. This can be handled using the unit vector components α_x and α_y as you will see in the outline below.

The program should provide for the possibility of different interval widths before and after burnout. Before burnout the acceleration may be great and both the speed and direction of travel may change rapidly. You will want to use a small interval. After burnout you will want to use a larger interval to save time. Here's an outline of a program.

> input burnout time: t_{bo}
> input firing angle: ϕ_0
> input interval widths: $(\Delta t)_b$ for $t < t_{bo}$, $(\Delta t)_a$ for $t > t_{bo}$
> set interval width: $\Delta t = (\Delta t)_b$
> calculate unit vector components: $\alpha_x = \cos\phi_0$, $\alpha_y = \sin\phi_0$
> set initial values: $t_b = 0$, $x_b = 0$, $v_{xb} = 0$, $y_b = 0$, $v_{yb} = 0$
> **begin loop** over intervals
>   calculate x component of acceleration at beginning of interval:
>   **if** $t_b < t_{bo}$ then $a_x = \alpha_x Ku/(M_0 - Kt_b)$
>   **if** $t_b \geq t_{bo}$ then $a_x = 0$
>   calculate y component of acceleration at beginning of interval:
>   **if** $t_b < t_{bo}$ then $a_y = -g + \alpha_y Ku/(M_0 - Kt_b)$
>   **if** $t_b \geq t_{bo}$ then $a_y = -9.8$
>   calculate velocity at end of interval: $v_{xe} = v_{xb} + a_x\Delta t$, $v_{ye} = v_{yb} + a_y\Delta t$
>   calculate "average" velocity: $v_x = (v_{xb} + v_{xe})/2$, $v_y = (v_{yb} + v_{ye})/2$
>   calculate x coordinates at end of interval: $x_e = x_b + v_x\Delta t$, $y_e = y_b + v_y\Delta t$
>   calculate time at end of interval: $t_e = t_b + \Delta t$
> *   **if** $t_e \geq t_f$ then
>   display or print t_b, x_b, y_b, v_{xb}, v_{yb}
>   display or print t_e, x_e, y_e, v_{xe}, v_{ye}
>   exit loop
>   **end if** statement

> set $t_b = t_e$, $x_b = x_e$, $v_{xb} = v_{xe}$, $y_b = y_e$, $v_{yb} = v_{ye}$
> if $t_b < t_{bo}$ set $\Delta t = (\Delta t)_b$, if $t_b \geq t_{bo}$ set $\Delta t = (\Delta t)_a$
> calculate unit vector components:
> $$v = \sqrt{v_{xe}^2 + v_{ye}^2}, \alpha_x = v_{xe}/v, \alpha_y = v_{ye}/v$$
> **end loop** over intervals when $y_e < 0$
> stop

The line marked with an asterisk will be different for different applications. To calculate the coordinates and velocity for the highest point on the trajectory, end the loop when $v_{ye} \leq 0$. To calculate the range over level ground, end the loop when $y_e \leq 0$.

PROJECT 3. A toy rocket has a mass (without fuel) of 250 g. It carries 10 g of fuel and is fired from rest at 45° above the horizontal. The fuel leaves the rocket at 4000 m/s and burns out after 0.200 s of flight. Use the program to find the coordinates and velocity components at burnout ($t = 0.20$ s). For 2 significant figure accuracy use $\Delta t = 0.001$ s before burnout and $\Delta t = 0.01$ s after burnout. [ans: $x = 12.0$ m, $y = 10.7$ m, $v_x = 121$ m/s, $v_y = 108$ m/s]

Use the program to find the time the rocket reaches the highest point on its trajectory. Find its coordinates and speed when it is there. [ans: $t = 11.2$ s, $x = 1.35 \times 10^3$ m, $y = 605$ m, $v = 121$ m/s]

Use the program to find the time the rocket lands if it is fired over level ground. Find its coordinates and speed just before it lands. [ans: $t = 22.3$ s, $x = 2.70 \times 10^3$ m, $v_x = 121$ m/s, $v_y = -109$ m/s]

IV. NOTES

Chapter 10
COLLISIONS

I. BASIC CONCEPTS

Here you will apply the principle of momentum conservation to collisions between objects. You should pay special attention to the role played by the impulses the objects exert on each other. You should also take notice of when energy conservation can be invoked to solve a problem and when it cannot.

Definitions. As succinctly as you can, tell in words what a collision is. Be sure to include the important characteristics. _____
_____

During a collision two objects exert forces on each other. What is important is not the force alone or its duration alone but a combination called the <u>impulse</u> of the force. If one body acts on the other with a force $\mathbf{F}(t)$ for a time interval from t_i to t_f, the impulse of the force is given by the integral

$$\mathbf{J} = $$

Don't confuse impulse and work. Impulse is an integral of force with respect to _____ and work is an integral of force with respect to _____. Impulse is a vector, work is a scalar. The SI unit of impulse is _____; the SI unit of work is _____.

In terms of the average force $\mathbf{F}_{ave}$ that acts during the collision, the impulse is given by $\mathbf{J} = $ _____, where Δt is _____. This expression is often useful for calculating the average force, which in turn is useful as an estimate of the strength of the interaction.

For most collisions the impulse of one colliding object on the other is usually much greater than any external impulse and we may neglect impulses exerted by the environment of the colliding bodies. We may, for example, neglect the effects of gravity and air resistance during the time a baseball is in contact with a bat.

Because Newton's second law is valid, the total impulse acting on an object gives the change in the _____ of the object. Because Newton's third law is valid the impulse of one object on the other is the negative of the impulse of the second object on the first. If external impulses can be neglected the total _____ of the two objects is conserved during a collision.

Collisions. Total momentum is conserved during the collisions we consider. Since momentum is conserved during a collision, the velocity of the center of mass of the colliding bodies is _____.

Kinetic energy may or may not be conserved in a collision. If it is, the collision is said to be _____. If it is not, the collision is said to be _____. The distinguishing characteristic of a two-body <u>completely</u> inelastic collision is _____
_____.

During a completely inelastic collision the loss in total kinetic energy is as large as conservation of momentum allows. That is, it is impossible for the bodies to lose a larger fraction of their original kinetic energy and together still retain all the original total momentum.

Explosions, in which an object splits into two or more parts as a result of internal forces, can be handled in exactly the same manner as a collision. During an explosion the total momentum is conserved but the total kinetic energy increases.

Elastic one-dimensional collisions. Consider an elastic two-body collision in one dimension and suppose both objects move along the x axis. Object 1 has mass m_1, moves with velocity v_{1i} before the collision, and moves with velocity v_{1f} after the collision. Object 2 has mass m_2, moves with velocity v_{2i} before the collision, and moves with velocity v_{2f} after the collision. The equation that expresses conservation of momentum during the collision is

The equation that expresses conservation of kinetic energy during the collision is

Using the equations of momentum and energy conservation you should be able to show that the relative velocity with which the objects approach each other before the collision is the same as the relative velocity with which they separate after the collision. In symbols, _____. The proof follows immediately when you divide the kinetic energy conservation equation, in the form $m_1(v_{1f}^2 - v_{1i}^2) = -m_2(v_{2f}^2 - v_{2i}^2)$, by the momentum conservation equation, in the form $m_1(v_{1f} - v_{1i}) = -m_2(v_{2f} - v_{2i})$. You must recognize that $A^2 - B^2 = (A - B)(A + B)$ for any quantities A and B. Carry out the steps in the space below:

In the space below solve $m_1v_{1i} + m_2v_{2i} = m_1v_{1f} + m_2v_{2f}$ and $(v_{1f} - v_{2f}) = -(v_{1i} - v_{2i})$ for v_{1f} and v_{2f} in terms of the other quantities. Draw a box around your answers and check them with the text.

Using these general expressions, you should be able to obtain the results for some special elastic collisions. Write the results below without looking them up and explain what has happened during the collision. Unless specifically stated otherwise do not assume either object is initially at rest.

a. If the two masses are the same, then $v_{1f} = $ _____ and $v_{2f} = $ _____.
   In words, what has happened is _____

b. If the two masses are the same and object 2 is initially at rest, then $v_{1f} = $ _____ and $v_{2f} = $ _____.
   In words, what has happened is _____

c. If the target object, object 2 say, is very massive compared to the incident object, object 1, then $v_{1f} = $ _____ and $v_{2f} = $ _____.
   In words, what has happened is _____

d. If the target object, object 2, is very massive compared to the incident object, object 1, and is initially at rest, then $v_{1f} = $ _____ and $v_{2f} = $ _____.
   In words, what has happened is _____

e. If object 1 is very massive compared to object 2 and object 2 is initially at rest, then $v_{1f} = $ _____ and $v_{2f} = $ _____.
   In words, what has happened is _____

For all these collisions the total kinetic energy (the sum of the individual kinetic energies of the colliding objects) is the same after the collision as before. However, kinetic energy is nearly always transferred from one object to the other. Sometimes you will want to know how much. The fractional loss of kinetic energy by object 1 is given by $(K_{1i} - K_{1f})/K_{1i}$ and, since $K = \frac{1}{2}mv^2$, this is $1 - v_{1f}^2/v_{1i}^2$.

Suppose object 2 is initially at rest and find an expression for the fractional energy loss of object 1, in terms of the masses of the two objects:

Consider some special cases, all with object 2 initially at rest. First suppose the masses of the two objects are the same. Then the fractional energy loss of object 1 is _____. Next suppose object 1 is much more massive than object 2. Its fractional energy loss is then _____. Finally suppose object 1 is much less massive than object 2. Its fractional energy loss is then _____.

Suppose object 2 is initially at rest and has a given mass m_2. How should you pick m_1 so object 1 loses a large fraction of its energy to object 2? _____ How should you pick m_1 so object 1 loses only a small fraction of its energy to object 2? _____

Completely inelastic one-dimensional collisions. Consider a completely inelastic two-body collision in one dimension. Both objects move along the x axis. Object 1 has mass m_1 and moves with velocity v_{1i} before the collision. Object 2 has mass m_2 and moves with velocity v_{2i} before the collision. Both objects move with velocity v_f after the collision. The equation than that expresses conservation of momentum during the collision is

If the masses and initial velocities are known, conservation of momentum is sufficient to determine the common final velocity. This velocity is given by

$v_f =$

An example of this type collision is _____ .

If object 2 is initially at rest then

$v_f =$

Notice that the final speed of the combination *must* be less than the initial speed of object 1. Explain why in words: _____

Kinetic energy is lost during a completely inelastic collision, transformed to internal energy, radiation, etc. Consider a completely inelastic collision with object 2 initially at rest. In terms of the masses and velocities the initial total kinetic energy is $K_i =$ _____ and the final total kinetic energy is $K_f =$ _____ . In the space below make the substitutions to express the kinetic energies in terms of the velocities, carry out the algebra, and obtain the result given below for the fractional energy loss of the two body system:

$$\left[\text{ans: } 1 - [(m_1 + m_2)/m_1](v_f^2/v_{1i}^2)\right]$$

Use conservation of momentum to substitute for v_f^2 in terms of v_{1i}^2 and find an expression for the fractional energy loss in terms of the masses alone:

$$\left[\text{ans: } m_2/(m_1 + m_2)\right]$$

To make the fractional energy loss large the mass of object _____ should be made much larger than the mass of the other object. To make the loss small the mass of _____ should be made much larger than the mass of the other object.

Two-dimensional collisions. Consider the two-dimensional elastic collision diagramed below.

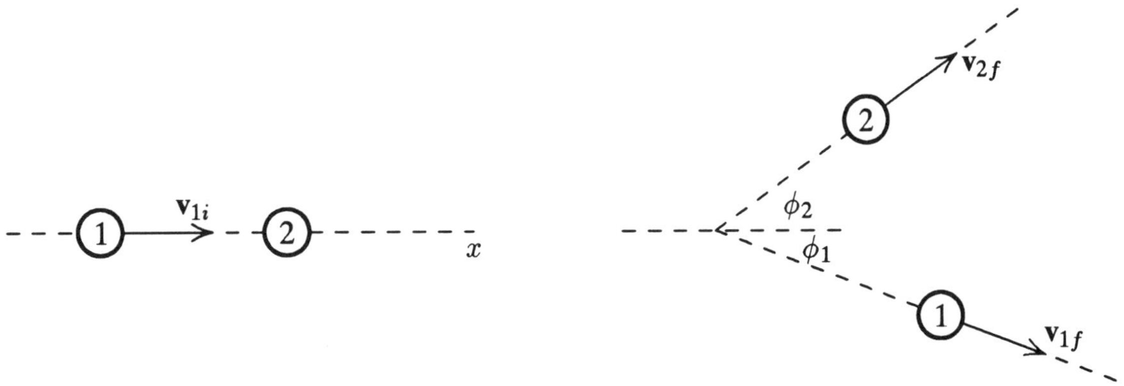

Before Collision   After Collision

Object 2 is initially at rest and object 1 impinges on it with speed v_{1i} along the x axis. Object 1 leaves the collision with speed v_{1f} along a line that is below the x axis and makes the angle ϕ_1 with that axis. Object 2 leaves the collision with speed v_{2f} along a line that is above the x axis and makes the angle ϕ_2 with that axis. Take the y axis to be upward in the diagram. In terms of the masses, speeds, and angles between the velocities and the x axis, conservation of total momentum **P** leads to the equations:

conservation of P_x:

conservation of P_y:

Suppose the collision is elastic and write the equation that expresses conservation of kinetic energy in terms of the masses and speeds:

conservation of K:

These equations can be solved for three unknowns. If the masses (or their ratio) are given then, of the five other quantities (v_{1i}, v_{1f}, v_{2f}, ϕ_1, and ϕ_2) that enter the equation, two must be given.

Note that the masses and the initial velocity of object 1 alone do not determine the outcome of a two-dimensional elastic collision. The magnitude and direction of the impulse acting on each object during the collision also play important roles. Often these are not known, so the outcome cannot be predicted. Experimenters usually measure the speed or the orientation of the line of motion of one of the objects after the collision then use the equations to calculate other quantities. You should know how to solve for any three of the quantities, given the others. Some hints are given in the Problem Solving section below.

If a two-dimensional collision is completely inelastic, the objects stick together after the collision and kinetic energy is not conserved. We now consider collisions for which both objects are initially moving. The geometry is shown below. Before the collision mass m_1 has velocity $\mathbf{v}_{1i}$ and mass m_2 has velocity $\mathbf{v}_{2i}$. After the collision both masses have velocity $\mathbf{v}_f$.

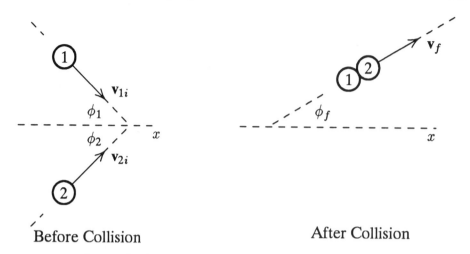

Before Collision After Collision

Since $v_{1f} = v_{2f}$, we write v_f for both of speeds. Take the y axis to be upward in the diagram. The conservation of momentum equations become

conservation of P_x:

conservation of P_y:

These two equations contain eight quantities (m_1, m_2, v_{1i}, v_{2i}, ϕ_1, ϕ_2, v_f, and ϕ_f) and can be solved for 2 unknowns. If the masses (or their ratio) are given, then 4 additional quantities must be given. Unlike an elastic collision, the masses and initial conditions completely determine the outcome. That is, knowledge of m_1, m_2, v_{1i}, v_{2i}, ϕ_1, and ϕ_2 allow us to solve for v_f and ϕ_f. Notice that the six given quantities allow us to find the velocity of the center of mass. Since momentum is conserved and the objects stick together this is the same as the final velocity of both objects.

Collisions in the center of mass frame. Measured in a reference frame that moves with the center of mass, the total momentum vanishes both before and after a collision. For a one-dimensional *elastic* collision the velocity of each particle is reversed by the collision so $v_{1f} = -v_{1i}$ and $v_{2f} = -v_{2i}$, where all these quantities are measured in the center of mass frame.

Suppose the masses and initial velocities are given for an elastic one-dimensional collision. One way to find the final velocities is to first find the velocity of the center of mass, then subtract it from each initial velocity to find the initial velocities in the center of mass frame. Now change the sign of each initial velocity in that frame to find the final velocities in the center of mass frame. Finally add the velocity of the center of mass to each of these to find the final velocities in the laboratory frame.

For a completely inelastic collision viewed in the center of mass frame, both objects are at rest after the collision. In the laboratory frame after the collision both objects move with the velocity of the center of mass.

Change in kinetic energy during an inelastic collision. Sometimes the gain or loss of kinetic energy during an inelastic collision is known. The energy equation is then useful. Suppose energy Q is gained by the system during a collision. Then, for either one- or two-dimensional collisions

$$\frac{1}{2}m_1v_{1f}^2 + \frac{1}{2}m_2v_{2f}^2 = \frac{1}{2}m_1v_{1i}^2 + \frac{1}{2}m_2v_{2i}^2 + Q.$$

This equation is also valid if kinetic energy is lost; Q is then negative. Q has a positive value, for example, when internal energy is converted to kinetic energy during the collision, perhaps through an explosion. On the other hand, Q is negative if kinetic energy is converted to internal energy or to another form. In a ballistic pendulum experiment, for example, Q is negative; most of the "lost" kinetic energy becomes internal energy or energy of deformation. In any event, the energy equation is solved simultaneously with the conservation of momentum equations.

The ideas discussed here are valid for atomic and subatomic collisions and for spontaneous decay of subatomic particles as well as for collisions of macroscopic objects. In many nuclear and subatomic collision processes the internal energy and mass energy may change, so they must be included in the conservation of energy equation. The number and type of particles may also change. Changes in particle identities are taken into account in the energy equation by including the energy associated with mass. In the energy equation a mass decrease is represented by a positive value for Q while a mass increase is represented by a negative value. To work problems involving changes in mass you must recall that the energy associated with mass m is _____, where c is the speed of light.

When the problem involves subatomic particles relativistically valid expressions for momentum and energy must usually be used rather than the non-relativistic expressions we have been working with. If a particle has mass m and velocity **v** its momentum is given by

$$\mathbf{p} =$$

and its kinetic energy is given by

$$K =$$

II. PROBLEM SOLVING

Most calculations of impulse are rather straightforward. You might be given the force acting on an object as a function of time and asked to find the impulse over a given time interval. You simply evaluate the integral that defines impulse. Remember that impulse is a vector quantity and you must, for example, use the x component of the force to find the x component of the impulse. For one-dimensional motion the force might be given graphically as a function of time. You must then recognize that the impulse is the area under the curve.

You may be asked to find the change in momentum during the interval the impulse acts. You must recognize this is equal to the impulse. Equate like components of impulse and momentum change. You might also be given the mass and asked for the change in velocity.

Lastly, you could be given the initial momentum or velocity and asked to find the final momentum or velocity. Simply add (vectorially) the change in momentum (i.e. the impulse) to the initial momentum.

PROBLEM 1. A 0.55-kg ball is dropped from rest. What is the impulse exerted by gravity on the ball during the first 1.7 s of its fall?

SOLUTION: Since the force is constant, the integral for the magnitude of the impulse reduces to $J = mg\Delta t$.

[ans: 9.16 N·s, downward]

The change in the momentum of the ball over this interval is 9.16 kg·m/s, downward. Calculate the velocity of the ball at the end of the interval and compare your result with the velocity you obtain using kinematics.

SOLUTION:

[ans: 16.7 m/s, downward]

What is the impulse of gravity on the ball during the first 20 m of its fall?

SOLUTION: The force is again mg, down, and the impulse is $mg\Delta t$, where Δt is the time interval. You must first calculate the time the ball takes to fall 20 m, then calculate the impulse.

[ans: 10.9 N·s, downward]

Be sure you understand what is meant by the *change* in momentum. It is the final momentum minus the initial momentum. Vector subtraction must be used, as illustrated by the following example.

PROBLEM 2. A 0.15-kg ball moving at 15 m/s hits the ground at an angle of 20° and bounces up with the same speed and at the same angle, as shown. What is the change in its momentum?

SOLUTION: Take the x axis to be horizontal and to the right; take the y axis to be vertical and upward. Let ϕ_i and ϕ_f be the angles the initial and final momenta, respectively, make with the x axis. They have the same values for this problem but may have different values for other problems. Fill in the following table by writing both algebraic expressions and numerical values for the components of the momentum before the collision, the momentum after the collision, and the change in momentum:

before: p_{ix} = _____ = _____ kg·m/s p_{iy} = _____ = _____ kg·m/s

after: p_{ix} = _____ = _____ kg·m/s p_{iy} = _____ = _____ kg·m/s

change: Δp_x = _____ = _____ kg·m/s Δp_x = _____ = _____ kg·m/s

Magnitude of the change in momentum: _____ kg·m/s

Angle the change in momentum makes with the x axis: _____

[ans: $\Delta p_x = 0$; $\Delta p_y = 1.54$ kg·m/s; $|\Delta \mathbf{p}| = 1.54$ kg·m/s; angle = 90°]

The inverse impulse-momentum problem gives information to compute the initial and final momenta and asks for the impulse. Find the change in momentum by subtracting (vectorially) the initial momentum from the final momentum, then equate the change to the impulse. Sample Problem 1 of the text is of this type. It emphasizes the vector nature of impulse. If the duration of the collision is known, you can use $\mathbf{J} = \mathbf{F}_{\text{ave}}\Delta t$ to find the average force acting.

PROBLEM 3. A 150-g baseball, traveling at 40 m/s and at an angle of 15° below the horizontal, is struck by a bat. It leaves the bat with a velocity of 55 m/s, 35° above the horizontal, on a trajectory over the pitcher's head. What were the horizontal and vertical components of the impulse of the bat on the ball? What was the magnitude of the impulse? If the ball was in contact with the bat for 2.3 ms what was the magnitude of the average force exerted by the bat on the ball?

SOLUTION: If p_{ix} and p_{iy} are the horizontal and vertical components, respectively, of the momentum of the ball before hitting the bat and p_{fx} and p_{fy} are the horizontal and vertical components, respectively, afterwards, then the components of the impulse are given by $J_x = p_{fx} - p_{ix}$ and $J_y = p_{fy} - p_{iy}$. Choose the x axis to be horizontal and positive in the direction from the pitcher toward the batter and the y axis to be positive in the upward direction. Then $p_{ix} = mv_0 \cos 15°$, $p_{iy} = -mv_0 \sin 15°$, $p_{fx} = -mv_0 \cos 35°$, and $p_{fy} = mv_0 \sin 35°$. Calculate J_x and J_y, then J. Use $J_x = F_{\text{ave}\,x}\Delta t$ and $J_y = F_{\text{ave}\,y}\Delta t$ to find the components of the force, then calculate the magnitude.

[ans: $J_x = -12.6\,\text{N·s}$; $J_y = 6.28\,\text{N·s}$; $J = 14.0\,\text{N·s}$; $F_{\text{ave}} = 6100\,\text{N}$]

Collision problems are more complicated. First decide if the collision is one- or two-dimensional. A head-on collision (one for which the impact parameter vanishes) is always one-dimensional. So is a completely inelastic two-body collision with one object initially at rest. If the objects move along different lines, whether initially or finally, the collision is two dimensional.

We consider one-dimensional collisions first. Since total momentum is always conserved in collisions, nearly every problem solution begins by writing the equation for momentum conservation. There is only one such equation for a one dimensional collision. To write it use the component, not the magnitude, of the momentum of each object and write the component as the product of the appropriate mass and velocity component. Always use symbols, not numbers, even for given quantities.

Make a list of the quantities given in the problem statement and a list of the unknowns. If there is only one unknown the momentum conservation equation can be solved immediately for it. Once the equation has been solved, the result can be used in other calculations. You might be asked to test for the conservation of kinetic energy, for example, and classify the collision as elastic or inelastic. You might also be asked to calculate the impulse exerted by one object on the other.

If more than one quantity is unknown, look for more information in the problem statement. If it specifies that the collision is elastic, write the equation for the conservation of kinetic energy in terms of the masses and speeds of the objects. If it specifies that the collision is completely inelastic, equate the final velocities of the objects to each other and use a single symbol to denote them. If it specifies that kinetic energy was gained or lost during the collision, write the energy equation in the form $K_f = K_i + Q$ (using masses and speeds to write K_i and K_f, of course).

Look for special conditions. If, for example, the problem statement tells you that the masses of the objects are the same, use the same symbol to represent both masses. If $Q = 0$ the masses then cancel from the equations and the algebra simplifies greatly. If the collision is elastic you may reduce your work by using $v_{1i} + v_{1f} = v_{2i} + v_{2f}$ instead of the energy equation.

PROBLEM 4. In an air track experiment a 0.35-kg glider impinges at 3.4 m/s on a 0.20-kg glider at rest. If the collision is elastic what are the velocities of the gliders afterwards?

SOLUTION: Solve $m_1 v_{1i} = m_1 v_{1f} + m_2 v_{2f}$ and $v_{1i} + v_{1f} = v_{2f}$ simultaneously for v_{1f} and v_{2f}.

[ans: 0.927 m/s, 4.33 m/s, both in the direction of motion of the incident glider]

How much energy was transferred from the incident glider to the glider initially at rest?

SOLUTION: Calculate the kinetic energy of the second glider after the collision. Since this glider was initially at rest the result is the energy transferred.

[ans: 1.87 J]

What impulse was exerted on the incident glider?

SOLUTION: Calculate the change in its momentum.

[ans: 0.866 N·s, opposite its initial direction of motion]

PROBLEM 5. In an air track experiment a 0.35-kg glider impinges at 3.4 m/s on a 0.20-kg glider at rest. If the collision is completely inelastic what are the velocities of the gliders afterwards?

SOLUTION: Solve $m_1 v_{1i} = (m_1 + m_2) v_f$ for the common final velocity v_f.

[ans: 2.16 m/s, in the direction of motion of the incident glider]

How much kinetic energy was transferred from one glider to the other?

SOLUTION:

[ans: 0.468 J]

How much kinetic energy was lost by the two-glider system?

SOLUTION: Write the energy equation and calculate Q.

[ans: 0.736 J]

What impulse was exerted on the incident glider?

SOLUTION:

[ans: 0.434 N·s, opposite its initial direction of motion]

PROBLEM 6. In an air track experiment a 0.35-kg glider and a 0.20-kg glider move toward each other with the heavy glider moving at 2.9 m/s and the light glider moving at 3.5 m/s. After they collide the heavy glider moves away in the opposite direction at 1.7 m/s. Was kinetic energy conserved? If it was not, was it gained or lost by the two-glider system? How much was gained or lost?

SOLUTION: Use conservation of momentum to calculate the velocity of the light glider after the collision, then calculate the energies.

[ans: 0.121 J was lost]

PROBLEM 7. In an air track experiment a 0.35-kg glider impinges at 2.9 m/s on a 0.20-kg glider at rest. The collision releases a compressed spring attached to the end of one of the gliders and, as a result, the kinetic energy of the two-glider system is increased by 2.3 J. What are the velocities of the gliders after the collision?

SOLUTION: Solve $m_1 v_{1i} = m_1 v_{1f} + m_2 v_{2f}$ and $\frac{1}{2} m_1 v_{1f}^2 + \frac{1}{2} m_2 v_{2f}^2 = \frac{1}{2} m_1 v_{1i}^2 + Q$ for v_{1f} and v_{2f}. First use the momentum equation to obtain an expression for v_{2f} in terms of v_{1f}, then substitute the expression into the energy equation. You will obtain a quadratic equation for v_{1f}. Solve it algebraically, then evaluate the result numerically. Take $Q = +2.3$ J. You will obtain two solutions, corresponding to the two possible signs of the square root. Only one makes physical sense. The other describes the passing of the gliders through each other.

[ans: heavy: 0.582 m/s opposite to $\mathbf{v}_{1i}$, light: 6.09 m/s in the direction of $\mathbf{v}_{1i}$]

You should be able to solve collision problems using the center of mass frame of reference. Let us practice with one of the problems above.

PROBLEM 8. In an air track experiment a 0.35-kg glider impinges at 3.4 m/s on a 0.20-kg glider at rest. If the collision is elastic what are the velocities of the gliders afterwards?

SOLUTION: First use the given masses and velocities to find the velocity of the center of mass.

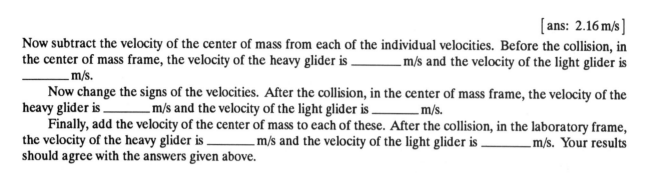

[ans: 2.16 m/s]

Now subtract the velocity of the center of mass from each of the individual velocities. Before the collision, in the center of mass frame, the velocity of the heavy glider is _____ m/s and the velocity of the light glider is _____ m/s.

Now change the signs of the velocities. After the collision, in the center of mass frame, the velocity of the heavy glider is _____ m/s and the velocity of the light glider is _____ m/s.

Finally, add the velocity of the center of mass to each of these. After the collision, in the laboratory frame, the velocity of the heavy glider is _____ m/s and the velocity of the light glider is _____ m/s. Your results should agree with the answers given above.

For two-dimensional collisions there are two momentum conservation equations, one for each component. The amount of algebraic manipulation can usually be reduced significantly if one of the coordinate axes is placed along the direction of the total momentum. For some problems, however, this direction cannot be found from the given data. In any event, select a coordinate system and write the momentum conservation equations in terms of the masses, speeds, and angles between the velocities and a coordinate axis.

If the line of motion of one of the objects, either before or after the collision, makes the angle ϕ with a coordinate axis, then you will find $\sin\phi$ in one of the components of the momentum conservation equation and $\cos\phi$ in the other. To eliminate ϕ, use the trigonometric identity $\cos^2\phi + \sin^2\phi = 1$. That is, solve one equation for $\sin\phi$ and the other for $\cos\phi$, square both results, add, and set the sum equal to 1. Sample Problem 4 of the text provides an example.

Sometimes the angle of the line of motion is sought. You then solve the momentum conservation equations for the sine or cosine of the angle and evaluate the inverse trigonometric function. This is also done in Sample Problem 4 of the text. Sometimes the equation for $\sin\phi$ or $\cos\phi$ contains an unknown and you must eliminate it before solving for the angle. An unknown can often be eliminated by dividing the expression for $\sin\phi$ by the expression for $\cos\phi$. The resulting equation for $\tan\phi$ is then solved.

Suppose the collision is elastic or Q is given for an inelastic collision and you wish to eliminate one of the speeds. Solve one of the momentum conservation equations for the speed to be eliminated and substitute the resulting expression in the energy equation, rather than vice versa. If you solve the energy equation first, the resulting expression will contain a square root and probably will be difficult to manipulate.

PROBLEM 9. On frictionless ice a block of mass m_1 impinges with speed v_{1i} on a block of mass m_2, initially at rest. The first block leaves the collision at an angle of ϕ_1 to the line of incidence while the target block leaves at an angle of ϕ_2, on the other side of the line of incidence. See the diagram in the Basic Concepts section. Find expressions for the speeds of the blocks after the collision, in terms of v_{1i}, ϕ_1, ϕ_1, m_1, and m_2.

SOLUTION: Solve $m_1 v_{1i} = m_1 v_{1f} \cos\phi_1 + m_2 v_{2f} \cos\phi_2$ and $0 = m_1 v_{1f} \sin\phi_1 - m_2 v_{2f} \sin\phi_2$ simultaneously for v_{1f} and v_{2f}: solve the second equation for v_{2f}, substitute the resulting expression into the first equation, and solve the result for v_{1f}. Use the second equation to solve for v_{2f}.

$$\left[\text{ans: } v_{1f} = v_{1i}\sin\phi_2/\sin(\phi_1+\phi_2),\ v_{2f} = (m_1/m_2)v_{1i}\sin\phi_1/\sin(\phi_1+\phi_1)\right]$$

The trigonometric identity $\sin\phi_1\cos\phi_2 + \sin\phi_2\cos\phi_2 = \sin(\phi_1+\phi_2)$ was used.

Take $m_1 = 400$ g, $m_2 = 700$ g, $v_{1i} = 50$ m/s, $\phi_1 = 45°$, and $\phi_2 = 25°$. What are the final speeds?

SOLUTION:

$$\left[\text{ans: } v_{1f} = 18.9\,\text{m/s},\ v_{2f} = 22.3\,\text{m/s}\right]$$

Notice that conservation of momentum was sufficient to solve the problem. We do not yet know if the collision was elastic or inelastic.

Did the kinetic energy of the two-block system change during the collision? If it did, by how much?

SOLUTION:

$$\left[\text{ans: decreased by } 73.5\,\text{J}\right]$$

Suppose the collision occurs again, with the same initial conditions and with the incident block leaving the collision at an angle of 45° to the line of incidence, as before. Now, however the target block leaves at an angle of 30° on the same side of the line of incidence as before. What now are the final speeds? How much kinetic energy was gained or lost by the two-block system?

SOLUTION:

$$\left[\text{ans: } v_{1f} = 25.9\,\text{m/s},\ v_{2f} = 20.9\,\text{m/s};\ 53.7\,\text{J gained}\right]$$

PROBLEM 10. On frictionless ice a block of mass m_1 impinges with speed v_{1i} on a block of mass m_2, initially at rest. The first block leaves the collision with a speed of v_{1f} and at an angle of ϕ_1 to the line of incidence. With what velocity does the target block leave the collision?

SOLUTION: Use the diagram given in the Basic Concepts section, with the x axis along the line of incidence and the y axis perpendicular to that line. The momentum conservation equations $m_1v_{1i} = m_1v_{1f}\cos\phi_1 + m_2v_{2f}\cos\phi_2$ and $0 = m_1v_{1f}\sin\phi_1 - m_2v_{2f}\sin\phi_2$ are obeyed. Solve for v_{2f} and ϕ_2: solve the first equation for $\sin\phi_2$ and the second for $\cos\phi_2$. Square the two results and sum them. Use the identity $\sin^2\phi_2 + \cos^2\phi_2 = 1$ to eliminate ϕ_2, then solve for v_{2f} in terms of the given quantities. Divide the expression for $\sin \phi_2$ by the expression for $\cos \phi_2$ to obtain an expression for $\tan \phi_2$ in terms of the given quantities. Solve for ϕ_2.

$$\left[\text{ans: } v_{2f} = (m_1/m_2)\sqrt{v_{1i}^2 + v_{1f}^2 - 2v_{1i}v_{1f}\cos\phi_1};\ \phi_2 = \arctan(v_{1f}\sin\phi_1)/(v_{1i} - v_{1f}\cos\phi_1)\right]$$

Take $m_1 = 400$ g, $m_2 = 700$ g, $v_{1i} = 50$ m/s, $v_{1f} = 20$ m/s, and $\phi_1 = 20°$. What are v_{2f} and ϕ_2?

SOLUTION:

[ans: 18.3 m/s; 12.4°]

How much kinetic energy was gained or lost by the two-block system?

SOLUTION:

[ans: 133 J lost]

PROBLEM 11. Suppose the initial situation is the same as in the previous problem: a block of mass 400 g is incident at 50 m/s on a 700-g block at rest. Now the first block leaves the collision with a speed of 30 m/s. At what angle to the line of incidence does it leave if the collision is elastic?

SOLUTION: The algebraic solution you obtained before is still valid. Substitute the expression you obtained for v_{2f} into the energy conservation equation $\frac{1}{2}m_1v_{1i}^2 = \frac{1}{2}m_1v_{1f}^2 + \frac{1}{2}m_2v_{2f}^2$ and solve for $\cos\phi_1$, then for ϕ_1.

[ans: 78.5°]

What is the velocity of the target block after the collision?

SOLUTION:

[ans: 30.2 m/s, 33.7° on the other side of the line of incidence]

PROBLEM 12. A car with mass m_1 is traveling at v_1, at an angle ϕ_1 south of east. A second car, with mass m_2 is traveling at v_2, at an angle ϕ_2 north of east. They collide and couple together. What is their velocity after the collision?

SOLUTION: The collision is completely inelastic so the conservation of momentum equations are

$$m_1 v_1 \cos\phi_1 + m_2 v_2 \cos\phi_2 = (m_1 + m_2) v_f \cos\phi_f$$

$$-m_1 v_1 \sin\phi_1 + m_2 v_2 \sin\phi_2 = (m_1 + m_2) v_f \sin\phi_f ,$$

where the x axis was taken to be toward the east and the y axis was taken to be toward the north. Solve the first equation for $\cos\phi_f$ and the second for $\sin\phi_f$. Square and sum the results, then after replacing $\sin^2\phi_f + \cos^2\phi_f$ with 1 solve for v_f. Divide the equation for $\sin\phi_f$ by the equation for $\cos\phi_f$ to obtain an expression for $\tan\phi_f$ in terms of known quantities.

[ans: $v_f = \sqrt{m_1^2 v_1^2 + m_2^2 v_2^2 + 2 m_1 m_2 v_1 v_2 \cos(\phi_1 + \phi_2)}/(m_1 + m_2)$;
$\phi_f = \arctan(-m_1 v_1 \sin\phi_1 + m_2 v_2 \sin\phi_2)/(m_1 v_1 \cos\phi_1 + m_2 v_2 \cos\phi_2)$]

Suppose the mass of the first car is 900 kg and it is initially traveling at 15 km/h at 35° south of east. The mass of the second car is 1200 kg and it is initially traveling at 55° north of east. What is the speed and direction of travel after their completely inelastic collision?

SOLUTION:

[ans: 15.9 m/s, 35.5° north of east]

What fraction of the initial total kinetic energy was lost in the collision?

SOLUTION:

[ans: 0.218]

III. COMPUTER PROJECTS

Many details of a two-body collision are controlled by the impulse exerted by each object on the other. Consider the situation in which an object of mass m_1, moving with velocity v_{1b}, impinges on an object of mass m_2, initially at rest. Suppose that after the collision both objects move along the line of incidence. Let J be the impulse object 1 exerts on object 2. Then the velocity of object 2 after the collision is given by $v_{2a} = J/m_2$. Object 2 exerts an impulse $-J$ on object 1 and its velocity after the collision is given by $v_{1a} = v_{1b} - J/m_1$. Both final velocities can be computed if the initial velocity of object 1 and the impulse J are known (along with the masses, of course).

Once the velocities are known the change in the kinetic energy of the system can be found. Before the collision the kinetic energy is $K_b = \frac{1}{2}m_1v_{1b}^2$ and after the collision it is $K_a = \frac{1}{2}m_1v_{1a}^2 + \frac{1}{2}m_2v_{2a}^2$. The kinetic energy gained during the collision is given by $Q = K_a - K_b$. You will use a computer program to investigate the outcome of a one-dimensional collision as a function of the impulse.

Write a program that calculates the final velocities and the change in kinetic energy for a given range of impulses. Input the masses and the initial velocity of object 1. Also input the limits of the range of impulses to be considered. The outline of a program is:

> input masses, initial velocity: m_1, m_2, v_{1b}
> calculate initial kinetic energy: $K_b = \frac{1}{2}m_1v_{1b}^2$
> input limits of range of impulse: J_i, J_f
> number of values of J: $N = 21$
> calculate impulse increment: $\Delta J = (J_f - J_i)/(N-1)$
> set $J = J_i$
> **begin loop** over impulse values; counter runs from 1 to N
> > calculate final velocities: $v_{1a} = v_{1b} - J/m_1$, $v_{2a} = J/m_2$
> > calculate final kinetic energy: $K_a = \frac{1}{2}m_1v_{1a}^2 + \frac{1}{2}m_2v_{2a}^2$
> > calculate kinetic energy gain: $Q = K_a - K_b$
> > print or display J, v_{1a}, v_{2a}, Q
> > increment J: replace J with $J + \Delta J$
>
> **end loop**
> stop

The number N was chosen to be 21 because 21 lines nicely fill the usual monitor screen. You may want to use a different value. Both J_i and J_f are included in the list of values considered.

PROJECT 1. An air-track cart of mass $m_1 = 250$ g impinges at 3.0 m/s on a cart of mass $m_2 = 600$ g, initially at rest. Use the computer program to list the final velocities and kinetic energy loss for values of the impulse J from 0 to 2 kg·m/s.

From the table you generated pick out the two values of the impulse that straddle the condition $v_{1a} = v_{2a}$. This is the completely inelastic collision. Note that Q is negative and has maximum magnitude in this region of the table. Rerun the program with a narrower range of impulse values. Repeat if necessary and find J, v_{1a}, and Q for a completely inelastic collision to 3 significant figure accuracy. [ans: 0.529 kg·m/s; 0.882 m/s; −0.794 J]

From the original table ($J = 0$ to 2 kg·m/s) pick out the two values that straddle the condition $v_{1a} = 0$. Find to 3 significant figures the impulse, final velocity of cart 2, and the kinetic energy loss. [ans: 0.750 kg·m/s; 1.250 m/s; 0.656 J]

From the original table pick out the two values that straddle the condition $Q = 0$. This is the elastic collision. Find to 3 significant figures the impulse and final velocities. [ans: 1.06 kg·m/s; −1.24 m/s; 1.77 m/s]

PROJECT 2. Now repeat the calculations for two carts with the same mass. Take $m_1 = m_2 = 250$ g and $v_{1b} = 3.0$ m/s. Notice that for this special case the impulse that stops the incident cart is the same as the impulse that produces an elastic collision. [ans: completely inelastic: $J = 0.375$ kg·m/s, $v_{1a} = v_{2a} = 1.50$ m/s, $Q = -0.563$ J; elastic: $J = 0.750$ kg·m/s, $v_{1a} = 0$, $v_{2a} = 3.00$ m/s, $Q = 0$]

Two-dimensional collisions are somewhat more difficult to analyze. Consider a collision between two pucks on an air table. One puck with mass m_1 and speed v_{1b} is incident along the

x axis on a puck with mass m_2, initially at rest. Suppose the impulse exerted by the moving puck on the puck at rest has magnitude J and makes the angle α with the x axis. Puck 2 leaves the collision along the line of the impulse; its velocity makes the angle $\theta_2 = \alpha$ with the x axis. Its speed after the collision is given by $v_{2a} = J/m_2$. The impulse-momentum equations for the incident puck are $-J\cos\alpha = m_1 v_{1a} \cos\theta_1 - m_1 v_{1b}$ and $-J\sin\alpha = m_1 v_{1a} \sin\theta_1$. These can be solved for v_{1a} and θ_1, with the result $v_{1a} = \sqrt{(J^2 + m_1^2 v_{1b}^2 - 2m_1 v_{1b} J \cos\alpha)/m_1}$ and $\theta_1 = -\arctan[J\sin\alpha/(m_1 v_{1b} - J\cos\alpha)]$. To conserve momentum the sign of θ_1 must be opposite that of θ_2; that is, the two pucks must leave the collision on opposite sides of the line of incidence. If the sign of θ_1 produced by the computer is not correct, add 180° to it or subtract 180° from it.

Write a program that accepts values for the masses, initial velocity of the incident puck, and angle of the impulse with the x axis, then calculates the magnitudes and directions of the velocities after the collision and the gain in kinetic energy, all as functions of the magnitude of the impulse. An outline might be:

> input masses, initial velocity: m_1, m_2, v_{1b}
> calculate initial kinetic energy: $K_a = \frac{1}{2} m_1 v_{1a}^2$
> input angle of impulse: α
> input limits of range of impulse: J_i, J_f
> number of values of J: $N = 21$
> calculate impulse increment: $\Delta J = (J_f - J_i)/(N-1)$
> set $J = J_i$
> **begin loop** over impulse values; counter runs from 1 to $N + 1$
> > calculate final speeds:
> > $$v_{1a} = \sqrt{(J^2 + m_1^2 v_{1b}^2 - 2m_1 v_{1b} J \cos\alpha)/m_1}$$
> > $$v_{2a} = J/m_2$$
> > calculate angles:
> > $$\theta_1 = -\arctan[J\sin\alpha/(m_1 v_{1b} - J\cos\alpha)]$$
> > $$\theta_2 = \alpha$$
> > check sign of θ_1 and correct if necessary
> > calculate final kinetic energy: $K_a = \frac{1}{2} m_1 v_{1a}^2 + \frac{1}{2} m_2 v_{2a}^2$
> > calculate kinetic energy gain: $Q = K_a - K_b$
> > print or display J, v_{1a}, v_{2a}, θ_1, θ_2, Q
> > increment J: replace J with $J + \Delta J$
> **end loop**
> stop

PROJECT 3. Puck 1 has a mass of 250 g and is incident at 3.0 m/s on puck 2, which has a mass of 600 g and is initially at rest. The impulse of puck 1 on puck 2 makes an angle of 20° with the axis of incidence (the x axis). Use the program to construct a table of the final speeds, angles of motion, and kinetic energy losses for impulses with magnitudes in the range from 0 to 2 kg·m/s.

Use the table to find two values of the impulse magnitude that straddle the value for an elastic collision. Rerun the program to find the impulse for an elastic collision to 3 significant figures. What are the final speeds and angle of motion then? [ans: $J = 0.995$ kg·m/s; $v_{1a} = 1.55$ m/s at $-119°$; $v_{2a} = 1.66$ m/s at $20°$]

Now find the magnitude of the impulse for which puck 2 is scattered through 90°. What are the velocities of the pucks after the collision and what kinetic energy is lost? [ans: $J = 0.798$ kg·m/s; $v_{1a} = 1.09$ m/s at $-90°$; $v_{2a} = 1.33$ m/s at $20°$; $Q = 1.33$ J]

Now suppose the impulse makes an angle of 60° with the line of incidence. What magnitude of impulse produces an elastic collision? What are the final velocities? [ans: $J = 0.529$ kg·m/s; $v_{1a} = 2.67$ m/s at $-43.4°$; $v_{2a} = 0.882$ m/s at $60°$]

For what magnitude of impulse is puck 1 scattered through 90°? What are the final velocities and the kinetic energy loss? [ans: $J = 1.50$ kg·m/s; $v_{1a} = 5.20$ m/s at $-90°$; $v_{2a} = 2.50$ m/s at $60°$; $Q = 4.13$ J]

The positive value of Q in the last case means kinetic energy must be *added* during the collision (in an explosion, for example) to obtain backscattering of the incident puck.

PROJECT 4. What happens if the incident puck has greater mass than the target puck? Take $m_1 = 600$ g, $v_{1b} = 3.0$ m/s, and $m_2 = 250$ g. For $\alpha = 20°$ and $60°$ calculate the magnitude of the impulse that will produce an elastic collision and find the velocities of the pucks after such a collision. [ans: 20°: $J = 0.995$ kg·m/s, $v_{1a} = 1.55$ m/s at $-21.5°$, $v_{2a} = 3.78$ m/s at $20°$; 60°: $J = 0.529$ kg·m/s, $v_{1a} = 2.67$ m/s at $-16.6°$, $v_{2a} = 2.12$ m/s at $60°$]

PROJECT 5. If the masses are equal the pucks always leave an elastic collision with their lines of motion making an angle of 90° with each other. Use the program to verify this for $m_1 = m_2 = 250$ g, $v_{1b} = 3.0$ m/s, and $\alpha = 20°$. Also verify it for $\alpha = 60°$. [ans: 20°: $J = 0.705$ kg·m/s, $v_{1a} = 1.03$ m/s at $-70.0°$, $v_{2a} = 2.82$ m/s at $20°$; 60°: $J = 0.375$ kg·m/s, $v_{1a} = 2.60$ m/s at $-30.0°$, $v_{2a} = 1.50$ m/s at $60°$]

IV. NOTES

Chapter 11
ROTATIONAL KINEMATICS

I. BASIC CONCEPTS

You now begin the study of rotational motion. The pattern is similar to that used for the study of linear motion: you first study kinematics, the description of the motion, and then, in the next chapter, you study the dynamics of the motion. When studying this chapter take special care to understand the definitions of the rotational variables and how they are related to linear variables.

Definitions. If a rigid body undergoes pure rotation about a fixed axis then the path followed by each point on the body is a _____, centered on the _____.

The <u>angular position</u> of the body is described by giving the angle between a <u>reference line</u>, fixed to the body, and a non-rotating coordinate axis. Suppose the body rotates around the z axis and describe how a reference line might be chosen. In particular, identify the plane it is parallel to and the angle it makes with the axis of rotation. _____

Illustrate the idea in the space to the right by drawing the outline of a rotating body, with the axis of rotation, the z axis, out of the page. Mark the axis with a circled dot ($\odot$). Show x and y axes and a reference line, all in the plane of the page. Usually the angle ϕ that gives the angular position is measured counterclockwise from a coordinate axis. Show ϕ on your diagram.

The angular position ϕ is often measured in radians. Define a radian: _____

You should know the relationship between radian, degree, and revolution measure: 1 radian = _____ degrees = _____ revolution. Some often used angles are:

$360°$ = _____ radians = _____ revolution
$270°$ = _____ radians = _____ revolution
$180°$ = _____ radians = _____ revolution
$90°$ = _____ radians = _____ revolution
$45°$ = _____ radians = _____ revolution

Since an angle in radians is the ratio of two lengths it is dimensionless.

If the body rotates from ϕ_1 to ϕ_2, then its <u>angular displacement</u> is $\Delta\phi$ = _____. If ϕ_2 is greater than ϕ_1 then the angular displacement is positive. According to the convention stated above for measuring ϕ, the body rotated in the _____ direction.

If ϕ_2 is less than ϕ_1 then the angular displacement is negative and the body rotated in the _____ direction.

If the body rotates through more than 1 revolution, ϕ continues to increase beyond 2π rad. Be sure you understand that an angle of 2π rad is NOT equivalent to 0. A rotation from $\phi = 0$ to $\phi = 2\pi$ rad represents an angular displacement of 2π rad, not 0.

If the body undergoes an angular displacement $\Delta\phi$ in time Δt, its average angular velocity during the interval is $\overline{\omega} =$ _____. Its instantaneous angular velocity at any time is given by the derivative $\omega =$ _____. According to the convention used above to measure the angular position, a positive value for ω indicates rotation in the _____ direction and a negative value for ω indicates rotation in the _____ direction. Instantaneous angular velocity is usually called simply angular velocity.

If the angular velocity of the body changes with time then the body has a non-vanishing angular acceleration. If the angular velocity changes from ω_1 to ω_2 in the time interval Δt, then the average angular acceleration in the interval is $\overline{\alpha} =$ _____. The instantaneous angular acceleration at any time t is given by the derivative $\alpha =$ _____. Instantaneous angular acceleration is usually called simply angular acceleration.

Common units of angular velocity are deg/s, rad/s, rev/s, and rev/min. Corresponding units of angular acceleration are _____, _____, _____, and _____.

You should understand that a positive angular acceleration does NOT necessarily mean the angular speed is increasing and a negative angular acceleration does NOT necessarily mean the angular speed is decreasing. Remember that the angular speed decreases if ω and α have opposite signs and increases if they have the same sign, no matter what the sign of α. If ω is negative and α is positive, for example, then ω becomes less negative as time goes on and its magnitude becomes smaller. If the positive acceleration continues beyond $\omega = 0$, of course, the magnitude of ω starts increasing with time.

The values of $\Delta\phi$, ω, and α are the same for every point in a rigid body. The angular positions of different points may be different, of course, but when one point rotates through any angle, all points rotate through the same angle and their angular velocities change at the same rate.

Angular velocity and angular acceleration, but NOT angular position or displacement, can be considered to be vectors along the axis of rotation. To determine the direction of ω, use the right hand rule: _____

The diagram on the right shows a rigid rod rotating around the z axis, with points to the right of the axis moving into the page and points to the left of the axis moving out of the page. Show the angular velocity vector ω on the diagram. If the body were rotating in the opposite direction the vector angular velocity would be in the _____ direction.

If the body rotates faster as time goes on, then α and ω are in the same direction. If it rotates more slowly, then α and ω are in opposite directions. Consider the situation illustrated above (right side moving into page, left side moving out of page) and suppose the body rotates

162   Chapter 11: Rotational Kinematics

faster as time goes on. Then the angular acceleration vector α is in the _____ direction. If the body were rotating slower as time goes on, α would be in the _____ direction.

The vector nature of the angular velocity is important for thinking about a body undergoing simultaneous rotations about two axes, as does the disk of Sample Problem 3 in the text. At any instant the motion is equivalent to rotation around an axis in the direction of the vector sum $\omega_1 + \omega_2$ and the magnitude of the angular velocity is the magnitude of this vector. In the sample problem one of the axes is fixed while the other rotates. Thus the direction of one of the angular velocities is constant while the other is not. This means the direction of the equivalent axis changes with time.

Although directions can be assigned arbitrarily to angular displacements, they do not add as vectors. The result of two successive angular displacements around different axes that are not parallel, for example, depends on the order in which the angular displacements are carried out. This means angular displacements are not vectors.

Rotation with constant angular acceleration. If a body is rotating around a fixed axis with constant angular acceleration α, then as functions of time its angular velocity is given by

$$\omega(t) =$$

and its angular position is given by

$$\phi(t) =$$

Here ω_0 is _____
and ϕ_0 is _____

These two equations are solved simultaneously to find answers to constant angular acceleration kinematics problems. You should memorize them or, better yet, learn to derive them quickly using techniques of the integral calculus.

Another useful equation can be obtained by using $\omega = \omega_0 + \alpha t$ to eliminate t from $\phi = \omega_0 t + \frac{1}{2}\alpha t^2$. It is: $\omega^2 - \omega_0^2 = 2\alpha\phi$, where phi_0 was taken to be 0. Table 1 of the text lists all possible results when one of the kinematic equations is used to eliminate a variable from the other. You may use these equations to solve rotational kinematics problems or else simultaneously solve the two you wrote above.

Any consistent set of units can be used. Thus ϕ in degrees, ω in degrees/s, and α in degrees/s^2 as well as ϕ in radians, ω in radians/s, and α in radians/s^2 or ϕ in revolutions, ω in revolutions/s, and α in revolutions/s^2 can be used. Be careful not to mix units, however.

Relationships between linear and angular quantities. Each point in a rotating body has a velocity and acceleration and these are related to the angular variables. Suppose the body is rotating around a fixed axis and consider a point in the body a perpendicular distance r from the axis. If the body turns through the angle ϕ (in radians) the point moves a distance $s =$ _____ along its circular path. The speed of the point is $v = ds/dt$ and the angular velocity of the body is $\omega = d\phi/dt$, so v is related to ω by $v =$ _____, where the units of ω MUST be _____. The tangential component of the acceleration of the point is $a_T = dv/dt$ and the angular acceleration is $\alpha = d\omega/dt$ so a_T is related to α by $a_T = $ _____, where the units of α MUST be _____.

Because the point is moving in a circular path its acceleration also has a radial component. You learned in Chapter 4 that $a_R = v^2/r$. Now write the relationship in terms of the angular velocity instead of the speed: $a_R = $ _____. In terms of the components a_R and a_T, the magnitude of the total acceleration is given by $a = $ _____ and, in terms of ω, α, and r, is given by $a = $ _____. You should know that since an angle in radians is the ratio of two lengths, it is really unitless. Check the units on the two sides of each equation you just wrote to be sure they are consistent.

For a point in a body rotating with constant angular acceleration, the tangential component of the acceleration is also constant. If s represents the distance traveled by the point along its circular path from time 0 to time t, then $s(t) = v_0 t + \frac{1}{2} a_T t^2$. The speed v of the point is given by $v(t) = v_0 + a_T t$. Divide these equations by the radius r of the orbit to obtain the kinematic equations for $\phi(t)$ and $\omega(t)$.

Be sure you understand the relationships between the angular and linear quantities. Every point on a rotating body turns through the same angle in the same time, has the same angular velocity, and has the same angular acceleration. This means that compared to a point on the rim of a rotating wheel, for example, a point halfway out travels _____ the distance, has _____ the speed, and has _____ the tangential acceleration.

The relationships between velocity and angular velocity and between acceleration and angular acceleration are actually vector relationships. Consider a point in a rotating body and suppose its position vector relative to an origin on the axis of rotation is **r**. In terms of the (vector) angular velocity, its (vector) velocity is given by the vector product **v** = _____. Be sure you have the order of the factors correct. Carefully note that the magnitude of **r** is NOT necessarily the radius of the orbit. This is because the origin is not necessarily at the center of the orbit.

To convince yourself that the vector product gives the correct direction for **v** (tangent to the circular path of the point, in the direction of motion) consider the diagram on the right. It shows the cross section of an object rotating about the z axis and two points in the object, one at $\mathbf{r}_A$ and the other at $\mathbf{r}_B$, both in the plane of the page. ω is in the positive z direction.

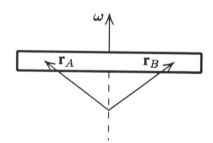

By writing either "into the page" or "out of the page" near the points, indicate on the diagram the velocity directions for the two points as predicted by the vector product $\boldsymbol{\omega} \times \mathbf{r}$. Now recall that ω in the positive z direction means the point to the right of the axis is moving into the page and the point to the left is moving out of the page. This should agree with the prediction of the vector product.

Also convince yourself that the vector product agrees with the scalar result for the speed: $v = $ (radius of orbit)$\times$(scalar angular velocity). According to the vector product equation for **v**, the speed is given by $v = $ _____, where θ is the angle between _____ and _____. Here r is the distance from the origin to the point and so is the radius of the orbit only if the origin happens to lie at the center of the orbit. If the origin is somewhere else on the axis, the radius of the orbit, in terms of r and θ, is given by _____. Label r, θ, and the orbit radius for the left point on the diagram above.

You should be able to differentiate the expression for **v** in terms of **r** and $\boldsymbol{\omega}$ to obtain the expression for **a**. Be careful to retain the original order in the vector product. In terms of the vectors $\boldsymbol{\alpha}$, $\boldsymbol{\omega}$, and **r** the acceleration of a point in the body is given by the sum of two vector products: **a** = _____ + _____. Of these, the radial acceleration is $\mathbf{a}_R$ = _____ and the tangential acceleration is $\mathbf{a}_T$ = _____.

II. PROBLEM SOLVING

Many problems of this chapter are mathematically identical to those of one-dimensional linear kinematics, discussed in Chapter 2: ϕ replaces x, ω replaces v, and α replaces a. For example, $\alpha = d\omega/dt$ and $\omega = d\phi/dt$.

PROBLEM 1. The angular position in radians for a certain wheel is given by $\phi(t) = 5.0t - 2.5t^3$, where t is in seconds. What are its angular velocity and angular acceleration as functions of time? What are its angular position, angular velocity, and angular acceleration at $t = 2.0\,\text{s}$?

SOLUTION:

[ans: $\omega = 5.0 - 7.5t^2$; $\alpha = -15t$; $\phi = -10.0\,\text{rad}$; $\omega = -25.0\,\text{rad/s}$; $\alpha = -30.0\,\text{rad/s}^2$]

According to the sign convention discussed above, the negative value for ϕ at $t = 2.0\,\text{s}$ indicates that the angular position is then 10.0 rad *clockwise* from the initial angular position. Notice that the wheel started turning in the counterclockwise direction, then stopped and began turning in the clockwise direction.

What is the average angular velocity of the wheel during the 2.0 s interval? What is its average angular acceleration?

SOLUTION: Use $\bar{\omega} = \Delta\phi/\Delta t$ and $\bar{\alpha} = \Delta\omega/\Delta t$.

[ans: $-5.0\,\text{rad/s}$; $-15\,\text{rad/s}^2$]

At what time does the wheel momentarily stop? What is its angular position and angular acceleration then?

SOLUTION: Solve $\omega = 5.0 - 7.5t^2 = 0$ for t, then evaluate $\phi(t)$ and $\alpha(t)$.

[ans: 0.816 s; 2.72 rad; $-12.2\,\text{rad/s}^2$]

The angular position is the integral with respect to time of the angular velocity and the angular velocity is the integral with respect to time of the angular acceleration. When you carry out these integrations, don't forget the constants of integration. They are evaluated using the initial conditions, as they are for linear kinematics problems.

PROBLEM 2. Initially a wheel has an angular velocity of 250 rad/s but it slows down with an angular acceleration that is given, in rad/s², by 2.5t, where t is in seconds. When does it stop and through what angle does it rotate while stopping?

SOLUTION: Integrate $\alpha(t)$ with respect to time to find an expression for the angular velocity $\omega(t)$ as a function of time, then solve $\omega(t) = 0$ for t. If you take ω_0 to be positive then α is negative. Integrate $\omega(t)$ to find an expression for $\phi(t)$ and evaluate it for the time the wheel stops. Since the direction of motion does not change during the interval from $t = 0$ to the time the wheel stops, the angle through which it turns is $\phi(t) - \phi_0$.

[ans: 14.1 s; 2360 rad]

If a problem involves rotation with constant angular acceleration first write the kinematic equations $\phi(t) = \phi_0 + \omega_0 t + \frac{1}{2}\alpha t^2$ and $\omega(t) = \omega_0 + \alpha t$, then solve them simultaneously for the unknowns. As an alternative, search Table 1 for the equation containing the quantities that are given in the problem statement.

Recall that when you solve one-dimensional kinematic problems you can always place the origin at any point you choose. In particular you might select the origin to be at the position of the particle when time $t = 0$ and thus take x_0 to be zero. Similarly, to describe rotational motion you can always orient the fixed reference frame and select the reference line in the body so the initial angular position ϕ_0 is zero. This is usually helpful because it reduces the number of algebraic symbols you must carry in the calculation.

Some problems require knowledge of the relative signs of ω and α. If the problem states or implies that the rotational motion is slowing down, ω and α have opposite signs. If the rotational motion is speeding up, they have the same sign. This is not important when ω, α, or both are among the unknowns. The answer will automatically include the sign. On the other hand, if numerical values are given then you will need to attach a sign to them. For example, if you are told a wheel has an angular velocity of 15 rad/s and is slowing down at a rate of 3.0 rad/s², then you might choose $\omega_0 = +15$ rad/s and $\alpha = -3.0$ rad/s². (You might also choose $\omega_0 = -15$ rad/s and $\alpha = +3.0$ rad/s².)

PROBLEM 3. After the driving motor is turned off, a heavy flywheel, originally rotating at 300 rev/min, slows down and stops in 5.0 min. If its decrease in speed is uniform, what is its angular acceleration? Through what angle does it turn as it slows down?

SOLUTION: Let $t = 0$ be the time the motor is turned off and take ϕ_0 to be 0. Let t be the time the wheel stops. Then $\phi(t)$ is the angle through which it rotates while stopping and $\omega(t)$ is its angular velocity when it stops; that is, $\omega(t) = 0$. Measure time in minutes, angle in revolutions, angular velocity in rev/min, and angular acceleration in rev/min². These are consistent with each other and can be used in the kinematic equations. Fill in the following table with values for the known quantities and question marks for the unknown quantities.

$\phi_0 = $ _____ $\omega_0 = $ _____ $t = $ _____

$\phi(t) = $ _____ $\omega(t) = $ _____ $\alpha = $ _____

Chapter 11: Rotational Kinematics

You should have assigned values to ϕ_0, ω_0, t, and $\omega(t)$. α and $\phi(t)$ are unknowns. Solve $\omega(t) = \omega_0 + \alpha t$ for α, then evaluate $\phi(t) = \phi_0 + \omega_0 t + \frac{1}{2}\alpha t^2$.

[ans: -60.0 rev/min^2; 750 rev]

The driving motor is now turned on and uniformly brings the flywheel from rest to a final angular velocity of 300 rev/min. During this time the wheel turns through 175 revolutions. How long does the process take and what is the angular acceleration of the wheel?

SOLUTION: Use the following table to identify the known and unknown quantities.

$\phi_0 = $ _____ $\omega_0 = $ _____ $t = $ _____

$\phi(t) = $ _____ $\omega(t) = $ _____ $\alpha = $ _____

Now solve for t and α.

[ans: 1.17 min; 257 rev/min^2]

PROBLEM 4. A wheel on an industrial machine has a constant angular acceleration of 9.7 rad/s^2, in the same direction as its angular velocity. During a 30 s time interval it turns through 7000 rad. What is its angular velocity at the beginning and end of the interval?

SOLUTION: Use the following table to identify the known and unknown quantities.

$\phi_0 = $ _____ $\omega_0 = $ _____ $t = $ _____

$\phi(t) = $ _____ $\omega(t) = $ _____ $\alpha = $ _____

Now solve for ω_0 and $\omega(t)$.

[ans: 87.8 rad/s; 379 rad/s]

PROBLEM 5. Initially a wheel is spinning at 250 rad/s. After 25 s its angular velocity has the same magnitude but the wheel is spinning in the opposite direction. If its angular acceleration is constant what is its value? What is its angular displacement and average angular velocity during this time?

SOLUTION:

[ans: -0.0828 rad/s^2; 0; 0]

To solve some problems you might need to convert angular units so all the data you use are mutually compatible. Here's an example that is similar to one above, but with different units.

Chapter 11: Rotational Kinematics

PROBLEM 6. A wheel on an industrial machine has a constant angular acceleration of 9.7 rad/s², in the same direction as its angular velocity. During a 30 s time interval it turns through 7000 rev. What is its angular velocity at the beginning and end of the interval?

SOLUTION: We choose to work the problem using radian measure. First convert 7000 rev to radians: since 1 rev is equivalent to 2π radians, 7000 rev is equivalent to $7000 \times 2\pi = 44,000$ rad. Now work the problem as before.

[ans: 1320 rad/s; 1610 rad/s]

Problems in another category deal with relationships between linear and angular variables. Given the radius of the orbit of a point in a rotating body you should be able to obtain values for the linear variables from values for the angular variables and vice versa. Use $\Delta s = r\Delta\phi$, $v = r\omega$, $a_T = r\alpha$, and $a_R = r\omega^2$. In some problems you may be given sufficient information to solve the kinematic equations for one or more of the angular variables but are asked for a related linear variable. Don't forget that the angular variables must be expressed in radian measure. This may require a change in units.

If you are asked to calculate the acceleration of a point in a spinning body, remember it has both tangential and radial components. You need to know ω to compute the radial component and α to compute the tangential component. These might be given or might be solutions to a kinematic problem. If you are asked for the magnitude of the acceleration use $a = \sqrt{a_R^2 + a_T^2}$.

PROBLEM 7. As a wheel starts rotating its angular velocity is given by $\omega(t) = 3.5t^2$. At time $t = 1.2$ s what are the radial and tangential components of the acceleration of a point on the wheel 3.0 cm from the axis of rotation? What is the magnitude of the acceleration of the point? What angle does the acceleration vector make with the radial line through the point?

SOLUTION:

[ans: 0.762 m/s²; 0.252 m/s²; 0.803 m/s²; 18.3°]

PROBLEM 8. A 1.2-m radius industrial flywheel is initially spinning at 420 rad/s. When the driving motor is shut off it slows down with constant angular acceleration and stops in 2.5 h. What is its angular acceleration? What are the tangential and radial components of the acceleration of a point on its rim when its angular velocity is half the initial value?

SOLUTION: Use $\omega(t) = \omega_0 + \alpha t = 0$ to calculate α, $a_T = r\alpha$ to calculate the tangential component of the acceleration, and $a_R = r\omega^2$ with $\omega = \omega_0/2$ to calculate the radial component.

[ans: $-0.0467\,\text{rad/s}^2$; $5.29 \times 10^4\,\text{m/s}^2$; $0.0560\,\text{m/s}^2$]

Some problems deal with two wheels that rotate about parallel axles, coupled by a belt around their rims, as shown. You must recognize that if the belt does not slip the speeds of points on the rims are the same. If the radii are different then the angular velocities must also be different. They are related by $R_1\omega_1 = R_2\omega_2$.

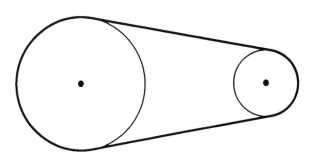

PROBLEM 9. Two wheels, one with a 12-cm radius and the other with a 30-cm radius, are coupled as shown above. The wheels start from rest and the smaller one is driven with an angular acceleration of $1.3\,\text{rad/s}^2$. Through what angle does the larger wheel turn during the first 10 s? What is its angular velocity and angular acceleration at the end of this interval?

SOLUTION: Use the kinematic equations to find the angular position ϕ_S and the angular velocity ω_S of the smaller wheel. If R_L is the radius of the larger wheel and R_S is the radius of the smaller, then $R_L\omega_L = R_S\omega_S$. Integration yields $R_L\phi_L = R_S\phi_S$ and differentiation yields $R_L\alpha_L = R_S\alpha_S$. Use these relationships to find ϕ_L, ω_L, and α_L.

[ans: $26.0\,\text{rad}$; $5.20\,\text{rad/s}$; $0.520\,\text{rad/s}^2$]

III. MATHEMATICAL SKILLS

You should know how to find the magnitude and direction of vector products, such as $\boldsymbol{\omega} \times \mathbf{r}$. Recall that the magnitude is $\omega r \sin\theta$, where θ is the smallest angle between $\boldsymbol{\omega}$ and $\mathbf{r}$ when they are drawn with their tails at the same point. You should also remember the right hand rule for finding the direction. The product is perpendicular to the plane of $\boldsymbol{\omega}$ and $\mathbf{r}$. Curl the fingers of your right hand so they rotate $\boldsymbol{\omega}$ toward $\mathbf{r}$ through the angle θ. Then the thumb will point in the direction of the vector product.

You may also need to know how to compute the vector product in terms of the components of the factors:

$$(\mathbf{a} \times \mathbf{b})_x = a_y b_z - a_z b_y$$

$$(\mathbf{a} \times \mathbf{b})_y = a_z b_x - a_x b_z$$
$$(\mathbf{a} \times \mathbf{b})_z = a_x b_y - a_y b_x$$

IV. NOTES

Chapter 12
ROTATIONAL DYNAMICS

I. BASIC CONCEPTS

Now that you have learned to describe rotational motion, you begin the study of the mechanism by which rotational motion is changed. The two most important new concepts are torque and rotational inertia. Study their definitions and learn the roles they play in rotational dynamics. You will also learn about rotational kinetic energy and the work-energy theorem for rotation.

Rotational inertia. Rotational inertia is a property of a rotating body that depends on the distribution of mass in the body and on the axis of rotation. It determines the angular acceleration and change in kinetic energy produced by the net torque acting. That is, the same torque acting on different bodies produces different angular accelerations and different changes in kinetic energy because the bodies have different rotational inertias.

The rotational inertia I of a rigid body consisting of N particles, rotating about a given fixed axis, is defined by the sum over all particles:

$$I =$$

where m_i is the mass of particle i and r_i is the distance of particle i from the axis of rotation. This distance is measured along a line that is perpendicular to the axis of rotation, from the axis to the particle.

The sum is difficult to carry out for a body with more than a few particles. Some bodies however, can be approximated by a continuous distribution of mass and techniques of the integral calculus can be used to find the rotational inertia. For a body with a continuous distribution of mass the rotational inertia is given by the integral

$$I =$$

In Section 12–3 of the text the integral is evaluated for a thin rod rotating about an axis through it center and for a plate rotating about an axis that is perpendicular to its face. Fig. 9 of the text gives the rotational inertias of several bodies, some for two different axes. You will need to refer to this figure as you work through problems.

The equation that defines the rotational inertia as a sum over the particles of a body is very useful when we need to think about the rotational inertia of an object. For example, the sum shows us that the rotational inertia depends on the square of the distance of each particle from the axis as well as on the mass of each particle. A particle that is far from the axis contributes more to the rotational inertia than a particle of equal mass close to the axis.

Look at the expressions for the rotational inertias of a hoop about the cylinder axis and for a solid cylinder, also about the cylinder axis. These are given in Fig. 9. The rotational inertia of a solid cylinder is less than the rotational inertia of a hoop with the same mass and radius because _____.

The defining equation also shows that the rotational inertia depends on the orientation and position of the axis of rotation. Look at Fig. 9 and compare the expression for the rotational inertia of a thin rod rotating about an axis through its center to the equation for the rotational inertia of the same rod but rotating about an axis through one end. The rotational inertia is smaller when the axis is through the center because _____

The defining equation for the rotational inertia is a sum over all particles in the body. If we like, we can consider the body to be composed of two or more parts and calculate the rotational inertia of each part about the same axis, then add the results to obtain the rotational inertia of the complete body. As an example, suppose two identical solid uniform balls, each of mass m and radius r, are attached to each other at a point on their surfaces and the composite body rotates around the line that joins their centers. Both balls rotate about a diameter so each contributes $2mr^2/5$ to the rotational inertia of the complete body. Thus $I = 4mr^2/5 = 2Mr^2/5$, where $M = 2m$ is the total mass.

The defining equation for the rotational inertia is used to prove the parallel axis theorem. Here we consider two identical bodies. One is rotating about an axis through the center of mass and the other is rotating about an axis that is parallel to the first axis but is displaced from the center of mass by a distance h (measured along a line that is perpendicular to the axis). The rotational inertia I for the second body is related to the rotational inertia I_{cm} for the first by $I = $ _____, where M is the total mass of the body.

An example is given in Fig. 9 of the text. Compare the equations for the two cases of a thin rod. The distance between the end and center of the rod is $L/2$; so for a rod rotating about one end $I = I_{cm} + ML^2/4 = ML^2/12 + ML^2/4 = ML^2/3$.

Now you try one. Consider a uniform sphere of radius r and mass M, rotating about an axis that is tangent to its surface. The distance between this axis and a parallel axis through the center is _____. Look up the expression for the rotational inertia of a uniform solid sphere about an axis through its center and apply the parallel axis theorem. The rotational inertia about the axis tangent to the surface is given by $I = $ _____ + _____ = _____. You should get $I = 7Mr^2/5$.

The parallel axis theorem tells us of all the places we can position the axis of rotation the one that leads to the smallest rotational inertia is the one through the center of mass and that the rotational inertia increases as the axis moves away from the center of mass (remaining parallel to its original orientation, of course).

Rotational kinetic energy. For a rigid body with rotational inertia I, rotating with angular velocity ω about a fixed axis, the total kinetic energy of all the particles in the body is given by $K = $ _____. For this expression to be valid ω must be in rad/s. You should realize this is simply the sum of the kinetic energies ($\frac{1}{2}mv^2$) of the individual particles.

According to this expression a particle far from the axis of rotation contributes more to the rotational kinetic energy than a particle of the same mass closer to the axis. Use $v = r\omega$ to explain: _____

Torque. The net torque acting on a rotating body gives the body an angular acceleration. It does work and changes the rotational kinetic energy of the body.

If a force **F** is applied to a particle with position vector **r** then the torque acting on the particle is given by the vector product $\tau = $ _____. Be sure you have the order of the factors correct. Reversing it reverses the sign of the product. The magnitude of the torque is given by $\tau = $ _____, where θ is the angle between **F** and **r** when they are drawn with their tails at the same point.

Carefully note that the torque depends on the choice of origin. Since a different position vector might mean a different torque the origin must be specified when torque is given or asked for. Note that only the component of **F** that is perpendicular to **r** contributes to the torque, not the component parallel to **r**. Thus $\tau = F_\perp r$, where $F_\perp = F \sin \theta$. The equation for the magnitude of the torque can also be written $\tau = Fr_\perp$, where $r_\perp = r \sin \theta$. $r_\perp$ is called the _____ of the force.

The diagram on the right shows a body subjected to three forces. The origin O, the forces, and their points of application are all in the plane of the page. For each force draw the vector **r** from the origin to the point of application of the force. For each, label the angle θ that appears in $\tau = Fr \sin \theta$.

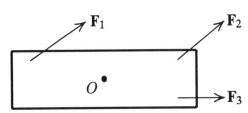

In this chapter we shall be concerned with bodies rotating about fixed axes. For them only the component of the torque along the axis of rotation is important. The only force components that contribute to this component of the torque are in a plane that is perpendicular to the axis of rotation. Each is also perpendicular to the line from the axis to the point of application. The component that is parallel to the axis does not contribute. The component that is along the line from the axis to the point of application also does not contribute. You can test these assertions for yourself by assuming the force is along one of these directions and showing that $\mathbf{r} \times \mathbf{F}$ does not have a component along the axis of rotation.

The diagram on the right shows a particle traveling counterclockwise around a circle of radius R. The resultant force **F** on it at some instant of time is also shown. On the diagram show the tangential and radial components of the force. The radial component holds it in the circle. The tangential component produces an angular acceleration. In terms of the magnitude F of the force and the angle θ shown, the radial component is _____ and the tangential component is _____. The magnitude of the torque is _____.

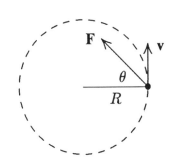

Notice that the magnitude of a torque depends on the distance from the axis of rotation to the point of application. A tangential force produces a greater angular acceleration if it is applied far from the axis than if it is applied closer.

Chapter 12: Rotational Dynamics   173

For rotation about a fixed axis we may take a torque to be positive if it tends to turn the object counterclockwise and negative if it tends to turn the body clockwise. This convention is consistent with the one introduced in the last chapter for the signs of the angular velocity and acceleration. For the situation illustrated in the diagram above, the sign of the torque is _____.

The sign convention is related to the vector direction of the torque by a right hand rule. If you point the thumb of your right hand in the direction of the torque vector, your fingers will curl in the direction the tangential component of the force is pushing the object. For the rotation illustrated above the torque vector is _____ the page.

Work done by a torque. Consider a particle traveling counterclockwise in a circular orbit, subjected to a force with tangential component F_t. As it moves through an infinitesimal angular displacement $d\phi$ the torque does work

$$W = $$

On the left hand diagram below, show a force that does positive work on the particle and give the signs of the torque and the angular velocity, using the convention described above (positive for counterclockwise, negative for clockwise). On the right hand diagram, do the same for a force that does negative work.

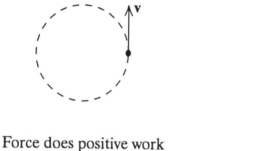
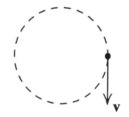

Force does positive work
Sign of torque: _____
Sign of angular velocity: _____

Force does negative work
Sign of torque: _____
Sign of angular velocity: _____

Note that a torque does positive work if it has the same sign as the angular velocity and does negative work if it has the opposite sign. In terms of the vector torque and angular velocity the torque does positive work if $\tau \cdot \omega > 0$ and does negative work if $\tau \cdot \omega < 0$.

If a torque τ is applied to a body that is rotating with angular velocity ω the rate at which it does work (the power supplied by the torque) is given by $P = $ _____.

A work-energy theorem is valid for rotational motion. The total work done by all torques acting on a body equals the change in its rotational kinetic energy. Suppose the net torque τ is constant. Further suppose that in a certain time interval the body turns through the angular displacement $\Delta\phi$ and its angular velocity changes from ω_0 to ω. In terms of these quantities the work-energy theorem becomes: _____ = _____ , where I is the rotational inertia of the body.

Newton's second law for rotation about a fixed axis. This law relates the net external torque $\sum \tau_{\text{ext}}$ acting on a rigid body to the angular acceleration α of the body. Quantitatively it is

$$\sum \tau_{\text{ext}} = \underline{\hspace{2cm}}$$

where I is the rotational inertia of the body about the axis of rotation. The relationship follows directly from the work-energy theorem for rotation. It is as important to the study of rotational motion about a fixed axis as $\sum F = ma$ is to the study of one-dimensional translational motion. Concentrate on learning to use it.

Be careful that you use the total torque in this equation. You must identify and sum all the torques that are acting and you must be careful about signs. If a torque tends to push the body in the counterclockwise direction it enters with a positive sign; if it tends to push the body in the clockwise direction it enters with a negative sign. The same direction (counterclockwise) is used for positive torque as is used for positive angular displacement, positive angular velocity, and positive angular acceleration.

Combined rotational and translational motion. In this section you study an object, such as a rolling wheel, that rotates as its center of mass moves along a line. The center of mass obeys Newton's second law: _____, where **F** is the total force acting on the object, M is the total mass of the object, and $\mathbf{a}_{\text{cm}}$ is the acceleration of the center of mass. Rotation about an axis through the center of mass is governed by Newton's second law for rotation: _____, where τ is the total torque acting on the object, I is the rotational inertia of the object, and α is the angular acceleration of the object.

You should understand that these two equations are related by the forces that act on the wheel. The total force accelerates the center of mass and the total torque (derived from the forces) produces an angular acceleration. If, in addition to the mass and rotational inertia of the body, you know all the forces acting and their points of application you can calculate the acceleration of the center of mass and the angular acceleration about the center of mass.

Rolling without slipping is a special case. We usually do not know the force of friction exerted by the surface on the body at the point of contact but we do know what that force does. If the object rolls without slipping, it causes the point on the wheel in contact with the surface to have a velocity of _____. This means that the speed v_{cm} of the center of mass and the angular speed ω around the center of mass are related by $v_{\text{cm}} = $ _____, where R is the radius of the wheel. Furthermore, since the wheel accelerates without slipping, the magnitude of the acceleration of the center of mass a_{cm} is related to the magnitude of the angular acceleration around the center of mass by $a_{\text{cm}} = $ _____.

You should understand the relationship between v_{cm} and $r\omega$ for a rolling wheel. The velocity of the point in contact with the ground is given by the sum of the velocity due to the motion of the center of mass and the velocity due to rotation: $v = v_{\text{cm}} - r\omega$, where the forward direction was taken to be positive. If the wheel slips then $v \neq 0$ and v_{cm} is different from $r\omega$. If it does not slip then $v = 0$ and $v_{\text{cm}} = r\omega$.

For a wheel rolling without slipping along a horizontal surface there are 3 equations containing the forces, torques, and accelerations. They are:

_____ for translation of the center of mass
_____ for rotation around the center of mass
_____ for the condition of no slipping.

These can be solved, for example, for the linear and angular accelerations and for the force of friction if the other quantities are known.

To test if the wheel slips or not, you will also need the vertical component of Newton's second law for the center of mass. This is used to solve for the normal force of the surface on the wheel. Imagine that the wheel does not slip and calculate the force of friction required to prevent slipping. Compare the magnitude of this force with $\mu_s N$, where μ_s is the coefficient of static friction. If it less, the wheel does not slip and the force of friction is as computed. If it is greater, the wheel does slip and the force of friction is $\mu_k N$, where μ_k is the coefficient of kinetic friction.

A rolling wheel, whether slipping or not, has both translational and rotational kinetic energy. If the center of mass has velocity v_{cm} and the wheel is rotating with angular velocity ω about an axis through the center of mass, then the total kinetic energy is given by $K =$ _____. If the wheel is not slipping then both terms of the kinetic energy can be written in terms of v_{cm}. Use $v_{cm} = r\omega$ to eliminate ω in favor of v_{cm}. The result is $K =$ _____ + _____. The kinetic energy can also be written in terms of ω: _____ + _____.

If an object is rolling without slipping the points in contact with the surface are instantaneously at rest. As a result, the frictional force acting there does zero work.

An object that is rolling without slipping on a surface may be considered to be in pure rotation about an axis through the point of contact with the surface. Consider a wheel of radius R, rolling on a horizontal surface. If its center of mass is at its center and its rotational inertia about an axis through that point is I_{cm}, then, according to the parallel axis theorem, its rotational inertia about the point of contact is given by $I =$ _____. The total kinetic energy of the wheel is now just its rotational kinetic energy $\frac{1}{2}I\omega^2$. Substitute for I in terms of I_{cm} and obtain the result you wrote in the last paragraph, $K =$ _____.

II. PROBLEM SOLVING

You should know how to compute the rotational inertia of a system composed of a small number of discrete particles. Given the masses and positions of the particles you must evaluate the sum $I = \sum m_i r_i^2$, where r_i is the perpendicular distance of particle i from the axis of rotation. You may need to use some geometry to obtain r_i. If, for example, the z coordinate axis is the axis of rotation then $r_i^2 = x_i^2 + y_i^2$. It is usually convenient to place one coordinate axis along the rotation axis.

Sample Problem 1 of the text is an example. It also shows that the rotational inertia may be different for different axes. Here is another example.

PROBLEM 1. An object consists of 4 small balls attached to a light rod with a length of 1.5 m. A 1.2-kg ball is at one end, a 0.75-kg ball is 0.50 m from that end, a 0.75-kg ball is 1.0 m from that end, and a 1.2-kg ball is at the other end. What is the rotational inertia of the object if it rotates about a fixed axis that is perpendicular to the rod through its center?

SOLUTION:

[ans: $1.44\,\text{kg}\cdot\text{m}^2$]

What is the rotational inertia of the object if it rotates about a fixed axis that is perpendicular to the rod through one end?

SOLUTION:

[ans: $3.64\,\text{kg}\cdot\text{m}^2$]

Notice that this result obeys the parallel axis theorem. The total mass is 3.9 kg and the center of mass is at the midpoint of the rod, 0.75 m from the end. Thus the rotational inertia when the axis is through an end should be $Mh^2 = 3.9 \times 0.75^2 = 2.19\,\text{kg}\cdot\text{m}^2$ greater than when the axis is through the center of mass. Do your results agree?

If the body is solid, you must evaluate the integral $I = \int r^2\,dm$. Place one coordinate axis along the rotation axis. Pick a region of the body extending from r to $r + dr$. All parts of the region are essentially the same distance from the axis. Calculate the mass in the region as a function of r. It will be proportional to dr. If we represent this mass by $f(r)\,dr$, then the integral you must evaluate becomes $I = \int r^2 f(r)\,dr$. An integral of this type is evaluated in the text for a thin rod. Use that calculation as a model.

PROBLEM 2. A disk has radius R and mass M, distributed so that the density increases from the center toward the rim in proportion to the distance from the center. If the disk rotates about an axis that is perpendicular to its faces and is through its center, what is its rotational inertia?

SOLUTION: Take the thickness of the disk to be T. Take the density to be Ar, where A is a constant of proportionality to be determined later. Divide the disk into concentric rings and consider the ring extending from r to $r + dr$. The volume of the ring is the product of its circumference, width, and thickness: $dV = 2\pi r T\,dr$. The mass contained in the ring is $\rho\,dV = 2\pi A T r^2\,dr$ and the contribution of the ring to the rotational inertia is $dI = r^2 \rho\,dV = 2\pi A T r^4\,dr$. The rotational inertia of the disk is the sum of the contributions of all rings; that is, it is the integral $I = 2\pi A T \int r^4\,dr$, from $r = 0$ to $r = R$.

The total mass in the disk is $M = T \int \rho\,dV = 2\pi A T \int r^2\,dr$, from $r = 0$ to $r = R$. This expression can be used to evaluate the constant of proportionality A.

[ans: $(3/5)MR^2$]

If the disk were uniform its rotational inertia would be $(1/2)MR^2$, somewhat less. The greater rotational inertia comes about because more of the mass is located near the rim and less near the center.

Clearly you must also know how to calculate the torque, given a force and its point of application. Some problems simply provide practice in calculating torque while others require you to find the torque as an intermediate step in answering a question about a rotating body. Since torque is a vector you must give both its magnitude and direction.

If a torque is required to investigate the motion of a body rotating around a *fixed* axis, all you will need is the component along the axis of rotation. If the force is in the plane of rotation, simply find the force component tangent to the circular orbit at the application point, then multiply by the perpendicular distance from the axis to the application point. For fixed axis rotation you may treat the torque as a scalar. Assign a sign to it according to whether it tends to cause rotation in the counterclockwise (+) or clockwise (−) direction.

PROBLEM 3. The object shown is pivoted at O so it can turn in the plane of the page. It is subjected to two forces, both in the plane of the page. Take $F_1 = 50$ N, $r_1 = 0.25$ m, $\theta_1 = 40°$, $F_2 = 75$ N, $r_2 = 0.45$ m, and $\theta_2 = 25°$. What is the net torque on the object?

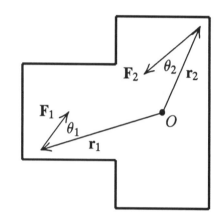

SOLUTION: The component of $\mathbf{F}_1$ tangent to the circle through its point of application is $F_1 \sin\theta_1$ so the magnitude of the torque it produces is $F_1 r_1 \sin\theta_1$. This torque tends to rotate the body in the clockwise direction so we shall take it to be negative. Alternatively, the vector torque is into the page, which we shall take to be the negative direction.

The component of $\mathbf{F}_2$ tangent to the circle through its point of application is $F_2 \sin\theta_2$ so the magnitude of the torque it produces is $F_2 r_2 \sin\theta_2$. This torque tends to rotate the body in the counterclockwise direction so we shall take it to be positive. Alternatively, the vector torque is out of the page and so is in the positive direction.

The net torque on the object is $\tau = F_2 r_2 \sin\theta_2 - F_1 r_1 \sin\theta_1$. Evaluate this expression.

[ans: $+6.23$ N·m]

The positive sign indicates that the net torque is counterclockwise or the vector torque is out of the page.

If the force is not parallel to the plane of rotation you will evaluate the component of the vector product $\mathbf{r} \times \mathbf{F}$ along the axis of rotation.

PROBLEM 4. A 0.55-m radius wheel rotates about a fixed axle, along the z axis. At some instant the external force, in newtons, is given by $\mathbf{F} = 23\mathbf{i} + 15\mathbf{j} - 18\mathbf{k}$, applied at the point on the rim that is crossing the x axis at that instant. What are the components of the torque about the origin due to this force?

SOLUTION: Take $\mathbf{r} = 0.55\mathbf{i}$ in meters and evaluate $\mathbf{r} \times \mathbf{F}$. See the Mathematical Skills section of Chapter 11 for the components of a cross product.

[ans: $\tau_x = 0$, $\tau_y = 9.90$ N·m, $\tau_z = 8.25$ N·m]

Both τ_y and τ_z tend to produce angular accelerations but only τ_z succeeds. For the axis of rotation to remain fixed the axle must exert a torque with a y component of -9.90 N·m. It may also contribute to the z component of the total torque. Friction in the bearings, for example, might make this contribution. In any event, the net torque, due to the applied force and to the axle itself, must be along the z axis since the vector angular acceleration must be along that axis. You should note that the point of application moves as the wheel rotates, so its coordinates are different for different times. The calculation is valid only for the instant the point of application crosses the x axis.

Many problems deal with the rotational motion of a rigid body about a fixed axis and can be solved using $\sum \tau_{ext} = I\alpha$. Don't forget that $\sum \tau_{ext}$ is the *net* torque on the body.

PROBLEM 5. A certain flywheel has a rotational inertia of 1.4 kg·m² and a radius of 0.45 m. To bring it from rest to its operating angular velocity of 150 rad/s a force of 45 N is applied tangentially to its rim. If the torque due to friction in the bearing is 2.0 N·m how long does the startup process take?

SOLUTION: Draw a free-body diagram for the wheel. Include the force applied at the rim and label it $\vec{F}$. Include the force of friction at the bearing and label it $\vec{f}$. You should also include the force of gravity acting downward at the center of the wheel and the supporting force of the axle, acting upward at the center.

Now sum the torques. The applied torque is FR. Use τ_f for the torque of friction. It is opposite the applied torque. Neither the force of gravity nor the supporting force exert torques about the rotation axis. So the net torque is $\sum \tau_{ext} = FR - \tau_f$ and $FR - \tau_f = I\alpha$. Calculate α, note that it is constant, then use $\omega(t) = \alpha t$ to find the time for which $\omega = 150$ rad/s.

[ans: 11.5 s]

In some cases you may need to use the kinematic equations to calculate the angular acceleration.

PROBLEM 6. A flywheel has a rotational inertia of 1.6 kg·m² and a radius of 0.35 m. When a 50-N force is applied tangentially to its rim it goes from rest to 150 rad/s in 22 s. What frictional torque acts?

SOLUTION: Use $\omega(t) = \omega_0 + \alpha t$ to find the angular acceleration, then $\sum \tau_{ext} = I\alpha$ to find the net torque. Assume only two torques act, the applied torque τ_a and the torque of friction τ_f, and write $\sum \tau_{ext} = \tau_a + \tau_f$. Use $\tau_a = RF$, where R is the radius of the wheel and F is the applied force. Thus $\tau_f = \sum \tau_{ext} - \tau_a = \sum \tau_{ext} - RF$.

[ans: 6.59 N·m]

Some problems involve two bodies, one in translation and one in rotation. Sample Problem 5 in the text is an example.

To solve this type problem draw two free-body diagrams, one for the translating body and one for the rotating body. Show all forces. For the rotating body, choose the direction of positive rotation (the counterclockwise direction, say) and write the sum of the torques, taking their directions into account. Set the sum equal to $I\alpha$. For the translating body, choose a coordinate system with one axis in the direction of the acceleration, if possible. Choose the direction of positive acceleration so that if the translating body has positive acceleration then

Chapter 12: Rotational Dynamics

the rotating body has positive angular acceleration. Write the sum of the force components for each coordinate direction and set each sum equal to the product of the mass of the body and the appropriate acceleration component. Write all equations in terms of the magnitudes of the forces and appropriate angles. Use algebraic symbols, not numbers.

Don't forget tensions in strings, forces of gravity, normal forces, and frictional forces, if they act. If a string is wrapped around a disk that is free to rotate and if the string does not slip on the disk, then the force exerted by the string on the disk is the tension T in the string. Since the string must be tangent to the disk, the torque it exerts has magnitude TR, where R is the radius of the disk. If a string passes over a pulley then the tension in the string will be different on different sides. If T_1 is the tension on one side and T_2 is the tension on the other side, then $\pm(T_1 - T_2)R$ is the net torque exerted by the string on the pulley. The sign, of course, depends on which force acts on the right side of the pulley and which acts on the left.

The acceleration of the translating body and the angular acceleration of the rotating body are usually related. Perhaps a string runs from the translating body and around the rotating body. Then, if the string does not slip or stretch, the magnitude of the tangential acceleration of a point on the rim of the rotating body must be the same as the magnitude of the translating body. You write $a = R\alpha$.

The second-law equations (for translation and rotation) and the equation that links the acceleration of the translating body and the angular acceleration of the rotating body are solved simultaneously for the unknowns.

PROBLEM 7. A block of mass M is on a frictionless inclined plane that makes an angle θ with the horizontal, as shown. A string is attached to the block and runs parallel to the plane to the top, where it is wrapped around a cylinder of radius R and rotational inertia I. The cylinder is free to rotate on a horizontal axis. Find expressions for the acceleration of the block, the angular acceleration of the cylinder, and the tension in the string.

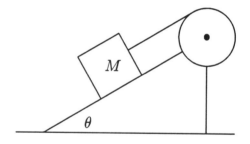

SOLUTION: Draw a free-body diagram for the cylinder. Label the tension in the string T. The torque on the cylinder is TR, so $TR = I\alpha$, if the counterclockwise direction is positive.

Draw a free-body diagram for the block. Take the positive x axis to be down the plane and the positive y axis to be in the direction of the normal force. The x component of Newton's second law is $Mg\sin\theta - T = Ma$.

The string does not slip on the cylinder so the acceleration of the block is the same as the tangential component of the acceleration of a point on the rim of the cylinder; that is $a = R\alpha$. A positive sign was used because we chose the coordinate systems in such a way that a positive acceleration of the block is associated with a positive angular acceleration of the cylinder.

Solve the three equations $TR = I\alpha$, $Mg\sin\theta - T = Ma$, and $a = R\alpha$ for a, α, and T in terms of the other quantities.

$$[\text{ans: } a = MR^2g\sin\theta/(MR^2 + I); \alpha = MRg\sin\theta/(MR^2 + I); T = IMg\sin\theta/(MR^2 + I)]$$

Examine the solution for a cylinder with a small rotational inertia. If $I \ll MR^2$ then the denominators of these expressions can be approximated by MR^2 and the acceleration of the block is given by $a = g\sin\theta$. You may recall that this is the acceleration of a block sliding down a frictionless inclined plane, without the string and cylinder. The influence of the cylinder on the motion of the block diminishes if its rotational inertia is diminished.

Take $\theta = 20°$, $M = 1.7$ kg, $R = 12$ cm, and $I = 0.19$ kg·m². Calculate a, α, and T.

SOLUTION:

[ans: 0.383 m/s²; 3.19 rad/s²; 5.05 N]

For a body that is simultaneously translating and rotating you should know how to calculate the speed of various points in the body.

PROBLEM 8. A 0.15-m radius wheel is rolling on a horizontal floor. It is rotating clockwise at 35 rad/s about its axis while the axis is moving to the right at 3.4 m/s. What is the speed of the point at the top? What is the speed of the point at the bottom? What is the speed of the right-most point?

SOLUTION: If the wheel were not translating the speed of every point on the rim would be $R\omega = 5.25$ m/s. The point at the top would be moving right, the point on the bottom would be moving left, and the right-most point would be moving down. We now superpose the velocity of translation on these velocities. The velocity of the point at the top is $(v_c + R\omega)\mathbf{i}$, where the x axis is to the right. The velocity of the point at the bottom is $(v_c - R\omega)\mathbf{i}$. The velocity of the right-most point is $v_c\mathbf{i} - R\omega\mathbf{j}$, where the y axis is upward. Its speed is $\sqrt{v_c^2 + R^2\omega^2}$.

[ans: 8.65 m/s; 1.85 m/s; 6.25 m/s]

Since the velocity of the point at the bottom is not zero this wheel is obviously slipping.

The dynamical solution of a problem involving a rolling body starts with a free-body diagram. See Fig. 26 and the paragraphs just below Sample Problem 8 of the text for an illustrative example. Show all forces and label them with algebraic symbols. Choose a coordinate system to describe the motion of the center of mass. If possible choose one axis to be in the direction of the acceleration. Sum the force components for each coordinate direction, again taking care with the signs. Equate the sum to the product of the mass and the corresponding component of the center of mass acceleration. Choose a direction for positive rotation. For each force write an expression that gives the torque about the axis of rotation. Be careful about the signs. Sum the torques and equate the sum to $I\alpha$.

PROBLEM 9. A wheel with a radius of 0.90 m, a mass of 12 kg, and a rotational inertia of 1.3 kg·m² stands vertically on a frictionless horizontal surface. A horizontal force of 25 N is then applied continuously to the top point on its rim. What is the acceleration of the center of mass and what is the angular acceleration about the center of mass?

SOLUTION: Draw the free-body diagram for the wheel. Take the applied force to be to the right and label it F. The torque it exerts is $-FR$, where the counterclockwise direction was chosen to be positive. Thus $-FR = I\alpha$. Newton's second law for the center of mass is $F = ma_{cm}$, where the coordinate axis was chosen to the positive to the right. Solve $-FR = I\alpha$ for α and $F = ma_{cm}$ for a_{cm}.

[ans: -17.3 rad/s²; 2.08 m/s²]

The applied force does two jobs: it causes the center of mass to accelerate and it causes the wheel to have an angular acceleration about its center of mass. Notice that a_{cm} and α are not related by $a_{cm} = \pm\alpha R$ because the wheel is slipping.

When a wheel accelerates without slipping a frictional force is present at the point of contact between the body and the surface on which it rolls. Don't forget to include it, even if it is not mentioned in the problem statement.

If the body rolls without slipping, the frictional force is usually an unknown, but we do know the result of its action: the acceleration of the center of mass is related to the angular acceleration about the center of mass by $a_{cm} = \pm R\alpha$. The sign depends on the choices made for the directions of positive translation and rotation.

PROBLEM 10. A horizontal force of 25 N is continuously applied to the top point on the rim of the wheel of the previous problem and the wheel rolls without slipping on a horizontal surface. What is the acceleration of the center of mass, the angular acceleration about the center of mass, and the force of friction?

SOLUTION: Draw a free-body diagram for the wheel. Take both the applied force F and the force of friction f to be to the right. The first acts at the top of the wheel; the second acts at the bottom. We don't know if the force of friction is to the right or left so either direction can be postulated. If f turns out to be negative the force is opposite to the direction we chose.

Choose the positive coordinate axis to be to the right and positive rotation to be counterclockwise. Take M to be the mass and I the rotational inertia of the wheel. Then Newton's second law for the center of mass is _____ and Newton's second law for rotation about the center of mass is _____.

The wheel is not slipping so the acceleration of the center of mass and the angular acceleration are related by _____. Be careful about the sign.

Solve these equations to obtain expressions for a_{cm}, α, and f in terms of given quantities.

You should obtain $a_{cm} = 2FR^2/(MR^2 + I)$, $\alpha = -2FR/(MR^2 + I)$, and $f = F(MR^2 - I)/(MR^2 + I)$. Since

$I < MR^2$, the force of friction is to the right. Now substitute numerical values.

[ans: $3.68\,\text{m/s}^2$; $-4.08\,\text{rad/s}^2$; $19.1\,\text{N}$]

If the body is slipping the magnitude of the force of friction is given by $f = \mu_k N$, where μ_k is the coefficient of kinetic friction for the body and surface and N is the magnitude of the normal force exerted by the surface on the body. In some cases you can determine the direction of the force before solving the problem and can take it into account in drawing the free-body diagram and writing Newton's second law. In other cases you must arbitrarily select a direction, then examine the solution to see if the direction of rotation is consistent with your selection. If it is not, rework the problem using the opposite direction.

PROBLEM 11. A horizontal force of 25 N is continuously applied to the top point on the rim of the wheel of the previous problem. If the coefficient of static friction between the wheel and the horizontal surface on which it rolls is 0.15, does the wheel slip?

SOLUTION: Assume the wheel does not slip and calculate the force of friction. See the last problem. The frictional force that keeps the wheel from slipping is $f_{\text{no slip}} =$ _____ N. Use the vertical component of Newton's second law to compute the normal force.

The maximum force of static friction that can be sustained is $\mu_s N =$ _____ N. Since $f_{\text{no slip}} > \mu_s N$, the wheel slips.

Assume the wheel starts from rest and that the coefficient of kinetic friction between the wheel and the surface is 0.12. What is the acceleration of the center of mass and what is the angular acceleration of the wheel?

SOLUTION: Again assume the force of friction f is toward the right. We will check on this later. The free-body diagram is as before. The horizontal component of Newton's second law for the center of mass is _____, the vertical component of Newton's second law for the center of mass is _____, and Newton's second law for rotation about the center of mass is _____. Set $f = \mu_k N$ and solve these equations for a and α.

[ans: $3.26\,\text{m/s}^2$; $-7.54\,\text{rad/s}^2$]

Now we must check to see if we made the right choice for the direction of the force of friction. The horizontal component of the acceleration of the point on the rim in contact with the surface is $a_{\text{cm}} + R\alpha = -3.26\,\text{m/s}^2$. The negative sign indicates it is toward the left. Since the wheel starts from rest, whatever point is in contact with the surface moves to the left, in the direction of the acceleration. The force of friction must therefore be to the right, as we originally assumed.

Many problems can be solved by the energy method. The fundamental equation is the work-energy theorem $W = \Delta K$, where W is the total work done by all forces and ΔK is the change in the kinetic energy. To this must be added any auxiliary equations (such as $v_{cm} = R\omega$) that apply. Notice that there are fewer equations than for the dynamical method. Less information is required and fewer unknowns can be evaluated. Directions of forces and torques enter only insofar as they determine the work. The energy method cannot be used to solve for directions except in special cases. If the work done by forces or torques is given, rather than the forces or torques themselves, the energy method *must* be used.

Consider first a single body rotating around a fixed axis. The work-energy theorem is $W = \frac{1}{2}I(\omega^2 - \omega_0^2)$. The ingredients of the method are the total work W done by the torques acting over some interval of time or angular displacement, the angular velocity ω_0 at the beginning of the interval, the angular velocity ω at the end of the interval, and the rotational inertia. All but one of these must be known, then the work-energy theorem for rotation can be used to compute the unknown.

PROBLEM 12. A flywheel with a rotational inertia of 1.3 kg·m², on a fixed frictionless axle, is brought from rest to an angular velocity of 300 rad/s. What is the total work done on the flywheel?

SOLUTION:

[ans: 58500 J]

If the angular acceleration of the wheel is constant and the wheel makes 700 revolutions as it is brought from rest up to speed, what is the net torque acting on it?

SOLUTION: Use $W = \tau \Delta\phi$, valid for a constant torque.

[ans: 13.3 N·m]

If the angular acceleration of the wheel is constant and the wheel is brought up to speed in 2.0 min, what is the net torque acting on it?

SOLUTION:

[ans: 6.50 N·m]

The energy method is also useful for some problems involving both rotating and translating bodies, such as Sample Problem 6 of the text. W is the sum of the works done by all forces acting on the translating body and all torques acting on the rotating body. K is the sum of the kinetic energies of the two bodies. Here is another example, involving three bodies.

184 *Chapter 12: Rotational Dynamics*

PROBLEM 13. Mass m_1 is on a frictionless inclined plane, at an angle θ above the horizontal. A string runs from m_1 over a frictionless pulley with radius R and rotational inertia I, to mass m_2, as shown. The system is released from rest and m_1 slides down the plane. Use the work-energy theorem to find an expression that can be solved for the speed of m_1 after it has gone a distance d down the plane.

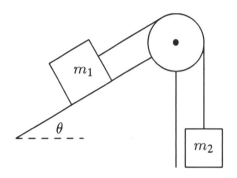

SOLUTION: As m_1 goes a distance d gravity does work $m_1 g d \sin\theta$ on it and $-m_2 g d$ on m_2. If the string exerts force T_1 on m_1 and T_2 on m_2 it does work $-T_1 d$ on m_1 and $T_2 d$ on m_2. The string applies torque $T_1 R$ clockwise and $T_2 R$ counterclockwise on the pulley and so does work $(T_2 R - T_1 R)\Delta\phi$ on the pulley. Now $R\Delta\phi = d$ so the work done on the pulley is $(T_2 - T_1)d$ and the total work done by the string is zero.

The kinetic energy is initially zero. After m_1 slides a distance d it has kinetic energy $\frac{1}{2}m_1 v^2$, m_2 has kinetic energy $\frac{1}{2}m_2 v^2$, and the pulley has kinetic energy $\frac{1}{2}I\omega^2 = \frac{1}{2}I(v/R)^2$. Note that the two masses and any point on the rim of the pulley have the same speed if the string does not slip on the pulley. We have used the same symbol v for the speed of all three.

The work-energy theorem, $m_1 g d \sin\theta - m_2 g d = \frac{1}{2}(m_1 + m_2 + I/R^2)v^2$, can be solved for v in terms of the given quantities.

Take $\theta = 20°$, $m_1 = 8.0$ kg, $m_2 = 2.0$ kg, $I = 1.75$ kg·m^2, and $R = 0.50$ m. What is the speed of m_1 after it has gone 2.5 m?

[ans: 1.46 m/s]

The work-energy theorem for an object that is rolling has the form $W = \frac{1}{2}m(v_{cm}^2 - v_{cm\,0}^2) + \frac{1}{2}I(\omega^2 - \omega_0^2)$, where v_{cm} is the speed of the center of mass, $v_{cm\,0}$ is its initial value, ω is the angular velocity, and ω_0 is its initial value. If the body rolls without slipping v_{cm} and ω are related by $v_{cm} = R\omega$. Thus the energy equation can be written in terms of either the translational speed or the angular speed alone: $W = \frac{1}{2}(m + I/R^2)(v_{cm}^2 - v_{cm}^2)$ or $W = \frac{1}{2}(mR^2 + I)(\omega^2 - \omega_0^2)$. Which you use depends on the problem statement. Since the point of application of the force of friction has zero velocity that force does not do any work. See Sample Problem 8 of the text for an example.

PROBLEM 14. A uniform sphere starts with an initial speed of 2.4 m/s and rolls up a 22° incline. If it rolls without slipping, how far up the incline does it go before momentarily stopping? Use the work-energy theorem.

SOLUTION: The only force that does work is the force of gravity, so $W = -mgd\sin\theta$, where m is the mass of the sphere, d is the distance it travels up the incline, and θ is the angle of the incline. The change in the kinetic energy is given by $\frac{1}{2}(m + I/R^2)(v_{cm}^2 - v_{cm\,0}^2)$, where $v_{cm} = 0$. Use $I = (2/5)mR^2$.

[ans: 1.10 m]

III. NOTES

Chapter 13
ANGULAR MOMENTUM

I. BASIC CONCEPTS

Here you will study rotational motion using a new concept, that of angular momentum. Learn the definition for a single particle and learn how to calculate the total angular momentum of a collection of particles. Pay particular attention to the result for a rigid body rotating about a fixed axis. You will also study the change in the angular momentum of a system brought about by the total external torque applied to the system. Angular momentum is at the heart of one of the great conservation laws of mechanics. Pay attention to the conditions for which angular momentum is conserved and learn to use to law to solve problems.

Definition. If a particle has momentum **p** and position vector **r** relative to some origin, then its <u>angular momentum</u> is defined by the vector product

$$\ell =$$

Notice that the angular momentum depends on the choice of origin. In particular, if the origin is picked so **r** and **p** are parallel at some instant of time then $\ell =$ _____ at that instant.

You should be familiar with some frequently occurring special cases. If a particle of mass m is traveling around a circle of radius R with speed v and the origin is placed at the center of the circle, then the magnitude of the angular momentum is $\ell =$ _____. If the orbit is in the plane of the page and the particle is traveling in the counterclockwise direction, the direction of its angular momentum is _____. If it is traveling in the clockwise direction, the direction is _____.

For rotation about a fixed axis the component of the angular momentum along the axis of rotation is, in most cases, the component of greatest interest. Nevertheless you should be able to compute the angular momentum vector and its component along any axis.

Suppose a particle of mass m is traveling around a circle of radius R parallel to the xy plane and centered on the z axis above the origin. The diagram on the right shows the xz plane as the particle crosses the plane going into the page. Draw the position vector **r** and the angular momentum vector ℓ. In terms of m, v, and r (the distance from the origin), the magnitude of ℓ is $\ell =$ _____.

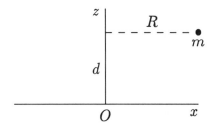

Let θ be the angle between **r** and the positive x axis. In terms of θ the z component of ℓ is $\ell_z =$ _____ and the x component is $\ell_x =$ _____. In terms of m, R, and v the z component of ℓ is given by $\ell_z =$ _____. In terms of m, d, and v the x component of ℓ is given by $\ell_x =$ _____. The component of ℓ in the plane of the orbit is radial

Chapter 13: Angular Momentum   187

and follows the particle around the circle, always pointing toward or away from the center of the circle, depending on the direction of travel. Notice that ℓ and ℓ_x depend on the position of the origin along the z axis but ℓ_z does not.

Even a particle moving in a straight line may have a non-zero angular momentum. Consider a particle of mass m traveling with velocity **v** in the negative y direction along the line $x = d$.

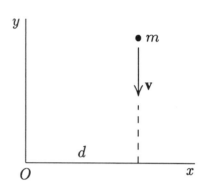

On the diagram draw the vector **r** from the origin to the particle. In terms of m, v, and d, the magnitude of its angular momentum about the origin is $\ell =$ _____ and the direction of its angular momentum is _____. Show the angular momentum vector on the diagram by drawing an × in a small circle (⊗) if it is into the page and a dot in a small circle (⊙) if it is out of the page. Label it ℓ. If the particle is traveling along the same line but in the positive y direction, the direction of its angular momentum is _____. If the particle is traveling along a line through the origin its angular momentum is _____.

The total angular momentum **L** of a system of particles is the vector sum of the individual angular momenta of all the particles in the system. For an object rotating with angular velocity ω about the z axis the z component of the angular momentum with the origin on the axis, written in terms of ω, the mass m_i, and distance r_i of each particle from the axis, is the sum

$$L_z =$$

In terms of the rotational inertia I of the object it is

$$L_z =$$

Notice that this equation is also valid for a single particle traveling in a circular orbit. Then the rotational inertia is mR^2.

If the mass is distributed symmetrically about the axis of rotation the total angular momentum is along the axis of rotation and the magnitude of L_z is also the magnitude of **L**. The direction of **L** is given by the right hand rule: curl the fingers in the direction of _____, then the thumb will point in the direction of _____.

A "dumbbell" consists of two masses connected by a light rod of length $2d$, spinning with angular velocity ω about the z axis, through the center of the rod and perpendicular to it. Take the plane of rotation to be a distance z from the origin. The diagram shows the situation when mass m_1 is moving into the page and mass m_2 is moving out of the page. At first suppose the masses are not equal. For mass 1, the magnitude of the angular momentum is _____, the z component is _____, and the x component is _____.

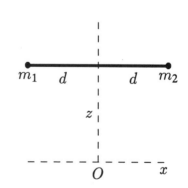

188   Chapter 13: Angular Momentum

For mass 2 the magnitude of the angular momentum is _____, the z component is _____, and the x component is _____. The z component of the total angular momentum is _____ and the x component is _____.

Now specialize the results to the symmetric case: the masses are equal. Then the z component of the total angular momentum is _____, while the x component vanishes. If the masses are unequal the rotational inertia of this object about the z axis is $I = $ _____. Notice that $L_z = I\omega$, whether or not the masses are equal.

Torque and the time rate of change of angular momentum. If a net torque acts on a particle, then its angular momentum changes with time. Recall that, in terms of the momentum of a particle, Newton's second law can be written $\mathbf{F} = d\mathbf{p}/dt$, where $\mathbf{F}$ is the net force acting on the particle. Take the vector product of this equation with $\mathbf{r}$ to obtain $\mathbf{r} \times \mathbf{F} = \mathbf{r} \times d\mathbf{p}/dt$. Note that $\mathbf{r} \times \mathbf{F} = \boldsymbol{\tau}$, the net torque acting on the particle. With a small amount of mathematical manipulation you can show that $\mathbf{r} \times (d\mathbf{p}/dt) = d(\mathbf{r} \times \mathbf{p})/dt - (d\mathbf{r}/dt) \times \mathbf{p} = d\boldsymbol{\ell}/dt$. The second term, $(d\mathbf{r}/dt) \times \mathbf{p}$, vanishes because _____ is parallel to _____. Thus, in terms of torque and angular momentum, Newton's second law becomes $\boldsymbol{\tau} = $ _____.

For a system of particles the time rate of change of the angular momentum of each particle equals the net torque acting on the particle. This means that the time rate of change of the total angular momentum for the system is the vector sum of all torques acting on all particles. The sum includes the torque exerted by any particle in the system on any other particle in the system as well as external torques, exerted on particles in the system by objects in the environment of the system.

According to Newton's third law (in the strong form), the sum of the torques exerted by particles of the system on each other is _____. Only torques exerted by objects outside the system can change the total angular momentum of the system. Be sure you understand this. One particle in the system can change the angular momentum of another particle in the system but the second particle simultaneously changes the angular momentum of the first and the two changes are equal in magnitude and opposite in direction. Internal interactions may change the angular momenta of individual particles in the system but they do not change the total angular momentum of the system as a whole.

Newton's second law for the angular momentum of the system is

$$\sum \boldsymbol{\tau}_{\text{ext}} = $$

where $\sum \boldsymbol{\tau}_{\text{ext}}$ is the total *external* torque exerted on all particles of the system and $\mathbf{L}$ is the total angular momentum of the system.

This is the fundamental equation for a system of particles. You should learn to apply it to as many situations as you can. Remember it is a three-dimensional equation. In component form $\sum \tau_{\text{ext }x} = $ _____, $\sum \tau_{\text{ext }y} = $ _____, and $\sum \tau_{\text{ext }z} = $ _____. You must use the same origin to compute the angular momentum and each torque.

Consider the spinning dumbbell with unequal masses. If the origin is on the axis of rotation but not at the rod that connects the masses, the angular momentum is not along the axis of rotation. Furthermore, the angular momentum vector rotates with the dumbbell and so is changing in time. The torque that causes this change in $\mathbf{L}$ is exerted by the rod on the masses

and is associated with the centripetal force that holds the masses in their circular orbits. When the origin is at the rod, then this torque vanishes because _____
_____.

If the masses are equal this torque vanishes, even if the origin is not at the rod, because _____
_____.

For a body rotating about the z coordinate axis, we are usually interested in only the z components of the torque and angular momentum. If an external torque with other components is applied, the bearings must apply whatever additional torque is required to hold the axis fixed. If the spinning body is symmetric about the axis of rotation then **L** is along that axis and the bearings do not need to exert a torque to hold the axis fixed.

Since $L_z = I\omega$ for a body rotating about the z axis, Newton's second law for fixed axis rotation becomes $\tau_{\text{ext } z} =$ _____. In some cases the body is rigid and I does not change with time. Then Newton's second law for fixed axis rotation becomes

$$\tau_{\text{ext } z} =$$

where α is the angular acceleration of the body. This relationship was derived in the last chapter using energy considerations. In this chapter you will encounter some situations for which I does change.

Precession. If the axis of rotation is not fixed a torque that has components perpendicular to the axis might change the orientation of the axis. One example is discussed in the text: a wheel spinning on a long horizontal axle. For an origin placed on the axle the angular momentum **L** of the wheel is along the axle. Now one end of the axle is placed on a support and the other end is released with the axle horizontal. The force of gravity, downward on the wheel, exerts a torque. Describe the direction of the torque relative to the vertical and to the axis of rotation:
_____
_____

Describe the change in the direction of **L** with time: _____
_____

If the wheel is spinning fast, the axle moves so as to remain in the direction of **L**.

Carefully note that τ remains perpendicular to **L**, so the magnitude of **L** is constant as **L** rotates. To convince yourself that L is constant, take the scalar product of **L** with both sides of $\tau = d\mathbf{L}/dt$ to obtain $\mathbf{L} \cdot \tau = \mathbf{L} \cdot d\mathbf{L}/dt$. Since τ remains perpendicular to **L**, $\mathbf{L} \cdot \tau = 0$. Furthermore $\mathbf{L} \cdot d\mathbf{L}/dt = \frac{1}{2}dL^2/dt$, so $dL^2/dt = 0$ and L^2 is constant. This means L is constant.

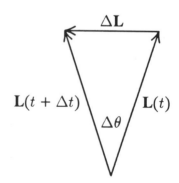

Suppose τ remains perpendicular to **L** and, over some time interval Δt, **L** rotates through the angle $\Delta\theta$, as shown on the right. The magnitude of the change in **L** is given by $|\Delta \mathbf{L}| = 2L\sin(\Delta\theta/2)$, as some simple trigonometry shows. In the limit as the angle becomes small the sine of any angle equals the angle itself in radians. Thus when $\Delta\theta$ is small $|\Delta \mathbf{L}| =$ _____ if $\Delta\theta$ is measured in radians. Divide by Δt, take the limit as Δt becomes small, and equate the result to the magnitude of τ.

190   Chapter 13: Angular Momentum

In terms of $d\theta/dt$, $\tau =$ _____. ω_p ($= d\theta/dt$) is called the <u>precessional angular velocity</u>. It is the angular velocity with which the angular momentum vector precesses. If the angular momentum of the wheel is large it is also the angular velocity with which the rotation axis precesses. For the equation that relates ω_p to L and τ to be valid, the units of ω_p must be _____.

The angular momentum changes magnitude if the applied torque has a non-vanishing component along a line that is _____; it changes direction if the applied torque has a non-vanishing component along a line that is _____.

Conservation of angular momentum. If the total external torque acting on a body is zero then the change over any time interval of the angular momentum of the body is _____. Angular momentum is then said to be conserved. This is a vector law. One component of the angular momentum may be conserved while other components are not. To see if angular momentum is conserved or not, calculate the total external torque acting on the body.

If the z component of the total external torque acting on a body rotating about a fixed axis vanishes then $I\omega$ is a constant. Consider an ice skater, initially rotating on the points of her skates with her arms extended. When she drops her arms to her side, her angular velocity increases. By dropping her arms she decreases her _____ and her angular velocity must increase for _____ to remain the same.

Suppose the skater is spinning in the counterclockwise direction when viewed from above, so her angular velocity vector points upward. Consider her torso and arms as two individual systems. As she drops her arms the angular momentum of her torso increases, so you know the torque exerted by her arms on her torso is in the _____ direction. Her torso exerts a torque of equal magnitude but opposite direction on her arms. This means the magnitude of the angular momentum of her arms _____ as they drop.

II. PROBLEM SOLVING

Some problems simply ask for a calculation of a torque or angular momentum using the definition or a closely related equation. Identify the point to be used as the origin, then evaluate the appropriate vector product: to find a torque take **r** to be the vector from the origin to the point of application of the force and evaluate $\mathbf{r} \times \mathbf{F}$; to find the angular momentum of a particle take **r** to be the vector from the origin to the particle, then evaluate $\mathbf{r} \times \mathbf{p}$. You must be given the momentum of the particle or else its mass and velocity.

PROBLEM 1. One end of a light rod, 1.5 m long, is attached to a small 5.0-kg ball and the ball is swung in a horizontal circle at 1.8 m/s with the center at the other end of the rod. Take the origin to be on a vertical line through the center of the circle, 2.0 m below the center, and calculate the magnitude of the angular momentum of the ball and the magnitude of its vertical component.

SOLUTION:

[ans: 22.5 kg·m²/s; 13.5 kg·m²/s]

Calculate the magnitude of the torque exerted by the rod on the ball and the magnitude of its vertical component.

SOLUTION: The force of the rod on the ball is mv^2/R, directed toward the center of the circle. Evaluate $\mathbf{r} \times \mathbf{F}$.

[ans: 21.6 N·m; 0]

A 12-N force, tangent to the circular path, is applied to the ball. What is the magnitude of the torque associated with this force and what is its vertical component?

SOLUTION:

[ans: 30.0 N·m; 18.0 N·m]

Now take the origin to be at the center of the circular orbit and calculate the magnitude of the angular momentum, the magnitude of the torque exerted by the rod, the magnitude of its vertical component, the magnitude of the external torque, and the magnitude of its vertical component.

SOLUTION:

[ans: 13.5 kg·m²/s; 13.5 kg·m²/s; 0; 0; 18.0 N·m; 18.0 N·m]

For a rigid body rotating about a fixed axis, the magnitude of the component of the angular momentum along the axis is given by $I\omega$. The direction is determined by a right hand rule. To calculate the angular momentum you must be given the rotational inertia and the angular velocity or else sufficient information to calculate them.

PROBLEM 2. A 3.5-kg uniform sphere with a radius of 20 cm is rotating about an axis that is tangent to its surface. If it starts from rest and has a constant angular acceleration of 1.1 rad/s² what is the component of its angular momentum along the axis of rotation as a function of time?

SOLUTION: Take the z axis to be along the axis of rotation. The angular velocity is given by $\omega(t) = \alpha t$, so the z component of the angular momentum is given by $L_z = I\omega = I\alpha t$. According to Fig. 9 of Chapter 12 the rotational inertia of a uniform sphere about a diameter is $I_{cm} = (2/5)MR^2$. The sphere of the problem is rotating about an axis that is displaced from a diameter by R so its rotational inertia is $I = (2/5)MR^2 + MR^2 = (7/5)MR^2$. Thus $L_z = (7/5)MR^2\alpha t$. Evaluate the expression for L_z.

[ans: 0.216 t, in SI units]

Dynamics problems for a body rotating about a fixed axis are based on $\tau_z = dL_z/dt = d(I\omega)/dt$. Here τ_z is the component of the total external torque along the axis of rotation and L_z is the component of the total angular momentum along that axis.

Some problems ask for L_z or τ_z themselves, others ask for related quantities, such as a force, speed, angular velocity, acceleration, or angular acceleration. Here are some examples.

PROBLEM 3. What torque is being applied to the sphere of the last problem?

SOLUTION:

[ans: 0.216 N·m]

PROBLEM 4. Consider the 12-N force applied to the ball of Problem 1. Suppose it is constant in magnitude and remains tangent to the orbit. Further suppose it increases the speed of the ball. Take the initial speed of the ball to be 1.8 m/s, assume the bearings are frictionless, and calculate the speed 2.5 s later.

SOLUTION: Use $\tau_z = dL_z/dt$ and $L_z = mRv$ to show that $\Delta v = \tau_z \Delta t/mR$, then evaluate this expression and add the result to the initial speed.

[ans: 7.8 m/s]

Note that since $\tau_z = FR$, the expression you derived for Δv reduces to $\Delta v = F\Delta t/m$, a result you can easily derive from Newton's second law without considering torque and angular momentum. The equivalence of the two methods should increase your confidence in $\tau_z = dL_z/dt$.

PROBLEM 5. A pulley of radius R and rotational inertia I is free to rotate on a fixed horizontal axis. A string runs over the pulley. Mass m_1 is hung on one end and mass m_2 is hung on the other. Treat the two masses and the pulley as the system and use $\tau_z = dL_z/dt$ to find an expression for the acceleration of m_1. Take $m_1 = 2.3$ kg, $m_2 = 1.4$ kg, $R = 15$ cm, and $I = 1.2$ kg·m², then evaluate the expression.

SOLUTION: Take the origin to be at the center of the pulley, take the y axis to be downward, and take the z axis to be out of the page. The external torques on the system are those due to the force of gravity acting on the masses. Gravity acts on the pulley and the pulley axle provides a supporting force, but these forces act through the center of the pulley and do not exert torques about that point. First show that the z component of the net external torque is $\tau_z = (m_1 - m_2)gR$, where m_1 was selected to be to the left of m_2 and counterclockwise rotation of the pulley was taken to be positive. Now show that the z component of the total angular momentum is $L_z = m_1 v_1 R - m_2 v_2 R + I\omega$, where ω is the angular velocity of the pulley. Since $v_2 = -v_1$ and $\omega = v_1/R$, $L_z = (m_1 + m_2)Rv_1 + I\omega/R$. Substitute into $\tau_z = dL_z/dt$ to obtain $(m_1 - m_2)gR = (m_1 + m_2)a_1 R + Ia_1/R$, where a_1 is the acceleration of m_1. Now solve for a_1.

[ans: 0.155 m/s²]

PROBLEM 6. An unbalanced wheel consists of a thin uniform 40-cm radius disk with an additional mass of 250 g at a point on the rim. It spins at 12 rad/s on the end of a fixed vertical axis, 80-cm long, through its center. What is the magnitude of the torque that must be exerted on the axle, about the end opposite the wheel, to keep it from wobbling?

SOLUTION: Place the origin at the end of the axle opposite the wheel. The angular momentum of the uniform disk is along the axle and does not change with time so you need to consider only the additional mass at the rim. The magnitude of its angular momentum about the origin is $\ell = mrR\omega$, where m is its mass, r is its distance from the origin, R is the radius of the wheel, and ω is the angular velocity of the wheel. The component along the axle is $\ell_z = mR^2\omega$ and the component along the wheel radius through the mass is $\ell_R = mdR\omega$, where d is the length of the axle. The former does not change with time but the latter does. It follows the mass around, always pointing toward or away from the wheel center, depending on the direction of rotation. Thus $|d\ell_R/dt| = \ell_R\omega$. This is the torque exerted by the axle on the wheel. The wheel exerts a torque of equal magnitude on the axle and a torque of magnitude $\tau = \ell_R\omega$ must be exerted on the axle by an external agent to hold it steady.

[ans: 11.5 N·m]

Notice that $\tau = |\mathbf{r} \times \mathbf{F}|$, where $\mathbf{F}$ is the centripetal force required to hold the additional mass on its circular orbit.

When the axis of rotation is not fixed you must use the vector equation $\boldsymbol{\tau} = d\mathbf{L}/dt$. If the net torque is given as a function of time, for example, this equation can be solved for the change in the angular momentum: $\Delta \mathbf{L} = \int \boldsymbol{\tau}\, dt$.

If $\boldsymbol{\tau}$ is always perpendicular to $\mathbf{L}$, then $\mathbf{L}$ rotates (or precesses) without change in magnitude. The magnitude τ of the net torque, the magnitude L of the angular momentum, and the precessional angular velocity ω_p are related by $\tau = L\omega_p$. Given any two of these quantities you can solve for the third.

PROBLEM 7. A body with an angular momentum of 85 kg·m²/s is subjected to a torque of 65 N·m, always perpendicular to L. What is its precessional angular velocity?

SOLUTION: Use $\omega_p = \tau/L$.

[ans: 0.765 rad/s]

PROBLEM 8. A gyroscope consists of a 35-cm radius uniform disk mounted at one end of a 1.2-m long horizontal axle, on which it is free to rotate. Only the other end of the axle is supported. If the disk is spinning at 1200 rad/s what is its precessional angular velocity?

SOLUTION: Take the origin to be at the supported end of the axle. The torque acting on the system is $\tau = mgd$, where d is the length of the axle and m is the mass of the disk. The angular momentum of the disk is $L = I\omega = \frac{1}{2}mR^2\omega$, where R is its radius. Use $\omega_p = \tau/L$ to compute the precessional angular velocity.

[ans: 0.160 rad/s]

Some problems can be solved using the principle of angular momentum conservation. To see if the principle can be used you must select a system and examine the external torque acting on it. If the total external torque vanishes then angular momentum is conserved. If one component of the total torque vanishes then that component of the angular momentum is conserved, regardless of whether or not the other components change. The following is an example in which angular momentum is conserved but angular velocity changes.

PROBLEM 9. A rope is looped over a pipe fastened vertically in the middle of a frozen pond. A 750-N ice skater holds the free end and uses the rope to skate at 3.5 m/s in a 5.5-m circle. The rope does not wrap around the pipe but simply slides around it. What is the angular momentum of the skater around the pipe?

SOLUTION: Use $L = mRv$, where m is the mass and v is the speed of the skater and R is the radius of the circle.

[ans: 1.47×10^3 kg·m²/s]

The skater now pulls the rope through her hands until she is skating in a 2.5-m radius circle. Neglect the force of the ice on her skates and calculate her new speed.

SOLUTION: Since the force of rope on the skater is toward the pipe, the torque it exerts is zero and her angular momentum is the same as when she was on the larger circle.

[ans: 7.70 m/s]

By how much did the kinetic energy of the skater change?

SOLUTION:

[ans: 1800 J]

As the skater is spiraling in her velocity, always tangent to her path, is not perpendicular to the force of the rope. In fact, it makes an angle of less than 90° with the force. This means the force increases her speed and does positive work, even though the torque associated with it is zero.

Some problems deal with an object whose rotational inertia changes at some time while it is spinning about a fixed axis. The problem statement gives information that can be used to calculate the initial values I_0 and ω_0, before the rotational inertia changes, and asks for I or ω at a later time, after it changes. If no external torques act, angular momentum is conserved and $I_0\omega_0 = I\omega$. This can be solved for one of the quantities in terms of the others.

PROBLEM 10. A girl stands on a rotating frictionless platform with arms outstretched, holding weights. The angular velocity of the platform and girl is 2.8 rad/s and the total rotational inertia of the girl, weights, and platform is 4.5 kg·m². When the girl drops her arms to her side the total rotational inertia decreases to 1.8 kg·m². What then is her angular velocity?

SOLUTION:

[ans: 7.00 rad/s]

What was the change in rotational kinetic energy?

SOLUTION: Use $K = \frac{1}{2}I\omega^2$ to calculate the rotational kinetic energy before and after the girl drops her arms.

[ans: 26.5 J]

PROBLEM 11. Let us examine the change in the angular velocity of the girl more closely. Suppose the torso of the girl and the platform have a combined rotational inertia of 1.56 kg·m² and she is carrying a 3.0 kg weight in each hand. Initially the weights are moving in a 0.70-m radius circle and later they are moving in a 0.20-m radius circle. Neglect the rotational inertia of her arms and hands. What was the change in the angular momentum of her torso?

SOLUTION: Use $\Delta L = I(\omega_f - \omega_i)$.

[ans: 6.55 kg·m²]

What was the change in the angular momentum of the weights?

SOLUTION: Use $\Delta L = 2m(R_f^2\omega_f - R_i^2\omega_i)$. The factor 2 appears because she is holding 2 weights.

[ans: −6.55 kg·m²]

Suppose her arms were brought down in such a way that they exerted a constant torque on her torso. What was that torque if the process took 2.0 s?

SOLUTION:

[ans: 3.28 N·m]

What was the torque exerted by her torso on the weights?

SOLUTION:

[ans: −3.28 N·m]

Note that the girl must exert a negative torque on the weights to reduce their angular momentum as she brings them in. If she did not the weights would be moving much faster than her torso. The weights exert an oppositely directed torque on her torso, increasing its angular momentum and its angular velocity.

To see why this is so, suppose the girl brought her arms in toward her torso by exerting a radial force on them, toward the axis of rotation. What then are the final angular velocities of the weights and torso?

SOLUTION: The torque is then zero and the angular momentum of the weights is conserved. Use $R_f^2 \omega_f = R_i^2 \omega_i$. The angular velocity of her torso does not change since no torque is exerted on it.

[ans: 34.3 rad/s; 2.8 rad/s]

The weights are moving much faster than the torso. An impossible situation!

Here's an example that requires you to know that an object moving along a straight line has an angular momentum about a point off the line.

PROBLEM 12. A 2.5-m radius frictionless playground merry-go-round has a rotational inertia of 800 kg·m². When it is at rest a 35-kg boy runs at 4.5 m/s along a line that is tangent to the rim and jumps on. What then is the angular velocity of the merry-go-round?

SOLUTION: No external torques act on the system consisting of the merry-go-round and the boy, so its total angular momentum is conserved. Take the origin to be at the center of the merry-go-round. Initially the boy is moving along a line that is offset from the center of the merry-go-round by a distance R, so his angular momentum is given by mRv, where m is his mass, v is his speed, and R is the radius of the merry-go-round. The merry-go-round has zero angular momentum. After the boy jumps on, the merry-go-round has angular momentum $I\omega$ and the boy has angular momentum $mR^2\omega$. Since angular momentum is conserved $mRv = (I + mR^2)\omega$. Solve for ω.

[ans: 0.387 rad/s]

PROBLEM 13. In a department store display a toy electric train runs on a circular track, mounted on a horizontal platform that is free to rotate on a nearly frictionless vertical axis. The train has a mass of 6.5 kg, the track has a radius of 1.6 m, and the platform has a rotational inertia of 5.0 kg·m². The display is started from rest by turning the train on so it runs at 1.3 m/s relative to the platform. What is the angular velocity of the platform?

SOLUTION: In writing an expression for the angular momentum of the train remember to use its speed relative to the earth, not relative to the rotating platform. If v' is its speed relative to the platform then $v = v' + R\omega$ is its speed relative to the earth. Here R is the radius of the track and ω is the angular velocity of the track. First show that conservation of angular momentum leads to $0 = I\omega + mR(v' + R\omega)$, where I is the rotational inertia of the platform alone. Then solve for ω.

SOLUTION:

[ans: 0.625 rad/s, opposite to the train]

In another type problem two objects spin independently about the same fixed axis. They are then coupled and eventually reach the same angular velocity. This is the rotational analog of a completely inelastic collision. If one object has rotational inertia I_1, initial angular velocity ω_{10}, and final angular velocity ω and the other has rotational inertia I_2, initial angular velocity ω_{20}, and final angular velocity ω, then conservation of angular momentum leads to $I_1\omega_{10} + I_2\omega_{20} = (I_1 + I_2)\omega$. This equation can be solved for one of the quantities, given the others.

PROBLEM 14. Two flywheels are free to rotate independently on the same frictionless axle. The first has a rotational inertia of $4.5\,\text{kg}\cdot\text{m}^2$ and is initially at rest. The second has a rotational inertia of $6.3\,\text{kg}\cdot\text{m}^2$ and is initially spinning at 150 rad/s. The first wheel is then pushed along the axle until it makes contact with the second. The force of friction between the wheels starts the first wheel spinning and slows the second. Eventually they spin together. What is the final angular velocity of the combination?

SOLUTION:

[ans: 87.5 rad/s]

How much rotational kinetic energy was lost?

SOLUTION:

[ans: 2.95×10^4 J]

This rotational energy was converted chiefly to internal energy by the forces of the flywheels on each other.

III. COMPUTER PROJECTS

Conservation of angular momentum problems usually involve a system of two or more parts that exert torques on each other. The net external torque, however, vanishes. The angular momentum of each part changes, but the sum of the changes vanishes. You can use a computer to examine details of the motion.

Consider two wheels that are free to rotate on the same axle and that exert torques on each other. If the torque on wheel 1 is τ, then the angular velocity ω_1 of that wheel obeys $\tau = I_1\,d\omega_1/dt$, where I_1 is the rotational inertia of the wheel. The torque on wheel 2 is $-\tau$ and the angular velocity of that wheel obeys $-\tau = I_2\,d\omega_2/dt$. If τ is given as a function of time a computer can be used to solve for θ_1, ω_1, θ_2, and ω_2, all as functions of time. Here's an outline, modeled after the program given in Chapter 2. The program also contains instructions to compute the angular momenta of the wheels and the total angular momentum of the system so you can check on angular momentum conservation.

> input initial angular velocities: ω_{10}, ω_{20}
> input final time, interval width: t_f, Δt
> input display interval: Δt_d

set $t_b = 0$, $t_d = \Delta t_d$, $\theta_{1b} = 0$, $\omega_{1b} = \omega_{10}$, $\theta_{2b} = 0$, $\omega_{2b} = \omega_{20}$
calculate torque on 1: $\tau_b = \tau(t_b)$
begin loop over intervals
 calculate time at end of interval: $t_e = t_b + \Delta t$
 calculate torque on 1 at end of interval: $\tau_e = \tau(t_e)$
 calculate "average" torque: $\tau = (\tau_b + \tau_e)/2$
 calculate angular velocities at end of interval:
  $\omega_{1e} = \omega_{1b} + \tau \Delta t / I_1$
  $\omega_{2e} = \omega_{2b} - \tau \Delta t / I_2$
 calculate "average" angular velocities:
  $\omega_1 = (\omega_{1b} + \omega_{1e})/2$
  $\omega_2 = (\omega_{2b} + \omega_{2e})/2$
 calculate angular positions at end of interval:
  $\theta_{1e} = \theta_{1b} + \omega_1 \Delta t$
  $\theta_{2e} = \theta_{2b} + \omega_2 \Delta t$
 if $t_e \geq t_d$ **then**
  calculate angular momenta: $L_1 = I_1 \omega_{1e}$, $L_2 = I_2 \omega_{2e}$, $L = L_1 + L_2$
  print or display t_e, θ_{1e}, ω_{1e}, θ_{2e}, ω_{2e}, L_1, L_2, L
  increment t_d by Δt_d
 end if statement
 if $t_e \geq t_f$ **then** exit loop
 set $t_b = t_e$, $\theta_{1b} = \theta_{1e}$, $\omega_{1b} = \omega_{1e}$, $\theta_{2b} = \theta_{2e}$, $\omega_{2b} = \omega_{2e}$, $\tau_b = \tau_e$
end loop over intervals
stop

PROJECT 1. Suppose wheel 1 has a rotational inertia of 2.5 kg·m² and is initially rotating at 100 rad/s. Wheel 2 has a rotational inertia of 1.5 kg·m² and is initially at rest. The wheels come in contact and exert torques on each other, the torque of wheel 2 on wheel 1 being $\tau(t) = -20te^{-t/2}$, in N·m for t in seconds. Use the program to plot the angular velocities of the wheels from $t = 0$ to $t = 15$ s. Use an integration interval of 0.005 s and a display interval of 0.5 s.

What are the initial values of the angular momenta of the wheels and the total angular momentum of the system? What are the values after 15 s? [ans: $L_1 = 250$ kg·m²/s, $L_2 = 0$, $L = 250$ kg·m²/s; $L_1 = 170$ kg·m²/s, $L_2 = 80$ kg·m²/s, $L = 250$ kg·m²/s]

Suppose that in addition to the torque that wheel 2 exerts on wheel 1 an external agent exerts a torque of 8.0 N·m on wheel 1, in the direction of its angular velocity. What now are the angular momenta of the wheels and the total angular momentum of the system after 15 s? The torque on wheel 1 is $8.0 - 20te^{-t/2}$ and the torque on wheel 2 is $20te^{-t/2}$. [ans: $L_1 = 290$ kg·m²/s, $L_2 = 80$ kg·m²/s, $L = 370$ kg·m²/s]

Notice that the angular momenta of the individual wheels change with time in both cases. The total angular momentum, however, is conserved in the first situation but not in the second. Since $dL/dt = \tau_{ext}$ and in the second situation $\tau_{ext} = 8.0$ N·m, the total angular momentum at the end of 15 s should be $8.0 \times 15 = 120$ kg·m²/s greater than the total angular momentum at $t = 0$. Is this result produced by the program?

PROJECT 2. Conservation of angular momentum is sometimes demonstrated by considering the inelastic "collision" between two wheels on the same axle. Suppose wheel 1 of the previous project starts with an angular velocity of 100 rad/s and exerts a constant torque of 8.0 N·m on wheel 2, which starts from rest. Wheel 2, of course, exerts a torque of -8.0 N·m on wheel 1. Use the program to find the time for which $\omega_1 = \omega_2$ and the value of the angular velocity then. Through what angle has each wheel rotated by the time they reach the same angular velocity? Compare the value obtained for the final angular velocity with that predicted by $I_1\omega_{10} = I_1\omega_1 + I_2\omega_2$. [ans: 11.7 s; 62.5 rad/s; 952 rad; 366 rad]

The time for the wheels to reach the same angular velocity and the angle through which they turn depend on the torques they exert on each other but the final value of the angular velocity does not. To verify this statement repeat the calculation for the same initial conditions but for a torque on wheel 1 that is given by $\tau = -20te^{-t/3}$, in N·m for t in seconds. [ans: 5.24 s; 62.5 rad/s; 438 rad; 144 rad]

PROJECT 3. The kinetic energy of the wheels is not conserved during the inelastic "collision" of the previous project. You can see this by calculating the initial and final kinetic energies, but you can also use a computer to calculate the work done by each torque. If a wheel turns through the small angle $\Delta\theta$ while the torque τ acts on it, the work done by the torque is given by $\Delta W = \tau\Delta\theta$. Modify the program so it computes the work done on each wheel. Just before the loop over intervals set $W_1 = 0$ and $W_2 = 0$. These variables will be used to sum the work done in the intervals. At the end of each interval have the computer increment W_1 by $\tau(\theta_{1e} - \theta_{1b})$ and W_2 by $-\tau(\theta_{2e} - \theta_{2b})$. Here τ is the average torque exerted by wheel 2 on wheel 1. Also have the computer calculate the total work $W = W_1 + W_2$ and display the result when it displays the angular momenta.

Take the initial conditions to be as before ($\omega_{10} = 100$ rad/s, $\omega_{20} = 0$). Take the torque of wheel 1 on wheel 2 to be a constant 8.0 N·m. Find the work done on each wheel and the total work done from $t = 0$ to the time when the wheels have the same angular velocity. Compare the total work done to the change in the total kinetic energy: $\Delta K = \frac{1}{2}(I_1 + I_2)\omega_f^2 - \frac{1}{2}I_1\omega_{10}^2$, where ω_f is the final angular velocity. [ans: -7.62×10^3 J; 2.93×10^3 J; -4.69×10^3 J]

Now repeat the calculation for a torque on wheel 1 that is given by $\tau = -20te^{-t/3}$, in N·m for t in seconds. You should get the same answers.

Can the "collision" be elastic? Take the torque on wheel 1 to be -8.0 N·m and calculate the total work done as a function of time. Search for a time such that the total work is zero. You might use the condition $W \geq 0$ to exit the loop over intervals. If the wheels no longer interact after this time the collision is elastic. What are the final angular velocities of the wheels? [ans: 23.4 s; 25.0 rad/s; 125 rad/s]

The text showed you how to analyze the change in angular velocity of a spinning ice skater as she brings her arms toward the center of her torso from an outstretched position. Before she brings them in her rotational inertia is I_i and her angular velocity is ω_i. After she brings them in her rotational inertia is I_f and her angular velocity is ω_f. Conservation of angular momentum leads to $I_i\omega_i = I_f\omega_f$, so $\omega_f = (I_i/I_f)\omega_i$.

Here you will use a computer program to investigate some of the details. In particular, you will be able to follow her angular velocity as she brings her arms in. Let I_T be the rotational inertia of her torso alone and let I_A be the rotational inertia of her arms, including any weights she might hold to heighten the effect. I_A is a function of time, determined by the rate with which she moves her arms. Angular momentum is conserved at each stage of the process so

$$\omega(t) = \frac{I_{Ai} + I_T}{I_A(t) + I_T}\omega_i,$$

where I_{Ai} is the initial value of the rotational inertia of her arms.

If $\omega(t)$ is known a computer program can be used to calculate the angle $\theta(t)$ through which the skater has rotated since $t = 0$. The angle θ_e at the end of any interval of width Δt is related to the angle θ_b at the beginning by $\theta_e = \theta_b + \omega\Delta t$, where ω is the average angular velocity in the interval. As usual, we approximate the average angular velocity by $(\omega_b + \omega_e)/2$. Here is an outline of a program.

    input final time and interval width: t_f, Δt
    input display interval: Δt_d

        set $t_b = 0$, $t_d = \Delta t_d$, $\theta_b = 0$
        calculate angular velocity at beginning of first interval: $\omega_b = \omega(t_b)$
        begin loop over intervals
                calculate time at end of interval: $t_e = t_b + \Delta t$
                calculate angular velocity at end of interval: $\omega_e = \omega(t_e)$
                calculate "average" angular velocity: $\omega = (\omega_b + \omega_e)/2$
                calculate angle at end of interval: $\theta_e = \theta_b + \omega \Delta t$
                if $t_e \geq t_d$ **then**
                        calculate torque at end of interval: τ_e
                        print or display t_e, θ_e, ω_e
                        increment t_d by Δt_d
                end if statement
                if $t_e \geq t_f$ **then** exit loop
                set $t_b = t_e$, $\theta_b = \theta_e$, $\omega_b = \omega_e$ in preparation for next interval
        end loop over intervals
        stop

Both the angular momentum of the skater's torso and the angular momentum of her arms change. Her torso exerts an torque on her arms and her arms exert a torque of equal magnitude but opposite direction on her torso. If τ represents the torque on her torso then $\tau = I_T\, d\omega/dt$. Differentiation of the expression above for $\omega(t)$ produces

$$\tau = -\frac{I_T \omega_i (I_{Ai} + I_T)}{(I_A + I_T)^2} \frac{dI_A}{dt}$$

Use this expression in the program to calculate the torque just before t_e, θ_e, and ω_e are displayed, then display it along with the other quantities. The instruction has already been included in the outline above.

PROJECT 4. A skater is initially turning at 12 rad/s. Her torso has a rotational inertia of 0.65 kg·m². She carries weights in her hands so that as she pulls her arms in their rotational inertia changes from 0.50 kg·m² to 0.04 kg·m². Suppose she moves her arms in such a way that their rotational inertia changes uniformly over 2.0 s. That is, $I_A(t) = I_{Ai} - Kt$, where $I_{Ai} = 0.50$ kg·m² and K is a constant chosen so $I_A = 0.040$ kg·m² when $t = 2.0$ s.

Use the program to plot her angular velocity and the torque acting on her torso as functions of time from $t = 0$ to $t = 2.0$ s. What is her final angular velocity? Through what angle did she rotate during the process? At what time did she exert the maximum torque on her arms and what was the magnitude of that torque? [ans: 20.0 rad/s; 30.7 rad (4.88 rev); 1.73 N·m at 2.0 s]

Suppose now that all of the rotational inertia of her arms is actually due to the weights she is carrying and that she pulls them in at a constant rate. If m is the mass of one weight and r is it distance from the axis of rotation then the rotational inertia of both weights together is $I_A = 2mr^2$. Since the weights are brought in at a constant rate, we may write $r = r_i + Kt$, where r_i is the initial distance and K is a constant. Take $m = 1.0$ kg and pick values of r_i and K so $I_{Ai} = 0.50$ kg·m² and $I_A = 0.05$ kg·m² at the end of 2.0 s. Then use the program to plot her angular velocity and the torque acting on her torso as functions of time from $t = 0$ to $t = 2.0$ s. What is her final angular velocity? Through what angle did she rotate during the process? At what time did she exert the maximum torque on her arms and what was the magnitude of that torque? [ans: 20.0 rad/s; 32.2 rad (5.12 rev); 2.82 N·m at 0.95 s]

IV. NOTES

Chapter 14
EQUILIBRIUM OF RIGID BODIES

I. BASIC CONCEPTS

This chapter consists largely of applications of Newton's second laws for center of mass motion and for rotation, specialized to situations in which the total force and total torque both vanish. However, several new ideas are introduced to deal with descriptions of the torque exerted by gravity and with elastic forces: the center of gravity and the elastic moduli. Pay careful attention to what they are and how they are used.

Equilibrium conditions. A rigid body is said to be in <u>static</u> <u>equilibrium</u> if its center of mass is not moving and if it is not rotating. Its total momentum and its total angular momentum both vanish. In terms of the external forces and torques acting on it, equilibrium means $\sum \mathbf{F} =$ _____ and $\sum \tau =$ _____, where $\sum \mathbf{F}$ is the vector sum of all external forces and $\sum \tau$ is the vector sum of all external torques.

You will use these equations in applications to determine the external forces and torques on static objects. In some cases you will be interested in the force or torque exerted by one part of an object on another part. In that event, simply consider the system to be the part of the object on which the force or torque acts and treat the force or torque as an external force or torque.

In nearly all examples and problems in this chapter the forces considered are in the same plane. If you pick an origin in this plane every torque is perpendicular to the plane and the number of equations to be solved is reduced to three. If the plane of the forces is the xy plane then the three equations are $\sum F_x =$ _____, $\sum F_y =$ _____, $\sum \tau_z =$ _____. They can be solved for three unknowns. Usually the last equation is written using τ to stand for τ_z.

Recall that the value of a torque depends on the origin used to specify the point of application of the force. When an object is in equilibrium, however, the torques sum to zero no matter which point is used as the origin. The choice is a matter of convenience, not necessity, but the same origin must be used to compute all torques. If an origin is not specified by the problem statement, place it in the plane of the forces.

Center of gravity. The force of gravity is often one of the forces acting on an object. Although each particle of the object experiences a gravitational force, for purposes of determining equilibrium these forces can be replaced by a single force acting at a point called _____. If the acceleration **g** due to gravity is uniform throughout the object, the magnitude of the replacement force is _____, where M is the total mass of the object, and the direction of the replacement force is _____. In addition, the point where the replacement force is applied then coincides with _____. In this case the position of the center of gravity does not depend on the orientation of the object.

As you read the sample problems of Section 14–3 in the text carefully note how gravitational forces are taken into account. In each case their sum is replaced by a single force of magnitude Mg, directed downward and applied at the center of mass.

Even if the acceleration due to gravity is not uniform over the object, the sum of the gravitational forces can still be replaced by a single force applied at the center of gravity. Then, however, the center of gravity does not coincide with the center of mass. Furthermore, the position of the center of gravity depends on the orientation of the object. Practically speaking, it is not as useful a concept then.

Types of equilibrium. Equilibrium conditions are classified according to whether the potential energy is a maximum, minimum, or constant as a function of the position and orientation of the object. The nature of the equilibrium can be tested by moving the object slightly away from equilibrium and finding if the potential energy increases, decreases, or stays the same. In this chapter you are concerned chiefly with gravitational potential energy.

If the potential energy is a maximum the equilibrium is said to be _____. A slight displacement from equilibrium results in forces or torques that _____
_____

If the potential energy is a minimum the equilibrium is said to be _____. A slight displacement from equilibrium results in forces or torques that _____
_____

If the potential energy is constant the equilibrium is said to be _____. A slight displacement from equilibrium does not give rise to any net force or torque.

If the potential energy is wholly gravitational, you need only look at what happens to the center of gravity as the object is displaced or rotated slightly. If the center of gravity moves downward, the equilibrium is _____, if the center of gravity moves upward, the equilibrium is _____, and if the center of gravity stays at the same height, the equilibrium is _____.

Carefully study the objects shown in Fig. 10 of the text. Assume each of them rotates about a point in contact with the table top.

The cube in (a) is in _____ equilibrium because _____
_____.

The cube in (b) is in _____ equilibrium because _____
_____.

The disk in (c) is in _____ equilibrium because _____
_____.

Elasticity. In many cases the conditions that the net force and torque both vanish are not sufficient to determine the forces acting on an object. From a mathematical viewpoint there are too many unknowns or not enough equations. As a result, the equations have an infinite number of solutions and additional information is needed to decide which one is physically correct.

Contact forces, like the normal force of a surface on an object, are <u>elastic</u> in nature. That is, they arise from slight deformations of the objects in contact. In many circumstances the

204 *Chapter 14: Equilibrium of Rigid Bodies*

additional information you need to understand equilibrium is how elastic forces depend on deformation.

You should know the special terminology that is used in the study of elasticity. A force per unit area is called a _____ and a fractional deformation is called a _____.

There are two regimes of deformation. If the stress is less than the yield strength then, when the stress is removed, _____
_____.

If the stress is greater than the yield strength but less than the ultimate strength then, when the stress is removed, _____
_____.

If the stress is greater than the ultimate strength, then _____
_____.

For stresses well below the yield strength, the strain is proportional to the stress and the material is said to be elastic. Elastic materials behave like springs.

Suppose forces that are equal in magnitude and opposite in direction are applied to the opposite ends of a rod. If the forces are perpendicular to the rod faces, the object will be elongated or compressed, depending on the directions of the forces. Suppose the forces are pulling outward and let ΔL be the elongation of the rod. Also suppose the magnitude of each force is F, the area of each face is A, and the undeformed rod length is L. Then the stress is given by _____ and the strain is given by _____. If the object remains elastic (small stress), then the stress is proportional to the strain. The proportionality is written $F/A = E\Delta L/L$, where the modulus of elasticity E is called _____. E is a property of the material and does not depend on F, A, L, or ΔL.

Values of E for some materials are listed in Table 1 of the text. A large value of E means a large force is required to compress or elongate a given length of sample by a given amount. If E is small and the length and area are the same, only a small force is required.

If the forces are opposite in direction but parallel to the rod faces, then shear occurs. Suppose one face moves a distance ΔL relative to the other, along a line that is parallel to the faces. Also suppose the magnitude of each force is F, the area of each face is A, and the rod length is L. The stress is still given by _____ and the strain is still given by _____. Stress is proportional to strain and the proportionality is written $F/A = G\Delta L/L$. The modulus of elasticity G is called _____. G is a property of the material. Materials with a small modulus of elasticity shear more easily than materials with a large modulus of elasticity provided, of course, the lengths and areas of the samples are the same.

II. PROBLEM SOLVING

Except for the ideas of center of gravity and of elasticity, this chapter chiefly covers applications of concepts discussed in earlier chapters and the problem solving techniques you will need are not new to you.

We consider situations in which all forces acting on the object under consideration are in the xy plane. Furthermore, in each case we select the origin for the computation of torques to be in that plane. All torques are then along the z axis, perpendicular to the plane. The method can easily be extended to situations in which forces with z components and torques with x and y components act.

Here are the steps you should use to solve problems. First select the object to be considered. Draw a free-body diagram showing as arrows all external forces acting on the object. Don't forget normal and frictional forces exerted by any objects in contact with the selected object. Use an algebraic symbol to designate the magnitude of each force, known or unknown. Use either Mg or W to label the force of gravity, depending on whether the weight or mass of the object is given.

Draw each force arrow in the correct direction, if you know it. Gravity acts downward, a string pulls along its length, a frictionless surface pushes along the perpendicular to the surface. Sometimes you will know the line of a force but not its direction. You might know, for example, that a rigid rod acts along its length but you may not know if it is pushing or pulling. In that event, draw the arrow in either direction. If, after you solve the problem, the value turns out to be positive then you picked the correct direction; if it turns out to be negative the force is actually opposite the direction you picked. In other cases you may know only that the force acts in the xy plane. Pick an arbitrary direction, not along one of the coordinate axes. You will then solve for both components.

Because you will need to compute torques, be sure you indicate the point of application of each force by placing the tail or head of the arrow at the appropriate place on the diagram. Gravity acts at the center of gravity. For the situations considered here this is the same as the center of mass. A string or rod acts at its point of attachment. In many cases an object makes contact only at a point and that is the point of application of the force it exerts.

Choose a coordinate system for calculating force components. If you know the direction of one of the unknown forces, you can usually simplify the algebra by selecting one of the coordinate axes to be in that direction. Write $\sum F_x = 0$ and $\sum F_y = 0$ in terms of the individual forces.

Choose an origin for calculating torques. The choice is completely arbitrary but you can usually reduce the algebra considerably by choosing an origin so that one or more torques vanish. Try to place the origin on the line of action of a torque associated with an unknown force. Write $\sum \tau = 0$ in terms of the forces and their lever arms.

Pay special attention to the signs of the torques. Pick one direction perpendicular to the plane of the forces (out of the page, for example) to be positive and the other to be negative. Then use the right hand rule for each force to find the direction of the corresponding torque.

To see how a problem is set up read the statement of Sample Problem 3 and examine Fig. 7 in the text. Note that the force of gravity on the ladder is applied at a point one third up the ladder because, according to the problem statement, that is where the center of gravity is. The force of gravity on the firefighter is applied at the position of the firefighter, halfway up the ladder. Also note that the wall exerts only a normal force, not a frictional force, while the ground exerts both normal and frictional forces. Point O, used to compute the torques, is selected at the intersection of the action lines of two forces and, as a result, these forces do not appear in the torque equation. Study Eq. 19 of the text carefully and note how easily the torques are written if you remember that a torque is the product of the magnitude of the force and the perpendicular distance from the action line to O.

PROBLEM 1. A 3.0-m long uniform plank weighing 120 N rests horizontally on two supports, one at one end and the other one quarter of the plank length from the other end. What force does each support exert on the plank?

SOLUTION: In the space to the right draw a free-body diagram for the plank. Label the force of the support at the end F_A and the force of the other support F_B. Take both to be upward. The force of gravity has magnitude W (= 120 N) and acts downward at the midpoint. Take the y axis to be positive in the upward direction and write the sum of the y components of the forces in terms of the magnitudes F_A, F_B, and W. Set the sum equal to zero: $F_A + F_B - W = 0$.

For purposes of calculating torques place the origin at the supported end of the plank and let d be the length of the plank. Then the torque due to gravity has magnitude $Wd/2$, the torque due to the support at the end is zero, and the torque due to the other support is $3F_B d/4$. Take the positive direction to be out of the page. Then the first torque is positive and the last is negative, so $Wd/2 - 3F_B d/4 = 0$. Solve this equation for F_B, then the force equation for F_A.

[ans: $F_A = 40.0$ N; $F_B = 80.0$ N]

How much weight can be placed on the plank at the unsupported end before the other end leaves the support?

SOLUTION: If weight W' is placed on the plank at the unsupported end it exerts a negative force W' and a positive torque $W'd$. The force and torque equations become $F_A + F_B - W - W' = 0$ and $Wd/2 + W'd - 3F_B d/4 = 0$. When the plank end is about the leave the support $F_A = 0$. Solve these equations for W'.

[ans: 120 N]

The plank is now balanced on the support located one fourth of its length from the unsupported end. The added weight is $d/4$ on one side of the support and the force of gravity acts at a point $d/4$ on the other side. At equilibrium these forces must be equal, so the added weight is the same as the weight of the plank, 120 N.

PROBLEM 2. A 3.0-m long uniform ladder weighing 25 N leans at 62° above the horizontal against a frictionless vertical wall. What is the normal force exerted by the wall, the normal force exerted by the floor, and the force of friction exerted by the floor?

SOLUTION: In the space to the right draw a free-body diagram for the ladder. Represent the normal force of the wall by N_W, the normal force of the floor by N_F, and the force of friction by f. The first force is horizontal and away from the wall, the second is vertical and upward, and the third is horizontal and toward the wall. The force of gravity is downward and is taken to act at the center of the ladder. Represent its magnitude by W.

Take the x axis to be horizontal and positive away from the wall; take the y axis to be vertical and positive upward. Newton's second law for the center of mass is then:

Chapter 14: Equilibrium of Rigid Bodies   **207**

x component:

y component:

If the positive direction for torque is taken to be out of the page and the origin is placed at the point of contact of the ladder with the floor, then the wall exerts a torque of $-N_W d \sin\theta$, gravity exerts a torque of $W(d/2)\cos\theta$, and the floor exerts a torque of zero. Here d is the length of the ladder and θ is the angle between the ladder and the horizontal. Newton's second law for rotation is:

Solve these equations for N_W, N_F, and f.

[ans: 6.65 N; 25.0 N; 6.65 N]

If the coefficient of static friction between the floor and the ladder is 0.30 does the ladder slip on the floor?

SOLUTION: The force of friction that holds the ladder at rest is 6.65 N, as calculated above. The maximum frictional force that the floor can exert on the ladder is $\mu_s N_F =$ _____ N. Since $f_{\rm rest} < \mu_s N_F$ the ladder does not slip.

If the coefficient of friction between the floor and the ladder is 0.30 what is the minimum angle the ladder can make with the horizontal without slipping on the floor?

SOLUTION: If the ladder leans at a smaller angle a greater force of friction is required to hold it. When the required force exceeds $\mu_s N_F$ the ladder slips. Substitute $f = \mu_s N_F$ for the force of friction in the force equations. Notice that the torque equation contains $\sin\theta$ in one term and $\cos\theta$ in the other, so it can be solved for $\tan\theta$ ($= \sin\theta/\cos\theta$) in terms of W and N_W. Do this, then use the force equations to write the result in terms of given quantities. You should obtain $\tan\theta = \frac{1}{2}\mu_s$. Evaluate this expression.

[ans: 59.0°]

Notice that the result depends only on the coefficient of static friction, not on the weight or length of the ladder.

Now try some by yourself.

PROBLEM 3. A 8.0-m long ladder weighing 30 N leans at 65° above the horizontal against a frictionless vertical wall. A 650-N man stands at a point 5.5 m up the ladder (as measured along the ladder). What is the normal force of the wall, the normal force of the floor, and the force of friction exerted by the floor?

SOLUTION:

[ans: $N_W = 215$ N; $N_F = 680$ N; $f = 215$ N]

Suppose the coefficient of static friction between the ladder and the floor is 0.40. How far up the ladder can the man climb before it slips?

208 *Chapter 14: Equilibrium of Rigid Bodies*

SOLUTION:

[ans: 6.99 m]

Many times you can predict the direction of an unknown force without actually solving the problem. For an example, read the statement of Sample Problem 2 and examine Fig. 6 in the text. With a little thought you should be able to see that the biceps must pull up on the arm at a point in front of the pivot. If they did not, the force of gravity on the ball and arm would produce an unbalanced clockwise torque around the pivot. The force of the biceps must be must larger than the combined weight of the ball and arm because it is applied much closer to the pivot. Now the forces are unbalanced. To balance them the bones must exert a downward force on the pivot point. This simple analysis has led us to the correct directions for the unknown forces.

PROBLEM 4. A strut is pinned to a vertical wall as shown. To hold the strut horizontal a wire runs from its midpoint to a point on the wall above the pin. A weight is placed on the unsupported end of the strut. In what direction is the vertical component of the force of the wall on the strut?

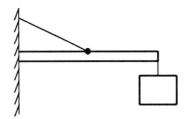

SOLUTION:

[ans: downward]

Take the length of the strut to be 10 m, the weight of the strut to be 200 N, and the weight placed on the end to be 700 N. Suppose the wire makes an angle of 25° with the strut. What are the forces exerted on the strut by the wall and wire?

SOLUTION:

[ans: horizontal component of force of wall = 3430 N, away from wall; vertical component of force of wall = 700 N, down; force of wire = 3790 N, along wire]

Now suppose the wire is fastened to the end of the strut, rather than to its midpoint. Without solving the problem find the direction of the vertical component of the force exerted by the wall on the strut.

SOLUTION:

[ans: upward]

Now calculate the forces exerted by the wall and the wire on the strut.

SOLUTION:

[ans: horizontal component of force of wall = 1720 N, away from wall; vertical component of force of wall = 100 N, up; force of wire = 1890 N, along wire]

Some elasticity problems are simply applications of Eq. 29 in the text or the analogous equation for shear. You may be asked to calculate the stress or the strain, given the other. In variants you may be asked for the change in length under compression or the relative displacement of the faces under shear. The appropriate modulus must be given or else listed in Table 1. In some cases you may be asked to calculate the modulus, given the other quantities in Eq. 29. These problems are all straightforward. Sample Problem 6 of the text is an example. Here are some others.

PROBLEM 5. A 0.65-cm radius steel rod (Young's modulus = 200×10^9 N/m^2) is compressed by applying an inward force of 5000 N uniformly to each end. What is the fractional change in its length?

SOLUTION: The equation that describes this situation is $F/A = -E\Delta L/L$, where F is the force applied to each face, A is the area of a face, E is Young's modulus, ΔL is the change in length, and L is the original length. The negative sign appears because the change in length is negative. The fractional change in length is $f = \Delta L/L$ so $f = -F/EA$. The area of a face, of course, is πR^2, where R is the radius.

[ans: -1.88×10^{-4}]

If the original length of the rod was 1.2 m what is the change in its length?

SOLUTION:

[ans: -2.26×10^{-4} m]

PROBLEM 6. What force must be applied to each end to change the length of the rod of the previous problem by 10 per cent? Assume the rod remains elastic.

SOLUTION:

[ans: 3.98×10^4 N]

Does the rod in fact remain elastic?

SOLUTION: Calculate the stress F/A.

[ans: 3.00×10^8 N/m^2]

According to Table 1 the yield strength for steel is 2.50×10^8 N/m^2. Since the applied stress is greater than the yield strength, the rod does not remain elastic but rather becomes permanently deformed.

PROBLEM 7. Each edge of a certain aluminum cube has a length of 5.5 cm. If an inward force of 5000 N is uniformly applied to each face what is the change in the volume of the cube? Aluminum has a Young's modulus of 7.0×10^{10} N/m^2.

SOLUTION: After compression each edge has a length of $a + \Delta a$, where a is the original length and $\Delta a/a = -F/EA$. The new volume is $(a + \Delta a)^3 = a^3 + 3a^2\Delta a + 3a(\Delta a)^2 + (\Delta a)^3$. Since Δa is much smaller than a this can be approximated by $a^3 + 3a^2\Delta a$. The change in volume is $\Delta V = 3a^2\Delta a$. Evaluate this expression.

[ans: -1.18×10^{-8} m^3]

What is the change in the area of any one face?

SOLUTION:

[ans: $-1.43 \times 10^{-7}\,\text{m}^2$]

Sample Problem 8 of the text shows how the equations of elasticity are used to determine forces at equilibrium. In this case you know that the table legs must all have the same length since the table does not wobble. This is sufficient to find the relationship between the forces on the legs: the force F_1 must be greater than the force F_3 by just enough to compress the leg that was originally longer by the extra amount d. Once this information is substituted into the equilibrium condition, the resulting equations can be solved for the forces.

For problems of this type you may think of the equilibrium equations for forces and torques and the elasticity equations as a set of simultaneous equations to be solved for the unknowns. For some cases, as for the sample problem, the strain of an object or a relationship between the strains of several objects, is given. In other cases the strain is an unknown.

PROBLEM 8. A 5.0-m long plank, used by a painter, is suspended by two cables alongside a building. The plank weighs 400 N. The cables have the same cross-sectional area and when unstretched have the same length, but one is made of steel (Young's modulus = $200 \times 10^9\,\text{N/m}^2$) and the other is made of aluminum (Young's modulus = $70 \times 10^9\,\text{N/m}^2$). Where should a 700-N painter stand so the plank is horizontal?

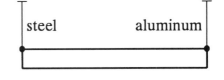

SOLUTION: Assume the painter is a distance x from the left edge of the plank and draw the free-body diagram for the plank. Include the force F_s of the steel cable, the force F_a of the aluminum cable, the force F_p of the painter (equal to his weight), and the weight W of the plank.

The vertical component of Newton's second law for the center of mass of the plank is

and Newton's second law for rotation is

The change in length of the steel cable is $\Delta L_s = LF_s/AE_s$ and the change in length of the aluminum cable is $\Delta L_a = LF_a/AE_a$, where L is the unstretched length of each cable and A is their cross-sectional areas. The painter must stand so the cable lengths are the same. Since the unstretched lengths are equal this means the changes in length must be equal and the forces F_s and F_a must be related by $F_s/E_s = F_a/E_a$. Solve this equation simultaneously with the second law equations. You should obtain $x = d[F_p + W(1 - E_s/E_a)/2]/[(E_s/E_a + 1)F_p]$, where d is the length of the plank. Evaluate this expression.

[ans: 0.608 m]

If the unstretched length of each cable is 40 m and each cable has a radius of 0.45 cm what is the elongation of the cables and what forces do they exert on the plank?

SOLUTION:

[ans: $\Delta L = 0.256\,\text{cm}$; $F_s = 815\,\text{N}$; $F_a = 285\,\text{N}$]

Suppose the painter stands at the midpoint of the plank. What then is the difference in the lengths of the two cables?

SOLUTION: An exact solution is difficult. For purposes of calculating the torque assume the plank is horizontal and write $(F_s - F_a)L/2$ for the total torque about the midpoint of the plank. Set this expression equal to 0 and solve it and the equation for the center of mass simultaneously for F_s and F_a, then use the relationship between stress and strain to calculate the elongations of the cables.

[ans: $0.321\,\text{cm}$]

III. NOTES

Chapter 15
OSCILLATIONS

I. BASIC CONCEPTS

This chapter is about periodic motion: the motion of an object that moves back and forth between two points, its motion being the same during every cycle. Pay attention to the meanings of the terms used to describe simple harmonic motion: amplitude, period, frequency, angular frequency, and phase constant; pay attention to the transfer of energy from kinetic to potential and back again as the object moves; and also pay attention to the form of the force law that leads to this type of motion.

Simple harmonic motion. Concentrate on simple harmonic motion, for which the position variable (angle or coordinate) is a sinusoidal function of time. For a model we take an object of mass m attached to a spring with force constant k, moving along the x axis with its equilibrium position at $x = 0$. Take the potential energy to be zero when the spring is neither stretched nor compressed. Then, as functions of the coordinate, the force on the mass is given by

$$F(x) =$$

and the potential energy of the system is given by

$$U(x) =$$

You should recognize that these equations are valid only if x = 0 is the equilibrium point. If the origin is placed elsewhere other terms must be added to the right sides.

The force and potential energy for the spring-mass system have certain distinguishing characteristics in common with all systems executing simple harmonic motion. The object, of course, does not move along a line for all SHM systems. For example, some objects have rotational motions and for them position is described by an angle rather than by a linear coordinate.

If we place the origin at the point for which the force or torque is zero then for SHM to occur the force or torque must always be proportional to _____ and if, in addition, we choose the potential energy to be zero when the position variable is zero then it will be proportional to _____.

The sign in the force law is important. When the coordinate is positive the force is _____ and when the coordinate is negative the force is _____. A force that always pushes an object toward its equilibrium position is called a _____ force.

The force law is substituted into Newton's second law to obtain the equation of motion. For the spring-mass system it is

$$\frac{d^2x}{dt^2} =$$

The most general solution to this equation is

$$x(t) = x_m \cos(\omega t + \phi),$$

where x_m, ω, and ϕ are constants. If the function given here is to obey the differential equation then the constant ω, in terms of m and k, must be $\omega = $ _____.

The maximum value of the coordinate x is denoted by _____. The object moves back and forth between $x = $ _____ and $x = $ _____. x_m is called the _____ of the oscillation.

The constant ω is called the _____ of the oscillation. Since ωt is measured in radians, the units of ω are _____.

The angular frequency is related to the frequency ν of the oscillation: $\omega = $_____. The physical significance of the frequency is: _____
_____

It has the dimension _____ and its SI unit is: _____.

The angular frequency and frequency are both related to the period T of the motion: $\omega = $ _____, and $\nu = $ _____. The physical significance of the period is _____
_____

It has the dimension _____ and its SI unit is _____.

As a fraction of the period T, the time taken by the mass to go from $x = 0$ to $x = x_m$ for the first time is _____ and the time taken to go from $x = -x_m$ to $x = +x_m$ for the first time is _____.

Suppose an oscillation has a period of 5.0 s. Then its frequency is _____ Hz and its angular frequency is _____ rad/s. The motion repeats every _____ s or, in other words, it repeats _____ times per s.

You should remember that the angular frequency is determined by the ratio of k to m. Consider two springs, one with force constant k and the other with force constant $2k$. An object of mass m is attached to the first spring and set into oscillation. We can obtain an oscillation of exactly the same angular frequency by attaching an object of mass _____ to the second spring. These two systems take the same time to complete a cycle of their motions.

An expression for the velocity of the object as a function of time can be found by differentiating the expression for $x(t)$ with respect to time. The result is

$$v(t) = $$

In terms of x_m and ω, the maximum speed of the object is _____. The object has maximum speed when its coordinate is $x = $ _____. The velocity of the object is zero when its coordinate is $x = $ _____ and also when its coordinate is $x = $ _____.

An expression for the acceleration of the object as a function of time can be found by differentiating $v(t)$ with respect to time. It is

$$a(t) = $$

In terms of x_m and ω, the magnitude of the maximum acceleration is _____. The mass has maximum acceleration when its coordinate is $x = $ _____ and also when its coordinate is $x = $ _____. This is when the force has maximum magnitude and the velocity vanishes. The acceleration is zero when the coordinate is _____. This is when the force vanishes and the speed is a maximum.

The argument of the trigonometric function, $\omega t + \phi$, is called the _____ of the oscillation and ϕ is called the _____.

For a given system, x_m and ϕ are determined by the initial conditions. Since $x(t) = x_m \cos(\omega t + \phi)$ the initial coordinate of the object is given by $x_0 =$ _____ and the initial velocity of the object is given by $v_0 =$ _____.

x_0 is positive for ϕ between _____ and _____.
x_0 is negative for ϕ between _____ and _____.
v_0 is positive for ϕ between _____ and _____.
v_0 is negative for ϕ between _____ and _____.

The equations $x_0 = x_m \cos \phi$ and $v_0 = -\omega x_m \sin \phi$ can be solved for x_m and ϕ. To obtain an expression for x_m solve the first equation for $\cos \phi$ and the second for $\sin \phi$, then use $\cos^2 \phi + \sin^2 \phi = 1$. To obtain an expression for ϕ, divide the second equation by the first and solve for $\tan \phi$. The results are

$x_m =$

$\phi =$

Be careful when you evaluate the expression for ϕ. There are always two angles that are the inverse tangent of any quantity but your calculator only gives the one closest to 0. The other is 180° or π radians away. Always check to be sure the values you obtain for x_m and ϕ give the correct initial coordinate and velocity. That is, $x_m \cos \phi$ must give the correct initial coordinate and $-\omega x_m \sin \phi$ must give the correct initial velocity. If they do not, add π to your answer for ϕ sand check again.

On the axes below draw a graph of $x(t) = x_m \cos(\omega t + \phi)$. Take ϕ to be 0 and x_m to have the value marked on the vertical axis. Draw the function so the period is T, as marked on the time axis. On your graph label the times for which magnitude of the velocity is a maximum, the times for which it is a minimum, the times for which the magnitude of the acceleration is a maximum, and the times for which it is a minimum.

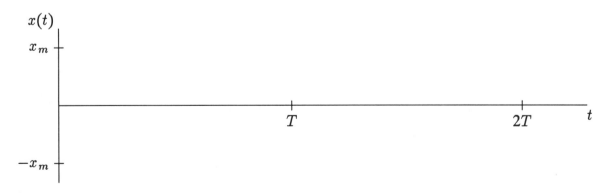

Energy considerations. An expression for the potential energy as a function of time can be found by substituting $x(t) = x_m \cos(\omega t + \phi)$ into $U = \frac{1}{2}kx^2$. The result is

$$U(t) =$$

An expression for the kinetic energy as a function of time can be found by substituting $v = -\omega x_m \sin(\omega t + \phi)$ into $K = \frac{1}{2}mv^2$. The result is

$$K(t) =$$

or, if $\omega^2 = k/m$ is used,

$$K(t) =$$

Both the potential and kinetic energies vary with time. As the object moves away from the equilibrium point it slows down; kinetic energy is converted to potential energy and stored in the extended spring. As the object moves toward the equilibrium point the stored potential energy is converted to kinetic energy and the object speeds up.

The potential energy is a maximum when the coordinate is $x =$ _____ and also when it is $x =$ _____. Then the speed of the object is $v =$ _____ and the kinetic energy vanishes. The kinetic energy is a maximum when the coordinate is $x =$ _____. Then the potential energy vanishes. Notice that the maximum kinetic energy has exactly the same value as the maximum potential energy.

Although both the potential and kinetic energies vary with time, the total mechanical energy $E = K + U$ is constant, as you can see by adding the two expressions you wrote above. If you know that the object has speed v when its coordinate is x, then you can use $E =$ _____ to compute the total mechanical energy. If you know the amplitude x_m of the oscillation then you can compute the total mechanical energy using $E =$ _____. If you know the maximum speed v_m of the object then you can compute the total mechanical energy using $E =$ _____. Since total mechanical energy is conserved these three expressions must have the same value.

Applications. Several other systems that oscillate in simple harmonic motion are discussed in the text. For each of them use one of the tables below to describe the displacement variable (linear or angular coordinate) that is oscillating and give its symbol. Then give an expression for the force or torque, as appropriate. For the torsional oscillator the torque is proportional to the angular displacement. For the pendula the force or torque is not proportional to the displacement unless the amplitude of the oscillation is small. Give both the exact expression and the small amplitude approximation. In all cases tell the physical meaning of the symbols that appear in the constant of proportionality that relates the force or torque and the displacement. Give the equation of motion obeyed by the displacement, in a form similar to that given above for a spring-mass system. Write the general form of the solution to the equation of motion and give the expression for the angular frequency ω and the period T in terms of properties of the system.

a. Torsional oscillator
   displacement variable: _____

   torque law: _____
   meaning of symbols: _____

   equation of motion: _____
   general solution: _____
   $\omega =$ _____ $T =$ _____

b. Simple pendulum
   displacement variable: _____

   torque law (exact): _____
   torque law (small amplitude): _____
   meaning of symbols: _____

   equation of motion: _____
   general solution: _____
   $\omega =$ _____ $T =$ _____

c. Physical pendulum
   displacement variable: _____

   torque law (exact): _____
   torque law (small amplitude): _____
   meaning of symbols: _____

   equation of motion: _____
   general solution: _____
   $\omega =$ _____ $T =$ _____

Remember that the small angle approximation $\sin\theta \approx \theta$ is valid only if θ is measured in radians. See the Mathematical Skills section for details.

Combinations of harmonic motions. When a particle moves around a circle with constant speed, the projection of its position vector on the x axis performs simple harmonic motion. Suppose the angular speed of the particle is ω and the radius of the circle is R. Put the origin of the coordinate system at the center of the circle and measure angles counterclockwise from the x axis. If the initial angular position of the particle is ϕ then the x component of its position vector is given by $x(t) =$ _____. You can identify the angular speed ω of the particle with the _____ of the oscillation and the radius of the circle with the _____ of the oscillation.

The projection of the particle's position vector on the y axis also performs simple harmonic motion. Since $y(t) = R\sin(\omega t + \phi) = R\cos(\omega t + \phi - \pi/2)$ the two coordinates are out of phase by $\pi/2$. Looked at another way, we can say that if a particle simultaneously performs simple harmonic motions with the same amplitude and frequency along two perpendicular axes and the motions are out of phase by $\pi/2$, then its trajectory is a _____. If the phases for the two motions are the same, the trajectory degenerates into a _____. If phases are not the same and the amplitudes are also unequal, the trajectory is an _____. If the frequencies are different the trajectory is more complicated. Look at Fig. 17 for some examples. These properties are often used to test for the equality of two frequencies, phases, or amplitudes.

Damped and forced harmonic motions. Many oscillating systems in nature are damped by a frictional force that is proportional to the velocity. Take the damping force to be $-b(\mathrm{d}x/\mathrm{d}t)$ and write the equation of motion obeyed by the displacement x of a mass m on a spring with force constant k:

$$\frac{\mathrm{d}^2 x}{\mathrm{d}t^2} =$$

If the damping coefficient b is small you may think of the motion as simple harmonic with an exponentially decreasing amplitude. The angular frequency is not quite the same as in the absence of damping and is now denoted by ω'. Give the solution in terms of the angular frequency ω' and the damping constant b:

$$x(t) =$$

Give an expression for the angular frequency in terms of k, m, and b:

$$\omega' =$$

Note that if b is small the angular frequency is nearly the same as the angular frequency for undamped motion, namely $\sqrt{k/m}$.

If $(b/2m)^2 < k/m$ then the mass oscillates with decreasing amplitude. If $(b/2m)^2 \geq k/m$ then the mass does not oscillate, but instead simply returns to its equilibrium point. The solution is then usually written in another form.

Sketch $x(t)$ on the graph below for $(b/2m)^2 < k/m$. Take the phase constant to be zero. Be sure to show both the oscillations and the decay of the amplitude.

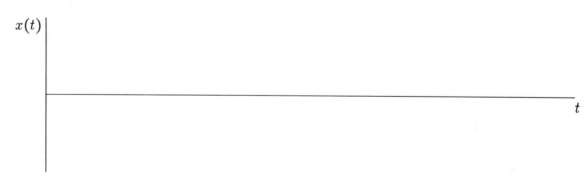

The total mechanical energy of a damped oscillator is not constant. As time goes on, energy is dissipated by _____.

Oscillations can be driven by applying an external sinusoidal force. Suppose the external force is given by $F_m \cos\omega''t$ and write the equation of motion for the displacement variable x, including a damping term proportional to dx/dt:

$$\frac{d^2x}{dt^2} =$$

After transients have been damped, the solution has the form

$$x(t) = \frac{F_m}{G}\sin(\omega''t - \phi),$$

where $G =$ _____ and $\phi =$ _____. Be careful to distinguish between the angular frequency ω'' of the external force and the natural angular frequency ω' of the oscillator. ω'' can be changed by changing the frequency of the driving force; ω' can be changed only by changing the oscillator itself: the mass, force constant or damping coefficient.

Note that the angular frequency of a forced oscillation is the same as the angular frequency of the driving force and is NOT necessarily the natural angular frequency of the oscillator. Also note that the amplitude of the oscillation depends on G and hence on the driving frequency. The driving frequency for which G is a minimum and the amplitude is a maximum is called the _____ frequency. If the damping coefficient b is small this frequency is very nearly _____.

The amplitude of the motion does not decay with time even though a resistive force is present and energy is being dissipated. The mechanism that drives the oscillator and provides the external force does work on the system and supplies the energy required to keep it going at constant amplitude.

Study Fig. 20 of the text to see how the amplitude depends on the driving frequency. For a given value of the damping coefficient b, the amplitude decreases as the driving frequency moves away from the resonance frequency in either direction. As b increases, the amplitude at resonance _____. Also the width of the curve, measured at an amplitude that is half the resonance amplitude, increases. Resonance becomes less sharp. This has important implications for the design of bridges and other structures that might resonate under the influence of cars, people, wind, or other external forces. It also has important implications for musical instruments, for tuning radios and television sets, for the emission of light from atoms, and in fact for almost every branch of science and engineering.

II. PROBLEM SOLVING

Some problems make use of the relationships between angular frequency, frequency, and period for simple harmonic motion: $\omega = 2\pi\nu$, $\nu = 1/T$, and $\omega = 2\pi/T$. Occasionally the period is given indirectly by describing a time interval. You must then know, for example, that the time the oscillator takes to go from maximum displacement in one direction to maximum displacement in the other direction is $T/2$ or the time it takes to go from maximum displacement to zero displacement is $T/4$. If these time intervals or others are given you should be able to calculate the period, frequency, and angular frequency.

PROBLEM 1. A block on the end of a spring is set into oscillation along a straight line. It takes 0.65 s to go from its greatest displacement in one direction to its greatest displacement in the opposite direction. What is the period, frequency, and angular frequency of the oscillation?

SOLUTION: The time for half a cycle is given so the period is $2 \times 0.65 = 1.30$ s. The frequency is given by $\nu = 1/T$ and the angular frequency by $\omega = 2\pi\nu$.

[ans: 1.30 s; 0.769 Hz; 4.83 rad/s]

Some problems depend on knowing the relationship between the angular frequency and the appropriate physical properties of the oscillating system: $\omega = \sqrt{k/m}$ for an undamped spring-mass system. All but one of quantities are given and you are asked to find the other.

PROBLEM 2. The mass of the block in the previous problem is 1.8 kg. What is the force constant of the spring?

SOLUTION:

[ans: 42.0 N/m]

A problem might give you an expression for the displacement as a function of time and ask for the amplitude, angular frequency, and phase constant (or related quantities). Simply identify the various constants in the given expression. It might also ask for the coordinate, velocity, and acceleration at some specific time. Simply evaluate the expression and its first and second derivatives.

PROBLEM 3. The coordinate of a mass on the end of a spring is given by $x(t) = 0.050\sin(30t + \pi/6)$, where x is in meters and t is in seconds. What is the amplitude, angular frequency, and phase constant of the oscillation?

SOLUTION:

[ans: 0.050 m; 30 rad/s; $\pi/6$ rad]

How long does it take the mass to travel from its greatest displacement to the equilibrium point?

SOLUTION: Calculate $T/4 = \pi/2\omega$.

[ans: .0524 s]

What is the maximum speed and maximum magnitude of the acceleration of the mass?

SOLUTION: Use $v_m = \omega x_m$ and $a_m = \omega^2 x_m$.

[ans: 1.5 m/s; 45 m/s^2]

What is the coordinate, velocity, and acceleration of the mass at $t = 0$?

SOLUTION: Use $x_0 = x_m \sin\phi$, $v_0 = \omega x_m \cos\phi$, and $a_0 = -\omega^2 x_m \sin\phi$.

[ans: 0.0433 m; -0.750 m/s; -39.0 m/s^2]

How long does it take the mass to travel from its starting position (at $t = 0$) to the equilibrium point?

SOLUTION: We cannot immediately tell how much time it takes in terms of the period so we must resort to a more general method: solve $x_m \sin(\omega t + \phi) = 0$ for t. There are many solutions, corresponding to values of $\omega t + \phi$ that differ by π radians. You want the smallest positive value of t.

[ans: 0.0873 s]

Some problems ask you to write an expression for $x(t)$. Since $x(t) = x_m \cos(\omega t + \phi)$, you first calculate x_m, ω, and ϕ from given information. Usually that information includes the initial conditions x_0 and v_0. Use $x_m^2 = x_0^2 + v_0^2/\omega^2$ to calculate x_m and $\tan\phi = -v_0/\omega x_0$ to calculate ϕ. Remember that x_m is positive and that you must choose one of the two solutions to $\tan\phi = -v_0/\omega x_0$. Check to be sure your choice for ϕ produces the given value for x_0 when substituted into $x_m \cos\phi$ and the given value for v_0 when substituted into $-\omega x_m \sin\phi$.

PROBLEM 4. A 1.8-kg block is attached to the end of a spring with a force constant of 200 N/m. It is set into oscillation by compressing the spring 2.3 cm from its equilibrium length and giving the block an initial velocity of 54 cm/s toward the equilibrium point. What is the angular frequency of the oscillation?

SOLUTION: Use $\omega = \sqrt{k/m}$.

[ans: 10.5 rad/s.]

What is the amplitude of the oscillation?

SOLUTION:

[ans: 5.62 cm]

If $x(t)$ is written $x_m \cos(\omega t + \phi)$ what is the phase constant ϕ?

SOLUTION: Substitute $x_0 = -2.3$ cm and $v_0 = 54$ cm/s into $\phi = \tan^{-1}(-v_0/\omega x_0)$. You will find two solutions, differing by π rad: -1.99 rad and $+1.15$ rad. If the first is used to compute x_0 and v_0 the results are $5.62 \cos(-1.99) = -2.3$ cm and $-10.5 \times 5.62 \sin(-1.99) = +54$ cm/s, in agreement with the given initial conditions. If the second is used the results are $5.62 \cos(1.15) = +2.3$ cm and $-10.5 \times 5.62 \sin(1.15) = -54$ cm/s, not in agreement.

[ans: -1.99 rad]

Write an expression for the coordinate of the block as a function of time.

SOLUTION:

[ans: $(5.62 \text{ cm}) \cos(10.5t - 1.99)$, for t in seconds]

Suppose the oscillation is started by compressing the spring 2.3 cm from its equilibrium length and giving the block an initial velocity of 54 cm/s away from the equilibrium point. What then is the coordinate as a function of time?

SOLUTION:

[ans: $(5.62 \text{ cm}) \cos(10.5t + 1.99)$, for t in seconds]

Some problems deal with the expressions for the maximum speed and acceleration.

PROBLEM 5. A block on a spring oscillates in such a way that its speed when it passes the equilibrium point is 2.6 m/s and its acceleration at the end points of its motion has a magnitude of 9.5 m/s². What is the angular frequency and amplitude of the oscillation?

SOLUTION: You should recognize that the speed given is the maximum speed and the acceleration given is the maximum acceleration. Solve $v_m = \omega x_m$ and $a_m = \omega^2 x_m$ simultaneously for ω and x_m.

[ans: 3.65 rad/s; 0.712 m]

If the maximum force of the spring on the block has a magnitude of 8.2 N what is the mass of the block and what is the force constant?

SOLUTION: Use $F_m = ma_m$ to find m and $F_m = kx_m$ or $\omega^2 = k/m$ to find k.

[ans: 0.863 kg; 11.5 N/m]

The total mechanical energy E, the speed v, and the coordinate x are related by $E = \frac{1}{2}mv^2 + \frac{1}{2}kx^2$. This equation can be solved for one of the quantities that appear in it. If the mass has speed v_1 when it is at x_1 and speed v_2 when it is at x_2 then conservation of energy yields $\frac{1}{2}mv_2^2 + \frac{1}{2}kx_2^2 = \frac{1}{2}mv_1^2 + \frac{1}{2}kx_1^2$.

PROBLEM 6. A 0.85 kg mass is attached to the end of a spring with a force constant of 280 N/m. It is started in oscillation by stretching the spring 12 cm from its equilibrium length and giving the mass a velocity of 2.2 m/s toward the equilibrium point. What is the total mechanical energy of the spring-mass system?

SOLUTION: Evaluate $E = \frac{1}{2}mv_0^2 + \frac{1}{2}kx_0^2$.

[ans: 4.07 J]

Once the total mechanical energy is known other quantities can be found. For example, what is the amplitude of the oscillation?

SOLUTION: When the mass is at its greatest displacement from the equilibrium point its speed is zero and the magnitude of its coordinate is the amplitude x_m. Thus $E = \frac{1}{2}kx_m^2$. Solve for x_m.

[ans: 17.1 cm]

What is the speed of the mass as it passes through the equilibrium point?

SOLUTION: At the equilibrium point $x = 0$ and $v = v_m$, the maximum speed. Thus $E = \frac{1}{2}mv_m^2$. Solve for v_m.

[ans: 3.10 m/s]

What is the period of oscillation?

SOLUTION: The angular frequency can be found directly from $v_m = \omega x_m$. Then use $T = 1/\nu = 2\pi/\omega$ to find the period.

[ans: 0.317 s]

Although we derived $v_m = \omega x_m$ using the expressions for $x(t)$ and $v(t)$, you should know that the same relationship is implied by the conservation of energy. Because energy is conserved, $\frac{1}{2}mv_m^2 = \frac{1}{2}kx_m^2$. Thus $k/m = v_m^2/x_m^2$. Since $k/m = \omega^2$, $\omega = v_m/x_m$.

Some problems deal with the other oscillating systems discussed in the text: the torsional oscillator, the simple pendulum, and the physical pendulum. In each case, you should know how the angular frequency depends on properties of the oscillating body: $\sqrt{\kappa/I}$ for a torsional oscillator, $\sqrt{g/\ell}$ for a simple pendulum, and $\sqrt{mgd/I}$ for a physical pendulum. If, for example, you are given the torque constant κ and the rotational inertia I for a torsional pendulum you should immediately realize you can calculate the angular frequency, frequency, and period.

PROBLEM 7. A uniform sphere with a radius of 10 cm is suspended from the ceiling by means of a 2.1-m long wire. If it swings as a small-amplitude pendulum, what is its angular frequency? What is its period?

SOLUTION: Use $\omega = \sqrt{mgd/I}$, where d is the distance from the center of mass to the point where the wire is attached to the ceiling. This distance is given by $d = \ell + R$, where ℓ is the length of the wire and R is the radius of the sphere. According to the parallel axis theorem the rotational inertia of the sphere about the point of attachment is given by $I = I_{cm} + md^2$, where $I_{cm} = (2/5)mR^2$ (see Fig. 9 of Chapter 12). Thus $\omega^2 = g(\ell + R)/[(2/5)R^2 + (\ell + R)^2]$, independently of the mass of the sphere. Evaluate this expression, then calculate the period.

[ans: 2.11 rad/s; 2.98 s]

Notice that the expression you derived above for ω^2 reduces to $\omega^2 = g/\ell$ when $\ell \gg R$. The system then behaves like a simple pendulum.

Suppose the same sphere, hung in the same way, oscillates as a torsional oscillator. What is the torque constant κ of the wire if the frequency is the same as when it oscillates as a physical pendulum?

SOLUTION: Use $\omega = \sqrt{\kappa/I}$. The rotational inertia is now $I = (2/5)mR^2$.

[ans: 0.0481 N·m/rad]

PROBLEM 8. Suppose the sphere of the last problem is started in torsional oscillation by giving it an initial angular velocity of 0.11 rad/s when it is at its equilibrium position. Write an expression for its angular position as a function of time.

SOLUTION: Write $\theta(t) = \theta_m \cos(\omega t + \phi)$ for the angular position of the sphere as a function of time. You already know ω (2.11 rad/s) but you need to find values for the angular amplitude θ_m and phase constant ϕ. The initial angular velocity is given by $d\theta/dt$, evaluated for $t = 0$. This is $-\omega\theta_m \sin\phi$. The initial angular position is $\theta_m \cos\phi$. Since this is zero ϕ is either $\pi/2$ or $3\pi/2$. Since $\sin(\pi/2) = 1$ and $\sin(3\pi/2) = -1$ the first choice leads to a negative initial angular velocity and the second leads to a positive initial angular velocity. Arbitrarily select the direction of positive rotation to be direction of the initial angular velocity. This means you take $\phi = 3\pi/2$.

Now the initial angular velocity is given by $\omega\theta_m$. Set this expression equal to 2.1 rad/s and solve for θ_m. Finally, substitute the values for θ_m, ω, and ϕ into the general expression for $\theta(t)$.

[ans: $\theta(t) = (0.0521 \text{ rad})\cos(2.11t + 3\pi/2)$ for t in seconds]

Some problems deal with oscillating systems that are not discussed in the text. Usually you are asked for the angular frequency or a related quantity. You must analyze the system using Newton's second law for translation or rotation, as appropriate. A generalization of the technique used in Section 15–5 can be employed. Go through the usual steps (free body diagram, identification of forces or torques, Newton's second law in component form) to find an expression for the second derivative of the displacement with respect to time. It should be proportional to the displacement itself, with a negative constant of proportionality that depends on properties of the system (force constant, rotational inertia, etc.). Equate the constant of proportionality to $-\omega^2$ and solve for ω. In some cases you may need to make the small angle approximation by replacing the sine of an angle with the angle in radians.

PROBLEM 9. Two springs, one with force constant k_1 and the other with force constant k_2, are fastened together to form a long composite spring. One end of the composite is attached to a wall and the other end is attached to a block of mass m on a frictionless horizontal surface, as shown. The block is pulled aside and released. In terms of m, k_1, and k_2, what is the angular frequency of the ensuing motion?

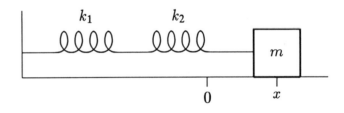

SOLUTION: We want to obtain an expression for the force acting on the block in terms of its displacement x from equilibrium. We hope the force is proportional to x and the constant of proportionality is negative. That is, if the mass oscillates in simple harmonic motion then $F = -Cx$ and we wish to find an expression for C in terms of k_1 and k_2. Then if C is positive the angular frequency of oscillation is $\sqrt{C/m}$.

Suppose spring 1 is extended by x_1 from its equilibrium length and spring 2 is extended by x_2 from its equilibrium length. Then the force on the block is $F = -k_2 x_2$ and the displacement of the block is $x = x_1 + x_2$. Now examine spring 2. Spring 1 pulls to the left with a force of magnitude $k_1 x_1$ and Newton's third law tells us that the block pulls to the right with a force of magnitude $k_2 x_2$. Since the mass of the spring is negligible, Newton's second law tells us that $k_1 x_1 = k_2 x_2$. Use $k_1 x_1 = k_2 x_2$ and $x = x_1 + x_2$ to eliminate x_2 from $F = -k_2 x_2$ in favor of x. You will find $F = -Cx$, where C is a positive constant that depends on k_1 and k_2. Now you can write the expression for the angular frequency.

[ans: $\sqrt{k_1 k_2/m(k_1 + k_2)}$]

PROBLEM 10. A wheel with radius R and rotational inertia I is free to rotate about a fixed horizontal axis through its center. A spring with force constant k is attached to the ceiling. A long inextensible wire is wrapped around the wheel and the free end is fastened to the spring. As the wheel rotates in either direction the force of the wire on the wheel is vertical and tangent to the wheel. What is the angular frequency of the spring-wheel system?

SOLUTION: The wheel is clearly a rotational oscillator. We wish to use Newton's second law for rotation to show that its angular acceleration is proportional to its angular displacement. If $\tau = -C\theta$, where C is a positive constant, then the angular frequency is $\sqrt{C/I}$. We need to find an expression for C in terms of R, I, and k.

Place the origin at the center of the wheel and take rotation in the counterclockwise direction to be positive. Measure the angular displacement of the wheel from its equilibrium orientation. Use Newton's second law for rotation: $\tau = I\alpha$. Suppose the wheel rotates through the angle θ, measured in radians. Then the spring expands by $R\theta$ and exerts the force $kR\theta$ upward. Thus $\tau = -kR^2\theta$. You should now be able to find expressions for the constant C and the angular frequency of oscillation.

[ans: $\sqrt{kR^2/I}$]

Take $R = 0.45$ m, $I = 0.75$ kg·m², $k = 200$ N/m, and evaluate this expression.

SOLUTION:

[ans: 7.35 rad/s]

III. MATHEMATICAL SKILLS

1. You need to know how to differentiate $\sin(\omega t + \phi)$ and $\cos(\omega t + \phi)$ to verify the solutions to several of the equations of motion in the chapter and to calculate the velocity and acceleration of an oscillating body. Remember how to use the chain rule. Let $\omega t + \phi = u$. Then $d\sin(\omega t + \phi)/dt = (d\sin u/du)(du/dt) = (\cos u)(\omega) = \omega\cos(\omega t + \phi)$. Similarly $d\cos(\omega t + \phi) = -\omega\sin(\omega t + \phi)$.

2. The derivative with respect to time of the function $e^{-bt/2m}\cos(\omega t + \phi)$ is required to verify the solution to the differential equation for a damped oscillator. Use the product rule:

$$\frac{d}{dt}\left[e^{-bt/2m}\cos(\omega t + \phi)\right] = \cos(\omega t + \phi)\frac{d}{dt}\left[e^{-bt/2m}\right] + e^{-bt/2m}\frac{d}{dt}\cos(\omega t + \phi)$$

$$= -\frac{b}{2m}e^{-bt/2m}\cos(\omega t + \phi) - \omega e^{-bt/2m}\sin(\omega t + \phi)$$

You will also need the second derivative:

$$\frac{d^2}{dt^2}\left[e^{-bt/2m}\cos(\omega t + \phi)\right] = e^{-bt/2m}\left\{\left[\frac{b^2}{4m^2} - \omega^2\right]\cos(\omega t + \phi) + \frac{\omega b}{m}\sin(\omega t + \phi)\right\}$$

3. Several of the oscillators discussed in this chapter are harmonic only if the amplitude is small. Simple and physical pendula are examples. For the motion to be considered harmonic the angle θ of swing must be sufficiently small that $\sin\theta$ may be replaced by θ in radians without generating unacceptable error.

The Maclaurin series for $\sin\theta$ is

$$\sin\theta = \sum_{n=0}^{\infty}(-1)^n \frac{\theta^{2n+1}}{(2n+1)!}$$

and its first three terms are

$$\sin\theta = \theta - \frac{\theta^3}{6} + \frac{\theta^5}{120} - \cdots$$

Notice that if θ is small each term in the series is less in magnitude than the previous term. The small angle approximation amounts to using only the first term of the series.

If θ is small the error generated by the small angle approximation is nearly the second term $\theta^3/6$ and the fractional error is roughly $\theta^2/6$. For example, the error is less than 1 per cent if $\theta^2/6 < 0.01$ or $\theta < 0.2$ radians. In fact the error for $\theta = 0.2$ radians is 0.7 per cent.

To check the approximation use your calculator to fill in the following table. Use $100(\theta - \sin\theta)/\sin\theta$ to compute the percent error.

θ (rad)	$\sin\theta$	percent error
1	_____	_____
0.1	_____	_____
0.01	_____	_____
0.001	_____	_____

IV. COMPUTER PROJECTS

Here you will use numerical integration to investigate the oscillation of a mass m on an ideal spring with force constant k. The acceleration of the mass is given by $a = -(k/m)x$, where x is the coordinate of the mass, measured from the equilibrium point. Recall that the motion is periodic with an angular frequency that is given by $\omega = \sqrt{k/m}$, a period that is given by $2\pi\sqrt{m/k}$, and an amplitude that is given by $x_m = \sqrt{x_0^2 + (v_0/\omega)^2}$, where x_0 is the initial coordinate and v_0 is the initial speed.

The first program discussed in the Computer Projects section of Chapter 6 can be used but the interval width Δt must be quite small and the number of intervals correspondingly large to avoid accumulating errors. The accuracy can be improved considerably by using more terms in the equations for the velocity and coordinate at the end of an interval. Including terms that are proportional to $(\Delta t)^2$, the velocity is given by $v_e = v_b + a_b\Delta t + \frac{1}{2}(da/dt)(\Delta t)^2$, where the derivative da/dt is evaluated for the beginning of the interval. Since $a = -(k/m)x$, $da/dt = -(k/m)v_b$. The derivative of the acceleration with respect to time is sometimes called the "jerk". Define $j_e = -(k/m)v_e$ and use $v_e = v_b + (\frac{1}{2}j_e\Delta t + a_e)\Delta t$ to calculate the velocity at the end of an interval. Some factoring has been carried out to save computational time and to reduce truncation errors. Similarly $x_e = x_b + v_b\Delta t + \frac{1}{2}a_b(\Delta t)^2$, which is best written $x_e = x_b + (\frac{1}{2}a_b\Delta t + v_b)\Delta t$. You must change appropriate lines of the program.

PROJECT 1. Consider a 2.0-kg mass attached to an ideal spring with force constant $k = 350\,\text{N/m}$. It is released from rest at $x = 0.070\,\text{m}$. Take the integration interval to be $\Delta t = 0.001\,\text{s}$ and plot the coordinate for every 0.05 s from $t = 0$ (the time of release) to $t = 1\,\text{s}$. For use later have the computer calculate the potential energy ($U = \frac{1}{2}kx^2$), the kinetic energy ($K = \frac{1}{2}mv^2$), and the total mechanical energy ($E = K + U$) for each point plotted.

Notice that the graph predicts an oscillatory motion. Use the graph to estimate the amplitude and compare the value with 0.070 m. Use the graph to estimate the period and compare the value with $2\pi\sqrt{m/k}$.

Also notice that at times when the spring has its greatest extension (0.070 m) the speed is 0, the kinetic energy is 0, the potential energy is maximum, and the magnitude of the acceleration is maximum. Also notice that when the spring is neither extended or compressed ($x = 0$) the speed is maximum, the kinetic energy is maximum, the potential energy is 0, and the acceleration is 0. Finally, notice that the total mechanical energy is constant. Your results may show some fluctuation after the third or fourth significant figure but this is due to computational errors.

The coordinate as a function of time can be written $x(t) = x_m \cos(\omega t + \phi)$, where ϕ is the phase constant. For any value of ϕ the initial coordinate is given by $x_0 = x_m \cos\phi$ and the initial velocity is given by $v_0 = -\omega x_m \sin\phi$. For the initial conditions you used above (x_0 positive and $v_0 = 0$) the phase constant is zero. Other initial conditions result in different values for the phase constant. If the amplitude is the same, a different phase constant produces a function $x(t)$ that is shifted along the time axis relative to the function for $\phi = 0$.

Suppose the phase constant is $\pi/6$ rad (30°). Take the amplitude to be $x_m = 0.070\,\text{m}$ and calculate the initial coordinate and velocity. Now use the program to plot $x(t)$ for the first second of the motion and find the time at which the first maximum ($x = 0.070\,\text{m}$) occurs. Since the first maximum occurs at $t = 0$ for $\phi = 0$, the result you obtain is the amount by which the plot is shifted.

A phase constant of 2π radians (360°) corresponds to a shift equal to one period of the motion. A phase constant of $\pi/6$ radians corresponds to one twelfth of a period. Does this agree with your result?

PROJECT 2. Suppose the mass is also subjected to a resistive force, proportional to its velocity. Then the acceleration is given by $a = -(k/m)x - (b/m)v$, where b is the drag coefficient. Take $m = 2.0\,\text{kg}$, $k = 350\,\text{N/m}$, $b = 2.8\,\text{kg/s}$, $x_0 = 0.070\,\text{m}$, and $v_0 = 0$. Use the computer program, with $\Delta t = 0.001\,\text{s}$, to plot the coordinate at intervals of 0.050 s from $t = 0$ (when the mass is released) to $t = 1.0\,\text{s}$. Also calculate the potential energy, the kinetic energy, and the total mechanical energy for these times. The rate of change of the acceleration is given by $j = -(k/m)v - (b/m)a$.

Notice that the amplitude decreases as time goes on. This decrease is related, of course, to the decrease in total mechanical energy. The resistive force does negative work on the mass. Plot the total mechanical energy as a function of time.

Does the drag force also change the period of the motion? Measure the period as the time between successive maxima and compare the result with $2\pi\sqrt{m/k}$.

PROJECT 3. The oscillator of the previous project is said to execute damped harmonic motion. The motion is said to be *underdamped* because the mass continues to oscillate. If the value of b is increased sufficiently no oscillations occur and the motion is said to be *overdamped*.

To see what underdamped motion is like, take $m = 2.0\,\text{kg}$, $k = 350\,\text{N/m}$, $b = 90\,\text{kg/s}$, $x_0 = 0.070\,\text{m}$, and $v_0 = 0$. Plot $x(t)$ every 0.050 s from $t = 0$ (when the mass is released) to $t = 1.0\,\text{s}$.

The motion of the mass is changed by an external force. Consider a force given by $F_m \cos(\omega'' t)$, where F_m and ω'' are constants. The angular frequency of the impressed force should not be confused with the natural angular frequency $\omega = \sqrt{k/m}$ of the mass and spring alone. The acceleration of the mass is now given by $a = -(k/m)x + (F_m/m)\cos(\omega''t)$ and its derivative is given by $j = -(k/m)v - (F_m\omega''/m)\sin(\omega''t)$. Change the program accordingly. The following project is designed to help you investigate forced oscillations without damping.

PROJECT 4. Take $m = 2.0$ kg, $k = 350$ N/m, $b = 0$, $x_0 = 0.070$ m, and $v_0 = 0$. The natural angular frequency is $\omega = \sqrt{350/2} = 13.2$ rad/s, corresponding to a period of 0.475 s. For each of the following impressed forces of the form $F_m \cos(\omega''t)$ plot $x(t)$ from $t = 0$ to $t = 1.0$ s. Use an integration interval of $\Delta t = 0.001$ s and a display interval of $\Delta t_d = 0.050$ s. (a.) $F_m = 18$ N, $\omega'' = 35$ rad/s (b.) $F_m = 18$ N, $\omega'' = 15$ rad/s.

For the conditions of part a, the motion is nearly sinusoidal with a period of about 0.47 s, the natural period. The influence of the impressed force is seen in deviations from a sinusoidal shape. When the impressed frequency is closer to the natural frequency, as in part b, the influence of the impressed force is more pronounced. The amplitude grows with time, then levels off. If the impressed frequency is made exactly equal to the natural frequency the amplitude grows without bound.

The increase in amplitude can easily be accounted for in terms of the work done by the impressed force. If ω'' is different from ω the impressed force is in the same direction as the velocity of the mass over some portions of the motion and in the opposite direction over other portions. If the frequencies are very different, as in part a, the net work done over a time period that is long compared with the period is almost zero. On the other hand, when the two frequencies are nearly the same, as in part b, the impressed force and velocity are in the same direction over a large portion of the motion and the amplitude grows with time. As time goes on the impressed force and velocity are in the same direction over smaller portions of the motion and in opposite directions over larger portions. Eventually the net work vanishes over a long time interval and the amplitude becomes constant.

You can easily verify these assertions by plotting the total mechanical energy as a function of time. Modify the program so it calculates and lists or plots the total mechanical energy every 0.05 s from $t = 0$ to $t = 2.0$ s. Note intervals during which the impressed force does positive work (the energy increases) and intervals during which it does negative work (the energy decreases).

PROJECT 5. Now investigate the motion when a damping force is present. Consider the same oscillator ($m = 2.0$ kg, $k = 350$ N/m) subjected to an impressed force with $F_m = 18$ N and $\omega'' = 35$ rad/s, but with the drag coefficient $b = 15$ kg/s. Use the same initial conditions ($x_0 = 0.070$ m, $v_0 = 0$) and integration interval ($\Delta t = 0.001$ s). Plot $x(t)$ every 0.050 s from $t = 0$ to $t = 2.0$ s.

Notice that motion starts out very much the same as when there is no damping. Its period is nearly the natural period. Now, however, this motion is quickly damped and what remains is a sinusoidal motion at the frequency of the impressed force. Verify that the period is about 0.18 s, corresponding to an angular frequency of 35 rad/s. The amplitude eventually becomes constant. The energy supplied by the impressed force is dissipated by damping.

Repeat the calculation for an impressed angular frequency of 15 rad/s, near the natural frequency. Note that the final amplitude is much larger than when the impressed frequency was far from the natural frequency.

A simple pendulum consists of a mass m at the end of a light rod of length ℓ, free to swing in a uniform gravitational field. The angular position is given by the angle $\theta(t)$, measured from the vertical. If g is the acceleration due to gravity then θ obeys the differential equation

$$\frac{d^2\theta}{dt^2} = -\frac{g}{\ell}\sin\theta.$$

You have learned that if θ is always small and is measured in radians then $\sin\theta$ may be replaced by θ itself and the solution to the resulting differential equation is $\theta(t) = \theta_m \sin(\omega t + \phi)$, where the angular frequency is given by $\omega = \sqrt{g/\ell}$.

Use a computer to investigate the motion when the amplitude is not small. You may want to change the program so it is written in terms of θ and its derivative Ω ($= d\theta/dt$): simply replace x with θ and v with Ω. The angular acceleration, which replaces a, is $\alpha = -(g/\ell)\sin\theta$ and its derivative is $j = -(g\Omega/\ell)\cos\theta$.

PROJECT 6. First use the program to check the small angle approximation. A simple pendulum with a length of 1.2 m is pulled aside 10° (0.175 rad) and released from rest. What is its period? Use an integration interval of 0.001 s and search for the second time after starting that Ω is zero. Compare the result with $2\pi\sqrt{\ell/g}$. [ans: 2.20 s]

Repeat the calculation for an initial angular displacement of 45° (0.785 rad). [ans: 2.29 s]

Repeat the calculation for an initial angular displacement of 75° (1.31 rad). [ans: 2.46 s]

Now check to see how close the motion is to simple harmonic. Take the initial angular displacement to be 75° and use the program to generate a table of $\theta(t)$ for the first 2.5 s of the motion. You might obtain values at intervals of 0.1 s. At the same time have the program generate values of $\theta_m \cos(\omega t)$, where the value of ω is calculated from the value you found for the period. Do you think the actual motion is simple harmonic?

The conversion of kinetic energy to potential energy and back again to kinetic energy is quite similar to the conversion that takes place for a spring-mass system, although the mechanism is the work done by the force of gravity, not the force of a spring. Suppose the pendulum is released from rest with an initial angular displacement of 75°. For each of the displayed points have the computer calculate the kinetic energy per unit mass ($\frac{1}{2}\ell^2\Omega^2$), the potential energy per unit mass [$g\ell(1-\cos\theta)$], and their sum.

V. NOTES

Chapter 16
GRAVITATION

1. BASIC CONCEPTS

You will learn about the force law that describes the gravitational attraction of two particles for each other. By considering the attraction of every particle in one object for every particle in a second object the law is extended to deal with objects containing many particles. Then the law is applied to the motion of planets and satellites. Here you will bring to bear some of the concepts you studied in earlier chapters, chiefly Newton's second law and the laws of energy and momentum conservation. The ideas of a force field and potential, discussed here, will be valuable to you not only in the study of gravitation but also in the study of electricity and magnetism, later.

Newton's law of gravity. According to Newton's law of gravity, two particles with masses m_1 and m_2, separated by a distance r, each attract the other with a force of magnitude

$$F = $$

where the universal gravitational constant G has the value $G = $ _____ N·m^2/kg^2. Let $\mathbf{r}_{12}$ be the displacement vector from particle 1 to particle 2. In vector notation the gravitational force exerted by particle 1 on particle 2 is given by

$$\mathbf{F}_{21} = $$

Note that the force obeys Newton's third law: the force of particle 1 on particle 2 has the same magnitude as the force of particle 2 on particle 1, but is in the opposite direction. Also note that the magnitude of the force is inversely proportional to the square of the particle separation. If that distance is doubled the gravitational force is one fourth as great.

Experimental evidence indicates that the masses in the law of gravity are identical to those in Newton's second law of motion. If the gravitational force exerted by particle 1 is the only force acting on particle 2 then the acceleration of particle 2 is given by $\mathbf{a} = $ _____, an expression that depends on m_1 but is independent of m_2.

Two very important consequences of Newton's law of gravity are proved in Section 16–5. Both are concerned with the force of gravity exerted by a uniform spherical shell of mass M on a particle of mass m. In the first case the particle is *outside* the shell, a distance r from the shell center. Then the magnitude of the force of the shell on the particle is given by

$$F = $$

The direction of the force is _____. In the second case the particle is *inside* the shell. Then the magnitude of the force of the shell on the particle is given by $F = $ _____, no matter where the particle is located inside.

The first of these theorems can be used to show that the force of attraction of two objects with spherically symmetric mass distributions is given by an equation that is identical to Newton's law of gravity for particles, provided we interpret r as _____.

The force of gravity of the earth on a particle of mass m at its surface can be written $\mathbf{F} = m\mathbf{g}$, where $\mathbf{g}$ is the acceleration due to gravity at the earth's surface. Equating this expression to the force given by Newton's law yields an expression for $\mathbf{g}$ in terms of the mass M_e and radius R_e of the earth:

$$\mathbf{g} = $$

This expression can be used to calculate the mass of the earth, once g, R_e, and G are known. A similar expression (with R_e replaced by the distance between the center of the earth and a particle) can be used to estimate the acceleration due to gravity at any given distance above the earth's surface. This calculation is an estimate because we have made the assumption that the earth has a spherically symmetric mass distribution, an assumption that is not precisely valid. If the mass of another spherical object is known it can be used to estimate the acceleration due to gravity a given distance from the center. See Sample Problem 3 of the text.

The theorems can also be used to find an expression for the gravitational force on a particle located *within* a spherically symmetric mass distribution. Consider a particle somewhere inside the earth, a distance r from the center. If the earth's mass distribution were spherically symmetric, we could calculate the force of the earth on the particle by considering only that part of the earth's mass that is less than r from the center. All parts of the earth attract the particle but the forces due to parts further from the center than the point mass sum to zero.

To test your understanding, assume the earth's mass is uniformly distributed and use the space below to show that the magnitude of the force on the particle is given by $(GM_em/R_e^3)r$, where M_e is the mass of the earth, m is the mass of particle, and r is the distance from the center of the earth to the particle. First show that the mass inside r is M_er^3/R_e^3, then substitute this expression for one of the masses in Newton's law of gravity.

Notice that the force on a particle at the center of the earth is zero.

In many cases the dimensions of an extended body are small enough that the acceleration due to gravity is very nearly uniform over the body. Then, for purposes of studying the motion of the body, we may treat it as a particle located at the center of mass. Thus objects, including satellites in the earth's gravitational field and planets in the sun's gravitational field, are discussed in the text as if they were particles.

Gravitational potential energy. The force of gravity is a conservative force. Go back to Chapter 8 and review the tests for conservative forces. In general, when a system changes configuration how does the work done by a conservative force depend on the paths taken by parts of the system? _____

Since the gravitational force is conservative a potential energy function is associated with it. If two point masses m and M are separated by a distance r, their gravitational potential energy is given by

$$U(r) =$$

where the potential energy was taken to be zero for infinite separation. This expression also gives the potential energy of two non-overlapping bodies with spherically symmetric mass distributions. Then r is interpreted as the separation of their _____. It is important to recognize that this energy is associated with the *pair* of masses, NOT with either mass alone.

As you know, the potential energy is a scalar. To calculate the gravitational potential energy of a system of particles, sum the potential energies of each *pair* of particles in the system. The total potential energy of a collection of particles is the work done by an external agent to assemble the particles from the reference configuration. The particles start from rest and are placed at rest in their final positions. If the particles are brought from infinite separation the external agent must do negative work since the mass already in place attracts any new mass being brought in and the agent must pull back on it.

If the particles of a system interact only via gravitational forces and no external forces act, then the total mechanical energy of the system is conserved. The sum of the kinetic energies of all the particles and the total gravitational potential energy remains the same as the particles move.

For a two-particle system the total mechanical energy consists of three terms, corresponding to the kinetic energy of each particle and the potential energy of their interaction. Suppose that at some instant particle 1 (with mass m_1) has speed v_1, particle 2 (with mass m_2) has speed v_2, and the particles are a distance r apart. Then the total mechanical energy of the system is given by

$$E =$$

At a later time v_1, v_2, and r may be different but the sum of the three terms will have the same value. You can equate the expressions for the energy in terms of the speeds and separation at two different times and use the resulting equation to solve for one of the quantities that appear in the equation.

If we consider a satellite in orbit around the earth or a planet in orbit around the sun, then one body is much more massive than the other. We may usually place the origin at the center of the more massive body and take its velocity to be zero. Then the total mechanical energy is the sum of two terms, the kinetic energy of the less massive body and the gravitational potential energy.

Suppose an earth-satellite with mass m (much less than that of the earth) has an initial speed v_0 when it is a distance r_0 from the earth's center and has speed v when it is a distance

r from the earth's center. Write the conservation of energy equation in terms of m, M_e, r_0, v_0, r, and v:

This equation can be used to solve for any one of the quantities in it, given the others.

The escape speed is the initial speed that an object must be given at the surface of the earth (or other large mass) in order to _____.

To use the conservation of energy equation to calculate the escape speed for an object on the earth's surface, set $r_0 =$ _____, $r =$ _____, and $v =$ _____, then solve for v_0. The result is: $v_0 =$ _____.

What other conservation principles are valid for the gravitational interaction of two spherically symmetric bodies? _____

Gravitational field and potential. We may think of a mass as creating a gravitational field in the space around it. This field then exerts a force on any other mass in the vicinity. Two important concepts are associated with a gravitational field: the field strength and the gravitational potential. Both are defined in terms of a test particle with mass m_0, which we can move around in the gravitational field. As we do, we measure the gravitational force on it and the gravitational potential energy.

The *gravitational field strength* **g** at any point in space is defined as the gravitational force per unit mass on the test particle when it is at the point. If m_0 is the mass of the test particle and **F** is the gravitational force on it then **g** = _____. This is, of course, just the acceleration due to gravity at the position of the test mass. Once the field at a point is known, we can calculate the gravitational force on any mass M using **F** = _____.

If the gravitational field is due to a single point mass m, then according to Newton's law of gravity the magnitude of the field strength at a point a distance r from it is given by

$$g =$$

You should note carefully that the field is due to m, NOT m_0.

Similarly, the *gravitational potential* at any point is the gravitational potential energy per unit test mass associated with the interaction between the gravitational field and the test mass when the test mass is at that point. If m_0 is the mass of the test particle and U is the gravitational potential energy then $V =$ _____ is the gravitational potential. If a mass M is placed at a point where the potential is V, the gravitational potential energy is $U =$ _____.

The gravitational potential a distance r from a point mass m is given by

$$V(r) =$$

where the potential was taken to be zero at points far from the mass.

Carefully note that the field strength is a vector and the gravitational potential is a scalar. To find the field strength of a collection of point masses you evaluate the vector sum of the individual field strengths. To find the gravitational potential of a collection of point masses you algebraically sum the individual potentials.

If the gravitational potential $V(x, y, z)$ is known at all points in an extended region of space, the field strength in that region can be calculated by evaluating its derivatives. More precisely,

$$g_x = \underline{\hspace{2cm}} \qquad g_y = \underline{\hspace{2cm}} \qquad g_z = \underline{\hspace{2cm}}$$

If a mass m is at the origin then the gravitational potential at a point with coordinate x on the x axis is $V = Gm/x$. The negative of the derivative is given by $dV/dx = \underline{\hspace{2cm}}$. The only component of the gravitational field that does not vanish is the x component and it is $g_x = \underline{\hspace{2cm}}$, just as Newton's law of gravity predicts.

Planetary motion. The motions of planets are controlled by gravity, due chiefly to the sun. Here we consider only the gravitational force of the sun and neglect the influence of other planets. The motion of a planet is then at least partially described by Kepler's three laws. State the laws in words and for each of them tell what property of the gravitational interaction (as described by Newton's law of gravity) is important for the validity of the law.

1. Law of orbits
   statement of law: _____

   property of the force law: _____

2. Law of areas
   statement of law: _____

   property of the force law: _____

3. Law of periods
   statement of law: _____

   property of the force law: _____

   A planetary orbit can be described by its semimajor axis, which is _____

   and its eccentricity, which is _____

The point of closest approach to the sun is called the _____ and, in terms of the semimajor axis a and eccentricity e, the distance of this point from the sun is given by $R_p = \underline{\hspace{2cm}}$. The point of maximum distance from the sun is called the _____ and this distance is given by $R_a = \underline{\hspace{2cm}}$. For circular orbits $e = \underline{\hspace{1cm}}$ and $R_a = R_p = a$. An eccentricity of nearly 1 corresponds to an ellipse that is much longer than it is wide. The same expressions are valid for a satellite in earth orbit but the point of closest approach is called the _____ and the point of maximum distance is called the _____.

Chapter 16: Gravitation

An elliptical orbit for a planet is shown to the right. Label the position of the sun, the semimajor axis, aphelion, and perihelion.

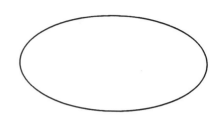

The law of areas provides us with a means of relating a planet's speed at one point to its speed at another point. The simplest relationship holds for points that are the greatest and least distances from the sun because at these points the velocity is perpendicular to the position vector from the sun.

In terms of the mass m of the planet, distance r from the sun, and speed v of the planet, the magnitude of the angular momentum at one of these points is given by $\ell = $ _____, where the origin was placed at the sun. Let R_p and v_p be its distance and speed at perihelion and let R_a and v_a be its distance and speed at aphelion. Then conservation of angular momentum leads to the equality: _____.

For a circular orbit the period T is related to the radius r by

$$T^2 = $$

where M is the mass of the central body (the sun). This expression can be used, for example, to calculate the mass of the sun. The equation is also valid for elliptical orbits if we replace r by the semimajor axis a.

You should recognize that asteroids and recurring comets in orbit around the sun and satellites (including the moon) in orbit around the earth or another planet also obey Kepler's laws. When the two bodies have comparable mass, as for example the two stars in a binary star system, each travels in an elliptical orbit around the _____.

Consider a body of mass m in a circular orbit about a much more massive sun (mass M), essentially at rest. The speed of the body and the energy of the system are closely related to the radius of the orbit. Gravity produces a centripetal force of magnitude $F = GmM/r^2$ and this must equal mv^2/r, so $v = $ _____. You can use this equation to find the speed of a body in a given circular orbit.

As a function of the orbit radius the kinetic energy is $K = \frac{1}{2}mv^2 = $ _____. If we take the potential energy to be zero for infinite separation then the potential energy for an orbit of radius r is $U = $ _____ and the total mechanical energy, as a function of r, is $E = $ _____. That the total energy is negative indicates that the system is bound. The orbiting body does not have enough kinetic energy to escape. This expression for the total energy is valid for elliptical orbits if we replace r with a, the semimajor axis. You should be aware that the kinetic energy is NOT given by $GmM/2a$ for non-circular orbits but the total energy is given by $-GmM/2a$.

To emphasize these ideas, consider a planet of mass m in an elliptical orbit with semimajor axis a, around a sun of mass M, essentially at rest. No matter where the planet is in its orbit, the total energy is given by $E = $ _____. When it is a distance r from the sun, the potential energy is given by $U = $ _____, and the kinetic energy is given by $K = E - U = GmM(1/r - 1/2a)$. This expression can be used to find the speed of the planet if M, r, and a are given.

As this discussion indicates, the speed of the orbiting body cannot be changed without changing the semimajor axis. Carefully review Sample Problem 11 of the text to see how a change in speed alters the orbit.

II. PROBLEM SOLVING

Some problems are straightforward applications of Newton's law of gravity. The law relates the force between two particles to their masses and separation. Three of these quantities are given and you are asked for the third. In some situations you are asked for the force of two or more masses on another. Then you must evaluate the vector sum of the individual forces. Find the cartesian components of each force and add corresponding components, with their correct signs. In some variants you are given the force and asked for the position of one of the masses. Then simply treat its coordinates as unknowns.

PROBLEM 1. What is the magnitude of the gravitational force exerted by the earth on the sun?

SOLUTION: According to data in an appendix of the text the mass of the sun is $M_s = 1.99 \times 10^{30}$ kg, the mass of the earth is $M_e = 5.98 \times 10^{24}$ kg, and their average separation is 149.9×10^6 km. Substitute these values into $F = GM_e M_s/r^2$.

[ans: 3.55×10^{22} N]

PROBLEM 2. Two 1.0-kg uniform balls are at diametrically opposite corners of a square with 1.0-m edges. What is the magnitude of the gravitational force exerted by one ball on the other?

SOLUTION:

[ans: 3.34×10^{-11} N]

Suppose a third identical ball is placed at another corner of the square. What is the gravitational force on it?

SOLUTION: Use vector addition.

[ans: 9.43×10^{-11} N, toward the center of the square]

Where must a fourth identical ball be placed so the total gravitational force on the third is zero?

SOLUTION: Clearly the fourth ball must be placed on the diagonal of the square through the third ball since the force it exerts must be along this line. Use $F = Gm^2/r^2$ to find the distance r for which the force of the fourth ball cancels the force of the other two balls.

[ans: on the diagonal that passes through the third ball, extended to a point outside the square, 0.897 m from the third ball]

Some problems ask for the gravitational force exerted on a point mass by a spherically symmetric distribution of charge. You use Newton's law, but for one of the masses you substitute the mass that is inside an imaginary spherical surface that passes through the point mass. Your first job is to decide what that mass is. For a shell, all the mass is either inside or all is outside the imaginary sphere. For a solid sphere of radius R, all the mass is inside if $r > R$, where r is the distance of the point mass from the sphere center. If $r < R$ and the mass density is uniform then the fraction of total mass that is inside is given by the ratio of the cubes of the radii: $M_{\text{inside}} = Mr^3/R^3$. If the mass density is not uniform you may need to resort to evaluating an integral to find the mass inside. See the Mathematical Skills section.

PROBLEM 3. A rocket is shot straight up from the surface of the earth. At what altitude is the gravitational force on it half the force at launch?

SOLUTION: The rocket is always outside the earth so the force on it is given by GM_eM_r/r^2, where M_e is the mass of the earth and M_r is the mass of the rocket. If R_e is the radius of the earth and h is the altitude of the rocket, then you want to solve $GM_eM_r/(R_e+h)^2 = GM_eM_r/2R_e^2$ for h. Take the value of R_e from an appendix of the text.

[ans: 2.64×10^6 m]

PROBLEM 4. In a science fiction story a narrow tunnel is drilled completely through the earth along a diameter. Assume the earth is a sphere with uniform mass density and show that a particle dropped in the tunnel will oscillate in simple harmonic motion. Find an expression for its angular frequency in terms of the mass and radius of the earth. What is its period?.

SOLUTION: Suppose the particle is a distance r from the center of the earth, in the tunnel. In the Basic Concepts section you showed that the magnitude of the gravitational force on the particle is given by GM_emr/R_e^3, where m is the mass of the particle, M_e is the mass of the earth, and R_e is the radius of the earth. The magnitude of the acceleration of the particle is given by $a = F/m = GM_er/R_e^3$ and **a** is toward the center of the earth, so r obeys the equation $d^2r/dt^2 = -GM_er/R_e^3$. This has the same form as the equation for a mass on a spring:

$d^2x/dt^2 = -(k/m)x$. We conclude that r is a sinusoidal function of the time and that the angular frequency ω is given by $\omega^2 = GM_e/R_e^3$.

Use data from an appendix of the text to evaluate this expression for ω. Use $T = 2\pi/\omega$ to calculate the period.

SOLUTION:

[ans: 1.24×10^{-3} rad/s; 5.06×10^3 s (1.40 h)]

PROBLEM 5. The mass density of a certain sphere increases in proportion to the distance from the sphere center. If the total mass of the sphere is M_s and its radius is R what is the gravitational force on a particle of mass m inside the sphere at a distance r from the sphere center?

SOLUTION: The volume of a spherical shell of width dr is $4\pi r^2\, dr$. The mass of such a shell is $dM = 4\pi r^2 \rho\, dr$, where ρ is the mass density. Write $\rho = Ar$, where A is a constant of proportionality. Then $dM = 4\pi A r^3\, dr$. If the shell is inside the particle it exerts a force of magnitude $dF = Gm\, dM/r^2 = 4\pi AGmr\, dr$; if it is outside it does not exert a force. Thus the total force on the particle has magnitude $F = 4\pi AGm \int_0^r r\, dr = 2\pi AGmr^2$. Notice that the integral extends only out to the position of the particle.

You now need to find an expression for the constant A in terms of given quantities. The total mass of the sphere is $M_s = 4\pi \int_0^R \rho r^2\, dr = 4\pi A \int_0^R r^3\, dr$. Evaluate the integral and solve for A. Substitute the result into the expression for F.

[ans: $2GM_s mr^2/R^4$]

Notice that the force is quite different from that of a uniform sphere. In particular, it is proportional to the square of r, rather than to r itself.

You can also find the force exerted on a point mass by a sphere with a spherical cavity, even if the cavity is off-center. First suppose the cavity is filled with material of the same mass density as the sphere and calculate the force exerted by the filled sphere. It is directed toward the center of the sphere. Then calculate the force exerted by the mass filling the cavity. It is directed toward the center of the cavity. Vectorially subtract the second force from the first. This technique is valid no matter what the location of the point mass: outside the sphere, inside the sphere, or inside the cavity. If the sphere is uniform (except for the cavity) its mass density can be found by dividing its mass by its volume, the difference between the volume of a solid sphere and the volume of the cavity.

Chapter 16: Gravitation **239**

PROBLEM 6. A 5.0-m radius ball has a 1.5-m radius spherical cavity with its center 2.0 m from the center of the ball, as shown in cross section. The ball has a mass of 25 kg, uniformly distributed. What is the gravitational force on a 1.0-kg particle located at P, 6.0 m from the center of the ball along the line joining the center of the ball and the center of the cavity?

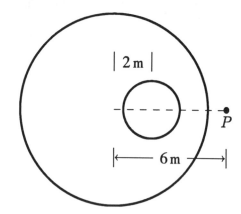

SOLUTION: Let R_b be the radius of the ball, R_c be the radius of the cavity, a be the distance from the center of the ball to the center of the cavity, and r be the distance from the center of the ball to the particle. Suppose the cavity is filled uniformly with material that has the same mass density as the ball. Let M_f be the mass the ball has with the cavity filled and let M_c be the mass of the material that fills the cavity.

The particle is outside the ball so the force exerted by the ball with its cavity filled is $F_f = GmM_f/r^2$. The distance from the center of the cavity to the particle is $r - a$ so the material in the cavity alone exerts the force $F_c = GmM_c/(r-a)^2$. Both these forces are toward the center of the ball so the force exerted by the ball when the cavity is not filled is the difference: $F = F_f - F_c = Gm\left[M_f/r^2 - M_c/(r-a)^2\right]$.

You must now find M_f and M_c in terms of the mass M of the ball when its cavity is empty. This is done most simply using ratios of volumes. The volume of the ball with its cavity empty is proportional to $R_b^3 - R_c^3$, the volume of the sphere with its cavity filled is proportional to R_b^3, and the volume of the cavity is proportional to R_c^3. Thus $M_f = MR_b^3/(R_b^3 - R_c^3)$ and $M_c = MR_c^3/(R_b^3 - R_c^3)$. Evaluate these expressions and use the results in the expression for the force.

[ans: 4.47×10^{-11} N]

What is the gravitational force on a 1.0-kg particle at the center of the ball?

SOLUTION:

[ans: 1.16×10^{-11} N, away from the center of the cavity]

Some problems deal with the gravitational field or, what is the same, the acceleration due to gravity. For a point mass m use $\mathbf{g} = -(Gm/r^3)\mathbf{r}$, where $\mathbf{r}$ gives the position of the field point relative to the mass. For a collection of point masses evaluate the vector sum of the fields due to the individual masses. For a spherically symmetric mass distribution find an expression for the gravitational force on a test mass and divide by the test mass.

PROBLEM 7. If the acceleration due to gravity at the surface of a certain planet is 10.4 m/s² what is it at a distance above the surface equal to one half the radius of the planet?

SOLUTION: Use the ratio $g_h/g_s = R^2/(R+h)^2$, where R is the radius of the planet and h is the distance above its surface. Set $h = R/2$ and solve for g_h.

[ans: 4.62 m/s²]

PROBLEM 8. What is the gravitational field inside a uniform sphere of mass M, at a point with position vector **r** relative to the center of the sphere?

SOLUTION: In the Basic Concepts section you derived an expression for the magnitude of the force on a particle of mass m. Now assume the particle is a test mass and divide by m to obtain the magnitude of the field. The direction is radially inward since the test mass is attracted toward the center of the sphere.

[ans: $-(GM_e/R_e^3)\mathbf{r}$]

What is the gravitational field at the point inside the sphere of Problem 5 with position vector **r** relative to the center of the sphere?

SOLUTION:

[ans: $-(2GM_s/R^4)r\mathbf{r}$]

The acceleration due to gravity enters into many of the calculations you performed while studying earlier chapters: the period of a pendulum, the normal force of a surface on an object, and the acceleration of a block sliding down an inclined plane, to mention a few instances. The settings of these and other problems may not be at the surface of the earth. Simply work these problems as you did before but use the value of g appropriate to the location. A simple example follows.

PROBLEM 9. Near the surface of the earth, where $g = 9.8$ m/s², a block suspended from the ceiling by means of a spring comes to rest with the spring extended 3.0 cm from its equilibrium length. If the same experiment were carried out at a location 2 earth radii above the earth's surface what would the spring extension be? Assume the ceiling at both locations is at rest relative to an inertial frame.

SOLUTION:

[ans: 0.333 cm]

Chapter 16: Gravitation

Some problems might require you to find the potential energy of a collection of point masses. The problem might be phrased in terms of the work required to assemble the masses from infinite separation or to remove them to infinite separation. In any case identify all the possible *pairs* of masses and sum the potential energies of the pairs. If there are 4 masses, for example, the pairs are 1–2, 1–3, 1–4, 2–3, 2–4, and 3–4. For each pair find the separation r_{ij} of the two masses, then evaluate $-Gm_im_j/r_{ij}$ for the potential energy of the pair. Finally, sum all the contributions to find the total potential energy U. As the system is assembled gravity does work $-U$ and an external agent does work $+U$. As the system is disassembled gravity does work $+U$ and the agent does work $-U$. U itself is negative.

PROBLEM 10. Four particles, each with mass m, are held at the corners of a square with edge length a. Derive an expression for the gravitational potential energy of the system, assuming it to be zero for infinite separation of the masses.

SOLUTION: There are six pairs of particles in the system. For four pairs the masses are a distance a apart and the contribution of each pair to the total potential energy is $-Gm^2/a$. For the other two pairs the masses are at diametrically opposite corners of the square, a distance $\sqrt{2}a$ apart, and each pair contributes $-Gm^2/\sqrt{2}a$. Sum all these contributions.

[ans: $-(Gm^2/a)(4 + \sqrt{2}) = -5.41 Gm^2/a$]

If $m = 2.5$ kg and $a = 3.2$ cm what is the potential energy of the system?

SOLUTION:

[ans: -7.05×10^{-8} J]

As the system is assembled from infinite separation the gravitational forces of the masses on each other do a total of $+7.05 \times 10^{-8}$ J of work and the external agent does -7.05×10^{-8} J of work.

You might be asked for the initial speed that one of the masses must be given so it barely escapes from the distribution, while the other masses remain at rest. After the mass is removed it is infinitely far from the other masses and its kinetic energy is zero. The initial energy is the potential energy of the system plus the kinetic energy of the mass to be removed. The final energy is simply the potential energy of the system with the mass removed. Use conservation of energy to find the initial kinetic energy of the mass that is removed.

PROBLEM 11. Consider the system composed of four identical masses held at the corners of a square, as described in the last problem, and suppose one of the masses is simultaneously released and given an initial speed. What should that speed be for the mass to escape from the gravitational attraction of the other three?

SOLUTION: The initial potential energy U_i is the potential energy when all four masses are at corners of the square and the initial kinetic energy K_i is $\frac{1}{2}mv^2$, where v is the initial speed of the mass that is released. The final potential energy U_f is the potential energy when there are only three masses at the corners. Calculate this quantity, then solve $U_i + \frac{1}{2}mv^2 = U_f$ for v.

[ans: 1.68×10^{-4} m/s]

Suppose that, when all four masses are at the corners of the square, one of them is released from rest. It will be accelerated toward the center of the square. What is its speed when it gets to the center?

SOLUTION: Use $U_i = U_f + \frac{1}{2}mv^2$, where U_i is the potential energy when all masses are at corners of the square and U_f is the potential energy when one mass is at the center of the square.

[ans: 1.27×10^{-4} m/s]

What work must be done by an external agent to carry one of masses from a corner to the center of the square, with identical masses at the other three corners?

SOLUTION: Use $W = U_f - U_i$.

[ans: -2.00×10^{-8} J]

PROBLEM 12. A rocket is launched straight up from the surface of the earth. When its altitude is one fourth an earth radius its fuel runs out and thereafter it coasts. What should it speed be when it starts to coast if it is to escape from the gravitational pull of the earth?

SOLUTION: Its distance from the center of the earth when its fuel runs out is $1.25R_e$, where R_e is the radius of the earth. The gravitational potential energy is then $-GmM_e/1.25R_e$, where m is the mass of the rocket and M_e is the mass of the earth. The kinetic energy of the rocket is $\frac{1}{2}mv^2$, where its speed v is the unknown in the problem. If the rocket barely escapes then its kinetic energy decreases to zero as its distance from the center of the earth increases without bound. Take $K_f = 0$ and $U_f = 0$. Now use conservation of energy to solve for v. Notice that the mass of the rocket does not influence the result.

[ans: 1.00×10^4 m/s]

How does this compare with the speed the rocket needs to escape if it starts coasting immediately after lift-off?
SOLUTION:

[ans: about 11% less]

You may be asked to compute the gravitational potential for a given configuration of mass. If the system consists of a collection of particles, simply sum the contributions of the individual particles. The contribution of a particle of mass m is $-Gm/r$, where r is the distance from the particle to the point for which the potential is sought. This calculation assumes the zero of the potential is infinitely far from all the particles.

For a continuous distribution of mass the calculation is more complicated. You calculate the work W done by the gravitational field as a test mass is brought from the reference point, where the potential is zero, to the point for which the potential is sought. This might require you to evaluate an integral. The potential is then given by $V = -W/m_0$. Here is an example.

PROBLEM 13. For the sphere of Problem 5 what is the gravitational potential at a point inside, a distance r from the center, if the potential is zero at the center?

SOLUTION: Place the origin at the center of the sphere and let $\mathbf{r}$ be the position vector of the test mass. According to the solution of Problem 5 the force on it is given by $\mathbf{F} = -(2GM_s m_0 r/R^4)\mathbf{r}$, where M_s is the mass of the sphere. Calculate the work done by the field as the test mass moves from the center of the sphere to a point a distance r from the center. Take the path to be along a radius and let $d\mathbf{s}$ be an infinitesimal displacement along this path, pointing outward from the sphere center. $\mathbf{F}$ and $d\mathbf{s}$ are in opposite directions so $\mathbf{F} \cdot d\mathbf{s} = -F\,dr$. Thus $W = \int \mathbf{F} \cdot d\mathbf{s} = -\int_0^r F\,dr = -(2GM_s m_0/R^4)\int_0^r r^2\,dr$. Evaluate the integral, then divide $-W$ by m_0.

[ans: $2GM_s r^3/3R^4$]

Check your answer by differentiating $U(r)$ with respect to r.
SOLUTION:

Many planetary motion problems can be solved using the relationship between the energy E and the semimajor axis a: $E = -GmM/2a$, where m is the mass of the satellite and M is the mass of the central body. In terms of the distance r from the central body and the speed v, $E = -GmM/r + \frac{1}{2}mv^2$. If the energy and mass are known you can calculate the semimajor axis. If, in addition, the distance from the central body is known you can calculate the speed.

You will also need to know the relationship between the apogee (or aphelion) distance, the perigee (or perihelion) distance, and the semimajor axis: $a = (R_a + R_p)/2$. Sometimes the eccentricity e is given or requested. Then you will need $R_a = a(1+e)$ and $R_p = a(1-e)$. When the period is given or requested use $T^2 = (4\pi^2/GM_e)a^3$.

PROBLEM 14. A 3000-kg artificial earth-satellite in an orbit with an eccentricity of 0.35 has an energy of -5.3×10^{10} J. What is its semimajor axis, apogee distance, and perigee distance?

SOLUTION: Solve $E = -GmM_e/2a$ for the semimajor axis a. Here m is the mass of the satellite and M_e is the mass of the earth. Then Use $R_a = a(1+e)$ and $R_p = a(1-e)$, where e is the eccentricity, to find its apogee and perigee distances.

[ans: 1.13×10^7 m; 1.52×10^7 m; 7.34×10^6 m]

What is the speed of the satellite at apogee and at perigee?

SOLUTION: Use $U = -GmM_e/r$ to find the potential energy, then use the conservation of energy to find the kinetic energy. Finally, solve for the speed.

[ans: 4.12×10^3 m/s; 1.04×10^4 m/s]

Verify that the angular momentum about the earth is the same at apogee and perigee.

SOLUTION: At these points **r** and **v** are perpendicular to each other so the magnitude of the angular momentum is given by mrv. Evaluate this quantity for the two points. You should get the same result.

What is the period of the motion?

SOLUTION: Use $T^2 = (4\pi^2/GM_e)a^3$.

[ans: 1.195×10^4 s ($= 3.32$ h)]

PROBLEM 15. An earth satellite is placed into orbit by taking it to a point 500 km above the surface of the earth and giving it an initial velocity of 8.0×10^3 m/s, perpendicular to its position vector from the center of the earth. What is the semimajor axis, eccentricity, apogee distance, and perigee distance of its orbit?

SOLUTION: Let r_0 be the initial distance from the center of the earth ($r_0 = 6.37 \times 10^6 + 0.500 \times 10^6 = 6.87 \times 10^6$ m). Let v_0 be the initial speed. Then the energy is $E = -GmM_e/r_0 + \frac{1}{2}mv_0^2$, where m is the mass of the satellite and M_e is the mass of the earth. In terms of the semimajor axis, $E = -GmM_e/2a$. Set these two expressions for the energy equal to each other and solve for a. Notice that the mass of the satellite cancels from the energy equation.

Chapter 16: Gravitation   245

If $a > r_0$ then r_0 must be the perigee distance. Use $r_0 = a(1-e)$ to find the eccentricity e and $R_a = a(1+e)$ to find the apogee distance. If $a < r_0$ then r_0 must be the apogee distance. Use $r_0 = a(1+e)$ to find the eccentricity and $R_p = a(1-e)$ to find the perigee distance.

[ans: 7.65×10^6 m; 0.102; 8.44×10^6 m; 6.87×10^6 m]

What is the period of the motion?

SOLUTION:

[ans: 6.66×10^3 s (= 1.85 h)]

What is the speed of the satellite at apogee?

SOLUTION: Use conservation of angular momentum in the form $mR_a v_a = mR_p v_p$ to find v_a.

[ans: 6.51×10^3 m/s]

PROBLEM 16. An earth satellite is placed in orbit by taking it to a point 500 km above the surface of the earth and giving it an initial velocity perpendicular to its position vector from the center of the earth. What should the initial speed be for the satellite to have a circular orbit?

SOLUTION: As for the previous problem the energy is $E = -GmM_e/r_0 + \frac{1}{2}mv_0^2$, where m is the mass of the satellite, M_e is the mass of the earth, r_0 is the initial distance from the center of the earth (6.87×10^6 m), and v_0 is the initial speed. Since the orbit is circular r_0 must be the radius of the orbit and is the same as the semimajor axis. So the energy is $-GmM_e/2r_0$. Equate the two expressions for the energy and solve for v_0.

[ans: 7.62×10^3 m/s]

What is the period of the motion?

SOLUTION:

[ans: 5.70×10^3 s (= 1.57 h)]

III. MATHEMATICAL SKILLS

1. The magnitude of the gravitational force exerted by the earth on a body a distance h above its surface is given by $GmM/(R+h)^2$, where M is the mass of the earth, m is the mass of the body, and R is the radius of the earth. You may need to calculate the force for h small compared to R. Use the binomial expansion:

$$(R+h)^{-2} = R^{-2} - 2R^{-3}h + 3R^{-4}h^2 + \ldots .$$

This expression finds practical use in some calculations. Suppose you wished to find the difference in the gravitational field at opposite ends of a vertical rod. In principle, you could use Newton's law for gravity directly, substituting $R+h_1$ to find the force at one end and $R+h_2$ to find the field at the other end. R, however, is generally so much greater than h_1 or h_2 that your calculator truncates both numbers to R. On the other hand, if you use the first two terms of the binomial expansion to write an expression for each force, then subtract the expressions, the first terms cancel and you are left with the terms that are proportional to h_1 and h_2. These can be calculated easily.

2. To work some problems you may need to find the mass enclosed by a spherical surface within which the mass density ρ depends on distance from the center of the sphere. To find the mass inside the surface you must then evaluate the integral $\int \rho \, dV$ over the volume inside the surface. If the mass distribution has spherical symmetry, evaluate the integral by considering spherical shells of thickness dr. If r is the radius of such a shell, its volume is $4\pi r^2 \, dr$ and the integral becomes $4\pi \int \rho(r) r^2 \, dr$. The lower limit is 0 or the inner boundary of the mass distribution, whichever is larger. The upper limit is the radius of the surface or the outer boundary of the mass distribution, whichever is smaller. Suppose the mass distribution has a concentric spherical cavity, so it extends from R_1 to R_2. Then if we wish to find the mass inside a surface of radius r ($> R_1$), the limits are R_1 and r if $r < R_2$, and R_1 and R_2 if $r > R_2$.

IV. COMPUTER PROJECTS

In this section you will use a computer to investigate satellite motion. A massive central body (the sun or earth, for example) is at the origin and is assumed to remain motionless. If the coordinates of the satellite are x and y then the force exerted by the central body on the satellite has components that are given by $F_x = -GMmx/r^3$ and $F_y = -GMmy/r^3$, where G is the universal gravitational constant (6.67×10^{-11} m^3/s^3·kg), M is the mass of the central body, m is the mass of the satellite, and $r = \sqrt{x^2 + y^2}$ is the center-to-center distance of the satellite from the central body. These expressions assume both the central body and satellite have spherically symmetric mass distributions. Notice that the force is directed radially from the center of the satellite toward the center of the central body.

The acceleration of the satellite has components that are given by $a_x = -GMx/r^3$ and $a_y = -GMy/r^3$. Since the force depends on the coordinates of the satellite, you cannot use $\mathbf{a} = (\mathbf{a}_b + \mathbf{a}_e)$ to approximate the average acceleration in an interval. The acceleration at the beginning of an interval is an extremely poor approximation to the average acceleration and if it is used the intervals must be quite narrow and the running time must be quite long.

In Chapter 15 you learned one trick, using the derivative of the acceleration. Here you will learn another.

The program outlined below calculates the coordinates and acceleration at the beginning of each interval and the velocity at the midpoint of each interval. The acceleration at the beginning of an interval and the velocity at the midpoint of the previous interval are used to compute the velocity at the midpoint of the interval, then that velocity and the coordinates at the beginning of the interval are used to compute the coordinates at the end of the interval.

To start, the velocity half an interval before t_0 must be computed using the acceleration and velocity at t_0. When the computer is asked to display results, in the loop over intervals, x_b and y_b are the coordinates at the beginning of the interval while v_{xm} and v_{ym} are the velocity components for the midpoint of the interval. We wish to display the velocity components for the beginning and end of the interval, so the acceleration for the beginning is used to calculate these quantities just before printing. Here's the outline.

> input initial conditions: x_0, y_0, v_{x0}, v_{y0}
> input final time and interval width: $t_f, \Delta t$
> input display interval: Δt_d
> calculate acceleration at t_0: a_x, a_y
> calculate velocity at $t_0 - \Delta t/2$:
> $\quad v_{xm} = v_{x0} - a_x \Delta t/2$
> $\quad v_{ym} = v_{y0} - a_y \Delta t/2$
> set $t_b = t_0$, $t_d = t_0 + \Delta t_d$, $x_b = x_0$, $y_b = y_0$
> **begin loop** over intervals
> $\quad$ calculate acceleration at beginning of interval: a_x, a_y
> $\quad$ calculate velocity at midpoint of interval:
> $\quad\quad$ replace v_{xm} with $v_{xm} + a_x \Delta t$
> $\quad\quad$ replace v_{ym} with $v_{ym} + a_y \Delta t$
> $\quad$ calculate coordinates at end of interval:
> $\quad\quad x_e = x_b + v_{xm} \Delta t$
> $\quad\quad y_e = y_b + v_{ym} \Delta t$
> $\quad$ calculate time at end of interval: $t_e = t_b + \Delta t$
> $\quad$ **if** $t \geq t_d$ **then**
> $\quad\quad$ calculate velocity at beginning of interval:
> $\quad\quad\quad v_{xb} = v_{xm} - a_x \Delta t/2$
> $\quad\quad\quad v_{yb} = v_{ym} - a_y \Delta t/2$
> $\quad\quad$ print or display $t_b, x_b, y_b, v_{xb}, v_{yb}$
> $\quad\quad$ calculate velocity at end of interval:
> $\quad\quad\quad v_{xe} = v_{xm} + a_x \Delta t/2$
> $\quad\quad\quad v_{ye} = v_{ym} + a_y \Delta t/2$
> $\quad\quad$ print or display $t_e, x_e, y_e, v_{xe}, v_{ye}$
> $\quad\quad$ set $t_d = t_d + \Delta t_d$
> $\quad$ **end if** statement
> $\quad$ **if** $t_e \geq t_f$ **then exit loop**

>   set $t_b = t_e$, $x_b = x_e$, $y_b = y_e$
>   **end of loop** over intervals
>   stop

To write an efficient program define the constant C by $C = -GM\Delta t$, then use $a_x = Cx_b/r^3$ and $a_y = Cy_b/r^3$ to compute the acceleration components.

In the following project you will use the program to verify Kepler's laws of planetary motion. First you will verify the conservation of energy and angular momentum.

PROJECT 1. Consider a satellite in orbit around the earth ($M = 5.98 \times 10^{24}$ kg). At $t = 0$ it is at $x = 7.2 \times 10^6$ m, $y = 0$ and has velocity components $v_{x0} = 0$, $v_{y0} = 9.0 \times 10^3$ m/s. Use the program with an integration interval of $\Delta t = 10$ s to plot the position for every 500 s from $t = 0$ to $t = 1.55 \times 10^4$ s.

Modify the program so it computes the total mechanical energy and angular momentum, both per unit satellite mass, for each point displayed. The total mechanical energy per unit mass is given by $E = \frac{1}{2}(v_{xe}^2 + v_{ye}^2) - GM/r$ and the angular momentum per unit mass is given by $\ell = x_e v_{ye} - y_e v_{xe}$. The results should be constant to 2 or 3 significant figures.

Now use the program to prove the orbit is an ellipse. An ellipse is a geometric figure such that the sum of the distances from any point on the figure to two fixed points, called foci, is the same as for any other point. For the orbit you generated above, one focus is at the origin and the other is on the x axis the same distance from the geometric center as the first. First locate the geometric center of the orbit. Use the coordinates generated by the program to find the x coordinates of the points where the orbit crosses the x axis, then calculate half their sum. Let this coordinate be $-x_c$. The second focus is at $x_F = -2x_c$.

Modify the program so it computes the sum of the distances from the point displayed to the foci. It is given by $\sqrt{x_e^2 + y_e^2} + \sqrt{(x_e - x_F)^2 + y_e^2}$. Check that this sum is the same for all displayed points to 3 or more significant figures.

Kepler's second law says that the vector from the central body to the satellite sweeps out equal areas in equal times. Let $\mathbf{r}_b$ be the position vector at the beginning of an interval and let $\mathbf{r}_e$ be the position vector at the end of the interval. The triangle defined by these two vectors has an area that is given by $A = \frac{1}{2}|\mathbf{r}_b \times \mathbf{r}_e| = \frac{1}{2}|x_b y_e - x_e y_b|$. This is the area swept out by the position vector in time Δt. Have the computer calculate A just before it displays results and ask it to display A along with the other quantities. Check to see if it is constant to a reasonable number of significant figures.

You should recognize that this calculation essentially repeats the calculation above to check on the constancy of the angular momentum. Since $\mathbf{r}_e = \mathbf{r}_b + \mathbf{v}\Delta t$, the area is given by $A = \frac{1}{2}|\mathbf{r}_b \times \mathbf{v}|\Delta t$. Since the angular momentum is given by $\boldsymbol{\ell} = m\mathbf{r} \times \mathbf{v}$, this becomes $A = \frac{1}{2}\ell\Delta t$. A is constant because ℓ is constant.

Kepler's third law tells us that the period T is related to the length a of the semimajor axis by $T^2 = (4\pi^2/GM)a^3$. For the orbit generated above the length of the semimajor axis is half the distance between the two points where the orbit crosses the x axis. Calculate this distance and compute the right side of the third law equation. Use the program to find the period. It is, for example, twice the time between two successive crossings of the x axis. Compute the left side of the third law equation and compare the value with the value you obtained for the right side.

PROJECT 2. The length of the semimajor axis of an elliptical orbit depends only on the total mechanical energy. Suppose the satellite of the previous project is started from $x = 5.5 \times 10^6$ m, $y = 0$ with a velocity in the positive y direction. Select the initial speed so its total mechanical energy is the same as before. Plot the trajectory and find the length of the semimajor axis. Compare with the length of the semimajor axis for the previous project.

The length of the semimajor axis is related to the total mechanical energy E by $E = -GMm/2a$. Suppose the satellite is started at $x = 7.2 \times 10^6$ m, $y = 0$ with a velocity in the positive y direction. What initial speed should it have if its orbit is to be circular? For a circular orbit the semimajor axis is the same as the radius and for an initial position on the x axis it must be the same as x_0. Equate $-GMm/2x_0$ to $\frac{1}{2}mv_{y0}^2 - GMm/x_0$ and solve for v_{y0}. Use the program to plot the orbit. You might have the computer calculate $r^2 = x_e^2 + y_e^2$ for every displayed point and see if this quantity remains constant. Compared to the orbit of the first project is the total mechanical energy greater or less for the circular orbit?

If the total mechanical energy is negative the satellite is bound and its orbit is an ellipse. This is so for the satellites studied above. If the total mechanical energy is zero the orbit is a parabola and if the total mechanical energy is positive the orbit is a hyperbola. In either of these cases the satellite is not bound and it eventually escapes from the gravitational pull of the central body. Here's an example.

PROJECT 3. Consider a spacecraft in the gravitational field of the earth, with initial coordinates $x_0 = 7.2 \times 10^6$ m, $y_0 = 0$. Take its initial velocity to be $v_{x0} = 0$, $v_{y0} = 1.2 \times 10^4$ m/s. Use the program to plot its position every 300 s from $t = 0$ to $t = 4500$ s. Also calculate the total mechanical energy for these times and note it is constant (within the limits of the calculation, of course) and positive. Notice that the trajectory tends to become a straight line as the spacecraft recedes from the central body. Find the angle this line makes with the x axis. [ans: 51°]

What happens to the line if the initial speed is increased? Use the program to verify your conjecture.

V. NOTES

Chapter 17
FLUID STATICS

I. BASIC CONCEPTS

In this chapter you will study gases and liquids at rest. The most important concepts for this study are those of pressure and density. Learn their definitions well and learn to calculate the pressure in various situations, paying particular attention to the variation of pressure with depth in a fluid. Then use the concepts to understand two of the most basic principles of fluid statics: Archimedes' and Pascal's principles.

Definitions. Describe some properties of solids, liquids, and gases that can be used to distinguish them from each other:

Solids: _____

Liquids: _____

Gases: _____

Both liquids and gases are fluids.

Every fluid exerts an outward force on the inner surface of the container holding it. The force on any small surface area ΔA is proportional to the area and is perpendicular to the surface. If ΔF is the magnitude of the force on the area ΔA then the pressure p exerted by the fluid at that place is defined by the scalar relationship

$$p =$$

You must take the limit as the area ΔA tends toward zero. Thus pressure is defined at each *point*. The pressure may vary from point to point on the surface.

Pressure ultimately arises from forces that must be exerted on particles of the fluid in order to contain them. Container walls must exert forces on particles that impinge on them in order to reverse the normal components of their velocities and keep the particles within the container. The particles, of course, exert forces with equal magnitude and opposite direction on the walls. If the container top is open, the atmosphere above exerts forces on fluid particles at the fluid surface.

Pressure also exists in the interior of a fluid. Fluid within any volume exerts an outward force on the fluid around it and fluid outside a volume exerts an inward force on the fluid it surrounds. The force is normal to the imaginary surface that bounds the volume and the pressure is defined in the same way as at the container walls.

The SI unit of pressure (N/m²) is called _____ and is abbreviated _____. Other units are: 1 atmosphere (atm) = _____ Pa, 1 bar = _____ Pa, 1 mm of Hg = _____ Pa, and 1 torr = _____ Pa. Pressure is a scalar quantity.

If a small volume ΔV of fluid has mass Δm then the density ρ at that place is given by

$$\rho =$$

The definition includes a limiting process in which the volume being considered shrinks to a point. Thus density is defined at each point in a fluid (or solid) and may vary from point to point. Note that density is a scalar. Sometimes specific gravity is given rather than density. The specific gravity of a fluid is _____
_____.

Table 1 of the text lists various pressures that exist in nature and Table 2 lists the densities of some materials. Note the wide range of values.

The density of a fluid depends on the pressure. If the pressure at any point is increased (by squeezing the container, for example) the density at that point increases. The fluid property that describes the change in the volume ΔV of a given quantity of fluid produced by a given change Δp in pressure is called the bulk modulus of the fluid and is defined by

$$B =$$

Its SI unit is _____.

Suppose that at a point in a fluid the density is ρ_0 when the pressure is p_0. If the pressure is increased to $p_0 + dp$, where dp is an infinitesimal, the density increases to $\rho_0 + d\rho$, where $d\rho = (\rho_0/B)\,dp$. This result tells us that if the bulk modulus is independent of the pressure then the change in density produced by a change in pressure is proportional to the pressure change and inversely proportional to the bulk modulus. If a fluid has a large bulk modulus then a given change in pressure produces a small change in density. For an incompressible fluid $B =$ _____. The density in an incompressible fluid is everywhere the _____, regardless of variations in the pressure.

Variation of pressure with depth in a fluid. Pressure varies with depth in a fluid subjected to gravitational forces. Consider an element of fluid at height y in a larger body of fluid. Suppose the upper and lower faces of the element each have area A and the element has thickness Δy, as shown. If the fluid has density ρ the mass of the element is _____ and the force of gravity on it is _____. The pressure at the upper face is $p(y + \Delta y)$ so the downward force of the fluid there is _____. The pressure at the lower face is $p(y)$ so the upward force of pressure there is _____. Since the element is in equilibrium the net force must vanish. In the limit as Δy becomes infinitesimal this means

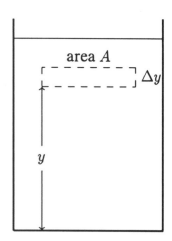

$$\frac{dp}{dy} =$$

The negative sign indicates that the pressure is _____ at points higher in the fluid than at lower points.

If the fluid is incompressible then the pressure difference between any two points in the fluid, at heights y_1 and y_2 respectively, is given by

$$p_2 - p_1 =$$

If p_0 is the pressure at the upper surface of an incompressible fluid then $p = $ _____ . is the pressure a distance h below the surface. Notice that the pressure is same at any points that are at the same height in a homogeneous fluid.

To derive an expression for the pressure as a function of height in a fluid, no matter how the density depends on pressure, start with $dp/dy = -\rho g$. This leads to $g\,dy = -dp/\rho$ and then to

$$g(y_2 - y_1) = -\int_{p_1}^{p_2} \frac{dp}{\rho}$$

where p_1 is the pressure at y_1 and p_2 is the pressure at y_2. If the density is a known function of pressure the integration can be carried out, at least in principle.

First assume the density does not depend on pressure, evaluate the integral, and show that $p_2 - p_1 = -\rho g(y_2 - y_1)$:

Now assume the density is proportional to the pressure, as it nearly is for the earth's atmosphere. Take $\rho = Cp$, where C is a constant. Evaluate the integral and show that $p_2 = p_1 e^{-Cg(y_2 - y_1)}$:

The constant C can be evaluated if the pressure and density for one position in the fluid are known. If, for example, you know the density is ρ_1 for a pressure of p_1 then you can use to $C = \rho_1/p_1$ to find a value for C.

If the fluid is inhomogeneous, as it is if it consists of layers of immiscible fluids with different densities, you must apply the expression for $p(y)$ to each layer separately. For practice, consider the stack of two incompressible fluids shown on the right. The upper surface is open to the atmosphere and the pressure there is atmospheric pressure p_0. Each layer has thickness $2h$. Write an expression for the pressure at each of the labelled points:

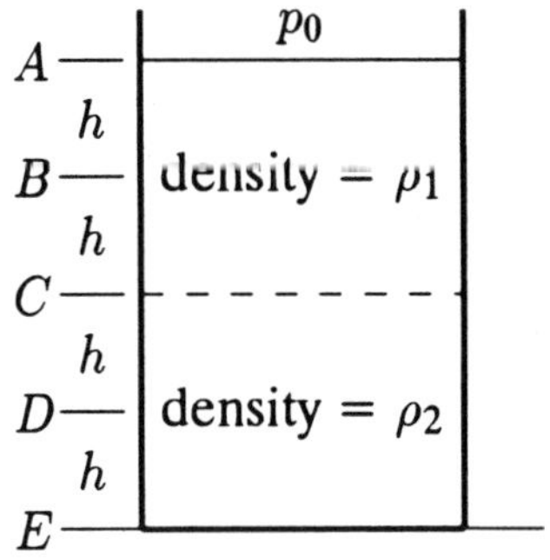

$$p_A =$$

Chapter 17: Fluid Statics

$p_B =$

$p_C =$

$p_D =$

$p_E =$

To work some problems you must also know when the pressure is the same at two points. Although atmospheric pressure varies with height, the variation is negligible over distances on the order of meters. The surfaces of two portions of a fluid that are both exposed to the atmosphere may be taken to be at the same pressure, atmospheric pressure. Thus the upper surface in each arm of an open U-tube may be taken to be at atmospheric pressure, even if the surfaces are at different heights.

If two points are at the same height *and* they can be joined by a line such that the fluid density does not change along the line, then the pressure is the same at the points. Thus the pressure is the same at the two points labeled C in Fig. 4b of the text. A line can be drawn from one to the other through the liquid in the bottom of the tube. Note that the pressure is not the same at the points labeled A and B, even though they are at the same height. A line joining them must pass through two liquids of different density.

Carefully go over Sample Problem 1 of the text and note how the dependence of pressure on fluid density and height is used to solve for one of the densities. Also note the other questions that can be asked about the same situation. If the densities are known, for example, the difference in heights can be calculated.

Pascal's principle. Suppose the external pressure applied to the surface of a fluid is changed by Δp_{ext}. Pascal's principle states that the pressure everywhere in the fluid changes by _____.

In the space to the right draw a diagram of a simple hydraulic system that can be used to lift a heavy object by applying a force that is much less than its weight. Clearly identify the input and output forces F_i and F_o. Suppose the input force is exerted over an area A_i and the output force is exerted over an area A_o. According to Pascal's principle these quantities are related by

$$\frac{F_i}{A_i} =$$

Suppose the fluid surface at the input force moves a distance d_i and the object moves a distance d_o. If the fluid is incompressible then the volume of fluid does not change and $d_i A_i = d_o A_o$. Substitute $A_o = d_i A_i / d_o$ into the equation for Pascal's principle to show that $F_i d_i = F_o d_o$:

This means that the work done by the external force on the fluid is the same as the work done by the fluid on the object. If the fluid is compressible the work done by _____ is larger than the work done by _____. The difference results in an increase in the internal energy of the compressed fluid.

An automobile braking system uses hydraulics to transmit the force you apply at the brake pedal to the brake shoes. You may think of a small diameter hose containing brake fluid running from the pedal to a shoe. Since a small force applied by you results in a huge force applied to the shoe, you know that the surface area of fluid being pushed on by the pedal is much _____ than the area of fluid pushing on the shoe. Furthermore, since the fluid is essentially incompressible, the distance the pedal moves is much _____ than the distance the shoe moves.

Archimedes' principle. State Archimedes' principle in your own words: _____
_____
_____

The buoyant force on an object wholly or partially immersed in a fluid is a direct result of the pressure exerted on it by the fluid and depends on the variation in pressure with height. The pressure at the bottom of the object is greater than the pressure at the top and the net buoyant force is upward.

You know that Archimedes' principle is valid because, if the submerged portion of an object is replaced by an equal volume of fluid, the fluid would be in equilibrium and the net force associated with the pressure of surrounding fluid would equal the _____ of the fluid that replaced the object.

To calculate the buoyant force acting on a given object, first find the submerged volume. If the object is completely surrounded by fluid, this is the volume of the object. If the object is floating on the surface, it is the volume that is beneath the surface. Suppose the submerged volume is V_s. Then the magnitude of the buoyant force is given by $F_b = $ _____, where ρ_f is the density of _____.

To predict if an object floats or sinks, first assume it is totally submerged and compare the buoyant force with the weight. If the weight is greater, the object _____. If the buoyant force is greater, it _____. For an object with uniform density in an incompressible fluid this procedure is the same as comparing the density of the object with the density of the fluid. If the density of the fluid is greater than the density of the object, the object _____. If the density of the object is greater than the density of the fluid, the object _____.

For purposes of calculating the torque on a floating object, the buoyant force of the fluid on the object can be treated as a single force acting at the center of gravity of _____. The force of gravity, on the other hand, can be treated as a single force acting at the center of gravity of _____. This point is called the center of _____. If buoyancy and gravity together produce a net torque the object tilts.

Buoyant torque is used to test the stability of a boat. Assume the boat is tilted slightly and calculate the buoyant torque about the center of gravity. If it tends to right the boat, the boat is stable. If it tends to tilt the boat more, the boat is unstable. Three cases are shown below, with the centers of buoyancy marked B and the centers of gravity marked G. For each, draw vectors to indicate the directions of the force of gravity and the buoyant force, give the direction of the buoyant torque about the center of gravity (into or out of the page), and label the boats "stable" and "unstable", as appropriate.

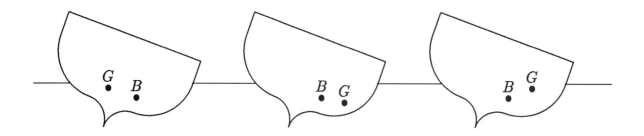

Measurement of pressure. This section serves two purposes. It describes two instruments used to measure pressure: the barometer and the open-tube manometer. It also provides some excellent applications of ideas discussed earlier in the chapter. You should study it with both these purposes in mind.

Some instruments measure <u>absolute pressure</u>, some measure <u>gauge pressure</u>. Distinguish between gauge and absolute pressure: _____
_____

In the space on the right draw a diagram of a mercury barometer. Clearly mark the mercury, the region of near vacuum, and the region where the pressure has the value indicated by the height of the mercury column.

You should be able to use what you know about the variation of pressure with height in an incompressible fluid to show that the pressure at the top of the mercury pool is given by $\rho g h$, where ρ is the density of mercury and h is the height of the top of the mercury column above the pool. Carry out the derivation in the space below:

Suppose a fluid with half the density of mercury is used in a barometer. The height of the column will be _____ the height of a mercury column.

You should understand the importance of the region above the mercury column. If this region is not a near vacuum but instead is filled with a gas, then the pressure at the surface of the mercury pool is NOT proportional to the height of the mercury column. In fact, if the pressure above the column is p_0 and the height of the column is h then the pressure above the pool is given by $p = $ _____.

In the space to the right draw a diagram of an open-tube manometer. Mark the fluid. Use p_0 to label the fluid surface where the pressure is atmospheric and p to label the fluid surface where the pressure has the value being measured.

Describe exactly what characteristic of the fluid is measured and tell how this is related to the pressure to be found:
_____
_____
_____

A barometer and an open-tube manometer measure different pressures. Tell what each measures:

barometer: _____

manometer: _____

Surface tension. Liquid surfaces resist stretching by generating a force that opposes any externally applied force. <u>Surface tension</u>, as this force is called, is parallel to the surface.

In the space to the right draw a diagram of the wire frame instrument used to measure surface tension and clearly label the fluid and the movable side of the rectangle. Assume the wire is moved so as to increase the area of the liquid and indicate with arrows the direction of the surface tension force exerted by the liquid on the wire.

Suppose the length of the movable wire is d and the distance between it and the opposite side of the rectangle is h. The force of the surface on the wire is proportional to _____ and does not depend on _____.

Carefully note that the force is NOT like the force of a spring. In particular, it is not proportional to the extension of the surface. The force behaves as it does because, when the surface is stretched, its area increases through the addition of liquid from the interior and not wholly through the elongation of bonds between atoms.

Surface tension γ is defined as the force per unit length exerted by the liquid across a line on its surface. The direction of the force is _____ to the line. If the surface tension of the liquid in the wire frame is γ then the total force exerted by the liquid on the sliding wire is _____. The factor 2 appears because _____.

The spherical shape of raindrops is due to surface tension. The radius of the sphere is determined by the balance of two forces: _____, acting at the surface of the drop, pulling neighboring portions together, and _____, acting to enlarge the drop.

II. PROBLEM SOLVING

In some problems the pressure in a fluid is given and you are asked for the force exerted by the fluid on its container. Remember that the force at every point is normal to the surface at that point. If the pressure is uniform and the surface is a plane the magnitude of the net force is simply the product of the pressure and area of the surface. When several planes are involved and you are asked for the resultant force you must add the forces vectorially. Resolve the force on each plane into its components.

We start with two situations in which the pressure is uniform. We expect the net force of the fluid on its container to be zero. This is because the condition of uniform pressure implies that the gravitational force on the fluid is zero or negligible. Since the weight of the fluid is zero and the fluid is at rest the net force of the container on it is zero and the net force of the fluid on the container must also be zero.

PROBLEM 1. Consider a container with cross section and dimensions as shown, filled with fluid at uniform pressure p. The sides, top, and bottom are rectangles that extend out of the page a distance W. Find expressions for the force of the fluid on each rectangle and the net upward force on the top and slanted sides.

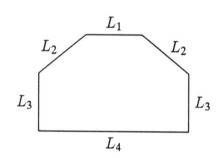

SOLUTION: On the diagram draw vector arrows to indicate the direction of the force of the fluid on each plane. Write expressions for the magnitudes of the forces in terms of given quantities:

$F_1 = $ _____ $F_2 = $ _____ $F_3 = $ _____ $F_4 = $ _____

Assume the slanted planes each make the angle α with the horizontal. The vertical component of the force on one of the slanted sides is then _____. From the diagram $\cos\alpha = (L_4 - L_1)/2L_2$, so in terms of the given dimensions this component is _____. The total upward force of the fluid on the top and slanted sides is _____. Note that this balances the downward force on the bottom of the container.

[ans: $F_1 = \rho L_1 W$; $F_2 = \rho L_2 W$; $F_3 = \rho L_3 W$; $F_4 = \rho L_4 W$; $F_{up} = \rho L_4 W$]

If the surface is curved you cannot find the total force of a fluid on it by multiplying the pressure by the surface area. You must take into account the direction of the force on each infinitesimal surface element and add these forces vectorially, by evaluating an integral.

PROBLEM 2. Suppose a hemisphere of radius R is filled with fluid at uniform pressure p. Find expressions for the force of the fluid on the curved portion and on the bottom.

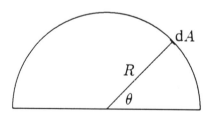

SOLUTION: Consider an infinitesimal area dA at the point shown on the diagram. In terms of the angle θ the vertical component of the force of the fluid there is _____. To find the net vertical component of the force on the curved portion of the hemisphere we must add the vertical compo-

258   *Chapter 17: Fluid Statics*

nents of the forces on all infinitesimal surface elements. Thus the vertical component of the force is given by the integral $F_v = \int p \sin\theta \, dA$ over the curved portion of the hemisphere.

All surface elements with the same area and at the same angular position θ experience the same vertical component of force, so we divide the hemisphere into narrow rings of angular width $d\theta$, each corresponding to a different value of θ. The radius of a ring is $R\cos\theta$ and its width is $Rd\theta$ so its area is $dA = R^2 \cos\theta \, d\theta$. The limits on θ are 0 and $\pi/2$ since each ring goes around the hemisphere. Thus

$$F_v = \int$$

Be sure to fill in the limits of integration. You should now be able to carry out the integration to obtain a result for F_v.

Finally, find the force on the bottom portion of the hemisphere.

[ans: $\pi R^2 p$, upward; $\pi R^2 p$, downward]

Notice that the two forces are equal in magnitude and opposite in direction. The net force of the fluid on the container is zero.

We now consider a fluid on which a gravitational force acts, so the pressure is not uniform.

PROBLEM 3. A trough has the cross-sectional shape and dimensions shown. It is 2.5 m in length and it is filled with water (density = 1.00×10^3 kg/m) to a height of 85 cm. What is the force of the water on the bottom? What is the force of the water on each side? What is the total force of the water on the trough? Assume water is incompressible.

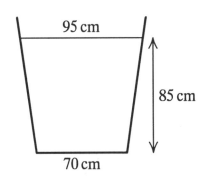

SOLUTION: Let W_b be the width of the bottom and L be the length of the trough. Then the area of the bottom is $A_b = W_b L$. The water pressure at the bottom is $p_0 + \rho gh$, where p_0 is atmospheric pressure, ρ is the density of water, and h is the height of the water in the trough. The force of the water on the bottom is $F_b = $ _____.

A distance y below the water surface the pressure is $p(y) = p_0 + \rho gy$. Consider an infinitesimal element of a side, with dimensions L and $d\ell$. Since its area is $dA = L\,d\ell$ the force of the water on it is given by $dF_s = p\,dA = L(p_0 + \rho gy)\,d\ell$. If θ is the angle a side makes with the horizontal then $d\ell = dy/\sin\theta$ and $dF_s = L(p_0 + \rho gy)\,dy/\sin\theta$. This expression is integrated from 0 to h to find the total force on the side: $F_s = L(p_0 h + \frac{1}{2}\rho gh^2)/\sin\theta$. Finally a little geometry shows that $\sin\theta = h/d$, where d is the length of one of the slanted sides. Thus $F_s = Ld(p_0 + \frac{1}{2}\rho gh)$. The length of a side can be computed using $d^2 = (W_t - W_b)^2/4 + h^2$, where W_t is the width of the trough at the upper water surface.

The total force is vertically downward and is given by $F_b + 2F_s \cos\theta$ or, since $\cos\theta = (W_t - W_b)/2d$, by $F_b + L(p_0 + \frac{1}{2}\rho gh)(W_t - W_b)$. Evaluate these expressions.

[ans: 1.90^5 N; 2.24×10^5 N; 2.55×10^5 N]

The last result is the force with which the surface under the trough must push up on the trough. It should equal the weight of the water in the trough plus the force of the air on the upper surface. The force of the air is $p_0 L W_t$. Now show that the volume of water in the trough is $(W_t - W_t)Lh/2$, then show that the weight of the water is $\rho g(W_t - W_b)Lh/2$. Evaluate these two expressions and add the results.

[ans: 2.55×10^5 N]

To calculate the variation of pressure with height in a static compressible fluid, start with $dp = -\rho g\, dy$, where the y axis was chosen to be positive in the upward direction. Here ρ is the fluid density and must be given as a function of the pressure. Recast the starting equation as $dp/\rho(p) = -g\, dy$, then integrate. After integrating, you may need to carry out some algebra to solve for the pressure at a given point. You have already considered fluids that are incompressible and those for which the density is proportional to the pressure. Here is another example.

PROBLEM 4. Over the pressure range of interest the density of a certain fluid is proportional to the square of the pressure: $\rho = Ap^2$, where A is a constant. Find an expression for the pressure as a function of depth when the fluid is subjected to a uniform gravitational field and the upper surface of the fluid is at atmospheric pressure p_0. To evaluate A suppose that the density is ρ_0 at atmospheric pressure.

SOLUTION: Take $y = 0$ to be at the upper surface, where $p = p_0$, and let h be a depth in the fluid. Integrate $dp/Ap^2 = -g\, dy$. The integral on the left side goes from p_0 to $p(h)$, the integral on the right side goes from 0 to $-h$.

[ans: $p(h) = p_0^2/(p_0 - \rho_0 g h)$]

A variation on this type problem concerns a fluid that is accelerating. Then the pressure differential across any fluid element provides some or all of the force that accelerates the element. Suppose the acceleration is horizontal, along the x axis. Then the pressure must vary with x. If A is the cross-sectional area of a fluid element, perpendicular to the acceleration, then the net force acting on the element is $Ap(x) - Ap(x + dx)$, where dx is the infinitesimal width of the element. This must equal the product of the acceleration a and mass dm of the element. Since the element has volume $A\, dx$, $dm = \rho A\, dx$ and $Ap(x) - Ap(x + dx) = \rho a A\, dx$. Thus $dp/dx = -\rho a$. For uniform acceleration of all parts of an incompressible fluid $p = p_0 - \rho a x$, where p_0 is the pressure at $x = 0$. Notice that the pressure must decrease in the direction of the acceleration.

If the acceleration is vertical then the force of gravity must be considered. If the y axis is taken to be positive in the upward direction then $Ap(y) - Ap(y + dy) - \rho g A\, dy = \rho a A\, dy$ and $dp/dy = -\rho g - \rho a$.

260 Chapter 17: Fluid Statics

PROBLEM 5. A closed tube is half filled with an incompressible liquid of density 1.5×10^3 kg/m^3, the region above the liquid being evacuated. The tube is carried vertically on an elevator. What is the pressure in the liquid a distance 10 cm from the upper surface when the elevator is moving at constant velocity? When it has an upward acceleration of 2.3 m/s^2? When it has a downward acceleration of 2.3 m/s^2?

SOLUTION: Start with $dp/dy = -\rho(g + a)$ and show that $p(h) = \rho h(g + a)$, where h is the depth measured from the liquid surface. You must use the condition that the pressure is zero at $h = 0$, true because the region above the liquid is evacuated. Evaluate the expression for $a = 0$, $a = +2.3$ m/s^2, and $a = -2.3$ m/s^2.

[ans: 1.47×10^3 Pa; 1.82×10^3 Pa; 1.13×10^3 Pa]

Some problems involve immiscible, incompressible fluids in U-tubes. Usually you know the pressure at the top surface of the fluid in each arm of the tube. If the end is open to the atmosphere it is atmospheric pressure. If the end is closed and a vacuum exists above the fluid it is zero. You can find a relationship between values of the pressure at the various fluid interfaces. Start at the fluid surface in one arm and follow a line in the tube to the surface in the other arm. Calculate the pressure at each interface and finally the pressure at the top of the second arm in terms of the pressure at the top of the first arm. Lastly, set the expression you obtain equal to the known value of the pressure at the top of the second arm. This gives you an expression to solve for the unknown in the problem.

PROBLEM 6. Both arms of a U-tube containing water are open to the atmosphere. When another incompressible liquid is poured in one arm, its upper surface is 1.70 cm above the water in that arm and 0.53 cm above the surface of the water in the other arm. Assume the liquids do not mix and calculate the density of the added liquid.

SOLUTION: In the space to the right draw a diagram showing the U-tube and the two liquids. Use ℓ_1 to label the length of the column of unknown liquid, ℓ_2 to label the length of the water column in the arm containing both liquids, and ℓ_3 to label the length of the water column in the arm containing only water. The pressure at the interface between the two liquids is $p_0 + \rho g \ell_1$ and the pressure at the bottom of the tube is $p_0 + \rho g \ell_1 + \rho_w g \ell_2$, where ρ_w is the density of water and ρ is the density of the other liquid. The pressure at the top of the water column in the second arm is $p_0 + \rho g \ell_1 + \rho_w g \ell_2 - \rho_w g \ell_3$. This must be atmospheric pressure, so $\rho \ell_1 + \rho_w \ell_2 - \rho_w \ell_3 = 0$.

If h is the difference in the heights of the upper liquid surfaces in the two arms then $h = \ell_1 + \ell_2 - \ell_3$. You can now solve for ρ in terms of ℓ_1 and h.

[ans: 687 kg/m^3]

The arm containing both liquids is attached to an air pressure line and the pressure is adjusted until the liquid level is the same in the two arms. What is the gauge pressure in the line?

SOLUTION: Now write $p_1 + \rho g \ell_1 + \rho_w g \ell_2 - \rho_w g \ell_3 = p_0$, where p_1 is the pressure in the line. The gauge pressure is $p_g = p_1 - p_0$ so $p_g + \rho g \ell_1 + \rho_w g \ell_2 - \rho_w g \ell_3 = 0$. As before $h = \ell_1 + \ell_2 - \ell_3$, but now $h = 0$. Solve for p_g in terms of ℓ_1 and the densities.

[ans: 51.8 Pa]

The air pressure line is disconnected from that arm and, without changing the pressure in the line, is connected to the arm containing only water. The arm containing both liquids is open to the atmosphere. What is the height of the upper liquid surface in the arm containing both liquids, relative to the height of the liquid in the other arm?

SOLUTION:

[ans: 1.06 cm above]

All Pascal's law problems are much the same. In every case a tube containing a fluid has a different cross section at different ends. A force F_1 is applied uniformly to an end with area A_1 and the fluid at the other end exerts a force F_2 on an object. If the area at the second end is A_2 then $F_1/A_1 = F_2/A_2$. Three of these quantities must be given, then the fourth can be evaluated.

Some problems may ask you to calculate the work done by each of the forces. These are $W_1 = F_1 d_1$ and $W_2 = F_2 d_2$, where d_1 and d_2 are the distances moved. If the fluid is incompressible then $d_1 A_1 = d_2 A_2$ and $W_1 = W_2$.

PROBLEM 7. A hydraulic system is designed to exert an output force of 3000 N on a square plate with an edge length of 2.0 cm when an input force of 100 N is applied. At the input end what is the radius of the tube containing the fluid? If the plate moves a distance of 1.5 mm, through what distance does the input piston move? Assume the fluid is incompressible.

SOLUTION:

[ans: 2.06 mm; 4.50 cm]

Two fundamental Archimedes' principle problems have been covered in the Basic Concepts section. They involve finding the buoyant force on an object, either floating or completely submersed in an incompressible fluid, and deciding if an object floats or sinks. These and many other Archimedes' law problems start with the equations $F_g = \rho g V$ for the force of gravity and $F_b = \rho_f g V_s$ for the buoyancy, where ρ is the density of the object, ρ_f is the density of the fluid in which it is immersed, V is the volume of the object, and V_s is the submerged volume. If the object is floating with no other forces acting then $\rho V = \rho_f V_s$.

In some cases other forces are present. For example, a string may be tied to the object, either holding it up from above or pulling it down from below, or weights may be placed on the object. In these instances simply include the additional force F in the equation for equilibrium: $\rho_f g V_s - \rho g V + F = 0$, where F is positive if the force is directed upward.

PROBLEM 8. A block of wood (density = 2.5×10^2 kg/m) with dimensions 3.0 m $\times$ 15 cm $\times$ 15 cm is floating in water (density = 1.0×10^3 kg/m^3) as shown. How deep in the water is the lower surface?

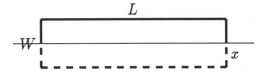

SOLUTION: Let L be the length, W be the width, and T be the thickness of the block and suppose x is the extent to which the block is submerged. The volume of the block is LWT and the submerged volume is LTx. The force of gravity is $F_g = \rho g LWT$ and the buoyant force is $F_b = \rho_w g LTx$, where ρ is the density of the wood and ρ_w is the density of water. Equilibrium occurs when these forces are equal, or $\rho W = \rho_w x$. So $x = \rho W/\rho_w$.

[ans: 3.75 cm]

What is the smallest mass that will submerge the block when it is placed on top of the block? The added mass is not submerged.

SOLUTION: Now $F_b - F_g - mg = 0$, where m is the added mass. In addition, $V = V_s$ so the condition for submersion is $\rho_w g LWT - \rho g LWT - mg = 0$.

[ans: 50.6 kg]

A variation on the standard Archimedes' principle problem involves a situation in which an object is in equilibrium at the interface between two immiscible fluids of different densities. In this case it is best to recall the origin of the buoyant force: the difference in pressure at the top and bottom of the object. Suppose the object has uniform cross section in planes that are parallel to a fluid surface. Suppose further that h_1 of the object's height is in the upper fluid, which has density ρ_1, and h_2 of the object's height is in the lower fluid, which has density ρ_2. Suppose the pressure at the upper surface is p_1. Then the pressure at the lower surface is $p_1 + \rho_1 g h_1 + \rho_2 g h_2$. The downward force on the upper surface is $p_1 A$ and the upward force on the lower surface is $p_1 A + \rho_1 g h_1 A + \rho_2 g h_2 A$, so the buoyant force is $\rho_1 g h_1 A + \rho_2 g h_2 A$, where A is the cross sectional area of the object. If no other forces act this must equal $\rho g V$, where ρ is the density and V is the total volume of the object. This condition can be used, for example, to find the extent h_2 to which the object sinks into fluid 2.

PROBLEM 9. A cube with an edge length of 5.5 cm and a density of 780 kg/m^3 floats at the interface between water (density = 1.0×10^3 kg/m^3) and oil (density = 570 kg/m^3). The oil covers its upper surface. How far into the water does its lower surface extend?

SOLUTION:

[ans: 2.69 cm]

Solutions to nearly all problems associated with this chapter rely on either the definition of pressure, the definition of density, the variation of pressure with depth, Archimedes' principle, or Pascal's principle. Read the problems in your homework assignment and classify them according to the definition or principle involved, then write the problem number on the appropriate line below. Some may belong in more than one category and some may belong in the category labelled "other".

Definition of pressure: _____
Definition of density: _____
Variation of pressure: _____
Archimedes' principle: _____
Pascal's principle: _____
Other: _____

III. NOTES

Chapter 18
FLUID DYNAMICS

I. BASIC CONCEPTS

This chapter is about fluids in motion. Two equations that are fundamental to the topic, the continuity and Bernoulli equations, relate the pressure, velocity, density, and height at points in a moving fluid. To understand these equations you must first understand the ideas of density and velocity fields, streamlines, and tubes of flow.

Density and velocity fields. A fluid in motion is described by giving the density and fluid velocity at every point in the fluid. That is, the functions $\rho(x, y, z, t)$ and $\mathbf{v}(x, y, z, t)$ are given. The first gives the fluid density at a point with coordinates x,y,z, at time t; the second gives the velocity of the fluid at that point and time. You should clearly understand that x, y, and z are the coordinates of a fixed point in space. They do not change with time.

Fluid flow can be categorized according to whether it is steady or nonsteady, compressible or incompressible, viscous or nonviscous, and rotational or irrotational. This chapter deals chiefly with steady, incompressible, nonviscous, irrotational flow.

When a fluid is in <u>steady flow</u>, the velocity and density fields do NOT depend on _____ but they may depend on _____. If we follow all the particles that eventually get to any selected point, we find they all have the same velocity as they pass that point, regardless of their velocities when they are elsewhere. In addition, particles flow into and out of any volume in such a way that the mass in the volume at any time is the same as at any other time.

Be very careful here. Steady flow does NOT imply that the velocity of any fluid particle is constant nor does it imply that the density is everywhere the same.

If the flow is <u>incompressible</u> the density does not depend on either the _____ or the _____. It is the same everywhere in the fluid and retains the same value through time. Steady flow does NOT imply incompressible flow. To be sure you are clear on this point, describe the density in a fluid that is flowing steadily but not incompressibly: _____

You should also know the meanings of the other terms. Qualitatively distinguish viscous from nonviscous flow: _____

If the fluid speed is sufficiently high in a viscous fluid, the flow becomes _____. Qualitatively distinguish rotational from irrotational flow: _____

<u>Streamlines</u> are a convenient and effective means of visualizing fluid flow. They are also important for several derivations carried out in this chapter. Define the term streamline: _____

In steady flow streamlines are fixed curves in space. Every fluid particle on the same streamline passes through the same sequence of points and at each point has the same velocity as other particles when they are at the point. At any point the fluid velocity is _____ to the streamline through the point.

In steady flow can the velocity of a particle be different when it is at different points on the same streamline? _____ Do different streamlines ever cross? _____ Give an argument to substantiate your claim: _____

Define the term <u>tube of flow</u>: _____

In steady flow do particles ever cross the boundary of a tube of flow? _____ Justify your answer: _____

Does a streamline ever cross the boundary of a tube of flow? _____ Why or why not? _____

We conclude from these statements that streamlines crowd closer together in narrow portions of tubes of flow than in wide portions. Consider a narrow tube of flow at a place where it has cross sectional area A. If the fluid there has density ρ and speed v, then the mass of fluid that passes the cross section per unit time is given by _____ and the volume of fluid that passes the cross section per unit time is given by _____. The former quantity is called the <u>mass flux</u> and the later is called the <u>volume flux</u> or <u>volume flow rate</u> and is denoted by R. The SI units of mass flux are _____ and the SI units of volume flow rate are _____. Other commonly used units for the volume flow rate are li/s, gal/s, and gal/min.

When you are finished studying this chapter, write a statement here to indicate why we choose to study fluids in tubes of flow. Why not pick some other volume of fluid? _____

Equation of continuity. The equation of continuity is derived from the condition that _____ is conserved in fluid flow. If there are no sources or sinks of fluid a change in the mass contained in any given volume can come about only because the mass that flows into the volume differs from the mass that flows out at any time. In steady flow the mass in any volume does not change, so in any time interval the mass flowing in equals the mass flowing out.

For steady flow the equation of continuity is a statement that the _____ is the same for every cross section along a tube of flow. Let the subscripts 1 and 2 label two places along a tube of flow with cross-sectional areas A_1 at 1 and A_2 at 2. Suppose the fluid density is ρ_1 at 1 and ρ_2 at 2 and the fluid speed is v_1 at 1 and v_2 at 2. Then for steady flow the equation of continuity is

$$\rho_1 A_1 v_1 =$$

If the flow is incompressible, then $\rho_1 = \rho_2$ and the equation of continuity becomes

$$A v_1 =$$

266   Chapter 18: Fluid Dynamics

For incompressible flow both the mass flux and volume flow rate are uniform along a tube of flow. This means that fluid particles have greater speed in a _____ portion of a tube than in a _____ portion. Because streamlines crowd closer together in narrow portions of tubes of flow than in wider portions we conclude that a high concentration of streamlines corresponds to a _____ fluid speed and a low concentration corresponds to a _____ fluid speed.

The Bernoulli equation. The Bernoulli equation for steady, incompressible, nonviscous, irrotational flow tells us that the quantity _____ has the same value at every point along a streamline. You should realize, however, that the value of the quantity may be different for different streamlines.

This equation arises directly from the work-energy theorem, applied to the fluid in a narrow tube of flow, in the limit as the tube becomes a streamline. The term p appears because the _____ exerted by neighboring portions of fluid do work, the term $\rho g y$ appears because _____ does work if the height of the tube varies, and the term $\frac{1}{2}\rho v^2$ appears because the _____ of the fluid changes if work is done on it.

If the tube does not change height, the gravitational terms cancel each other and the Bernoulli equation becomes _____ = constant along a streamline. This equation indicates that the pressure is large where the fluid speed is _____ and is small where the fluid speed is _____. Combining this result with the continuity equation, we can conclude that for steady, incompressible, horizontal flow the pressure is _____ where a tube of flow is narrow and _____ where it is wide.

Because the pressure is different in wide and narrow portions of a tube a net force acts on a fluid element as it moves from one region to another. This changes its velocity. The pressure is _____ where the fluid moves slowly and _____ where it moves faster.

To fix these ideas, consider the horizontal tube of flow shown on the right. Draw several streamlines within the tube, clearly illustrating where they crowd close together and where they spread apart. Label one end of the tube "high fluid velocity" and the other end "low fluid velocity", as appropriate. Label one end "high pressure" and the other end "low pressure", as appropriate.

Applications A number of applications of the Bernoulli equation depend on the lowering of pressure that accompanies an increase in fluid velocity in a horizontal tube of flow. Read the text carefully and for each of the two applications listed below tell what brings about the change in velocity. Draw a diagram. Indicate the direction of fluid flow, label regions of high and low velocity, and label regions of high and low pressure.

Venturi meter:
Mechanism of velocity change: _____   Diagram:

Chapter 18: Fluid Dynamics   **267**

Airplane wing:
Mechanism of velocity change: _____   Diagram:
_____
_____
_____

A rocket engine is somewhat different. Here high pressure is utilized to force gas through a small opening, thereby increasing its speed.

Viscosity. Viscosity is a measure of the drag force that one layer of fluid exerts on a neighboring layer. It results in a loss of mechanical energy. Two important situations are discussed in the text. For each of them, briefly describe the velocity field associated with the fluid.

In the first, the fluid is initially at rest between two stationary parallel plates. One plate is then moved at constant velocity parallel to itself while the other plate remains stationary. Qualitatively describe the variation in fluid velocity between the plates: _____
_____

This situation is used to measure the <u>coefficient of viscosity</u>. If F is magnitude of the force required to maintain the constant velocity of the plate, A is the area of a fluid layer, and dv/dy is the velocity gradient, then the coefficient of viscosity η is given by

$$\eta = $$

Here the y axis is _____ to the plates.

In the second situation, fluid is flowing through a circular pipe. Qualitatively describe the variation in velocity from the pipe wall to its center: _____
_____

In this situation fluid flow is maintained by _____
_____

II. PROBLEM SOLVING

You should be able to relate the volume flow rate and mass flux to the dimensions of a tube of flow and to the fluid velocity. The volume flow rate gives the volume of fluid that passes a cross section per unit time and is given by Av, where A is the cross-sectional area of the tube and v is the fluid speed. The mass flux is the mass of fluid that passes a cross section per unit time and is given by ρAv, where ρ is the fluid density. Notice that the mass flux is the product of the density and the volume flow rate.

PROBLEM 1. A 12-m wide canal with a horizontal bottom and vertical walls is filled with water (density $= 1.0 \times 10^3$ kg/m^3) to a depth of 2.8 m. The water flows steadily at 1.3 m/s. What is the volume flow rate and what volume of water passes a point on the shore in 5.0 min?

SOLUTION: The volume flow rate is given by $R = Av = Whv$, where W is the width of the canal and h is the water depth. This is the volume of water that passes in a second. Multiply by the number of seconds in 5.0 min.

[ans: $43.7\,\text{m}^3/\text{s}$; $1.31 \times 10^4\,\text{m}^3$]

What is the mass flux and how much mass passes a point on the shore in 5.0 min?

SOLUTION:

[ans: $4.37 \times 10^4\,\text{kg/s}$; $1.31 \times 10^7\,\text{kg}$]

PROBLEM 2. Water flowing steadily from a kitchen faucet fills a 1 gallon container in 1.2 min. What is the volume flow rate in m^3/s; what is the mass flux in kg/s? The density of water is $1.0 \times 10^3\,\text{kg/m}^3$.

SOLUTION: Since water is essentially incompressible a gallon of fluid passes through the faucet in 1.2 min so the volume flow rate is $R = 1/1.2 = 0.833$ gal/min. Use the conversion tables in the appendix to help convert gal/min to m^3/s. Use ρR to find the mass flux.

[ans: $5.26 \times 10^{-5}\,\text{m}^3/\text{s}$; $0.0526\,\text{kg/s}$]

Some problems deal with the continuity equation alone. In each instance you should be able to draw the tube of flow to which it is applied. The fluid velocity must be uniform over the cross section at the points where the equation is applied. For steady, nonviscous flow in a pipe you may usually take the whole pipe as a tube of flow and assume the velocity is uniform throughout a cross section unless the problem states the contrary. The continuity equation is a statement that for steady flow the mass flux is the same in all parts of the tube: $\rho_1 A_1 v_1 = \rho_2 A_2 v_2$. Nearly all these problems deal with incompressible fluids, for which the volume flow rate is the same in all parts of the tube and $v_1 A_1 = v_2 A_2$. A typical problem gives three of the quantities that appear in this equation and asks for the third. Here is an example in which the mass flux is given.

PROBLEM 3. A long pipe with a radius of 2.5 cm is joined smoothly to another long pipe with a radius of 1.6 cm. If the mass flux is a steady 6.8 kg/s what is the volume flow rate and what is the fluid speed in each of the pipes? Water is essentially incompressible and its density is $1.0 \times 10^3\,\text{kg/m}^3$.

SOLUTION:

[ans: $6.80 \times 10^{-3}\,\text{m}^3/\text{s}$; 3.46 m/s in the wide pipe; 8.46 m/s in the narrow pipe]

Sometimes you must use kinematics to obtain one of the speeds. An example is water streaming steadily from a faucet and falling under the influence of gravity. The trajectories of the outer fluid particles define a tube of flow and since the particles speed up as they fall the tube is more narrow at the bottom than at the top. If you know the speed at the top you can use free fall kinematics to compute the speed at any point along the path. If you know the area of the faucet you can use the equation of continuity to compute the area of the stream at any point.

PROBLEM 4. Water flows steadily at a rate of 1.2 gal/min from a circular faucet with a radius of 3.6 mm. What is the radius of the stream 12 cm below the faucet?

SOLUTION: The volume flow rate is $R = A_0 v_0$, where A_0 is the area of the faucet and v_0 is the water speed as it leaves the faucet. Use the conversion tables in the appendix to convert gal/min to m³/s and solve for v_0. The water speed v a distance h below the faucet is given by $v^2 = v_0^2 + 2gh$. Solve for v. Finally, the area A of the stream at the lower point can be found from $R = Av$. The radius r, of course, is found using $A = \pi r^2$.

[ans: 3.16 mm]

Sample Problem 2 in the text is a nice example of how the equation of continuity and the Bernoulli equation are used together. In part a the definition of volume flow rate and the equation of continuity are used to find the fluid velocity at two points in a tube of flow, then these are substituted into the Bernoulli equation to find the difference in pressure. Here are some other examples.

PROBLEM 5. A garden hose has a radius of 6.3 mm and its nozzle has a radius of 2.0 mm. If water is to leave the nozzle at 15 m/s how fast must it be going in the hose?

SOLUTION: Use $v_1 A_1 = v_2 A_2$.

[ans: 1.51 m/s]

If the hose is horizontal what is the difference in pressure between the nozzle and the hose? The density of water is 1.0×10^3 kg/m³.

SOLUTION: Use $p_1 + \frac{1}{2}\rho v_1^2 = p_2 + \frac{1}{2}\rho v_2^2$. Take point 1 to be in the hose and point 2 to be in the nozzle, then solve for $p_1 - p_2$.

[ans: 1.11×10^5 Pa]

The positive answer indicates that the pressure is higher in the hose than in the nozzle.

PROBLEM 6. Suppose the joined pipes of Problem 3 are horizontal. If the pressure in the wide pipe is 1.6×10^5 Pa what is the pressure in the narrow pipe?

SOLUTION:

[ans: 1.30×10^5 Pa]

Sometimes you must solve the continuity and Bernoulli equations simultaneously.

PROBLEM 7. A horizontal pipe carrying water narrows from a radius of 3.1 cm to a radius of 1.8 cm. A pressure difference of 2.3×10^5 Pa is maintained between the two ends. What is the water speed in the two sections?

SOLUTION: Solve $A_1 v_1 = A_2 v_2$ and $p_1 + \frac{1}{2}\rho v_1^2 = p_2 + \frac{1}{2}\rho v_2^2$ simultaneously for v_1 and v_2. Substitute $v_2 = (A_1/A_2)v_1$, from the first equation, into the second and show that $p_1 - p_2 = \frac{1}{2}\rho(A_1^2/A_2^2 - 1)v_1^2$. Solve for v_1, then solve $A_1 v_1 = A_2 v_2$ for v_2. Don't forget that the pressure is higher in the wide portion of the tube than in the narrow portion.

[ans: 7.68 m/s; 22.8 m/s]

As fluid passes from the narrow portion to the wide portion of a tube its speed increases. This means that the momentum and kinetic energy of any given quantity of fluid both increase. A net force acts on the fluid and net work is done on the fluid. For a horizontal tube the pressure is responsible for the work. Both the pressure and the walls of the tube are responsible for the net force.

PROBLEM 8. Consider a segment of fluid that initially straddles the junction between the two pipes of Problem 3, one boundary of the segment being in the wide pipe and the other boundary being in the narrow pipe, as shown by the dotted lines. The fluid is moving from left to right in the diagram. Take the pressure in the wide pipe to be 1.6×10^5 Pa, as in Problem 6. During the time that the boundary in the wide pipe moves 10 cm toward the junction what is the work done on the fluid segment by the pressure in the wide pipe? What is the work done on the fluid segment by the pressure in the narrow pipe? What is the net work done on the fluid segment? Assume the boundary in the wide pipe started more than 10 cm from the junction.

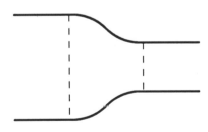

SOLUTION: The force exerted on the boundary in the wide pipe is $F_1 = p_1 A_1$, where p_1 is the pressure and A_1 is the cross-sectional area. The work done by this force is $W_1 = F_1 d_1 = p_1 A_1 d_1$, where d_1 is the distance the boundary moves.

You now need to compute the distance d_2 traveled by the boundary of the segment in the narrow pipe. Since the fluid is incompressible the volume of fluid that leaves the wide pipe is the same as the volume of fluid that enters the narrow pipe. Thus $A_1 d_1 = A_2 d_2$ and $d_2 = (A_1/A_2) d_1$. The work done by the pressure in the narrow pipe is $W_2 = -p_2 A_2 d_2$ and the net work is $W_1 + W_2$. Note that W_2 is negative since the fluid is moving in the direction opposite to that of the force.

The fluid speeds were calculated in Problem 3 and the pressure in the narrow pipe was calculated in Problem 6.

[ans: 31.4 J; −25.6 J; 5.84 J]

What is the change in the total kinetic energy of the fluid segment as it moves 10 cm?

SOLUTION: The same change would occur if the kinetic energy of a part of the segment with volume $V = A_1 d_1 (= A_2 d_2)$ changes from $\frac{1}{2}\rho v_1^2 A_1 d_1$ to $\frac{1}{2}\rho v_2^2 A_2 d_2$, so $\Delta K = \frac{1}{2}\rho(v_2^2 - v_1^2) A_1 d_1$.

[ans: 5.84 J]

Notice that the change in kinetic energy equals the net work done on the segment. The work-energy theorem is obeyed. But, of course, the Bernoulli equation guarantees that. In the space below use the continuity and Bernoulli equations to show that $p_1 A_1 d_1 - p_2 A_2 d_2 = \frac{1}{2}\rho(v_2^2 - v_1^2) A_1 d_1$, regardless of the particular values of any of the quantities.

PROBLEM 9. What is the change in the total momentum of the fluid segment described in the last problem as its boundary in the wide pipe moves 10 cm toward the junction?

SOLUTION: The change in momentum is the same as the change that would occur if the momentum of a part of the fluid segment with volume $A_1 d_1$ changes from $\rho A_1 d_1 v_1$ to $\rho A_2 d_2 v_2$, so the change in momentum is given by $\Delta P = \rho A_1 d_1 (v_2 - v_1)$.

[ans: 0.980 kg·m/s]

According to Newton's second law this should be Ft, where F is the net force on the segment and t is the time it takes to move the prescribed distance. Calculate $F_1 t$ and $F_2 t$, where F_1 is the force of neighboring fluid in the wide pipe on the segment and F_2 is the force of neighboring fluid in the narrow pipe on the segment. You will find that the two contributions do not sum to 0.980 kg·m/s. There is an additional force, the net force of the pipe walls at the constriction. Calculate the horizontal component F_{wall} of this force.

272   Chapter 18: Fluid Dynamics

SOLUTION: First calculate the time for the fluid segment to move the prescribed distance. If d_1 is the distance moved by the boundary in the wide pipe then $t = d_1/v_1$. If d_2 is the distance moved by the boundary in the narrow pipe then $t = d_2/v_2$. Both these equations yield the same result.

Since $F_1 = p_1 A_1$, $F_1 t = p_1 A_1 d_1/v_1$. Similarly, $F_2 t = -p_2 A_2 d_2/v_2$. Use $\Delta P = F_1 t + F_2 t + F_{wall} t$ to calculate the horizontal component F_{wall} of the net force of the walls on the fluid segment.

[ans: 9.07 kg·m/s; −3.02 kg·m/s; −175 N]

The negative sign indicates that the horizontal component of the force opposes the fluid flow. The force of a pipe wall on the fluid is perpendicular to the wall. On the diagram above indicate with arrows the direction of the forces of the walls in the constriction and verify that the horizontal components oppose the fluid flow.

Because the forces of the walls are perpendicular to the fluid velocity they do zero work and so were not considered when you calculated the work done on the fluid. They do change the directions of the streamlines, however.

In some cases the tubes of flow are not horizontal. Then the gravitational terms must be included in the Bernoulli equation.

PROBLEM 10. A utility company uses power generated during off-peak hours to pump water to a reservoir on the top of a hill. During peak hours the water is allowed to run downhill to a hydroelectric plant where it is used to generate electrical power. Suppose a pipe with a uniform radius is used to carry water from a pumping station to the top of a hill, 1000 m above the station. If the pipe outlet is at atmospheric pressure, what pressure should be maintained at the pumping station?

SOLUTION: Since the radius of the pipe is everywhere the same the fluid speed does not change along the pipe and the Bernoulli equation becomes $p_1 = p_2 + \rho g h$, where p_1 is the pressure at the station, p_2 is atmospheric pressure, and h is the height of the hill. Calculate p_1.

[ans: 9.90×10^6 Pa]

Suppose a volume flow rate of 4.0 m³/s is desired. Can the input pressure be lowered if the pipe is narrowed at the top? If it is widened? Take the radius of the pipe to be 15.0 cm at the pumping station and 5.00 cm at the top of the hill and calculate the pressure difference from end to end of the pipe. Repeat the calculation for a pipe radius of 45.0 cm at the top of the hill.

SOLUTION:

[ans: 1.08×10^7 Pa; 9.80×10^6 Pa]

Some problems deal with an incompressible fluid flowing from a hole in a tank. If the tank has a large cross-sectional area and the hole has a small area the fluid velocity at the upper surface of the fluid is much smaller than the velocity at the hole and may be neglected. If the tank is open then the pressure at the top and at the hole is the same, atmospheric pressure. The Bernoulli equation becomes $\frac{1}{2}v^2 = gy$, where y is the height of the upper fluid surface above the hole. Thus $v = \sqrt{2gy}$. This is the speed a particle has after falling a distance y from rest. If there is more than one hole, each can be treated independently if their areas are much smaller than the area of the tank. Here's an example for which the speed of the upper surface cannot be neglected.

PROBLEM 11. A cylindrical tank, open at the top, has a radius of 2.4 m and is initially filled with water to a height of 8.0 m. If it has a hole of radius 1.2 m in the bottom what is the speed with which water initially flows from the hole? What is the speed with which the surface of the water in the tank is initially descending?

SOLUTION: Since the pressure is atmospheric at the top and at the hole, the Bernoulli equation becomes $\frac{1}{2}v_1^2 + gh = \frac{1}{2}v_2^2$, where the subscript 1 refers to the water surface in the tank, the subscript 2 refers to the hole, and the water depth is denoted by h. The equation of continuity is $A_1v_1 = A_2v_2$. Show that $2gh = v_2^2(1 - A_2^2/A_1^2)$ and solve for v_2. Show that $2gh = v_1^2(A_1^2/A_2^2 - 1)$ and solve for v_1.

[ans: 12.9 m/s; 3.23 m/s]

The fluid speed at the hole is not the same as $\sqrt{2gh} = 12.5$ m/s. The initial speed v_1 is significant when compared to the speed v_2 at the hole. Now repeat the calculation for a hole with a radius of 0.25 m.

SOLUTION:

[ans: 12.5 m/s; 0.136 m/s]

Now the speed v_1 is much smaller than v_2 and the water at the top of the tank moves slowly.

Study of the Venturi meter is an excellent way to review many of the concepts of this and the previous chapter. To derive Eq. 11 of the text you must make use of both the continuity and Bernoulli equations as well as the variation of pressure with depth in a fluid.

PROBLEM 12. Derive Eq. 11 of the text for the velocity of a fluid in a Venturi meter. Suppose the density of the moving fluid is ρ, the area of the main part of the tube is A, the area of the constricted part of the tube is a, the density of the manometer fluid is ρ', and the difference in height of the manometer fluid in the two arms is h. Show that the speed of the fluid in the main part of the tube is given by

$$v = a\sqrt{\frac{2(\rho' - \rho)gh}{\rho(A^2 - a^2)}}$$

SOLUTION: Since the Venturi tube is horizontal, you need to consider only the pressure and velocity terms in the Bernoulli equation. If you were given the pressure difference between points 1 and 2 in Fig. 9 you could solve the Bernoulli and continuity equations simultaneously for the fluid speed. Do this:

You should have obtained

$$v = a\sqrt{\frac{2(p_1 - p_2)}{\rho(A^2 - a^2)}}$$

The manometer supplies the pressure difference in terms of the difference in heights of the fluid column in the two arms. Use what you know about the variation of pressure with depth to find an expression for the pressure difference in terms of h, ρ, and ρ'. In the space below show that the pressure difference between points 1 and 2 in Fig. 9 is given by $p_1 - p_2 = (\rho' - \rho)gh$. Don't forget to take into account the fluid of density ρ that fills the upper parts of the manometer tube.

Now substitute for $p_1 - p_2$ in the expression for v. You should obtain the desired result.

The continuity or Bernoulli equation alone can be used to solve some of the problems of this chapter; for others they must be used in conjunction with each other. Read the problems in your homework assignment and write the number of each problem in the space provided below. Next to the number of the problem write either "continuity", Bernoulli", "both", or "neither", as appropriate.

PROBLEM	EQUATION USED FOR SOLUTION
_____	_____
_____	_____
_____	_____
_____	_____
_____	_____
_____	_____

III. NOTES

Chapter 19
WAVE MOTION

I. BASIC CONCEPTS

In this chapter you study wave motion, a mechanism by which a disturbance (or distortion) created at one place in a medium propagates to other places. A wave may carry energy and momentum with it, but not matter.

The general ideas are specialized to a mechanical wave on a taut string. Pay attention to those characteristics of a string that determine the speed of a wave. You will also learn about sinusoidal waves, for which the disturbance has the shape of a sine or cosine function. This type wave has a special terminology associated with it; you will need to know the meaning of the terms amplitude, frequency, period, angular frequency, wavelength, wave number, and phase. You will also need to understand what determines the values of each of these quantities.

Two or more waves present at the same time and place give rise to what are called interference effects. Special combinations of traveling waves result in standing waves. Learn what these phenomena are and how to analyze them.

General characteristics of traveling waves. For each of the following types of waves tell precisely what is distorted and how a distortion might be created:

mechanical wave on a taut string
what is distorted: _____
how wave is generated: _____

sound wave in air
what is distorted: _____
how wave is generated: _____

water wave
what is distorted: _____
how wave is generated: _____

This chapter concentrates on mechanical waves on taut strings.

A distortion in one region of space causes a disturbance in neighboring regions and so the disturbance moves from place to place. After a taut string is distorted and released, for example, the distorted section pulls on neighboring sections to distort them and the neighboring sections pull the originally distorted portion back to its undistorted position.

When a mechanical wave is present, both the waveform (the disturbance) and the medium (string, water, air) move. These motions are related to each other but they are NOT the same. While studying this chapter you should be continually aware of which of these motions is being discussed.

Waves are sometimes classified as transverse or longitudinal, according to the direction the medium moves relative to the direction the wave moves. In a <u>longitudinal</u> wave, the medium is displaced in a direction that is _____ to the direction the wave travels and in a <u>transverse</u> wave it is displaced in a direction that is _____ to the direction the wave travels. Many waves, water waves among them, are neither transverse nor longitudinal.

A wave may also be classified as one-, two-, or three-dimensional, according to the number of dimensions in which it propagates. For example, a wave on a taut string is _____ dimensional, a water wave produced by a dropped pebble is _____ dimensional, and a sound wave emitted from a point source is _____ dimensional.

Waves may also be classified as <u>pulses</u> or <u>trains</u>. Distinguish between a wave pulse and a wave train: _____
_____

Some wave trains are <u>periodic</u>. A periodic wave train is one for which _____
_____

Some periodic wave trains are <u>harmonic</u>. A harmonic wave train is one for which _____
_____

The concept of <u>wavefront</u> is important for waves moving in two and three dimensions. A wavefront is a line (for two-dimensional waves) or a surface (for three-dimensional waves) such that all parts of the medium on it _____
_____

A <u>ray</u> is a line that is _____
_____

The <u>wave speed</u> is the speed with which the distortion moves. The graph below shows a distorted string at time $t = 0$. The string stretches along the x axis and its displacement is denoted by y. Suppose the distortion travels without change in shape to the right at 2.0 m/s. On the coordinates provided sketch the string at $t = 1$ s and $t = 2$ s.

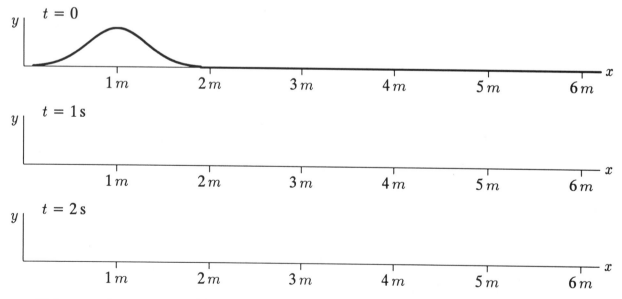

If the waveform moves with constant speed v and without change in shape then any particular point on the waveform (the maximum displacement, for example) moves so its coordinate

x at time t is given by $x - vt =$ constant for waveforms traveling in the *positive* x direction and by $x + vt =$ constant for waveforms traveling in the *negative* x direction. Carefully note the signs in these expressions.

For the waveform shown on the three graphs above take x to be the coordinate at the maximum string displacement and fill in the following table.

t (s)	x (m)	$x - vt$ (m)
0		
1		
2		

If you did not get the same answer for $x - vt$ each time, you probably did not draw the graphs correctly.

A wave on a string is described by giving the string displacement $y(x, t)$ as a function of position and time. For example, the graphs above show $y(x, t)$ for three different values of t. If you know the function $y(x, t)$ you can find the displacement of any point on the string at any time. Simply substitute the value of the coordinate x of the point and the value of the time t into the function.

If you are given the wave speed and the function $f(x)$ that describes the string displacement at $t = 0$, then you can find the displacement $y(x, t)$ for later times. If the waveform is moving in the positive x direction you substitute _____ for x in the function $f(x)$. If the waveform is traveling in the negative x direction you substitute _____ for x.

Sinusoidal waves. A sinusoidal wave is shown below for time $t = 0$.

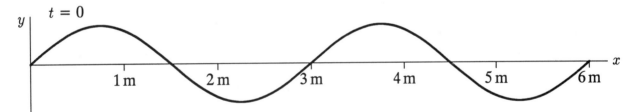

The function that describes the string displacement is $f(x) = y_m \sin(kx)$, where y_m and k are constants. y_m is called the _____ of the wave. Indicate it on the graph. k is called the _____ of the wave. It is related to the wavelength λ by $k =$ _____. Indicate the wavelength on the graph by marking an appropriate distance with the symbol λ. Any two points that are separated by a multiple of λ have identical displacements at every instant. The wavelength of the wave shown is about _____ m.

If the wave travels with speed v in the positive x direction, the displacement $y(x, t)$ at any time t is given by

$$y(x, t) = y_m \sin[k(x - vt)] = y_m \sin(kx - \omega t),$$

where $\omega = kv$.

Every point of the string moves in simple harmonic motion with angular frequency _____. In terms of the angular frequency the frequency of its motion is given by $\nu =$ _____ and the period of its motion is given by $T =$ _____. The period of an oscillation tells us _____ and the frequency tells us _____

The graph below shows the displacement of a point on the string as a function of time. On the time axis use the symbol T to label a time interval equal to a period.

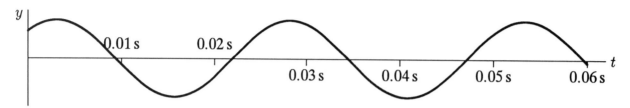

The period of the wave shown is about _____ s and the frequency is about _____ Hz.

When ω is written in terms of the period T and k is written in terms of the wavelength λ, $\omega = kv$ becomes $\lambda = vT$. That is, the wave moves a distance equal to one wavelength in a time equal to _____. This result should be obvious to you. Consider a point on the string that has maximum displacement at $t = 0$. There is another point with the same displacement a distance λ away. By the time this part of the wave gets to the first point that point has moved through one complete cycle.

The most general functional form for a sinusoidal wave traveling in the positive x direction is $y(x,t) = y_m \sin(kx - \omega t - \phi)$. The quantity $kx - \omega t - \phi$ is called the _____ and ϕ is called the _____. If $\phi = 0$ then at $t = 0$ the point at $x = 0$ has displacement $y =$ _____, if $\phi = \pi/2$ radians it has displacement $y =$ _____, and if $\phi = 3\pi/2$ radians it has displacement $y =$ _____.

At any time, waves with different phase constants are shifted relative to each other along the x axis. Different values of ϕ result in different shifts. In fact, the displacement $y_1(x,t) = y_m \sin(kx - \omega t)$ evaluated at $x = x_0$ is the same as the displacement $y_2(x,t) = y_m \sin(kx - \omega t - \phi)$ evaluated at $x = x_0 + \Delta x$, where $\Delta x = \phi/k = \lambda\phi/2\pi$. For example, x_0 might be the coordinate of a maximum of $y_1(x,t)$ at some time t. Then $x_0 + \lambda\phi/2\pi$ is the coordinate of a maximum of $y_2(x,t)$ at the same time. If $\phi = \pi/2$ the wave y_2 is shifted by $\Delta x =$ _____ from the wave y_1; if $\phi = \pi$ it is shifted by $\Delta x =$ _____. If ϕ is negative the shift is in the negative x direction.

On the axes below use different colors to draw two waves traveling in the same direction with the same wavelength and frequency but with different phase constants: $y_1(x,t) = y_m \sin(kx - \omega t)$ and $y_2(x,t) = y_m \sin(kx - \omega t - \pi/2)$ for any time. Take the wavelength of each wave to be about 2 m. Indicate the amount of the shift by marking a maximum of y_1 and the nearest maximum of y_2. Check to be sure it is the fraction of a wavelength that you calculated above.

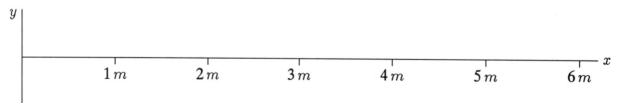

Wave speed. Carefully read Section 4 of the text. There you are shown that Newton's second law leads directly to an expression for the wave speed in terms of the tension F in the string and the linear mass density μ of the string. Specifically,

$$v =$$

The tension in the string is usually determined by the external forces applied at its ends to hold it taut.

You should carefully note that the wave speed does NOT depend on the frequency or wavelength. If the frequency is increased the wavelength must decrease so that $v = \lambda \nu$ has the same value. Suppose, for example, the wave speed on a certain string is 5.0 m/s. Then the wavelength of a 100 Hz sinusoidal wave is _____ m. If a 200 Hz sinusoidal wave travels on the same string with the same tension, its speed is _____ m/s and its wavelength is _____ m.

Particle velocity The transverse velocity $u(x,t)$ of the string at the position with coordinate x can be found as a function of time by differentiating _____ with respect to time. If the displacement is given by $y_m \sin(kx - \omega t - \phi)$ then

$$u(x,t) =$$

You should distinguish between wave and particle velocities. The wave velocity is associated with the motion of _____, while the particle velocity is associated with the motion of _____. Consider a transverse wave on a string and list some characteristics that are different for these two velocities: _____

_____

The acceleration of a point on a string can be found by differentiating _____ with respect to time. For a sinusoidal wave the result is

$$a(x,t) =$$

Notice that it is proportional to the displacement $y(x,t)$ and that the constant of proportionality is negative, a result you should expect since the string at any point is in simple harmonic motion.

The wave equation. As a wave moves past a point on a string, both the displacement and string velocity there vary with time. As pointed out in Section 4 of the text the forces exerted by neighboring portions of string produce the acceleration of the string. Carefully read Section 5 of the text.

The end result is the wave equation for one-dimensional waves, labelled Eq. 25 in the text. This equation is a differential equation that relates the second derivative of the displacement with respect to time to its second derivative with respect to coordinate. For reference copy it here:

$$\frac{\partial^2 y}{\partial x^2} =$$

Solutions to this equation *must* be functions of $x - vt$ or $x + vt$. The coordinate x and time t cannot enter a solution in any other combination. Thus waves that obey this equation travel with speed v and do not change shape as they move.

Power and Intensity. A mechanical wave on a taut string, as well as other waves, carries energy from place to place. Each segment of a string does work on the neighboring section and thereby transfers energy. For a transverse wave the power transmitted is the product of the string velocity and the *transverse* component of the tension. Since $u = \partial y/\partial t$ is the string velocity and $F_t = F\,\partial y/\partial x$ is the transverse component of the tension, the power transmitted is given by $P =$ _____. This is a general expression that can be used for any transverse waveform. For a sinusoidal wave the power transmitted, written as a function of position and time, is given by

$$P(x,t) =$$

and the average over a cycle is given by

$$P_{\text{ave}} =$$

Note that averaging removes the dependence on position and time. Also note that the power transmitted is proportional to the square of the amplitude. This is true for other types of waves as well.

Intensity is used to discuss the power transmitted by three-dimensional waves. Tell what this term means: _____

_____

Consider a point source of light or sound. If no energy is lost in transmission, then during a cycle the energy crossing every sphere centered at the source is the same. Since the area of a sphere is proportional to the square of its radius r, the intensity is proportional to _____, where r is the distance from the source. Thus the amplitude as a function of r is proportional to _____. Suppose, for example, the intensity is $20\,\text{W/m}^2$ at a point 2 m from a point source of light. Then the intensity 4 m from the source is _____ W/m².

Superposition and interference. To derive some results in this and following sections you should know the trigonometric identity

$$\sin A + \sin B = 2\sin\left(\frac{A+B}{2}\right)\cos\left(\frac{A-B}{2}\right),$$

valid for any angles A and B. It is proved in the Mathematical Skills section.

When two waves, one with displacement $y_1(x,t)$ and the other with displacement $y_2(x,t)$ are simultaneously on the same string, the displacement of the string is given by their sum: $y(x,t) = y_1(x,t) + y_2(x,t)$, provided the amplitudes are small. The addition of waveforms is called superposition. Carefully note that displacements, not intensities, add. If the amplitudes are not small the presence of one wave changes the shape of a second wave. This situation is not considered here.

The addition of waveforms leads to some interesting and important phenomena, called interference phenomena. Consider two sinusoidal waves with the same frequency, traveling in the same direction on the same string. Suppose the waves have the same amplitude but different phase constants and let $y_1(x,t) = y_m \sin(kx - \omega t - \phi_1)$ and $y_2(x,t) = y_m \sin(kx -$

$\omega t - \phi_2$). The waves are identical except that at every instant the second is shifted along the x axis from the first by an amount that depends on the value of ϕ. Eq. 36 gives the equation for the resultant string displacement. Copy it here:

$$y(x,t) =$$

where $\Delta\phi = \phi_2 - \phi_1$ and $\phi' = (\phi_1 + \phi_2)/2$. Now use the trigonometric identity given above to prove the result:

Notice that the composite wave is sinusoidal with the same frequency and wavelength as either of the constituent waves.

Concentrate on the amplitude of the composite wave. It is a function of $\Delta\phi$ and, in particular, is given by _____. The maximum amplitude is given by _____ and the amplitude is a maximum when $\Delta\phi$ has any of the values _____ Maximum amplitude occurs when $\Delta\phi$ is adjusted so the crests of one of the constituent waves fall exactly on the _____ of the other. This condition is called _____ interference. The minimum amplitude is zero and the amplitude is a minimum when $\Delta\phi$ has any one of the values _____. Minimum amplitude occurs when $\Delta\phi$ is adjusted so the crests of one of the constituent waves fall exactly on the _____ of the other. This condition is called _____ interference. For other values of $\Delta\phi$ the interference is destructive but it is not complete.

Standing waves and resonance. In a standing wave each part of the string oscillates back and forth but the waveform does not move, as it does in a traveling wave. No energy is transmitted from place to place.

A standing wave can be constructed as the superposition of two traveling waves with the same amplitude and frequency, but moving in opposite directions. Let $y_1(x,t) = y_m \sin(kx - \omega t)$ and $y_2(x,t) = y_m \sin(kx + \omega t)$ represent the two traveling waves. The sum is given by Eq. 40 of the text. Copy it here:

$$y(x,t) =$$

Now use the trigonometric identity given above to prove the result:

Note that x and t do NOT enter the expression for $y(x,t)$ in either of the forms $x - vt$ or $x + vt$. The wave is NOT a traveling wave. Also note that each point on the string vibrates in simple harmonic motion with an amplitude that varies with position along the string. In fact, the amplitude of the oscillation of the point at x is given by _____.

At certain points, called <u>nodes</u>, the amplitude is zero. Since their displacements are always zero, these points on the string do not vibrate at all. For the standing wave given above nodes occur at positions for which kx is a multiple of _____ or, what is the same thing, x is a multiple of _____. Nodes are _____ wavelength apart.

At other points, called _____, the amplitude is a maximum. In terms of y_m it is _____. For the standing wave given above, these points occur at positions for which kx is an odd multiple of _____ or, what is the same thing, x is an odd multiple of _____.

The phase constant for both traveling waves was chosen to be 0 but that is not a necessity. If the phase constants of the constituent traveling waves are different, the nodes and antinodes are still the same distances apart but they are shifted relative to those that occur when the phase constants are the same.

One way to generate a standing wave is simply to allow a sinusoidal wave train to be reflected from an end of the string. The reflection, of course, travels in the opposite direction. If such a wave is reflected from a *fixed* end the reflected and incident waves at the end have phase constants that differ by _____ radians. Specifically, if the incident wave is $y_m \sin(kx - \omega t)$ and the fixed end is at $x = L$ then the reflected wave is _____. This phase change must occur so that the fixed end of the string is a node of the standing wave pattern.

On the other hand, if the same wave is reflected from a free end at $x = L$, the reflected wave is given by _____. There is no phase change at the end. The free end of the string is an antinode of the standing wave pattern.

For a string with both ends fixed, both ends are nodes of a standing wave pattern. This means that the traveling waves producing the pattern may have only certain wavelengths. Possible wavelengths are determined by the condition that the length of the string must be a multiple of _____, where λ is the wavelength. If v is the wave speed for traveling waves on the string, then the standing wave frequencies are given in terms of the string length L by $\nu =$ _____, where n is a positive integer.

The lines below represent a string with fixed ends. Draw the amplitude as a function of position for the standing waves with the three lowest frequencies.

|———————| |———————| |———————|

Standing wave frequencies are called the _____ frequencies of the string. If the string is driven at one of its standing wave frequencies by an applied sinusoidal force, the amplitude at the antinodes becomes large. The driving force is said to be in <u>resonance</u> with the string. Of course, the string can be driven at another frequency but then the amplitude remains small.

When an applied driving force is in resonance with a string the energy supplied by the applied force is dissipated in internal friction. In the absence of friction the standing wave amplitude would grow without bound. This is to be contrasted with the situation in which the driving force is not in resonance with the string. Then, in addition to frictional dissipation, the string loses energy by doing work on _____ .

II. PROBLEM SOLVING

You can "watch" a wave pulse move past a point by evaluating the displacement function $y(x,t)$ for a given value of x and a succession of values of t.

PROBLEM 1. A string extends along the x axis from large negative values of x to large positive values of x. At time $t = 0$ the shape of the string is given by the function

$$f(x) = Ae^{-x^2/a},$$

where A and a are constants. This function describes a bell-shaped curve with a peak of height A located at $x = 0$. Suppose the distortion moves in the positive x direction with speed v. Write the expression for the string displacement $y(x,t)$, valid for any later time t.

SOLUTION: Substitute $x - vt$ for x.

[ans: $y(x,t) = Ae^{-(x-vt)^2/a}$]

Take $A = 1.2$ cm, $a = 1.5$ m^2, and $v = 6.0$ m/s. Calculate the displacement at the point $x = 6.0$ m for $t = 0, 1, 2, 3, 4, 5$, and 6 s. Fill in the following table.

t(s)	y(cm)
0	
1	
2	
3	
4	
5	
6	

Notice that the peak, where y = 1.2 cm, arrives at $x = 6.0$ m at time $t = 3.0$ s. This is consistent with a wave moving at 2.0 m/s.

Many of the problems involving sinusoidal waves on a string deal with the relationships $v = \lambda \nu = \lambda/T = \omega/k$, where v is the wave speed, λ is the wavelength, ν is the frequency, T is the period, ω is the angular frequency, and k is the wave number. Typical problems might give you the wavelength and frequency, then ask for the wave speed or might give you the wave speed and period, then ask for the wavelength or wave number.

Sometimes the quantities are given by describing the motion. For example, a problem might tell you that the string at one point takes a certain time to go from its equilibrium position to maximum displacement. This, of course, is one-fourth the period. In other problems the frequency of the source (a person's hand or a mechanical vibrator, for example) might be given. You must then recognize that the frequency of the wave is the same as the frequency of the source.

PROBLEM 2. A sinusoidal wave is traveling on a string. At one instant the point on the string at $x = 6.5$ cm has its maximum displacement, the point at $x = 4.5$ cm has zero displacement, and no part of the string between these points has zero displacement. The displacement at $x = 6.5$ cm drops to zero in 1.5 ms. What are the period, frequency, angular frequency, wavelength, and speed of the wave?

SOLUTION: The 1.5 ms interval represents one quarter of a period. Calculate the period, then take the reciprocal to find the frequency and use $\omega = 2\pi\nu$ to find the angular frequency. The two points are one quarter of a wavelength apart. Use their coordinates to calculate the wavelength. Finally use $v = \lambda\nu$ or $v = \lambda/T$ to calculate the wave speed.

[ans: 6.00 ms; 167 Hz; 1050 rad/s; 8.0 cm; 13.3 m/s]

Instead of the wave speed, the tension F in the string and the linear mass density μ of the string might be given. You must then use $v = \sqrt{F/\mu}$ to find the wave speed. In other instances you find the wave speed from the wavelength and frequency or other related quantities, then use $v = \sqrt{F/\mu}$ to find F or μ, given the other.

PROBLEM 3. The linear mass density of the string in the previous problem is 2.0 g/m. What is the tension in the string?

SOLUTION:

[ans: 0.356 N]

If a sinusoidal wave with a frequency of 500 Hz is generated in the same string, with the same tension, what are the wave speed, wavelength, and wave number of the wave?

SOLUTION: The wave speed is the same as before. Use $v = \lambda\nu$ to find the wavelength, then $k = 2\pi/\lambda$ to find the wave number.

[ans: 13.3 m/s; 2.67 cm; 236 m^{-1}]

The tension may not be given directly but sufficient information may be given to compute it. For example, one end of a horizontal string may be attached to a wall and the other end may go around a pulley so it hangs vertically. If mass m is attached to the hanging end, the tension may be taken to be mg. If, in another example, a string with linear mass density μ is hung from the ceiling, the tension must increase linearly with distance along the string, so the change in tension over a segment of length Δd is just right to cancel the downward pull of gravity on the segment.

PROBLEM 4. One end of a horizontal string with a linear mass density of 25 g/m is fixed to a wall. The other end passes over a light frictionless pulley and a mass of 200 g is attached to it. What is the tension in the string? What is the speed of a wave on the string? What is the wavelength of a 100 Hz sinusoidal wave on the string?

SOLUTION: The tension is given by $F = mg$, the speed by $v = \sqrt{F/\mu}$, and the wavelength by $\lambda = v/\nu$.

[ans: 1.96 N; 8.85 m/s; 8.85 cm]

A string with linear mass density μ is hung vertically from the ceiling and mass m is attached to its free end. If the length of the string is L, what is the tension in the string a distance d from the ceiling?

SOLUTION: The tension decreases linearly with d. It must be $mg + \mu g L$ for $d = 0$ and mg for $d = L$.

[ans: $mg + \mu g(L - d)$]

If $\mu = 25$ g/m, $m = 200$ g, and $L = 2.5$ m what is the wave speed near the ceiling? near the hanging mass?

SOLUTION:

[ans: 10.2 m/s; 8.85 m/s]

If two strings are joined together and the composite is held taut, the tension is the same in both segments. As a sinusoidal wave passes from one segment to the other its frequency does not change. If, however, the segments have different linear densities then the wave speed and wavelength change.

PROBLEM 5. Suppose the ends of two strings, one with a linear mass density of 5.0×10^{-4} kg/m and the other with a linear mass density of 4.0×10^{-3} kg/m, are joined and held taut with a tension of 1.25×10^{-2} N. A 100 Hz sinusoidal wave travels from the lighter to the heavier string. What are the wave speed and wavelength of the wave when it is in each of the segments?

SOLUTION:

[ans: lighter: 5.00 m/s, 5.00 cm; heavier: 1.77 m/s, 1.77 cm]

In some problems the displacement is given as a function of x and t and you must be able to recognize the various parameters that appear in the function.

PROBLEM 6. Suppose $y(x,t) = 0.020 \sin(450x + 4000t - 2.0)$, where y and x are in meters and t is in seconds, describes a sinusoidal wave on a taut string. What are the amplitude, angular frequency, wave number, and phase constant of this wave? In what direction is the wave traveling?

SOLUTION:

[ans: 0.020 m; 4000 rad/s; 450 m^{-1}; 2.0 rad; in the negative x direction]

What are the frequency, period, wavelength, and wave speed of this wave?

SOLUTION:

[ans: 637 Hz; 1.57 ms; 1.40 cm; 8.89 m/s]

The function $y(x,t)$ can be used to find the displacement and string velocity at any point and any time.

PROBLEM 7. For the sinusoidal wave of the previous problem what are the displacement and string velocity at $x = 1.5$ cm and $t = 2.5$ s?

SOLUTION: Evaluate $y = y_m \sin(kx + \omega t - \phi)$ and $\partial y/\partial t = \omega y_m \cos(kx + \omega t - \phi)$.

[ans: 1.88 cm; −27.3 m/s]

Find the coordinates of all points on the string that, at $t = 0$, are moving in the positive y direction with a speed that is half the maximum speed. Find the displacements of the string at these points.

SOLUTION: These are the points for which $\cos(kx - \phi) = 1/2$. Remember that if α is the solution to $\cos \alpha = 1/2$ then so is $-\alpha$. Thus $\alpha + 2\pi n$ and $-\alpha + 2\pi n$, where n is any integer, are all solutions. Be sure to include both sets of points.

[ans: at $0.677 \pm 1.40n$ cm with displacement $+1.73$ cm, at $0.212 \pm 1.40n$ cm with displacement -1.73 cm, where n is any integer]

For some problems you must be able to write an expression for the displacement as a function of position and time for a sinusoidal wave. Since it has the form $y(x, t) = y_m \sin(kx \pm \omega t - \phi)$, you must be able to determine k, ω and ϕ from data given in the problem statement. The sign in front of ωt is determined by the direction of travel.

PROBLEM 8. A string with a linear mass density of 3.0 g/m lies along the x axis and is held at a uniform tension of 0.15 N. It carries a sinusoidal wave with a frequency of 85 Hz. At time $t = 0$ the point on the string at $x = 0$ has a displacement of $+2.5$ cm (in the positive y direction) and a velocity of -10 m/s (in the negative y direction). The wave is traveling in the positive x direction. Write an expression for the displacement y of the string as a function of x and t.

SOLUTION: For a sinusoidal wave traveling in the positive x direction the displacement has the form $y(x, t) = y_m \sin(kx - \omega t - \phi)$. The angular frequency can be found from $\omega = 2\pi \nu$, the wave speed can be found from $v = \sqrt{F/\mu}$, and the wave number can be found from $\omega = kv$.

We still need the amplitude and phase constant. The initial displacement of the point at $x = 0$ is given by $y_0 = -y_m \sin \phi$ and its initial velocity is given by $u_0 = -\omega y_m \cos \phi$. These equations can be solved for y_m and ϕ. Divide one by the other to obtain $\tan \phi = \omega y_0/u_0$. Use $\sin^2 \phi + \cos^2 \phi = 1$ to obtain that $y_m^2 = y_0^2 + u_0^2/\omega^2$. Don't forget to check your solution for ϕ to be sure $-\sin \phi$ gives the correct sign for the initial displacement and $-\cos \phi$ gives the correct sign for the initial velocity.

[ans: $3.12 \sin(75.5x - 534t + 0.928)$ cm, for x in meters and t in seconds]

PROBLEM 9. A 75 Hz sinusoidal wave is traveling in the positive x direction along a string with a linear mass density of 3.5×10^{-3} kg/m and a tension of 35 N. At time $t = 0$ the point at $x = 0$ has maximum displacement in the positive y direction. When this point next has zero displacement the slope of the string there is 0.15. Write an expression for the displacement of the string as a function of x and t.

SOLUTION: Take the form of the displacement to be $y(x,t) = y_m \sin(kx - \omega t - \phi)$. Use $\omega = 2\pi\nu$ to find the angular frequency, use $v = \sqrt{F/\mu}$ to find the wave speed, and use $\omega = kv$ to find the wave number. Since $y(0,0) = y_m$, $-\sin\phi = 1$ and $\phi = 3\pi/2$. The slope of the string is given by $\partial y/\partial x = ky_m \cos(kx - \omega t - \phi)$ in general and by $S = ky_m \cos(\omega t + \phi)$ at $x = 0$. Since this must be $+0.15$ when $\sin(\omega t + \phi) = 0$ (and $\cos(\omega t + \phi) = \pm 1$), solve $ky_m = 0.15$ for y_m.

[ans: $3.18\sin(4.71x - 471t - 3\pi/2)$ cm, for x in meters and t in seconds]

The equations $P = y_m^2 \mu v \omega^2 \cos^2(kx - \omega t - \phi)$ for the power transmitted at time t past the point at x and $P_{\text{ave}} = \frac{1}{2}y_m^2 \mu v \omega^2$ for the average power transmitted at any time past any point are typically used in a straightforward manner to compute the power, given the parameters of the wave. You may also be asked to find the value of one of the parameters (μ, ω, v, F, y_m, or a related quantity) to achieve a given level of power transmission.

PROBLEM 10. What is the average power transmitted by the wave of Problem 9?
SOLUTION:

[ans: 39.3 W]

The tension in the string is changed to increase the average power transmitted to 50 W without changing the frequency or amplitude. What is the new tension in the string?

SOLUTION: Solve the power equation for v, then use $v = \sqrt{F/\mu}$ to calculate F.

[ans: 56.8 N]

If the frequency is doubled to 150 Hz, keeping the tension at 56.8 N and the amplitude at 3.18 cm, what is the average transmitted power?

SOLUTION:

[ans: 200 W]

The concept of intensity can be used to solve several types of related problems, most dealing with spherical waves emitted from isotropic point sources. Recall that the power transmitted across every sphere centered at the source is the same and is the same as the power emitted by the source. The intensity is the same at all places on any given sphere and is proportional to the inverse of the square of the distance from the source. In one type problem you are given the intensity I_1 at some distance r_1 and asked for the intensity at another distance r_2. Use the relationship $I_1/I_2 = r_2^2/r_1^2$ to solve for I_2. In another case you might be given the power P emitted by a source and asked for the intensity at a point r distant. Use

$I = P/4\pi r^2$. You might also be given the intensity at some point and asked for the power emitted by the source.

Variations deal with the amplitude instead of the intensity. Since the amplitude is proportional to the square root of the intensity, the relationship you will use is $y_{1m}/y_{2m} = r_2/r_1$.

PROBLEM 11. The intensity 500 m from a point source of radio waves is 6.4×10^{-4} W/m². What is the intensity 1500 m from the source?

SOLUTION:

[ans: 7.11×10^{-5} W/m²]

What is the power output of the source?

SOLUTION:

[ans: 2.01×10^3 W]

Interference problems may give the phase difference of two waves and ask for the resultant wave. In variations the resultant is given and you are asked for the phase difference.

PROBLEM 12. Two sinusoidal waves of equal amplitude and frequency are traveling in the same direction on the same string. Suppose the waves are 30° out of phase with each other. What is the ratio of the resultant amplitude to the amplitude of either wave?

SOLUTION:

[ans: 1.81]

For what phase difference does the resultant wave have the same amplitude as either wave? How then is the phase constant of the resultant related to the phase constants of the constituent waves?

SOLUTION:

[ans: 2.09 rad or 120°; 60° greater than one, 60° less than the other]

A phase difference at a point of observation may arise from a phase difference at the sources or from a difference in the distances traveled by the waves to get to the point of observation. Suppose two waves start in phase. If wave 1 travels distance x_1 and wave 2 travels distance x_2, then their respective phases at the observation point are $kx_1 - \omega t$ and $kx_2 - \omega t$. Their phase difference is $k|x_1 - x_2| = 2\pi|x_1 - x_2|/\lambda$, where λ is the wavelength of either wave. If $|x_1 - x_2|$ is a multiple of λ then the phase difference is a multiple of 2π radians and the amplitude of the composite wave at the observation point is a maximum. If $|x_1 - x_2|$ is an odd multiple of $\lambda/2$ then the phase difference is an odd multiple of π radians and the amplitude of the composite wave at the observation point is a minimum.

PROBLEM 13. Identical sinusoidal waves, each with a wavelength of 30 cm, are produced by speakers S_1 and S_2, a distance 1.6 m apart, as shown. The waves have the same phase at the speakers. What is their phase difference at point A, a distance of 2.5 m from the line joining the speakers?

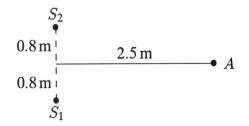

SOLUTION: Notice that the waves travel the same distance.

[ans: 0]

Suppose S_1 is moved along the dotted line an additional 25 cm away from S_2. What then is the phase difference at A?

SOLUTION: The wave from S_1 travels $x_1 = \sqrt{2.5^2 + 1.05^2} = 2.71$ m and the wave from S_2 travels $x_2 = \sqrt{2.5^2 + 0.80^2} = 2.62$ m. Calculate $2\pi|x_1 - x_2|/\lambda$.

[ans: 1.82 rad or 104°]

How far should S_1 be moved along the dotted line from its original position so the speaker separation is the smallest for which the phase difference at A is 180°?

SOLUTION: If d is the distance moved then the wave from S_1 travels $x_1 = \sqrt{2.5^2 + (0.80 + d)^2}$. Solve for the value of d for which $|x_1 - x_2| = \lambda/2$.

[ans: 0.404 m]

How far should S_1 be moved along the dotted line from its original position so the waves are again in phase at A?

SOLUTION: Solve for d so that $|x_1 - x_2| = \lambda$.

[ans: 0.718 m]

Some problems dealing with standing waves on a string give you the amplitude, frequency, and wave number (or related quantities) for the traveling waves and ask for the standing wave pattern. The inverse problem gives the standing wave in the form $y = A\sin(kx)\cos(\omega t)$ and asks for the component traveling waves. They have the form $y_m \sin(kx \pm \omega t)$ with $y_m = A/2$.

Instead of the wavelength or wave number of the traveling waves you might be told the distance between successive nodes or successive antinodes. Double it to find the wavelength. If you are told the distance between a node and a neighboring antinode, multiply it by 4 to find the wavelength.

PROBLEM 14. A certain string has a linear mass density of 4.5×10^{-2} kg/m and is held taut with a uniform tension of 3.0 N. When one end is set into oscillation with a frequency of 95 Hz a standing wave pattern with a maximum amplitude of 1.2 cm results. Write expressions for two traveling waves that combine to form the standing wave.

SOLUTION: The waves have the form $y_m \sin(kx \pm \omega t)$. The angular frequency is given by $\omega = 2\pi\nu$, the wave speed is given by $v = \sqrt{F/\mu}$, the wavelength is given by $\lambda = v/\nu$, and the wave number is given by $k = 2\pi/\lambda$. The amplitude of each traveling wave is half the maximum amplitude of the standing wave.

SOLUTION:

[ans: $0.60\sin(42.3x \pm 346t)$ cm, for x in meters and t in seconds]

How far apart are the nodes?

SOLUTION:

[ans: 29.6 cm]

If a standing wave is generated in a string with both ends fixed, the wave pattern must have a node at each end of the string. This means the length L of the string and the wavelength λ of the traveling waves must be related by $L = n\lambda/2$, where n is an integer.

PROBLEM 15. What are the four longest wavelengths associated with standing waves on a 2.0-m long string?

SOLUTION:

[ans: 4.0 m; 2.0 m; 1.33 m; 1.0 m]

If the linear mass density of the string is 4.5×10^{-3} kg/m and the tension in the string is 8.0 N what are the frequencies associated with these standing wave patterns?

SOLUTION:

[ans: 10.5 Hz; 21.1 Hz; 31.6 Hz; 42.2 Hz]

PROBLEM 16. A 2.0-m long string with a linear mass density of 5.6×10^{-3} kg/m and a tension of 6.5 N has both ends fixed and vibrates in a standing wave pattern with three antinodes. What is the frequency of vibration?

SOLUTION: Draw a graph of the amplitude as a function of position and note that the string is divided into three half-wavelength segments. Thus $\lambda/2 = L/3$, where L is the length of the string. Use this condition to calculate λ. Use $v = \sqrt{F/\mu}$ to calculate the wave speed and $v = \lambda\nu$ to calculate the frequency.

[ans: 25.6 Hz]

III. MATHEMATICAL SKILLS

1. Partial derivatives are important for understanding much of this chapter. The displacement $y(x, t)$ of a string carrying a wave is a function of two variables, the coordinate x of a point on the string and the time t. You may differentiate with respect to either variable.

The notation $\partial y(x, t)/\partial x$ stands for the derivative of y with respect to x, with t treated as a constant. Similarly, $\partial y(x, t)/\partial t$ means the derivative of y with respect to t, with x treated as a constant. The result of either differentiation may again be a function of x and t. For example, $\partial \sin(kx - \omega t)/\partial x = k\cos(kx - \omega t)$ and $\partial \sin(kx - \omega t)/\partial t = -\omega \cos(kx - \omega t)$.

You should understand the physical significance of a partial derivative as well as be able to evaluate it, given the function. The partial derivative $\partial y(x, t)/\partial x$ gives the slope of the string at the point x and time t; here you are evaluating the rate at which the displacement changes with *distance* along the string at some instant of time. For this to be meaningful the definition must make use of displacements for slightly separated points on the string *at the same time*, in the limit as the separation tends to zero. That is why t is treated as a constant.

The partial derivative $\partial y(x, t)/\partial t$ gives the string velocity at the point x and time t; here you are evaluating the rate at which the displacement changes with time at a given point. Clearly this is associated with the displacement *of the same point* but at slightly different times, in the limit as the time interval approaches zero. That is why x is treated as a constant.

Your instructor may require you to carry out partial differentiation of functions other than sine and cosine functions. For example, the string displacement associated with a bell-shaped pulse traveling in the positive x direction is given by

$$y(x, t) = A e^{-(x-vt)^2/a},$$

where A and a are constants. The partial derivatives are

$$\frac{\partial y}{\partial x} = -\frac{2A(x - vt)}{a} e^{-(x-vt)^2/a}$$

and

$$\frac{\partial v}{\partial t} = \frac{2Av(x - vt)}{a} e^{-(x-vt)^2/a}$$

Check these yourself.

2. You should also be able to use the chain rule to show that *any* function of $x - vt$ or of $x + vt$ satisfies the wave equation $\partial^2 y/\partial x^2 = (1/v^2)\partial^2 y/\partial t^2$. Consider a wave traveling in the positive x direction and let $y(x, t) = f(u)$, where $u = x - vt$. Then $\partial y/\partial x = (df/du)(\partial u/\partial x) = df/du$ and $\partial^2 y/\partial x^2 = (d^2 f/du^2)(\partial u/\partial x) = d^2 f/du^2$. In a similar manner, $\partial y/\partial t = (df/du)(\partial u/\partial t) = -v df/dt$ and $\partial^2 y/\partial t^2 = -v(d^2 f/du^2)(\partial u/\partial t) = v^2 d^2 f/du^2$. Thus $(1/v^2)\partial^2 y/\partial t^2$ is clearly the same as $\partial^2 y/\partial x^2$.

3. The trigonometric identity

$$\sin A + \sin B = 2\sin\left(\frac{A+B}{2}\right)\cos\left(\frac{A-B}{2}\right)$$

plays an important role in this chapter. You should be able to show its validity. Write $\sin A + \sin B = \sin\left(\frac{A+B}{2} + \frac{A-B}{2}\right) + \sin\left(\frac{A+B}{2} - \frac{A-B}{2}\right)$ and use the rules for expanding the sine of the sum and the sine of the difference of two angles: $\sin(\theta_1 + \theta_2) = \sin\theta_1\cos\theta_2 + \cos\theta_1\sin\theta_2$ and $\sin(\theta_1 - \theta_2) = \sin\theta_1\cos\theta_2 - \cos\theta_1\sin\theta_2$. These give

$$\sin A + \sin B =$$
$$\sin\left(\frac{A+B}{2}\right)\cos\left(\frac{A-B}{2}\right) + \cos\left(\frac{A+B}{2}\right)\sin\left(\frac{A-B}{2}\right)$$
$$+ \sin\left(\frac{A+B}{2}\right)\cos\left(\frac{A-B}{2}\right) - \cos\left(\frac{A+B}{2}\right)\sin\left(\frac{A-B}{2}\right)$$
$$= 2\sin\left(\frac{A+B}{2}\right)\cos\left(\frac{A-B}{2}\right)$$

IV. COMPUTER PROJECTS

Start with a rather simple program to plot the displacement $y(x,t) = y_m \sin(kx - \omega t)$ for a sinusoidal traveling wave at a given time. Input the amplitude y_m, wavelength λ, frequency ν, and time t. Have the computer calculate values of $y(x,t)$ for x from 0 to some final value x_f, at intervals of width Δx. Here's an outline.

> input amplitude, wavelength, frequency: y_m, λ, ν
> input final value of x, interval width: x_f, Δx
> input time: t
> calculate angular wave number, angular frequency: $k = 2\pi/\lambda$, $\omega = 2\pi\nu$
> set $x = 0$
> **begin loop** over intervals
>   calculate $y(x,t)$: $y = y_m \sin(kx - \omega t)$
>   display or plot x, y
>   increment x by Δx
> if $x > x_f$ then **end loop** over intervals
> stop

A reasonably good graph can be obtained if Δx is chosen to be about $\lambda/20$ and x_f is chosen to be about 2λ. If you program the computer to plot the wave on your monitor you might have the program calculate values of Δx and x_f rather than read them. If you are plotting by hand you will want values that are 1, 2, or 5 times a power of ten.

PROJECT 1. Take $y_m = 2.0$ mm, $\lambda = 5.0$ cm, and $\nu = 10$ Hz. Make a graph of the wave at $t = 0$ and note the position of the maximum nearest the origin.

Now make a graph of the wave at $t = 4.0 \times 10^{-2}$ s and again note the position of the maximum nearest the origin. Measure the distance this maximum has moved in 4.0×10^{-2} s and use the value you obtain to calculate the wave speed. Compare your answer with $v = \lambda\nu$. The time was chosen so that no other maximum appears between the origin and the maximum you are following.

Two sinusoidal waves with the same frequency and wavelength, traveling in the same direction, sum to form another sinusoidal wave. You can use a computer program to plot the sum of the waves, then read the resultant amplitude and phase from the graph. Suppose $y_1(x,t) = y_{1m} \sin(kx - \omega t)$ and $y_2(x,t) = y_{2m} \sin(kx - \omega t - \phi)$. Then the resultant wave is given by $y(x,t) = y_{1m} \sin(kx - \omega t) + y_{2m} \sin(kx - \omega t - \phi)$. Here's the outline of a program.

> input amplitudes, wavelength, frequency, phase constant: $y_{1m}, y_{2m}, \lambda, \nu, \phi$
> input final value of x, interval width: $x_f, \Delta x$
> input time: t
> calculate angular wave number, angular frequency: $k = 2\pi/\lambda, \omega = 2\pi\nu$
> set $x = 0$
> **begin loop** over intervals
>   calculate $y(x,t)$: $y = y_{1m} \sin(kx - \omega t) + y_{2m} \sin(kx - \omega t - \phi)$
>   display or plot x, y
>   increment x by Δx
> if $x > x_f$ then **end loop** over intervals
> stop

PROJECT 2. No matter what the amplitudes and the phase difference, the resultant wave is sinusoidal and moves with same speed as the constituent waves. Take $y_{1m} = 2.0$ mm, $y_{2m} = 4.0$ mm, $\lambda = 5.0$ cm, $\nu = 10$ Hz, and $\phi = 65°$. Plot the resultant wave from $x = 0$ to $x = 10$ cm for $t = 0$. Then plot the resultant wave for $t = 4.0 \times 10^{-2}$ s and calculate the wave speed. You should get the same answer as you obtained for the first project.

PROJECT 3. The amplitude of the resultant wave depends on the phase difference ϕ. First consider two waves with the same amplitude. Then the resultant amplitude is given by $y_m = 2y_{1m} \cos(\phi/2)$. Take $y_{1m} = y_{2m} = 2.0$ mm, $\lambda = 5.0$ cm, $\nu = 10$ Hz, and $\phi = 65°$. Plot the resultant wave from $x = 0$ to $x = 10$ cm for $t = 0$. Measure the amplitude and compare the result with the calculated value.

The phase constant α of the resultant wave also depends on the phases of the constituent waves. For equal amplitudes $\alpha = \phi/2$. Use your graph to find α. At $t = 0$ the first constituent wave y_1 is zero and has positive slope at the origin. Use your graph to find the coordinate where the resultant wave is zero and has positive slope. The ratio of this coordinate to a wavelength is the same as the ratio of the phase constant to 360° or 2π rad. The coordinate should be $(32.5/360)\lambda = 0.45$ cm. Is it?

What happens when the amplitudes are not equal? Take $y_{1m} = 2.0$ mm, $y_{2m} = 4.0$ mm, $\lambda = 5.0$ cm, and $\nu = 10$ Hz. For each of the following values of ϕ find the amplitude and phase constant of the resultant wave: (a.) 0; (b.) 30°; (c.) 45°; (d.) 90°; (e.) 180°. [ans: 6.0 mm, 0; 5.8 mm, 20°; 5.6 mm, 30°; 4.5 mm, 63°; 2.0 mm, 180°]

Your program can also be used to investigate standing waves. A standing wave is composed of two traveling waves with the same amplitude, frequency, and wavelength, but traveling in opposite directions. Revise the program so only one amplitude is read and so $y = y_m \sin(kx - \omega t) + y_m \sin(kx + \omega t - \phi)$ is used to calculate the resultant wave.

PROJECT 4. Take $y_m = 2.0$ mm, $\lambda = 5.0$ cm, and $\nu = 10$ Hz. Make a graph of the wave at $t = 0$ and note the position of the maximum nearest the origin. Now make a graph of the wave at $t = 0.020$ s and again note the position of the maximum nearest the origin. During this time each of the traveling waves moved 1.0 cm, one fifth of a wavelength. But the standing wave maximum did not move.

Notice that the maximum displacement is less than at $t = 0$. In fact, all parts of the string move together toward $y = 0$. At $t = .025$ s the displacement is everywhere 0. Use the program to verify this. Also plot the wave for $t = 0.05$ s, one half period after the start.

PROJECT 5. Identify the nodes on the graph you made as part of the last project for $t = 0$ and verify that they are half a wavelength apart. How do the phases of the traveling waves affect the positions of the nodes? For the same wavelength and frequency as the last project and for each of the following values of ϕ find the coordinate of the node nearest the origin and verify that the node separation is the same: (a.) 0; (b.) 45°; (c.) 90°. [ans: 0; 0.31 cm; 0.63 cm]

V. NOTES

Chapter 20
SOUND WAVES

I. BASIC CONCEPTS

In this chapter the ideas of wave motion introduced in the last chapter are applied to sound waves. Pay particular attention to the dependence of the wave speed on the properties of the medium in which sound is propagating. Although the idea is the same, the properties are different from those that determine the speed of a wave on a taut string. In addition, you will be dealing chiefly with variations in pressure rather than particle displacement. Completely new concepts include beats and the Doppler shift. Be sure you understand what these phenomena are and how they originate.

General description. Sound waves are propagating distortions of a material medium. In fluids they are longitudinal: particles of the medium move back and forth along the line of _____. In solids they may be longitudinal, transverse, or neither.

Since particles at slightly different positions move different amounts the medium becomes compressed or rarefied as a sound wave passes. That is, we can consider a sound wave to be the propagation of a local increase or decrease in density. Since a change in pressure is associated with a change in density a sound wave may also be considered to be the propagation of a deviation in local pressure from the ambient pressure.

To understand this chapter you must know the relationships between density, volume, and pressure discussed in Chapter 17. Consider a tube of fluid with cross-sectional area A, density ρ, and bulk modulus B, at pressure p. Suppose an element of fluid originally of length ℓ is uniformly elongated by $\Delta\ell$ so its new length is $\ell + \Delta\ell$ and its new volume is $A(\ell + \Delta\ell)$. The mass of the fluid in the element is given by $m = \rho A \ell$, so the density after elongation is $\rho' = m/A(\ell + \Delta\ell) = \rho\ell/(\ell + \Delta\ell)$. This expression for ρ' can be written $\rho' = \rho + \Delta\rho$, where $\Delta\rho$ is the change in density. In the space below show that if $\Delta\ell$ is much smaller than ℓ then $\Delta\rho = -\rho\Delta\ell/\ell$:

Notice that $\Delta\rho$ is negative if $\Delta\ell$ is positive. The density decreases because the mass occupies a larger volume after elongation. The change in pressure that accompanies the elongation is given by $\Delta p = -B\Delta V/V = -B\Delta\ell/\ell$. It is also proportional to $\Delta\ell$. The same expressions are also valid for a compression. Then $\Delta\ell$ is negative.

The relationships $\Delta\rho = -\rho\Delta\ell/\ell$ and $\Delta p = -B\Delta\ell/\ell$ are used to derive an expression for the speed of sound in terms of properties of the medium and to relate displacement, density, and pressure waves to each other.

Speed of sound. The speed of sound in a fluid is determined by the density ρ and bulk modulus B of the fluid. Specifically, it is given by

$$v = $$

To understand this expression, apply Newton's second law to an element of fluid in which a sound pulse is traveling and make use of the relations between density, volume, and pressure given above. The derivation follows closely the derivation of the expression for the speed of a wave on a string, given in Section 4 of Chapter 19. You may wish to review that derivation.

We view a compressional pulse in a fluid from a reference frame that moves with the speed of sound v, so the pulse is stationary and the fluid moves with speed v into the right end and out of the left end of the pulse, as shown in Fig. 2 of the text. An element of fluid with cross-sectional area A and length ℓ (when just outside the pulse) takes time $\Delta t = \ell/v$ to completely enter the pulse. In terms of v and Δt the volume of the element when it is outside is _____ and if ρ is the fluid density outside then the mass of the element is $m = $ _____.

Now consider the fluid element when it is partly inside and partly outside the pulse. If the pressure outside the pulse is p and the pressure inside the pulse is $p + \Delta p$ then the net force on the element is $F = $ _____. Since the pressure inside the pulse is greater than the pressure outside, this force slows the element as it enters the pulse, so its velocity changes from v to $v + \Delta v$, where Δv is negative. The acceleration of the element is $a = \Delta v/\Delta t$. In $F = ma$ replace F with $-A\Delta p$, m with $\rho v A \Delta t$ and a with $\Delta v/\Delta t$, then solve for v:

Multiply the result by v to obtain $v^2 = -v\Delta p/\rho\Delta v$.

The key to the remainder of the derivation is to observe that the volume of the fluid element changes as it enters the pulse and that the fractional change in volume is the same as the fractional change in speed: that is, $\Delta V/V = \Delta v/v$. The leading edge of the element, inside the pulse, travels slower than the trailing edge, outside the pulse, by Δv. In time Δt it goes a distance $\Delta v \Delta t$ less, and this must be the amount by which the element is shortened. Thus $\Delta V = A\Delta v\Delta t$. Now $V = Av\Delta t$ is the volume V of the element when it is completely outside. Divide one of these results by the other to obtain $\Delta V/V = \Delta v/v$. Substitute $\Delta V/V$ for $\Delta v/v$ in the expression for ρv^2, then replace $-\Delta p/(\Delta V/V)$ with the bulk modulus B and solve for v:

You should have obtained $v = \sqrt{B/\rho}$. Thus the expression for the speed of sound can be derived straightforwardly from Newton's second law and is a direct result of the forces neighboring portions of the fluid exert on each other. You should know that the speed of sound is about _____ m/s in air, about _____ m/s in water, and about _____ m/s in solids.

Pressure and density waves. To describe a traveling sound wave you can use any of three quantities: the particle displacement, the deviation of the density from its ambient value, and the deviation of the pressure from its ambient value. All propagate as waves. They are related to each other and you should understand the relationships.

Suppose a sound wave is traveling along the x axis through a fluid and suppose further that the displacement of the fluid at coordinate x and time t is given by a known function $s(x,t)$. Consider an element of fluid that in the absence of the wave extends from x to $x + \Delta x$. Its length is Δx. In the presence of the wave the same fluid element extends from $x + s(x,t)$ to $x + \Delta x + s(x + \Delta x, t)$ so its change in length is $\Delta \ell = s(x + \Delta x, t) - s(x,t)$. In the limit as Δx becomes small $\Delta \ell$ becomes $[\partial s(x,t)/\partial x]\Delta x$. Thus the density in the presence of a wave is $\rho'(x,t) = \rho + \Delta\rho(x,t)$, where $\Delta\rho(x,t) = -\rho\Delta\ell/\Delta x = -\rho\partial s(x,t)/\partial x$.

Similarly, the pressure in the presence of a wave is $p'(x,t) = p + \Delta p(x,t)$ where $\Delta p = -B\Delta\ell/\Delta x = -B\partial s(x,t)/\partial x$ and B is the bulk modulus.

Suppose the sound wave is sinusoidal and the fluid displacement is given by $s(x,t) = s_m \cos(kx - \omega t)$, where k is the wave number and ω is the angular frequency. In terms of these quantities the deviation of the density from its ambient value is given by $\Delta\rho(x,t) = $ _____ and the deviation of the pressure from its ambient value is given by $\Delta p(x,t) = $ _____.

These expressions can be written

$$\Delta\rho = \Delta\rho_m \sin(kx - \omega t)$$

and

$$\Delta p = \Delta p_m \sin(kx - \omega t),$$

where the density amplitude $\Delta\rho_m$ and the pressure amplitude Δp_m are given in terms of the displacement amplitude s_m, the wave speed v, and the wave number k by $\Delta\rho_m = $ _____ and $\Delta p_m = $ _____.

Notice that the pressure and density waves are not in phase with the displacement wave. At points where the displacement is a maximum the deviation of the pressure from its ambient value is a _____ and the deviation of the density from its ambient value is a _____. These results make physical sense because the fluid displacement is nearly uniform in the neighborhood of a displacement maximum. An element of fluid is neither compressed or elongated there.

At points where the displacement is zero the deviation of the pressure from its ambient value is a _____ and the deviation of the density from its ambient value is a _____. Here the rate of change of the displacement with distance has its greatest magnitude. The elongation or compression of the fluid is greater here than anywhere else. Thus we expect deviations of the pressure and density to be the greatest at these points.

Particle displacements are vectors and when two or more sound waves are simultaneously present the total displacement is calculated as the vector sum of the displacements due to the individual waves. On the other hand, pressure is a scalar and the total pressure is calculated as the algebraic sum of the individual pressures. It is possible to have total destructive interference of two pressure waves without having total destructive interference of the corresponding displacement waves. This occurs when the pressure waves are 180° out of phase with each other. The displacement waves are also 180° out of phase but if the displacements are not along the same line they do not cancel each other. Ears and microphones are sensitive to pressure, not particle displacements, so pressure waves are usually considered in studies of sound. We concentrate on pressure waves.

Power and intensity. The power P transmitted from one element of fluid to a neighboring element is the product of the force exerted by the element on its neighbor and the velocity of the neighbor. For a sound wave in a tube with cross-sectional area A the force is $A\Delta p$ and the velocity is $\partial s/\partial t$, where s is the particle displacement. Thus $P(x,t) = $ _____. For a sinusoidal wave with $s = s_m \cos(kx - \omega t)$, $P(x,t) = $ _____. This is usually written in terms of the pressure amplitude rather than the displacement amplitude. Use $s_m = $ _____ Δp_m to eliminate s_m and obtain

$$P(x,t) = $$

The average power over a cycle is given by

$$\overline{P} = $$

a quantity that does not depend on x or t. The intensity of a sound wave is the average power transmitted per unit _____ and, for the sinusoidal wave discussed above, is given by

$$I = $$

Its SI units are _____.

Sound level is often used instead of intensity. The sound level associated with intensity I is defined by $SL = $ _____, where I_0 is the standard reference intensity (_____ W/m^2). Sound level is measured in units of _____, abbreviated _____.

Carefully note that the sound level is defined in terms of the *logarithm to the base* 10 of I/I_0. If $I = I_0$ the sound level is _____ db. If the intensity is increased by a factor of 10 the sound level increases by _____ db.

The standard reference intensity is roughly the threshold of human hearing. The intensity at the upper end of the human hearing range (called the threshold of pain) is about _____ W/m^2 and the sound level is about _____ db. Look at Table 2 of the text for some other sound level values.

Standing sound waves. Two sinusoidal sound waves with the same frequency and amplitude but traveling in opposite directions combine to form a standing wave. The text considers pressure waves. If p_0 is the ambient pressure then at a pressure node the pressure is always _____. At a pressure antinode the pressure oscillates between _____ and _____, where Δp_m is the pressure amplitude. At other points the pressure oscillates with an amplitude that is given by _____, where k is the wave number of either of the traveling waves and x is the coordinate of the point. In writing the expression for the amplitude, the pressure amplitude at $x = 0$ was assumed to be _____.

300   Chapter 20: Sound Waves

If there are no losses, standing waves are created in pipes by the superposition of a sinusoidal wave and its reflection from the end of the pipe. A pressure _____ exists at a closed end of a pipe. A pressure _____ exists at an open end. If the sound wave is generated by a speaker at one end of the pipe the point at the speaker is very nearly _____ of the standing pressure wave.

If sound is produced by a speaker at one end of a pipe of length L and the other end is open, the wavelengths associated with possible standing waves are such that the pipe length is a multiple of _____. In terms of L the standing wave wavelengths are given by $\lambda_n =$ _____ and the standing wave frequencies are given by $\nu_n =$ _____, where v is the speed of sound for the fluid that fills the pipe and n is _____.

For the lowest three standing wave frequencies use the coordinates below to plot the pressure (relative to the ambient pressure) as a function of position along the pipe. Take the speaker to be at the left end.

If sound is produced by a speaker at one end of a pipe of length L and the other end is closed, the wavelengths associated with possible standing waves are such that the pipe length is an odd multiple of _____. In terms of L the standing wave wavelengths are given by $\lambda_n =$ _____ and the standing wave frequencies are given by $\nu_n =$ _____, where v is the speed of sound for the fluid that fills the pipe and n is _____.

For the lowest three standing wave frequencies use the coordinates below to plot the pressure amplitude as a function of position along the pipe. Take the speaker to be at the left end.

A string of a stringed instrument or the air in an organ pipe can vibrate at any one of its standing wave (or natural) frequencies. Which sound is produced depends on how the instrument is excited. Usually the lowest frequency dominates but higher frequency sound is mixed in. The admixture of higher frequencies gives any instrument the quality of sound peculiar to that instrument and, for example, allows us to distinguish a violin from a piano.

The lowest frequency is called the _____ frequency, while higher frequencies are called _____ and are numbered in order of increasing frequency. If the higher frequencies are multiples of the lowest frequency they are called _____.

Chapter 20: Sound Waves   301

Beats. Beats are created when two sound waves with nearly the same _____ are simultaneously present. We can view the resultant wave as one with a frequency that is the average of the frequencies of the constituent waves and an amplitude that varies with time, but much more slowly than either of the constituent waves.

Since we are interested in the time dependence of the wave, we can study the pressure at a single point in space and ignore variations with position. Let $\Delta p_1 = \Delta p_m \sin(\omega_1 t)$ represent the pressure variation at the point due to one of the waves and $\Delta p_2 = \Delta p_m \sin(\omega_2 t)$ represent the pressure variation at the same point due to the other wave. The resultant pressure variation is the sum of the two and, once the trigonometric identity given in the last chapter is used, it can be written as Eq. 31 of the text:

$$\Delta p(t) =$$

Notice that there are two time dependent factors, both periodic. One has an angular frequency $\overline{\omega} =$ _____, the average of the two constituent angular frequencies. This is the greater of the angular frequencies associated with the two factors and if the two constituent frequencies are nearly the same it is essentially equal to either of them.

The angular frequency of the second time dependent factor is $\omega_{amp} =$ _____. Note that it depends on the difference in the two constituent frequencies. If ω_1 and ω_2 are nearly the same the factor associated with this angular frequency is slowly varying. We may think of it as a slow variation in the amplitude of the faster vibration. The effect can be produced, for example, by blowing a note on a horn at the angular frequency $\overline{\omega}$, but modulating it so it is periodically loud and soft.

A <u>beat</u> is one cycle in the *intensity*, from loud to soft and back to loud again, for example, and occurs as $\cos(\omega_{amp}t)$ goes from $+1$ to _____. Thus the beat angular frequency ω_{beat} is NOT the same as ω_{amp}. In fact $\omega_{beat} =$ _____ ω_{amp}, or in terms of ω_1 and ω_2, $\omega_{beat} =$ _____. If one constituent wave has a frequency of 1000 Hz and the other has a frequency of 1005 Hz, then the beat frequency is $\nu_{beat} =$ _____ Hz and the beat angular frequency is $\omega_{beat} =$ _____ rad/s.

Doppler effect. Suppose a sustained note with a well-defined frequency ν is played by a stationary trumpeter. If you move rapidly *toward* the trumpeter you will hear a note with a _____ frequency. If you move rapidly *away* from the trumpeter you will hear a note with a _____ frequency. Similar effects occur if you are stationary and the trumpeter is moving. The note has a higher frequency if the trumpeter is moving _____ and a lower frequency if the trumpeter is moving _____. These are examples of the <u>Doppler</u> <u>effect</u>. The next time you hear a police or ambulance siren on the highway listen carefully as the vehicle approaches and then recedes from you. If it is going sufficiently fast you should hear the Doppler shift in frequency.

To understand how the Doppler effect comes about, suppose you are moving with speed v_o away from a source of sound with frequency ν and wavelength λ ($= v/\nu$, where v is the speed of sound). If you were at rest you would receive _____ wave crests in time t. Because you are moving away from the source you receive _____ fewer crests in the same time. Thus the number of crests you receive in time t is _____ and the frequency you hear is this number divided by t, or $\nu' =$ _____.

Write down the equations for the other possibilities:

> If you are moving with speed v_o *toward* a stationary source then the frequency you hear is $\nu' = $ _____ .
>
> If you are stationary and the source is moving with speed v_s *toward* you then the frequency you hear is $\nu' = $ _____ .
>
> If you are stationary and the source is moving *away* from you with speed v_s then the frequency you hear is $\nu' = $ _____ .

Eq. 40 in the text covers all possibilities. Write it here:

$$\nu' = $$

You can easily determine which signs to use in any particular situation by remembering that motion of the source toward the observer or the observer toward the source results in hearing a higher frequency while motion of the source away from the observer or the observer away from the source results in hearing a lower frequency than would be heard if both were stationary.

You should understand that the velocities in the Doppler effect equation are measured relative to the medium in which the wave is propagating (the air, for example). The Doppler shift is different for these two cases: (1) the source is stationary and the observer is moving toward it with a speed of 40 m/s; (2) the observer is stationary and the source is moving toward it with a speed of 40 m/s. What counts is not the motion of the source relative to the observer but the motions of both the source and observer relative to the medium of propagation.

You should also understand that the Doppler effect equations given above are valid only if the motion is along the line joining the source and observer. For motion in other directions v_o and v_s must be interpreted as components of the velocities along that line.

To obtain an understanding of the speeds involved, estimate the speed with which you would have to move toward a stationary source of sound in still air to hear a 10% increase in frequency: _____ m/s. Can this speed be achieved by walking slowly, walking fast, riding a bicycle, driving a car at a moderate speed, or driving a car at high speed?

The Doppler effect also occurs for electromagnetic radiation. Light from stars that are moving at high speeds away from the earth is shifted toward the red and the extent of the shift is used to calculate the speeds of the stars. Doppler shifts of radar waves reflected from moving objects can be used to find their speeds. Police use the effect to detect speeders and TV technicians use it to find the speeds of thrown baseballs.

If a source of sound is moving through a medium faster than the speed of sound in the medium, a shock wave is produced. Then a wavefront has the shape of a _____, with the source at its _____, as shown in Fig. 14 of the text. The half angle θ of the cone is given by $\sin\theta = $ _____ , where v is the speed of _____ and v_s is the speed of _____ . Note that no shock wave is produced if $v < v_s$.

Shock waves are responsible for sonic booms. A listener hears a sonic boom when

In the space to the right draw a plane flying horizontally over level ground at an altitude h at the time when the envelope of the shock wave intersects the ground at point A. Show the envelope of the shock and label the half angle of the cone θ. Let d be the horizontal distance from A to the point on the ground under the plane. The relationship between θ, d, and h is $\tan\theta =$ _____ .

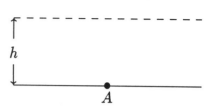

II. PROBLEM SOLVING

Since many of the problems are similar to those of the last chapter, you might want to review Chapter 19 of this manual.

One difference is that the speed of sound in an isotropic medium is given by $\sqrt{B/\rho}$, where B is the bulk modulus and ρ is the density of the medium. If you are given the bulk modulus and density you can calculate the speed of sound. Alternatively you might use $v = \lambda \nu$ (or a related expression) to find the speed of sound, then use $v = \sqrt{B/\rho}$ to solve for B or ρ, given the other.

PROBLEM 1. At a pressure of 1.0 atm and a temperature of 20° C air has a density of 1.21 kg/m³ and a bulk modulus of 1.4×10^5 N/m². Water has a density of 0.998×10^2 kg/m³ and a bulk modulus of 2.2×10^9 N/m². Compare the speed of sound in air to the speed of sound in water.
SOLUTION:

[ans: $v_a = 340$ m/s, $v_w = 1480$ m/s]

What are the wavelengths of 500-Hz sound waves in these two media?
SOLUTION:

[ans: $\lambda_a = 0.680$ m, $\lambda_w = 2.97$ m]

Remember that if the frequency is changed then the wavelength changes but the wave speed does not. To change the wave speed you must change the bulk modulus or density of the medium.

Sound waves are usually discussed in terms of deviations of the pressure from its ambient value rather than in terms of particle displacement. Some problems give the displacement wave $s(x,t)$ and ask for the pressure wave $\Delta p(x,t)$. For a one-dimensional wave use $\Delta p(x,t) = -B\partial s(x,t)/\partial x$. If $s(x,t) = s_m \cos(kx - \omega t)$, for example, then $\Delta p(x,t) = kBs_m \sin(kx - \omega t)$ and the pressure amplitude Δp_m is related to the displacement amplitude by $\Delta p_m = kBs_m$ or by $\Delta p_m = k\rho v^2 s_m$, where $B = v^2 \rho$ was used. Similarly the deviation of the density from its ambient value is given by $\Delta \rho(x,t) = -\rho \partial s(x,t)/\partial x$. For $s = s_m \cos(kx - \omega t)$, $\Delta \rho(x,t) = \rho k s_m \sin(kx - \omega t)$ and the density amplitude is given by $\Delta \rho_m = \rho k s_m$.

PROBLEM 2. A 500-Hz sound wave with a pressure amplitude of 5.0 Pa travels in air (density = 1.21 kg/m^3 and bulk modulus = 1.4×10^5 N/m^2). What are the density and displacement amplitudes?

SOLUTION:

[ans: 4.32×10^{-5} kg/m^3; 3.87×10^{-6} m]

What is the maximum particle speed?

SOLUTION: The particle speed is given by $u(x,t) = \partial s(x,t)/\partial t = \omega s_m \sin(kx - \omega t)$ so the maximum speed is given by $u_m = \omega s_m$.

[ans: 1.21 cm/s]

Carefully note that you had to find the *displacement* amplitude to calculate the maximum particle speed.

To solve power, intensity, and sound level problems use the equations $\overline{P} = \frac{1}{2}A(\Delta p_m)^2/\rho v$ for the average power transmitted through area A, $I = \frac{1}{2}(\Delta p_m)^2/\rho v$ for the intensity, and $SL = 10\log(I/I_0)$ for the sound level. These are all related.

PROBLEM 3. A sound wave with a pressure amplitude of 0.20 Pa is incident in air on a human ear with an area of 1.0 cm^2. Take the speed of sound to be 340 m/s and the density of air to be 1.21 kg/m^3, then calculate the sound intensity, the sound level, and the average power transmitted into the ear.

SOLUTION:

[ans: 4.86×10^{-5} W/m^2; 76.9 db; 4.86×10^{-9} W]

If the sound level is increased by 10 db what is the new pressure amplitude?

SOLUTION:

[ans: 0.632 Pa]

Look at Table 2 of the text and decide if these sound levels are high or low. They both fall between the sound levels of _____ and _____ .

The pressure amplitude for sound in a given medium depends only on the intensity or sound level. The displacement amplitude, on the other hand, also depends on the frequency.

PROBLEM 4. 100- and 1000-Hz sound waves in air each have sound levels of 45 db. What are the pressure amplitudes of these waves? What are the displacement amplitudes? Take the speed of sound in the pipe to be 340 m/s, the density of air to be 1.21 kg/m³, and the bulk modulus of air to be 1.4×10^5 N/m².

SOLUTION:

[ans: 100 Hz: 0.20 Pa, 7.73×10^{-7} m; 1000 Hz: 0.20 Pa, 7.73×10^{-8} m]

Many problems deal with the production of sound by pipes with either both ends closed or one end open and one end closed. In some cases you are given the separation of adjacent nodes or adjacent antinodes. Equate this distance to $\lambda/2$ to find the wavelength λ. In other cases you are given the distance from the pipe end to the first node or antinode. If the end is open a pressure node occurs there and the distance to the next node is $\lambda/2$ while the distance to the first antinode is $\lambda/4$. If the end is closed a pressure antinode occurs there and the distance to the first node is $\lambda/4$ while the distance to the next antinode is $\lambda/2$.

In any event the frequency can be found using $v = \lambda \nu$. The speed of sound may be given directly or B and ρ may be given and you must calculate v.

PROBLEM 5. A 2.0-m long pipe containing air is open at both ends. What are the three longest wavelengths associated with standing waves in the pipe?

SOLUTION: Pressure nodes occur at both ends, so $L = n\lambda/2$, where L is the length of the pipe, λ is the wavelength, and n is an integer. Calculate λ for $n = 1, 2,$ and 3.

[ans: 4.0 m, 2.0 m, 1.33 m]

What frequency sound must be generated to produce these standing waves? Take the speed of sound to be 340 m/s.

SOLUTION:

[ans: 85 Hz; 170 Hz; 255 Hz]

PROBLEM 6. Suppose the pipe of the previous problem is open at one end and closed at the other. What then are the three longest wavelengths and the frequencies associated with them?

SOLUTION: Now a pressure node occurs at one end (the open end) and a pressure antinode occurs at the other, so $L = n\lambda/4$, where n is an odd integer. Calculate λ for $n = 1$, 3, and 5. Use $v = \lambda \nu$ to calculate the frequencies.

[ans: 8.0 m, 42.5 Hz; 2.67 m, 127.5 Hz; 1.6 m, 212.5 Hz]

Some problems deal with the production of beats by two sinusoidal sound waves with nearly the same frequency. You may be given the frequency ν_1 of one of the waves and the beat frequency ν_{beat}, then asked for the frequency ν_2 of the other wave. Since $\nu_{beat} = |\nu_1 - \nu_2|$, it is given by $\nu_2 = \nu_1 \pm \nu_{beat}$. You require more information to determine which sign to use in this equation. One way to give this information is to tell you what happens to the beat frequency if ν_1 is increased (or decreased). If the beat frequency increases when ν_1 increases, then ν_1 must be greater than ν_2 and $\nu_2 = \nu_1 - \nu_{beat}$. If the beat frequency decreases then ν_1 must be less than ν_2 and $\nu_2 = \nu_1 + \nu_{beat}$.

PROBLEM 7. A certain violin string emits sound with a frequency of 600 Hz. When it is played simultaneously with a second string 6 beats per second occur and when the tension in the second string is increased slightly without changing its length, the beat frequency decreases. What was the original frequency of the second string?

SOLUTION: Since the beat frequency before tightening was 6 Hz, the original frequency of the second string was either 594 Hz or 606 Hz. Tightening the string increases the speed of waves on the string and, since the wavelength does not change, it increases the frequency.

[ans: 594 Hz]

PROBLEM 8. Two identical strings have lengths of 2.0 m and linear mass densities of 2.5×10^{-4} kg/m. One has a tension of 475 N and the other has a tension of 450 N. When they are both vibrating in their fundamental modes with both ends fixed what beat frequency do they produce?

SOLUTION: The wavelength associated with the standing waves in each string is twice the length of the string. Use $v = \sqrt{F/\mu}$ to calculate the speed of waves on each string, then use $v = \lambda \nu$ to calculate the frequencies of the sound produced. Finally use $\nu_{beat} = |\nu_1 - \nu_2|$.

[ans: 9.19 Hz]

To what value should the tension in the second string be increased to reduce the beat frequency to 1 Hz?

SOLUTION:

[ans: 472 N]

Nearly all Doppler shift problems can be solved using

$$\nu' = \nu \left[\frac{v \pm v_o}{v \mp v_s} \right],$$

where v is speed of sound, v_s is the speed of the source, v_o is the speed of the observer, ν is the frequency of the source, and ν' is the frequency detected by the observer. The upper sign in the numerator refers to a situation in which the observer is moving toward the source; the lower sign to a situation in which the observer is moving away from the source. The upper sign in the denominator refers to a situation in which the source is moving toward the observer; the lower sign to a situation in which the source is moving away from the observer. Remember that all speeds are measured relative to the medium in which the sound is propagating. You might be given the velocities and one of the frequencies, then asked for the other frequency. In other situations you might be given the two frequencies and one of the velocities, then asked for the other velocity. In all cases, simple algebraic manipulation of the equation will produce the desired expression.

PROBLEM 9. A source emits a 300-Hz sound wave in air (speed of sound = 340 m/s). What frequency does an observer hear if the source is stationary and the observer is moving toward it at 50 m/s? if the observer is stationary and the source is moving toward him at 50 m/s? if the observer is moving toward the source at 25 m/s as the source is moving toward him at 25 m/s?

SOLUTION:

[ans: 344 Hz; 352 Hz; 348 Hz]

PROBLEM 10. A 0.75-m long string with a linear density of 3.7×10^{-4} kg/m and a tension of 150 N, fixed at both ends, is vibrating in its fundamental mode as it moves along the line between it and a stationary observer. If the observer hears sound with a frequency of 380 Hz how fast is the source moving?

SOLUTION: First calculate the frequency of vibration of the string. The wavelength associated with the standing wave is twice the length of the string or 1.5 m. The speed of a wave on the string is given by $V = \sqrt{F/\mu}$, where F is the tension in the string and μ is the linear mass density. Use $\nu = V/\lambda$ to calculate the frequency. Carefully note that the speed of a wave *on the string* is used, not the speed of sound in air.

You will find that the observer hears a lower frequency than if the string were stationary. This means $\nu' = \nu v/(v + v_s)$, where v is the speed of sound in air, ν is the frequency of vibration of the string, and ν' is the frequency heard by the observer. Solve for v_s.

[ans: 39.8 m/s]

One set of problems deals with a source of sound that is moving with speed v_s away from a stationary observer and toward a stationary reflecting surface. The two waves reaching the observer, one directly and one after reflection, have different frequencies and produce beats. The direct wave comes from a source that is moving away from the observer. Its frequency is lowered and is given by $\nu_d = \nu v/(v + v_s)$. The wave reaching the reflecting surface comes from a source that is moving toward the surface. Its frequency is raised and is given by $\nu_r = \nu v/(v - v_s)$. This the also the frequency of the reflected wave reaching the observer. Now try an example in which the source is moving away from the wall.

PROBLEM 11. A source of 400-Hz sound waves is moving at 10 m/s directly away from a reflecting wall. A stationary detector is in front of the source. What are the frequencies of the direct and reflected waves at the detector? What is the beat frequency? Take the speed of sound to be 340 m/s.

SOLUTION:

[ans: 412 Hz; 389 Hz; 23 Hz]

Suppose the speed of the source is changed so the beat frequency is 15 Hz. What then is its speed?

SOLUTION: The beat frequency is given by $\nu_{beat} = \nu v[1/(v - v_s) - 1/(v + v_s)] = 2\nu v v_s/(v^2 - v_s^2)$. Solve for v_s.

[ans: 6.37 m/s]

PROBLEM 12. Two cars are traveling on a highway, one behind the other. When the leading car is going 20 m/s and the trailing car is going 15 m/s, the horn of the trailing car emits a 500-Hz sound wave that is reflected from the leading car. What is the frequency of the reflected wave as detected by a stationary detector behind the trailing car? Take the speed of sound to be 340 m/s.

SOLUTION: An observer riding with the leading car is traveling at $v_o = 20$ m/s away from the source. The source is traveling at $v_s = 15$ m/s toward the observer. The frequency heard by the observer is $\nu' = (v-v_o)/(v-v_s)$. We may now think of the leading car as a source traveling with speed v'_s away from the stationary detector and emitting a wave with frequency ν'. The frequency detected is $\nu'' = \nu'v/(v+v'_s)$.

[ans: 465 Hz]

Solutions to most shock wave problems start with $\sin\theta = v/v_s$, where θ is the half angle of the shock envelope, v is the speed of sound, and v_s is the speed of the source. For an airplane flying horizontally at altitude h you may also need to make use of $\tan\theta = h/d$, where d is the horizontal distance from the intersection of the shock wave with the ground to the point on the ground under the plane. Some problems give the time interval t from when the plane is overhead to when the sonic boom is heard. Use $d = v_s t$.

PROBLEM 13. A supersonic airplane flys horizontally with constant speed over an observer on the ground. The observer hears the sonic boom 12 s after the plane is directly overhead. At the instant the boom is heard the line of sight from the observer to the plane is 40° above the horizontal. What is the speed and altitude of the plane? Take the speed of sound to be 340 m/s.

SOLUTION: The half angle of the shock envelope is $\theta = 40°$. Use $\sin = v/v_s$ to find the plane speed v_s. Use $\tan\theta = h/d = h/v_s t$ to find the altitude h.

[ans: 529 m/s; 5.33 km]

Here is a mathematically more complicated example.

PROBLEM 14. A supersonic airplane flys horizontally with constant speed at an altitude of 6.5 km. The sonic boom is heard on the ground 10 s after the plane is directly overhead. What is the speed of the plane and what is the half angle of the shock cone? Take the speed of sound to be 340 m/s.

SOLUTION: A little geometry shows that $\sin\theta = h/(h^2+d^2)^{1/2} = h/(h^2+v_s^2 t^2)^{1/2}$, where $d = v_s t$ was used. Thus $v/v_s = h/(h^2+v_s^2 t^2)^{1/2}$. Solve for v_s, then use $\sin\theta = v/v_s$ to find θ.

[ans: 399 m/s; 58.5° or 1.02 rad]

III. COMPUTER PROJECTS

Modify the program of the last chapter to investigate beats. Two waves with slightly different frequencies are summed: $y(x,t) = y_{1m}\sin(k_1 x - \omega_1 t) + y_{2m}\sin(k_2 x - \omega_2 t - \phi)$. Once a value is chosen for x this can be written $y = y_{1m}\sin(\omega_1 t - \phi_1) + y_{2m}\sin(\omega_2 t - \phi_2)$, where $\phi_1 = k_1 x - \pi$ and $\phi_2 = k_2 x + \phi - \pi$. For the first few projects choose x and ϕ so both ϕ_1 and ϕ_2 vanish. Then $y = y_{1m}\sin(\omega_1 t) + y_{2m}\sin(\omega_2 t)$. Input the amplitudes and frequencies, then have the computer generate a plot of y for times from $t = 0$ to $t = t_f$, with an interval of Δt. Here's an outline.

> input amplitudes, frequencies: $y_{1m}, y_{2m}, \nu_1, \nu_2$
> input final value of t, interval width: $t_f, \Delta t$
> calculate angular frequencies: $\omega_1 = 2\pi\nu_1, \omega_2 = 2\pi\nu_2$
> set $t = 0$
> **begin loop** over intervals
>   calculate y: $y = y_{1m}\sin(\omega_1 t) + y_{2m}\sin(\omega_2 t)$
>   plot y
>   increment t by Δt
> if $t > t_f$ then **end loop** over intervals
> stop

Reasonable graphs are produced if Δt is taken to be about $1/50(\nu_1 + \nu_2)$ and t_f is taken to be about $2/|\nu_1 - \nu_2|$. The displacement y oscillates with a frequency of $(\nu_1 + \nu_2)/2$ so this value of Δt produces 100 points per period of oscillation. The beat frequency is $|\nu_1 - \nu_2|$ so this value of t_f lets you see two periods of the beat. Since a large number of points are generated you should program the computer to produce the graph on the monitor screen.

PROJECT 1. First take $y_1 = y_2 = 2.0$ mm, $\nu_1 = 100$ Hz, and $\nu_2 = 110$ Hz. The graph shows a rapidly varying oscillation inside a more slowly varying envelope. Measure the period of the rapid oscillation and verify that it is $2/(\nu_1 + \nu_2)$. Measure the period of the envelope and verify that it is $1/|\nu_1 - \nu_2|$.

Now try $y_{1m} = y_{2m} = 2.0$ mm, $\nu_1 = 100$ Hz, and $\nu_2 = 105$ Hz. Verify that the period of the rapid oscillation is nearly the same and that the period of the envelope has doubled.

Finally try $y_{1m} = y_{2m} = 2.0$ mm, $\nu_1 = 500$ Hz, and $\nu_2 = 510$ Hz. Verify that the period of the rapid oscillation has decreased by a factor of about 5 while the beat period is the same as for the first case above ($\nu_1 = 100$ Hz, $\nu_2 = 110$ Hz).

PROJECT 2. What happens if the amplitudes are different? Try $y_{1m} = 2.0$ mm, $y_{2m} = 4.0$ mm, $\nu_1 = 100$ Hz, and $\nu_2 = 110$ Hz. Pay special attention to the regions between beats and tell how these compare to the same regions when the amplitudes are the same: _____

Now increase the amplitude of the second wave to $y_{2m} = 6.0$ mm. Does the result substantiate your statement?

PROJECT 3. Now investigate the influence of a phase difference. Rewrite the program instruction for the calculation of y so it reads $y = y_{1m}\sin(\omega_1 t) + y_{2m}\sin(\omega_2 t - \phi)$ and add an input statement for ϕ near the beginning of the program. Take $y_{1m} = y_{2m} = 2.0$ mm, $\nu_1 = 100$ Hz, and $\nu_2 = 110$ Hz. Plot the displacement as a function of time for $\phi = 30°, 60°, 90°$, and $180°$. You already have a graph for $\phi = 0$. As you look at the graphs pay particular attention to the position of the beat maxima along the time axis. What is the influence of the phase? _____

IV. NOTES

Chapter 21
THE SPECIAL THEORY OF RELATIVITY

I. BASIC CONCEPTS

When two observers who are moving relative to each other measure the same physical quantity, they may obtain different values. The theory of special relativity tells how the values are related to each other when both observers are at rest in different inertial frames. Although the complete theory deals with all physical quantities, the ones you consider here are the coordinates and time of an event and the velocity, momentum, and energy of a particle.

All the equations you will use are linear algebraic equations, so the mathematics is quite simple. The concepts, however, are difficult for some students because they run counter to experience. You must get used to the idea, for example, that the time interval and spatial distance between two events depend on the velocity of the observer carrying out the measurements. The distance between New York and Los Angeles is different for a motorist traveling between those cities at 55 mph and a rocket doing the same at close to the speed of light. The lifetime of a human being on earth is different for that person and for an alien who watches the person from a fast-traveling spaceship.

The basis of special relativity. The theory is based on two postulates. The first deals with the laws of physics. It is: _____
_____

Keep in mind that the laws of physics are relationships between physical quantities, not the quantities themselves. Newton's second law and the conservation principles are examples of laws. The momentum of a system has a different value for different reference frames but if it is conserved for one inertial frame it is conserved for all inertial frames, according to the postulate.

The second postulate deals with the speed of light. It is: _____
_____

Suppose a light source sends a pulse of light toward you. If you are stationary with respect to the source, the speed of the pulse relative to you is _____. If you are moving at $c/2$ toward the source the speed of the pulse relative to you is _____. If you are moving away from the source the speed of the pulse relative to you is _____.

You should understand the phenomena of relativity not only qualitatively but also from the mathematical viewpoint of the Lorentz transformation. Suppose an event is viewed by two observers, one at rest in inertial frame S and another at rest in inertial frame S'. From the viewpoint of S, S' is moving in the positive x direction with velocity u. The coordinate systems and clocks are arranged so that the origins coincide at time $t = 0$ and the clocks of S' are synchronized to read 0 when the origins coincide. The coordinates of the event are S are

x, y, z and time of the event is t, all as measured in S. The same event has coordinates x', y', z' and occurs at time t', as measured in S'. Then, according to the Lorentz transformation, the coordinates and times are related by

$$x' =$$

$$y' =$$

$$z' =$$

$$t' =$$

where the <u>Lorentz factor</u> γ is defined by $\gamma =$ _____ .

You will often deal with the time and spatial intervals between two events. Suppose the spatial interval has components Δx, Δy, and Δz in S and has components $\Delta x'$, $\Delta y'$, and $\Delta z'$ in S'. The time interval is Δt in S and $\Delta t'$ in S'. Then the intervals are related by

$$\Delta x' =$$

$$\Delta y' =$$

$$\Delta z' =$$

$$\Delta t' =$$

You should recognize that the same equations can be used if S' is moving in the negative x direction relative to S. The velocity u is then negative.

You might be given the intervals in S' and asked for those in S. You then use

$$\Delta x =$$

$$\Delta y =$$

$$\Delta z =$$

$$\Delta t =$$

Carefully note that distance measurements along the y or z axis produce the same values in all inertial frames that move relative to each other along their x axes. Different values are obtained only when the measurement is along the direction of relative motion.

Whether you must take special relativity into account or not is controlled by the value of γ. If γ is close to 1 you can safely ignore relativity. But if γ is much greater than 1 relativity is important. If $u = 0.1c$, $\gamma =$ _____; if $u = 0.5c$, $\gamma =$ _____; if $u = 0.9c$, $\gamma =$ _____; if $u = 0.95c$, $\gamma =$ _____; and if $u = 0.99c$, $\gamma =$ _____.

Time dilation. The digram to the right shows a clock. The flash unit F emits a light pulse that travels to mirror M and is reflected back to the flash unit. It is detected there and immediately triggers the next flash. If the flash unit and mirror are separated by a distance L_0 then the time interval between flashes is given by $\Delta t_0 = $ _____.

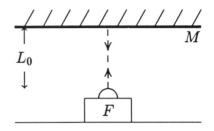

Suppose S' carries a clock like this at speed u past observer S. Complete the diagram on the right by drawing the path of one pulse as seen by S. If Δt is the time interval between emission and detection of the pulse, as measured by S, then during this interval the flash unit moves a distance $\ell = $ _____ and the light pulse moves a distance $2L = $ _____. This follows because L is the hypotenuse of a right triangle with sides of length L_0 and $u\Delta t/2$. Substitute $2L = c\Delta t$ and solve for Δt:

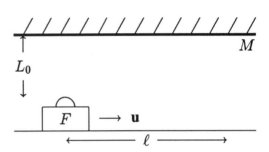

You should obtain $\Delta t = \gamma \Delta t_0$. Notice that the derivation of this result depends strongly on the second postulate of relativity. An observer comparing a moving clock with his clocks concludes that the moving clock ticks at a slower rate. It is not important that the clock utilize light as does the clock you used to derive the relationship above. Any clock will do, even a heartbeat.

You should realize that there is perfect symmetry between the two reference frames. An observer in S' watching a clock in S sees that clock tick at a slower rate.

The concept of <u>proper time</u> is important for understanding the relativity of time. Suppose two events, such as the emission and detection of a light pulse, occur at the same coordinate in one frame. The time interval between them, as measured in that frame, is the proper time between the events. The time interval between the same two events, as measured in a frame that is moving relative to the first, is longer. The relationship is $\Delta t = \gamma \Delta t'$, where $\Delta t'$ is the proper time interval, Δt is the interval in the second frame, and $\gamma = 1/\sqrt{1 - u^2/c^2}$. Here u is the speed of the second frame relative to the first.

Use the Lorentz transformation equations to obtain this result. Take $\Delta x' = 0$ (the events occur at the same coordinate in S') and solve for Δt:

Now suppose the two events occur at the same coordinate in the S frame. Take $\Delta x = 0$ and solve for $\Delta t'$:

You should obtain $\Delta t' = \gamma \Delta t$.

Notice that it is possible to give the time interval in one frame and ask for it in another such that the events do not occur at the same coordinate in either frame. Then the clocks in neither frame measure the proper time between the events.

Length contraction. A light pulse can be used to measure a distance. Place a flash unit at one end of the distance to be measured and a mirror at the other end. Then time the interval between the flash and its detection back at the flash unit. If Δt_0 is the time interval then $L_0 =$ _____ is the length. The diagram on the right illustrates the process.

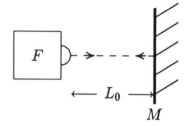

Suppose this process is viewed by an observer moving with speed u to the left and suppose the time interval in this frame from the emission of the light pulse to its reflection at the mirror is Δt_1. While the pulse is in transit the mirror moves a distance $u\Delta t_1$ to the right and the total distance traveled by the pulse is $L + u\Delta t_1$, where L is the distance between the flash unit and the mirror. Since the pulse travels at speed c, $(L + u\Delta t_1)/c =$ _____. Solve this to find $\Delta t_1 =$ _____. Let Δt_2 be the time taken by the pulse to go from the mirror back to the flash unit. During this interval the flash unit moves a distance $u\Delta t_2$ to the right and the distance traveled by the pulse is $L - u\Delta t_2$. Thus $(L - u\Delta t_2)/c =$ _____. Solve this to find $\Delta t_2 =$ _____.

The total time for the trip is $\Delta t_1 + \Delta t_2 =$ _____ and since Δt_0 is the proper time interval between the emission and detection of the flash this must be equal to $\gamma \Delta t_0$. Replace Δt_0 with $2L_0/c$ and solve for L in terms of L_0. The result is $L =$ _____.

The distance between the flash unit and the mirror as measured in the frame in which they are at rest is called the _____ length of the spatial interval. The length as measured in any frame moving parallel to the length is less by the factor _____. The length as measured in any frame moving perpendicularly to the length is the same as the rest length.

The same results can be derived by means of the Lorentz transformation. Suppose an object of rest length L_0 is at rest along the x axis of frame S and its length is measured in S', a frame that is moving with speed u in positive x direction relative to S'. One way of measuring the length is to place two marks on the x' axis of S', one at the front of the object and one at the back, then measure the distance between the marks. The marks, of course, must be made simultaneously. Set $\Delta x' = L$, $\Delta t' = 0$, and $\Delta x = L_0$. Use the transformation equations to solve for $\Delta x'$ in terms of Δx:

You should obtain $\Delta x' = \Delta x/\gamma$, where $\gamma = 1/\sqrt{1 - u^2/c^2}$.

Now suppose the object is at rest on the x' axis of S' and its length is measured in S. Set $\Delta x' = L_0$, $\Delta x = L$, and $\Delta t = 0$. Solve for Δx in terms of $\Delta x'$:

You should obtain $\Delta x = \Delta x'/\gamma$.

Simultaneity. Notice from the derivation of the length contraction equation that two events that are simultaneous in one frame are not simultaneous in another that is moving relative to the first. Suppose the events are simultaneous in S and are spatially separated by Δx. The time interval between them, as measured in S', is given by $\Delta t' =$ _____ if S' is moving with speed u in the positive x direction relative to S.

Now suppose two events are simultaneous in S' and are spatially separated by x'. Give an expression for the time interval between them, as measured in S: $\Delta t =$ _____ .

Relativistic velocities. Here you deal with the relationship between the velocity of an object as measured in one frame and its velocity as measured in another, moving with speed u in the positive x direction.

Suppose the velocity components as measured in S are v_x, v_y, and v_z. Then the components in S' are given by

$$v'_x =$$

$$v'_y =$$

$$v'_z =$$

These expressions can be derived easily from the Lorentz transformation equations. Divide $\Delta x' = \gamma(\Delta x - u\Delta t)$ by $\Delta t' = \gamma(\Delta t - u\Delta x/c^2)$ to obtain $\Delta x'/\Delta t' =$ _____ . Now divide both the numerator and the denominator by Δt, replace $\Delta x'/\Delta t'$ with v'_x, and replace $\Delta x/\Delta t$ with v_x:

Divide $\Delta y' = \Delta y$ by $\Delta t' = \gamma(\Delta t - u\Delta x/c^2)$ to obtain $\Delta y'/\Delta t' =$ _____ . Divide both the numerator and denominator by $\Delta t'$, replace $\Delta y'/\Delta t'$ with v'_y, replace $\Delta x/\Delta t$ with v_x, and replace $\Delta y/\Delta t$ with v_y:

The derivation of the expression for v'_z follows in exactly the same manner.

When $u \ll c$ the quantity $1 - uv_x/c^2$ in the denominators can be approximated by 1 and the Galilean velocity transformation equations are obtained. They are $v'_x =$ _____ , $v'_y =$ _____ , and $v'_z =$ _____ .

One of the most important consequences of the relativistic velocity transformation equations is: if the speed of an object, as measured in one frame, equals the speed of light then its speed, as measured in any other frame, also equals the speed of light. No matter how fast you travel away from an approaching light pulse its speed relative to you will be c. The same is true no matter how fast you travel toward it. Prove this for an object moving in the positive x direction: replace v_x with c in the expression for v'_x and show that the result is $v'_x = c$.

A corollary is: if the speed of an object, as measured in one frame, is less than the speed of light, then its speed, as measured in any frame, is less than the speed of light. If an object is moving toward you then its speed relative to you will be less than the speed of light no matter how fast you travel toward or away from it

The one exception occurs if you travel at the speed of light. Suppose both the object is traveling in the positive x direction and you are traveling in the negative x direction. Replace u with $-c$ and show that $v'_x = c$:

Relativistic momentum and energy. Relativity theory requires that the definitions of momentum and kinetic energy be revised if these quantities are to obey the familiar conservation laws. More precisely, if measurements taken in one inertial frame show that momentum is conserved then measurements taken in any other inertial frame should also show that momentum is conserved. Similarly if energy is conserved in one inertial frame then it should be conserved in all inertial frames.

If a particle has mass m and travels with velocity **v** then its momentum **p** is given by

$$\mathbf{p} = $$

If $u \ll c$ this expression reduces to $\mathbf{p} = $ _____, the non-relativistic definition.

If a system consists of several particles the total momentum is the vector sum of the individual momenta. If the net external force on particles of a system vanishes then the total momentum of the system is conserved.

Suppose a particle with mass m is moving in the positive x direction with velocity v. Its momentum, of course, is in the positive x direction and is given by $p = mv/\sqrt{1 - v^2/c^2}$. The same particle is observed in frame S', moving with speed u in the positive x direction. The momentum of the particle, as measured in this frame, is given by $p' = mv'/\sqrt{1 - (v')^2/c^2}$, where $v' = $ _____.

The kinetic energy of a particle with mass m moving with speed v is given by

$$K = $$

The total kinetic energy of a system of particles is the scalar sum of the individual kinetic energies. This quantity is conserved in collisions provided the masses of the particles do not change.

In many nuclear scattering and decay processes the masses do change. You must then take into the account the <u>rest energies</u> of the particles. The rest energy of a particle of mass

m is given by $E_0 = $ _____. The total energy, the sum of the rest and kinetic energies, of a particle is given in terms of the mass and speed of the particle by

$$E = $$

The total energy and the magnitude of the momentum of any particle are related by

$$E^2 = $$

This expression replaces $E = p^2/2m = mv^2/2$, the non-relativistic relationship between E and p or between E and v. Note that the energy E in the relativistic relationship includes both rest and kinetic energies.

If a particle has zero mass, as do photons and perhaps neutrinos, the its energy is given in terms of its momentum by $E = $ _____.

III. PROBLEM SOLVING

Many problems give the coordinate and time of an event in one reference frame and ask for the coordinate and time in another. The problems can usually be set up so that each frame is moving along the x axis of the other and so that clocks at the origins read $t = 0$ and $t' = 0$ when the origins coincide. Then the Lorentz transformation equations are $x' = \gamma(x - ut)$, $y' = y$, $z' = z$, and $t' = \gamma(t - ux/c^2)$, where $\gamma = 1/\sqrt{1 - u^2/c^2}$. Here u is the velocity of the primed frame relative to the unprimed frame; it is positive if the primed frame is moving in the positive x direction and negative if it is moving in the negative x direction.

Sometimes you are given the space and time intervals between two events rather than the coordinates and times of the events. Then you use $\Delta x' = \gamma(\Delta x - u\Delta t)$, $\Delta y' = \Delta y$, $\Delta z' = \Delta z$, and $\Delta t' = \gamma(\Delta t - u\Delta x/c^2)$.

Two special cases are of interest. If two events occur simultaneously in reference frame S, at places separated by Δx along the x axis, then the time interval between them in frame S' is given by $\Delta t' = -\gamma u\Delta x/c^2$ and their spatial separation is given by $\Delta x' = \gamma\Delta x$. You should recognize that a measurement of the length of a moving object consists of simultaneously noting the coordinates of the front and back of the object. Thus the length of an object moving parallel to its length is shorter than its rest length by the factor γ.

If two events occur at the same place in reference frame S and the time interval between them is Δt, then in frame S' they are separated by $\Delta x' = -\gamma u\Delta t$ and the time interval between them is $\Delta t' = \gamma\Delta t$. Δt is the proper time interval and the interval in all frames moving with respect to S is longer by a factor of γ.

PROBLEM 1. According to clocks at rest on the earth a space ship enters a galaxy at $t = 0$ and exits the other side at $t = 6.4 \times 10^7$ s. The galaxy is 1.8×10^{16} m wide. What is the proper time between these two events? What clocks measure the proper time?

SOLUTION: The proper time is measured by a clock on board the space ship, traveling at $u = \Delta x/\Delta t$ relative to the earth. The proper time is given by $\Delta t' = \Delta t/\gamma$.

[ans: 2.23×10^7 s]

PROBLEM 2. A rocket traveling at 0.96c crosses a galaxy in 1.7 yr (5.36×10^7 s), as measured by clocks on board. Frame S is attached to the rocket and frame S' is attached to the galaxy. What is the time of the trip as measured in S'?

SOLUTION: The two events occur at the same coordinate in S so these clocks measure the proper time interval between them. The time interval in S' is given by $\Delta t' = \gamma \Delta t$.

[ans: 1.91×10^8 s]

How wide is the galaxy as measured by observers on the rocket? How wide is it as measured by observers at rest relative to the galaxy?

SOLUTION: The galaxy goes by the windows of the rocket in the time interval $\Delta t = 5.36 \times 10^7$ s, at a speed of 0.96c. The width of the galaxy in S is given by $\Delta x = u\Delta t$. The width of the galaxy in S' is the rest width and is given by $\Delta x' = \gamma \Delta x$.

[ans: 1.54×10^{16} m; 5.51×10^{16} m]

As a check divide the width of the galaxy by the time interval, both as measured in S'. This should yield the speed of the rocket.

PROBLEM 3. A muon is created in a nuclear reaction 3000 m above the surface of the earth. It travels at 0.99c toward the surface and, in its rest frame, lives for 1.5×10^{-6} s before it decays into other particles. Note that if the lifetime were the same in the frame of the earth the muon would travel about 446 m before decaying and so would not reach the earth. In the rest frame of the earth what is the muon lifetime? How far does the muon travel in that time if it is not impeded? Does it reach the earth?

SOLUTION: The proper time interval between the creation and decay of the muon is $\Delta t' = 1.5 \times 10^{-6}$ s. In the frame of the earth the muon lifetime is $\Delta t = \gamma \Delta t'$. The muon travels a distance $d = u\Delta t$ in that frame.

[ans: 1.06×10^{-5} s; 3.16×10^3 m; yes]

The same conclusion must be reached if the problem is worked using times and distances measured in the rest frame of the muon. As measured in that frame how far away is the earth when the muon is created? In that frame the earth is moving toward the muon with a speed of 0.99c. How far does the earth move during the lifetime of the muon? Does the earth reach the muon before it decays?

SOLUTION: In the frame of the earth the distance between the earth and the point where the muon is created is $\Delta x = 3000$ m. This is a rest length. The distance in the rest frame of the muon is $\Delta x' = \Delta x/\gamma$. The earth moves a distance $d' = u\Delta t' = 0.99c\Delta t'$.

[ans: 423 m; 446 m; yes]

The two events considered may not occur at the same coordinate in either S or S'. Then neither Δt nor $\Delta t'$ is the proper time interval between them and these two quantities are not related to each other by a factor γ. The two events may not occur simultaneously in either S or S'. Then Δx and $\Delta x'$ are not related by the factor γ either.

PROBLEM 4. A space ship travels at 0.96c directly away from the earth. Event 1 occurs at the back of the ship and event 2 occurs 3.2×10^{-6} s later at the front, 200 m away. This time and distance are measured by instruments at rest with respect to the ship. Find the time and distance intervals between the two events as measured by instruments at rest with respect to the earth. Show by direct comparison that neither the ratio of the distance intervals nor the ratio of the time intervals equals γ.

SOLUTION: If the primed frame is attached to the space ship then $\Delta x' = 200$ m and $\Delta t' = 3.2 \times 10^{-6}$ s. Use the Lorentz transformation equations to find Δx and Δt.

[ans: 2.58×10^3 m; 9.14×10^{-6} s]

Find the speed of a reference frame for which the two events occur at the same time. Is such a frame physically realizable (is its speed less than the speed of light)?

SOLUTION: If the unprimed frame is at rest with respect to the space ship and the events occur at the same time in the primed frame, then $\Delta t' = \gamma(\Delta t - \Delta x u/c^2) = 0$, so $u = c^2 \Delta t/\Delta x$.

[ans: 1.44×10^9 m/s; no]

Find a frame for which the two events occur at the same coordinate. Is such a frame physically realizable?

SOLUTION:

[ans: 6.25×10^7 m/s; yes]

PROBLEM 5. A barn is 30 m long and has doors in both the front and back walls. Initially both doors are open. A 40-m rod, traveling parallel to its length at $u = 0.95c$, enters the front door and, when its back end is inside the barn, both doors are closed. Since, in the rest frame of the barn, the length of the rod is $40/\gamma$ m long, and therefore shorter than 30 m, this can clearly be done with the rod entirely within the barn. Shortly thereafter, of course, the front end of the rod slams through the closed rear door of the barn. For a short time, however, the rod is entirely in the barn with both doors shut. From the point of view of an observer moving with the rod, however, the rod is 40 m long and barn is shorter, $30/\gamma$ m. It is clearly impossible for the rod to fit in the barn.

Consider the following three events: the front of the rod enters the barn, the back of the rod enters the barn, and the front of the rod reaches the back of the barn. Let frame S be at rest relative to the barn and suppose the front door of the barn is at $x = 0$, the rear door is at $x = 30$ m, and the front end of the rod enters the barn at $t = 0$. Let frame S' be at rest with respect to the rod and suppose the front end is at $x' = 0$ and it enters the barn at $t' = 0$. For both frames, calculate the time when the back end of the rod enters the barn and the time when the front end of the rod reaches the back door of the barn.

SOLUTION: First consider the frame of the barn. The back end of the rod enters at time $t_1 = \ell_{\rm rod}/u$, where $\ell_{\rm rod}$ is the length of the rod in S. The front end of the rod hits the back of the barn at time $t_2 = \ell_{\rm barn}/u$.

[ans: $t'_1 = 4.38 \times 10^{-8}$ s; $t'_2 = 1.05 \times 10^{-8}$ s]

Use the Lorentz transformation to find the position and time of each of the events in the rest frame of the rod.

SOLUTION:

[ans: $t_1 = 1.40 \times 10^{-7}$ s; $t_2 = 3.29 \times 10^{-8}$ s]

In the frame of the barn the events occur in this order: front end of rod enters barn, back end of rod enters barn, front end of rod hits back of barn. In the frame of the rod the order of events is: front end of rod enters barn, front end of rod hits back of barn, back end of rod enters barn.

Some problems deal with length measurements. Be sure to remember that the length of a moving object must be measured by *simultaneously* noting the coordinates of the ends.

PROBLEM 6. Suppose frame S' is attached to a rocket ship traveling at $0.97c$ in the positive x direction of S. At time $t = 0$ an observer in S simultaneously makes two marks on his x coordinate axis. One, at $x_1 = 0$, is at the back of the rocket and the other, at $x_2 = 85$ m, is at the front. The length of the rocket, as measured in S, is $x_2 - x_1 = 85$ m. Find the coordinates and times of the two events as measured in S'.

SOLUTION: Since $\Delta t = 0$, $x' = \gamma x$ and $t' = -\gamma x u/c^2$.

[ans: back: $x_1' = 0$, $t_1' = 0$; front: $x_2' = 350$ m, $t_2' = -1.13 \times 10^{-6}$ s]

$x_2' - x_1'$ is the length of the rocket as measured in S'. It is the rest length of the rocket since it is measured with a meter stick that is at rest relative to the rocket.

According to instruments in S' the mark at the front of the rocket is made before the mark at the back. During the time interval between the making of the marks, the mark at the front moves backward toward the rear with speed $u = 0.97c$. Find the x' coordinate of the mark on the x axis originally at the front of the rocket, but find it at the time the mark at the back of the rocket is made.

SOLUTION: Set $t' = $ the time the mark at the back is made, $x = 85$ m, and solve the Lorentz transformation equations for x'.

[ans: 20.7 m]

Now add $|ut'|$ to the result and compare the answer to the rest length of the rocket.

Another way of working the problem: the mark will move a distance $|ut'| = 329$ m before the mark at the back is made. Its coordinate can be found by subtracting this from the rest length of the rocket.

PROBLEM 7. Here's an example of how the length of a moving object *cannot* be measured. Frame S' is attached to a rocket traveling at $0.97c$ in the positive x direction of frame S. At time $t' = 0$ an observer in S' simultaneously marks both the x and x' axes. One mark, at $x_1' = 0$ is at the back of the rocket and the other, at $x_2' = 300$ m, at the front. The rest length of the rocket is $x_2' - x_1' = 300$ m. Find the coordinates of the two events, as measured in S. What is the distance between the marks on the x axis, as measured in S? What is the length of rocket as measured in S?

SOLUTION:

[ans: $x_1 = 0$, $x_2 = 1.23 \times 10^3$ m; $d = 1.23 \times 10^3$ m; $\ell = 72.9$ m]

The length of the rocket in S is not d because the marks were not made simultaneously in S.

Here is a problem that will help you analyze the relativity of simultaneity.

PROBLEM 8. According to clocks in reference frame S two events occur simultaneously at time $t = 0$. Event 1 is the explosion of a firecracker at the origin and event 2 is the explosion of another firecracker at $x = 5.0 \times 10^7$ m. These events are viewed by an observer at rest in reference frame S', which moves relative to S in the positive x direction at 2.4×10^8 m/s. What are the coordinates and times of the two events, as measured in the primed reference frame?

SOLUTION: Use $x' = \gamma x$ and $t' = -\gamma x u/c^2$, where $\gamma = 1/\sqrt{1 - u^2/c^2}$. For the first event $x = 0$; for the second $x = 5.0 \times 10^7$ m.

[ans: event 1: $x'_1 = 0$, $t'_1 = 0$; event 2: $x'_2 = 8.33 \times 10^7$ m, $t'_2 = -0.222$ s]

Notice that the two events do not occur simultaneously in the primed frame. The minus sign indicates that event 2 occurs before event 1. Also notice that the distance between the events as measured in that frame is not the same as the distance between them as measured in S.

The leading edges of the flashes meet at $x = 2.5 \times 10^7$ m, halfway between the places where the events occurred. They meet at time $t = 2.5 \times 10^7/c = 8.33 \times 10^{-2}$ s. Where and when do they meet as measured by instruments at rest in the primed frame?

SOLUTION: The meeting of the flashes is not simultaneous in either S or S' with either firecracker explosion. Nor does it occur at the same coordinate as either explosion. You must use the full Lorentz transformation: $x' = \gamma(x - ut)$, $t' = \gamma(t - xu/c^2)$.

[ans: 8.33×10^6 m; 2.78×10^{-2} s]

In the unprimed frame the fact that the flashes meet halfway between the sources is a clear indication that they started at the same time. In the primed frame, the flash that started first travels the greater distance and the meeting place is closer to the source of the second flash. Do your results substantiate this conclusion?

The following diagrams may help you organize the various distance and time measurements described in this problem.

The diagram on the right shows the view from frame S when the origins of the two frames coincide. The points $x = 0$ and $x = 5.0 \times 10^7$ m, where the events occur, are marked. The lower clocks are at rest in S and both show $t = 0$. The upper clocks are at rest in S' and move to the right with speed u relative to S. On each S' clock record the time of the event near it, as calculated above. If either observer looks at the nearest S' clock when either of the events occurs these are the readings he sees. They are different.

Also record on the diagram the coordinates x'_1 and x'_2 of the events. If either observer looks at the x' axis when either of the events occurs these are the coordinates he sees at the position of the event.

The diagram on the right shows the view from S' when event 2 occurs. Frame S and its clocks are moving to the left with speed u and the clocks in S' show the same time, the time of event 2. Fill in the values of t'_2 and x'_2. Notice that the origins of the frames do not yet coincide but they will in a short while. The event at the origin has not yet occurred but it will when the origins coincide. The clock at the origin of S does not read 0 and we wish to find its reading. To do this, first find the position of the origin of S, as measured in S'. This is just $x' = -ut'_2$, a positive number. Now pretend there is an event at this value of x' and at time t'_2 and use the inverse Lorentz transformation to find x and t for this event. Record the values on the diagram.

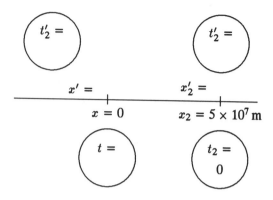

SOLUTION:

[ans: $x' = 5.33 \times 10^7$ m; $t = -0.329$ s]

Notice the symmetry in these diagrams. An observer always finds the clocks at rest with respect to him to be synchronized and always finds the clocks moving with respect to him to be out of synchronization.

Given the velocity of an object as measured in one frame of reference you should be able to calculate its velocity in another frame. If the second frame is moving with speed u along the x axis of the first and if the components of the particle velocity in the first frame are v_x, v_y, v_z, then the components in the second are given by $v'_x = (v_x - u)/(1 - uv_x/c^2)$, $v'_y = v_y/\gamma(1 - uv_x/c^2)$, and $v'_z = v_z/\gamma(1 - uv_x/c^2)$, where $\gamma = 1/\sqrt{1 - u^2/c^2}$.

PROBLEM 9. Two rocket ships are moving relative to an observer at rest in frame S. The first moves in the positive x direction with velocity $v_1 = 0.750c$ and the second moves in the negative x direction with velocity $v_2 = -0.850c$. They move toward each other.

What is the velocity of the second rocket as measured by an observer on board the first?

SOLUTION: Take $u = v_1 = 0.750c$ and find v'_2.

[ans: $-0.977c$ (-2.93×10^8 m/s)]

What is the velocity of the first rocket as measured by an observer on board the second?

SOLUTION: Take $u = v_2 = -0.850c$ and find v'_1.

[ans: $0.977c$ (2.93×10^8 m/s)]

The crew of the first rocket sends a message capsule toward the second rocket. The capsule travels at $0.800c$ relative to the first rocket. At what speed does the second observer see the capsule approach?

SOLUTION: Take $u = -.977c$ and $v = 0.800c$.

[ans: $0.997c$ (2.99×10^8 m/s)]

What is the speed of the capsule relative to S?

SOLUTION: S has a velocity of $-0.750c$ relative to the first rocket.

[ans: $0.969c$ $(2.91 \times 10^8$ m/s$)$]

PROBLEM 10. Two rocket ships are moving relative to an observer at rest in S. The first moves with velocity $v_{1x} = 0.750c$, $v_{1y} = 0$ and the second moves with velocity $v_{2x} = 0$, $v_{2y} = 0.850c$. Both rockets start at the origin at $t = 0$. What are the velocity components of the second rocket, as measured by an observer on board the first? What is the speed of the second rocket according to this observer?

SOLUTION:

[ans: $v'_{2x} = -0.750c$ $(-2.25 \times 10^8$ m/s$)$; $v'_{2y} = 0.562c$ $(1.69 \times 10^8$ m/s$)$; $v'_2 = 0.937c$ $(2.81 \times 10^8$ m/s$)$]

At $t = 0.50$ s, as measured in S, the observer on the first rocket beams a light signal to the second rocket. If the beam makes an angle of $7.2°$ with the y axis of S, does it hit the second rocket? If it hits at what time does it hit according to clocks in S?

SOLUTION: The coordinates of the leading edge of the signal are given by $x(t) = x_0 - c(t - t_0) \sin \alpha$ and $y(t) = c(t - t_0) \cos \alpha$, where $\alpha = 7.2°$, $t_0 = 0.50$ s, and $x_0 = v_{1x} t_0$ is the coordinate of the first rocket when the signal is sent. The coordinates of the second rocket are given by $x = 0$, $y(t) = v_{2y} t$. The question reduces to: does $x_0 - c(t - t_0) \sin \alpha = 0$ when $c(t - t_0) \cos \alpha = v_{2y} t$?

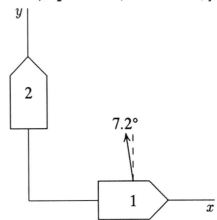

[ans: yes; 3.49 s]

At what angle to the y axis of his rest frame must the sender aim the beam?

SOLUTION: In S the velocity components of the light signal are given by $c_x = -c \sin \alpha$, $c_y = c \cos \alpha$. Use the velocity transformation equations to find the components in the rest frame of the sending rocket. Then use $\tan \alpha' = -c'_x/c'_y$.

[ans: $53.1°$]

Some problems deal with the relativistic definitions of momentum and energy. If $\mathbf{v}$ is the velocity of a particle with mass m then $\mathbf{p} = m\mathbf{v}/\sqrt{1 - v^2/c^2}$ gives its momentum and $E = mc^2/\sqrt{1 - v^2/c^2}$ gives its energy. E is the sum of its rest energy $E_0 = mc^2$ and its kinetic energy. The energy and momentum are related by $E^2 = (pc)^2 + (mc^2)^2$.

Chapter 21: The Special Theory of Relativity

PROBLEM 11. A 2.7×10^{-27}-kg particle has a speed of $0.85c$. What is the magnitude of its momentum? What is its total energy, rest energy, and kinetic energy?

SOLUTION:

[ans: 1.31×10^{-18} kg·m/s; 4.61×10^{-10} J; 2.43×10^{-10} J; 2.18×10^{-10} J]

A 2.7×10^{-27}-kg particle has a momentum of 1.4×10^{-18} kg·m/s. What is its total energy, rest energy, and kinetic energy?

SOLUTION: Solve $E^2 = (pc)^2 + (mc^2)^2$ for its energy, $E_0 = mc^2$ for its rest energy, and $K = E - E_0$ for its kinetic energy.

[ans: 4.85×10^{-10} J; 2.43×10^{-10} J; 2.42×10^{-10} J]

PROBLEM 12. A particle moves in the positive x direction of frame S. It has momentum 2.7×10^{-18} kg·m/s and energy 8.6×10^{-10} J. What is its mass? What is its velocity?

SOLUTION: Solve $E^2 = (pc)^2 + (mc^2)^2$ for m and $p = mv/\sqrt{1 - v^2/c^2}$ for v.

[ans: 3.21×10^{-27} kg; $0.887c$ (2.66×10^8 m/s)]

PROBLEM 13. For what speed is the kinetic energy of any particle equal to its rest energy?

SOLUTION: The total energy must be twice the rest energy. Solve $mc^2/\sqrt{1 - v^2/c^2} = 2mc^2$ for v.

[ans: $0.707c$]

If you know the momentum and energy in one reference frame you can find them in another.

PROBLEM 14. In the laboratory, an electron ($m = 9.1096 \times 10^{-31}$ kg) travels at $0.93c$ in the x, y plane. Its path makes an angle of $30°$ with the positive x axis. Find the x and y components of its momentum and its energy as measured in the laboratory.

SOLUTION:

[ans: $p_x = 5.99 \times 10^{-22}$ kg·m/s; $p_y = 3.46 \times 10^{-22}$ kg·m/s; $E = 2.23 \times 10^{-13}$ J]

Its motion is also viewed from the S' frame, which moves in the positive x direction with a velocity of $0.96c$, relative to the laboratory. Find the velocity components of the electron velocity in S'.

SOLUTION: The components in S are $v_x = 0.93c\cos 30°$, $v_y = 0.93c\sin 30°$. Substitute into $v'_x = (v_x - u)/(1 - uv_x/c^2)$ and $v'_y = v_y/\gamma(1 - uv_x/c^2)$.

[ans: $v'_x = -0.682c$ (-2.04×10^8 m/s); $v'_y = 0.574c$ (1.72×10^8 m/s)]

What are the momentum components and the energy as measured in S'?

SOLUTION: Use $p'_x = mv'_x/[1 - (v')^2/c^2]$ and $p'_y = mv_y/[1 - (v')^2/c^2]$, where $v' = \sqrt{(v'_x)^2 + (v'_y)^2}$. The magnitude of the momentum is given by $p' = \sqrt{(p'_x)^2 + (p'_y)^2}$ and the energy is given by $(E')^2 = (p'c)^2 + (mc^2)^2$.

[ans: $p'_x = -4.10 \times 10^{-22}$ kg·m/s; $p'_y = 3.45 \times 10^{-22}$ kg·m/s; $E' = 1.80 \times 10^{-13}$ J]

For a system composed of more than one particle, the total momentum is the vector sum of the individual momenta and the total energy is the sum of the individual energies, all measured in the same frame. $E^2 = (pc)^2 + (mc^2)^2$ is used to define what is meant by the mass m of the system. Given the total energy E and the magnitude of the total momentum p it is solved for m. The result is not necessarily the sum of the masses of the individual particles in the system.

PROBLEM 15. Two particles move toward each other along the x axis of frame S. The first particle has a mass of 7.6×10^{-25} kg and moves to the right with a speed of $0.96c$ while the second particle has a mass of 5.4×10^{-25} kg and moves to the left with a speed of $0.91c$. Find the individual momenta, the individual energies, the total momentum, and the total energy, as measured in S.

SOLUTION:

[ans: $p_1 = 7.82 \times 10^{-16}$ kg·m/s; $E_1 = 2.44 \times 10^{-7}$ J; $p_2 = -3.56 \times 10^{-16}$ kg·m/s; $E_2 = 1.17 \times 10^{-7}$ J; $p = 4.26 \times 10^{-16}$ kg·m/s; $E = 3.62 \times 10^{-7}$ J]

What is the mass of the system?

SOLUTION: Solve $E^2 = (pc)^2 + (mc^2)^2$ for m.

[ans: 3.76×10^{-24} kg]

Notice that the mass of the system is greater than the sum of the masses of the individual particles. The mass of the system depends on the kinetic energy of the constituent particles as well as on their masses.

III. COMPUTER PROJECTS

First write a program that carries out the Lorentz transformation for a succession of values of the relative velocity u of the reference frames. Input the coordinate x and time t of the event, as measured in frame S. Then input the first value of u to be considered, the last, and the interval Δu. Positive values of u mean S' is moving in the positive x direction; negative values mean it is moving in the negative x direction. Use $x' = \gamma(x - ut)$ and $t' = \gamma(t - ux/c^2)$, where $\gamma = 1/\sqrt{1 - u^2/c^2}$ and c is the speed of light, to calculate the coordinate and time in the primed frame. Here's an outline.

> input coordinate and time in S: x, t
> input first velocity, last velocity, and increment: $u_i, u_f, \Delta u$
> set $u = u_i$
> **begin loop** over intervals
>   exit loop if $|c - u|/c < 10^{-5}$
>   calculate γ: $\gamma = 1/\sqrt{1 - u^2/c^2}$
>   calculate x': $x' = \gamma(x - ut)$
>   calculate t': $t' = \gamma(t - ux/c^2)$
>   display u, x, t, x', t'
>   increment u by Δu
> **end loop** if $u > u_f$
> stop

The computer will overflow if it tries to calculate γ for u nearly equal to c so these values are rejected. If $u_f = u_i$ then results for only one velocity are produced.

Use the program to investigate simultaneity and length measurements.

PROJECT 1. Two events occur at time $t = 0$ in reference frame S, one at the origin and the other at $x = 5.0 \times 10^7$ m. Plot the distance $|\Delta x'|$ between the events as measured in frame S', as a function of the velocity u of that frame relative to frame S. Consider values from $-0.95c$ to $+0.95c$.

For the range considered which value of u results in the greatest value of $|\Delta x'|$? Which results in the smallest value? [ans: $\pm 0.95c$; 0]

Plot the time interval $|\Delta t'|$ between the events as measured in S', as a function of u. For the range considered which value of u results in the greatest value of $|\Delta t'|$? Which results in the smallest value? [ans: $\pm 0.95c$, 0]

For what range of u does the second event occur before the first ($\Delta t' < 0$)? For what range does it occur after the first ($\Delta t' > 0$)? [ans: $u > 0$; $u < 0$]

You may think of the values of $|\Delta x'|$ as the results of a series of length measurements, with the object being measured traveling at the various velocities considered. Its length is S is measured by simultaneously making marks at the front and back ends on the x axis and measuring the distance between them. Since the marks are made simultaneously $\Delta t = 0$ and $\Delta x' = \gamma \Delta x$, where $\Delta x'$ is the length in the rest frame of the object. Note that $|\Delta x'| > |\Delta x|$ and the discrepancy becomes greater as the speed becomes greater.

PROJECT 2. Two events at separated places that are simultaneous in one frame are not simultaneous in any other frame that is moving with respect to the first along the line joining the positions of the events. The closer together the two events are in space, however, the smaller is the time interval between them in S'. Verify this statement by repeating the calculations of the last project, but take $x = 2.5 \times 10^7$ m. Compare the results you obtain with those of the last project.

If two events occur at the same coordinate then the time between the events is the proper time interval. The time between the events, as measured in another frame, is longer than the proper time by the factor γ. This is the phenomenon of time dilation.

PROJECT 3. Two events occur at $x = 0$ in reference frame S, one at $t = 0$ and the other at $t = 8.0 \times 10^{-5}$ s. Use the program to plot the time interval $|\Delta t'|$ between the events as measured in S', as a function of the velocity u of that frame relative to S. Consider values of u from $-0.95c$ to $+0.95c$.

For the range considered which value of u results in the greatest value of $|\Delta t'|$? Which results in the smallest value? [ans: $\pm 0.95c, 0$]

The proper time interval for these events is measured in frame S, where both events occur at the same place. Notice that the time interval, as measured in any other frame, is longer.

Plot the distance $|\Delta x'|$ in frame S' as a function of the velocity of that frame. For what range of u is $\Delta x'$ negative? For what range is it positive? [ans: $u < 0; u > 0$]

The two events considered may not occur at the same coordinate in either S or S'. Then neither Δt nor $\Delta t'$ is the proper time interval between them and these two quantities are not related to each other by a factor of γ. If $\Delta x/\Delta t < c$ then a frame, traveling at less than the speed of light, exists for which the two events occur at the same coordinate. The time interval, as measured in that frame, *is* the proper time interval between the events.

PROJECT 4. One event occurs at $x = 0$, $t = 0$ and another occurs at $x = 200$ m, $t = 9.5 \times 10^{-7}$ s. Use the program and a trial and error technique to find the velocity of a frame for which the events occur at the same place. Check your answer by a direction calculation: $\Delta x' = 0$ means $\gamma(\Delta x - u\Delta t) = 0$, or $u = \Delta x/\Delta t$. [ans: 2.11×10^8 m/s $(0.702c)$]

The next project gives a nice demonstration of time dilation. Each observer signals the other, giving the time read by his clocks. Each finds that the other's clocks run slowly. You may wish to revise the program so it will calculate x' and t' for various values of x and t, all for the same value of the relative velocity u. Here's an outline.

> input velocity of S' relative to S: u
> calculate γ: $\gamma = 1/\sqrt{1 - u^2/c^2}$
> input coordinate and time in S: x, t
> calculate x': $x' = \gamma(x - ut)$
> calculate t': $t' = \gamma(t - ux/c^2)$
> display u, x, t, x', t'
> another calculation?
> > if yes go back to third line
> > if no stop

PROJECT 5. A rocket starts on earth and travels away at $0.97c$. Every hour for the first five hours an earth bound transmitter at the launch pad sends a radio signal to the rocket. These signals are electromagnetic and travel at the speed of light. Assume the rocket starts at time $t = 0$ and the first signal is sent at $t_{s1} = 3600$ s. Signal n is sent at $t_{sn} = 3600n$, where $n = 1, 2, 3, 4, 5$. Take the launch pad to be at the origin of the earth's rest frame S and the rocket to be at the origin of its rest frame S'.

Chapter 21: The Special Theory of Relativity

Fill in the first 3 columns of the following table (t_{sn} is the time signal n is sent, t_{rn} is the time signal n is received, x_{rn} is the coordinate of the rocket when signal n is received, all in the rest frame of the earth). You will need to show that signal n is received at time $t_{rn} = t_{sn}c/(c-u)$, where u is the speed of the rocket. This is easy since the coordinate of the rocket is given by $x = ut$ and the coordinate of signal n is given by $x = c(t - t_{sn})$. The signal is received when these are equal. The distance from the earth to the rocket when it receives signal n is given by $x_{rn} = ut_{rn}$.

n	t_{sn}	t_{rn}	x_{rn}	t'_{rn}	t'_{sn}
1					
2					
3					
4					
5					

Now use the program to find the time t'_{rn} when signal n is received, as measured by a clock on board the rocket. Fill in the fourth column of the table.

An observer on the rocket can calculate the time, according to his clock, when each signal was sent. The signal starts from the coordinate of the launch pad at time t'_{sn}. This is $x' = -ut'_{sn}$, where the negative sign appears because in the frame of the rocket the earth is moving in the negative x direction. The coordinate of the signal is given by $x' = -ut'_{sn} + c(t' - t'_{sn})$. This must be zero when $t' = t'_{rn}$ (the rocket receives the signal at the origin of S'. So $t'_{sn} = t'_{rn}c/(c+u)$. Fill in the last column of the table. According to clock on the rocket is the earth clock slow or fast? _____

Now suppose the rocket emits a signal every hour, according to on-board clocks. Let t'_{sn} be the time signal n is sent from the rocket (at $x' = 0$), let t'_{rn} be the time it is received on earth, and let x'_{rn} be the coordinate of the earth when the signal is received. Fill out the following table for those signals.

n	t'_{sn}	t'_{rn}	x'_{rn}	t_{rn}	t_{sn}
1					
2					
3					
4					
5					

According to earth clocks is the clock on the rocket slow or fast? _____

The first program above can be used to investigate the velocity of a particle as measured by observers moving at various velocities. Suppose the particle has constant velocity v along the x axis of S. If reference frame S is placed so its origin is at the position of the particle at $t = 0$, then the coordinate of the particle is given by $x = vt$. Pick a value for t and calculate x. Now use the program to find x' and t', the coordinate and time in another frame. Finally use $v' = x'/t'$ to compute the velocity in the primed frame.

PROJECT 6. A particle moves along the x axis of reference frame S with a velocity of $0.10c$. Plot the velocity in frame S' as a function of the velocity u of that frame. Consider values from $-0.95c$ to $+0.95c$. Notice that for u close to zero the Galilean transformation is nearly correct. That is, v' is nearly $v - u$. But for u close to the speed of light v' is also close to the speed of light.

If the particle speed is c in frame S then it is c in every frame. Take $v = c$ and find the particle velocity for values of u in the range from $-0.95c$ to $+0.95c$. The answer should be c for every value of u.

Now consider a particle moving along the y axis of S. The y component of its velocity in S' is not the same as the y component in S because $\Delta t' \ne \Delta t$. In addition, the velocity in S' has an x component equal to $-u$. Test these assertions. Take the y component of the velocity in S to be $v_y = 0.10c$ and the x component to be $v_x = 0$.

Modify the program to find the components of the velocity and the speed in S' as function of the velocity of that frame. Let $x = 0$ and $y = v_y t$. Use $v'_x = x'/t'$ and $v'_y = y'/t'$ to calculate the components of the particle velocity in S'. Consider values of u from $-0.95c$ to $+0.95c$.

Notice that the y' component becomes smaller as the speed of S' approaches the speed of light. This is because $\Delta t'$ becomes larger. In the limit as $u \to c$, $v'_y \to 0$ and $v'_x \to c$. The particle moves along the x' axis at the speed of light.

As the speed of a particle with mass increases toward the speed of light its kinetic energy increases and, for speeds near the speed of light, the increase is dramatic. According to the defining equations, $\mathbf{p} = m\mathbf{v}/\sqrt{1 - v^2/c^2}$ and $E = mc^2/\sqrt{1 - v^2/c^2}$, both the energy and momentum become infinite as v approaches c. The ratio of the magnitude of the momentum to the energy, however, does not blow up but approaches $1/c$ in the limit as the speed approaches the speed of light. When pc is much greater than mc^2 then $E \approx pc$.

The first program of this section can be modified to plot the energy and the magnitude of the momentum as functions of the particle velocity. Consider a particle that moves along the x axis. Here's an outline.

> input mass, first velocity, last velocity, and increment: $m, v_i, v_f, \Delta v$
> set $v = v_i$
> **begin loop** over intervals
>   exit loop if $|c - v|/c < 10^{-5}$
>   calculate the constant α: $\alpha = \sqrt{1 - v^2/c^2}$
>   calculate p: $p = mv/\alpha$
>   calculate E: $E = mc^2/\alpha$
>   display $v, p, E, p/E$
>   increment v by Δv
> **end loop** if $v > v_f$
> stop

PROJECT 7. A 6.5×10^{-29}-kg particle travels along the x axis. Use the program to make separate plots of the energy, momentum, and ratio p/E as function of the particle speed. Plot points every $0.05c$ from $v = 0$ to $v = 0.95c$. Mark the value of the rest energy on the energy graph.

At slow speeds the energy is nearly the rest energy. Added to this is the kinetic energy, which increases as the square of the speed. Close to the speed of light the energy increases more rapidly with the speed of the particle and becomes infinite at $v = c$. The momentum starts at zero and increases linearly with the speed of the particle. At relativistic speeds it increases more rapidly and becomes infinite at $v = c$. The ratio p/E is small near $v = 0$ since p is small. It increases since E is nearly constant and p increases. In the relativistic region, p/E approaches $1/c$ as a limiting value.

IV. NOTES

OVERVIEW II
THERMODYNAMICS

Thermodynamic phenomena are familiar to everyone. During a long trip the tires of your car get hot and the air pressure in them increases; the car engine and coolant also warm but the fan blows cooler outside air over them and keeps them at appropriate operating temperatures; perhaps the car has an air conditioner to cool the air inside and you notice that you use fuel at a slightly greater rate when it is on; the brake shoe and disks become hot when you brake, but after the car has stopped for sufficient time they have the same temperature as the surrounding air.

Thermodynamics deals chiefly with the transfer of energy between large systems of particles and the changes that thereby occur in the systems. Simple gases are used to illustrate thermodynamic phenomena and, for them, the quantities considered are the volume, pressure, and temperature. The list is longer for other systems. Chapter 22 deals with the concept of temperature and when you study that chapter you will learn its precise definition and how it is measured.

An energy transfer can change both the internal energy of the system and the kinetic energy associated with the motion of the system as a whole. As you may recall from your study of Chapter 9, the internal energy is the sum of the kinetic energies of the particles, as measured in a reference frame that moves with the center of mass, and the potential energies of their interactions with each other. A kinetic energy is associated with the motion of the system as a whole but for simplicity we treat situations in which the center of mass remains at rest and only the internal energy changes. You will study ideal gases, for which the internal energy is simply the sum of the kinetic energies of the particles, including their rotational kinetic energies. These gases have no internal potential energy.

To complete the study of changes in a system brought about by the transfer of energy you must know how the internal energy depends on the other thermodynamic variables. The internal energy of an ideal gas depends only on the temperature and changes in these quantities are directly proportional to each other. This important result is obtained in Chapter 23.

Before you can understand the connection between temperature and internal energy you must understand the connection between the pressure and internal energy on the one hand and the speeds of the particles on the other. These relations, which link the microscopic and macroscopic descriptions of a gas, are developed in Chapter 23. The story is continued in Chapter 24 where you will see how the speeds of the particles can be described without listing them individually and how this information can be used to obtain the pressure, internal energy, and temperature from the distribution of speeds.

The pressure, volume, and temperature of a gas are related to each other: a change in one brings about changes in one or both of the others. The relationship is different for different systems but such a relationship, called an equation of state, exists for every system. Every equation of state contains at least three variables and so does not by itself determine the changes that accompany a change in temperature. More information about the process that changed the temperature must be known. For example, information about the change in volume, along with the equation of state, allows us to determine the change in pressure. Similarly, information about the change in pressure allows us to determine the change in volume. You will also study processes in which the temperature does not change. At the beginning of Chapter 23 you will learn about the equation of state for an ideal gas. The end of the chapter includes a discussion of an equation of state that is useful for other gases.

The study of energy transfers themselves form an important part of thermodynamics, apart from the changes they bring about in a system. You will learn that energy can be transferred to or from a system in two distinct ways, through work done on or by the system and through heat exchanged between the system and its environment. You are already familiar with work; heat is probably new to you. For processes involving a system with its center of mass at rest, work involves a change in volume. Heat, on the other hand, is energy that is transferred because the system and its environment are at different temperatures and does not involve a change in volume. Calculations of the work done on or by a gas during various thermodynamic processes are discussed in Chapter 23. Calculations of the heat transferred are discussed in Chapter 25.

Work done on a system and heat entering the system both increase the internal energy and temperature of the system. In fact, the principle of energy conservation, modified for a system with its center of mass at rest, tells us that the sum of the work done on a system and the heat entering the system is equal to the change in internal energy of the system. This principle, known as the first law of thermodynamics, is the central theme of Chapter 25. You will use it to calculate any of the three quantities, work, heat, and change in internal energy, in terms of the other two.

There are certain natural limitations on thermodynamic processes. For example, if no work is done on a system then heat flows out of the system if its temperature is higher than that of its environment and flows into the system if its temperature is lower. Only if work is done can heat flow be from the colder to the hotter body. Another limitation concerns processes in which heat enters a system and the system does work. No process can convert all the heat into work; at some point in the process heat must flow out of the system or else the internal energy must change. These limitations and others are codified in what is known as the second law of thermodynamics.

The limitations described by the second law have important practical consequences for the design of refrigerators, air conditioners, engines, electrical power generators, and many other devices. More importantly for our purposes, they also give us a glimpse at the way nature operates.

Chapter 26 is devoted to the second law of thermodynamics. You will learn several equivalent statements of the law and how to apply them. At the end of the chapter you will study another important thermodynamic property of systems, called the entropy. You will learn to calculate changes in entropy for various processes and will learn an important fact about it: the total entropy of a system and its environment never decreases. This is a direct consequence of the second law and is important for understanding many thermodynamic processes.

Chapter 22
TEMPERATURE

I. BASIC CONCEPTS

You now begin the study of thermodynamics. Temperature, the most important concept introduced in this chapter, is familiar to you but you have probably not thought about it in detail. Be sure you understand its definition and how it is measured. Pay particular attention to the zeroth law of thermodynamics, which codifies the characteristic of nature that allows temperature to be defined. You will also study the phenomenon of thermal expansion, the familiar increase in the dimensions of an object when its temperature is raised.

Thermodynamic quantities. Thermodynamics is formulated in terms of *macroscopic*, not microscopic, quantities. For a gas the important quantities are:

_____
_____
_____
_____
_____

There are relatively few of them. You should carefully distinguish between a macroscopic description of a system, using these and similar quantities, and a microscopic description. What quantities might be specified in a microscopic description of a gas? _____
_____

As you read this and succeeding chapters you should be on the lookout for discussions that show how macroscopic and microscopic quantities are related.

Temperature. Temperature is intimately related to the idea of <u>thermal equilibrium</u>. If two objects that are not in thermal equilibrium with each other are allowed to exchange energy they will do so and, as a result, some or all of their macroscopic properties will change. When the properties stop changing there is no longer a net flow of energy and the objects are said to be in thermal equilibrium with each other. They are then at the same temperature.
  Temperature is the property of an object that _____

  Note carefully that, when two systems are in thermal equilibrium with each other, all macroscopic quantities except temperature might have different values for the two systems. Suppose the systems are gases. When in thermal equilibrium with each other their pressures may be different, their volumes may be different, their particle numbers may be different, and their internal energies may be different. But their temperatures are the same.

To reach thermal equilibrium two gases must be placed in thermal contact in such a way that they do not mix. A <u>diathermal wall</u> is used to separate them. The important properties of such a wall are: _____

_____

A diathermal wall must be distinguished from an <u>adiabatic wall</u>, the important properties of which are: _____

_____

The law of nature that legitimizes the temperature as a property of a body in thermal equilibrium is the <u>zeroth law of thermodynamics</u>. The text gives two statements of the law. They are

_____

_____

_____

and

_____

_____

_____

Suppose the law were not valid and suppose further that body A and body B are in thermal equilibrium and therefore at the same temperature. If body C is in equilibrium with A but not with B, then the temperature of C is not a well defined quantity and cannot be considered a property of C.

Another important consequence of the zeroth law is that it allows us to select some object for use as a thermometer. Suppose that when the thermometer is in thermal equilibrium with body A it is also in thermal equilibrium with body B. Then, because the law is valid, we know that A and B have the same temperature. We do not need to place A and B in thermal contact to test if they are equilibrium with each other.

Measurement of temperature. In general, temperature is measured by measuring a property, called the _____ property, of some substance, called the _____ substance. The property and substance are chosen for convenience and accuracy of measurement. For an ordinary mercury thermometer the thermometric substance is _____ and the thermometric property is _____.

Three types of temperature scales are in common use: Celsius, Fahrenheit, and Kelvin (or absolute). For the first two the temperature T and thermometric property X are related by $T =$ _____, where a and b are constants. For the <u>Fahrenheit</u> scale a and b are selected so the freezing point of water is $T_F =$ _____° F and the boiling point of water is _____° F. For the Celsius scale a and b are selected to the freezing point of water is _____° C and the boiling point of water is _____° C.

Temperature on the Kelvin scale is defined by the relationship $T = aX$, where the constant a is selected so $T =$ _____ K at the triple point of water. The triple point of water is the temperature for which _____

_____.

If X is the value of the thermometric property when the thermometer is in thermal equilibrium with some object and X_{tr} is its value when the thermometer is in equilibrium with water at the triple point, then the temperature of the object is given by $T = $ _____. The unit of temperature on the Kelvin scale is called the _____ and is abbreviated _____. The definition automatically assures that the temperature of water at its triple point is _____ K.

When different thermometric properties are used to measure the temperature of the same substance, different values may be obtained. You will learn of two techniques that are independent of the thermometric property and substance. One uses a constant-volume gas thermometer and yields what is known as the ideal gas temperature. The other, to be studied in Chapter 26, yields what is known as the thermodynamic temperature. These two temperatures agree with each other over the entire range for which both techniques can be used. This temperature is implied if a value is given in Kelvins without mentioning a thermometric substance and property.

The Celsius temperature T_C (independent of thermometric substance and property) is defined in terms of the ideal gas temperature T by $T_C = $ _____. The Fahrenheit temperature T_F is given in terms of the Celsius temperature by $T_F = $ _____. A degree Celsius is the same as a Kelvin and is _____ as large as a degree Fahrenheit.

For constant-volume gas thermometer the thermometric substance is _____ and the thermometric property is _____. Look at Fig. 4 of the text. The gas is placed in thermal contact with the object whose temperature is to be measured. The position of the gas reservoir R is adjusted so the _____ of gas is a certain value selected for the thermometer, then the _____ is measured by means of a _____. The temperature is given by $T = $ _____, where p is the pressure when the gas is in thermal contact with the object and p_{tr} is the pressure when the gas is in thermal contact with water at its triple point. It is important that the _____ of the gas be the same for these two pressure measurements.

Different gases used in constant-volume thermometers yield different numerical values of the temperature when in contact with the same object. However, the temperature readings for all gases approach the same value in the limit as _____. This value is called the ideal gas temperature. Temperatures measured on the ideal gas scale have units of _____, abbreviated _____.

Thermal expansion. Most solids and liquids expand when the temperature is increased and contract when it is decreased. If the temperature is changed by ΔT, a rod originally of length L changes length by $\Delta L = $ _____, where α is the coefficient of linear expansion. Over small temperature ranges α may be considered to be independent of temperature. Be careful. For a situation in which the temperature is raised by ΔT, then lowered by the same amount, repeated application of the equation predicts the rod is shorter than its original length, whereas it actually returns to its original length. Consult Table 3 of the text for values of some coefficients of linear expansion. You may need them to solve some problems.

When the temperature changes, the length of every line in an isotropic material changes by the same *fractional* amount, given by $\Delta L/L = \alpha \Delta T$. The length of a scratch on the surface of an isotropic solid also changes by the same fractional amount. The fractional change in the diameter of a round hole in an object, for example, is given by $\Delta D/D = $ _____, where α is

Chapter 22: Temperature 337

the coefficient of linear expansion of the object.

When the temperature changes by a small amount ΔT the area A of a face of a solid changes by $\Delta A = $ _____ and the volume V of the solid changes by $\Delta V = $ _____, where α is the coefficient of linear expansion. For a fluid the change in volume is given by the same expression. In the space below prove the first result for yourself. Consider a rectangular face with sides L_1 and L_2. Its area before a temperature change is $A = L_1 L_2$ and its area after the change is $A + \Delta A = (L_1 + \Delta L_1)(L_2 + \Delta L_2)$. Substitute $\Delta L_1 = \alpha L_1 \Delta T$ and $\Delta L_2 = \alpha L_2 \Delta T$ for the changes in length and retain only terms that are proportional to ΔT when you multiply the two factors in parentheses:

You should have obtained $\Delta A = 2\alpha A \Delta T$. The expression for the change in volume can be obtained in a similar manner.

Water near 4°C and a few other substances decrease in volume when the temperature is increased. These materials have negative coefficients of thermal expansion in the temperature range for which such a contraction occurs.

Qualitatively explain the microscopic basis of thermal expansion. As the temperature is raised the atoms of the material have greater vibrational energy and vibrate with larger _____.

In the space to the right draw a diagram showing the potential energy of interaction between two atoms as a function of their separation and show two possible energies. Mark the turning points and the average separation for each. For a given energy the average separation is halfway between the turning points. If you have drawn the curve correctly the average separation will be greater for greater vibrational energy than for smaller.

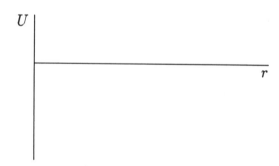

II. PROBLEM SOLVING

Many of the problems of this chapter deal with the definition of temperature. You will need to manipulate the Kelvin scale relationship $T_1/T_2 = X_1/X_2$, where X_1 is the value of a thermodynamic property at temperature T_1 and X_2 is its value at temperature T_2. You must know three of the quantities that appear in the equation, then you can solve for the fourth. In some cases one of the temperatures is the triple point of water, 273.16° on the temperature scale defined by the thermometric property and substance.

PROBLEM 1. The electrical resistance of a certain resistor is used to measure the temperature. If the resistance is 50.15 ohms at the triple point of water, what is the temperature on the scale defined by the resistor when the resistance is 50.93 ohms?

SOLUTION: Use $T/T_{tr} = R/R_{tr}$, with $T_{tr} = 273.16°$, $R_{tr} = 50.15$ ohms, and $R = 50.93$ ohms. Solve for T.

[ans: 277.4 K]

The value of the temperature you found in this problem is on the scale defined by the resistor. To convert to the ideal gas scale, for example, you need to know the thermometric property as a function of the ideal gas temperature. Here is a problem to show you how to convert.

PROBLEM 2. The resistance of a certain resistor is proportional to the square of the ideal gas temperature. What is the relationship between the value of the temperature on the scale defined by this resistor and the ideal gas temperature?

SOLUTION: Let Θ represent the temperature on the scale defined by the resistor and T be the ideal gas temperature. The resistance as a function of T is given by $R = AT^2$, where A is a constant of proportionality. Substitute this expression into $R/R_{tr} = \Theta/\Theta_{tr}$, use the numerical equality $T_{tr} = \Theta_{tr}$, and solve for T or Θ.

[ans: $\Theta = T^2/T_{tr}$]

What is the ideal gas temperature when $\Theta = 5.0°$? when $\Theta = 100°$? when $\Theta = 300°$?

SOLUTION:

[ans: 37.0 K; 165 K; 286 K]

If the resistance is 75 ohms at the triple point of water what is the ideal gas temperature when the resistance is 45 ohms?

SOLUTION:

[ans: 212 K]

You should know how to convert from one of the common temperature scales to another. The relevant equations are $T_C = T - 273.15$ (Kelvin to Celsius) and $T_F = 32 + (9/5)T_C$ (Celsius to Fahrenheit). The most straightforward problems give the temperature on one scale and ask for the temperature on another.

PROBLEM 3. On a warm summer day the temperature is 90°F. What is the temperature on the Celsius and Kelvin scales?

SOLUTION: Solve $T_F = 32 + (9/5)T_C$ for T_C, then add 273.15 to obtain T.

[ans: 32.2°C; 305 K]

Some problems are slightly more complicated.

PROBLEM 4. At what temperature is the numerical value the same on both the Celsius and Fahrenheit scales?

SOLUTION: Use $T_F = 32 + (9/5)T_C$ with $T_F = T_C$.

[ans: $-40°F$; $-40°C$]

Another set of problems is concerned with the coefficient of linear expansion. The important relationship is $\Delta \ell = \alpha \ell \Delta T$. You might be given the original length, the coefficient α, and the temperature change. You would then be asked for the change in length $\Delta \ell$ or else the new length $\ell + \Delta \ell$. In other problems you might be asked for the temperature change required to achieve a given change in length or a given final length.

PROBLEM 5. A steel rod has a length of 2.4 m at 25°C. What is the increase in its length when the temperature is raised to 50°C? The coefficient of linear expansion for steel is 11×10^{-6} per C°.

SOLUTION:

[ans: 0.66 mm]

Sometimes the lengths for two different temperatures are given and you must calculate the coefficient, assumed to be independent of temperature. The fractional change $\Delta \ell / \ell$ might be given instead of the initial and final lengths.

PROBLEM 6. When the temperature is increased from 10°C to 50°C the fractional change in the length of a certain rod is 7.2×10^{-4}. What is the coefficient of linear expansion of the rod?

SOLUTION:

[ans: 1.8×10^{-5} per C°]

In some cases you are asked to compare the expansions of two objects. Suppose, for example, the original lengths of two rods are ℓ_{10} and ℓ_{20} and their coefficients of expansion are α_1 and α_2 respectively. At what temperature will they have the same length? Since $\ell_1 = \ell_{10} + \alpha_1 \ell_{10} \Delta T$ and $\ell_2 = \ell_{20} + \alpha_2 \ell_{20} \Delta T$, they have the same length when $\ell_{10} + \alpha_1 \ell_{10} \Delta T = \ell_{20} + \alpha_2 \ell_{20} \Delta T$ or $\Delta T = -(\ell_{10} - \ell_{20})/(\alpha_1 \ell_{10} - \alpha_2 \ell_{20})$. To find the final temperature add the initial temperature to ΔT.

In a variation of this problem, suppose that at a certain temperature the diameter of a circular hole in a flat plate is D_{10} and the diameter of a sphere made of a different material is D_{20}. If $D_{20} > D_{10}$ the sphere will not fit through the hole, but a change in temperature might change both diameters so the sphere will fit.

PROBLEM 7. At 25°C a hole in an aluminum plate has a diameter of 2.140 cm and a steel ball has a diameter of 2.150 cm. What is the lowest temperature for which the ball will fit through the hole? The coefficients of linear expansion are 23×10^{-6} per C° for aluminum and 11×10^{-6} per C° for steel. Assume they are independent of temperature.

SOLUTION: Start with $D_1 = D_{10} + \alpha_a D_{10} \Delta T$ and $D_2 = D_{20} + \alpha_s D_{20} \Delta T$, set $D_1 = D_2$, and solve for ΔT. Finally, add 25°C to ΔT.

[ans: 416°C]

Some problems deal with areas and volumes. These are essentially the same as the problems dealing with linear dimensions but the coefficient of expansion is different. If α is the coefficient of linear expansion then 2α is the coefficient of area expansion and 3α is the coefficient of volume expansion. The fractional change in area is twice the fractional change in a linear dimension and the fractional change in volume is three times the fractional change in a linear dimension.

PROBLEM 8. At 25°C a certain steel plate has a face with dimensions that are exactly 35 cm × 45 cm. The plate is exactly 0.015 cm thick. If the temperature is raised to 75°C what is the change in the area of a face? what is the change in the volume of the plate? The coefficient of linear expansion for steel is 11×10^{-6} per C°.

SOLUTION:

[ans: 1.73 cm²; 3.90×10^{-2} cm³]

Thermal expansion has consequences for the motion of an object, as the following example illustrates.

PROBLEM 9. At 20°C a 3.5-kg uniform steel sphere has a radius of exactly 15 cm. What is its rotational inertia for rotation about a diameter? If the temperature is raised to 45°C what is the change in the rotational inertia? The coefficient of linear expansion for steel is 11×10^{-6} per C°.

SOLUTION: The rotational inertia of a sphere spinning about a diameter is given by $I = (2/5)MR^2$, where M is its mass and R is its radius. The radius increases with the temperature and the increase in I is given by $\Delta I = 2(2/5)MR\Delta R$. Since $\Delta R = \alpha R \Delta T$, $\Delta I = 2(2/5)MR^2 \alpha \Delta T$, where α is the coefficient of linear expansion.

[ans: 3.15×10^{-2}; 1.73×10^{-5} kg·m²]

While at 20°C the sphere is set spinning with an angular velocity of 100 rad/s around a diameter. Then the temperature is raised to 45°C without exerting any torque. By how much is the angular velocity changed? By how much is the rotational kinetic energy changed?

SOLUTION: Since no torque is applied the angular momentum, given by $I\omega$, does not change. Thus $\omega \Delta I + I \Delta \omega = 0$ and $\Delta \omega = -\omega \Delta I/I$. Since the kinetic energy is given by $K = \frac{1}{2}I\omega^2$, its change is $\Delta K = \frac{1}{2}\omega^2 \Delta I + I\omega \Delta \omega = -\frac{1}{2}\omega^2 \Delta I$. The last expression is obtained by substituting $-\omega \Delta I/I$ for $\Delta \omega$.

[ans: -5.50×10^{-2} rad/s; -8.66×10^{-2} J]

III. NOTES

Chapter 23
KINETIC THEORY
AND THE IDEAL GAS

I. BASIC CONCEPTS

Kinetic theory is used to relate macroscopic quantities like pressure, temperature, and internal energy to the energies and momenta of the particles that comprise a gas. Here you will gain an understanding of the relationship between the microscopic and macroscopic descriptions of a gas.

Ideal gas law. Since you will be dealing with ideal gases you should know what one is. Describe the properties that distinguish an ideal gas from a real gas:
_____
_____
_____

The quantity of matter in the system of interest is important for most of the discussions of this chapter. It might be given as the number of particles or as the number of moles. The number of molecules in a mole of molecules is _____. This is the Avogadro number, denoted by N_A. If a gas contains N molecules then it contains _____ moles of molecules. If the mass of one mole is M, then the mass of one molecule is $m = $ _____. If m is the molecular mass and M is the molar mass, then a gas containing N molecules has a total mass of _____ and a gas containing n moles has a total mass of _____.

To understand the derivations of this chapter you must first understand the <u>ideal gas law</u> (or ideal gas equation of state), which expresses a relationship between the temperature, pressure, volume, and number of particles for an ideal gas.

The ideal gas law is given in two equivalent forms. One form is in terms of the number of molecules N in the gas and the Boltzmann constant k. It is

$$pV = $$

The second form is in terms of the number of moles n of gas and the universal gas constant R. It is

$$pV = $$

In both of these forms p is the _____, V is the _____, and T is the _____ of the gas. It is important to recognize that for these equations to be valid, the Kelvin or ideal gas temperature scale must be used.

The SI values of k and R are $k = $ _____ J/K and $R = $ _____ J/mol·K. You should know the relationship between k and R: $R = $ _____. Since $N = nN_A$ the relationship between k and R can be used to obtain one form of the ideal gas law from the other.

Real gases obey the ideal gas law only approximately; the approximation being better when _____
_____.

This is chiefly because the ideal gas law assumes the gas molecules are point particles that do not interact with each other except for elastic collisions of extremely short duration. In a real gas molecules have extent in space and interact continuously with each other. Molecular sizes and interactions play less of a role as the number of molecules per unit volume decreases.

Fig. 1 of the text shows a device that allows us, in principle, to vary individually the macroscopic properties of a gas. The device can be used to experimentally verify the ideal gas law. For each of the following processes tell how you would manipulate the device to achieve the desired goal. In all cases you must change two quantities, only one of which is given. Name the second quantity and tell whether it should be increased or decreased.

1. Increase the pressure, holding the volume and number of particles constant.
   Second quantity: _____   Increase or decrease? _____
   How to manipulate device: _____
   _____

2. Increase the pressure, holding the volume and temperature constant.
   Second quantity: _____   Increase or decrease? _____
   How to manipulate device: _____
   _____

3. Increase the volume, holding the pressure and temperature constant.
   Second quantity: _____   Increase or decrease? _____
   How to manipulate device: _____
   _____

4. Increase the number of particles, holding the pressure and volume constant.
   Second quantity: _____   Increase or decrease? _____
   How to manipulate device: _____
   _____

Pressure. The pressure in a gas is intimately related to the average of the squares of the speeds of the molecules. The relationship is

$$p =$$

where ρ is the mass density.

Pressure can be computed as the average force per unit area exerted by the molecules of a gas on the container walls. We assume that at each collision between a molecule and a wall the normal component of the molecule's _____ is reversed. By computing the change per unit time we can find the force exerted by the molecule on the wall. When this is divided by the area of the wall, the result is the contribution of the molecule to the pressure.

We consider a gas in a cubical container with edge L. If m is the mass of a molecule, v_x is the component of its velocity normal to the wall, and Δt is the time from one collision to the next with the same wall, then the average force exerted by the molecule on the wall is given by _____. In terms of L and v_x, the time between collisions is $\Delta t =$ _____. Thus the average force of the molecule is _____ and its contribution to the pressure is _____. This is now summed over all molecules. Replace the sum of v_x^2 with the product of its average value and the number of molecules N. You obtain $p =$ _____. Finally, replace the average of v_x^2 with one third the average of v^2 and m/L^3 with the density ρ. These substitutions complete the derivation. You obtain $p =$ _____.

Explain why the average of v_x^2 is one third the average of v^2:

_____

_____

_____

The root-mean-square speed of the molecules in a gas is not quite the same as the average speed but it gives us a rough idea of the average. Its definition, in terms of the average of the squares of the molecular speeds, is $v_\text{rms} =$ _____ and, in terms of the pressure and density, it is given by

$$v_\text{rms} =$$

You can use this equation in conjunction with the ideal gas law to estimate the root-mean-square speed of a given gas at a given temperature. You need to know the molar mass M (or the molecular mass m). Substitute $p = nRT/V$ into the expression for v_rms and use $\rho = nM/V$ to obtain

$$v_\text{rms} =$$

Carefully note that v_rms depends only on the temperature and the molar (or molecular) mass. Thus v_rms remains the same if the temperature does not change as the pressure and volume are changed; that is, if the product pV does not change. v_rms^2 is directly proportional to the temperature and inversely proportional to the molecular or molar mass.

Suppose molecules in an ideal gas have a root-mean-square speed v_0. If the temperature is doubled the root-mean-square speed becomes _____.

Suppose two ideal gases are in thermal equilibrium with each other. Molecules in the first gas have mass M and root-mean-square speed v_0. If molecules in the second gas have mass $2M$ then their root-mean-square speed is _____.

Some values of v_rms are listed in Table 1. For gases at room temperature they run from a low of _____ m/s (for _____) to a high of _____ m/s (for _____). You should know that the root-mean-square speed is roughly the same as the speed of sound for any given gas.

Temperature. For an ideal gas the translational kinetic energy per mole is proportional to the temperature. In terms of the universal gas constant R the exact relationship is

$$\tfrac{1}{2} M v_\text{rms}^2 =$$

Clearly the kinetic energy per molecule is also proportional to the temperature. In terms of the Boltzmann constant the exact relationship is

$$\tfrac{1}{2} m v_\text{rms}^2 =$$

Both of these equations follow from $p = \frac{1}{3}\rho v_{rms}^2$. To obtain the first, use $pV = nRT$ and recognize that ρV is the total mass of the gas: $\rho V = nM$. To obtain the second, use $pV = NkT$ and recognize that $\rho V = Nm$. You use the first form when you know the molar mass and the second when you know the molecular mass.

Carefully note that the molecules of two ideal gases in thermal equilibrium with each other have exactly the same average translational kinetic energy. The root-mean-square speeds, however, are different if the molecular masses are different.

Work. When the volume of a gas changes from V_i to V_f the work that is done *on* it by external forces is given by the integral

$$W = $$

Processes exist for which work is done and the volume does not change but these necessarily involve an acceleration of the center of mass or an angular acceleration about the center of mass and we do not consider them.

The process that changes the volume must be carried out so the gas is nearly in thermal equilibrium at all times. Only then is the pressure well defined throughout the process. The process can be plotted as a curve on a graph with p as the vertical axis and V as the horizontal axis. Then the magnitude of W is the _____ under the curve. If $V_f > V_i$ then the work done on the system is _____; if $V_i > V_f$ then the work done on the system is _____.

The work done as the system goes from V_i to V_f depends on the functional dependence of p on V. Different dependencies are represented by different paths on a p-V graph. You should know how to compute the work for the four special processes described below, all involving n moles of an ideal gas. Write out the function $p(V)$, carry out the integration, and write the expression for the work done on the gas. Give expressions for the changes requested. On the p-V axes to the right, draw a curve that represents the process. Finally describe how you would manipulate the device of Fig. 1 of the text to carry out the work.

1. Change in pressure from p_i to p_f at constant volume V
   $p(V)$ is immaterial
   $W = $

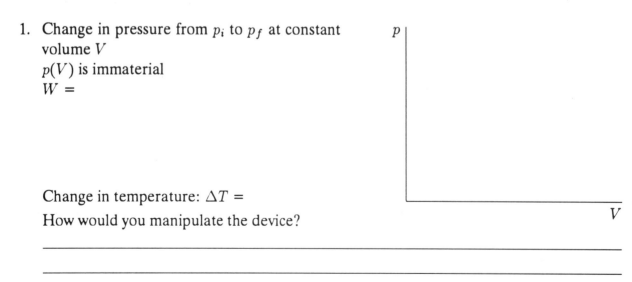

Change in temperature: $\Delta T = $

How would you manipulate the device?

_____

_____

2. Change in volume from V_i to V_f at constant pressure p
   $p(V)$ = constant
   $W =$

   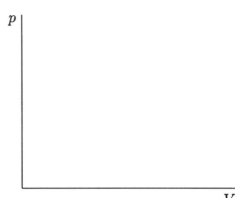

   Change in temperature: $\Delta T =$
   How would you manipulate the device?
   _____
   _____

3. Change in volume from V_i to V_f at constant temperature T
   $p(V) =$
   $W =$

   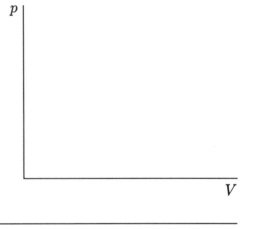

   Change in pressure: $\Delta p =$
   How would you manipulate the device?
   _____
   _____

4. Adiabatic (in thermal isolation) change in volume from V_i to V_f
   $p(V) =$
   $W =$

   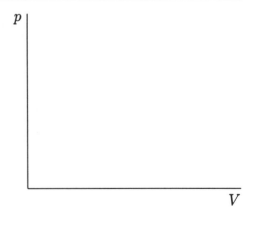

   Change in pressure: $\Delta p =$
   Change in temperature: $\Delta T =$
   How would you manipulate the device?
   _____
   _____

Internal energy. For a monatomic ideal gas the internal energy is the total kinetic energy of the molecules. Since the average kinetic energy of a molecule is $\overline{K} = \tfrac{3}{2}kT$, $E_{\text{int}} = $ _____, where N is the number of molecules. This is the same as $E_{\text{int}} = $ _____, where n is the number of moles.

Chapter 23: Kinetic Theory and the Ideal Gas 347

Note that the internal energy of an ideal gas depends only on the temperature and the amount of gas. This information will prove useful in discussing the thermodynamic temperature scale as well as in solving ideal gas problems. For other systems the internal energy may also depend on the pressure, volume, or other macroscopic quantities.

The <u>equipartition theorem</u> tells us that, in thermal equilibrium, the internal energy is equally divided among the degrees of freedom of a system, with the energy associated with each degree of freedom being _____ kT. One degree of freedom is associated with each independent energy term. For example, _____ degrees of freedom are associated with the translational kinetic energy of a molecule since it can move in any of three independent directions. To find the total number of degrees of freedom for the gas we multiply the number of degrees of freedom for a molecule by the number of molecules. The internal energy is the product of the result and $\frac{1}{2}kT$.

Consider three gases, each with N molecules. For the first the molecules are monatomic, for the second they are diatomic, and for the third they are polyatomic. Fill in the following table giving the translational, rotational, and total internal energies in terms of kT.

	MONATOMIC	DIATOMIC	POLYATOMIC
Number of translational degrees of freedom	_____	_____	_____
Translational energy	_____	_____	_____
Number of rotational degrees of freedom	_____	_____	_____
Total internal energy	_____	_____	_____

You should remember that the internal energy and molecular speeds are measured in a reference frame for which the center of mass is at rest. The internal energy does NOT include the translational kinetic energy of the system as a whole.

Knowing how the internal energy depends on the temperature allows us to compute the change in temperature when work is done on the gas or heat is absorbed by the gas. If work W is done on n moles of a monatomic ideal gas in thermal isolation then the increase in internal energy is $\Delta E_{\text{int}} =$ _____ and the increase in temperature is $\Delta T =$ _____.

Van der Waals equation of state. For many gases the van der Waals equation of state describes the relationship between p, V, n, and T more accurately than the ideal gas equation of state. It is

$$\left[p + \frac{an^2}{V^2}\right](V - nb) = nRT$$

where a and b are constants that are different for different gases. This equation clearly differs from the ideal gas equation of state. One way of looking at the difference is to think of the effective volume as being less than the true volume V of the gas and the effective pressure as being greater than the true pressure p. The quantity nb is subtracted from the volume of the gas because _____

The quantity an^2/V^2 is added to the pressure because _____

The van der Waals equation of state reduces to the ideal gas equation of state when the volume per mole V/n is much greater than _____ and also much greater than _____.

The van der Waals equation of state leads to isothermal and adiabatic curves that are different from those of an ideal gas. On the p-V diagram to the right sketch and label isotherms for an ideal gas and a van der Waals gas at the same temperature. These need not be quantitatively correct but they should show qualitative differences. In particular, they should show where the two curves are nearly the same and where they are not.

II. PROBLEM SOLVING

You should be adept at manipulating the ideal gas equation of state ($pV = nRT$ or $pV = NkT$). These equations each contain four thermodynamic variables. Given any three you should be able to solve for the fourth. If the quantity of matter is given in moles, use nRT on the right side; if it is given in number of molecules, use NkT. For some problems the mass density ρ might be given. Use $\rho = nM/V$ or $\rho = Nm/V$, where M is the mass of 1 mole and m is the mass of 1 molecule, to eliminate either V, n, or N from the equation of state.

PROBLEM 1. Suppose 2.0 moles of an ideal gas occupy $15 \times 10^{-4}\,\mathrm{m}^3$ at 250 K. What is the pressure?
SOLUTION:

[ans: $2.8 \times 10^6\,\mathrm{Pa}$]

Suppose 3.0×10^{23} molecules of an ideal gas are held at a pressure of 2.5×10^5 Pa and a temperature of 320 K. What is the volume of the container?
SOLUTION:

[ans: $5.3 \times 10^{-3}\,\mathrm{m}^3$]

Some problems ask you to compare the values of a thermodynamic quantity for two different states. You might be asked, for example, to tell what happens to the temperature of an ideal gas when both the pressure and volume are doubled. The equation of state is written twice, once for each state, and the two equations are solved simultaneously. Usually the easiest way to do this is to divide the equation for one state by the equation for the other to obtain $p_1V_1/p_2V_2 = n_1T_1/n_2T_2$ (or $p_1V_1/p_2V_2 = N_1T_1/N_2T_2$). For example, suppose you know the pressure p_1, volume V_1, and temperature T_1 for one state and the pressure p_2 and volume V_2 for another. You may then be asked for the temperature of the second state. In this case $n_2 = n_1$ so $p_1V_1/p_2V_2 = T_1/T_2$ or $T_2 = (p_2V_2/p_1V_1)T_1$. Notice that the final expression contains ratios of the pressures and volumes. Only these need be given, not the initial and final values themselves.

PROBLEM 2. Suppose the temperature of an ideal gas is 325 K. If the number of molecules is tripled and the volume is doubled without changing the pressure, what is its new temperature?

SOLUTION:

[ans: 217 K]

In some cases the pressure is not given directly. You might, for example, be given the height of a fluid column in a barometer, the difference in fluid height on the two sides of a U-tube, or the depth in a fluid. To find the pressure you will need to recall information from Chapter 17.

As a bubble of air rises through water, the pressure in the bubble decreases and the volume of the bubble increases. If the rise is not swift the pressure in the bubble is the same as the water pressure at the height of the bubble and can be computed using $p = p_o + \rho_w g d$, where p_o is the pressure at the water surface (usually atmospheric pressure), ρ_w is the density of water, and d is the depth of the bubble in the water. You can use this relationship to calculate the ratio of the pressure at two different depths. If you know the ratio of the temperatures you can then compute the ratio of the bubble volumes.

PROBLEM 3. A bubble starts at the bottom of a 60-m deep lake with a volume of 5.4×10^{-9} m^3. Neglect the difference in temperature between the bottom and the surface and calculate its volume when it reaches the surface. The density of water is 1.000×10^3 kg/m^3. Atmospheric pressure is 1.013×10^5 Pa.

SOLUTION: Use $p_b = p_o + \rho_w g d$ to calculate the pressure p_b at the bottom of the lake, then use $p_b V_b = p_o V_t$ to compute the bubble volume V_t at the top.

[ans: 3.67×10^{-8} m^3]

The relationship $p = \frac{1}{3}\rho v_{rms}^2$ between the pressure and the mean-square molecular speed is important for some problems. Given the pressure and mass density you should be able to compute the mean-square speed and its square root, the root-mean-square speed. For some problems the volume, temperature, and number of moles or number of molecules are given and the equation of state must be used to find the pressure. When using the equation of state you may need to recognize that the density, volume, and amount of matter are related by $\rho = Nm/V$ or by $\rho = nM/V$.

PROBLEM 4. Find the mean-square speed and root-mean-square speed of molecules in an ideal gas at 400 K if the molecular mass is 4.5×10^{-26} kg.

SOLUTION: Since $p = \frac{1}{3}\rho v_{rms}^2$, $pV = \frac{1}{3}\rho V v_{rms}^2 = \frac{1}{3}Nmv_{rms}^2$. Use $pV = NkT$ to show that $kT = \frac{1}{3}mv_{rms}^2$. Solve for v_{rms}^2, then take the square root.

[ans: 3.7×10^5 m^2/s^2; 6.1×10^2 m/s]

Other problems depend on the relationship between the internal energy and the temperature for an ideal gas. You should recognize that the average translational kinetic energy of a molecule is given by $\frac{1}{2}mv_{rms}^2$ and the total translational kinetic energy is given by $\frac{1}{2}Nmv_{rms}^2$ (or $\frac{1}{2}nMv_{rms}^2$). Thus you can calculate the average kinetic energy if you know the root-mean-square speed and the molecular mass and you can calculate the total translational energy if, in addition, you know the number of molecules (or number of moles). The temperature and average translational kinetic energy are related by $\overline{K} = \frac{3}{2}kT$, regardless of the number of degrees of rotational freedom.

PROBLEM 5. What is the average translational kinetic energy of the molecules of an ideal gas at 350 K?

SOLUTION: Use $\overline{K} = \frac{3}{2}kT$.

[ans: 7.2×10^{-21} J]

The mass of each molecule is 8.7×10^{-26} kg. What is the root-mean-square speed?

SOLUTION: Use $\overline{K} = \frac{1}{2}mv_{rms}^2$.

[ans: 4.1×10^2 m/s]

If the gas consists of 5.9×10^{23} molecules and occupies a volume of 3.6×10^{-3} m^3 what is the pressure?

SOLUTION: Use either $p = \frac{1}{3}\rho v_{rms}^2 = \frac{1}{3}(Nm/V)v_{rms}^2$ or $pV = NkT$.

[ans: 7.9×10^5 Pa]

The expressions given above for the translational energy are valid for all ideal gases, regardless of the number of atoms in each molecule. If the molecules are monatomic the total translational energy is the same as the internal energy. If they are diatomic there are two additional degrees of freedom per molecule, both rotational. On average each has energy $\frac{1}{2}kT$, so the rotational energy per molecule is kT and the total energy per molecule is $\frac{5}{2}kT$. If the molecules are polyatomic there are three rotational degrees of freedom per molecule. On average each has energy $\frac{1}{2}kT$, so the rotational energy per molecule is $\frac{3}{2}kT$ and the total energy per molecule is $3kT$. The total internal energy is found by multiplying the total energy per molecule by the number of molecules.

PROBLEM 6. Suppose the gas described in Problem 5 is diatomic and calculate the rotational energy per molecule, the total energy per molecule, and the total internal energy of the gas.

SOLUTION:

[ans: 4.8×10^{-21} J; 1.2×10^{-20} J; 7.1×10^{3} J]

PROBLEM 7. Suppose the gas described in Problem 5 is polyatomic and calculate the rotational energy per molecule, the total energy per molecule, and the total internal energy of the gas.

SOLUTION:

[ans: 7.2×10^{-21} J; 1.4×10^{-20} J; 8.5×10^{3} J]

You should be able to compute the work done on an ideal gas by its environment for various quasi-static processes. In each case the work is given by $W = -\int p\,dV$, taken from the initial state to the final state. You derived the appropriate equations in the previous section but you should be able to rederive them quickly. To work each of the following examples, start with $W = -\int p\,dV$, make the appropriate substitution for p as a function of V, carry out the integration, and evaluate the result to obtain a numerical answer. If a problem asks for the work done by the gas, evaluate $+\int p\,dV$ rather than $-\int p\,dV$.

PROBLEM 8. As the volume of an ideal gas containing 5.2×10^{23} diatomic molecules is increased from 3.5×10^{-3} m^3 to 9.5×10^{-3} m^3 the pressure is held at 5.2×10^{5} Pa. What work is done *on* the gas?

SOLUTION: Use $W = -\int_{V_i}^{V_f} p\,dV = -p\int_{V_i}^{V_f} dV = -p(V_f - V_i)$.

[ans: -3.1×10^{3} J]

What work is done *by* the gas?

SOLUTION:

[ans: $+3.1 \times 10^{3}$ J]

Use $pV = NkT$ to calculate the change in the temperature of the gas.

SOLUTION:

[ans: 4.3×10^{2} K]

By how much does the internal energy of the gas change?

SOLUTION: Use $\Delta E_{\text{int}} = \frac{5}{2}nk\Delta T$.

[ans: 7.8×10^{3} J]

The change in the internal energy was not the same as the work done because heat was absorbed by the gas. You will learn more about this in Chapter 25. For now remember that you cannot always equate the work done on the gas to the change in its internal energy.

PROBLEM 9. As the temperature of an ideal gas containing 5.2×10^{23} diatomic molecules is increased from 250 K to 400 K the pressure is held at 5.2×10^5 Pa. What work is done on the gas?

SOLUTION: Substitute from $pV = NkT$ into $W = -p\Delta V$ to obtain $W = -Nk\Delta T$.

[ans: -1.1×10^3 J]

By how much does the internal energy of the gas change?

SOLUTION:

[ans: 2.7×10^3 J]

By how much does the volume of the gas change?

SOLUTION:

[ans: 2.1×10^{-3} m^3]

PROBLEM 10. As the volume of an ideal gas containing 5.2×10^{23} diatomic molecules is increased from 3.5×10^{-3} m^3 to 9.5×10^{-3} m^3 the temperature is held at 350 K. What work is done on the gas?

SOLUTION: The process is isothermal. Use $W = -\int_{V_i}^{V_f} p\,dV = -NkT \int_{V_i}^{V_f} (1/V)\,dV = -NkT \ln(V_f/V_i)$.

[ans: -2.5×10^3 J]

What is the change in pressure?

SOLUTION:

[ans: -4.5×10^5 Pa]

What is the change in internal energy?

SOLUTION: The answer is 0. Why?

PROBLEM 11. As the pressure of an ideal gas containing 5.2×10^{23} diatomic molecules is increased from 5.1×10^5 Pa to 1.1×10^6 Pa the temperature is held at 350 K. What work is done on the gas?

SOLUTION:

[ans: 1.9×10^3 J]

What is the change in volume of the gas?

SOLUTION:

[ans: -2.6×10^{-3} m^3]

For an ideal gas undergoing an adiabatic process, $p_f V_f^\gamma = p_i V_i^\gamma$, where γ is the ratio of the heat capacities. This relationship can be used to find any one of p_i, V_i, p_f, and V_f if the other three are given. Once the initial and final volumes and pressures are known, $W = (p_f V_f - p_i V_i)/(\gamma - 1)$ is used to compute the work done on the gas during the process. If the number of molecules or moles is known the equation of state can be used to find the initial and final temperatures.

PROBLEM 12. A monatomic ideal gas ($\gamma = 1.67$) containing 5.2×10^{23} molecules has a pressure of 5.1×10^5 Pa and a volume of 3.5×10^{-3} m^3. While in thermal isolation its volume is increased to 9.5×10^{-3} m^3. What is the final pressure?

SOLUTION: Use $p_f V_f^\gamma = p_i V_i^\gamma$.

[ans: 9.6×10^4 Pa]

What is the work done on the gas?

SOLUTION:

[ans: -1.3×10^3 J]

What are the initial and final temperatures?

SOLUTION: Use $pV = NkT$.

[ans: 249 K, 127 K]

What is the change in internal energy?

SOLUTION: Use $\Delta E_{\text{int}} = \frac{3}{2} N \Delta T$.

[ans: -1.3×10^3 J]

PROBLEM 13. A monatomic ideal gas ($\gamma = 1.67$) containing 5.2×10^{23} molecules has a pressure of 5.1×10^5 Pa and a volume of 3.5×10^{-3} m^3. While in thermal isolation its pressure is increased to 1.1×10^6 Pa. What is the final volume?

SOLUTION:

[ans: 2.2×10^{-3} m^3]

What is the work done on the gas?

SOLUTION:

[ans: 9.6×10^2 J]

What are the initial and final temperatures?

SOLUTION:

[ans: 249 K, 339 K]

What is the change in internal energy?

SOLUTION:

[ans: 9.7×10^2 J]

Since $p_i V_i = NkT_i$ and $p_f V_f = NkT_f$, the expression for the work done on an ideal gas during an adiabatic process can also be written $Nk(T_f - T_i)/(\gamma - 1)$.

PROBLEM 14. A monatomic ideal gas ($\gamma = 1.67$) containing 5.2×10^{23} molecules is in thermal isolation while the temperature is changed from 250 K to 400 K. What work is done on the gas?

SOLUTION:

[ans: 1.6×10^3 J]

III. MATHEMATICAL SKILLS

This chapter introduces some of the ideas of statistics. Most important are the average and root-mean-square of a collection of numbers. To find the average, add the numbers and divide the result by the number of terms in the sum. To find the root-mean-square, add the squares of the numbers, divide the sum by the number of terms, and take the square root of the result. The root-mean-square is the square root of the average of the squares of the numbers.

The speeds of 5 molecules are 350 m/s, 225 m/s, 432 m/s, 375 m/s, and 450 m/s. Find their average and root-mean-square speeds. The following table might help.

PARTICLE	SPEED (m/s)	SPEED SQUARED (m^2/s^2)
1	_____	_____
2	_____	_____
3	_____	_____
4	_____	_____
5	_____	_____
sum	_____	_____
average	_____	_____

$\bar{v} =$ _____ m/s $v_{rms} =$ _____ m/s

Your answers should be $\bar{v} = 366$ m/s and $v_{\text{rms}} = 375$ m/s. Notice that the average and root-mean-square values are different. In fact, the root-mean-square speed is greater than the average speed.

You should be able to show algebraically that if all the molecules have the same speed, their average and root-mean-square speeds are the same. If there are N molecules and each has speed v, then $\sum v = Nv$ and $\sum v^2 = Nv^2$. Complete the proof in the space below:

IV. NOTES

Chapter 24
STATISTICAL MECHANICS

I. BASIC CONCEPTS

In this chapter you will apply some concepts from the field of statistics to the molecules of a gas. There are three main topics: distances traveled by molecules between collisions, molecular speeds, and molecular kinetic energies. Instead of dealing with average or root-mean-square values, you will consider a much more detailed description, in terms of distributions of values. Pay attention to the definition of a distribution function and learn how to use such a function to compute average and root-mean-square values.

Distributions and average values. A distribution of values can be represented on a histogram. Suppose we are dealing with the speeds of molecules in a gas and explain how a histogram of their values might be plotted. Be sure to tell what the vertical and horizontal axes represent and how the horizontal axis is divided into bins.

_____
_____
_____
_____

Let v_i be the representative speed for bin i and suppose there are $n(v_i)$ molecules with speeds in that bin. Then the average speed is estimated by evaluating the sum over bins:

$$\overline{v} = $$

The mean-square speed is estimated by evaluating

$$\overline{v^2} = $$

The relative frequency of occurrence of speeds in bin i is defined by $f(v_i) = $ _____, where N is the total number of molecules. In terms of relative frequencies, the expression for the average speed can be written as the sum

$$\overline{v} = $$

and the expression for the mean-square speed can be written

$$\overline{v^2} = $$

Some histograms are plotted with the relative frequency on the vertical axis, rather than the number of molecules.

Mean free path. The mean free path of a molecule in a gas is the average distance it travels between _____. Fig. 4 of the text shows a histogram of the relative frequency of the various intercollision distances traveled for a molecule in a typical gas. You should examine it and verify that a molecule has a relatively high probability of traveling a short distance before suffering a collision and the probability decreases for longer distances.

The distances traveled by molecules between collisions are distributed continuously. We take $f(r)\,dr$ to be the relative frequency for intercollision distances; that is, $f(r)\,dr$ gives the fraction of intercollision distances that are between r and $r + dr$. In terms of integrals containing $f(r)$ the mean free path λ is given by

$$\lambda = \frac{\int_0^\infty r f(r)\,dr}{\int_0^\infty f(r)\,dr}$$

The functional dependence of the relative frequency on the distance traveled between collisions can be found. We imagine that a beam of molecules is incident on a thin layer of the system and compare the incident beam intensity (the number of particles entering the layer per unit time, for example) with the exiting beam intensity. The particles in the exiting beam are those that did not suffer collisions.

When we increase the thickness by dr the number of particles that exit decreases because more particles suffer collisions. The change in the intensity of the exiting beam is proportional to the intensity of the beam because _____

_____

It is proportional to dr because _____

_____

Thus the change in the intensity I is given by $dI = -cI\,dr$, where c is a positive constant. Integration yields $I(r) = $ _____, where I_o is _____. We take $f(r)$ to be Ce^{-cr}, where C is another positive constant.

When the ratio of integrals given above for the mean free path is evaluated using $f(r) = Ce^{-cr}$ the result is

$$\lambda =$$

The constant C is usually evaluated using the condition $\int_0^\infty f(r)\,dr = $ _____. In the space below show that this condition leads to $C = 1/\lambda$:

Thus, in terms of λ,

$$f(r) =$$

The probability that between collisions a molecule goes a distance in the range from r_1 to r_2 is given by the integral

$$P(r_1, r_2) = \int_{r_1}^{r_2} f(r)\,dr = \frac{1}{\lambda} \int_{r_1}^{r_2} e^{-r/\lambda}\,dr$$

In the space below show that $P(r_1, r_2) = e^{-r_1/\lambda} - e^{-r_2/\lambda}$:

The mean free path (not the distribution of distances between collisions) can be estimated theoretically by considering the number of molecules that are encountered by any given molecule as it moves through the gas. Assume the molecules each have a diameter d. Concentrate on one that is moving with speed v and assume all others are stationary. The moving molecule has a collision with another molecule if the distance between their centers is less than _____. In terms of the diameter d and speed v, the number of collisions it has in time t equals the number of molecules in the volume of a cylinder with cross-sectional area $A = $ _____ and length $\ell = $ _____. If ρ_n is the number of molecules per unit volume in the gas, this is _____. The mean free path is the distance vt divided by the number of collisions in time t or, in terms of d and ρ_n, $\lambda = $ _____.

Because each molecule collides with moving rather than stationary molecules the mean free path is actually somewhat less. It is, in fact, given by

$$\lambda =$$

Carefully distinguish between the molecular concentration ρ_n and the mass density ρ. The first quantity gives the number of molecules per unit volume, the second gives the mass per unit volume. They are related by $\rho = $ _____ ρ_n, where m is the molecular mass.

The mean free path of air molecules at sea level is about _____; the mean free path in a reasonably good laboratory vacuum is about _____.

The distribution of molecular speeds. Not all molecules in a gas have the same speed. The distribution of speeds is described by the <u>Maxwellian</u> <u>distribution</u> <u>function</u> $n(v)$, defined so that $n(v)\,dv$ gives the number of molecules with speed in the range from v to $v + dv$. If a gas at temperature T contains N molecules, all of mass m, then the Maxwellian distribution function is given by

$$n(v) =$$

The factor that multiplies the exponential has been chosen so that the integral $\int_0^\infty n(v)\,dv$ gives N. See the Mathematical Skills section for a proof. Sketch the function $n(v)$ on the graph below:

To find the number of molecules with speed between v_1 and v_2 evaluate the definite integral

$$P(v_1, v_2) = \int_{v_1}^{v_2} n(v)\,dv$$

If the interval $v_2 - v_1$ is small you can approximate this by $(v_2 - v_1)n(v_1)$. See Sample Problem 5 in the text.

The Maxwellian speed distribution can be used to calculate the most probable speed, the average speed, and the root-mean-square speed. The <u>most probable speed</u> v_p is the one for which $n(v)$ is _____. In terms of the mass of a molecule and the temperature, it is given by

$$v_p =$$

The <u>average speed</u> is given by

$$\bar{v} = \frac{1}{N}\int_0^\infty v n(v)\,dv =$$

where the explicit expression for the Maxwellian distribution function was substituted for $n(v)$ and the integral was evaluated. The <u>mean-square speed</u> is given by

$$\overline{v^2} = \frac{1}{N}\int_0^\infty v^2 n(v)\,dv =$$

where the explicit expression for the Maxwellian distribution function was again substituted for $n(v)$ and the integral was evaluated. The root-mean-square speed v_{rms} is the square root of this. When the square root is taken, the result is

$$v_{rms} =$$

Note that all three speeds (v_p, $\bar{v}$, and v_{rms}) depend on the temperature as well as on the molecular or molar mass. When the temperature increases, they all _____. For a given gas at a given temperature the greatest of the three characteristic speeds is _____ and the least is _____.

The distribution of molecular kinetic energies. We can also consider the distribution of translational kinetic energies in an ideal gas. If $n(E)$ is the energy distribution function then $n(E)\,dE$ gives the number of particles with energy between E and $E + dE$. $n(E)$ is related to $n(v)$ by $n(E) = n(v)\,dv/dE$. Since $v = \sqrt{2E/m}$, $dv/dE = $ _____ and the energy distribution function is

$$n(E) = $$

where the explicit expression for $n(v)$ has been used. $n(E)$ is called the Maxwell-Boltzmann energy distribution function for the translational energy of molecules in an ideal gas. Carefully note that the molecular mass does not appear explicitly in the expression for $n(E)$. The function depends only on the temperature T and, of course, on the energy E. It is therefore exactly the same function for every ideal gas, regardless of the mass of the molecules.

In terms of the distribution function $n(E)$ the total translational kinetic energy of an ideal gas is given by the integral $K = $ _____. This integral can be evaluated, with the result

$$K = $$

in agreement with the result of kinetic theory. If the molecules are monatomic this is also the internal energy. If they are not, rotational energy terms must be added to it to obtain the internal energy.

Quantum mechanical distribution functions. The Maxwell-Boltzmann energy distribution is valid only for a gas of classical particles. Different distributions apply when quantum mechanics must be used. There are actually two quantum distribution functions. Both can be written in the form $n(E) = Cf_i(E)$, where f_i is a different function of the energy for each distribution.

The <u>Fermi</u>-<u>Dirac</u> distribution function applies to a collection of particles each of which have spin angular momentum given by $nh/4\pi$, where h is the Planck constant and n is an _____ integer. Electrons, protons, and neutrons are all particles of this type.

The characteristic function for fermions, as these particles are called, is

$$f_{FD}(E) = $$

where E_0 is a parameter that depends on the particle concentration and on the temperature but is independent of the energy. Use the axes below to make a rough graph of $f_{FD}(E)$ and label E_0.

Chapter 24: Statistical Mechanics

The Bose-underline{Einstein} distribution function applies to a collection of particles each of which have spin angular momentum given by $nh/2\pi$, where h is the Planck constant and n is an _____. Helium atoms are examples of this type particle.

The characteristic function for bosons, as these particles are called, is

$$f_{BE}(E) =$$

Use the axes below to make a rough graph of $f_B E$.

f_{BE} vs E

One important difference between the Fermi-Dirac and Bose-Einstein distributions is that only one Fermi-Dirac particle can occupy any quantum state whereas _____ number of Bose-Einstein particles can occupy any state. As the temperature approaches $T = 0\,\text{K}$, the total internal energy of each type of gas approaches the lowest possible value. For a Fermi-Dirac gas many particles have fairly high energy. The lowest energy states are occupied and since only one particle may occupy these states other particles must have higher energies. For a Bose-Einstein gas, on the other hand, all particles may occupy the lowest energy state. The prohibition against more than one particle in any quantum state of a Fermi-Dirac gas is called the _____ principle.

Brownian motion. Brownian motion originally referred to the motion of pollen grains suspended in water. The most important characteristic of the motion is _____.

The importance of this phenomenon is that it allows us to observe the motion of "large molecules" and confirm by direct observation the results of statistical mechanics. These observations are of historical importance because they convinced many doubters of the existence of molecules.

Compare the pollen grains with the water molecules. Place an S next to all the following quantities that are the same for both; place a D next to all that are different:

average kinetic energy	_____
root-mean-square speed	_____
average speed	_____
temperature	_____
distribution of speeds	_____
distribution of energies	_____

II. PROBLEM SOLVING

Many of the problems dealing with mean free path can be solved by manipulation of the expression $\lambda = 1/\sqrt{2}\pi d^2 \rho_n$, where d is the molecular diameter and ρ_n is the molecular concentration (number of molecules per unit volume). There are three variables in this expression; given any two you can solve for the third.

PROBLEM 1. For a certain gas the mean free path is 5.0×10^{-7} m and the molecular concentration is 4.7×10^{25} molecules/m³. What is the diameter of the molecules in this gas?

SOLUTION: Solve $\lambda = 1/\sqrt{2}\pi d^2 \rho_n$ for d.

[ans: 9.8×10^{-11} m]

If the mean free path and average speed are known the mean time between collisions (collision rate) can be computed. The relationship you use is rate $= \bar{v}/\lambda$. If molecules in the gas have an average speed of 500 m/s what is the collision rate?

SOLUTION:

[ans: 1.0×10^{9} collisions/s]

For some problems you may need to use the root-mean-square speed as an approximation to the average speed.

In some cases the mean free path may not be given directly. Instead the collision rate and average speed are given. Use $\lambda = v/\text{rate}$ to find the mean free path, then substitute into $\lambda = 1/\sqrt{2}2\pi d^2 \rho_n$ and solve for either d or ρ_n, depending on which is given.

PROBLEM 2. Molecules in a certain gas have an average speed of 450 m/s and a collision rate of 2.8×10^8 s^{-1}. If the molecular concentration is 9.1×10^{25} m^{-3} what is the molecular diameter?

SOLUTION:

[ans: 3.9×10^{-11} m]

Sometimes the molecular concentration must be computed from the equation of state. If the gas is ideal $\rho_n = p/kT$, a relationship that follows from the equation of state once N/V is replaced by ρ_n.

PROBLEM 3. Molecules in a nearly ideal gas at 3.4×10^6 Pa and 450 K have masses of 7.5×10^{-26} kg. What is the molecular concentration and the root-mean-square speed?

SOLUTION: Use the equation of state to obtain ρ_n and $p = \frac{1}{3}\rho v_{\text{rms}}^2$ to obtain v_{rms}.

[ans: 5.5×10^{26} m^{-3}; 5.0×10^2 m/s]

Suppose the molecules suffer collisions at the rate 4.4×10^9 s^{-1}. Approximate the average speed by the root-mean-square speed, then estimate the mean free path and the molecular diameter.

SOLUTION:

[ans: 1.1×10^{-7} m; 6.0×10^{-11} m]

Some problems deal with the distribution function for intercollision distances: $f(r) = (1/\lambda)e^{-r/\lambda}$. Given the mean free path you can use this expression to find the probability that a molecule will travel a given range of distances between collisions. Use $P(r_1, r_2) = e^{-r_1/\lambda} - e^{-r_2/\lambda}$.

PROBLEM 4. What is the probability a molecule travels a distance less than a mean free path before suffering a collision? Greater than a mean free path?

SOLUTION: In the first case the range is from $r_1 = 0$ to $r_2 = \lambda$. In the second it is from $r_1 = \lambda$ to $r_2 = \infty$.

[ans: 0.632; 0.368]

PROBLEM 5. Suppose molecules in a certain gas have a mean free path of 5.6×10^{-6} m. What is the probability a molecule travels a distance between 1.5×10^{-6} m and 2.0×10^{-6} m before suffering a collision?

SOLUTION:

[ans: 0.0653]

Many problems deal with the Maxwellian speed distribution. You should be able to use the expressions for the most probable, average, and root-mean-square speeds: $v_p = \sqrt{2RT/M} = \sqrt{2kT/m}$, $\bar{v} = \sqrt{8RT/\pi M} = \sqrt{8kT/\pi m}$, and $v_{\text{rms}} = \sqrt{3RT/M} = \sqrt{3kT/m}$. Here T is the temperature, M is the molar mass, and m is the molecular mass.

PROBLEM 6. A gas consisting of molecules with masses of 3.5×10^{-26} kg is in thermal equilibrium at 350 K. If the distribution of speeds is Maxwellian what are the most probable speed, the average speed, and the root-mean-square speed?

SOLUTION:

[ans: 5.3×10^2 m/s; 5.9×10^2 m/s; 6.4×10^2 m/s]

You should recognize that all three characteristic speeds are proportional to the square root of the temperature and inversely proportional to the square root of the molecular or molar mass. For example, the most probable speed at temperature T_1 is related to the most probable speed at temperature T_2 by $v_{p1}/v_{p2} = \sqrt{T_1/T_2}$. Similar expressions hold for the other characteristic speeds. If two gases are in thermal equilibrium the characteristic speeds are less for the gas containing the most massive molecules. For example, if the molecular mass for one gas is m_1 and the molecular mass for the other is m_2, then the ratio of the most probable speeds for the two gases is given by $v_{p1}/v_{p2} = \sqrt{m_2/m_1}$. Similar expressions hold for the other characteristic speeds.

PROBLEM 7. When the temperature of a certain gas is 250 K the average molecular speed is 450 m/s. What is the average molecular speed when the temperature is raised to 350 K?

SOLUTION:

[ans: 532 m/s]

PROBLEM 8. Two ideal gases are in thermal equilibrium with each other. Molecules of the first gas have masses of 7.3×10^{-26} kg and a root-mean-square speed of 300 m/s. Molecules of the second gas have masses of 2.8×10^{-26} kg. What is their root-mean-square speed?

SOLUTION:

[ans: 484 m/s]

The characteristic speeds are related to each other: notice that all three expressions contain the combination kT/m (or RT/M). Thus if you know one of them you can easily calculate the other two.

PROBLEM 9. Suppose the speeds of molecules in a gas are Maxwellian and the most probable speed is 500 m/s. What then are the average and root-mean-square speeds?

SOLUTION:

[ans: 564 m/s; 612 m/s]

Some problems deal with the Maxwell-Boltzmann energy distribution function $n(E) = 2\pi N(1/\pi kT)^{3/2} E^{1/2} e^{-E/kT}$

PROBLEM 10. Find an expression for the most probable kinetic energy of a molecule in an ideal gas at temperature T.

SOLUTION: Find the value of E for which $n(E)$ is a maximum. $dn(E)/dE = 0$ for this value.

[ans: $kT/2$]

Chapter 24: Statistical Mechanics

PROBLEM 10. An ideal gas is at a temperature of 25°C. Estimate the fraction of molecules that have kinetic energy in the range from $0.49kT$ to $0.51kT$. This is a small range centered at the most probable energy.

SOLUTION: Since the energy range being considered is narrow we may take the fraction to be $n(E)\Delta E/N$, evaluated for $E = kT/2$. Don't forget to convert the temperature value to the ideal gas scale.

[ans: 9.68×10^{-3}]

Repeat the calculation for the same interval width, but with the interval centered at $E = kT/4$ and with the interval centered at $E = kT$.

SOLUTION:

[ans: 8.79×10^{-3}; 8.30×10^{-3}]

III. MATHEMATICAL SKILLS

1. In this chapter the statistical ideas introduced in the last chapter are extended to apply to a continuous distribution of values. The ideas are the same but sums are replaced by integrals. Let x denote one number of the continuous set to be considered. It might be a molecular speed or the distance a molecule travels between collisions. Since possible values form a continuum we cannot speak meaningfully about the frequency with which any one value occurs. It is essentially zero for all values. On the other hand, we can speak about the frequency with which values in a given *range* occur. If the range is small the frequency is proportional to its width and we may write $f(x)\,dx$ for the frequency. The function $f(x)$ is the distribution function for the values of x. A distribution function might be determined from some theoretical argument, as it is for the Maxwellian distribution of molecular speeds, or it might be determined experimentally by taking a large number of measurements of the quantity x.

In what follows we assume that x can have any value from 0 to ∞. This is consistent with an interpretation of x as a molecular speed or intercollision distance.

If we take a large number of measurements we expect the fraction of readings that lie between x_1 and x_2 to be given by

$$P(x_1, x_2) = \int_{x_1}^{x_2} f(x)\,dx$$

The fraction of measurements that lie between 0 and ∞ is 1 since, according to our assumption about x, all measurements must lie in that range. Thus $\int_0^\infty f(x)\,dx = 1$. This is a restriction placed on the function $f(x)$.

To find the average value we multiply each possible value by the frequency of its occurrence and sum the results. For a continuous distribution this prescription leads to

$$\bar{x} = \int_0^\infty x f(x)\,dx$$

To find the mean-square value we multiply the square of each possible value by the frequency of its occurrence and sum the results. For a continuous distribution this prescription leads to

$$\overline{x^2} = \int_0^\infty x^2 f(x)\,dx$$

The root-mean-square is the square root of this result.

2. This chapter contains several integrals that are difficult to evaluate using only a knowledge of introductory calculus. They are associated with the average and mean-square speed for molecules with a Maxwellian speed distribution. The average speed is given by

$$\overline{v} = 4\pi \left[\frac{M}{2\pi RT}\right]^{3/2} \int_0^\infty v^3 e^{-Mv^2/2RT}\,dv$$

and the mean-square speed is given by

$$\overline{v^2} = 4\pi \left[\frac{M}{2\pi RT}\right]^{3/2} \int_0^\infty v^4 e^{-Mv^2/2RT}\,dv$$

Most standard integral tables list the following integrals:

$$\int_0^\infty x^{2a} e^{-px^2}\,dx = \frac{(2a-1)!!}{2(2p)^a}\sqrt{\frac{\pi}{p}}$$

and

$$\int_0^\infty x^{2a+1} e^{-px^2}\,dx = \frac{a!}{2p^{a+1}}$$

Here p is any positive number and a is any integer. $a!$ is the factorial of a; that is, $a! = 1 \cdot 2 \cdot 3 \cdot 4 \ldots a$. $(2a-1)!!$ is the product of all odd integers from 1 to $2a-1$; that is, $(2a-1)!! = 1 \cdot 3 \cdot 5 \cdot 7 \ldots (2a-1)$. We use the first integral when x to an *even* power multiplies the exponential in the integrand and the second when x to an *odd* power multiplies the exponential.

To evaluate the integral for the average speed of a Maxwellian distribution substitute $v = x$ and $m/2RT = p$ into the equation for $\overline{v}$. You should obtain

$$\overline{v} = 4\pi \left[\frac{M}{2\pi RT}\right]^{3/2} \int_0^\infty x^3 e^{-px^2}\,dx$$

The integral is the same as the second integral taken from the tables, with $a = 1$. So

$$\overline{v} = 4\pi \left[\frac{M}{2\pi RT}\right]^{3/2} \frac{1}{2p^2} = 4\pi \left[\frac{M}{2\pi RT}\right]^{3/2} \frac{2R^2 T^2}{M^2} = \sqrt{\frac{8RT}{\pi M}}$$

To evaluate the integral for the mean-square speed of a Maxwellian distribution make the same substitutions to obtain

$$\overline{v^2} = 4\pi \left[\frac{M}{2\pi RT}\right]^{3/2} \int_0^\infty x^4 e^{-px^2}\,dx$$

The integral is the same as the first one taken from the tables, with $a = 2$. So

$$\overline{v^2} = 4\pi \left[\frac{M}{2\pi RT}\right]^{3/2} \frac{3}{2(2p)^2}\sqrt{\frac{\pi}{p}} = 4\pi \left[\frac{M}{2\pi RT}\right]^{3/2} \frac{3R^2T^2}{2M^2}\sqrt{\frac{2\pi RT}{M}} = \frac{3RT}{M}$$

Thus the root-mean-square speed is $\sqrt{3RT/M}$.

IV. NOTES

Chapter 25
HEAT AND THE FIRST LAW OF THERMODYNAMICS

I. BASIC CONCEPTS

Having learned something about the microscopic basis of thermodynamics you now return to the main theme: the macroscopic behavior of a system as it exchanges energy with its environment. Here you are introduced to the concept of heat and you learn about changes in the internal energy, temperature, pressure, and volume that occur when heat is absorbed by a system or when work is done on it.

Heat. Recall from your study of Chapter 8 that the total work done on an object changes the sum of its translational kinetic and internal energies: $W = \Delta K + \Delta E_{\text{int}}$, where we have assumed that the objects of the system do not interact and so have omitted the potential energy term. Now another term is added to the left side of the energy equation. The energy of an object may also be changed by its absorbing or emitting heat.

Heat is the energy that flows between a system and its environment because _____

_____

When heat Q is included in the energy equation it becomes $W + Q = \Delta K + \Delta E_{\text{int}}$. In the thermodynamics chapters we deal with systems whose centers of mass remain at rest. So $\Delta K = 0$ and
$$W + Q = \Delta E_{\text{int}}$$
Heat and work are alternate ways of transferring energy.

A sign convention is adopted in writing the energy equation. W is taken to be positive when _____ and Q is taken to be positive when _____.
Heat flows from the environment to the system when the temperature of the environment is _____ than the temperature of the system.

The SI units of heat are _____. Other units in common use and their SI equivalents are 1 cal = _____ J, 1 Btu (British thermal unit) = _____ J, and 1 Cal = _____ J. The Calorie unit (with a capital C) is commonly used in nutrition.

Heat capacity and specific heat. The heat capacity relates the heat flowing into or out of the system to the change in temperature. If during some process heat Q is absorbed and the temperature increases by a small increment ΔT, then the heat capacity of the system for that process is given by $C' =$ _____. The heat capacity depends not only on the kind of material in the system but also the amount of amount of material. A related property that depends only on the kind of material and not the amount is the specific heat, defined by $c =$ _____, where m is the mass of the system. Another is the molar heat capacity (or heat capacity per mole), defined by $C =$ _____, where n is the number of moles in the system.

Suppose a certain system has molar heat capacity C for the process being considered. Then the heat capacity of n moles is $C' =$ _____ and the heat capacity of N molecules is $C' =$ _____, where N_a is the Avogadro constant. Suppose a certain system has specific heat c for a process and each of its molecules has mass m. Then the heat capacity of N molecules is $C' =$ _____ and the heat capacity of n moles is $C' =$ _____. The molar heat capacity for this system and process is $C =$ _____.

Consider an object of mass m, composed of material with specific heat c. If its temperature is changed from T_i to T_f by some process, then the heat exchanged is given by the integral $Q =$ _____. If the specific heat is independent of the temperature then the integral can easily be evaluated, with the result $Q =$ _____. Strictly speaking, the heat capacity and specific heat are usually temperature dependent, but they may usually be approximated by constants if the temperature change is small.

Carefully note that $Q = C'(T_f - T_i)$ is valid regardless of which temperature is higher. If $T_f > T_i$ then Q is positive, indicating that heat enters the object. If $T_f < T_i$ then Q is negative, indicating that heat leaves the object.

When two objects, initially at different temperatures, are placed in thermal contact with each other and isolated from their surroundings, the hotter substance cools, and the cooler substance warms until they reach the same temperature. The heat capacities of the objects can be used to find the final common temperature. The principle used is that the heat leaving the hotter object has the same magnitude as the heat entering the cooler object. The algebraic relationship is $Q_A + Q_B = 0$. If object A has mass m_A, specific heat c_A, and initial temperature T_A, then $Q_A =$ _____, where T_f is the final temperature. If object B has mass m_B, specific heat c_B, and initial temperature T_B, then $Q_B =$ _____. Note that the heat corresponding to the hotter object is negative, indicating that heat actually flows *from* it. Since $Q_A + Q_B = 0$, the final temperature is given by $T_f =$ _____. You should use the specific heats (for constant volume, constant pressure, or some other process) appropriate to the actual process. If more than 2 objects are involved, use $\sum Q = 0$.

Both heat and work tend to change the internal energy of the system and thus tend to change the temperature. The change in temperature that accompanies an exchange of heat may be greater or less than it would be if no work were done and the same heat were exchanged. For example, if heat Q is absorbed by an object and positive work W is done on it, then the heat capacity is _____ than it would be if same amount of heat is absorbed with no work being done.

Two special cases are important. In the first the volume is held constant while heat is supplied; in the second the pressure is held constant. The heat capacities are distinguished from each other by subscripts: the specific heat *at constant volume* is denoted by _____ and the specific heat *at constant pressure* is denoted by _____. These two specific heats have different values. Similar statements are true for the heat capacity and the molar heat capacity.

For a constant volume process the environment does _____ work on the system and, according to the energy equation, $Q =$ _____. Thus the molar heat capacity at constant volume can be calculated using $C_v = \Delta E_{\text{int}}/n\Delta T$, where n is the number of moles. For a monatomic ideal gas the result is $C_v = \frac{3}{2}R =$ _____ J/mol·K, for a diatomic ideal gas $C_v = \frac{5}{2}R =$ _____ J/mol·K, and for a polyatomic ideal gas $C_v = 3R =$ _____ J/mol·K.

Carefully note that no matter what process is used to change the internal energy (constant volume or not), changes in the temperature and internal energy are related by $\Delta T = \Delta E_{\text{int}}/nC_v$ or its equivalent $\Delta T = \Delta E_{\text{int}}/mc_v$. On the other hand, $\Delta T = Q/nC_v$ is valid only if the process is at constant volume. Suppose that in a certain process work W is done on the system and heat Q is absorbed by the system. Then, in terms of the molar heat capacity at constant volume, the change in temperature of the system is given by $\Delta T = $ _____.

For n moles of an ideal gas undergoing a constant pressure process, the ideal gas law gives $W = -p\Delta V = $ _____ ΔT. Now $\Delta E_{\text{int}} = nC_v\Delta T$, so $Q = \Delta E_{\text{int}} - W = ($ _____$)\Delta T$. Thus the molar heat capacity at constant pressure is $C_p = Q/n\Delta T = C_v + $ _____. For a monatomic ideal gas $C_p = \frac{5}{2}R = $ _____ J/mol·K, for diatomic ideal gas $C_p = \frac{7}{2}R = $ _____ J/mol·K, and for a polyatomic ideal gas $C_p = 4R = $ _____ J/mol·K.

Why do you expect the heat capacity at constant pressure to be greater than the heat capacity at constant volume? In particular, why is the heat required to obtain a given temperature change greater when it is supplied at constant pressure than when it is supplied at constant volume? _____

As you learned in the last chapter the ratio of the heat capacities $\gamma = C_p/C_v$ is important for calculating the work done during an adiabatic process. For a monatomic gas $\gamma = $ _____, for a diatomic gas $\gamma = $ _____, and for a polyatomic gas $\gamma = $ _____. Look at Table 3 to see how close the molar heat capacities and their ratios for real gases at room temperature are to the ideal gas values.

For a solid we can consider each atom to have _____ degrees of freedom, all vibrational. The equipartition theorem therefore predicts that the internal energy per mole is $E_{\text{int}}/n = $ _____ and that the molar heat capacity at constant volume is $C_v = \Delta E_{\text{int}}/n\Delta T = $ _____. Many solids have this value of the heat capacity at high temperatures but at low temperatures the heat capacity is significantly less. Look at Fig. 3 of the text to see how the heat capacities of several solids vary with temperature. These experimentally generated curves can be explained by quantum mechanics.

If the temperature is scaled by dividing it by the Debye temperature T_D, a constant that is different for different solids, the resulting heat capacities for nearly all solids lie on the same curve. See Fig. 4.

Heats of transformation. A substance accepts or rejects heat when it changes phase (melts, freezes, vaporizes, or condenses, for example), even though the temperature remains constant during a phase change. In words, a <u>heat of transformation</u> is _____

If L is the heat of transformation for a system of mass m then the magnitude of the heat accompanying a phase change is given by $|Q| = $ _____. Q is positive (heat absorbed) for melting and vaporization; it is negative (heat rejected) for freezing and condensing. The <u>heat of fusion</u> L_f is the heat per unit mass transferred during _____ or _____. The <u>heat of vaporization</u> L_v is the heat per unit mass transferred during _____ or _____.

The first law of thermodynamics. The first law of thermodynamics postulates the existence of the internal energy and states that its change as the system goes from any initial thermal equilibrium state to any final thermal equilibrium state is independent of how the change is brought about. It then equates the change in the internal energy to $Q + W$, evaluated for any path between the initial and final equilibrium states. For later review write the statement of the first law as given in the text:

_____

_____

As you learned previously, the internal energy is actually the sum of the kinetic energies of all the particles in the system and the potential energy of their interactions. We consider only systems for which the center of mass is at rest so the kinetic energy of translation of the system as a whole is zero. Then the mathematical statement of the first law, in differential form, is

$$dE_{\text{int}} = $$

Carefully contrast the internal energy with heat and work. When a system undergoes a change from one equilibrium state to another the work done on the system and the heat absorbed by the system depend on the actual process by which the state is changed and may be different for different processes. The change in the internal energy, like the temperature change, does NOT depend on the process but only on the initial and final states.

Because a change in the internal energy is a function of only the initial and final equilibrium states, the process used to change states is immaterial for its calculation. At intermediate stages of the process the system might not even be in thermal equilibrium. If n moles of a monatomic ideal gas in thermal equilibrium with a thermal reservoir at temperature T_i is removed from the reservoir and placed in contact with another reservoir at temperature T_f, then when it reaches equilibrium its internal energy will have changed by $\Delta E_{\text{int}} = $ _____ . If the difference between the two temperatures is not infinitesimal the gas is in thermal equilibrium only at the ends points of the process, NOT during the process. Note that the work done on the gas and the heat absorbed by the gas cannot be computed for this case.

The free expansion of a gas is another important example of a process for which the system is not in thermal equilibrium during intermediate stages. Initially a partition confines the gas to one part of its container, which is thermally insulated. Then the partition is removed and the gas expands into the other part. For this process, $W = $ _____ and $Q = $ _____ . Thus, according to the first law, $\Delta E_{\text{int}} = $ _____ . If the gas is an ideal gas then $\Delta T = $ _____ .

An important consequence of the first law is that the net change in the internal energy is zero over a cycle of a cyclic process, for which the initial and final equilibrium states are identical.

Another important consequence is that pV^γ is constant for an ideal gas undergoing an adiabatic quasi-static process. Here γ is the ratio of the heat capacities: $\gamma = C_p/C_v$. To prove the statement start with $dE_{\text{int}} = dW$, an equality that is valid for adiabatic processes. Use $dW = -p\,dV$ and $dE_{\text{int}} = $ _____ dT to show that $p\,dV = -nC_v\,dT$. Now the equation of state for an ideal gas, in differential form, is $p\,dV + V\,dp = nR\,dT$. Substitute $-nC_v\,dT$ for

372   *Chapter 25: Heat and the First Law of Thermodynamics*

$p\,dV$ and transfer this term to the right side of the equation to get $V\,dp = $ _____. Now use $C_p = C_v + R$ to obtain $V\,dp = nC_p\,dT$. The ratio $V\,dp/p\,dV$ is thus $-C_p/C_v$ or $-\gamma$. Rearrange $V\,dp/p\,dV = -\gamma$ to obtain $dp/p = -\gamma\,dV/V$. Integrate from an initial to a final state to obtain $\ln(p_f/p_i) = \ln(V_f/V_i)^{-\gamma}$, or $p_f/p_i = (V_i/V_f)^\gamma$. Rearrange to get the final result: $p_i V_i^\gamma = p_f V_f^\gamma$.

Transfer of heat. There are three mechanisms by which heat flows from one place to another. Briefly describe each of them in words, being sure the descriptions distinguish between them:

conduction: _____

convection: _____

radiation: _____

The important quantity is the *rate* of heat transfer, with symbol H. If heat Q is transferred in a short time Δt then the rate of transfer over that time is given by $H = $ _____.

To discuss conduction we consider a homogeneous bar of material with one end held at temperature T_H (high) and the opposite end held at temperature T_L (low). Let the x axis be along the bar, with $x = 0$ at the hot face and $x = L$ at the cold face. The temperature in the bar varies from point to point along its length and so is a function of x.

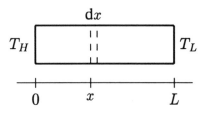

Consider a slab with infinitesimal thickness dx, parallel to the ends of the bar and located at an arbitrary point x along the bar. Its faces are at different temperatures and heat is transferred from the hotter side to the colder side. The rate at which heat passes through the infinitesimal slab is proportional to the temperature gradient dT/dx at its position and to its cross sectional area A. The constant of proportionality is a property of the material and is called the _____. Its symbol is k.

Mathematically the rate of heat flow through the slab at x is given by

$$H = $$

Be sure you have the sign correct. Positive H indicates flow in the positive x direction; negative H indicates flow in the negative x direction. According to the heat transfer equation positive H results if the temperature gradient is negative. Describe in words how the temperature depends on x if the gradient is negative: _____

Your answer should be consistent with a flow of heat from hot to cold regions.

If the bar is homogeneous the temperature increases uniformly from the cold face to the hot face or, what is the same, the temperature gradient is uniform. Since the bar has length

L and its ends are at temperatures T_H and T_L, respectively, then $dT/dx = $ _____ and $H = $ _____ .

The thermal conductivity can be considered to be independent of the temperature over small temperature ranges but does, in fact, vary slightly with T.

Building materials are often characterized by their thermal resistances or R values rather than their thermal conductivities. The R value is related to the thermal conductivity k by $R = $ _____ .

II. PROBLEM SOLVING

The heat capacity for a given substance and process is proportional to the amount of matter, usually specified by giving the mass or number of moles. The specific heat is the heat capacity per unit mass and the molar heat capacity is the heat capacity per mole.

PROBLEM 1. For a certain process a gas has a specific heat of 0.65 J/g·K. A 2.5-mol sample has a mass of 0.20 kg. What is the heat capacity and the molar heat capacity?

SOLUTION: Use $C' = mc$, where m is the mass of the gas, to calculate the heat capacity and $C = C'/n$, where n is the number of moles, to calculate the molar heat capacity.

[ans: 130 J/K; 52 J/mol·K]

PROBLEM 2. A certain gas consists of 2.5×10^{18} molecules, each with a mass of 5.2×10^{-26} kg. The molar heat capacity for a certain process is 15 J/mol·K. What is the specific heat? What is the heat capacity?

SOLUTION: Use $c = C/M$, where M is the molar mass. M can be found using $M = mN_a$, where m is the molecular mass and N_a is the Avogadro constant. The heat capacity is given by NC/N_a, where N is the number of molecules.

[ans: 479 J/kg·K; 6.23×10^{-5} J/K]

What is the heat capacity of 5.0 moles of this gas?

SOLUTION:

[ans: 75 J/K]

What is the heat capacity of 100 g of this gas?

SOLUTION:

[ans: 47.9 J/K]

Some problems are simple manipulations of the energy equation $\Delta E_{int} = W + Q$.

PROBLEM 3. During a certain process the environment does 4.7×10^2 J of work on a system and the system absorbs 3.1×10^2 J of heat. What is the change in its internal energy?

SOLUTION: Evaluate $\Delta E_{int} = W + Q$. Both W and Q are positive.

[ans: 7.8×10^2 J]

PROBLEM 4. During a certain process a system does 4.7×10^2 J of work on its environment and the system absorbs 3.1×10^2 J of heat. What is the change in the internal energy?

SOLUTION: Evaluate $\Delta E_{int} = W + Q$. W is negative and Q is positive.

[ans: -1.6×10^2 J]

The following problems deal with processes during which the system is always near thermal equilibrium, the heat capacity is constant, and no phase changes take place. Each of the terms in the first law can be computed from information about the system and process. Use $\Delta E_{int} = C'_v \Delta T$ for the change in the internal energy, $Q = C' \Delta T$ for the heat, and $W = -\int p\, dV$ for the work. In most cases two of the three quantities Q, W, and ΔE_{int} are known or can be computed and the first law can be used to compute the third. The result might be used to compute a heat capacity or temperature change, for example.

PROBLEM 5. A certain sample has a heat capacity at constant volume of 350 J/K and a heat capacity at constant pressure of 400 J/K. It undergoes a constant pressure process in which its temperature increases by 50 K. How much heat did it absorb?

SOLUTION: Use $Q = C'_p \Delta T$.

[ans: 2.0×10^4 J]

By how much did its internal energy increase?

SOLUTION: Use $\Delta E_{int} = C'_v \Delta T$.

[ans: 1.75×10^4 J]

How much work was done on it?

SOLUTION: Use $W + Q = \Delta E_{int}$.

[ans: -2.5×10^3 J]

If the pressure was 5.0×10^6 Pa, by how much did its volume change?

SOLUTION: Use $W = -p\Delta V$.

[ans: 5.0×10^{-4} m^3]

ΔV is positive, indicating an increase in volume.

PROBLEM 6. The heat capacity at constant volume for a piece of lead is 75 J/K. If 350 J of work is done adiabatically on the lead, what is its increase in temperature?

SOLUTION: Use $W + Q = \Delta E_{int}$, with $Q = 0$, to find the change in the internal energy. Use $\Delta E_{int} = C'_v \Delta T$ to find ΔT.

[ans: 4.7 K]

PROBLEM 7. During a certain process the environment does 3.7×10^3 J of work on a system and the temperature of the system increases by 25 K. The heat capacity of the system for this process is constant and has a value of 180 J/K. What heat is absorbed and what is the change in the internal energy of the system?

SOLUTION: Use $Q = C' \Delta T$ to compute Q, then the first law to compute ΔE_{int}.

[ans: 4.5×10^3 J; 8.2×10^3 J]

The heat capacities for ideal gases undergoing constant volume or constant pressure processes are known and so may not be given in the problem. Be careful to distinguish between monatomic, diatomic, and polyatomic ideal gases.

PROBLEM 8. In a certain process 1000 J of heat is absorbed by 2.0 moles of a diatomic ideal gas and the gas does 500 J of work on its environment. What is the change in the internal energy of the gas?

SOLUTION: Use $Q + W = \Delta E_{int}$.

[ans: 500 J]

What is the change in temperature of the gas?

SOLUTION: Use $\Delta E_{int} = nC_v \Delta T$. The molar heat capacity at constant volume of a diatomic gas is $C_v = 5R/2$.

[ans: 12 K]

What is the molar heat capacity of the gas for this process?

SOLUTION: Use $Q = nC\Delta T$.

[ans: $5R = 42$ J/mol·K]

PROBLEM 9. As the volume of an ideal gas containing 5.2×10^{23} diatomic molecules is slowly increased from 3.5×10^{-3} m^3 to 9.5×10^{-3} m^3 the pressure is held at 5.2×10^5 Pa. Calculate the work done on the gas, the change in the internal energy, and the heat absorbed by the gas.

SOLUTION: Use $W = -p(V_f - V_i)$ to calculate the work done on the gas, $p\Delta V = Nk\Delta T$ to calculate the change in temperature, and $\Delta E_{\text{int}} = C'_v \Delta T$ to calculate the change in the internal energy. The heat absorbed can be computed using either $Q = C'_p \Delta T$ or $Q = \Delta E_{\text{int}} - W$. Check your answer by computing it both ways. The heat capacities, of course, are given by $C'_v = 5Nk/2$ and $C'_p = 7Nk/2$.

[ans: -3.1×10^3 J; 7.8×10^3 J; 1.1×10^4 J]

PROBLEM 10. As the temperature of an ideal gas containing 5.2×10^{23} diatomic molecules is slowly increased from 250 K to 400 K the pressure is held at 5.2×10^5 Pa. Calculate the work done on the gas, the change in the internal energy, and the heat absorbed by the gas.

SOLUTION:

[ans: -1.1×10^3 J; 2.7×10^3 J; 3.8×10^3 J]

PROBLEM 11. As the volume of an ideal gas containing 5.2×10^{23} diatomic molecules is slowly increased from 3.5×10^{-3} m^3 to 9.5×10^{-3} m^3 the temperature is held at 350 K. Calculate the work done on the gas, the change in the internal energy, and the heat absorbed by the gas.

SOLUTION: Since $pV = NkT$, $p = NkT/V$ and $W = -\int p\,dV = -\int (1/V)\,dV = -NkT \ln(V_f/V_i)$. The change in the internal energy is zero. Why? Use the first law to calculate Q.

[ans: -2.5×10^3 J; 0; 2.5×10^3 J]

If the heat capacity depends on the temperature you must use $Q = \int C'(T)\,dT$, where the lower limit is the initial temperature and the upper limit is the final temperature.

PROBLEM 12. Suppose for a certain process the heat capacity of a substance varies with temperature T according to $C' = \alpha + \beta T^2$, where α and β are constants. Derive an expression for the heat absorbed when the temperature increases from T_i to T_f. If $\alpha = 130$ J/K and $\beta = 8.5 \times 10^{-4}$ J/K^3, how much heat is absorbed as the temperature increases from 250 K to 350 K?

SOLUTION:

[ans: $Q = \alpha(T_f - T_i) + \frac{1}{3}\beta(T_f^3 - T_i^3)$; 2.1×10^4 J]

Some problems deal with calorimetry. If two objects, initially at different temperatures, are placed in thermal contact with each other and the combined system is thermally isolated then $Q_A + Q_B = 0$. If no phase changes take place $Q_A = C'_A(T_f - T_A)$ and $Q_B = C'_B(T_f - T_B)$, so $T_f = (C'_A T_A + C'_B T_B)/(C'_A + C'_B)$. Here T_A and T_B are the initial temperatures and T_f is the common final temperature. You should be aware that only differences in initial and final temperatures enter the heat balance equation. Thus, if heat capacities are given in J/K, values entered for the temperatures may be on either the Kelvin or Celsius scales.

PROBLEM 13. Body A, initially at 250 K, and body B, initially at 375 K, are placed in thermal contact and the combination is thermally isolated. If the heat capacity of A is 220 J/K and the heat capacity of B is 145 J/K, what is the final common temperature and how much heat is exchanged by the bodies?

SOLUTION:

[ans: 300 K; 1.1×10^4 J]

PROBLEM 14. 2.0 moles of a monatomic ideal gas at 150 K and 1.5 moles of a polyatomic ideal gas at 250 K, both in rigid containers, are placed in thermal contact and the combination is isolated. What is the common final temperature?

SOLUTION: Since the containers are rigid the process is at constant volume. The monatomic gas absorbs heat $n_1 C_{v1} \Delta T_1$ and the diatomic gas delivers heat $n_2 C_{v2} \Delta T_2$. Equate the magnitudes of these heats and solve for the final temperature.

[ans: 210 K]

Initial and final temperatures measured in a calorimetry experiment are often used to find the specific heat of a substance.

PROBLEM 15. A 200-g block initially at 200°C is placed in 500 g of water at 25°C and the combination is thermally isolated. The final common temperature is 150°C. If the specific heat of water is 4.2×10^3 J/kg·K, what is the specific heat of the block?

SOLUTION:

[ans: 2.6×10^4 J/kg·K]

During a phase change the system absorbs or rejects heat without a change in temperature. The amount is given by $|Q| = mL$, where L is the heat of transformation and m is the mass of the system.

PROBLEM 16. How much heat is required to melt 0.50 kg of ice at its freezing point? Take the heat of fusion to be 3.33×10^5 J/kg.

SOLUTION:

[ans: 1.7×10^5 J]

PROBLEM 17. A 0.50-kg block of ice is at $-30°$C. How much heat is required to turn it to water at $+30°$C? Take the specific heat of ice to be 2.2×10^3 J/kg·K, the specific heat of water to be 4.2×10^3 J/kg·K, and the heat of fusion to be 3.33×10^5 J/kg.

SOLUTION: In this situation the temperature of the ice must be raised to the melting point, $0°$C, then the ice must be melted, and finally the temperature of the water must be raised to $+30°$C. During the first part of the process the heat absorbed is $Q_1 = mc_I \Delta T$, where m is the mass of the ice, c_I is the specific heat of ice, and ΔT is 30 C°. During the second part the heat absorbed is $Q_2 = mL_f$, where L_f is the heat of fusion. During the third part the heat absorbed is $Q_3 = mc_w \Delta T$, where c_w is the specific heat of water and ΔT is 30 C°.

[ans: 2.6×10^5 J]

Some calorimetry problems might involve a change in phase. Suppose a block of ice below its freezing point is placed in thermal contact with another object at a higher temperature. Any of three outcomes might result. The final common temperature might be reached without the ice melting, the ice might melt completely and the resulting water might reach a temperature above the melting point, or only part of the ice might melt and the final temperature is at melting point.

PROBLEM 18. Suppose a 0.50-kg block of ice at $-30°$C is placed inside a metal container at $100°$C and the ice-container system is isolated. The specific heat of ice is 2.2×10^3 J/kg·K, the specific heat of water is 4.2×10^3 J/kg·K, the heat of fusion of water is 3.33×10^5 J/kg, and the heat capacity of the container is 5.8×10^3 J/K. What is the final common temperature of the ice (and/or water) and container?

SOLUTION: You must test to determine the actual outcome. First assume the ice does not melt and calculate the final temperature: $T_f = (mc_I T_I + C'_c T_c)/(mc_I + C'_c)$, where c_I is the specific heat of ice, m is the mass of ice, C'_c is the heat capacity of the container, T_I is the initial temperature of the ice, and T_c is the initial temperature of the container.

[ans: $79.3°$C]

This result is above the melting point of ice ($0°$C) so at least some of the ice melts and the temperature found is NOT the correct final temperature. Had the answer turned out to be below the melting point of ice then it would have been the correct answer for the final temperature of the ice and container. Had the answer turned out to be exactly $0°$C, then both the ice and container would end up at $0°$C and none of the ice would be melted.

For the conditions of the problem at hand we now know that some or all of the ice melts. Assume all the ice melts and the water reaches the common final temperature T_f. Calculate T_f.

SOLUTION: The total heat absorbed by the ice and water is $mc_I(T_M - T_I) + mL_f + mc_W(T_f - T_M)$, where L_f is the heat of fusion of water and T_M is the melting point of ice (0°C). This must equal $-C'_c(T_f - T_c)$. Solve for T_f.

[ans: 48.4°C]

This result is above the melting point of ice so it is the correct answer. All the ice melts and the final temperature of the water and container is 48.4°C.

Suppose the answer had turned out to be below the melting point of ice (with the answer to the first part above the melting point, as it was). Then you know that only a part of the ice melts. The final temperature for the remaining ice, water, and container is then 0°C. You can now calculate the mass m' of the ice that melted. The heat balance equation becomes $mc_I(T_M - T_I) + m'L_f + C'_c(T_M - T_c) = 0$, so $m' = [C'_c(T_c - T_M) - mc_I(T_M - T_I)]/L_f$.

PROBLEM 19. Suppose the situation is the same as in the previous problem except that the heat capacity of the container is 9.0×10^2 J/K. First check to be sure that some, but not all, of the ice melts, then calculate the mass of ice that melts.

SOLUTION:

[ans: 0.173 kg]

Solutions to many heat conduction problems start with the equation for the rate of heat flow: $H = -kA\, dT/dx$, where k is the thermal conductivity and A is the cross-sectional area of the sample. For a bar in steady state with one end at temperature T_H and the other at temperature T_L, $H = kA(T_H - T_L)/L$, where L is the length of the slab.

PROBLEM 20. Aluminum has a thermal conductivity of 235 W/m·K. One end of a 2.0-m long aluminum pipe with a radius of 3.5 cm is held at 300 K and the other end is held at 400 K. What is the rate of heat flow along the pipe?

SOLUTION:

[ans: 45 W]

When two rods are joined end-to-end and the ends of the combination are held at different temperatures, the steady state rate of heat flow is the same in the two rods. Thus $k_1 A_1(T_{H1} - T_{L1})/L_1 = k_2 A_2(T_{H2} - T_{L2})/L_2$. Furthermore, the temperature at the junction is the same for both rods. Suppose the free end of rod 1 is held at temperature T_H and the free end of rod 2 is held at temperature T_L. Then $T_{H1} = T_H$, $T_{L2} = T_L$, and $T_{L1} = T_{H2} = T_J$, where T_J is the temperature at the junction. Thus $k_1 A_1(T_H - T_J)/L_1 = k_2 A_2(T_J - T_L)/L_2$. If the thermal conductivities, the geometry, and the temperatures of the ends are known, this can be solved for T_J. Once T_J is known either $H = k_1 A_1(T_H - T_J)/L_1$ or $H = k_2 A_2(T_J - T_L)/L_2$ can be solved for the rate of heat flow.

PROBLEM 21. Aluminum has a thermal conductivity of 235 W/m·K and copper has a thermal conductivity of 401 W/m·K. A rod of aluminum and a rod of copper, each 2.5 m long and each with a radius of 2.1 cm, are joined end-to-end. The free end of the aluminum is held at 300 K while the free end of the copper is held at 450 K. What is the temperature at the junction?

SOLUTION:

[ans: 394 K]

What is the rate of heat flow?

SOLUTION:

[ans: 12.3 W]

When two rods are joined in parallel along their lengths and the ends of the combination are held at different temperatures, the total rate of heat flow is the sum of the rates for the individual rods.

PROBLEM 22. Suppose the two rods of the previous example are joined along their lengths. If one end of the combination is held at 300 K and the other is held at 450 K, what is the total rate of heat flow?

SOLUTION:

[ans: 52.9 W]

In some problems heat passing out of a rod is used to change the phase of a substance in contact with the end. For example, the rod might be in contact with a quantity of ice at its melting point. Since the heat required to melt mass m is given by $Q = mL_f$, where L_f is the heat of fusion, the rate dm/dt at which the ice is melting is proportional to the rate of heat flow into the ice: $L_f \, dm/dt = H$. Once H is found using $H = -kA \, dT/dx$, dm/dt can be computed.

PROBLEM 23. Suppose one end of a 2.5-m long, 2.1-cm radius aluminum rod ($k = 235$ W/m·K) is held at 20°C and the other end is in contact with a block of ice (heat of fusion = 3.33×10^5 J/kg) at its melting point. At what rate is the ice melting? Assume it does not exchange energy with any other object.

SOLUTION:

[ans: 7.82×10^{-6} kg/s]

III. NOTES

Chapter 26
ENTROPY AND THE SECOND LAW OF THERMODYNAMICS

I. BASIC CONCEPTS

Here you study the second law of thermodynamics and a closely associated property of any macroscopic system, its entropy. The law is formulated in three equivalent forms: in terms of heat engines, refrigerators, and entropy. Be sure you understand all three formulations and the far-reaching implications of the law. The third and final law of thermodynamics is also discussed.

The second law of thermodynamics. The concepts of reversible and irreversible processes are important for understanding the second law. Explain what each is, being sure to distinguish between them.

reversible: _____

irreversible: _____

Give an example of each. You may wish to use the device pictured in Fig. 1 of the text.

reversible: _____

irreversible: _____

Reversible processes must be carried out quasi-statically but not all quasi-static processes are reversible. Tell what is meant by a quasi-static process: _____

Give an example of a quasi-static process that is not reversible: _____

If two different processes are used to take a system from the same initial equilibrium state to the same final equilibrium state, one reversible and one irreversible, the work and heat may be different but changes in the _____, _____, _____, and _____ will be the same. These quantities are called state variables. Their changes during any process that begins and ends at an equilibrium state depends only on the initial and final states and not on the process. The entropy, a quantity to be defined later, is also a state variable and its change also depends only on the initial and final states. Over a cycle the net change in any state variable is zero.

Although the second law has universal applicability, it is discussed in terms of two types of devices that act in cycles: a <u>heat engine</u> and a <u>refrigerator</u>. In terms of work and heat describe what these two types of devices do:

heat engine: _____
_____
refrigerator: _____
_____

Both make use of a working substance whose thermodynamic state varies during the process but which periodically returns to the same state. The same sequence of steps is performed on it during every cycle.

A possible reversible heat engine consists of an increase in the pressure of a gas at constant volume, followed by an expansion at constant pressure, followed by a decrease in pressure at constant volume, and finally a contraction at constant pressure. It is shown on the p-V diagram to the right. Use an arrow to show the direction in which the cycle is traversed. Also label on the diagram the parts of the cycle during which work is done by the system, work is done by the environment, heat is received by the system, and heat is rejected by the system.

The <u>efficiency</u> of a heat engine is defined by $e =$ _____, where W is _____ and Q_{in} is _____. The efficiency is positive. The negative sign appears because _____ is negative and _____ is positive.

A perfect heat engine takes in heat and does an identical amount of work. No heat is rejected. Since $W = -Q_{in}$ for a perfect heat engine, its efficiency is _____. Such an engine does not violate the first law of thermodynamics but it does violate the second law.

For the cycle depicted above let $|Q_{in}|$ be the magnitude of the heat absorbed and $|Q_{out}|$ be the magnitude of the heat rejected. According to the first law the magnitude of the work done by the engine is given by $|W| =$ _____. In terms of $|Q_{in}|$ and $|Q_{out}|$ the efficiency is given by $e =$ _____. Note that the efficiency can be increased to 1 if $Q_{out} = 0$. But such an engine violates the second law of thermodynamics and cannot be built.

Quote the <u>Kelvin-Planck</u> statement of the second law from the text: _____
_____
_____

This statement tells us that Q_{out} in the above example CANNOT be 0 and that there CANNOT be a 100% efficient heat engine. The first law tells us that in a cycle of a heat engine heat input is turned to work and heat output. The second law tells us it cannot be turned completely to work; some must become heat output.

The cycle depicted above, when run in reverse, becomes the cycle for a refrigerator. Sketch the cycle on the p-V axes to the right. Use an arrow to show the direction of traversal. Also label on the diagram the parts of the cycle during which work is done by the system, work is done by the environment, heat is received by the system, and heat is rejected by the system.

The <u>coefficient</u> <u>of</u> <u>performance</u> of a refrigerator is defined by $K = $ _____, where Q_L is the heat extracted from the low temperature reservoir and W is the work done on the working substance. For a perfect refrigerator $K = $ _____ because $W = $ _____. A perfect refrigerator does not violate the first law of thermodynamics but it does violate the second law and cannot be built.

Quote the <u>Clausius</u> statement of the second law from the text: _____

_____

_____

This statement tells us that work must be done to extract heat from a low temperature reservoir and deliver it to a high temperature reservoir cyclically. The statement does NOT preclude heat flowing from a high temperature reservoir to a low temperature reservoir without work being done. This is a natural occurrence.

The two statements of the second law are equivalent to each other. If the Kelvin-Planck statement were not valid, then a perfect heat engine could be combined with an imperfect refrigerator in such a way that the combination would be a perfect refrigerator and thus would violate the Clausius statement. Similarly, if the Clausius statement were not valid then a perfect refrigerator could be combined with an imperfect heat engine in such a way that the combination would be a perfect heat engine and thus would violate the Kelvin-Planck statement.

Carnot cycles. A Carnot cycle consists of two isothermal processes at different temperatures linked by two adiabatic processes. Diagram a Carnot cycle on the p-V axes to the right. Label the isotherms and adiabats. Assume the cycle is run as a heat engine and mark the portion of the cycle during which heat flows from a reservoir to the working substance and the portion during which heat flows from the working substance to a reservoir.

In terms of the heat $|Q_H|$ absorbed at the higher temperature and the heat $|Q_L|$ rejected at the lower temperature the efficiency of a Carnot heat engine is given by $e = $ _____. If an ideal gas is used as the working substance then, in terms of the reservoir temperatures

Chapter 26: Entropy and the Second Law of Thermodynamics **385**

T_H and T_L, $|Q_H|/|Q_L| =$ _____ and the efficiency is given by $e =$ _____. Notice the efficiency increases as the temperature of the hot reservoir increases but cannot reach 1 unless $T_L = 0\,$K. A reservoir at $T_L = 0\,$K does not violate either the first or second laws of thermodynamics, but it does violate the third.

If a Carnot cycle is run as a refrigerator the coefficient of performance, in terms of the temperatures, is given by $K =$ _____.

The efficiency of a Carnot cycle is an upper limit for the efficiency of all heat engines operating between the same temperatures. This is stated precisely by Carnot's theorem, which is: _____
_____
_____

The theorem can be proven by assuming an engine exists with a greater efficiency than a Carnot engine. Work done by such an engine is used to drive the Carnot cycle as a refrigerator and the combination is shown to be a perfect refrigerator, in violation of the Clausius statement of the second law. Notice that the heat engine need not be reversible for the theorem to be valid.

Every *reversible* heat engine operating between two thermal reservoirs must have the same efficiency as a Carnot engine operating between the same reservoirs. If it did not we could combine the engine with a Carnot engine and build either an engine or a refrigerator that violates the second law.

These two theorems, taken together, tell us that a heat engine has the same efficiency as a Carnot engine operating between the same temperatures if the engine is reversible and a lower efficiency if it is not. Another important consequence is that the efficiency of a Carnot heat engine is given by $e = (T_H - T_L)/T_H$, regardless of the working substance.

Thermodynamic (Kelvin) temperature scale. A Carnot heat engine can be used to define a temperature scale. In this section the symbol Θ is used temporarily to represent a temperature on this scale. One reservoir is taken to be at a standard temperature. By convention this is the triple point of water and is assigned the value $\Theta_{tr} = 273.16\,$K. The other reservoir is at the temperature Θ to be found. The quantity measured is _____ and the temperature Θ is taken to be $\Theta =$ _____ Θ_{tr}. The advantage of this scale is that the value assigned to the temperature is independent of the working substance.

Recall that if an ideal gas is used as the working substance then $|Q_H|/|Q_L| = T_H/T_L$, where the temperatures T_H and T_L are measured on the ideal gas scale. Thus the ideal gas and thermodynamic scales agree identically at all temperatures for which an ideal gas can be used as the working substance. From now on the symbol T will be used for both thermodynamic and ideal gas temperatures.

The third law of thermodynamics. State the third law: _____
_____
_____

This law tells us that a heat engine with an efficiency of _____ and a refrigerator with a coefficient of performance of _____ cannot be constructed by using a low temperature

386   *Chapter 26: Entropy and the Second Law of Thermodynamics*

reservoir at 0 K. This is distinct from the second law, which tells us that a heat engine with an efficiency of 1 cannot be constructed using reservoirs at non-zero temperatures because Q_{in} cannot be _____.

Entropy. Entropy is another state variable. If heat dQ is received by the system when its state changes from an equilibrium state to a nearby equilibrium state, then the entropy of the system changes by $dS =$ _____, where T is the temperature. This defines entropy. If the state changes reversibly from some initial equilibrium state i to some final equilibrium state f, the change in entropy can be computed by evaluating the integral $\Delta S =$ _____. Notice that the integral $\int_i^f dQ$ depends on the process but the integral $\int_i^f dQ/T$ does not.

To evaluate the integral for ΔS you must pick some path that connects the initial and final states but because entropy is a state variable it need not be the one actually followed by the system. For an irreversible process pick a *reversible* process with the same initial and final states. Sample Problem 5 and examples in Section 26–7 of the text show how to calculate the entropy changes for various situations. Here are some other examples.

Change in phase. A reversible change of phase provides a simple example. Suppose mass m of a substance with heat of fusion L_f melts at temperature T. For this process T is constant and $\Delta S = Q/T$. Since $Q = mL_f$, $\Delta S =$ _____. On freezing the change in entropy is $\Delta S =$ _____. Note that the entropy of the substance increases on melting and decreases on freezing.

Constant volume process. Suppose the heat capacity at constant volume C'_v for a gas is independent of T, V, and p. If the gas undergoes an infinitesimal reversible change of state at constant volume, so its temperature changes by dT, then the heat absorbed is $dQ =$ _____ dT and the change in entropy is $dS = dQ/T =$ _____. If the temperature changes from T_i to T_f then the change in entropy is the integral of this expression. In the space below carry out the integration to show that $\Delta S = C'_v \ln(T_f/T_i)$:

Constant pressure process. Suppose the heat capacity at constant pressure C'_p for a gas is independent of T, V, and p. If the gas undergoes an infinitesimal reversible change of state at constant pressure, so its temperature changes by dT, then the heat absorbed is $dQ =$ _____ dT and the change in entropy is $dS = dQ/T =$ _____. If the temperature changes from T_i to T_f then the change in entropy is the integral of this expression. In the space below carry out the integration to show that $\Delta S = C'_p \ln(T_f/T_i)$:

Isothermal process. If the gas undergoes an infinitesimal reversible change of state at constant temperature, then $dE_{\text{int}} = C'_v dT = 0$ and $dQ = -dW = +p\,dV$. Thus $dS = (p/T)\,dV$ and $\Delta S = \int (p/T)\,dV$, where the limits of integration are the initial volume V_i and the final volume V_f. To evaluate this integral p must be known as a function of volume and temperature. The equation of state gives this information.

Consider n moles of an ideal gas, with heat capacity at constant volume C'_v and heat capacity at constant pressure C'_p. Its pressure, volume, and temperature obey the equation of state $pV = nRT$. For a reversible change in temperature from T_i to T_f at constant volume, $\Delta S =$ _____. For a reversible change in pressure from p_i to p_f at constant volume, $\Delta S =$ _____. For a reversible change in temperature from T_i to T_f at constant pressure, $\Delta S =$ _____. For a reversible change in volume from V_i to V_f at constant pressure, $\Delta S =$ _____. For a reversible change in pressure from p_i to p_f at constant temperature, $\Delta S =$ _____. For a reversible change in volume from V_i to V_f at constant temperature, $\Delta S =$ _____.

For a cycle $\Delta S =$ _____. This is true because the entropy is a thermodynamic state variable and has the same value for a given state no matter how that state is reached.

The second law of thermodynamics, in its most general form, is expressed in terms of entropy changes. It is: _____
_____
_____

If a change of state is reversible then the change in the total entropy of the system and its environment is _____. If the entropy of the system decreases in a reversible process then the entropy of the environment must _____. If the system changes state through an irreversible process then the change in the total entropy of the system and its environment is _____.

Suppose two identical systems go from the same initial state to the same final state. For system 1 the process is carried out reversibly while for system 2 it is carried out irreversibly. Suppose the change in the entropy for system 1 is ΔS_1. Then the change in the entropy of the environment of system 1 is _____. The change in entropy of system 2 is _____. The change in the magnitude of the entropy of the environment of system 2 is _____ than $|\Delta S_1|$.

You should be able to show that perfect heat engines and perfect refrigerators operating between non-zero temperatures violate this statement of the second law.

For example, consider a heat engine that absorbs heat Q_H at temperature T_H and rejects heat $-Q_L$ at temperature T_L (Q_L is negative). Over a cycle the change in entropy of the working substance is _____. The change in the entropy of the reservoir at T_H is _____ and the change in the entropy of the reservoir at T_L is _____. (Be careful about signs here: if heat is positive for the working substance it is negative for the reservoir.) The change in the total entropy of the working substance and the reservoirs is $\Delta S_{\text{total}} =$ _____. Note that Q_L cannot be zero since that would make ΔS_{total} negative.

If the engine is reversible then over a cycle $\Delta S_{\text{total}} =$ _____ and, in terms of the temperatures, $|Q_L|/|Q_H| =$ _____, regardless of the working substance. If the engine is irreversible then ΔS is positive and $|Q_L| >$ _____.

All real (non-ideal) processes are irreversible to some extent. We therefore expect the total entropy of every real system and its environment to _____ as time goes on. This has often been used to assign a "direction" to time.

At the microscopic level entropy is related to a certain probability. That probability is:

The relationship is $S =$ _____, where P is _____.

II. PROBLEM SOLVING

Many problems of this chapter are concerned with reversible thermodynamic cycles, in which a working substance is repeatedly carried through a series of equilibrium states. If the working substance is an ideal gas you should be able to compute the work done and the heat absorbed during each portion of the cycle. Review the calculation of work and heat for these gases. The following is a summary:

constant volume
$W = 0$
$Q = nC_v \Delta T = C_v V \Delta p / R$

constant pressure
$W = -p\Delta V = -nR\Delta T$
$Q = nC_p \Delta T = C_p p \Delta V / R$

isothermal
$W = -nRT \ln(V_f/V_i) = -nRT \ln(p_i/p_f)$
$Q = -W = nRT \ln(V_f/V_i) = nRT \ln(p_i/p_f)$

adiabatic
$W = (p_f V_f - p_i V_i)/(\gamma - 1) = nR\Delta T/(\gamma - 1)$
$Q = 0$

The ideal gas law was used to obtain the alternate forms given. You must supply the values for C_v and C_p appropriate to the gas being considered.

You should also be able to calculate the net work done on the system and the net heat delivered to the system during a cycle. Simply sum the works for all portions of the cycle to find the net work and sum the heats for all portions to find the net heat. You must be careful of signs here. Work and heat are positive when energy is entering the system and negative when energy is leaving.

PROBLEM 1. Starting with a volume of 1.5×10^{-4} m³ and a temperature of 250 K, 2.5 moles of a monatomic ideal gas is carried through the following reversible cycle:

A. The temperature is raised at constant volume to 400 K.
B. The gas is isothermally expanded to 3.4×10^{-4} m³.
C. The temperature is decreased at constant volume to 250 K.
D. The gas is isothermally compressed to 1.5×10^{-4} m³.

Sketch the cycle on the p-V axes to the right. Use arrows to show the direction of traversal. Mark portions of the cycle for which positive work is done by the gas, negative work is done by the gas, heat is absorbed by the gas, and heat is rejected by the gas.

What work is done by the environment and what heat is absorbed by the gas during portion A of the cycle?

SOLUTION: Use $Q = nC_v \Delta T$, with $C_v = \frac{3}{2}R$ and $\Delta T = +150$ K, to calculate the heat.

[ans: 0; 4.67×10^3 J]

What work is done by the gas and what heat is absorbed by the gas during portion B of the cycle?

SOLUTION: Use $W = -nRT \ln(V_f/V_i)$ and $Q = nRT \ln(V_f/V_i)$, with $T = 400\,\text{K}$, $V_i = 1.5 \times 10^{-4}\,\text{m}^3$, and $V_f = 3.4 \times 10^{-4}\,\text{m}^3$.

[ans: 6.80×10^3 J; 6.80×10^3 J]

What work is done by the environment and what heat is rejected by the gas during portion C of the cycle?

SOLUTION: Use $Q = nC_v \Delta T$, with $\Delta T = -150\,\text{K}$.

[ans: 0; 4.67×10^3 J]

What work is done by the environment and what heat is rejected by the gas during portion D of the cycle?

SOLUTION: Use $W = -nRT \ln(V_f/V_i)$ and $Q = nRT \ln(V_f/V_i)$, with $T = 250\,\text{K}$, $V_i = 3.4 \times 10^{-4}\,\text{m}^3$, and $V_f = 1.5 \times 10^{-4}\,\text{m}^3$.

[ans: 4.25×10^3 J; 4.25×10^3 J]

Over one complete cycle what is the net work done by the gas on the its environment, what heat is absorbed by the gas, and what heat is rejected by the gas?

SOLUTION: Sum the work done during parts B and D of the cycle. Be careful of signs. W is negative for part B and positive for part D. Q is positive for parts A and B and the sum of these heats gives Q_{in}. Q is negative for parts C and D and the sum of these heats gives Q_{out}.

[ans: 2.55×10^3 J; 11.47×10^3 J; 8.92×10^3 J]

In some cases you must calculate the efficiency e of a cycle run as a heat engine or the coefficient of performance K of a cycle run as a refrigerator. Use $e = -W/Q_{in}$ and $K = Q_{in}/W$, where Q_{in} is the heat entering and W is the net work done on the working substance during a cycle. You must identify the portions of a cycle during which heat enters and sum only those heats to find Q_{in}. Carefully note that Q_{in} is not the same when a cycle is run as an engine and when it is run as a refrigerator. In fact Q_{in} for the refrigerator is $-Q_{out}$ for the engine. Q_{in} is always positive; W is negative for a heat engine and positive for a refrigerator.

PROBLEM 2. What is the efficiency of the heat engine of the previous problem?

SOLUTION: Use $e = -W/Q_{in}$, with $W = -2.55 \times 10^3$ J and $Q_{in} = 11.47 \times 10^3$ J.

[ans: 0.222]

What is the coefficient of performance of the cycle if it is run in the opposite direction, as a refrigerator?

SOLUTION: Use $K = Q_{in}/W$, with $W = 2.55 \times 10^3$ J and $Q_{in} = 8.92 \times 10^3$ J.

[ans: 3.50]

A Carnot cycle consists of two isotherms and two adiabats. If the working substance is an ideal gas then pV is constant along each isotherm and pV^γ is constant along each adiabat. These relations and the ideal gas equation of state can be used to find p, V, and T at the "corners" of the cycle, where an isotherm and an adiabat intersect.

PROBLEM 3. Consider a Carnot cycle with points 1 and 2 on the same isotherm, points 3 and 4 on the other isotherm, points 1 and 4 on an adiabat, and points 2 and 3 on the other adiabat. The working substance is 2.5 moles of a monatomic ideal gas. Suppose $p_1 = 1.1 \times 10^6$ Pa and $V_1 = 7.3 \times 10^{-3}$ m^3. What is T_1?

SOLUTION: Use $p_1V_1 = nRT_1$.

[ans: 387 K]

Suppose $V_2 = 1.2 \times 10^{-2}$ m^3. What is p_2?

SOLUTION: Since points 1 and 2 are on the same isotherm, $T_2 = T_1$ and $p_1V_1 = p_2V_2$.

[ans: 6.69×10^5 Pa]

Suppose $V_3 = 3.8 \times 10^{-2}$ m^3. What are p_3 and T_3?

SOLUTION: Use $p_2V_2^\gamma = p_3V_3^\gamma$ to find p_3, then use $p_3V_3 = nRT_3$ to find T_3.

[ans: 9.80×10^4 Pa; 179 K]

What are V_4 and p_4?

SOLUTION: Solve $p_4V_4 = p_3V_3$ and $p_4V_4^\gamma = p_1V_1^\gamma$ simultaneously for V_4 and p_4.

[ans: 2.31×10^{-2} m^3; 1.61×10^5 Pa]

The cycle is traversed in the order 1 → 2 → 3 → 4 → 1. During the process 1 → 2 what work is done on the gas and what heat is absorbed by the gas?

SOLUTION:

[ans: -4.00×10^3 J; $+4.00 \times 10^3$ J]

During the process 2 → 3 what work is done on the gas and what heat is absorbed by the gas?

SOLUTION:

[ans: -6.48×10^3 J; 0]

During the process 3 → 4 what work is done on the gas and what heat is absorbed by the gas?

SOLUTION:

[ans: $+1.85 \times 10^3$ J; -1.85×10^3 J]

During the process 4 → 1 what work is done on the gas and what heat is absorbed by the gas?

SOLUTION:

[ans: $+6.48 \times 10^3$ J; 0]

Use $e = -W/Q_{in}$ to compute the efficiency and compare the result with $1 - T_L/T_H$.

SOLUTION:

[ans: 0.537]

If this Carnot cycle is run in the opposite direction, as a refrigerator, what is its coefficient of performance?

SOLUTION:

[ans: 0.861]

If a reversible engine operates between a high temperature reservoir at T_H and a low temperature reservoir at T_L then its efficiency is given by $e = 1 - T_L/T_H$. Thus you can calculate the efficiency if the temperatures are given. Once the efficiency is known, other quantities can be calculated, as the following examples show.

PROBLEM 4. In each cycle a reversible heat engine absorbs 1500 J of heat at 450 K and rejects heat to a reservoir at 325 K. What is the efficiency of this engine?

SOLUTION: Use $e = 1 - T_L/T_H$, with $T_L = 325$ K and $T_H = 450$ K.

[ans: 0.278]

What is the net work done by the engine during each cycle?

SOLUTION: Use $e = -W/Q_{in}$, with $Q_{in} = 1500$ J.

[ans: 417 J]

What heat is rejected during each cycle?

SOLUTION: Use the first law, $W + Q_{in} + Q_{out} = 0$, with $W = -417$ J and $Q_{in} = 1500$ J.

[ans: 1083 J]

PROBLEM 5. A reversible heat engine operates between 200 K and 500 K. What is its efficiency?

SOLUTION:

[ans: 0.60]

If it does a net work of 600 J each cycle, how much heat is absorbed during each cycle?

SOLUTION:

[ans: 1000 J]

How much heat is rejected during each cycle?

SOLUTION:

[ans: 400 J]

Similar questions can be asked about a refrigerator cycle.

PROBLEM 6. The coefficient of performance of a certain refrigerator is 5.5. During each cycle it absorbs 1200 J of heat. During each cycle how much work is required and how much heat is delivered to the high temperature reservoir?

SOLUTION:

[ans: 218 J; 1418 J]

The efficiency of a cycle operating as a heat engine is related to the coefficient of performance of the same cycle operating as a refrigerator.

PROBLEM 7. If a reversible heat engine has an efficiency of 0.68, what is its coefficient of performance when it is run as a refrigerator?

SOLUTION: Use lower case letters to designate quantities for the engine and upper case quantities to designate quantities for the refrigerator. Thus $e = -w/q_{in}$ and $K = Q_{in}/W$. Now $Q_{in} = -q_{out} = w + q_{in}$ and $W = -w$, so $K = -(w + q_{in})/w$, or, since $w = -eq_{in}$, $K = q_{in}(1-e)/eq_{in} = (1-e)/e$.

[ans: 0.47]

You should know how to compute the change in entropy for various processes. It is given by $\Delta S = \int dQ/T$, where the integral is over any reversible path that connects the actual initial and final states.

PROBLEM 8. What is the change in entropy of 3.0 moles of a monatomic ideal gas when its temperature changes from 450 K to 300 K at constant volume? at constant pressure?

SOLUTION: For a reversible constant volume process $dQ = nC_v\, dT$, so $\Delta S = \int dQ/T = \int (nC_v/T)\, dT = nC_v \ln(T_f/T_i)$. For a reversible constant pressure process $dQ = nC_p\, dT$, so $\Delta S = nC_p \ln(T_f/T_i)$. The molar heat capacities are $C_v = \frac{3}{2}R$ and $C_p = \frac{5}{2}R$.

[ans: −15.2 J/K; −25.3 J/K]

What is the change in entropy of 3.0 moles of a monatomic ideal gas when its volume changes from 2.1×10^{-2} m³ to 4.5×10^{-2} m³ at constant pressure? at constant temperature?

SOLUTION: For the constant pressure process $\Delta S = nC_p \ln(T_f/T_i) = nC_p \ln(V_f/V_i)$, where the ideal gas law was used to obtain the second form. For the constant temperature process $\Delta S = Q/T = nR \ln(V_f/V_i)$.

[ans: 47.5 J/K; 19.0 J/K]

Most processes are not at constant volume, constant pressure, or constant temperature. In fact, most are not strictly reversible. Nevertheless a reversible process, starting at the same initial state and ending at the same final state, can be constructed as a sequence of reversible constant volume, constant pressure, and constant temperature processes. This sequence is used to calculate the change in entropy.

PROBLEM 9. What is the change in entropy of 2.5 moles of a monatomic ideal gas when it is taken from a pressure of 1.5×10^5 Pa and volume of 3.1×10^{-3} m^3 to a pressure of 2.4×10^5 Pa and volume of 4.5×10^{-3} m^3?

SOLUTION: Let us take a two segment path: in the first segment the gas is taken at constant volume (3.1×10^{-3} m^3) from a pressure of 1.5×10^5 Pa to a pressure of 2.4×10^5 Pa; in the second segment it is taken at constant pressure (2.4×10^5 Pa) from a volume of 3.1×10^{-3} m^3 to a volume of 4.5×10^{-3} m^3.

What is the change in entropy for the first segment?

[ans: 14.6 J/K]

What is the change in entropy for the second segment?

[ans: 19.4 J/K]

What is the change in entropy for the entire path?

[ans: 34.0 J/K]

This is the change in entropy regardless of whether or not the actual process is reversible, and, if it is reversible, regardless of whether or not the actual path is the same as the one used for the calculation.

Changes in entropy are also associated with phase changes.

PROBLEM 10. A 0.25-kg block of ice at $-50°$C is placed in thermal contact with a thermal reservoir at $50°$C. Eventually it becomes water at $50°$C. Take the specific heat of ice to be 2.22×10^3 J/kg·K, the specific heat of water to be 4.19×10^3 J/kg·K, and the heat of fusion to be 3.33×10^5 J/kg. Calculate the change in entropy of the ice/water.

SOLUTION: First, the temperature of the ice increases from $-50°$C to $0°$C (223 K to 273 K). For this process $dQ = mc_I\, dT$ and $\Delta S = \int dQ/T = \int (mc_I/T)\, dT = mc_I \ln(T_f/T_i)$, where m is the mass of the ice and c_I is the specific heat of ice. Second, the ice melts. For this process $\Delta S = Q/T = mL_f/T$, where L_f is the heat of fusion. Lastly, the temperature of the water increases from $0°$C to $50°$C (273 K to 323 K). The change in entropy for this process is $\Delta S = mc_w \ln(T_f/T_i)$, where c_w is the specific heat of water. Calculate each contribution and sum the results.

[ans: 593 J/K]

What is the change in entropy of the reservoir?

SOLUTION: The reservoir remains at a constant temperature of 50°C (323 K) so $\Delta S = Q/T$. Q is found by summing the heats exchanged during the three portions of the process. While the temperature of the ice is increasing the reservoir supplies heat $mc_I \Delta T$, while the ice is melting it supplies heat mL_f, and while the temperature of the water is increasing it supplies heat $mc_w \Delta T$. In each case heat leaves the reservoir so Q is negative.

[ans: −506 J/K]

Notice that the sum of the entropy changes for the ice/water system and the reservoir is positive, indicating that the process is irreversible.

Sometimes you must use what you learned in previous chapters to calculate values for some of the thermodynamic variables before you can compute the change in entropy. Consider the following example.

PROBLEM 11. Gas A, with a heat capacity at constant volume of $C'_{vA} = 85$ J/K, is initially at −30°C. Gas B, with a heat capacity at constant volume of $C'_{vB} = 55$ J/K, is initially at +100°C. Each gas is in a rigid diathermal container, the containers are in thermal contact, and the combination is thermally isolated. From the time thermal contact is first made to the time thermal equilibrium is reached, what are the changes in entropy for each gas separately and for the isolated system consisting of the two gases?

SOLUTION: First find the final common temperature. Since the containers are rigid the process takes place at constant volume. Use $T_f = (C'_{vA} T_A + C'_{vB} T_B)/(C'_{vA} + C'_{vB})$.

[ans: 21.1°C]

Gas A goes from −30°C to +21.1°C (243 K to 294 K) at constant volume. What is the change in its entropy?

SOLUTION: The actual process is irreversible but we can find a reversible process that carries the gas from the same initial state to the same final state. For this process $dQ = C'_{vA} \, dT$ and $\Delta S = C'_{vA} \ln(T_f/T_i)$.

[ans: 16.2 J/K]

Gas B goes from +100°C to +21.1°C (373 K to 294 K). What is the change in its entropy?

SOLUTION:

[ans: −13.1 J/K]

What is the total change in entropy for the two-gas system?

SOLUTION:

[ans: +3.1 J/K]

This process is irreversible. Note that no heat enters or leaves the composite system but the total entropy increases.

To solve some problems you will need to calculate the change in entropy of the environment of a system as well as that of the system itself. The last problem is an example: gas A can be considered to be the environment of gas B and vice versa.

Often one or more thermal reservoirs form the environment of a system. A reservoir can exchange unlimited heat with a system without a change in its temperature. If T is the temperature of the reservoir and Q is the heat exchanged then the change in entropy of the reservoir is $\Delta S = Q/T$. Q and ΔS are positive if heat flows from the system to the reservoir; Q and ΔS are negative if heat flows from the reservoir to the system.

PROBLEM 12. A Carnot heat engine operating between 200 K and 450 K does 600 J of work during each cycle. What is its efficiency?

SOLUTION:

[ans: 0.556]

During each cycle how much heat is absorbed by the working substance from the high temperature reservoir and what is the change in entropy of this reservoir?

SOLUTION:

[ans: 1080 J; −2.40 J/K]

During each cycle how much heat is rejected by the working substance to the low temperature reservoir and what is the change in entropy of this reservoir?

SOLUTION:

[ans: 480 J; +2.40 J/K]

This is a reversible process so the change in the total entropy of the working substance and its environment (the reservoirs) is zero. The entropy of the working substance is the same at the beginning and end of a cycle so the entropy of the environment must also be the same. It is: the gain in entropy by one reservoir equals the loss by the other.

PROBLEM 13. A 0.250-kg piece of ice at 0°C is melted by placing it in contact with a reservoir only infinitesimally warmer than 0°C. Take the heat of fusion of water to be 3.33×10^5 J/kg and compute the changes in the entropies of the ice and the reservoir.

SOLUTION:

[ans: +305 J/K; −305 J/K]

This process is reversible and the total entropy of the ice and reservoir does not change.

An identical piece of ice is melted by placing it in contact with a reservoir at 50°C. Consider only the melting process, not the warming of the water after melting. That is, take the initial state to be ice at 0°C and the final state to be water at 0°C. Calculate the changes in the entropies of the ice and reservoir.

SOLUTION:

[ans: +305 J/K; −258 J/K]

This process is irreversible and the total entropy of the ice and reservoir increases by 47 J/K.

III. NOTES

OVERVIEW III
ELECTRICITY AND MAGNETISM

Electric and magnetic phenomena pervade our lives. We use electric power to heat and light our homes, schools, and places of work. We use it to carry out a vast number of industrial and manufacturing tasks. We use it, via our radios, television sets, and movie projectors, to provide entertainment. Magnetism provides the force used to run all electric motors and in every electric generating plant a coil of wire is rotated near a large magnet to produce electric power.

As important as all these uses are, electricity is even more vital to us in another area: electric forces bind particles together to form atoms and atoms together to form all materials. We literally cannot exist without electrical forces! Nor can any solid or liquid, or for that matter any atom.

Visible light, microwaves, x rays, and radio waves are all electromagnetic in nature and differ only in their frequency. Optics, x-ray imaging, radio transmission and reception, and microwave production and detection are all logically part of the field of electricity and magnetism.

Particles such as electrons and protons exert electric and magnetic forces on each other; other particles do not. The property of a particle that gives rise to these forces is called charge. Charged particles at rest interact only via electric forces; charged particles in motion exert both electric and magnetic forces on each other.

Electric forces between charged particles at rest are described in Chapter 27. You will learn there are two types of charge, called positive and negative, and that like charges repel each other while unlike charges attract each other. In each case the force is along the line that joins the charges and is proportional to the reciprocal of the square of the distance between them. Electric forces, of course, are vectors and, when two or more charges act on another, the resultant force is the vector sum of the individual forces.

We may think of a charge creating electric and magnetic fields in all of space; the fields then exert forces on other charges. The problem of finding the force exerted by a collection of charges is then broken down into two somewhat simpler problems: *what fields are produced by the collection?* and *what force do the fields exert on the other charge?* Electric and magnetic interactions can be described much more easily in terms of fields than in terms of forces. When charges are moving or produce electromagnetic radiation the field viewpoint is almost a necessity.

You start your study of electric fields, in Chapter 28, by learning about the force exerted by a field on a charge. If no counterbalancing forces act the charge accelerates. You will apply Newton's second law and your knowledge of kinematics to examine the motions of charges in electric fields.

The relationship between a collection of charges and the electric field it produces is discussed in Chapters 28 and 29. It can be formulated in two equivalent, but quite different, ways. In the first chapter it is given in terms of the charges and their positions while in the second it is given in terms of an integral of the field over a surface. You should understand both formulations. The first is useful when you are given the charges and their positions, then asked to find the electric field; the second, known as Gauss' law, is useful when you know the field and want to find the charge distribution. It can also be used to find the field in certain situations of high symmetry.

Electrostatic forces, those of charges at rest, are conservative, so a potential energy can be associated them. Chapter 30 deals with electric potential energy and a closely related quantity, electric potential. You also will learn to calculate the work that must be done to assemble a collection of charges.

In Chapter 31 the ideas of electric field and potential, as well as Gauss' law, are put to practical use in a study of capacitance. Capacitors are extremely useful devices for the storage of electrical energy, which can then be recovered in short intense bursts. Small capacitors are used in electronic flash units for cameras. Large capacitors are used to power the extremely intense lasers used in an attempt to produce controlled nuclear fusion. In a more fundamental vein you will learn here that you may think of the potential energy of a collection of charges as being stored in the electric field and you will learn how to calculate the energy per unit volume stored in any field.

An electric current is simply a collection of moving charges. Currents are studied for two reasons. First, electrical circuits carrying current have many important practical applications. Think of the miniature circuits that operate computers and the house and industrial circuits that operate lights, appliances, and machinery. Secondly, currents produce magnetic fields. Chapters 32 and 33, in which currents are studied, form a bridge to later chapters on the magnetic field.

In Chapter 32 you will learn that the current in any non-superconducting material is determined by the electric field. You will also learn about resistivity, the property of materials that determines the current for a given electric field. It is closely related to the transfer of energy from moving charges to the material through which they move. Chapter 33 deals with batteries and other devices used to produce currents. Here you will apply the ideas of this and the previous chapter to the analysis of simple electrical circuits carrying time-independent currents. Energy balance in electrical circuits, important because current is often used to carry energy from one place to another, is also studied in this chapter.

In Chapter 34 you start the study of magnetic fields by learning about the force exerted by a magnetic field on a moving charge. This is extended to a discussion of the force and torque exerted on a current-carrying circuit.

The relationship between a current and the magnetic field it produces is discussed in Chapter 35. Just as for electric fields, two equivalent views are presented. In the first, known as the Biot-Savart law, the field produced by an infinitesimal element of current is postulated and this expression is integrated over the entire current. In the second, known as Ampère's law, the integral of the magnetic field around a closed loop is related to the current passing through the loop. The first viewpoint is of value when the current circuit is known and the field is sought; the second is of value when the field is known and the current is sought. Ampère's law can also be used to find the magnetic fields of highly symmetric current distributions.

A time-varying current produces an electric as well as a magnetic field, even if the net charge is zero in every region of space.

The relationship between the fields is given by Faraday's law and is discussed in Chapter 36. The law also deals with situations in which an object moves in a constant magnetic field. The result is exactly as if an infinitesimal battery were placed at each point in the object. Current is induced if the object forms part of an electrical circuit. Faraday's law is applied to electrical circuits in Chapter 37. There you will learn about inductors: circuit elements that are used to control the rate of change of current in a circuit and that store energy in magnetic fields just as capacitors store energy in electric fields.

When an externally generated magnetic field is applied to a substance it produces a field of its own. You may think of currents circulating within the material. For paramagnetic materials the induced field is in the same direction as the applied field and the total field is greater than the applied field. For diamagnetic materials the induced field is in the opposite direction and the total field is smaller than the applied field. Certain materials, such as iron, are ferromagnetic and can be permanently magnetized. They produce a magnetic field even in the absence of an applied field. Magnetic properties of materials are examined in Chapter 38. Gauss' law for magnetism is also presented in this chapter. It is the mathematical equivalent of the statement that no particle exists that produces a magnetic field when at rest.

Most electrical power is transmitted by means of alternating currents, for which the direction of motion of the charges changes periodically. Current in an AC circuit containing capacitors and inductors has a natural frequency of oscillation and this property is used, for example, to tune radios and television sets. You will study this type circuit in Chapter 39.

Electromagnetic theory is completed by an additional relationship, described in Section 40–1: a magnetic field is always associated with a time-dependent electric field, just as an electric field is always associated with a time-dependent magnetic field. You may think of a time-varying magnetic field as the source of an electric field and a time-varying electric field as the source of an magnetic field. You will learn later that the relationship is such that changes in the fields propagate through space as waves.

Gauss' law for electricity, Gauss' law for magnetism, Faraday's law, and Ampère's law provide a complete description of the electric and magnetic fields produced by any given charge and current distribution. Collectively they are known as Maxwell's equations and, taken with the equations for the electric and magnetic forces on a charge, they provide a complete classical description of electromagnetic phenomena. They are collected and reviewed in Chapter 40.

The force that one charge exerts on another, whether electric or magnetic, depends on the relative positions of the charges. If one charge moves the force it exerts on the other charge does not change immediately. On the contrary, the electric and magnetic fields first change in the near vicinity of the moving charge and these changes propagate as waves. The force on the second charge is not altered until the wave reaches it. Electromagnetic waves travel with the speed of light and, in fact, visible light is such a wave with frequency in a certain range. In Chapter 40 you will learn that the wave-like propagation of electric and magnetic fields is predicted by Maxwell's equations. A detailed description of electromagnetic waves is given in Chapters 41 and 42.

Once you understand the nature of electromagnetic waves you are ready for Chapters 43 through 48, the optics chapters. First you study two fundamental phenomena: reflection and refraction. Two new waves are generated when an electromagnetic wave impinges on the boundary between two regions with different optical properties. One, the reflected wave, propagates back into the region of the incident wave and the other, the refracted wave, continues into the other region. The propagation direction of the refracted wave is different from that of the incident wave and depends on the optical properties of both regions. In Chapter 44 you will learn how the reflecting properties of mirrors and the refracting properties of lenses are used to form images. The ideas are applied to microscopes and telescopes.

Electromagnetic waves obey a superposition principle like other waves: if two or more waves are simultaneously present the total electric field, for example, is the vector sum of the electric fields associated with the individual waves. Superposition gives rise to interference effects, the topic of Chapter 45. If two waves are in phase then they interfere constructively and the resultant amplitude is larger than either constituent amplitude. If their phases differ by 180° they interfere destructively and tend to cancel each other. If the difference in phase comes about because the waves travel different distances the effect is wavelength dependent; destructive interference of red light, for example, occurs at different places than the destructive interference of blue light. Interference is responsible for the colors on soap bubbles and oil films.

Waves passing by the edge of an object travel into the shadow region and produce interference-like fringes in that region: alternating dark and bright bands appear if only a single wavelength is present. Diffraction effects such as this are studied in Chapter 46. Diffraction gratings, which consist of closely spaced, extremely narrow, transparent lines ruled on glass or plastic, are used to analyze the spectrum of light from materials. X rays are diffracted from crystals and the pattern they produce can be used to find the atomic structure of the crystal. Diffraction by gratings and crystals is discussed in Chapter 47.

If the electric field of an electromagnetic wave is always parallel to the same line the wave is said to be linearly polarized. By way of contrast, the field in other waves, said to be circularly polarized, rotates as the wave travels. Some materials, such as are used in many sunglasses, transmit only light that is linearly polarized in a certain direction. Polarization is studied in Chapter 48.

Chapter 27
ELECTRIC CHARGE AND COULOMB'S LAW

I. BASIC CONCEPTS

You start your study of electromagnetism with Coulomb's law, which describes the force that one stationary charged particle exerts on another. Pay attention to both the magnitude and direction of the force and learn how to calculate the total force when more than one charge acts.

Electric charge. Electric charge is defined in terms of electric current, which in turn is defined in terms of the magnetic force exerted by two current carrying wires on each other. You must delay a precise understanding of charge until you study magnetism. For now you should know that charge is a property possessed by some particles and that two charged particles exert electric forces on each other. The SI unit for charge is _Coulomb_ (abbreviated _C_).

There are two types of charge, called <u>positive</u> and <u>negative</u>. In the equations of electromagnetism a positive number is substituted for the value of a positive charge and a negative number is substituted for the value of a negative charge.

The charge on an electron is __−__ C; the charge on a proton is __+__ C. Charge is quantized: the charge on all particles detected so far is a positive or negative multiple of a fundamental unit of charge e. The value of e is _1.60217733 × 10⁻¹⁹ C_.

All macroscopic materials contain enormous numbers of electrons and protons but if the number of electrons in an object equals the number of protons then the net charge is zero and the object is said to be <u>neutral</u> (or uncharged). If there are more electrons than protons the object has a net negative charge; if there are more protons than electrons it has a net positive charge. In either case it said to be <u>charged</u>. Notice that the net charge is computed by algebraically adding the charges of all particles in the object, taking their signs into account. Normally all macroscopic bodies are neutral.

Objects may be given net charges by rubbing them together. When a glass rod is rubbed with silk the rod becomes __+__ charged. When a plastic rod is rubbed with fur the rod becomes __−__ charged.

Charge is conserved. The net charge of a closed system (with no particles entering or leaving) is always the same. After rubbing a glass rod silk becomes __−__ charged and, if no particles leak on or off, the magnitude of its charge is the same as the magnitude of the charge on the rod. Carefully note that it is the algebraic sum of the charges that remains the same. Before rubbing the net charge was zero. After rubbing it is still zero, the sum of the negative charge on one object and the positive charge on the other.

The force that one point charge exerts on another is along the line that joins the charges. If the charges have the same sign then the force is one of repulsion; if the charges have different signs then the force is one of attraction. The diagram below shows three possible pairings.

At each charge draw a vector that indicates the force exerted on it by the other charge in the pair.

Conductors, insulators, and semiconductors. The *electrons* in materials move and are transferred to or from other objects by rubbing; the atomic nuclei (containing protons) are nearly immobile. Materials are often classified according the freedom with which electrons can move. Electrons in an insulator are not free to move far; any charge placed on an insulator remains where it is placed. On the other hand, conductors contain many electrons that are free to move throughout the conductor. When you touch a charged conductor to an uncharged conductor charge is transferred and, as a result, both conductors are charged. When you touch a charged insulator to a neutral conductor very little, if any, charge is transferred. Scraping or rubbing is required to transfer charge to or from an insulator.

Give some examples of conductors and insulators.
conductors: _Copper, metal, water, human body_
insulators: _Rubber, plastic, glass_

Your body is a moderately good conductor, particularly if there is moisture on it. Although you can charge a conductor by rubbing it, you cannot touch it while you do or else the charge will move to your body. Conductors in the laboratory are usually equipped with insulating handles.

Conductors have large numbers of nearly free electrons, typically __10^{23}__ electrons/m^3. Most insulators also have nearly free electrons but at room temperature the concentration is small, about __1__ electrons/m^3. Semiconductors are intermediate between insulators and conductors, with a nearly free electron concentration of about __$10^{10} - 10^{12}$__ electrons/m^3 at room temperature.

Electrical force. Two point particles, one with charge q_1 and the other with charge q_2, are a distance r apart. The magnitude of the force exerted by either of the charges on the other is given by

$$F = k \frac{q_1 q_2}{r^2} \qquad k = \frac{1}{4\pi\epsilon_0} = 8.99 \times 10^9 \ \text{N}\cdot\text{m}^2/\text{C}^2$$

You should write this equation with the factor $1/4\pi\epsilon_0$. The constant of nature ϵ_0 is called _permittivity constant_ and has the value $\epsilon_0 = $ _8.85×10^{-12}_ C^2/N·m^2 (to three significant figures). The factor $1/4\pi\epsilon_0$ has the value _8.99×10^9_ N·m^2/C^2 (to three significant figures). When you use this force equation you must remember to enter the magnitudes of the charges. If you also enter their signs you might get a negative result, which is not correct for the magnitude of a vector.

Notice that the magnitude of the force is proportional to the inverse square of the distance between the charges and is also proportional to the product of the magnitudes of the charges. If the distance between the charges is doubled the force is reduced by a factor of __$1/2$__; if one of the charges is doubled the force increases by a factor of __2__; if both charges are doubled the force increases by a factor of __4__.

You must know how to express the direction of the force in terms of a unit vector in the direction from the charge that is exerting the force toward the charge that is experiencing the force. Suppose you wish to write an equation for the force $\mathbf{F}_{12}$ of charge 2 on charge 1. Let $\hat{\mathbf{r}}_{12}$ be a unit vector directed from charge 2 toward charge 1. Then

$$\mathbf{F}_{12} = \frac{1}{4\pi\epsilon_0} \frac{q_1 q_2}{r_{12}^2} \hat{\mathbf{r}}_{12}$$

If $\mathbf{r}_{12}$ is the displacement of charge 1 from charge 2 then $\hat{\mathbf{r}}_{12} = \mathbf{r}_{12}/r_{12}$. The unit vector $\hat{\mathbf{r}}_{12}$ is dimensionless. Now write the equation for the force of charge 1 on charge 2. Let $\hat{\mathbf{r}}_{21}$ be a unit vector directed from charge 1 toward charge 2. Then

$$\mathbf{F}_{21} = \frac{1}{4\pi\epsilon_0} \frac{q_1 q_2}{r_{21}^2} \hat{\mathbf{r}}_{21}$$

These two equations express Coulomb's law for the interaction of two point charges. Here you substitute the signed values of the charges. Then the equations correctly predict that charges of like sign repel each other and charge of unlike sign attract each other.

Electrostatic forces between two charges obey Newton's third law: they are equal in magnitude and opposite in direction. The equations predict this result since $\hat{\mathbf{r}}_{12} = -\hat{\mathbf{r}}_{21}$.

When more than two charges are present the force on any one of them is the vector sum of the forces due to the others. Each force is computed using Coulomb's law. This is the <u>principle of superposition</u> for electric forces.

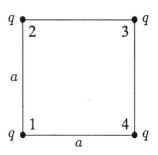

To illustrate the principle of superposition, consider four identical charges q at the corners of the square with edge a, as shown. Develop an expression for the total force on the charge labelled 1 in the diagram. Fill in the following table, giving the magnitude of each force, the x component, and the y component, all in terms of q and a:

	magnitude	x component	y component
force of 2 on 1	$F_{12} = \frac{1}{4\pi\epsilon_0} \frac{q_1 q_2}{r_{12}^2}$		F_{12y}
force of 3 on 1	$F_{13} = \frac{1}{4\pi\epsilon_0} \frac{q_1 q_3}{r_{13}^2}$	F_{13x}	F_{13y}
force of 4 on 1	$F_{14} = \frac{1}{4\pi\epsilon_0} \frac{q_1 q_4}{r_{14}^2}$	F_{14x}	F_{14y}

The x component of the total force on 1 is given by

$$F_x = F_{12x} + F_{13x} + 0$$

and the y component is given by

$$F_y = 0 + F_{13y} + F_{14y}$$

Forces on neutral objects. Charged objects attract neutral macroscopic objects. A neutral conductor is attracted to a charged rod because the electrons within the inductor are redistributed, making some regions positively charged and others negatively charged. The forces on the negative and positive regions are, of course, in opposite directions but the force is greater on the region nearer to the charged object.

The diagram on the right shows a positively charged insulating rod held near a neutral conductor. Indicate the distribution of charge on the conductor and draw arrows of different length to indicate the forces on the positive and negative regions of the conductor. The net force should be toward the rod.

The diagram on the right shows a negatively charged insulating rod held near a neutral conductor. Indicate the distribution of charge on the conductor and draw arrows of different length to indicate the forces on the positive and negative regions of the conductor. The net force should be toward the rod.

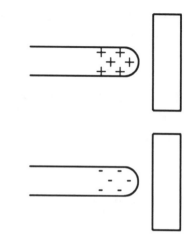

You touch the conductor with the insulating rod nearby. Electrons flow from you to the conductor if the insulating rod is positive or from the conductor to you if the rod is negative. Remove your finger, then remove the rod. The conductor is now charged. It is _____ if the insulating rod was positive and _____ if the rod was negative. This process is called charging by induction.

Charged objects also attract neutral insulators but the force is much smaller than the force of attraction for a conductor. Under the influence of the external charge electrons in an insulator move slightly from their normal orbits so the center of the negative charge is slightly displaced from the center of the positive charge. This results in a net force. A charged rod can be used to pick up small pieces of paper and a rubbed balloon "sticks" to a wall.

II. PROBLEM SOLVING

Some problems ask you to compute the total number of electrons or protons in an object, given the mass of the object. We suppose the object has only one type of atom — it is a chemical element. First calculate how many atoms are in the object. This is given by $N = M/m$, where M is the mass of the object and m is the mass of an atom. The mass of an atom in grams is given by $m = A/N_a$, where A is the atomic mass number and N_a is the Avogadro constant (6.022×10^{23} mol^{-1}). The atomic mass number of an atom is the larger of the two numbers printed in the appropriate box of the periodic table in the appendix of the text. When you evaluate $N = M/m$ be sure the two masses are expressed in the same units.

Now you can calculate the number of electrons and protons in the object. Each atom has of Z electrons and Z protons, where Z is the atomic number (the smaller of the two numbers in the appropriate box of the periodic table), so the number of electrons in the object is ZN. The number of protons is the same. Finally, the total negative charge in the object is $-eZN$ and the total positive charge is $+eZN$.

PROBLEM 1. What is the mass of an iron atom? How many atoms are in 2.50 kg of iron? How many electrons? What is the total charge on the electrons?

SOLUTION: According to the periodic table, iron (chemical symbol Fe) has atomic mass number $A = 55.874$ and atomic number $Z = 26$. The mass of a mole of iron atoms is 55.874×10^{-3} kg so the mass of a single iron atom is $m = 55.874 \times 10^{-3}/6.022 \times 10^{23} = 9.28 \times 10^{-26}$ kg. Use $N = M/m$ to compute the number of atoms, ZN to compute the number of electrons, and $-eZN$ to compute the total charge on the electrons.

[ans: 9.28×10^{-26} kg; 2.69×10^{25}; 7.01×10^{26}; 1.12×10^8 C]

To solve some problems you must manipulate the scalar form of Coulomb's law. In other problems the direction of the force is important.

PROBLEM 2. What is the magnitude of the force exerted by a proton on an electron 1.0 m away?

SOLUTION: Use $F = (1/4\pi\epsilon_0)e^2/r^2$.

[ans: 2.30×10^{-28} N]

How near to the proton should the electron be brought so the force on it has a magnitude of 2.0 N?

SOLUTION: Solve $F = (1/4\pi\epsilon_0)e^2/r^2$ for r.

[ans: 1.07×10^{-14} m]

PROBLEM 3. A positive charge q_1 is at the origin and a second charge q_2 ($= +2.3 \times 10^{-9}$ C) is on the x axis at $x = 1.3$ m. Where can a third charge q_3 ($= -6.2 \times 10^{-9}$ C) be placed so the net force on q_1 is zero?

SOLUTION: Since the force of the second charge is in the negative x direction the third charge must be on the x axis and it must be positioned so the force it exerts on q_1 is in the positive x direction. Since the third charge is negative and therefore attracts the charge at the origin its x coordinate must be positive. If x_2 is the coordinate of the second charge and x_3 is the coordinate of the third then the net force on the charge at the origin is

$$F = \frac{q_1}{4\pi\epsilon_0}\left[\frac{q_2}{x_2^2} + \frac{q_3}{x_3^2}\right]$$

Set this equal to zero and solve for x_3. Notice that the answer does not depend on the value of q_1.

[ans: 2.13 m]

PROBLEM 4. A 3.8×10^{-9}-C charge is at the origin and a 8.7×10^{-9}-C charge is on the x axis at $x = 0.25$ m. A third charge is placed on the x axis so that the force on each of the three charges is zero. What is the third charge and where is it located?

SOLUTION: Clearly, the third charge is negative and lies between the other two. Try the other possibilities to convince yourself they can't do the job. Draw a diagram to indicate the positions of the charges.

Let q_1 represent the charge at the origin, q_2 the second charge, and q_3 the third. Let x_2 be the coordinate of the second charge and x_3 be the coordinate of the third. For the force on the first charge to vanish $[q_3/x_3^2] + [q_2/x_2^2] = 0$; for the force on the second to vanish $[q_3/(x_2 - x_3)^2] + [q_1/x_2^2] = 0$; and for the force on the third to vanish $[q_1/x_3^2] - [q_2/(x_2 - x_3)^2] = 0$. Signed values for the charges are substituted into these equations. The third equation can be solved for x_3. You will get two solutions, but only one makes physical sense. Either of the other equations can be solved for q_3. Use the remaining equation to check that the conditions of the problem are met by your solution.

$[\text{ans: } x_3 = 0.0995 \text{ m}; q_3 = -1.38 \times 10^{-9} \text{ C}]$

III. NOTES

Chapter 28
THE ELECTRIC FIELD

I. BASIC CONCEPTS

Here the idea of an electric field is introduced and used to describe electrical interactions between charges. It is fundamental to our understanding of electromagnetic phenomena and is used extensively throughout the rest of this course. Build a firm foundation by learning well the material of this chapter.

The electric field. A <u>field</u> is an important concept in physics. It gives some property of a region as a function of position. An example of a scalar field is _____ _____ ; an example of a vector field is _____ .

To describe the electric force that one charge exerts on another, we associate an <u>electric field</u> with the first charge and say that this field exerts a force on the second charge. Consider two charges Q and q, for example. Charge Q creates an electric field and that field exerts a force on q. Charge q also creates an electric field and *that* field exerts a force on Q. The electric field of a charge exists throughout all space, so anywhere a second charge is placed it will experience a force.

The electric field at any point in space is defined as the electric force per unit charge on a stationary positive test charge placed at that point, in the limit as the test charge becomes vanishingly small. If the test charge is q_0 and **F** is the electric force on it then the electric field at the position of the test charge is

$$\mathbf{E} =$$

You must be aware that this field is *not* the field created by q_0; it is the total field created by all *other* charges. The test charge does not exert a force on itself. The SI units of the electric field are _____ .

Explain why the limit as q_0 becomes vanishingly small must be included in the definition: _____
_____

Suppose a collection of stationary charges creates an electric field **E** and a charge q is placed in this field. Then the force exerted on q is given by

$$\mathbf{F} =$$

If **E** varies from place to place, as it usually does, then you use the value at the position of q.

Lines of force. Electric fields are often represented in diagrams by <u>lines of force</u> (also called electric field lines). These lines graphically depict both the direction and magnitude of the electric field throughout a region of space. Tell how the direction of the electric field at a point is related to the line of force through that point: _____

Arrows are placed on field lines to indicate the direction of the field but electric field lines themselves are *not* vectors.

Tell how the magnitude of the electric field at a point is related to the lines of force in the vicinity of the point: _____

Lines of force emanate from _____ charge and terminate on _____ charge.

To see the lines of force for some charge distributions look carefully at Figs. 5, 6, 7, and 8 of the text. Be aware that the configuration of lines is different for different charge distributions: the electric field of a point charge, for example, is *not* the same as the electric field of a dipole. For each figure notice the directions of the lines, where they are close together, and where they are far apart. You should recognize that the number of field lines is proportional to the charge. If all charge is doubled then the number of lines doubles in every region.

The diagram to the right shows some lines of force. Indicate on it a region where the magnitude of the electric field is relatively large and a region where it is relatively small. Suppose an electron is at the point marked •. Draw an arrow to indicate the direction of the force on it and label the arrow **F**.

Point charges. The magnitude of the electric field at a point a distance r from a single point charge q is given by

$$E = $$

The field is along the line that joins the charge q and the point. If q is positive it points away from q, if q is negative it points toward q.

If more than one point charge is responsible for the electric field you calculate the total field by summing the fields due to the individual charges. Don't forget that electric fields are vectors and the sum is a vector sum. You can sum the magnitudes only if the fields are in the same direction.

In any given situation you must distinguish between the total electric field and the field that acts on a given charge. Suppose, for example, there are three charges q_1, q_2, and q_3. The total field at any given point is the vector sum of the fields due to the three charges. The field that acts on q_1, however, is the vector sum of the fields due to q_2 and q_3 only.

Electric fields of continuous distributions. A continuous charge distribution is characterized by a <u>charge</u> <u>density</u>. If the charge is on a line (straight or curved) the appropriate charge density is the _____ charge density, denoted by _____ and defined so that the charge dq in an infinitesimal segment ds of the line is given by $dq = $ _____. If the charge is distributed on a surface the appropriate charge density is the _____ charge density, denoted by _____ and defined so that the charge dq in an infinitesimal element of area dA is given by $dq = $ _____. If the charge is distributed in a volume the appropriate charge density is the _____ charge density, denoted by _____ and defined so that the charge dq in an infinitesimal element of volume dV is given by dq = _____. A uniform charge distribution means that the appropriate charge density is the same everywhere along the line, on the area, or in the volume.

Carefully study the derivations of expressions for the electric field on the axis of a uniform ring of charge and on the axis of a uniform disk of charge. You should understand why the field is parallel to the axis in each case.

The diagram on the right shows an infinitesimal segment ds of the ring and the field it produces at point P on the axis, a distance z from the ring center. On the diagram show the location of another infinitesimal segment such that the sum of its field and the field shown is along the axis. Draw the electric field vector at P for this segment. Since every infinitesimal segment of the ring can be paired in this way with another segment, the total field must be along the axis.

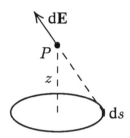

If the ring has radius R and total charge q then the linear charge density is λ = _____, the charge in an infinitesimal segment ds is dq = _____, the magnitude of the field produced by the segment at P is

$$dE =$$

and the component along the axis is

$$dE_\parallel =$$

You must now integrate around the ring. The integration, of course, sums the components along the axis of the fields due to the ring segments. The result for the total field is

$$E =$$

To calculate the field produced by a uniform disk of charge, the disk is divided into rings, each having infinitesimal width ds. The area of a ring of radius s is dA = _____ and if σ is the area charge density then dq = _____ is the charge in a such a ring. The magnitude of the field it produces at a point on the axis a distance z from the ring center is

$$dE =$$

To find an expression for the total field this expression is integrated from $s = 0$ to $s = R$. The result is:

$$E =$$

By taking R to be much larger than z you can find an expression for the field produced by a uniform plane of charge. Notice that if $R \gg z$ then $z/\sqrt{x^2 + R^2}$ is much smaller than 1 and

$$E =$$

Chapter 28: The Electric Field

You will find this expression extremely useful in your study of later chapters. It is a good approximation to the field of a finite plane of charge at points that are close to the plane and far from the edges. Notice that the field does not depend on distance from the plane.

Note that at points near a ring, disk, or plane superposition produces total fields that are quite different from each other and quite different from the field of a point charge. At points far from a charge distribution, however, the electric field tends to become like that of a point particle with charge equal to the net charge in the distribution. For a ring of charge with uniform linear density λ the net charge is $q =$ _____ and the field far away ($z \gg R$) is given by $E =$ _____. For a disk with uniform area density σ the net charge is $q =$ _____ and the field far away ($z \gg R$) is given by $E =$ _____.

Electric dipoles. The geometry is shown in Fig. 4 of the text. An electric dipole consists of _____

It is characterized by a dipole moment **p**, a vector with magnitude given by $p =$ _____, where q is _____
and d is _____.
The direction of **p** is _____.
Look carefully at Fig. 8 of the text to see the lines of force produced by a dipole.

Go over the derivation of Eq. 9 in the text for the electric field at point P in Fig. 4. Be sure you understand why the field there is in the negative y direction. In terms of q, x, and d the magnitude of the field produced at P by the positive charge is given by $E_+ =$ _____ and the magnitude of the field produced there by the negative charge is given by $E_- =$ _____. Both these expressions must be multiplied by $\cos\theta$ to find the y components. The results are then summed to produce an expression for the total field. It is

$$E =$$

We shall often be interested in atomic dipoles, for which the separation of the charges is considerably less than an angstrom and much less than x. Use the binomial theorem to expand $[x^2 + (d/2)^2]^{3/2}$ and retain only those terms that are proportional to d. In terms of the dipole moment the field is

$$E =$$

Notice that the field is inversely proportional to x^3, not x^2.

Consider a dipole in a uniform electric field **E** (created by other charges). The net force on the dipole is _____ because _____.

The dipole does, however, experience a torque. If **p** is the dipole moment and **E** is the electric field then the torque is given by

$$\tau =$$

This torque tends to rotate the dipole so that **p** _____.

When an electric dipole rotates in an electric field the field does work on it. For a rotation during which the angle between the dipole and field changes from θ_0 to θ the work done by the field is $W = $ _____. The sign of the work is _____ if the angle increases and _____ if the angle decreases. During a rotation the potential energy of the field and dipole changes, the change being given by $\Delta U = $ _____. If we take the potential energy to be zero when the dipole moment is perpendicular to the field the potential energy for any orientation can be written as the scalar product $U = $ _____. The potential energy is a maximum when _____
and is a minimum when _____.

Suppose a dipole is initially oriented with its moment perpendicular to an electric field. It is then rotated through 90° so it is in the same direction as the field. During this process the sign of the work done by the field is _____ and the sign of the change in the potential energy is _____. Suppose, instead, that the dipole is rotated so its direction becomes opposite to that of the field. During this process the sign of the work done by the field is _____ and the sign of the change in the potential energy is _____.

II. PROBLEM SOLVING

Some problems involve simple manipulation of the expression for the electric field of a point charge. In other cases you must evaluate the vector sum of two or more fields.

PROBLEM 1. What is the magnitude of the electric field 0.50 m from a 3.1×10^{-9} C point charge?

SOLUTION: Use $E = (1/4\pi\epsilon_o)q/r^2$.

[ans: 1.11×10^2 N/C]

How far from the charge is the field half this value?

SOLUTION: Solve $E = (1/4\pi\epsilon_o)q/r^2$ for r.

[ans: 0.707 m]

PROBLEM 2. Charge $q_1 = 7.9 \times 10^{-9}$ C is at the origin and charge $q_2 = -7.9 \times 10^{-9}$ C is on the x axis at $x = 3.8$ cm. What is the electric field midway between the charges?

SOLUTION: On the axis to the right locate the charges and the point at $x = 1.9$ cm. Draw the individual electric field vectors at the point. Both fields are in the positive x direction and they have the same magnitude. Calculate the magnitudes of the individual fields, then sum the x components.

[ans: 3.93×10^5 N/C, in the positive x direction]

What is the total electric field on the x axis at $x = 5.7$ cm?

SOLUTION: Locate the point on the diagram and draw the individual electric field vectors. Note that at the point the field due to q_1 is in the positive x direction while the field due to q_2 is in the negative x direction. Find the magnitudes and sum the x components.

[ans: 1.75×10^5 N/C, in the negative x direction]

What is the force on an electron placed on the x axis at $x = 5.7$ cm?

SOLUTION: Use $\mathbf{F} = q\mathbf{E} = -e\mathbf{E}$.

[ans: 2.80×10^{-14} N, in the positive x direction]

The force is opposite the electric field because the electron has negative charge.

PROBLEM 3. Charges q_1 and q_2 are a distance L apart, as shown. Find an expression for the electric field at point P on the perpendicular bisector of the line joining them, a distance x from the line.

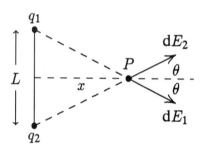

SOLUTION: First find the magnitude of the field due to q_1. Its distance from the point is $[x^2+(L/2)^2]^{1/2}$. Multiply the magnitude by $\cos\theta = x/[x^2+(L/2)^2]^{1/2}$ to find the x component and by $-\sin\theta = -(L/2)/[x^2+(L/2)^2]^{1/2}$ to find the y component. Similarly, find the x and y components of the field due to q_2. Add the x components to find the x component of the total field and add the y components to find the y component of the total field.

[ans: $E_x = (1/4\pi\epsilon_0)(q_1 + q_2)x/[x^2 + (L/2)^2]^{3/2}$; $E_y = (1/4\pi\epsilon_0)(q_2 - q_1)(L/2)/[x^2 + (L/2)^2]^{3/2}$]

What relationship between q_1 and q_2 causes the total field to be parallel to the x axis?

SOLUTION:

[ans: $q_2 = q_1$]

What relationship between q_1 and q_2 causes the total field to be parallel to the y axis?

SOLUTION:

[ans: $q_2 = -q_1$]

What relationship between q_1 and q_2 causes the total field to make an angle of 45° with the x axis?

SOLUTION:

[ans: $q_2 = \pm q_1(x + L/2)/(x - L/2)$]

If $q_1 = 7.8 \times 10^{-9}$ C, $q_2 = -3.4 \times 10^{-9}$ C, $L = 2.2$ cm, and $x = 4.6$ cm, what is the magnitude of the total field and what angle does it make with the x axis?

SOLUTION: Evaluate the expressions for E_x and E_y, then use $E^2 = E_x^2 + E_y^2$ and $\tan\theta = E_y/E_x$ to find the magnitude E and angle θ. When you solve the last equation you must choose between the angle produced by your calculator and that angle plus 180°. Draw a diagram showing the individual fields and use it to determine roughly the direction of the total field. Pick the angle that most closely corresponds to that direction.

[ans: 2.60×10^4 N/C; $-31.3°$]

You should also know how to compute the electric field produced by a continuous distribution of charge.

PROBLEM 4. Charge is distributed with uniform linear density λ along a line of length L, as shown. What is the electric field at point P, a distance x from the line and a distance ℓ_1 above the bottom of the line?

SOLUTION: Place the coordinate system so the line of charge is on the y axis and P is on the x axis. The line of charge then extends from $y = -\ell_1$ to $y = +\ell_2$, where $\ell_1 + \ell_2 = L$.

Let ds be an infinitesimal segment of the line a distance s from the origin. This segment contains charge $dq = \lambda\,ds$ and is $r = \sqrt{s^2 + x^2}$ distant from P. Thus the magnitude of the field it produces at P is $dE = (1/4\pi\epsilon_0)\lambda\,ds/(s^2 + x^2)$. To find the x component multiply this by $\cos\theta$ and to find the y component multiply it by $-\sin\theta$, where θ is the angle between the field and the x axis. You will need to express the trigonometric functions in terms of s and x. A little geometry shows that $\cos\theta = x/\sqrt{s^2 + x^2}$ and $\sin\theta = s/\sqrt{s^2 + x^2}$. Thus

$$E_x = \frac{1}{4\pi\epsilon_0}\int_{-\ell_1}^{\ell_2}\frac{\lambda\cos\theta\,ds}{s^2 + x^2} = \frac{1}{4\pi\epsilon_0}\int_{-\ell_1}^{\ell_2}\frac{\lambda x\,ds}{(s^2 + x^2)^{3/2}}$$

and

$$E_y = -\frac{1}{4\pi\epsilon_0}\int_{-\ell_1}^{\ell_2}\frac{\lambda\sin\theta\,ds}{s^2 + x^2} = -\frac{1}{4\pi\epsilon_0}\int_{-\ell_1}^{\ell_2}\frac{\lambda s\,ds}{(s^2 + x^2)^{3/2}}$$

The indefinite integrals $\int (s^2 + x^2)^{-3/2} ds = (1/x^2) s (s^2 + x^2)^{-1/2}$ and $\int s(s^2 + x^2)^{-3/2} ds = -(s^2 + x^2)^{-1/2}$ can be used to evaluate the expressions.

$$\left[\text{ans: } E_x = \frac{\lambda}{4\pi\epsilon_0 x} \left[\frac{\ell_1}{[x^2 + \ell_1^2]^{1/2}} + \frac{\ell_2}{[x^2 + \ell_2^2]^{1/2}} \right]; E_y = \frac{\lambda}{4\pi\epsilon_0} \left[\frac{1}{[x^2 + \ell_2^2]^{1/2}} - \frac{1}{[x^2 + \ell_1^2]^{1/2}} \right] \right]$$

Find expressions for the field components on the perpendicular bisector of the line joining the charges, a distance x from the line.

SOLUTION: Take $\ell_1 = \ell_2 = L/2$.

$$\left[\text{ans: } E_x = (1/4\pi\epsilon_0)(\lambda L/x)[x^2 + (L/2)^2]^{1/2}; E_y = 0\right]$$

PROBLEM 5. Charge is distributed around the perimeter of a square with side a, as shown. The linear charge density on the left and top edges is λ_1 and the linear charge density on the right and bottom edges is λ_2. What is the electric field at the center of the square?

SOLUTION: The square center is on the perpendicular bisector of all four edges, so the results of the last problem can be used. If λ_1 and λ_2 are positive the left edge contributes a field to the right, the right edge contributes a field to the left, the upper edge contributes a field that points downward, and the lower edge contributes a field that points upward. The center is $a/2$ distant from each edge. Superpose the fields. Take the x axis to point to the right and the y axis to point upward.

$$\left[\text{ans: } E_x = (1/4\pi\epsilon_0 a)\sqrt{2}(\lambda_1 - \lambda_2); E_y = (1/4\pi\epsilon_0 a)\sqrt{2}(\lambda_2 - \lambda_1) \right]$$

Some geometries involve arcs of circles.

PROBLEM 6. Charge is distributed with uniform linear charge density λ around a semicircle of radius R, as shown. What is the electric field at point P, the center of the semicircle?

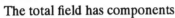

SOLUTION: Consider the line element ds, with charge $dq = \lambda\, ds$. The field it produces at P has magnitude $dE = (1/4\pi\epsilon_0)\lambda\, ds/R^2$, x component $dE_x = (1/4\pi\epsilon_0)\lambda \cos\theta\, ds/R^2$, and y component $E_y = -(1/4\pi\epsilon_0)\lambda \sin\theta\, ds/R^2$. For geometries involving arcs of circles integration is usually simpler if angular variables are used. The segment subtends the angle $d\theta = ds/R$, in radians, at the center of the semicircle, so we replace ds with $R\,d\theta$ and take the limits of integration to be 0 and π radians.

The total field has components
$$E_x = \frac{1}{4\pi\epsilon_0 R}\int_0^\pi \lambda \cos\theta\, d\theta$$

and
$$E_y = -\frac{1}{4\pi\epsilon_0 R}\int_0^\pi \lambda \sin\theta\, d\theta$$

Evaluate these integrals.

[ans: $E_x = 0;\ E_y = -\lambda/2\pi\epsilon_0 R$]

You might have used symmetry to convince yourself that $E_x = 0$ before setting up and evaluating the integral.

If the charge on the semicircle is given rather than the linear charge density you would work the problem in exactly the same way but you would substitute $Q/\pi R$ for λ. Think about similar problems with slightly different geometries. What, for example, would you change in the solution if the charge were distributed along a $90°$ arc? What would you do if the linear charge density is not uniform but is given as a function of θ?

PROBLEM 7. Charge Q is distributed on the semicircle of the last problem in such a way that the linear charge density increases in proportion to the angle θ shown in the diagram. What now is the electric field at P?

SOLUTION: Take $\lambda = A\theta$, where A is a constant, and follow the steps of the last problem to obtain expressions for E_x and E_y in terms of A. You will need the indefinite integrals $\int \theta \cos\theta\, d\theta = \cos\theta + \theta\sin\theta$ and $\int \theta \sin\theta\, d\theta = \sin\theta - \theta\cos\theta$. To complete the solution you will need an expression for A in terms of the total charge Q. Since $Q = \int \lambda\, ds = \int_0^\pi A\theta R\, d\theta = \frac{1}{2}AR\theta^2|_0^\pi = \frac{1}{2}AR\pi^2$, $A = 2Q/\pi^2 R$. Substitute this result into your expressions for E_x and E_y.

[ans: $E_x = -Q/\pi^3\epsilon_0 R^2;\ E_y = -Q/2\pi^2\epsilon_0 R^2$]

PROBLEM 8. Total charge $Q_1 = 7.5 \times 10^{-9}$ C is distributed uniformly over the upper half of a circle with a 3.5-cm radius and total charge $Q_2 = -3.6 \times 10^{-9}$ C is distributed over the lower half. What is the magnitude of the electric field at the center of the circle?

SOLUTION: Use the results found above for a uniform distribution of charge on a semicircle and superpose the fields of two semicircles. Be careful about the directions of the two fields.

[ans: 5.19×10^4 N/C]

PROBLEM 9. The two large parallel rectangular plates, shown edge-on, each have an area of 0.75 m². Charge is distributed uniformly on them, $+4.5 \times 10^{-9}$ C on the left plate and -4.5×10^{-9} C on the right plate. What is the electric field at point A between the plates? What is the electric field at point B?

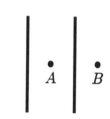

SOLUTION: The electric field due to a uniform distribution of charge on a plane is given by $E = \sigma/2\epsilon_0$, where σ is the area charge density. The field points away from a plane of positive charge and toward a plane of negative charge. Use $\sigma = Q/A$ to calculate the area charge density and superpose the fields due to the two planes, being careful about field directions.

[ans: 339 N/C at A; 0 at B]

Suppose each plate has a charge of $+4.5 \times 10^{-9}$ C. What is the field at A now? at B?

SOLUTION:

[ans: 0 at A; 339 N/C at B]

You should know how to solve problems involving electric dipoles in external fields. The torque on a dipole is given by $\tau = \mathbf{p} \times \mathbf{E}$ and this torque tends to produce an angular acceleration. For rotation about a fixed axis $\tau = I\alpha$, where τ is the component of the torque along the axis, I is the rotational inertia of the dipole, and α is its angular acceleration. The potential energy of a dipole in an external field is given by $U = -\mathbf{p} \cdot \mathbf{E}$. When the dipole rotates the work done by the external field is given by $W = -\Delta U$.

PROBLEM 10. A certain electric dipole consists of two point charges, $+3.5 \times 10^{-6}$ C and -3.5×10^{-6} C, separated by 0.24 mm. The dipole is oriented so the line from the negative charge to the positive charge makes an angle of 25° with a uniform electric field of 25 N/C. What is the magnitude of the dipole moment? What is the magnitude of the torque exerted by the field on the dipole? What is the potential energy of the dipole, relative to the potential energy when the dipole moment is perpendicular to the field?

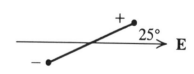

SOLUTION: Use $p = qd$ to calculate the magnitude of the dipole moment, use $\tau = |\mathbf{p} \times \mathbf{E}| = pE \sin\theta$ to calculate the torque, and use $U = -\mathbf{p} \cdot \mathbf{E} = -pE \cos\theta$ to calculate the potential energy.

[ans: 8.4×10^{-10} C·m; 8.87×10^{-9} N·m; -1.90×10^{-8} J]

Suppose each charge has a mass of 140 g. What is the angular acceleration of the dipole?

SOLUTION: The rotational inertia about the center of mass is given by $I = 2m(d/2)^2 = md^2/2$ and the angular acceleration is given by $\alpha = \tau/I$.

[ans: 2.20 rad/s^2]

PROBLEM 11. If the torque of the electric field is the only torque acting a dipole will oscillate about the direction of the field. Suppose the dipole has a moment with magnitude p and a rotational inertia I. Develop an expression for the angular frequency of oscillation in an electric field of magnitude E. Assume the amplitude is sufficiently small that $\sin\theta$ can be approximated by θ in radians.

SOLUTION: The magnitude of the torque is given by $pE \sin\theta$ and can be approximated by $pE\theta$ for small angles, so $-pE\theta = I\, d^2\theta/dt^2$. The negative sign appears because the torque is a restoring torque: if θ is positive then the angular acceleration is negative. Compare this equation to Newton's second law for a mass on a spring: $-kx = m\, d^2x/dt^2$. The angular frequency for the spring-mass system is given by $\omega = \sqrt{k/m}$.

[ans: $\sqrt{pE/I}$]

What is the angular frequency of oscillation for the dipole of the previous problem?

SOLUTION:

[ans: 2.28 rad/s]

If the dipole is released from rest at $\theta = 25°$, what is its angular velocity as rotates past the direction of the field ($\theta = 0$)?

SOLUTION: Note that the initial angle is not small enough to make the small angle approximation valid. Use an energy method rather than Newton's second law. The initial potential energy is $-pE \cos\theta$, the initial kinetic energy is 0. The final potential energy is $-pE$ and the final kinetic energy is $\frac{1}{2}I\omega^2$, where ω is the angular speed. Write the expression for the conservation of energy and solve for ω.

[ans: 0.910 rad/s]

Some problems of this chapter deal with the motion of a charge in an electric field. If the field is uniform then the force it exerts is constant. In many cases the electric field exerts the only force. Then the acceleration of the charge is constant and the kinematic equations for constant acceleration motion are valid. The force is given by $\mathbf{F} = q\mathbf{E}$ and the acceleration by $\mathbf{a} = \mathbf{F}/m = q\mathbf{E}/m$.

PROBLEM 12. A particle with positive charge q and mass m, initially traveling with speed v_0 enters the region between two parallel plates, as shown. A uniform electric field E, perpendicular to the initial velocity, exists between the plates. Take $q = 5.5 \times 10^{-5}$ C, $m = 3.5 \times 10^{-6}$ kg, $d = 2.5$ cm, $\ell = 75$ cm, $v_0 = 4.0 \times 10^2$ m/s, and $E = 850$ N/C. Does the charge strike the upper plate? If it does where does it strike?

SOLUTION: Place the coordinate system with the x axis parallel to the initial velocity of the charge, the y axis along the electric field, and the origin at the left edge of the region, where the charge enters. The coordinates of the charge are given as functions of time by $x(t) = v_0 t$ and $y(t) = \frac{1}{2}(qE/m)t^2$. Set $y(t) = d/2$ and solve for t. Use the result to evaluate $x(t)$. If $x(t)$ is greater than ℓ the charge does not strike the upper plate. If $x(t)$ is less than or equal to ℓ it does and the coordinates of the point where it strikes are $x(t), d/2$.

[ans: it strikes at $x = 54.7$ cm, $y = 1.25$ cm]

What is the minimum initial speed the charge must have to get through the plates without striking them?

SOLUTION: Set $x(t) = \ell$ and $y(t) = d/2$, then solve for v_0.

[ans: 5.48×10^2 m/s]

If it has this initial speed what is its velocity as it leaves the region between the plates?

SOLUTION: Solve for the time it leaves the region then use $v_x = v_0$ and $v_y = at = qEt/m$.

[ans: $v_x = 5.48 \times 10^2$ m/s, $v_y = 18.3$ m/s]

Since no forces act on the charge after it leaves the region between the plates it continues with this velocity.

III. COMPUTER PROJECTS

A computer can be programmed to calculate the electric field due to a line of charge if you know how the charge is distributed along the line. If the charge is on the x axis from x_0 to x_f and the charge density is given by the known function $\lambda(x)$ then the electric field at the point in the xy plane with coordinates x and y is given by

$$E_x = \frac{1}{4\pi\epsilon_0} \int_{x_0}^{x_f} \frac{\lambda(x')(x-x')\,dx'}{\left[(x-x')^2 + y^2\right]^{3/2}},$$

$$E_y = \frac{y}{4\pi\epsilon_0} \int_{x_0}^{x_f} \frac{\lambda(x')\,dx'}{\left[(x-x')^2 + y^2\right]^{3/2}}.$$

The z component is zero. A little thought should lead you to the conclusion that you do not need to consider points with non-vanishing z components.

The Simpson's rule program discussed in the Computer Projects section of Chapter 7 can be used to carry out the integrations. For most problems, however, this technique requires a large number of intervals because the functions $(x-x')/\left[(x-x')^2 + y^2\right]^{3/2}$ and $1/\left[(x-x')^2 + y^2\right]^{3/2}$ in the integrands vary rapidly with x'. On the other hand, λ usually varies much more slowly. In order to avoid this problem divide the x axis from x_0 to x_f into N segments, each of width $\Delta x'$, and approximate λ by a constant in each segment. The integral over each segment can be evaluated analytically. The indefinite integrals are $\int (x-x')\left[(x-x')^2 + y^2\right]^{-3/2}\,dx' = \left[(x-x')^2 + y^2\right]^{-1/2}$ and $\int \left[(x-x')^2 + y^2\right]^{-3/2}\,dx' = -(1/y^2)(x-x')\left[(x-x')^2 + y^2\right]^{-1/2}$. Let an interval start at x_b and evaluate λ at the midpoint of the interval: $\lambda_m = \lambda(x_b + \Delta x'/2)$. Then the components of the electric field are given approximately by

$$E_x = \frac{1}{4\pi\epsilon_0} \sum_{i=1}^{N} \lambda_m \left\{ \frac{1}{\left[(x-x_b-\Delta x')^2 + y^2\right]^{1/2}} - \frac{1}{\left[(x-x_b)^2 + y^2\right]^{1/2}} \right\},$$

$$E_y = -\frac{1}{4\pi\epsilon_0}\frac{1}{y} \sum_{i=1}^{N} \lambda_m \left\{ \frac{(x-x_b-\Delta x')}{\left[(x-x_b-\Delta x')^2 + y^2\right]^{1/2}} - \frac{(x-x_b)}{\left[(x-x_b)^2 + y^2\right]^{-1/2}} \right\}.$$

Write a program to calculate the components of the electric field produced by a line of charge. Input the coordinates x and y and the limits of integration x_0 and x_f, then have the program evaluate the sums and multiply the results by $1/4\pi\epsilon_0$ or $(1/4\pi\epsilon_0)(1/y)$, as appropriate. You will need to supply a programming line to define the function $\lambda(x)$. Here's an outline.

> input number of intervals, limits of integration: N, x_0, x_f
> calculate segment length: $\Delta x' = (x_f - x_0)/N$
> input coordinates: x, y
> set $S_x = 0$, $S_y = 0$
> set $x_b = x_0$

calculate $f_{1x} = 1/\left[(x-x_b)^2 + y^2\right]^{1/2}$
calculate $f_{1y} = (x-x_b)/\left[(x-x_b)^2 + y^2\right]^{1/2}$
begin loop over intervals: counter runs from 1 to N
  set $x_e = x_b + \Delta x'$
  calculate $f_{2x} = 1/\left[(x-x_e)^2 + y^2\right]^{1/2}$
  calculate $f_{2y} = (x-x_e)/\left[(x-x_e)^2 + y^2\right]^{1/2}$
  calculate linear charge density at center of segment: λ_m
  $S_x = S_x + \lambda_m(f_{2x} - f_{1x})$
  $S_y = S_y + \lambda_m(f_{2y} - f_{1y})$
  replace x_b with x_e, f_{1x} with f_{2x}, f_{1y} with f_{2y}
end loop
calculate field components: $E_x = (1/4\pi\epsilon_0)S_x$, $E_y = -(1/4\pi\epsilon_0)(1/y)S_y$
display E_x, E_y
go back to enter another set of coordinates or quit

You will need to run the program several times to select an appropriate value for N. Start with $N = 10$, say, and double it on successive runs. Continue until the results agree to 3 significant figures.

PROJECT 1. Test the program by considering a line of uniform charge density. A line of charge runs along the x axis from the origin to $x = 0.10$ m. Suppose the line contains 5.5×10^{-9} C of charge, distributed uniformly (λ = constant). Use the program to calculate the electric field components at $x = 0$, $y = 0.05$ m. Evaluate the analytic expressions and compare answers.

Now suppose the same total charge is distributed on the same line with a linear density that is given by $\lambda = Ax^2$, where A is a constant. First show that $\lambda = 1.65 \times 10^{-5}x^2$ C/m for x in meters. Then use the program to find the electric field components at points along the line $y = 0.050$ m. Take points every 0.020 m from $x = -0.060$ m to $x = 0.100$ m.

Estimate the value of x for which the electric field is in the y direction. Explain why it is not on the center line of the wire, $x = 0.050$ m.

At points far away from the line of charge the field tends to become like that of a point charge. Use the program to calculate the field components along the line $x = 0$. Take $y = 0.10, 1.0, 10, 100$, and 1000 m. Modify the program so it also calculates the electric field at the same points for a point charge $q = 5.5 \times 10^{-9}$ C at the origin. Compare the fields of the line and point charge. Does the field of the line become more like that of the point charge at far away points?

Given a charge distribution that creates an electric field, a computer can be used to plot the electric field lines. We consider a distribution of point charges. It is usual to start at a point close to one of the charges, where the field line is along the line that joins the point and the charge. The electric field at the point is calculated and the result is used to locate a neighboring point on the same field line. It is a short distance away in the direction of the electric field vector. The process is then continued to locate other points on the same field line.

For simplicity we will deal with charges, fields, and field lines in the x, y plane. Suppose the electric field at a point with coordinates x and y has components E_x and E_y and we wish to find another point on the field line through x, y. $(E_x/E)\mathbf{i} + (E_y/E)\mathbf{j}$ is a unit vector tangent to the field line and $x + (E_x/E)\Delta s$, $y + (E_y/E)\Delta s$ are the coordinates of a point on the line

Δs distant from x, y. The following is the outline of a program that calculates a sequence of points on a single field line. You must supply the coordinates x_0, y_0 of the first point, the distance Δs between points, and the program instructions to calculate the electric field.

> input distance between points, number of points: $\Delta s, N$
> input the coordinates of the first point: x_0, y_0
> set $x_b = x_0, y_b = y_0$
> plot x_b, y_b
> **begin loop** over points; counter runs from 1 to N
>   calculate field components at x_b, y_b: E_x, E_y
>   calculate magnitude of field at x_b, y_b: $E = \sqrt{E_x^2 + E_y^2}$
>   calculate coordinates of new point:
> $x_e = x_b + (E_x/E)\Delta s$
> $y_e = y_b + (E_y/E)\Delta s$
>   plot x_e, y_e
>   set $x_b = x_e, y_b = y_e$
> **end loop** over points
> go back to get another starting point or quit

You may want to provide instructions so the field lines are plotted on the monitor screen. Alternatively, you may have the computer display the coordinates of the points so you can plot the line by hand. You should realize that the lines generated are approximate but the approximation becomes more accurate as Δs is made smaller. Do not make Δs so small that significance is lost in the calculation. If you plot by hand you will not want to display and plot every point, particularly if Δs is small. Add program instructions so that only every 10 or 20 calculated points are displayed.

You might also want to stop plotting when the line gets close to a charge. One way to do this is to save the coordinates of the charges, then check x_e and y_e to see if they are near a corresponding charge coordinate. If they are, have the computer go to the last line of the program.

PROJECT 2. Check the program by considering a dipole. Charge $q_1 = 7.1 \times 10^{-9}$ C is located at the origin and charge $q_2 = -7.1 \times 10^{-9}$ C is located on the y axis at $y = -0.40$ m. Use the program to plot 4 field lines. Start one at $x = 1.0 \times 10^{-3}$ m, $y = 1.0 \times 10^{-3}$ m, the second at $x = 1.0 \times 10^{-3}$ m, $y = -1.0 \times 10^{-3}$ m, the third at $x = -1.0 \times 10^{-3}$ m, $y = -1.0 \times 10^{-3}$ m, and the fourth at $x = -1.0 \times 10^{-3}$ m, $y = 1.0 \times 10^{-3}$ m. Take $\Delta s = 0.005$ m and plot about 200 points on each line. Do these lines look like the lines of a dipole?

PROJECT 3. Four identical charges, each with $q = 5.6 \times 10^{-9}$ C, are placed at the corners of a square with edge length $a = 0.36$ m. The square is centered at the origin and its sides are parallel to the coordinate axes. Take $\Delta s = 0.005$ m and use the program to plot 6 field lines, each starting on a circle of radius 0.10 m centered on the charge at $x = 0.18$ m, $y = 0.18$ m. One line starts parallel to the x axis and the others start at intervals of $\pi/3$ radians around the circle. When a line is within 0.10 m of any charge or more than 0.70 m away from all charges, do not continue it. One of the lines goes toward the center of the square, where the electric field vanishes. The program does not properly evaluate E_x/E and E_y/E in the limit of vanishing field. Stop plotting when the field becomes less than 10^{-4} times the field at the initial points.

Plot 6 field lines emanating from each of the other charges. This can be done using the data generated for the first charge and a symmetry argument. It is not necessary to calculate new points.

Far away from all charges the distribution has the same electric field as that of a single charge equal to the net charge in the distribution and located at the origin. The field lines are then radially outward if the net charge does not vanish and is positive. Look at your plot and notice that the lines tend to become more nearly in the radial direction as the distance from the origin increases. Because of the choice of starting points for the lines, they also tend to be arranged with equal angles between adjacent lines. You should see 24 lines, all nearly in a radial direction, with adjacent lines separated by $\pi/12$ radians. Do you?

Now change the sign of the charges at $x = -0.18\,\text{m}$, $y = 0.18\,\text{m}$ and at $x = 0.18\,\text{m}$, $y = -0.18\,\text{m}$ and use the program to plot lines starting at the same points as before. Notice that now the net charge in the distribution is zero and that all lines start and stop at charges within the distribution.

IV. NOTES

Chapter 29
GAUSS' LAW

I. BASIC CONCEPTS

Gauss' law is one of the four fundamental laws of electromagnetism. It relates an electric field to the charges that create it and is therefore closely akin to Coulomb's law. Unlike Coulomb's law, however, it is valid when the charges are moving, even at relativistic speeds. The central concept for an understanding of Gauss' law is that of electric flux, a quantity that is proportional to the number of field lines penetrating a given surface. You should pay careful attention to the definition of flux and learn how to compute it for various fields and surfaces. You should then learn how to use Gauss' law to compute the charge in any region if the electric field is known on the boundary and also how to use the law to compute the electric field in certain highly symmetric situations.

Electric flux. Electric flux is associated with any surface (real or imaginary) through which electric field lines pass. It is defined by the integral

$$\Phi_E = $$

where $\mathbf{E}$ is the electric field on the surface and $d\mathbf{A}$ is an infinitesimal element of surface area. This integral tells us to evaluate the scalar product $\mathbf{E} \cdot d\mathbf{A}$ for each element of the surface and sum the results for all elements. Notice that the infinitesimal area element is written as a vector. What is its direction? _____
For an open surface, such as the top of a table, $d\mathbf{A}$ may be in either of the two directions that are perpendicular to the surface (up or down for a horizontal surface). Which one you pick determines the sign but not the magnitude of the flux. For a closed surface, one that completely surrounds a volume, $d\mathbf{A}$ is always chosen to be _____.

The scalar product that appears as the integrand may be interpreted as the product of the infinitesimal area and the component of the field normal to the surface. Thus only that part of the field that pierces the surface contributes to the flux. Consider two identical surfaces and suppose a uniform electric field exists at each of them. If the fields have the same normal components then the flux is the same even if they have different tangential components. A field that is parallel to a surface does not contribute to the flux through that surface.

The flux through a surface is proportional to the number of field lines that penetrate the surface. Recall that the number of lines per unit area through an infinitesimal area dA perpendicular to the field is proportional to the magnitude of the field. Thus if $d\mathbf{A}$ is in the same direction as the field the number of lines that penetrate it is proportional to $E\,dA$. If the area is rotated so $d\mathbf{A}$ makes the angle ϕ with the field the number of lines that penetrate the area decreases to $E\,dA\cos\phi$. Carefully study Fig. 1 of the text.

When studying this and subsequent chapters you may think of the electric flux through a surface as giving the net number of lines crossing the surface. There are, however, two important distinctions you should make. First, Φ_E is not necessarily an integer. Second, Φ_E is negative if the field makes an angle of more than 90° with dA or, what is the same thing, the field lines cross the surface in a direction roughly opposite to dA. Field lines that cross a closed surface from inside to outside make a _____ contribution to Φ_E while lines that cross the surface from outside to inside make a _____ contribution. Since some parts of a closed surface may make positive contributions to the flux while other parts make negative contributions, the total flux through a surface may vanish even though an electric field exists at every point on the surface.

Gauss' law. Gauss' law states that for any closed surface the total flux through the surface is proportional to the net charge enclosed by the surface. The constant of proportionality involves ϵ_0, the permittivity constant. Copy Eq. 9 of the text here:

When you use the law be sure to evaluate the integral for a *closed* surface. The small circle on the integral sign is a reminder. The surface is known as a Gaussian surface. When you use Gauss' law to solve a problem you must identify the Gaussian surface you are using. It might be the physical surface of an object or it might be a purely imaginary construction. Also be very careful that the charge you use in the Gauss' law equation is the net charge *enclosed* by the Gaussian surface. Charge outside does not contribute to the total flux through a closed surface, although it does contribute to the electric field at the surface.

The law should not surprise you since the magnitude of the electric field and the number of field lines associated with any charge are both proportional to the charge. All lines from a single positive charge within a closed surface penetrate the surface from inside to outside, so the total flux is positive and proportional to the charge. All lines associated with a single negative charge within a closed surface penetrate from outside to inside so the total flux is negative and again proportional to the charge. Some lines associated with a charge outside a surface penetrate the surface but those that do penetrate twice, once from outside to inside and once from inside to outside, so this charge does not contribute to the total flux through the surface. When more than one charge is present we add the individual flux contributions, with their appropriate signs.

You should be aware that the electric field at points on a closed surface containing a net charge of zero is not necessarily zero. Each charge, whether in the interior or exterior, creates a field that does not vanish on the surface. Nevertheless, a net charge of zero inside means a total flux of zero. Zero total flux simply means there are just as many field lines entering the volume enclosed by the surface as there are leaving.

Gauss' law can be used to find an expression for the electric field of a point charge. Imagine a sphere of radius r with a positive point charge q at its center. Since the electric field is radially outward from the charge, the normal component of the field at any point on the surface of the sphere is the same as the magnitude E of the field. In terms of E and r the total flux through the sphere is $\Phi_E =$ _____. Equate this to q/ϵ_0 and solve for E. The result is: $E =$ _____, in agreement with Coulomb's law.

Gauss's law can be written in differential form and used to solve for the electric field of any distribution of charge. In its integral form it is useful in several ways. If the electric field is known at all points on a surface the law can be used to calculate the net charge enclosed by the surface. If the charge is known the law can be used to find the total electric flux through the surface. If the charge distribution is highly symmetric so a symmetry argument can be used to show that $E\cos\theta$ has the same value at all points on the surface then Gauss' law can be used to solve for the electric field at points on the surface.

Gauss' law and conductors. The electric field vanishes at all points in the interior of a conductor in electrostatic equilibrium, with all charge stationary. What would happen if this were not true? _____

_____

Remember that $\mathbf{E} = 0$ inside a conductor even when excess charge is placed on it or when an external field is applied to it.

At electrostatic equilibrium the electric field at all points just outside a conductor is perpendicular to the surface. It has only a normal, not a tangential component. The charge on the surface can bring about a change in the normal component of the field so it is zero inside and non-zero outside but it cannot change the tangential component. Since the tangential component must be zero inside it must also be zero just outside.

The condition that the electric field is zero leads, via Gauss' law, to an interesting property of conductors: in an electrostatic situation any excess charge on a conductor must reside on its surface; there can be no net charge in its interior. Imagine a Gaussian surface that is completely within the conductor. The electric field is _____ at every point on the Gaussian surface, so the total flux through the surface is _____ and, according to Gauss' law, the net charge enclosed by the surface is _____. Any net charge in the conductor must lie outside the Gaussian surface. Since this result is true for *every* Gaussian surface that can be drawn completely within the conductor, no matter how close to its surface, we conclude that any excess charge must be on the surface of the conductor.

You should understand conductors with cavities. The diagram on the right shows the cross section of a conductor with a cavity containing a point charge q_1. The net charge within the Gaussian surface indicated by the dotted line is _____ so if q_2 is the charge on the inner surface of the conductor $q_1 + q_2 =$ _____ and $q_2 =$ _____. Furthermore if Q is the total excess charge on the conductor then $q_2 + q_3 =$ _____ and $q_3 =$ _____.

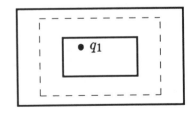

Note that when there is charge in the cavity there must be charge on the inner surface of the conductor. If the net charge on the conductor is zero the charge on its outer surface is _____. If $Q =$ _____ there is no net charge on the outer surface of the conductor. Notice that Gauss' law cannot tell us the distribution of charge on a surface, only the net charge there.

You should realize that the area charge density may not be the same at every point on the surface of a conductor. Whether it is or not the magnitude of the electric field at any point just outside is directly proportional to the area charge density at the corresponding point on the surface. The exact relationship is $E = $ _____, where σ is the area charge density.

Gauss' law can be used to prove this result. Consider a small portion of the surface of a conductor and suppose its area is A and its area charge density is σ. The electric field is normal to the surface. Describe the Gaussian surface you will use: _____

_____

One end of the Gaussian surface should be inside the conductor. The flux through this end is _____. Another end should be outside the conductor and perpendicular to the field. If the magnitude of the field is E the flux through this end is _____. The sides are parallel to the field so the flux through them is _____. In terms of E and A the total flux through the Gaussian surface is _____.

In terms of σ and A the net charge enclosed by the surface is _____. Gauss' law yields _____ = _____. $E = \sigma/\epsilon_0$ follows immediately. You should realize that this field is produced by the charge on *all* parts of the surface and by external sources if they are present.

Calculating the electric field. Successful use of Gauss' law to solve for the electric field depends greatly on choosing the right surface. First of all, it must pass through the point at which you want the field. Second, either the electric field must have constant magnitude over the entire surface or else it must have constant magnitude over part of the surface and the flux through the other parts must be zero. A symmetry argument to justify the use of the Gaussian surface you have chosen should be included in the solution of every problem of this type.

To understand how such an argument is made consider a long straight wire, with positive charge distributed uniformly on it. Take the wire to be infinitely long and use symmetry to show that the electric field at any point is radially outward from the wire. The outline of the argument is: pretend that the field is *not* radially outward and show that this leads to a contradiction of the principle that identical charge distributions produce identical electric fields.

Assume the field at point P is in the direction shown in the upper diagram, a non-radial direction. Now turn the wire over end-for-end, through a 180° rotation about the dotted line through P. If the direction chosen for the field is correct, then after turning the wire the field will be in the direction shown in the lower diagram. This is impossible since the charge distribution is exactly the same in the two cases. A radial field is the only one that does not change when the wire is turned.

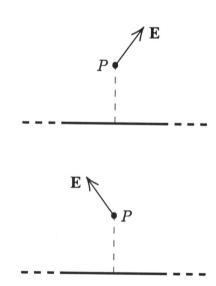

Note that if the wire is not infinite the argument works only if P is opposite the midpoint of the wire. Similarly the argument does not work (and the field is not radial) if the charge is not distributed uniformly on the wire. In those cases turning the wire changes the distribution of charge.

For an infinite wire with a uniform charge distribution the electric field must have the same magnitude at all points that are the same distance from the wire. The charge distribution looks exactly the same from point P as it does from a point 2 m (or 2 km) further along the wire, the same distance from the wire.

The points where the magnitude of the field are the same as at P form the rounded portion of the surface of a cylinder, so take a cylinder of length h and radius r, with P on the surface, for your Gaussian surface. In the space to the right draw a diagram of the wire and the Gaussian surface. Label dA and **E** at point P.

The completion of the problem is now straightforward. First evaluate the integral $\int \mathbf{E} \cdot d\mathbf{A}$ over the rounded portion of the cylinder. Since **E** and d**A** are parallel at all points on the surface $\mathbf{E} \cdot d\mathbf{A} = $ _____ and since E is the same for all points on the surface $\int \mathbf{E} \cdot d\mathbf{A} = E \int dA = $ _____, in terms of r and h. Since a Gaussian surface must be closed you must also consider the circles that form the ends of the cylinder. No flux passes through either end because _____.

Solve the Gauss' law equation. If q is the net charge enclosed by the cylinder, then the magnitude of the electric field is given by $E = $ _____. Usually the linear charge density λ of the wire is given. Since this is q/h, $E = $ _____.

Now consider some of the other charge distributions discussed in the text. Think of these derivations in two ways. First, they will give you a chance to practice solving problems using Gauss' law. Second, in some cases the results are useful for later work. You should particularly remember the expression for the electric field of a large plane sheet of charge. You will use it several times when you study capacitors.

Consider an infinite sheet with uniform area charge density σ. In the space to the right sketch a portion of the sheet and the Gaussian surface used to find the electric field at a point away from the sheet. The field is perpendicular to the sheet and points away from it if it is positive. Draw a vector indicating the electric field at a point on the Gaussian surface. Also label the parts of the surface through which the flux is zero.

In terms of the magnitude E of the field and area A of one end of the Gaussian surface, the total flux through the surface is $\Phi_E = $ _____. In terms of σ and A the total charge enclosed by the surface is _____. Gauss' law thus gives the following expression for the electric field in terms of the area charge density: $E = $ _____.

You might think this result is inconsistent with the expression for the magnitude of the electric field just outside a conductor, $E = \sigma/\epsilon_0$. It is not. Consider an infinite plane conducting sheet with a uniform charge density σ on one surface. This charge produces an electric field with magnitude $\sigma/2\epsilon_0$, just like any other large uniform sheet of charge. The field exists on both sides of the sheet, in the interior of the conductor as well as in the exterior, and on each side it points away from the surface if σ is positive.

To obtain the layer of charge on the surface of the conductor another electric field must be present, produced perhaps by charge on another portion of the conductor or by external charge. In the diagram E_1 is the magnitude of the field due to the charge layer and E_2 is the magnitude of the second field. In the interior the second field exactly cancels the field due to the charge layer to produce a total field of zero and in the exterior it augments the field due to the charge layer to produce a total field with magnitude σ/ϵ_0.

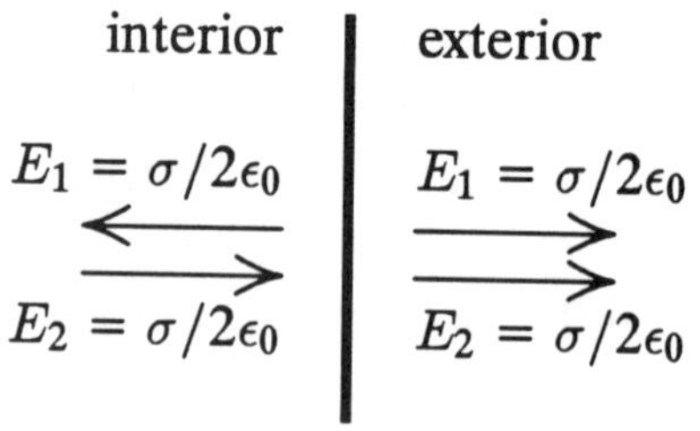

interior	exterior
$E_1 = \sigma/2\epsilon_0$ ←	$E_1 = \sigma/2\epsilon_0$ →
$E_2 = \sigma/2\epsilon_0$ →	$E_2 = \sigma/2\epsilon_0$ →

Another important charge configuration is a uniform spherical shell carrying total charge Q. Symmetry leads to us to conclude that the electric field must be radial so we use a Gaussian surface in the form of a sphere, concentric with the shell. If the radius of the Gaussian surface is r and the electric field has magnitude E at points on it then the flux through the surface is $\Phi_E =$ _____. If the Gaussian surface is inside the shell the charge enclosed is _____, so the electric field is $E =$ _____ at points inside the shell. If the Gaussian surface is outside the shell the charge enclosed is _____, so the electric field is _____ at points outside the shell.

II. PROBLEM SOLVING

You should know how to compute the flux associated with a given electric field and surface.

PROBLEM 1. What is the flux through $2.0\,\text{m}^2$ of the xy plane if the electric field in N/C is $\mathbf{E} = 23\mathbf{i} + 47\mathbf{k}$?

SOLUTION: Since the electric field is uniform $\Phi_E = \int \mathbf{E} \cdot d\mathbf{A} = \mathbf{E} \cdot \mathbf{A}$, where $\mathbf{A}$ is the vector area. Since the area being considered is in the xy plane $\mathbf{A}$ is in the z direction and we may write $\mathbf{A} = 2.0\,\mathbf{k}$. Thus $\Phi_E = (23\mathbf{i} + 47\mathbf{k}) \cdot 2.0\,\mathbf{k}$. Complete the evaluation of the scalar product. Notice that the x component of the field is immaterial.

[ans: $94\,\text{N·m}^2/\text{C}$]

The problem might also be worded in terms of an angle rather than vector components. Suppose a 75-N/C electric field makes an angle of 25° with the z axis. What is the flux through a $2.0\,\text{m}^2$ portion of the xy plane?

SOLUTION:

[ans: $136\,\text{N·m}^2/\text{C}$]

In both cases the vector area might have been in either the positive or negative z direction. Thus the negatives of the answers given are equally correct.

PROBLEM 2. A uniform electric field with magnitude E passes through a hemisphere of radius R as shown. Derive an expression for the flux through the hemisphere.

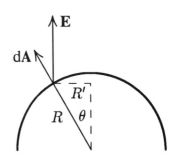

SOLUTION: The direction of dA is shown for an infinitesimal element of surface area. We wish to evaluate $\Phi_E = \int \mathbf{E} \cdot d\mathbf{A} = E \int \cos\theta \, dA$. Take a ring of width $R\,d\theta$, parallel to the bottom face of the hemisphere, as the element of area. It has radius $R' = R\sin\theta$ and area $dA = 2\pi R^2 \sin\theta \, d\theta$, so $\Phi_E = 2\pi R^2 E \int_0^{\pi/2} \sin\theta \cos\theta \, d\theta$. Evaluate the integral and obtain an expression for Φ_E.

[ans: $\pi R^2 E$]

Notice that the magnitude of the flux through the hemisphere is the same as the magnitude of the flux through the flat base. The fluxes have opposite signs so the total flux through the hemisphere and its base is zero and, according to Gauss's law, the net charge enclosed by the hemisphere and base is zero. This is an example of a general consequence of Gauss' law: the net charge is zero in any region of space containing a uniform electric field throughout.

Gauss' law is sometimes used to calculate the flux through a closed surface with known charge inside. Remember that the precise location of the charge is immaterial as long as it is inside.

PROBLEM 3. A 7.0×10^{-9} C point charge is at the center of a 5.0-cm radius sphere. What is the electric flux through the sphere surface?

SOLUTION: Use $\Phi_E = q/\epsilon_0$.

[ans: 7.91×10^2 N·m^2/C]

The charge is moved to a point 3.0 cm from the sphere center. What then is the total flux through the sphere surface?

SOLUTION:

[ans: 7.91×10^2 N·m^2/C]

The charge is moved to a point 6.0 cm from the sphere center. What then is the total flux through the sphere surface?

SOLUTION:

[ans: 0]

Gauss' law can be used to find the net charge in any region if the electric field is known for every point on the boundary of the region.

PROBLEM 3. A student measures the electric field at the walls, ceiling, and floor of a small room and finds it is zero everywhere except at the front and back walls. At the front wall the field is uniform, perpendicular to the wall, points into the room, and has a magnitude of 85 N/C. At the rear wall the field is also uniform, perpendicular to the wall, and points into the room but it has a magnitude of 115 N/C. If the area of each wall is 12 m^2 what is the sign and magnitude of the net charge in the room?

SOLUTION:

[ans: -2.13×10^{-8} C]

The next day the measurements are taken again. The field at the front wall is the same but the field at the rear wall is now in the opposite direction, pointing out of the room. Its magnitude is the same. What is the sign and magnitude of the net charge in the room now?

SOLUTION:

[ans: $+3.19 \times 10^{-9}$ C]

PROBLEM 4. A cube with edges of length 2.0 m is positioned as shown. In the region occupied by the cube the electric field in SI units is given by $\mathbf{E} = Bx\mathbf{i} + By\mathbf{j}$, where $B = 5.0 \times 10^2$ N/C·m. What is the net charge in the cube?

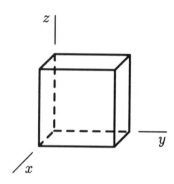

SOLUTION: Calculate the total flux through the surface of the cube. Consider first the right face and let a be the length of a cube edge. Notice that the y component of the field is uniform over this face and has the value $E_y = Ba$, since y has the value a at every point on the face. For this face $d\mathbf{A}$ is in the positive y direction, $\mathbf{E} \cdot d\mathbf{A} = E_y\, dA = Ba\, dA$, and its contribution to the flux is $\int Ba\, dA = Ba^3$.

The flux through the left face is 0 since $E_y = 0$ there. Show that the flux through the front face is Ba^3, the flux through the back face is 0, and the fluxes through the top and bottom faces are also 0. Calculate the total flux through the cube surface and evaluate $\epsilon_0 \Phi_E$.

[ans: 7.08×10^{-8} C]

PROBLEM 5. The electric field in a certain spherical charge distribution is radially outward and has magnitude given in N/C by $E(r) = 15 \times 10^5 r^2$, where r is the distance from the center in meters. What is the net charge inside a sphere with a radius of 2.5 cm, concentric with the distribution? Assume the radius of the charge distribution is larger than 2.5 cm.

SOLUTION: Since the electric field is everywhere normal to the surface of the 2.5-cm radius sphere and since the magnitude of the field is uniform over the surface, the flux through the surface is $\int \mathbf{E} \cdot d\mathbf{A} = EA = 15 \times 10^5 r^2 \times 4\pi r^2 = 60\pi \times 10^5 r^4$, where $r = 0.025$ m. Multiply by ϵ_0 to obtain the charge enclosed.

[ans: 6.52×10^{-11} C]

Note that the expression for the field must be evaluated for points on the Gaussian surface.

Some problems are designed to help you investigate the consequences of zero electric field in a conductor.

PROBLEM 6. A neutral conducting plate is placed in a uniform electric field with magnitude E, perpendicular to its faces, as shown. What is the charge density on each face?

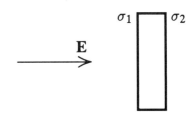

SOLUTION: Assume the area charge density on the left face is σ_1 and the area charge density on the right face is σ_2. Since the electric field must vanish in the interior of the conductor, $E + (\sigma_1/2\epsilon_0) - (\sigma_2/2\epsilon_0) = 0$. Since the net charge on the conductor is zero, $\sigma_1 + \sigma_2 = 0$. Solve for σ_1 and σ_2.

[ans: $\sigma_1 = -\epsilon_0 E$, $\sigma_2 = +\epsilon_0 E$]

Suppose now that excess charge Q is placed on the plate. Take the plate area to be A and find an expression for the charge density on each face.

SOLUTION:

[ans: $\sigma_1 = (Q - 2\epsilon_0 AE)/2A$; $\sigma_2 = (Q + 2\epsilon_0 AE)/2A$]

PROBLEM 7. A conducting sheet with an area of $1.3 \, m^2$ has an excess charge of 7.5×10^{-5} C. An external electric field is applied perpendicularly to the sheet. For what magnitude of the external field does all the excess charge lie on one face of the sheet?

SOLUTION: Use the solution to the previous problem. Find the value of E for which $\sigma_1 = 0$.

[ans: 3.26×10^2 N/C]

PROBLEM 8. Two large conducting plates a and b, each with face area A, are parallel to each other. Charge q_a is placed on the left hand plate and charge q_b is placed on the right hand plate. What is the charge density on each surface of each plate?

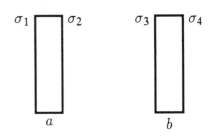

SOLUTION: Suppose the area charge densities are σ_1, σ_2, σ_3, and σ_4, as labelled in the diagram. The electric field is everywhere perpendicular to the plates. Draw a Gaussian surface in the form of a rectangular solid with one end in each of the conductors and with four sides parallel to the field. Use it to show that $\sigma_2 = -\sigma_3$.

The area charge densities for plate a are related to the total charge on the plate by $(\sigma_1 + \sigma_2)A = q_a$. Similarly $(\sigma_3 + \sigma_4)A = q_b$. The sum of the electric fields due to the four planes of charge vanishes inside each conductor. Express this statement in equation form, using $\sigma/2\epsilon_0$ for the magnitude of the field due to a large plane with area charge density σ. You now have four equations to solve for the four charge densities.

[ans: $\sigma_1 = -\sigma_3 = \sigma_4 = (q_a + q_b)/2A$; $\sigma_2 = (q_a - q_b)/2A$]

Now look at some examples of the use of Gauss' law to calculate the electric field.

PROBLEM 9. Charge Q is distributed uniformly throughout a thick spherical shell with inner radius R_1 and outer radius R_2. Find an expression for the electric field in each of the three regions $r < R_1$, $R_1 < r < R_2$, and $R_2 < r$, where r is the distance from the sphere center.

SOLUTION: You should be able to apply the shell theorems to answer easily for the first and last regions. To answer the second part consider a Gaussian surface in the form of sphere with radius r, concentric with the shell. If Q is positive the field is radially outward, in the same direction as dA, so $\Phi_E = 4\pi r^2 E$. Since the shell volume that lies inside the Gaussian surface is $(4\pi/3)r^3 - (4\pi/3)R_1^3$, the charge enclosed is given by $q = (4\pi/3)\rho(r^3 - R_1^3)$, where ρ is the volume charge density. You can now use Gauss' law to solve for E in terms of ρ. The volume charge density is the total charge divided by the volume of the shell, or $\rho = (3/4\pi)Q/(R_2^3 - R_1^3)$. Make the substitution to obtain E in terms of Q.

[ans: 0; $(Q/4\pi\epsilon_0 r^2)(r^3 - R_1^3)/(R_2^3 - R_1^3)$; $Q/4\pi\epsilon_0 r^2$]

Show that the expressions for the outer two regions yield the same result for $r = R_2$.

SOLUTION:

PROBLEM 10. Charge Q is distributed inside a sphere of radius R in such a way that the volume charge density is proportional to the distance from the center of the sphere. Find an expression for the electric field inside the sphere, a distance r from its center.

SOLUTION: Take the volume charge density to be $\rho = Br$, where B is a constant. For the Gaussian surface use a sphere of radius r, concentric with the sphere of charge. If Q is positive the field at the Gaussian surface is radially outward and the total flux through the surface is $4\pi r^2 E$. The charge enclosed is given by $q = \int_0^r \rho 4\pi r^2 \, dr = 4\pi B \int_0^r r^3 \, dr = \pi B r^4$. Use Gauss' law to solve for E. You will need to know B in terms of Q. The total charge in the sphere is given by $Q = \int_0^R \rho 4\pi r^2 \, dr = \pi B R^4$. Use this expression to eliminate B in favor of Q in the equation you developed for E.

[ans: $Qr^2/4\pi\epsilon_0 R^4$]

PROBLEM 11. Charge is distributed with a uniform charge density ρ throughout a long cylindrical shell with inner radius R_1 and outer radius R_2. What is the magnitude of the electric field in each of the three regions $r < R_1$, $R_1 < r < R_2$, and $R_2 < r$, where r is the distance from the cylinder axis?

SOLUTION: If ρ is positive the field is everywhere radially outward from the cylinder axis. Take the Gaussian surface to be that of a cylinder with radius r and length h, concentric with the shell. Show that if E is the magnitude of the field a distance r from the axis then the flux through the Gaussian surface is given by $2\pi E r h$. For $r < R_1$ the charge enclosed is zero, for $R_1 < r < R_2$ the charge enclosed is $\pi\rho(r^2 - R_1^2)h$, and for $R_2 < r$ the charge enclosed is $\pi\rho(R_2^2 - R_1^2)h$. Use Gauss' law to solve for E in each case.

[ans: 0; $(\rho/2\epsilon_0)(r^2 - R_1^2)/r$; $(\rho/2\epsilon_0)(R_2^2 - R_1^2)/r$]

III. MATHEMATICAL SKILLS

1. To calculate the electric flux you should be familiar with the evaluation of simple area integrals. If the region of integration is a rectangle with sides a and b you will probably want to position the coordinate system so that the rectangle is in the x, y plane with two of its sides along the axes. The infinitesimal element of area can be taken to be a rectangle with sides dx and dy. The flux integral then becomes

$$\Phi_E = \int_{x=0}^{a} \int_{y=0}^{b} E_z(x, y)\, dx\, dy$$

Integrations in x and y are carried out independently. If, for example, the z component of the electric field is given by $E_z = 9x^2y + 2y$ then

$$\Phi_E = \int_{x=0}^{a} \int_{y=0}^{b} (9x^2y^2 + 2y)\, dx\, dy = \int_{y=0}^{b} (3x^3y^2 + 2xy)\Big|_{x=0}^{a} dy$$

$$= \int_{y=0}^{b} (3a^3y^2 + 2ay)\, dy = (a^3y^3 + ay^2)\Big|_{y=0}^{b} = a^3b^3 + ab^2$$

where the integration over x was carried out first, then the integration over y.

Another example, the flux of a uniform electric field through a hemispherical surface, was given in the Problem Solving section. Study that problem carefully to be sure you understand the steps.

For many problems of this chapter a Gaussian surface can be chosen so that the normal component of the electric field has the same value at all points on a portion of it and is zero on the other portions. The integral for the flux then reduces to $\Phi_E = EA$, where A is the area of the region over which the normal component of the field is not zero and E is the magnitude of the normal component.

To evaluate the flux you must know how to calculate the areas of various surfaces. The area of a rectangle with sides of length a and b is ab; the area of a circle with radius R is πR^2; the surface area of a sphere with radius R is $4\pi R^2$; and the surface area of the rounded portion of a cylinder with radius R and length h is $2\pi Rh$.

If you have trouble remembering that $2\pi rh$ gives the area of a cylindrical surface, imagine a paper towel roll wrapped exactly once around with a towel. The surface area of the roll is the same as the area of the towel. Since the towel is a rectangle with sides of length $2\pi r$ and h, its area is $2\pi rh$.

2. You must also be able to calculate the charge enclosed by a Gaussian surface when the volume, area, or linear charge density is given. If an object has a uniform volume charge density ρ, for example, the charge enclosed is given by ρV, where V is the volume of the part of the object that lies within the Gaussian surface. If the Gaussian surface is completely within the object and the object does not have any cavities then the region enclosed by the Gaussian surface is completely filled with charge and V is the volume enclosed by the Gaussian surface. If the object has a cavity that is wholly within the Gaussian surface then V is the volume enclosed by the Gaussian surface minus the volume of the cavity.

Suppose, for example, a sphere of radius R has a uniform volume charge density ρ and the Gaussian surface is a concentric sphere of radius r. If $r < R$ then the Gaussian sphere is filled with charge and the charge enclosed is $4\pi\rho r^3/3$. If, on the other hand, $R < r$ then Gaussian surface is only partially filled with charge. The charge enclosed is the total charge, or $4\pi\rho R^3/3$.

If a spherical object has a spherical cavity with radius R_c and the Gaussian surface is within the object but outside the cavity then the charge enclosed is $\rho[(4\pi r^3/3) - (4\pi R_c^3/3)]$. The first term in the brackets is the volume enclosed by the Gaussian surface and the second is the volume of the cavity. If $R < r$ the charge enclosed is $\rho[(4\pi R^3/3) - (4\pi R_c^3/3)]$.

Now consider a solid cylinder with radius R and length h, having a uniform volume charge density ρ. Suppose the Gaussian surface is a concentric cylinder with radius r and the same length as the cylinder of charge. If $r < R$ the charge enclosed is $2\pi\rho rh$ and if $R < r$ it is $2\pi\rho Rh$. If the cylinder has a cavity that is inside the Gaussian surface its volume must be subtracted from the volumes in these expressions.

If the volume charge density varies from point to point in the object you must evaluate the integral $\int \rho \, dV$ over the volume of that part of the object lying within the Gaussian surface. The most common example is a sphere with a charge density that varies only with distance r from the center. Carry out the integration by dividing the sphere into spherical shells with infinitesimal thickness dr. A typical shell extends from r to $r + dr$ and has a volume of $4\pi r^2 \, dr$. Notice that this is the product of the surface area of the shell and its thickness. The charge in the shell is $4\pi\rho(r) r^2 \, dr$. If the Gaussian surface is sphere of radius r, concentric with the sphere of charge and entirely within it, the charge enclosed is $\int_0^r 4\pi\rho(r) r^2 \, dr$. If the Gaussian surface is entirely outside the charge distribution the charge enclosed is $\int_0^R 4\pi\rho(r) r^2 \, dr$. Notice the upper limits of integration for these two cases. If the sphere contains a concentric spherical cavity with radius R_c the lower limits of integration for the two cases are both R_c.

For a cylinder with a charge density that depends only on the distance r from the axis, divide the cylinder into concentric cylindrical shells with thickness dr. The volume of a shell is $2\pi rh\,dr$, where h is the length of the cylinder. If the Gaussian surface is a concentric cylinder with radius r and is inside the charge distribution then the charge enclosed is $\int_0^r 2\pi rh\rho(r)\,dr$. If the Gaussian surface is outside the distribution the charge enclosed is $\int_0^R 2\pi rh\rho(r)\,dr$. The lower limits must be changed for a distribution with a concentric cylindrical cavity.

IV. COMPUTER PROJECTS

Gauss's law can be verified directly by using a computer to evaluate the integral $\oint \mathbf{E}\cdot d\mathbf{A}$ over a closed surface. Consider the special case of a cube bounded by the 6 planes $x = 0$, $x = a$, $y = 0$, $y = a$, $z = 0$, and $z = a$, as shown in the figure. Place a single charge q on the line $x = a/2$, $z = a/2$, through the cube center, and carry out the integration one face at a time.

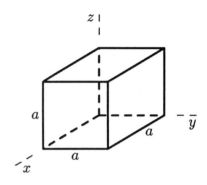

First consider the top face. Divide it into small rectangles of width Δx and length Δy, evaluate the electric field at the center of each rectangle, and calculate $E_z\Delta x\Delta y$. Finally, sum the results. Then carry out similar calculations for the other faces.

If the y coordinate of the charge is y' then its electric field at x, y, z is given by

$$\mathbf{E} = \frac{q}{4\pi\epsilon_0} \frac{(x-a/2)\mathbf{i} + (y-y')\mathbf{j} + (z-a/2)\mathbf{k}}{\left[(x-a/2)^2 + (y-y')^2 + (z-a/2)^2\right]^{3/2}}.$$

By symmetry the flux through the top, bottom, front, and back faces are all the same. Calculate the flux through the top face, say, and multiply by 4. On the top face $z = a$ and the quantity to be summed is

$$E_z\Delta x\Delta y = \frac{q}{4\pi\epsilon_0} \frac{a/2}{\left[(x-a/2)^2 + (y-y')^2 + a^2/4\right]^{3/2}}.$$

For the left face $y = 0$ and the quantity to be summed is

$$-E_y\Delta x\Delta z = \frac{q}{4\pi\epsilon_0} \frac{y'}{\left[(x-a/2)^2 + (y')^2 + (z-a/2)^2\right]^{3/2}}.$$

For the right face $y = a$ and the quantity to be summed is

$$E_y\Delta x\Delta z = \frac{q}{4\pi\epsilon_0} \frac{a-y'}{\left[(x-a/2)^2 + (a-y')^2 + (z-a/2)^2\right]^{3/2}}.$$

Here is the outline of a program to calculate the flux through the upper face. The x axis and the y axis, both from 0 to a, are each divided into N segments and $\Delta x = \Delta y = a/N$ is computed. In the program this quantity is called $\Delta\ell$. There are two loops: an outer loop over x and an inner loop over y. The first value of x is $\Delta\ell/2$, at the center of the first segment, and x is incremented by $\Delta\ell$ each time around the loop over x. For each value of x, y starts at $\Delta\ell/2$ and is incremented by $\Delta\ell$ each time around the loop over y. The sum is saved in S. The factor $q(a/2)(\Delta\ell)^2/4\pi\epsilon_0$ appears in every term so it is not included until after the sum is completed.

```
input charge and its coordinate: q, y'
input edge of cube, number of segments: a, N
calculate segment length: Δℓ = a/N
set S = 0
set x = Δℓ/2
begin loop over x
        set A = (x − a/2)² + a/4
        set y = Δℓ/2
        begin loop over y
                replace S with S = S + [A + (y − y')²]⁻³/²
                increment y by Δℓ
        end loop over y
        increment x by Δℓ
end loop over x
multiply S by q(a/2)(Δℓ)²/4πε₀ and display result
stop
```

Program lines for other faces are similar. For the left face the loop lines are:

```
set S = 0
set x = Δℓ/2
begin loop over x
        set A = (x − a/2)² + (y')²
        set z = Δℓ/2
        begin loop over z
                replace S with S = S + [A + (z − a/2)²]⁻³/²
                increment z by Δℓ
        end loop over z
        increment x by Δℓ
end loop over x
multiply S by qy'(Δℓ)²/4πε₀ and display result
```

For the right face the loop lines are:

```
set S = 0
set x = Δℓ/2
begin loop over x
        set A = (x − a/2)² + (a − y')²
        set z = Δℓ/2
        begin loop over z
                replace S with S = S + [A + (z − a/2)²]⁻³/²
                increment z by Δℓ
        end loop over z
        increment x by Δℓ
```

end loop over x

multiply S by $q(a - y')(\Delta \ell)^2 / 4\pi\epsilon_0$ and display result

PROJECT 1. Use the program to evaluate $\oint \mathbf{E} \cdot d\mathbf{A}$ for the surface of a cube with edge $a = 10$ m and with charge $q = 3.7 \times 10^{-9}$ C inside. Use $N = 15$ for 3 significant figure accuracy. First place the charge at the center of the cube: $x' = 5$ m, $y' = 5$ m, $z' = 5$ m. The flux is the same through each face so you need run the program for only one face, then multiply by 6. Compare the result with q/ϵ_0.

Now place the charge at $x' = 5$ m, $y' = 7.5$ m, $z' = 5$ m. Notice that the value of the flux through each face has changed from the previous situation. Also notice that the flux through the left, right, and top faces differ from each other. Nevertheless, the total flux is the same to within the accuracy of the calculation.

Finally, place the charge at $x' = 5$ m, $y' = 12.5$ m, $z' = 5$ m. It is outside the cube and the total flux through the cube should be zero. The result of the program may not be exactly zero because the flux through each face was computed to only about 3 significant figures. The total, however, should be significantly less than the flux through any individual face and should be still less if the calculation is done with a larger value of N.

Compare the calculations by filling out the following table with values of the flux:

y'	5 m	7.5 m	12.5 m
top	_____	_____	_____
bottom	_____	_____	_____
right	_____	_____	_____
left	_____	_____	_____
back	_____	_____	_____
front	_____	_____	_____
total	_____	_____	_____

PROJECT 2. If the net charge inside the cube is zero then, according to Gauss' law, the total flux through the surface of the cube is zero. The flux through any particular face, however, does not necessarily vanish. Suppose $q_1 = -3.7 \times 10^{-9}$ C is at $y' = 2.5$ m on the line through the center of the cube and $q_2 = +3.7 \times 10^{-9}$ C is at $y' = 7.5$ m on the same line. Use the program to find the flux through each face of the cube due to each charge separately, then fill out the following table with values of the flux. If you completed the first project you already have values for q_1.

	q_1	q_2
top	_____	_____
bottom	_____	_____
right	_____	_____
left	_____	_____
back	_____	_____
front	_____	_____
total	_____	_____

The two total fluxes may not sum to zero because of errors in the calculation. However, the sum should be considerably less than the flux through any of the faces.

V. NOTES

Chapter 30
ELECTRIC POTENTIAL

I. BASIC CONCEPTS

Because charges exert forces on each other work, (perhaps negative) must be done to assemble a collection of charges and because the forces are conservative a potential energy is associated with the assembled collection. If the charges are released, potential energy is converted to kinetic energy as each charge moves in response to the forces of the other charges. In this chapter you will learn to calculate and use the potential energy of a system of charges.

Electric potential is closely related to potential energy and plays a vital role in most succeeding discussions of electricity. Play careful attention to its definition and learn how to compute it for collections of point charges and for continuous charge distributions.

Electric potential energy. Since the electric force is conservative a potential energy function can be defined. Consider a collection of interacting charged particles. If one of them, with charge q, moves through an infinitesimal displacement ds the electric field $\mathbf{E}$ acting on it does work $dW = $ _____ and the potential energy of the collection changes by $\Delta U = -dW = $ _____. If the displacement of q is not infinitesimal, but instead is from point a to point b, the work done by the field is given by the integral $W_{ab} = $ _____ and the change in the potential energy is given by $U_b - U_a = -W_{ab} = $ _____.

You should be familiar with the sign of the change in potential energy. If, for example, a positive charge moves in the direction of the electric field acting on it then the sign of the work done by the field is _____ and the sign of the change in the potential energy of the system is _____. If a positive charge moves in the direction opposite to the field acting on it then the sign of the work done by the field is _____ and the sign of the change in the potential energy is _____. These signs are reversed if the charge is negative.

Coulomb's law can be used to write the work and potential energy in terms of the charges and their separation. The simplest case is that of two point charges. Suppose charges q_1 and q_2 start a distance r_a apart and move so they end a distance r_b apart. Then the work done by the electric field is $W_{ab} = $ _____ and the change in the potential energy of the two-charge system is _____. The result is the same if either charge remains stationary while the other moves or if both move. All that matters is their initial and final separations.

If the total charge in the collection is finite the potential energy is usually taken to be zero for infinite charge separation. In the expression you wrote above for the change in the potential energy of two point charges, let the initial separation r_a become large without bound and write the potential energy as a function of the separation r:

$$U(r) = $$

This expression is valid no matter what the signs of the charges. If they have the same sign the potential energy is positive; it decreases if their separation becomes greater and increases if their separation becomes less. Note that the electrical force is repulsive so the field does positive work in the first case and negative work in the second. If the two charges have opposite signs the potential energy is negative; it increases (becomes less negative) if their separation becomes greater and decreases (becomes more negative) if their separation becomes less. The field now does negative work in the first case and positive work in the second.

To calculate the potential energy of a collection of point charges, sum the potential energies of all *pairs* of charges. If the system consists of four charges, for example, add the potential energies of q_1 and q_2, q_1 and q_3, q_1 and q_4, q_2 and q_3, q_2 and q_4, and q_3 and q_4. If r_{12} is the separation of q_1 and q_2, r_{13} is the separation of q_1 and q_3, etc., then the potential energy of this system is

$$U = \underline{}$$

If one or more of the charges move then some or all of the separations change and a change in the potential energy results. To find the change calculate the initial and final potential energies, then subtract the initial potential energy from the final.

You may view the electric potential energy of the system as the work that must be done by an external agent to assemble the collection, bringing the charges from infinite separation to their final positions. The charges are at rest at the beginning and end of the process, so there is no change in kinetic energy.

You should recall how the principle of energy conservation is used. If the charges are released from some initial configuration their mutual attractions and repulsions might cause them to move. The potential energy of the system will decrease as potential energy is converted to kinetic energy. You write $\Delta U + \Delta K = 0$, where K is the total kinetic energy of the charges. Thus if you know the initial and final configurations of the charge you can compute the change in the total kinetic energy.

Electric potential. Electric potential and electric potential energy are closely related but they are not the same. Be careful to distinguish the two concepts. To find the electric potential of a collection of point charges, a reference point is chosen and the electric potential at that point is set equal to zero. Then a positive *test* charge, not one of the charges in the collection, is moved from the reference point to any point a and the work done by the electric field on the test charge is calculated. The electric potential at a is the negative of this work divided by the test charge. Note that this is also the change in the electric potential energy per unit test charge of the system consisting of the original collection of charges and the test charge. The charges of the collection must remain in fixed positions as the test charge is moved. If the total charge is finite the reference point is usually selected to be infinitely far removed from the charge collection. Note that a value for the electric potential is associated with each point in space. To find the value for a particular point the test charge is moved from the reference point to that point.

Since electric potential is an energy divided by a charge its SI unit is J/C. This unit is called a _____ (abbreviation: _____).

Suppose you wish to find the electric potential a distance r from a single isolated point charge q. Move a test charge q_0 from infinitely far away to a point r distant from q. As you do this the work done by the electric field of q is $W = $ _____ and the potential energy of the system consisting of q and q_0 changes from zero to $U = $ _____. The electric potential at the final position of the test charge is given by

$$V = $$

This expression is also valid if q is negative. The potential then has a negative value. Be sure to note that the point charge q does not have a potential energy (only collections of more than one charge have electric potential energies) but it does create an electric potential in the space around it.

The electric potential of a collection of charges can be used to compute changes in the potential energy when another charge, not belonging to the collection, is moved from one point to another. If charge Q is placed at point a, where the electric potential due to charges in the collection is V_a, the potential energy of the system consisting of the collection and Q is $U = $ _____ and if Q is moved from that point to a point where the potential due to the collection is V_b the change in the potential energy of the system is $\Delta U = $ _____.

The electric potential from the charge distribution. The electric potential of a collection of point charges is the sum of the individual potentials of the charges in the collection. Since the electric potential is a scalar the sum is algebraic. Suppose the system consists of charges q_1, q_2, and q_3. If you want to compute the potential at some point a you must know the distance from each charge to a. Suppose q_1 is a distance r_1 from a, q_2 is a distance r_2, and q_3 is a distance r_3. Then the potential at a is given by

$$V = $$

When you substitute values for the charges be sure you include the appropriate signs.

Charge may be distributed continuously along a line, on a surface, or throughout a volume. If it is you divide the region into infinitesimal elements, each containing charge dq, and sum (integrate) the potential due to the regions. If an infinitesimal region is a distance r from point a then the contribution of that region to the potential at a is $dV = $ _____ and the potential due to the whole distribution is the integral

$$V = $$

In practice dq is replaced by $\lambda\, ds$ for a line distribution, by $\sigma\, dA$ for a surface distribution, and by $\rho\, dV$ for a volume distribution. Be careful here. The symbol V is used sometimes to denote a potential and sometimes to denote a volume. To carry out the integration you will normally write ds, dA, or dV in terms of variables that are appropriate to the geometry of the problem. Don't forget to write the distances in terms of these quantities also.

To see how the integration is done, study the calculations in the text of the potentials at points on the axes of a uniformly charged ring and a uniformly charged disk. In these calculations z is the distance along the axis from the center of the ring or disk to the point.

All charge on a ring is the same distance from the point; in particular, $r = $ _____, where s is the radius of the ring. Thus the potential produced by the ring at z is $V = $ _____.

A disk is divided into infinitesimal rings. If a ring has radius s and width ds then its area is $dA = $ _____ and its charge is $dq = $ _____, where λ is the area charge density. The contribution of this ring to the potential at z is $dV = $ _____ and the total potential is given by the integral _____. Since $\int (s^2 + z^2)^{-1/2} s \, ds = (s^2 + z^2)^{1/2}$ the integral can be evaluated easily. Carefully note the limits of integration. The result is $V = $
_____.

The electric potential from the field. An expression for the potential difference between two points can be written as an integral involving the electric field at points along any path that connects the points. If **E** is the electric field then the potential difference between the points a and b is given by an integral:

$$V_b - V_a = $$

You should understand that this integral is a line integral: the path is divided into infinitesimal segments, the integrand is evaluated for each segment, and the results are summed. Because the electric field is conservative every path will give the same value for the potential difference.

You should also understand that the expression you wrote follows immediately from the definition of the potential difference as the negative of the work per unit test charge done by the field on a test charge as the test charge moves from a to b. The force of the electric field on the test charge q_0 is _____ and the work done by the field is _____.

The integral expression for $V_b - V_a$ in terms of the electric field is valid for any situation, whether the field is produced by a single point charge or by a collection of charges. Specialize the expression for a situation in which the field is uniform, as it is outside a large sheet with uniform charge density, for example. If **E** is the electric field and $\Delta \mathbf{r}$ is the displacement from point a to point b then $V_b - V_a = $ _____.

For some situations you must use different expressions for the electric field as the test charge passes through different regions. For example, the electric field of a uniform distribution of charge is given by $E = Q/4\pi\epsilon_0 r^2$ for r outside the distribution and by $Qr/4\pi\epsilon_0 R^3$ for r inside. Here Q is the total charge and R is the radius of the distribution. The integral for the electric potential at a point inside must be broken into two parts. Use the first expression for the field as you integrate from far away to the surface of the charge distribution and the second expression as you integrate from the surface to a point inside.

Roughly speaking, if the potential varies from place to place, being high in one region and lower in a neighboring region, the electric field points from the region where the potential is _____ toward the region where the potential is _____. A positive charge is accelerated toward the _____ potential region; a negative charge is accelerated toward the _____ potential region.

Since an electric potential is the product of an electric field and a distance the SI unit for an electric field is often taken to be V/m.

You have been considering the problem of finding the potential when the field is given. The inverse problem can also be solved: if the electric potential is a known function of position in a region of space then to find the component of **E** in any direction you calculate the rate of change of V with position along a line in that direction. Mathematically, if ds is the magnitude of an infinitesimal displacement in a given direction, then the component of **E** in that direction is given by the derivative $E_s = $ _____. Specialize this relationship to each of the three coordinate directions. Suppose the potential is given as some function $V(x, y, z)$. Then the components of the electric field at the point with coordinates x, y, z are given by

$$E_x = \qquad\qquad E_y = \qquad\qquad E_z = $$

When you differentiate with respect to x to find E_x you treat y and z as constants. Similar statements hold for E_y and E_z. The partial derivative symbol ∂ is used to remind you. When you are finished differentiating, but not before, you evaluate the result for the coordinates of interest. As an example, suppose the potential in volts is given by $V(x, y, z) = 3x^2y - 5xy^2$, with the coordinates in meters. Then $E_x = -\partial V/\partial x = -6xy + 5y^2$ and $E_y = -\partial V/\partial y = -3x^2 + 10xy$. At $x = 2\,\text{m}$, $y = 3\,\text{m}$, $E_x = -6 \times 2 \times 3 + 5 \times 3^2 = 9\,\text{V/m}$ and $E_y = -3 \times 2^2 + 10 \times 2 \times 3 = 48\,\text{V/m}$.

Equipotential surfaces. An equipotential surface is a surface (imaginary or real) such that the potential _____ at all points on it. An equipotential surface can be drawn by connecting neighboring points at which the potential has the same value and the surface can be labelled by giving the value of the potential on it. Equipotential surfaces do not cross each other. One and only one of them goes through any point in space.

The electric field line through any point is _____ to the equipotential surface through that point. If the field has a non-vanishing component tangent to a surface then a potential difference must exist between points on the surface and the surface cannot be an equipotential surface.

The equipotential surfaces of an isolated point charge are _____, centered on the charge. The equipotential surfaces of a uniform field are _____, perpendicular to the field. Equipotential surfaces associated with other charge distributions are more complicated but if the distribution has a net charge they are _____ far from the distribution. Look at Fig. 15 of the text to see the surfaces associated with a dipole, for which the net charge is zero. All conductors are equipotential volumes and their boundaries are equipotential surfaces.

You can approximate the magnitude of the electric field in any region by calculating $\Delta V/\Delta s$, where Δs is the perpendicular distance between two equipotential surfaces that differ in potential by ΔV. Equipotential surfaces are close together in regions of high field and far apart in regions of low field.

If the equipotential surfaces are known for some charge distribution they can be used to calculate changes in the potential energy and the work done by the electric field or an external agent as an additional charge is moved from one place to another. The diagram to the right shows a family of equipotential surfaces where they cut the plane of the page. Several points, labelled a, b, c, d, and e are also shown.

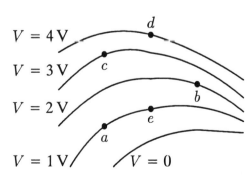

Calculate the change in potential energy as a particle is carried from one of these points to another, then calculate the work done by the electric field and the work done by the agent carrying the particle. Use the following table to record your answers.

PROCESS	CHANGE IN P.E.	WORK DONE BY FIELD	WORK DONE BY AGENT
electron from a to b			
electron from a to c			
electron from a to e			
proton from a to d			
proton from a to e			

II. PROBLEM SOLVING

Some problems are designed to help you understand and interpret the potential energy of a distribution of point charges. Here are some examples.

PROBLEM 1. Four identical 3.7×10^{-7} C charges are to be assembled one at a time at the corners of a square with an edge length of 6.7 cm. What work must the agent do to bring the first charge to a corner?

SOLUTION: No charges are yet in position so no force acts on the charge being brought in.

[ans: 0]

What work must the agent do to bring the second charge to the diametrically opposite corner?

SOLUTION: In their final positions the two charges are $\sqrt{2}a$ apart, where a is the length of a cube edge. The agent must do work $(1/4\pi\epsilon_0)q^2/\sqrt{2}a$ against the force exerted by the first charge.

[ans: 0.0130 J]

This is the potential energy of the two-charge system.
What work must the agent do to bring the third charge to one of the empty corners?

SOLUTION: The agent must do work against forces exerted by the two charges in place, both of them a distance a from the final position of the third charge.

[ans: 0.0367 J]

This is the potential energy that is added to the system when the third charge is brought in. The total potential energy of the three-charge system is $0.0130 + 0.0367 = 0.0497$ J.
Finally, what work is done to bring the last charge to the remaining corner?

SOLUTION:

[ans: 0.0497 J]

This is the potential energy that is added to the system when the fourth charge is brought in. The total potential energy of the four-charge system is $0.0497 + 0.0497 = 0.0994$ J.

PROBLEM 2. Three charges are placed at the corners of a right triangle with two equal sides, as shown. Take $a = 2.5$ cm, $q_1 = -3.7 \times 10^{-8}$ C, $q_2 = 5.5 \times 10^{-8}$ C, and $q_3 = 8.9 \times 10^{-8}$ C. Calculate the potential energy of the system, relative to the potential energy at infinite separation.

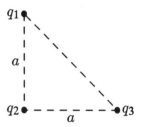

SOLUTION: The potential energy is the sum of three terms, corresponding to the interactions between q_1 and q_2, q_1 and q_3, and q_2 and q_3. The first term is $(1/4\pi\epsilon_0)q_1 q_2/a$ and the others have a similar form. Note that the distance between q_1 and q_2 is a, the distance between q_2 and q_3 is a, and the distance between q_1 and q_3 is $\sqrt{2}a$.

[ans: 1.91×10^{-4} J]

This is also the work that must be done by an external agent to assemble the charges from infinite separation. How much additional work must be done by the agent to move q_2 from the vertex of the triangle to the midpoint of the line joining the other two charges?

SOLUTION: Calculate the potential energy of the system for the second configuration, then the change in the potential energy.

[ans: 4.26×10^{-4} J]

PROBLEM 3. Now use the concept of electric potential to calculate the work done by the external agent in moving q_2 as described above. First calculate the potential due to q_1 and q_3 at the initial and final positions of q_2. Then calculate the potential difference.

SOLUTION: Use $q_1/4\pi\epsilon_0 a + q_3/4\pi\epsilon_0 a$ to calculate the potential at the initial position of q_2 and a similar expression to calculate the potential at the final position.

$$\left[\text{ans: } V_i = 1.87 \times 10^4 \text{ V}; V_f = 2.64 \times 10^4 \text{ V}; \Delta V = 7.74 \times 10^3 \text{ V}\right]$$

Multiply the potential difference by q_2 to obtain the work done by the agent. The answer should be the same as above.

SOLUTION:

One of the advantages of using the potential method is that the potential due to q_1 and q_3 is independent of q_2. If q_2 has another value you simply multiply it by the same potential difference to find the work done by the agent. What work must the agent do to move a -3.5×10^{-8}-C charge from the right angle vertex to the center of the line joining the other two charges?

SOLUTION: Use $W = q\Delta V$, with $q = -3.5 \times 10^{-8}$ C and $\Delta V = 7.74 \times 10^3$ V.

$$\left[\text{ans: } -4.34 \times 10^{-4} \text{ J}\right]$$

You should know how to compute the electric potential produced by a collection of point charges. Simply add terms of the form $q/4\pi\epsilon_0 r$, where r is the distance from the charge q to the point of interest.

PROBLEM 4. Point charges q_1 and q_2 are on the x axis, q_1 at $x = -d$ and q_2 at $x = +d$. Derive an expression for the electric potential at points on the y axis.

SOLUTION: Each charge is $\sqrt{d^2 + y^2}$ from the point and contributes $q/4\pi\epsilon_0\sqrt{d^2 + y^2}$ to the potential at y.

$$\left[\text{ans: } (q_1 + q_2)/4\pi\epsilon_0\sqrt{d^2 + y^2}\right]$$

You should know how to derive an expression for the electric potential if you are given the electric field. Sometimes an expression for the field has been given earlier, perhaps in another chapter. Sometimes you must use Gauss' law to find an expression for the field.

PROBLEM 5. A 3.5-cm radius thin conducting shell has 3.8×10^{-8} C of charge distributed uniformly on its surface. What is the electric potential at its surface, relative to the potential infinitely far away? What is the potential at its center?

SOLUTION: At points outside the shell the electric field is like that of a point charge: radially outward with magnitude given by $E(r) = Q/4\pi\epsilon_0 r^2$, where Q is the charge on the shell and r is the distance from its center. This means the potential is given by $V(r) = Q/4\pi\epsilon_0 r$. Evaluate this expression for $r = R$. The field inside the shell vanishes so the potential has the same value at all points in the interior and this is the same as the value on the shell surface.

[ans: 9.76×10^3 V; 9.76×10^3 V]

PROBLEM 6. A sphere of radius R is uniformly charged throughout with a charge density ρ. Find an expression for the electric potential at a point a distance r from its center, relative to the potential infinitely far away.

SOLUTION: The electric field outside the sphere is radial and has a magnitude given by $E(r) = Q/4\pi\epsilon_0 r^2 = \rho R^3/3\epsilon_0 r^2$. Here $4\pi R^3 \rho/3$ was substituted for Q. The potential is therefore $V(r) = \rho R^3/3\epsilon_0 r$. Inside the sphere the electric field is radial and has a magnitude given by $E(r) = \rho r/3\epsilon_0$, a result previously derived using Gauss' law. Break the integral from ∞ to r into two parts. The first is from ∞ to R and yields $V(R) = \rho R^2/3\epsilon_0$. The second is from R to r. Thus $V(r) = V(R) - \int_R^r \rho r/3\epsilon_0 \, dr$. Evaluate the integral and calculate $V(r)$.

[ans: $\rho R^3/3\epsilon_0 r$ for $r > R$; $(\rho/6\epsilon_0)(3R^2 - r^2)$ for $r < R$]

What is the potential at the center of a sphere with a radius of 3.0 cm and 5.8×10^{-10} C of charge distributed uniformly throughout? What is the potential at the rim?

SOLUTION: Use the expressions derived above and $\rho = 3Q/4\pi R^3$. In the first case $r = 0$, in the second $r = R$.

[ans: 2.61×10^2 V; 1.74×10^2 V]

An electron is carried from the center to the rim. What work is done by the electric field? What work is done by the carrying agent?

SOLUTION:

[ans: -1.39×10^{-17} J; $+1.39 \times 10^{-17}$ J]

With what minimum speed must an electron be started at the center of the sphere in order to become free of the sphere (without falling back toward it)? Assume the non-relativistic expression for the kinetic energy is valid.

SOLUTION: Use conservation of energy. As the electron moves away from the sphere the potential energy tends toward zero. If it has the minimum energy required to get far away its kinetic energy also tends toward zero. Thus $K_c + U_c = 0$, where the subscript denotes values at the sphere center. Replace K_c with $\frac{1}{2}mv_c^2$ and U_c with $-eV_c$, where $m = 9.11 \times 10^{-31}$ kg is the mass of an electron and V_c is the electric potential at the sphere center.

[ans: 9.57×10^5 m/s]

If the initial speed of the electron is 6.8×10^6 m/s how far does it travel from the sphere center before turning around?

SOLUTION: Let U_f be the potential energy when the electron turns around. Since its kinetic energy is then zero, $K_c + U_c = U_f$, or $K_c + U_c = -eV_f$. Solve for V_f and compare the value with the electric potential at the sphere surface. If it is greater the electron turns around inside the sphere and you solve $V_f = (\rho/6\epsilon_0)(3R^2 - r^2)$ for r. If it is less the electron turns around outside the sphere and you solve $\rho R^3/3\epsilon_0 r^2$ for r.

[ans: 4.04 cm]

Some problems deal with the relationships between the charge, area charge density, potential, and field for a conductor. If the conductor is a sphere $V = Q/4\pi\epsilon_0 R$ gives the potential at the surface and in the interior, $\sigma = Q/4\pi R^2$ gives the charge density on the surface, and $E = \sigma/\epsilon_0$ gives the magnitude of the electric field just outside the surface.

PROBLEM 7. A 3.7-cm radius conducting sphere is charged until its potential is 100 V relative to the potential far away. What is the charge on the sphere?

SOLUTION: Use $V = Q/4\pi\epsilon_0 R$. Set $V = 100$ V and $R = 0.037$ m. Solve for Q.

[ans: 4.12×10^{-10} C]

What is the area charge density on the surface of the sphere? What is the electric field just outside the sphere?

SOLUTION: Use $\sigma = Q/4\pi R^2$ and $E = \sigma/\epsilon_0$.

[ans: 2.39×10^{-8} C/m^2; 2.70×10^3 V/m]

PROBLEM 8. Two isolated conducting spheres are far apart. Sphere A has a radius of 3.5 cm and a charge of 8.6×10^{-8} C. Sphere B has a radius of 6.1 cm and a charge of 3.3×10^{-8} C. After the spheres are connected by a conducting wire what is the charge and potential of each sphere?

SOLUTION: Let Q_A be the final charge on sphere A and Q_B be the final charge on sphere B. Since the spheres and wire form a single conductor the potential is the same everywhere in them. This means $Q_A/R_A = Q_B/R_B$. Charge is conserved so $Q_A + Q_B = Q$, where Q is the total charge, $8.6 \times 10^{-8} + 3.3 \times 10^{-8} = 11.9 \times 10^{-8}$ C. Solve these equations simultaneously for Q_A and Q_B, then use $V = Q_A/4\pi\epsilon_0 R_A$ or $V = Q_B/4\pi\epsilon_0 R_B$ to find the potential.

[ans: $Q_A = 4.34 \times 10^{-8}$ C; $Q_B = 7.56 \times 10^{-8}$ C; $V = 1.11 \times 10^4$ V]

What is the charge density on the surface of each sphere?

SOLUTION:

[ans: $\sigma_A = 2.82 \times 10^{-6}$ C/m^2; $\sigma_B = 1.62 \times 10^{-6}$ C/m^2]

What is the electric field just outside each sphere?

SOLUTION:

[ans: $E_A = 3.19 \times 10^5$ V/m; $E_B = 1.83 \times 10^5$ V/m]

Notice that the smaller sphere ends up with less charge than the larger sphere but its charge density is greater and so is the electric field it produces just outside its surface.

PROBLEM 9. Three identical raindrops are charged to the same potential V. If they coalesce into one large drop, what is potential at its surface?

SOLUTION: Since $V = Q/4\pi\epsilon_0 R$ at the surface, the charge on each small drop is $Q = 4\pi\epsilon_0 RV$ and the total charge is $Q' = 12\pi\epsilon_0 RV$. Now find the radius R' of the large drop. Its volume must be 3 times the volume of one of the small drops. Finally use $V' = Q'/4\pi\epsilon_0 R'$ to find the potential at the surface of the large drop.

[ans: $\sqrt[3]{9}V$]

Here are some problems that deal with equipotential surfaces.

PROBLEM 10. The equipotential surfaces of a point charge are spheres. For a 7.3×10^{-9}-C charge what is the radius of the equipotential sphere corresponding to 1 V? 2 V? 30 V? 31 V?

SOLUTION: Solve $V = Q/4\pi\epsilon_0 r$ for r.

[ans: 65.6 m; 32.8 m; 2.19 m; 2.12 m]

Notice that equipotential spheres differing by 1 V are close together at high potential, near the charge, and far apart at low potential, far from the charge. The electric field is greater near the charge than far away.

PROBLEM 11. A large plane sheet has a uniform charge density of 5.8×10^{-6} C/m². What is the distance between equipotential surfaces that differ by 1 V?

SOLUTION: The electric field is perpendicular to the sheet and has a magnitude given by $E = \sigma/2\epsilon_0$ so the potential is given by $V = -\sigma x/2\epsilon_0$ if the x axis is parallel to the field. Solve $\Delta V = -\sigma \Delta x/2\epsilon_0$ for Δx.

[ans: 3.05×10^{-6} m]

PROBLEM 12. A 3.0-cm radius sphere is uniformly charged with a charge density of 4.6×10^{-6} C/m³. What is the radius of the equipotential surface corresponding to 100 V? to 200 V?

SOLUTION: Recall that $V(r) = \rho R^3/3\epsilon_0 r$ for $r > R$ and $V(r) = (\rho/6\epsilon_0)(3R^2 - r^2)$ for $r < R$. First find the potential at the surface of the sphere by substituting $r = R$ in either of these expressions. The answer is

234 V. The 100 V equipotential surface is outside the sphere so you solve $V(r) = \rho R^3/3\epsilon_0 r$ for r. The 200 V equipotential surface is inside the sphere so you solve $V(r) = (\rho/6\epsilon_0)(3R^2 - r^2)$ for r.

[ans: 4.68 cm; 1.98 cm]

III. COMPUTER PROJECTS

If charge is distributed along the x axis with linear charge density $\lambda(x')$, then the electric potential at a point in the xy plane is given by

$$V(x,y) = \frac{1}{4\pi\epsilon_0} \int \frac{\lambda(x')}{\left[(x-x')^2 + y^2\right]^{1/2}} \, dx',$$

where the integral extends over the charge distribution. You can use the Simpson's rule program of Chapter 7 to evaluate the integral. After evaluating the integral, multiply by $1/4\pi\epsilon_0$. Here is an outline of the program.

    input limits of integral: x'_i, x'_f
    input number of intervals: N
    replace N with nearest even integer
    calculate interval width: $\Delta x' = (x'_f - x'_i)/N$
    input coordinates of field point: x, y
    initialize quantity to hold sum of values with even labels: $S_e = 0$
    initialize quantity to hold sum of values with odd labels: $S_o = 0$
    set $x' = x'_i$
    begin loop over intervals: counter runs from 1 to $N/2$
        calculate integrand: $I(x') = \lambda(x')\left[(x-x')^2 + y^2\right]^{-1/2}$
        add it sum of values with even labels: replace S_e with $S_e + I(x')$
        increment x' by $\Delta x'$
        calculate integrand: $I(x') = \lambda(x')\left[(x-x')^2 + y^2\right]^{-1/2}$
        add it to the sum of values with odd labels: replace S_o with $S_o + I(x')$
        increment x' by $\Delta x'$
    end loop
    calculate integrand at upper and lower limits:
    $I_0 = \lambda(x'_i)\left[(x-x'_i)^2 + y^2\right]^{-1/2}$
    $I_N = \lambda(x'_f)\left[(x-x'_f) + y^2\right]^{-1/2}$
    evaluate integral: $(\Delta x/3)(I_N - I_0 + 2S_e + 4S_o)$
    multiply by $1/4\pi\epsilon_0$ and display result
    go back for coordinates of another field point or stop

First use the program to investigate the relationship between the electric potential and the electric field: $E_x = -\partial V/\partial x$, $E_y = -\partial V/\partial y$.

PROJECT 1. Charge is distributed from $x' = 0$ to $x' = 0.10$ m along the x axis with a linear charge density given by $\lambda(x') = 1.83 \times 10^{-5}\sqrt{x'}$ C/m, where x' is in meters. Use the Simpson's rule program to find values for the electric potential at $x = -0.005$ m and at $x = +0.005$ m on the line $y = 0.20$ m. Start with $N = 20$ and repeat the calculation with N doubled each time until you get the same results to 3 significant figures.

Estimate the x component of the electric field at $x = 0$, $y = 0.2$ m by evaluating $-(V_2 - V_1)/\Delta x$, where V_1 is the potential at $x = -0.005$ m, V_2 is the potential at $x = +0.005$ m, and $\Delta x = 0.01$ m. Check your answer by using the program of Chapter 28 to compute the electric field directly. To how many figures do you obtain agreement? Some significance is lost when you subtract the two values of the potential.

Use the Simpson's rule program to calculate the electric potential at $y = 0.195$ m and at $y = 0.205$ m on the y axis. Use the results to estimate the y component of the electric field at $x = 0$, $y = 0.2$ m. Check your result by using the program of Chapter 28.

PROJECT 2. You can use the program to plot equipotential surfaces. In this project you will consider the line charge of the previous project and plot a line in the xy plane along which the electric potential has a given value V. Start with $x = -0.01$ m and use trial and error to find the y coordinate of the point for which the potential has the value V. You might start with $y = .01$ m and increment y by 0.1 m until you find two points with potentials that straddle V, then narrow the gap until the coordinates of the two points at its ends are the same to 2 significant figures. Increment x by 0.01 m and repeat. Continue until you reach $x = +0.11$ m. Once you have found the first few points a pattern should emerge and later points should be easier to find. Try potentials of 3, 5, and 10 V.

IV. NOTES

Chapter 31
CAPACITORS AND DIELECTRICS

I. BASIC CONCEPTS

Capacitors are electrical devices that are used to store charge and electrical energy and, as you will see in a later chapter, are important for the generation of electromagnetic oscillations. This chapter is an excellent review of the principles you have learned in previous chapters: you will make extensive use of Gauss' law and the concepts of electric potential and electrical potential energy.

Capacitors and capacitance. A capacitor is simply two _____, isolated from each other. Each is called a <u>plate</u>. The symbol used to represent a capacitor in an electrical circuit is: _____.

In normal use one plate holds positive charge and the other holds negative charge of the same magnitude. The charge creates an electric field, pointing roughly from the positive plate toward the negative plate, and because an electric field exists in the region between the plates, the plates are at different electric potentials. The positive plate, of course, is at a higher potential than the negative.

The potential difference of the plates and the charge on either one are proportional to each other. If the magnitude of the potential difference is V and the magnitude of the charge on either plate is q, then

$$q = $$

where _____ is the <u>capacitance</u> of the capacitor. Because q and V are proportional the capacitance is independent of both q and V. If q is doubled _____ doubles, but _____ remains the same. A capacitor can be charged by transferring charge from one plate to the other. The capacitance is a measure of how much is transferred for a given _____.

A battery connected to a capacitor transfers electrons from the plate at its positive terminal to the plate at its negative terminal. When fully charged, the potential difference of the plates is the same as the terminal voltage of the battery. That is, for example, a 9 V battery will transfer charge until the potential difference of the plates is 9 V. Remember, however, that a potential difference exists between the plates of a charged capacitor whether or not a battery is connected.

The SI unit for capacitance is a coulomb/volt. This unit is called a _____ and is abbreviated _____. Microfarad (abbreviated _____) and picofarad (abbreviated _____) capacitors are commonly used in electronic circuits. A 1 F parallel plate capacitor must have a huge plate area or else an extremely small plate separation. Until recently the technology was not available to construct such a capacitor in a reasonable volume.

Chapter 31: Capacitors and Dielectrics

Calculation of capacitance. To calculate capacitance, first imagine charge q is placed on one plate and charge $-q$ is placed on the other. Use Gauss' law to find an expression for _____ in the region between the plates. In every case you will find that the field is proportional to q. Once an expression for the electric field is known, use _____ to compute the magnitude V of the potential difference of the plates. V is also proportional to q. Finally use _____ to find an expression for the capacitance. It will not depend on q or V but in general terms it will depend on: _____
_____.

Carefully study the examples given in the text. They clearly show that the electric field and potential difference used in the calculation are due to the charge on the plates and, in each case, they explicitly give the geometric quantities on which the capacitance depends.

Suppose a capacitor consists of two parallel metal plates, each of area A, separated by a distance d. If charge q is paced on one and charge $-q$ is placed on the other, the electric field between the plates is given by $E =$ _____ and the potential difference of the plates is given by $V =$ _____. Thus the capacitance is given by $C =$ _____. Carefully note that the capacitance depends only on the permittivity constant ϵ_0, the plate area A, and the plate separation d, not on q or V.

Remember the expression for the capacitance of a parallel plate capacitor. It will be used many times. Also remember that C is proportional to A and inversely proportional to d. If the plate separation is halved without changing the plate area, the capacitance is _____. If the plate area is halved without changing the separation, the capacitance is _____.

Suppose a capacitor consists of two coaxial cylinders of length L, the inner one with radius a and the outer one with radius b. Charge q is placed on the inner cylinder and charge $-q$ is placed on the outer cylinder. The electric field between the cylinders is given by $E =$ _____, the potential difference of the cylinders is given by $V =$ _____, and the capacitance is given by $C =$ _____. Note that C depends only on ϵ_0, a, b, and L.

Suppose a capacitor consists of two concentric spherical shells, with radii a and b. Charge q is placed on the inner shell and charge $-q$ is placed on the outer shell. The electric field between the shells is given by $E =$ _____, the potential difference of the shells is given by $V =$ _____, and the capacitance is given by $C =$ _____.

Capacitance is also defined for a single conductor. The other plate is assumed to be infinitely far away. To find an expression for the capacitance imagine charge q is on the conductor, find an expression for the electric field outside the conductor, calculate the potential V of the conductor relative to the potential at infinity, and use $q = CV$ to find the capacitance. For example, the capacitance of a spherical conductor of radius R is given by $C =$ _____.

Capacitors in series and parallel. Be sure you can distinguish between <u>parallel</u> and <u>series</u> combinations of two or more capacitors. For a _____ combination the potential difference is the same for all the capacitors and for a _____ combination the charge is the same. If neither the potential difference nor the charge is the same then the capacitors do not form either a parallel or series combination.

In the space below draw two capacitors connected in parallel and two connected in series.

Assume a potential difference V is applied to each combination and indicate on the diagrams where it is applied. Label the charge on each plate to show which plates have the same charge and which have different charges.

PARALLEL   SERIES

In each case the capacitors can be replaced by a single capacitor with capacitance C_{eq} such that the charge transferred is the same as the total charge transferred for the original combination when the potential difference is the same as the potential difference across the original combination. If the capacitances are C_1 and C_2 then the value of C_{eq} for a parallel combination is given by $C_{eq} =$ _____ and the value of C_{eq} for a series combination is given by $C_{eq} =$ _____. For a parallel combination the equivalent capacitance is greater than the greatest capacitance in the combination; for a series combination the equivalent capacitance is less than the smallest capacitance in the combination.

Energy storage. As a capacitor is being charged, energy must be supplied by an external source, such as a battery, and energy is stored by the capacitor. You may think of the stored energy in either one of two ways: as the potential energy of the charge on the plates or as an energy associated with the electric field produced by the charges.

Suppose that, at one stage in the charging process, the positive plate has charge q' and the potential difference between the plates is V'. If an additional infinitesimal charge dq' is taken from the negative plate and placed on the positive plate, the energy is increased by $dU =$ _____ or, what is the same, by $dU = (q'/C)\, dq'$, where C is the capacitance. This expression is integrated from 0 to the final charge q. In terms of the final charge the total energy required to charge an originally uncharged capacitor is $U =$ _____. In terms of the final potential difference V it is $U =$ _____.

Instead of associating the energy of a charged capacitor with the mutual interactions of the charges on its plates, it may be associated with the electric field created by those charges. The text shows that the energy density (or energy per unit volume) in a parallel plate capacitor is given by

$$u =$$

where E is the magnitude of the electric field between the plates.

This expression is quite generally valid for any electric field, whether in a capacitor or not. Most fields are functions of position so a volume integral must be evaluated to calculate the energy required to produce the field: $U = \frac{1}{2}\epsilon_0 \int E^2\, dV$. Part c of Sample Problem 7 in the text shows how to evaluate this integral for a charged conducting sphere.

If, after charging a capacitor, the plates are connected by a conducting wire, then electrons will flow from the negative to the positive plate until both plates are neutral and the electric field vanishes. The stored energy is converted to kinetic energy of motion and eventually to internal energy in resistive elements of the circuit.

Dielectrics. If the space between the plates of a capacitor is filled with insulating material the capacitance is greater than if the space is a vacuum. In fact, the capacitance is given by $C =$ _____, where C_0 is the capacitance of the unfilled capacitor and κ_e is the _____ of the insulator. This last quantity is a property of the insulator, is unitless, and is always greater than 1. See Table 1 of the text for some values.

Consider two capacitors that are identical except capacitor A has a dielectric between its plates while capacitor B does not. If the capacitors have the same charge on their plates, the potential difference across capacitor _____ is greater than the potential difference across the other capacitor. If the capacitors have the same potential difference, the charge on the positive plate of capacitor _____ is greater than the charge on the positive plate of the other capacitor.

Polarization of the dielectric by the electric field brings about an increase in capacitance. Suppose the dielectric is composed of polar molecules and describe what happens when it is polarized: _____
_____

Now suppose the dielectric is composed of non-polar molecules and describe what happens:
_____
_____

The electric dipoles in a polarized dielectric produce an electric field that is directed opposite to the field produced by the charge on the conducting plates. Thus the total field is weaker than the field produced by charge on the plates alone. It is, in fact, weaker by the factor κ_e. That is, if $\mathbf{E}_0$ is the field produced by the charge on the plates then the total field is given by $\mathbf{E} =$ _____. If the electric field produced by charge on the plates is uniform, as it essentially is between the plates of a parallel plate capacitor, then the dipole field and the total field are also uniform.

Since the electric field for a given charge is weaker if the capacitor is filled with a dielectric than if it is not, the potential difference is _____ when the dielectric is present. Since $q = CV$, this means the capacitance is larger when the dielectric is present.

For a parallel plate capacitor the effect of the dielectric is exactly the same as a uniform distribution of positive charge q' on the surface nearest the negative plate and a uniform distribution of negative charge $-q'$ on the surface nearest the positive plate. If the dielectric constant of the dielectric is κ_e and the charge on the positive plate is q then $q' =$ _____. If the plates have area A then the electric field due to the charge on the plates is _____, the field due to the dipoles is _____, and the total field is _____. In terms of q, A, and d the potential difference is given by _____.

The energy stored in a capacitor is still given by $U = \frac{1}{2}q^2/C = \frac{1}{2}CV^2$, where q is the magnitude of the charge on either plate and V is the potential difference. As a dielectric is inserted between the plates of a capacitor with the potential difference held constant (by a battery, say), the charge on the positive plate _____ and the stored energy _____. For the potential difference to remain the same the battery must transfer charge as the dielectric is inserted. As a dielectric is inserted with the capacitor in isolation, so the charge cannot change, the potential difference _____ and the stored energy _____. If the dielectric is inserted by an external agent so it is at rest, the difference in energy is associated with _____. If no external agent acts the difference in energy is associated with _____.

Although a capacitor filled with a dielectric holds more charge than an identical one at the same potential difference, but without a dielectric, the dielectric places an upper limit on the charge and potential difference. The <u>dielectric strength</u> of an insulator is _____

If a parallel plate capacitor has capacitance C and plate separation d and the insulator filling it has dielectric strength E_{max} then the maximum potential difference it can sustain is $V_{max} = $ _____ and the maximum charge than can be placed on the positive plate is $q_{max} = $ _____.

The relationship $\mathbf{E} = \mathbf{E}_0/\kappa_e$ between the field $\mathbf{E}_0$ in the absence of a dielectric and the field $\mathbf{E}$ in the presence of a dielectric is valid no matter what the source of $\mathbf{E}_0$. If a point charge is imbedded in a dielectric with dielectric constant κ_e the magnitude of the total field a distance r from the charge is $E = $ _____. The charge itself produces a field with magnitude $E_0 = $ _____ and the dipoles of the dielectric produce a field with magnitude _____.

II. PROBLEM SOLVING

Many problems are based on the definition of capacitance $C = q/V$ and on the expression for the capacitance of a parallel plate capacitor $C = \epsilon_0 A/d$.

PROBLEM 1. A $3.7\,\mu$F capacitor is connected to a variable source of electric potential. What is the charge on the positive plate when the potential difference is 5.0 V? 10 V? 15 V?

SOLUTION:

[ans: 1.85×10^{-5} C; 3.7×10^{-5} C; 5.55×10^{-5} C]

PROBLEM 2. The charge on a parallel plate capacitor is 4.4×10^{-9} C when the potential difference is 30 V. What is the capacitance?

SOLUTION:

[ans: 147 pF]

The capacitor is electrically isolated and the separation between its plates is doubled. What is the capacitance then? What is the potential difference across its plates?

SOLUTION:

[ans: 73.3 pF; 60 V]

The plate separation is doubled from its original value while the capacitor is connected to the 30 V battery. What then is charge on its plates?

SOLUTION:

[ans: 2.2×10^{-9} C]

You should understand series and parallel connections well enough to calculate the charge and potential difference for each capacitor as well as the equivalent capacitance of the combination.

PROBLEM 3. Two capacitors, $C_1 = 2.4\,\mu\text{F}$ and $C_2 = 4.5\,\mu\text{F}$, are connected in parallel. What is the equivalent capacitance?

SOLUTION: Use $C_{eq} = C_1 + C_2$.

[ans: $6.9\,\mu\text{F}$]

A 100-V potential difference is applied across the combination. What is the potential difference across each capacitor and what is the charge on the plates of the each capacitor?

SOLUTION: The potential difference is the same for both capacitors and is 100 V. Use $q_1 = C_1 V$ and $q_2 = C_2 V$ to find the charges.

[ans: $V_1 = 100$ V; $V_2 = 100$ V; $q_1 = 2.4 \times 10^{-4}$ C; $q_2 = 4.5 \times 10^{-4}$ C]

The capacitors are now connected in series. What is the equivalent capacitance?

SOLUTION:

[ans: $1.57\,\mu\text{F}$]

A 100-V potential difference is applied across the combination. What is the potential difference across each capacitor and what is the charge on the plates of each capacitor?

SOLUTION: The charge is the same for both capacitors and is $q = C_{eq} V$. Use $q = CV$ to calculate the potential difference across each capacitor.

[ans: $V_1 = 65.2$ V; $V_2 = 34.8$ V; $q_1 = 1.57 \times 10^{-4}$ C; $q_2 = 1.57 \times 10^{-4}$ C]

PROBLEM 4. Capacitor 1 has capacitance $C_1 = 5.4\,\mu\text{F}$ and charge $q_1 = 6.7 \times 10^{-6}$ C on its plates. Capacitor 2 has capacitance $C_2 = 7.3\,\mu\text{F}$ and charge $q_2 = 3.6 \times 10^{-6}$ C on its plates. The positive plates are then connected by a conducting wire and the negative plates are connected by another conducting wire. What then is the charge and potential difference for each capacitor?

SOLUTION: Since the potential differences of the capacitors are the same the capacitors form a parallel combination. The equivalent capacitor has potential difference V and charge $q = q_1 + q_2$, so $V = q/C_{eq}$, where $C_{eq} = C_1 + C_2$. This is also the potential difference across each of the original capacitors. The charge on capacitor 1 is $C_1 V$ and the charge on capacitor 2 is $C_2 V$.

[ans: $q_1 = 6.87 \times 10^{-6}$ C; $V_1 = 1.27$ V; $q_2 = 3.43 \times 10^{-6}$ C; $V_2 = 1.27$ V]

Now suppose that, starting with the same charge as before, the negative plate of each capacitor is connected by a conducting wire to the positive plate of the other capacitor. What are the charges and potential differences?

SOLUTION: The problem is exactly the same except that the charge on the equivalent capacitor is $q = q_1 - q_2 = 3.1 \times 10^{-6}$ C.

[ans: $q_1 = 2.07 \times 10^{-6}$ C; $V_1 = 0.383$ V; $q_2 = 1.03 \times 10^{-6}$ C; $V_2 = 0.383$ V]

PROBLEM 5. Consider the combination of capacitors shown on the right. A potential difference V is applied between points a and b. In terms of C_1, C_2, C_3, and V what is the charge on the plates of each capacitor?

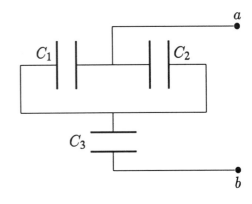

SOLUTION: Note that C_1 and C_2 are in parallel since their potential differences are the same. They can be replaced by an equivalent capacitor with capacitance $C_{12} =$ _____. Redraw the circuit in the space provided. Now notice that C_{12} and C_3 are in series since their charges are the same. They can be replaced by an equivalent capacitor with capacitance $C_{123} =$ _____.

The charge on each plate of C_{123} has magnitude $q_{123} = C_{123}V =$ _____, where the expression for C_{123} has been substituted. This charge is the same as the charge on both C_{12} and C_3 of the previous circuit. Thus $q_3 =$ _____ and $q_{12} =$ _____.

In terms of the original circuit q_{12} is divided between C_1 and C_2. We must calculate the potential differences across these capacitors to find out how it is divided. The potential difference for C_3 is $V_3 = q_3/C_3 =$ _____ and the potential differences for C_1 and C_2 are both $V - V_3 =$ _____. Thus $q_1 = C_1V_1 =$ _____ and $q_2 = C_2V_2 =$ _____.

[ans: $q_1 = C_1C_3V/(C_1 + C_2 + C_3)$; $q_2 = C_2C_3V/(C_1 + C_2 + C_3)$; $q_3 = C_3(C_1 + C_2)V/(C_1 + C_2 + C_3)$]

Some problems deal with the capacitances of parallel plate, cylindrical, and spherical capacitors in terms of their geometries.

PROBLEM 6. Two identical parallel plate capacitors each have plate area A and plate separation d. What is the plate area of a single parallel plate capacitor with plate separation d and capacitance equal to the equivalent capacitance of the two in series?

SOLUTION: If C is the capacitance of one of the original capacitors then $C_{eq} = C/2 = \epsilon_0 A/2d$. This is the same as the capacitance of a parallel plate capacitor with plate separation d and a different area A'. Write $C_{eq} = \epsilon_0 A'/d$ and solve for A'.

[ans: $A/2$]

What is the plate area of a single parallel plate capacitor with plate separation d and capacitance equal to the equivalent capacitance of the two in parallel?

SOLUTION:

[ans: $2A$]

Chapter 31: Capacitors and Dielectrics

PROBLEM 7. A spherical capacitor has an inner radius of 1.500 cm and an outer radius of 1.520 cm. What is its capacitance?

SOLUTION: Use $C = 4\pi\epsilon_0 ab/(b-a)$, where a is the inner radius and b is the outer radius.

[ans: 1.27×10^{-10} F]

A cylindrical capacitor has an inner radius of 1.500 cm and an outer radius of 1.520 cm. It length is such that the inner cylinder has the same area (exclusive of the ends) as the inner shell of the spherical capacitor considered above. What is its capacitance?

SOLUTION: Use $4\pi a^2 = 2\pi aL$ to calculate the length L of the cylinder and $C = 2\pi\epsilon_0 L/\ln(b/a)$ to calculate the capacitance.

[ans: 1.26×10^{-10} F]

A parallel plate capacitor has a plate separation of 0.020 cm and a plate area that is the same as the area of the inner sphere above. What is its capacitance?

SOLUTION: Use $A = 4\pi a^2$ to calculate the area and $C = \epsilon_0 A/d$ to calculate the capacitance.

[ans: 1.25×10^{-10} F]

As these results suggest, capacitances of all capacitors with the same area and plate separation are nearly the same, provided the separation is much smaller than any of other dimensions. If these conditions hold you can use the expression for a parallel plate capacitor to calculate the capacitance.

Start with the equation $C = 4\pi\epsilon_0 ab/(b-a)$ for the capacitance of a spherical capacitor. The quantity $b-a$ in the denominator is the plate separation; replace it with d. If the plate separation is small then $b \approx a$ and $ab \approx a^2$. Replace $4\pi ab$ in the numerator with A, the area of the sphere. You should obtain $C = \epsilon_0 A/d$.

Start with the equation $C = 2\pi\epsilon_0 L/\ln(b/a)$ for a cylindrical capacitor. If b is nearly the same as a then $\ln(a/b)$ can be approximated by $(b-a)/a$ and this is d/a, where d is the plate separation. Make this substitution and replace $2\pi aL$ with the plate area A to obtain $C = \epsilon_0 A/d$.

You should know how to derive an expression for the capacitance. Follow the steps used in the text to obtain expressions for parallel plate, spherical, and cylindrical capacitors.

PROBLEM 8. What is the capacitance of the two-conductor system consisting of two well-separated spherical conductors with radii R_1 and R_2?

SOLUTION: Imagine sphere 1 has charge Q and sphere 2 has charge $-Q$. Since they are far apart the charge is distributed uniformly on each sphere surface. The potential relative to infinity of sphere 1 is $V_1 = (1/4\pi\epsilon_0)Q/R_1$ and the potential relative to infinity of sphere 2 is $V_2 = -(1/4\pi\epsilon_0)Q/R_2$. Find the potential difference $V = V_1 - V_2$ and use $Q = CV$ to find an expression for the capacitance.

[ans: $4\pi\epsilon_0 R_1 R_2/(R_1 + R_2)$]

You should also know how to find the capacitance of a single conductor. Suppose the spheres are connected by a thin conducting wire to form a single conductor. What then is the capacitance?

SOLUTION: Suppose charge Q is placed on the system, with sphere 1 obtaining q_1 and sphere 2 obtaining q_2. The potential of sphere 1 is $V = (1/4\pi\epsilon_0)q_1/R_1$ and the potential of sphere 2 is $V = (1/4\pi\epsilon_0)q_1/R_2$. Since the spheres are connected by a conductor these must be equal. In addition, $q_1 + q_2 = Q$. Eliminate q_1 and q_2 and solve for V in terms of Q. Finally use $Q = CV$ to find an expression for C.

[ans: $4\pi\epsilon_0(R_1 + R_2)$]

Since the two spheres are at the same potential they may be considered to be two single-plate capacitors in parallel. Notice that the capacitance of the system is the sum of the capacitances of the spheres separately.

You should be able to work problems dealing with force, work, and energy, particularly in connection with a parallel plate capacitor.

PROBLEM 9. A 2.9-μF and a 4.6-μF capacitor are connected in parallel and a 150-V potential difference is applied across the combination. What energy is stored in each capacitor?

SOLUTION: Since they are connected in parallel the potential difference is the same across each capacitor and is 150 V. Use $U = \frac{1}{2}CV^2$.

[ans: 3.26×10^{-2} J; 5.18×10^{-2} J]

Notice that the capacitor with the larger capacitance stores more energy than the other.

If the capacitors are connected in series and the same potential difference is applied to the combination, what energy is stored in each?

SOLUTION: The capacitors now carry the same charge. Use $U = \frac{1}{2}q^2/C$. The charge can be found using $q = C_{eq}V$, where $C_{eq} = C_1C_2/(C_1 + C_2)$.

[ans: 1.23×10^{-2} J; 7.74×10^{-3} J]

PROBLEM 10. A parallel plate capacitor with a plate area of 0.85 m^2 and a plate separation of 0.15 mm is charged to 150 V. What is the charge on the positive plate?

SOLUTION:

[ans: 7.53×10^{-6} C]

What is the electric field produced by this charge at the position of the other plate?

SOLUTION: Use the expression $E = \sigma/2\epsilon_0$ for the field produced by a uniform plane of charge. Here σ is the area charge density on the plate. Notice that E is not the total field between the plates, only the field produced by charge on one of the plates.

[ans: 5.00×10^5 V/m]

What is the magnitude of the electric force exerted by one plate on the other?

SOLUTION: Use $F = qE$, where E is the magnitude of the field of one plate and q is the magnitude of the charge on the other. The force is attractive.

[ans: 3.76 N]

The battery is now removed and the plate separation is doubled. What work does the external agent do in pulling the plates apart?

SOLUTION: The agent must exert a force F equal in magnitude to the electric force and must pull one plate a distance d. The agent does work $W = Fd$.

[ans: 0.564 J]

Calculate the energy stored in the capacitor before and after the agent pulls the pates apart.

SOLUTION: Use $U = \frac{1}{2}q^2/C = \frac{1}{2}q^2 d/\epsilon_0 A$.

[ans: initial: 0.564 J, final: 1.13 J]

Notice that the change in energy equals the work done by the agent.

PROBLEM 11. A 15-μF capacitor is charged to 150 V and a 25-μF capacitor is separately charged to 100 V. The batteries are removed, the positive plates are connected with a conducting wire, and the negative plates are connected with another conducting wire. Calculate the energy stored in the capacitors both before and after the connections are made.

SOLUTION: Use $C = \frac{1}{2}V^2$ to compute the energy stored in each capacitor before the connections are made. Also compute the sum of the charges on the two positive plates: $q = q_1 + q_2 = C_1 V_1 + C_2 V_2$. After the connections are made the capacitors form a parallel combination. Calculate the equivalent capacitance, then use $U = \frac{1}{2}q^2/C_{eq}$ to calculate the energy.

[ans: before: 0.294 J; after: 0.282 J]

Notice that the energy stored decreases when the connections are made. Energy was changed to internal energy in the wires as charge was transferred from one capacitor to the other.

Some problems deal with capacitors filled with insulating material. Most of these are like the problems above except that the capacitance is increased by the factor κ_e, the dielectric constant of the insulating material.

PROBLEM 12. A cylindrical capacitor has an inner radius of 6.0 cm, an outer radius of 6.2 cm, and a length of 7.5 cm. You want to use it to hold a charge of 3.5×10^{-8} C when a potential difference of 150 V is applied. What should be the dielectric constant of the insulator between its plates?

SOLUTION: Use $C = 2\pi\epsilon_0 \kappa_e L / \ln(b/a)$ and $q = CV$ to solve for κ_e.

[ans: 1.83]

You want the capacitor to store an energy of 3.0×10^{-6} J when charged to 150 V. What then should the dielectric constant be?

SOLUTION: Use $C = 2\pi\epsilon_0 \kappa_e L / \ln(b/a)$ and $U = \frac{1}{2}CV^2$ to solve for κ_e.

[ans: 2.10]

PROBLEM 13. Two identical capacitors, each with capacitance C, are connected in parallel and a potential difference V is applied to the combination. The battery is then removed but the parallel connection is maintained. An insulator with dielectric constant κ_e is inserted into one capacitor and fills the gap. What is the charge and potential difference for each capacitor?

SOLUTION: While the battery is connected each capacitor has charge CV so the total charge on the positive plates is $Q = 2CV$. Let q_1 be the charge on the unfilled capacitor and q_2 be the charge on the filled capacitor after the battery is removed and the insulator is inserted. Then the potential difference across the unfilled capacitor is $V' = q_1/C$ and the potential difference across the filled capacitor is $V' = q_2/\kappa_e C$. These must be the same since the capacitors remain in parallel. In addition, $q_1 + q_2 = 2CV$. Solve $q_1/C = q_2/\kappa_e C$ and $q_1 + q_2 = 2CV$ for q_1 and q_2, then solve $V' = q_1/C$ for V'.

[ans: $q_1 = 2CV/(\kappa_e + 1)$; $q_2 = 2CV\kappa_e/(\kappa_e + 1)$; $V' = 2V/(\kappa_e + 1)$]

You may be asked to integrate the energy density of an electric field over a volume to obtain the energy stored in the field.

PROBLEM 14. Evaluate the volume integral of the energy density between the plates of a cylindrical capacitor to find an expression for the total energy when the charge on the positive plate is q. The capacitor has length L, inner radius a, and outer radius b.

SOLUTION: If charge q is placed on one plate and $-q$ is placed on the other, the magnitude of the electric field between the plates is given by $E = q/2\pi\epsilon_0 L r$, where r is the distance from the cylinder axis. Thus the energy density is $u = q^2/8\pi^2\epsilon_0 L^2 r^2$. Divide the cylinder into cylindrical shells, one of which has radius r and thickness dr. Its volume is $dV = 2\pi L r \, dr$. Thus the energy stored is given by

$$U = \int u \, dV = \frac{q^2}{4\pi\epsilon_0 L} \int_a^b \frac{dr}{r}.$$

Evaluate the integral.

[ans: $(q^2/4\pi\epsilon_0 L)\ln(b/a)$]

Check the answer using $U = q^2/2C$.

SOLUTION:

III. NOTES

Chapter 32
CURRENT AND RESISTANCE

I. BASIC CONCEPTS

Here you will begin the study of electric current, the flow of charge. Pay close attention to the definitions of current and current density and understand how they depend on the concentration and speed of the charged particles. For most materials a current is established and maintained only when an electric field is present. You will learn about the properties of materials that determine the magnitude and direction of the current for any given field.

Electric current and current density. As precisely as you can, tell in words what it means for a conducting wire to contain an electric current: _____
_____

The current i through any area is given by

$$i = dq/dt,$$

where dq is _____ that passes through _____ in the time interval _____. Although current is a scalar, not a vector, it is assigned a direction. Positive charge moving to the right and negative charge moving to the left, for example, are both currents in the same direction, namely to the _____. Positive charge moving to the right and negative charge moving in the same direction are currents in _____ directions: the current associated with the positive charge is to the _____ while the current associated with the negative charge is to the _____.

Explain why the flow of neutral atoms or molecules is not an electric current: _____
_____

Explain why electric current is not a vector: _____
_____

For the steady flow of charge through a conductor the current is the same for all cross sections along the conductor. Explain why: _____
_____

The SI unit for current is coulomb/second. This unit is called _____ and abbreviated _____.

Current is associated with an area, like the cross-sectional area of a conductor. On the other hand, <u>current density</u> **j** is a related quantity that can be associated with each point in a conductor. At any point it is the current per unit area passing through an area at the point, oriented perpendicularly to the charge flow, in the limit as the area shrinks to the point. Current density is a vector. If positive charge is flowing **j** is in the direction of the velocity; if negative charge is flowing **j** is in the direction opposite the velocity.

Consider any surface within a current-carrying conductor and let d**A** be an infinitesimal vector area, with direction perpendicular to the surface. If **j** is the current density, then the current i through the surface is given by the integral over the surface:

$$i = \underline{}$$

The scalar product indicates that only the component of **j** normal to the surface contributes to the current. This is consistent with the statement that charge must flow through the surface for there to be a current through the surface. If the surface is a plane with area A and is perpendicular to the particle velocity and if the current density is uniform over the surface, then $i = \underline{}$.

For a collection of electrons with uniform concentration n, each with charge $-e$ and each moving with velocity $\mathbf{v}_d$, the current density is given in terms of e, n, and $\mathbf{v}_d$ by the vector relationship

$$\mathbf{j} = \underline{}$$

This expression is derived by computing the number of electrons per unit time that pass through an area perpendicular to their velocity. Note that **j** and $\mathbf{v}_d$ are in opposite directions for negative charge carriers. As Sample Problem 2 shows, the expression above can be used to compute the drift speed v_d for a given current in a given conductor.

All electrons in a current-carrying conductor do not actually move with the same velocity, but each of their velocities can be considered to be the sum of two velocities, one of which changes direction often as the electron collides with atoms. In addition, when all the electrons are taken into account this component averages to zero. The second component, the drift velocity, is much smaller in magnitude than the first and is the same for all electrons. Except for producing extremely small, rapid fluctuations the first does not contribute to the current. Thus the velocity that enters the expression for the current density is the drift velocity $\mathbf{v}_d$, not the total velocity. Be careful to distinguish between drift and total velocities.

In an ordinary conductor an electric field is required to produce a net drift and hence an electric current. Explain what a field does: _____

_____

For electrons how is the direction of the drift velocity related to the direction of the electric field? _____ This means the current is in the same direction as the field and it is directed from a region of high electric potential toward a region of lower electric potential.

That an electric field can be maintained in a conductor does not contradict the statement proved earlier that the electrostatic field in a conductor is zero. Describe how the situation considered in this chapter differs from that considered in Chapter 29: _____

_____

Resistance and resistivity. The current in any material (except a superconductor) is zero unless an electric field exists in the material. <u>Resistance</u> is a measure of the current generated in a given conductor by a given potential difference. To determine resistance a potential difference V is applied between two points in the material and the current i is measured. The resistance for that material and those points of application is defined by

$$R = $$

Resistance has an SI unit of volt/ampere, a unit that is called _____ and abbreviated _____. In drawing an electrical circuit an element whose function is to provide resistance, called a resistor, is indicated by the symbol _____.

Resistance is intimately related to a property of the material called its resistivity. If at some point in the material the electric field is **E** and the current density is **j** then the resistivity ρ at that point is given by

$$\mathbf{E} = $$

We consider materials that are homogeneous and isotropic; the resistivity is the same at every point in the material and is the same for every orientation of the electric field.

The SI unit for resistivity is _____. Values for some materials are given in Table 1 of the text. Note that typical semiconductors have resistivities that are greater than those of metals by factors of 10^5 to 10^{11} and insulators have resistivities that are greater than those of metals by factors of 10^{18} to 10^{24}. The conductivity σ of any substance is related to the resistivity by $\sigma = $ _____.

Given the resistivity of the material and the points of application of a potential difference the resistance can, in principle, be calculated. The simplest case is that of a homogeneous wire with uniform cross section. Let L be the length of the wire, A be its cross-sectional area, and ρ be its resistivity. If one end of the wire is held at potential 0 and the other is held at potential V the electric field in the wire is given by $E = $ _____. If the current is i then, since the current density is uniform, it is given by $j = $ _____. Substitute $E = V/L$ and $j = i/A$ into $E = \rho j$ and solve for V/i ($= R$). The result is $R = $ _____. Note that the resistance depends on a property of the material (ρ) and on the geometry of the sample (L and A). Resistivity, on the other hand, is a property of the material alone.

Ohm's law. Ohm's law describes an important characteristic of the resistance of certain samples, called ohmic samples (or devices). It is: _____

Carefully note that $V = iR$ defines the resistance R and so holds for every sample. For an ohmic sample V is a linear function of i or, what is the same, R does not depend on V or i. In addition, for ohmic samples a reversal of the potential difference simply reverses the direction of the current without changing its magnitude. For some non-ohmic samples, such as a *pn* junction, a reversal of the potential difference not only changes the direction of the current but also its magnitude.

In the space below sketch graphs of the current as a function of potential difference for an ohmic and for a non-ohmic sample. Include both positive and negative values of V.

Chapter 32: Current and Resistance

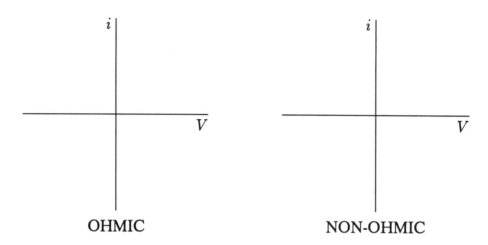

OHMIC   NON-OHMIC

For an ohmic substance the resistivity is independent of the electric field and the current density is therefore proportional to the electric field. This implies that the drift velocity is also proportional to the field. Since an electric field accelerates charges you should be surprised.

To show why Ohm's law is valid for some materials the text considers the collection of so-called free electrons in a conducting sample. These electrons are accelerated by the electric field applied to the sample and suffer collisions with atoms of the material. We suppose that the effect of a collision is to stop the drift of an electron so that, as far as drift is concerned, an electron starts from rest at a collision and is accelerated by the field until the next collision. The collision stops it and the process is repeated. The time from the last collision, averaged over all electrons, is designated by τ. Because the collisions occur at random times τ is independent of the time; its value is the same no matter when the averaging is done. In addition, τ gives the average time between collisions and is, in fact, called the <u>mean time between collisions</u> or the <u>mean free time</u>.

Since the magnitude of the acceleration is $a =$ _____, where E is the magnitude of the electric field and m is the mass of an electron, the average electron drift speed is given by $v_d =$ _____. Although each electron accelerates between collisions v_d does not vary with time if the field is constant. Also note that v_d is proportional to E if τ does not depend on E. Explain why you expect τ to be independent of E: _____

In terms of E, m, e, and τ, $j = env_d =$ _____. This expression immediately gives $\rho =$ _____ for the resistivity. Because τ is independent of E the resistivity is also independent of E and the material obeys Ohm's law.

Give a qualitative argument to convince yourself that a long mean free time leads to a small resistivity. In particular, explain why a long mean free time leads to a larger current than a short mean free time if the electric field is the same: _____

Because the mean time between collisions is determined to a large extent by thermal vibrations of the atoms, it is temperature dependent. As the temperature increases we expect an electron to suffer more collisions per unit time so τ _____ as the temperature increases.

Temperature dependence of the resistivity. The resistivity of a metal increases with increasing temperature, chiefly because the mean free time decreases. The temperature dependence is characterized by a quantity $\overline{\alpha}$, called the mean temperature coefficient of resistivity, which is a measure of the deviation of the resistivity from its value at a reference temperature. Let ρ_0 be the resistivity at the reference temperature T_0 and let ρ be the resistivity at a nearby temperature T. Then, in terms of $\overline{\alpha}$,

$$\rho - \rho_0 =$$

Temperatures are given in K or C°.

Table 1 lists values for some materials. Notice that $\overline{\alpha}$ is negative for semiconductors. For them the resistivity decreases as the temperature increases because the concentration of nearly free electrons increases much more rapidly than the mean free time decreases. For metals, on the other hand, n is essentially independent of temperature.

Energy transfers. If current i passes through a potential difference V, from high to low potential, the moving charge loses potential energy at a rate given by

$$P = dU/dt =$$

For a resistor the energy is transferred to atoms of the material in collisions. The result is usually an increase in the vibrational energy of the atoms and is accompanied by an increase in the temperature of the resistor. The phenomenon, known as Joule heating, finds practical application in toasters and electrical heaters.

Since $V = iR$ for a resistor, the expression for the rate of energy loss can be written in terms of the current i and resistance R as $P =$ _____ or in terms of the potential difference V and resistance R as $P =$ _____.

Suppose the resistance of resistor A is greater than that of resistor B. If the same potential difference is applied to them then the internal energy of resistor _____ will increase at the greater rate. If they have the same current then the internal energy of resistor _____ will increase at the greater rate.

II. PROBLEM SOLVING

To carry out calculations involving the relationship $j = env_d$ between the current density and the drift velocity you may need to compute a value for n, the concentration of nearly free electrons. Review the calculation of the electron concentration for a chemical element as presented in the Problem Solving section for Chapter 27. Recall that the mass m in grams of one atom is given by $m = A/N_a$, where A is the atomic mass number of the element and N_a is the Avogadro constant (6.022×10^{23} mol^{-1}). The number of atoms per unit volume is given by ρ_m/m, where ρ_m is the mass density. This is multiplied by the number of nearly free electrons per atom to obtain the number of nearly free electrons per unit volume in the substance. Be sure m is in kg if ρ is in kg/m^2. Here is an example to try.

PROBLEM 1. Find the nearly free electron concentration in silver, which has a mass density of 10.5×10^3 kg/m^3 and one free electron per atom. The atomic mass number can be found in an appendix of the text.

SOLUTION:

[ans: 5.86×10^{28} electrons/m^{-3}]

In general, $i = \int \mathbf{j} \cdot d\mathbf{A}$ relates the current density $\mathbf{j}$ and current i. If $\mathbf{j}$ is uniform and normal to the area being considered then $i = jA$.

PROBLEM 2. Two wires, one of aluminum and one of copper, are joined end-to-end and carry a current of 3.0 A. If each wire has a radius of 0.20 mm what is the current density in each wire? Assume it is uniform.

SOLUTION: Use $j = i/A = i/\pi r^2$. The current is the same in each wire and the wires have the same cross-sectional area so the current densities are the same.

[ans: 2.39×10^7 A/m^2]

PROBLEM 3. Two wires, one of copper and one of silver, each carry a current of 3.0 A. The copper wire has a radius of 0.20 mm and the silver wire has a radius of 0.15 mm. What is the current density in each wire? Assume it is uniform.

SOLUTION:

[ans: copper: 2.39×10^7 A/m^2; silver: 4.24×10^7 A/m^2]

Notice that the currents are equal but the current densities are not, because the cross-sectional areas are not. What is the electron drift speed in each wire?

SOLUTION: Use $j = env_d$. The nearly free electron concentration for copper is calculated in Sample Problem 2 of the text and is 8.49×10^{28} electrons/m^{-3}. You calculated the nearly free electron concentration for silver above.

[ans: copper: 1.76×10^{-3} m/s; silver: 4.52×10^{-3} m/s]

In most materials current is driven by an electric field. The resistance of the sample tells how large the current is for a given potential difference. If the resistance is large the current is small and vice versa. It is defined by $V = iR$. For ohmic samples the resistance is independent of V and i; for non-ohmic samples it is not.

PROBLEM 4. The current i in a certain device is given as a function of the potential difference V by the expression $i = 0.65V + 0.17V^2$, where i is in amperes and V is in volts. What is the resistance of the device for $V = 5.0$ V? for $V = 10$ V?

SOLUTION: Use $R = V/i$ and compute i for each value of V.

[ans: $0.667\,\Omega$; $0.426\,\Omega$]

If the current through an ohmic device for a potential difference of 5.0 V is the same as the current through this non-ohmic device for the same potential difference, what is the current through the ohmic device for a potential difference of 10 V?

SOLUTION: Calculate the resistance of the ohmic device using the current through the non-ohmic device and a potential difference of 5.0 V. Then use the same value of the resistance to calculate the current when the potential difference is 10 V.

[ans: 15 A]

You should understand how to calculate the resistance of a wire, given its length and cross-sectional area and the resistivity of the material.

PROBLEM 5. A wire 1.2 m long and 0.13 mm in radius is made of conducting material with a resistivity of 2.5×10^{-8} Ω·m. What is its end-to-end resistance if its ends are equipotential surfaces?

SOLUTION: Use $R = \rho L/A$, where ρ is the resistivity, L is the length, and A is the cross-sectional area.

[ans: 0.565 Ω]

A 1.3-V potential difference is maintained between its ends. What is the current? What is the current density?

SOLUTION:

[ans: 2.30 A; 4.33×10^7 A/m^2]

What is the electric field in the wire?

SOLUTION: For the information given this can be calculated in two ways. First use $E = V/L$, then use $E = \rho j$. You should get the same answer.

[ans: 1.08 V/m]

PROBLEM 6. A wire with resistance R is pulled through a die, thereby reducing its radius to one third its original value. What then is the resistance of the wire?

SOLUTION: Neither the resistivity or the volume of the material changes but the radius and length does. If r is the original radius then the final radius is $r' = r/3$. Since the volume does not change $\pi r^2 L = \pi r'^2 L'$ and $L' = (r/r')^2 L$. You should now be able to find an expression for the new resistance in terms of the old.

[ans: 81R]

PROBLEM 7. Each of two wires is 1.5 m long and has a radius of 0.17 mm. Wire 1 has resistivity $\rho_1 = 1.8 \times 10^{-6}$ Ω·m and wire 2 has resistivity $\rho_2 = 5.6 \times 10^{-8}$ Ω·m. They are joined end-to-end and a 2.5-V potential difference is maintained across the ends of the combination. What is the current in the wires?

SOLUTION: The current is the same in both wires so $V_1 = iR_1$ and $V_2 = iR_2$, where R_1 and R_2 are the resistances ($R_1 = \rho_1 L_1/A_1$ and $R_2 = \rho_2 L_2/A_2$). Furthermore, $V_1 + V_2 = V$, the potential difference across the combination. Thus $i(R_1 + R_2) = V$ and $i = V/(R_1 + R_2)$.

[ans: 2.04 A]

What is the potential difference across each wire?

SOLUTION: Use $V = iR$.

[ans: $V_1 = 0.608$ V; $V_2 = 1.89$ V]

What is the electric field in each wire?

SOLUTION: Use $E = V/L$.

[ans: $E_1 = 0.405$ V/m; $V_2 = 1.26$ V/m]

What is the current density in each wire?

SOLUTION: Use $j = i/A$ or $E = \rho j$.

[ans: $j_1 = j_2 = 2.25 \times 10^7$ A/m²]

If the cross-sectional areas had been different, the current densities would also have been different.

PROBLEM 8. The same two wires are isolated from each other and a potential difference of 2.5 V is maintained across each of them. What is the current in each wire?

SOLUTION: In each case use $V = iR$, with $R = \rho L/A$.

[ans: $i_1 = 8.41$ A; $i_2 = 2.70$ A]

What is the electric field in each wire?

SOLUTION: Use $E = V/L$.

[ans: $E_1 = E_2 = 1.67$ V/m]

If the lengths of the wires had been different, the field would also have been different.

What is the current density in each wire?

SOLUTION: Use $j = i/A$.

[ans: $j_1 = 9.26 \times 10^7$ A/m²; $j_2 = 2.98 \times 10^7$ A/m²]

Many problems deal with energy dissipation in a resistor. Use $P = i^2 R$ or $P = V^2/R$ to calculate the rate at which electrical energy is being changed to thermal energy.

PROBLEM 9. Heater 1 has a resistance R_1 and heater 2 has a greater resistance R_2. If the current is the same which heater delivers the greater heat?

SOLUTION: Compare $P = i^2 R$ for the two heaters.

[ans: heater 2]

If the potential difference is the same which heater delivers the greater heat?

SOLUTION: Compare $P = V^2 R$ for the two heaters.

[ans: heater 1]

PROBLEM 10. A certain light bulb converts electrical energy to internal energy at a rate of 25 W when the potential difference is 100 V. What is the rate of energy conversion if the potential difference is 50 V? Assume the resistance does not change.

SOLUTION: Use $P_1 = V_1^2/R$ to find the resistance, then $P_2 = V_2^2/R$ to find the power at the new potential difference.

[ans: 6.25 W]

PROBLEM 11. A heater has a resistance of 3.7 Ω and is operated at 100 V. At what potential difference should a second heater, with a resistance of 7.8 Ω, be operated so its thermal output is the same as the first?

SOLUTION:

[ans: 145 V]

A new potential difference is applied to the first heater and the current is then 8.0 A. What should the current be in the second heater to achieve the same thermal output as the first?

SOLUTION:

[ans: 5.51 A]

If the current density is not uniform you will need to evaluate the integral $i = \int \mathbf{j} \cdot d\mathbf{A}$ to obtain an expression for the current. Here is an example.

PROBLEM 12. A certain cylindrical conductor of radius R carries current i. The current density is not uniform but, instead, is proportional to the square of the distance from the axis. Find an expression for the current density in terms of i, R, and the distance r from the axis.

SOLUTION: Let r be the distance from the axis. Since j is proportional to r^2 you may write $j = Br^2$, where B is a constant of proportionality. You must now find an expression for B. The current is given by $i = \int_0^R j \, dA$, where dA is an element of the cross-sectional area of the conductor. Divide the circular cross section into rings, one of which has radius r and width dr. Its area is $dA = 2\pi r \, dr$, so $i = 2\pi \int_0^R r j \, dr$. Carry out the integration, solve for B in terms of i and R, then substitute the expression for B into $j = Br^2$.

[ans: $2ir^2/\pi R^4$]

What is the current density at the axis? at the rim?

SOLUTION:

[ans: $j(0) = 0$; $j(R) = 2i/\pi R^2$]

III. NOTES

Chapter 33
DC CIRCUITS

I. BASIC CONCEPTS

The concept of electromotive force (emf) is introduced in this chapter. Seats (or sources) of emf are used to maintain potential differences and drive currents in electrical circuits. You will use this concept, along with those of electric potential, current, resistance, and capacitance, learned earlier, to solve both simple and complicated circuit problems. You will also learn about energy balance in electrical circuits.

Electromotive force. A seat of electromotive force performs two functions: it maintains a potential difference and, to do that, it moves charge from one terminal to the other inside the seat. A battery contains a seat of emf as well as internal resistance.

In electrical circuits a seat of emf is indicated by the symbol _____. The arrow with the small circle at its tail points from the _____ terminal toward the _____ terminal. A seat tends to drive current in the direction of the arrow but you should remember that, for any circuit, the actual direction of the current through a seat depends on other elements in the circuit.

The emf of a seat, denoted by $\mathcal{E}$, is defined in terms of the work it does as it transports charge. If the seat does work dW on charge dq then

$$\mathcal{E} =$$

Electromotive force has an SI unit of _____.

A seat of emf does positive work if positive charge moves from the _____ to the _____ terminal and negative work if positive charge moves in the other direction. It does positive work if negative charge moves from the _____ to the _____ terminal and negative work if negative charge moves in the other direction. This can be summarized by saying a seat does positive work on the charge if the current through it is from the negative to the positive terminal and negative work if the current is in the other direction.

Positive work done by a seat of emf results in a decrease in the store of energy of the seat, chemical energy in the case of a battery. Negative work results in an increase in the energy of the seat. If the seat is a battery, then in the first case it is said to be discharging and in the second it is said to be charging.

The potential difference across a seat of emf $\mathcal{E}$ is _____, with the positive terminal being at the higher potential. This statement is true no matter what the direction of the current through the seat.

Single loop circuits. To solve single loop circuits you must be able to express the potential differences across seats of emf and resistors in terms of the emf's, resistances, and currents.

For each of the circuit elements shown below assume the potential is V_a at the left side and write an expression for the potential V_b at the right side in terms of V_a, $\mathcal{E}$, R, and i, as appropriate.

$V_b =$ $V_b =$ $V_b =$ $V_b =$

You will use Kirchhoff's second rule or loop rule to solve both single-loop and multiple-loop circuits. Write this rule in words: _____

_____

In general, the rules for solving a single loop circuit are: pick a direction for the current and draw a current arrow on the circuit diagram. For a single-loop circuit the current is the same in every circuit element. Go around the circuit and, as you do, add the changes in electric potential for the various circuit elements, with appropriate signs. For a resistor assume the current is in the direction of the arrow you drew. The diagrams above should help you. Then equate the sum to zero and solve for the unknown quantity.

Consider the single-loop circuit shown on the right and use Kirchhoff's loop rule to develop an equation that relates the emf $\mathcal{E}$, resistance R, and current i. In terms of these quantities the potential difference $V_b - V_a$ is _____ and the potential difference $V_c - V_b$ is _____. Now sum the potential differences around the loop. Since $V_c = V_a$, $(V_b - V_a) + (V_c - V_b) = 0$ or, in terms of $\mathcal{E}$, R, and i, $\mathcal{E} - iR = 0$. This equation can be solved for any one of the quantities appearing in it. If, for example, $\mathcal{E}$ and R are known, then $i = \mathcal{E}/R$.

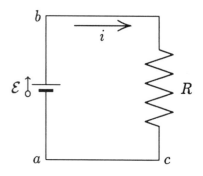

Notice that the value of i is positive, no matter what the values of $\mathcal{E}$ and R. This is because the current arrow was picked to be in the direction of the actual current. In other circuits the direction of the current might not be so obvious and you might not pick the proper direction for the current arrow. If you pick the wrong direction the value you obtain for i will be negative. Suppose, for example, the current arrow for the circuit above was picked in the opposite direction, upward through the resistor. Then the loop equation would be _____ and the expression for the current in terms of $\mathcal{E}$ and R would be $i =$ _____. The value for i is obviously negative now.

A real battery contains an internal resistance in addition to a seat of emf. Although the internal resistance is distributed inside the battery you may think of it as being in series with the seat. You must distinguish between the emf and terminal potential difference of a real battery. The latter includes the potential difference of the internal resistance. To give an example, suppose a battery is characterized by emf $\mathcal{E}$ and internal resistance r. If the current in the battery is zero then the terminal potential difference is $V =$ _____; if the battery is discharging, with current i in the direction of the emf, then $V =$ _____; and if the battery is charging, with current i opposite the emf, then $V =$ _____.

Suppose a single-loop circuit consists of a battery with emf $\mathcal{E}$ and internal resistance r, connected to an external resistance R. Then the current in the circuit is given by $i =$ _____. Study Sample Problem 2 of the text for an example of a circuit with two batteries. Note that the battery with the larger emf is discharging and determines the direction of the current while the battery with the smaller emf is charging.

Energy considerations. A seat of emf with current in the direction of the emf supplies energy to the moving charges by increasing their potential energy. If the emf is $\mathcal{E}$ and the current is i then the rate at which energy is supplied is given by $P =$ _____. Seats with current opposite the direction of the emf remove energy from the moving charges by decreasing their potential energy. The rate at which energy is removed is _____. A resistor removes energy from moving charges in collisions between the charges and atoms of the resistor. If the resistance is R and the current is i then the rate at which energy is removed is given by _____. A discharging battery, being a combination of a seat of emf and a resistor, both supplies and removes energy.

For any circuit in steady state the rate at which energy is supplied by discharging batteries exactly equals the rate at which energy is removed by resistors and charging batteries.

Multiloop circuits. To analyze multiloop circuits another rule, called <u>Kirchhoff's first rule</u> or <u>junction rule</u>, is needed in addition to Kirchhoff's loop rule. State the junction rule in words:
_____
_____

Successful analysis of a multiloop circuit depends on your ability to identify <u>junctions</u> and <u>branches</u>. A junction is _____
_____

A branch is _____
_____

The diagrams below show two junctions and the current arrows associated with them. For each write the equation that follows from the junction rule.

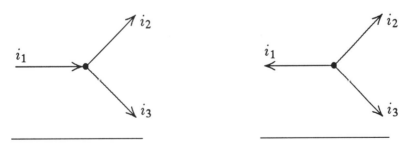

The form of the junction equation does not depend on the directions of the actual currents, only on the directions of the current arrows. Note that all of the currents in the second example above cannot have the same sign.

Chapter 33: DC Circuits   479

Consider the circuit shown to the right and assume the battery has negligible internal resistance. Identify the 4 junctions by writing the letters associated with them: _____
Identify the 6 branches by writing the letters corresponding to their end points: _____
_____

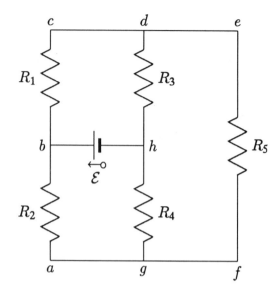

The currents in the various branches may be different. On the diagram draw a current arrow for each branch and label them i_1 through i_6, with i_j being the current in resistor R_j and i_6 being the current in the battery. The directions of the arrows are immaterial. In general for any circuit with N junctions there will be $N-1$ independent junction equations. Pick any 3 junctions for the circuit shown and write the junction equations here:

The total number of independent equations equals the number of branches so the number of independent loop equations you will use equals the number of branches minus the number of junction equations. For the circuit shown you will need $6 - 3 = 3$ loop equations. Pick 3 loops. You must be a little careful here. Every branch of the circuit must contribute to at least one of the loop equations. You cannot pick $abhg$, $bcdh$, and $acdg$, for example, since this set leaves out the branch deg. Substitute a loop containing R_5 for one of the 3 loops just mentioned. Now write the corresponding loop equations:

The 6 equations can be solved for 6 unknowns. If the emf and resistances are known then the equations can be solved for the currents, for example. If the numerical value of a current is positive the actual current is in the direction of the corresponding arrow on the diagram and electrons move opposite to the direction of the arrow. If the numerical value of a current is negative the actual current is opposite the direction of the arrow and electrons move in the direction of the arrow.

Resistors in series and parallel. Be sure you can distinguish between parallel and series combinations of two or more resistors. As for any circuit elements, the potential difference is the same for all the resistors in a _____ combination and the current is the same for all resistors in a _____ combination. If neither the potential difference nor the current is the same then the resistors do not form either a parallel or series combination.

In the space below draw two resistors connected in parallel and two connected in series. Assume a potential difference V is applied to each combination and indicate on the digrams where it is applied. Draw a current arrow for each resistor and label them to show which resistors have the same current and which have different currents.

PARALLEL   SERIES

In each case the resistors can be replaced by a single resistor with resistance R_{eq} such that the current is the same as the total current into the original combination when the potential difference is the same as the potential difference across the original combination. If the resistances are R_1 and R_2 then the value of R_{eq} for a parallel combination is given by $R_{eq} =$ _____ and the value of R_{eq} for a series combination is given by $R_{eq} =$ _____. For a parallel combination the equivalent resistance is less than the least resistance in the combination; for a series combination the equivalent resistance is greater than the greatest resistance in the combination.

RC circuits. You will now study a circuit with a time dependent current: a single-loop containing a capacitor, a resistor, and a seat of emf. Kirchhoff's loop rule is still valid and the circuit equation is derived by summing the potential differences around the loop.

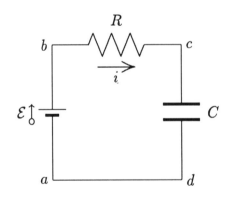

The diagram on the right shows the circuit; it might be used to charge the capacitor. $\mathcal{E}$ is the emf, R the resistance, and C the capacitance. Assume a switch (not shown) is closed at time $t = 0$, when the capacitor is uncharged. Let $i(t)$ be the current at time t, positive in the direction of the arrow; let $q(t)$ be the charge on the upper plate of the capacitor at time t. The potential difference from a to b is $V_b - V_a =$ _____; the potential difference from b to c is $V_c - V_b =$ _____; and the potential difference from c to d is $V_d - V_c =$ _____. Thus $\mathcal{E} - iR - q/C = 0$.

Both i and q are unknown, but since they are related you can eliminate one in favor of the other to obtain an equation in one unknown. Use $i = dq/dt$ to eliminate i and show that

$$R\frac{dq}{dt} + \frac{q}{C} - \mathcal{E} = 0.$$

This is a differential equation to be solved for $q(t)$.

The relationship between i and q depends in sign on the direction of the current arrow and the selection of q to represent the charge on the *upper* plate. For the choices made above positive i is equivalent to the statement that positive charge is flowing onto the upper plate and this, in turn, means dq/dt is positive. For a current arrow in the other direction $i = -dq/dt$.

The solution to the differential equation that obeys the initial condition $q(0) = 0$ is

$$q(t) =$$

This expression is differentiated with respect to time to obtain an expression for the current:

$$i(t) =$$

On the axes below sketch graphs of $q(t)$ and $i(t)$:

According to this result the current just after the switch is closed, at $t = 0$, is given by $i(0) = $ _____, just as if there were no capacitor. The results also predict that after a long time the charge on the capacitor is $q(\infty) = $ _____ and the current is $i(\infty) = $ _____. These conclusions make physical sense. At $t = 0$ the charge on the capacitor is zero so the potential difference across that element is _____. Since the circuit equation reduces to $\mathcal{E} = iR$ we expect $i = $ _____. As charge builds up on the capacitor the potential difference across it increases and, as a consequence, the potential difference across the resistor _____ (they must always sum to $\mathcal{E}$). Since the potential difference across the resistor is iR this means the current _____ with time. When the capacitor is fully charged $dq/dt = 0$ and the circuit equation becomes $\mathcal{E} = q/C$. Thus the current is _____ and the charge on the capacitor is _____.

The quantity $\tau_C = RC$ is called the <u>capacitive time constant</u> of the circuit. It has units of _____ and controls the time for the charge on the capacitor to reach any given value. Describe what happens to the curves you drew above if τ_C is made longer (by increasing C or R): _____

Suppose now that a capacitor C, initially with charge q_0, is discharged by connecting it to a resistor R. In the space to the right draw a circuit diagram with a current arrow. In terms of q and i the circuit equation is _____. Use the relationship between q and i to eliminate i and write the resulting differential equation for q:

The solution to the differential equation is

$$q(t) =$$

Note that $q(0) = q_0$ and that after a long time q becomes zero. The current as a function of time is

$$i(t) =$$

Initially the current is $i(0) = $ _____; after along time the current is _____. On the axes below sketch the behavior of $q(t)$ and $i(t)$:

II. PROBLEM SOLVING

Here is an example to help you understand potential differences and energy transfers in a circuit.

PROBLEM 1. Consider the circuit branch shown on the right, consisting of a seat of emf and a resistor. A current arrow is also shown. First suppose $\mathcal{E} = 10$ V, $R = 2.0\,\Omega$, and $i = 2.0$ A. What is the potential difference $V_a - V_b$, the rate of energy dissipation in the resistor, and the rate of energy conversion in the seat?

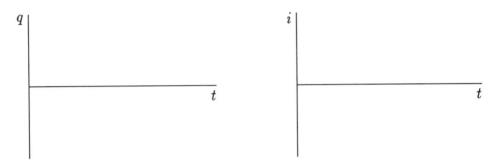

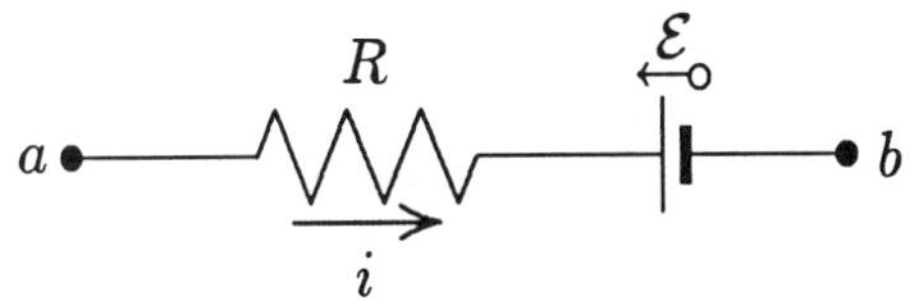

SOLUTION: Use $V_a - V_b = \mathcal{E} + iR$, i^2R and $i\mathcal{E}$. Be careful about signs.

[ans: 14 V; 8.0 W; −20 W]

The current is opposite the emf so the seat is charging and the rate of energy conversion is negative.
Now suppose $i = -2.0$ A and calculate the same quantities.

SOLUTION:

[ans: 6.0 V; 8.0 W; +20 W]

Chapter 33: DC Circuits

Suppose $V_a - V_b = 10\,\text{V}$, $R = 5.0\,\Omega$, and $\mathcal{E} = 5.0\,\text{V}$. What then is the current?

SOLUTION: Use $V_a - V_b = \mathcal{E} + iR$.

[ans: 1.0 A]

It is in the direction of the arrow.

Suppose $V_a - V_b = 1.0\,\text{V}$, $R = 5.0\,\Omega$, and $\mathcal{E} = 5.0\,\text{V}$. What then is the current?

SOLUTION:

[ans: −0.80 A]

It is opposite the direction of the arrow.

Some problems deal with various aspects of single-loop circuits. You should know how to use the loop rule to derive an equation that relates the current, emf's, and resistances. You should also know the energy relationships $P = i\mathcal{E}$ for the rate of energy exchange with a seat of emf (either charging or discharging) and $P = i^2R = V^2/R$ for the rate of energy dissipation in a resistor.

PROBLEM 2. Consider a single-loop circuit consisting of a battery with emf $\mathcal{E}$ and internal resistance r connected to an external resistor with resistance R. In terms of these quantities what is the current in the loop?

SOLUTION: In the space to the right draw a diagram of the circuit. Draw a current arrow pointing in the actual direction of the current, out of the positive terminal of the seat of emf and into the negative terminal. Write the loop equation and solve for the current i.

[ans: $\mathcal{E}/(r + R)$]

In terms of $\mathcal{E}$, r, and R what is the rate of energy dissipation in the external resistor?

SOLUTION: Use $P = i^2R$ and the expression you found for i.

[ans: $\mathcal{E}^2R/(r + R)^2$]

For what value of R is the energy dissipation a maximum?

SOLUTION: Notice that for $R \ll r$ the rate of energy dissipation increases in proportion to R and for $R \gg r$ the rate decreases in proportion to $1/R$. There is indeed a maximum somewhere. To find it solve $dP/dR = 0$ for R.

[ans: r]

For $\mathcal{E} = 5.0$ V and $r = 2\,\Omega$, what is the value of the current in the circuit at maximum dissipation? What is the rate of energy dissipation in R? What is the rate at which the seat of emf is supplying energy?

SOLUTION:

[ans: 1.25 A; 3.13 W; 6.25 W]

PROBLEM 3. A battery with an emf of 10 V and an internal resistance of $2.0\,\Omega$ is used to charge a battery with an emf of 7.0 V and an internal resistance of $1.5\,\Omega$. Assume the external resistance is negligible. What is the rate at which the smaller battery is charging?

SOLUTION: You must first find the current. Draw the circuit diagram in the space to the right, write the loop equation, and solve for i. Then use $P = i\mathcal{E}$, where $\mathcal{E} = 7.0$ V.

[ans: 6.0 W]

You can use energy balance in the circuit to check your solution for the current. What is the total rate of energy dissipation of the internal resistances of the batteries? What is the rate at which the larger battery is supplying energy?

SOLUTION:

[ans: 2.57 W; 8.57 W]

The rate at which energy is being supplied should equal the sum of the rate at which energy is entering the charging emf and the rate of at which it is being dissipated.

Chapter 33: DC Circuits

Multiloop circuits require you to use both the junction and loop rules. Here are some examples.

PROBLEM 4. Consider the circuit shown to the right. Take $i_1 = 2.0\,\text{A}$, $i_2 = 4.0\,\text{A}$, and $i_3 = 5.0\,\text{A}$. Find the values of i_4 and i_5.

SOLUTION: Apply the junction rule to the junctions at b and e (or d).

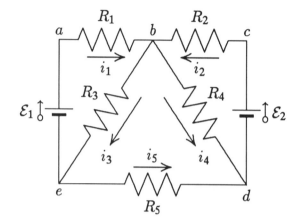

[ans: 1.0 A; 3.0 A]

Take $\mathcal{E}_1 = 20\,\text{V}$ and $R_1 = 5.0\,\Omega$. Find the value of R_3.

SOLUTION: Write and solve the equation for the loop $abea$.

[ans: $2.0\,\Omega$]

Take $R_4 = 13\,\Omega$ and find the value of R_5.

SOLUTION:

[ans: $1.0\,\Omega$]

Take $R_2 = 2.0\,\Omega$ and find the value for $\mathcal{E}_2$.

SOLUTION:

[ans: 21 V]

PROBLEM 5. Two batteries are wired in parallel and a resistor R is connected across the combination, as shown. Each battery has emf $\mathcal{E}$ and internal resistance r. What is the current in each battery and in R?

SOLUTION: Write a set of independent junction and loop equations for the circuit, then solve for the currents. There are _____ junctions and _____ branches, so you must find _____ equations, of which _____ are junction equations and _____ are loop equations. Take the current arrows in the directions shown and write the junction equation:

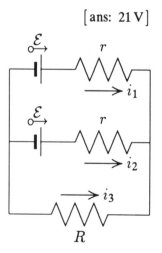

Write 2 loop equations:

486 Chapter 33: DC Circuits

Now solve for i_1, i_2, and i_3. You can simplify the algebra considerably if you recognize that i_1 must be the same as i_2 since the two branches have identical elements and the potential difference across them must be the same.

$$\left[\text{ans: } i_1 = i_2 = \mathcal{E}/(r+2R); i_3 = -2\mathcal{E}/(r+2R)\right]$$

The value for i_3 is negative, of course, because the current is actually from right to left in R.

Take $\mathcal{E} = 10$ V, $r = 1.0\,\Omega$, and $R = 10\,\Omega$. What are the currents?

SOLUTION:

$$\left[\text{ans: } i_1 = i_2 = 0.476\,\text{A}; i_3 = -0.952\,\text{A}\right]$$

What is the terminal potential difference of each battery?

SOLUTION: Use $V = \mathcal{E} - i_1 r$.

$$\left[\text{ans: } 9.52\,\text{V}\right]$$

You should know how to calculate the currents and potential differences for resistors in parallel and series and how to use equivalent resistances to solve problems.

PROBLEM 6. A 10-Ω resistor is connected in series with a 5.0-Ω resistor and a 25-V potential difference is maintained across the combination. What is the current in each resistor?

SOLUTION: The equivalent resistance is given by $R_{eq} = R_1 + R_2$, where $R_1 = 10\,\Omega$ and $R_2 = 5.0\,\Omega$. The current is the same in the two resistors and is given by $i = V/R_{eq}$, where V is the potential difference across the combination.

$$\left[\text{ans: } 1.67\,\text{A}\right]$$

What is the potential difference across each resistor?

SOLUTION: Use $V_1 = iR_1$ and $V_2 = iR_2$.

$$\left[\text{ans: } 16.7\,\text{V}; 8.33\,\text{V}\right]$$

Notice that the sum is 25 V, the potential difference across the combination.

PROBLEM 7. A 10-Ω resistor is connected in parallel with a 5.0-Ω resistor and a 25-V potential difference is maintained across the combination. What is the current in each resistor?

SOLUTION: The potential difference across each resistor is $V = 25\,\Omega$. Use $V = i_1 R_1$ and $V = i_2 R_2$.

[ans: 2.50 A; 5.0 A]

What is the total current into the combination?

SOLUTION: You should know two ways to compute the total current. First, it is the sum of the currents in the resistors: $i = i_1 + i_2$. Second, the equivalent resistance is given by $R_{eq} = R_1 R_2 / (R_1 + R_2)$ and the total current is given by $i = V/R_{eq}$. Verify that these give the same result.

[ans: 7.50 A]

PROBLEM 8. Heater 1 delivers 1200 W at a potential difference of 100 V while heater 2 delivers 1800 W. What power does each deliver if they are wired in series and a potential difference of 100 V is maintained across the combination? Assume the resistances do not change.

SOLUTION:

[ans: $P_1 = 432$ W; $P_2 = 288$ W]

PROBLEM 9. For the circuit shown, $R_1 = 7\,\Omega$, $R_2 = 15\,\Omega$, $R_3 = 20\,\Omega$, and $\mathcal{E} = 10$ V. Find the value of the current in each resistor.

SOLUTION: Use the ideas of series and parallel resistors to solve the problem. R_2 and R_3 are in parallel. Their equivalent resistance has the value $R_{23} = R_2 R_3/(R_2 + R_3) =$ _____. In the space provided redraw the circuit, replacing R_2 and R_3 with R_{23}. R_1 and R_{23} are in series. Their equivalent resistance is $R_{123} = R_1 + R_{23} =$ _____. In the space provided redraw the circuit, replacing R_1 and R_{12} with R_{123}.

The current in the circuit of the last diagram is $i_{123} = \mathcal{E}/R_{123} =$ _____. This is the same as the current i_1 in R_1 and the current i_{23} in R_{23} of the previous diagram. Now find the potential difference across R_{23}. It is $V_{23} = i_{23} R_{23} =$ _____ and is the same as the potential difference across each of R_2 and R_3 of the original circuit. Finally find the current in R_2 and R_3. Use $i_2 = V_{23}/R_1$ and $i_3 = V_{23}/R_2$.

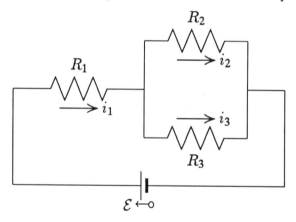

[ans: $i_1 = 0.642$ A; $i_2 = 0.367$ A; $i_3 = 0.275$ A]

488 Chapter 33: DC Circuits

Now solve the problem using the junction and loop rules.

SOLUTION: The circuit has _____ junctions and _____ branches so you should have _____ junction equations and _____ loop equations. Write them here:

Solve the equations simultaneously for i_1, i_2, and i_3. You should get the same answers as above.

Some problems deal with circuits containing capacitors. The circuit in problems 10 and 11 below consists of a capacitor in series with a resistor, just like the circuit discussed in the Basic Concepts section above. There you wrote the solutions for the charge on the capacitor and the current in the circuit as functions of time. The circuit in problem 12 is different but the junction and loop rules lead to a differential equation for the charge that is similar. It is a good test of your understanding of capacitor circuits.

PROBLEM 10. A 7.5-μF capacitor is connected in series to a 5.5-kΩ resistor and a 10-V seat of emf. What is the final charge on the capacitor?

SOLUTION: When the capacitor is fully charged the current is zero, so the potential difference across the resistor is also zero. The potential difference across the capacitor is $\mathcal{E}$, so $q = C\mathcal{E}$.

[ans: 7.5×10^{-5} C]

If the capacitor is uncharged at time $t = 0$, when the connection is made, at what time is its charge half the final charge? Express your answer in units of seconds and in units of the capacitive time constant for the circuit.

SOLUTION: Use $q = C\mathcal{E}(1 - e^{-t/RC})$. When the capacitor is half charged, $q/C\mathcal{E} = 0.5$ so $0.5 = 1 - e^{-t/RC}$, or $0.5 = e^{-t/RC}$. Take the natural logarithm of both sides and multiply by RC to obtain $t = -RC \ln 0.5$.

[ans: 2.86×10^{-2} s; 0.693]

What is the current then?

SOLUTION: Substitute into $i(t) = (\mathcal{E}/R)e^{-t/RC}$.

[ans: 9.09×10^{-4} A]

How much additional time is required for the capacitor to reach three-fourths its final charge?

SOLUTION:

$$\left[\text{ans: } 2.86 \times 10^{-2}\,\text{s}\right]$$

Notice this is the same as the time required for the capacitor to become half charged. In fact, the time to go halfway from *any* initial charge to the final charge is the same, no matter what the starting point. This is a property of the exponential form of the solution for $q(t)$.

What is the current when the capacitor is three-fourths charged?

SOLUTION:

$$\left[\text{ans: } 4.55 \times 10^{-4}\,\text{A}\right]$$

Notice that the capacitor is now charging at exactly half the rate it was charging when it was half charged.

PROBLEM 11. Consider the circuit of the last problem and suppose the current starts at time $t = 0$. How much energy is supplied by the seat of emf during the first 0.020 s?

SOLUTION: The rate at which the seat supplies energy is given by $P_\mathcal{E} = i\mathcal{E} = (\mathcal{E}^2/R)e^{-t/RC}$, where $i = (\mathcal{E}/R)e^{-t/RC}$ was used. The energy supplied is given by $E_\mathcal{E} = \int_0^t P_\mathcal{E}\,dt = \int_0^t (\mathcal{E}^2/R)e^{-t/RC}\,dt$. Evaluate the integral.

$$\left[\text{ans: } 2.88 \times 10^{-4}\,\text{J}\right]$$

How much energy is dissipated in the resistor during the first 0.020 s?

SOLUTION: The rate of dissipation is $P_R = i^2 R = (\mathcal{E}^2/R)e^{-2t/RC}$ and the energy dissipated is $E_R = \int_0^t P_R\,dt = \int_0^t (\mathcal{E}^2/R)e^{-2t/RC}\,dt$. Evaluate the integral.

$$\left[\text{ans: } 2.33 \times 10^{-4}\,\text{J}\right]$$

How much energy is stored in the capacitor during the first 0.020 s?

SOLUTION: Use $E_C = q^2/2C = (\mathcal{E}^2 C/2)[1 - e^{-t/RC}]^2$.

$$\left[\text{ans: } 5.54 \times 10^{-5}\,\text{J}\right]$$

Note that $E_\mathcal{E} = E_R + E_C$.

PROBLEM 12. Consider the circuit shown to the right. Resistor R_2 and capacitor C are in parallel and the combination is in series with a seat of emf $\mathcal{E}$ and another resistor R_1. At time $t = 0$ the capacitor is uncharged. What is the charge on the capacitor as a function of time?

SOLUTION: Use the current arrows shown and let q be the charge on the upper plate of the capacitor. Write one junction and two loop equations.

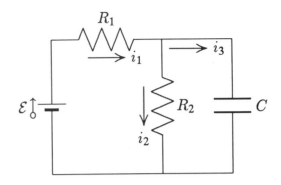

[ans: $i_1 = i_2 + i_3$; $\mathcal{E} - i_1 R_1 - i_2 R_2 = 0$; $\mathcal{E} - i_1 R_1 - q/C = 0$]

Another set of equations is possible: either of the loop equations can be replaced with $i_2 R_2 - q/C = 0$.

Use two of the equations to eliminate i_1 and i_2 from the third and replace i_3 with dq/dt to obtain a differential equation for $q(t)$.

[ans: $dq/dt + (q/C)(R_1 + R_2)/R_1 R_2 = \mathcal{E}/R_1$]

Compare this equation with Eq. 29 of the text and use Eq. 31 as a model to write the solution, an expression for $q(t)$. Notice that Eq. 29 matches the equation you derived if R is replaced by $R_1 R_2/(R_1 + R_2)$ and $\mathcal{E}$ is replaced by $\mathcal{E} R_2/(R_1 + R_2)$.

SOLUTION:

[ans: $[CR_2\mathcal{E}/(R_1 + R_2)][1 - e^{-t/\tau}]$, where $\tau = CR_1 R_2/(R_1 + R_2)$]

III. COMPUTER PROJECTS

Many circuit problems involve the solution of simultaneous linear equations. They can be solved on a computer. We describe what is known as the Gauss-Seidel iteration scheme, in which a solution is guessed and the given equations are used to improve the guess.

Suppose there are N equations and the unknowns are $i_1, i_2, \ldots, i_N$. For many problems they are the currents in the various branches of a circuit. Equation number j may be written

$$\sum_{k=1}^{N} A_{jk} i_k = B_j$$

where A_{jk} is the coefficient of unknown i_k in equation j and B_j is a term that contains no unknown. If, for example, the set of equations to be solved is

$$3i_1 + 2i_2 + 2i_3 = 8$$
$$3i_1 + 4i_2 + 3i_3 = 5$$
$$7i_1 + 5i_2 + 3i_3 = 3,$$

then $A_{11} = 3$, $A_{12} = 2$, $A_{13} = 3$, $A_{21} = 3$, $A_{22} = 4$, $A_{23} = 3$, $A_{31} = 7$, $A_{32} = 5$, $A_{33} = 3$, $B_1 = 8$, $B_2 = 5$, and $B_3 = 3$.

Solve the first equation for i_1 in terms of the other unknowns, the second equation for i_2 in terms of the other unknowns, etc. The result is

$$i_j = \frac{\left[B_j - \sum_{\substack{k=1 \\ k \neq j}}^{N} A_{jk} i_k\right]}{A_{jj}}.$$

Notice that the sum contains all terms except the one for unknown j. For the set of equations above you would write $i_1 = (8 - 2i_2 - 2i_3)/3$, $i_2 = (5 - 3i_1 - 3i_3)/4$, and $i_3 = (3 - 7i_1 - 5i_2)/3$.

The first step is to guess values for $i_1, i_2, \ldots, i_N$. These guesses are used in the above equations to calculate new values. The process is then carried out again using the results of the first run. Iteration is continued until two successive runs yield the same results to within an acceptable error. The most current values are used on the right side of the equations as soon as they are calculated.

Care must be taken to arrange the equations so A_{jj} is not zero for any equation in the set. Even so, the results do not converge for some sets of equations. After many iterations successive results may differ greatly and may show no sign of getting closer in value. When this occurs, the original set of equations must be modified by adding (repeatedly, perhaps) some equations to others or subtracting some equations from others and using the resulting equation to replace one of the originals. Such manipulations do not change the solution.

We state without proof that the Gauss-Seidel iteration scheme converges toward the correct solution if, for every equation in the set, the so-called diagonal term (A_{jj} for equation j) is larger in magnitude than the sum of the magnitudes of the other coefficients in the equation. That is,

$$|A_{jj}| > \sum_{\substack{k=1 \\ k \neq j}}^{N} |A_{jk}|.$$

The goal of any modifications made to the original set is to obtain a new set for which this inequality holds.

The first step in the modification process is to put the equations in optimal order. Search for the largest coefficient and arrange the equations so this coefficient becomes a diagonal coefficient. Now search for the largest coefficient that is not in the same equation and does not multiply the same unknown as the first one found, then arrange the equations so this one is also diagonal. Continue until all equations have been considered. For example, in the set of equations given above the largest coefficient is 7. It multiplies i_1 in the third equation, so this equation should be the first. Of the coefficients that multiply other unknowns in other equations, the largest is 4, which multiplies i_2 in the second equation, so this equation should remain the second. The optimal order is

$$7i_1 + 5i_2 + 3i_3 = 3$$
$$3i_1 + 4i_2 + 3i_3 = 5$$
$$3i_1 + 2i_2 + 2i_3 = 8.$$

None of these equations obey the inequality. Subtract the second from the first and use the result to replace the first. The new set is

$$4i_1 + i_2 = -2$$
$$3i_1 + 4i_2 + 3i_3 = 5$$
$$3i_1 + 2i_2 + 2i_3 = 8.$$

Now the first equation obeys the inequality but the others do not. Subtract the third from the second and use the result to replace the second. The equations are now

$$4i_1 + i_2 = -2$$
$$2i_2 + i_3 = -3$$
$$3i_1 + 2i_2 + 2i_3 = 8.$$

Both the first and second equations obey the inequality. To bring the third into line, multiply the first by 3, the third by 4, and subtract. Replace the third equation with the result. The new set is

$$4i_1 + i_2 = -2$$
$$2i_2 + i_3 = -3$$
$$5i_2 + 8i_3 = 38.$$

All three equations now satisfy the inequality and we expect the Gauss-Seidel iteration scheme to work. The scheme may work even if the inequality is not satisfied, so you may want to run through a few iterations before spending time modifying the equations.

Write a program to solve a set of linear simultaneous equations. Store the coefficients in a subscripted variable $A(j,k)$, the constant terms in the subscripted variable $B(j)$, and the unknowns in the subscripted variable $I(j)$. Take the initial guesses to all be zero. Here's an outline.

> input number of equations: N
> **begin loop** over equations; counter j runs from 1 to N
>   **begin loop** over variables; counter k runs from 1 to N
>   input coefficients: $A(j,k)$
>   **end loop** over variables
>   input constant term: $B(j)$

                set $I(j) = 0$
            end loop over equations
        * begin loop over equations; counter j runs from 1 to N
                set $S = 0$ in preparation for computing sum
                begin loop over variables; counter k runs from 1 to N
                    if $k \neq j$ replace S with $S + A(j,k)I(k)$
                end loop over variables
                replace $I(j)$ by $[B(j) - S]/A(j,j)$
            end loop over equations
            display solution
            go back to starred instruction for another iteration or stop

When the solution is displayed you must judge whether convergence has been reached or not. If values of $I(j)$ have not changed much from the last iteration you will want to stop the program. If they have changed you will want the program to perform another iteration. You might arrange the display so that results of two or more iterations are on the monitor screen simultaneously.

PROJECT 1. To test the program, use it to solve the following set of 4 simultaneous equations. Obtain 3 significant figure accuracy.

$$2i_1 - i_2 = 5$$
$$2i_2 - i_3 = 7$$
$$2i_3 - i_4 = 9$$
$$2i_4 - i_1 = 11.$$

[ans: $i_1 = 6.47$, $i_2 = 7.93$, $i_3 = 8.87$, $i_4 = 8.73$]

PROJECT 2. Consider the circuit shown to the right with current arrows and labels. Write down 2 junction and 3 loop equations. Modify the set of equations so the convergence conditions are met. Take $\mathcal{E}_1 = 10$ V, $R_1 = R_2 = 5\,\Omega$, $R_3 = R_4 = 8\,\Omega$, and $R_5 = 12\,\Omega$. Use the program to find values of the 5 currents for each of the following values of $\mathcal{E}_2$: 5, 7.5, 10, 12, and 15 V. Assume the given values are exact and find the solutions with an accuracy of at least 3 significant figures.

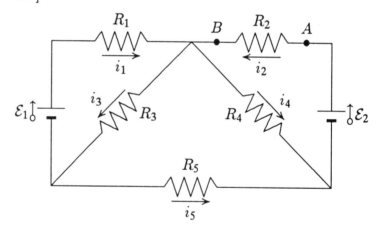

For each value of $\mathcal{E}_2$ tell if the seats of emf are charging or discharging.

PROJECT 3. Suppose the ends of resistor R_2 in the circuit of the previous project are also connected to an external circuit. Current i_6 enters the external circuit at A and returns to the circuit of the diagram at B.

Take $\mathcal{E}_2 = 15$ V and the other quantities as given in the previous project. For each of the following values of i_6 solve for the values of the other currents: 0, 2, 4, 6, 8, and 10 A. Obtain 3 significant figure accuracy.

For each of the cases considered calculate the potential difference ΔV across R_2. Plot ΔV vs. i_6. Notice that it is a straight line. As far as the external circuit is concerned the circuit shown in the diagram above can be replaced by a seat of emf and a resistor in series. The emf has the value of ΔV for $i_6 = 0$ and the value of the resistance is the slope of the line. These values are $\mathcal{E} = $ _____ V and $R = $ _____ Ω.

We now use the program to investigate the operation of a measuring instrument. In this case the instrument is a Wheatstone bridge and is used to measure resistance. It is typical of many different bridge circuits used for various electrical measurements.

PROJECT 4. The circuit for a Wheatstone bridge is shown to the right. The symbol G stands for a galvanometer, an instrument that can detect small currents. We suppose the unknown resistor is R_3 and the other resistors are variable. They are set so the galvanometer reads 0 and $i_6 = 0$. Then $R_3 = R_1 R_4 / R_2$. The resistances R_1, R_2, and R_4 are read and their values are used to calculate R_3.

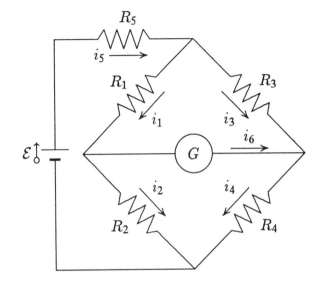

First verify that $i_6 = 0$ when $R_3 = R_1 R_4 / R_2$. Take $R_1 = R_2 = 12\,\Omega$, $R_3 = R_4 = 18\,\Omega$, and $R_5 = 3.2\,\Omega$. The resistance of the galvanometer is $2.0\,\Omega$ and $\mathcal{E} = 10$ V. Note that the balance condition is met. Write 3 junction and 3 loop equations, then use the program to solve for the 6 currents. Obtain 3 significant figure accuracy. Don't forget to modify the equations so the convergence conditions are met. You should find that $i_6 = 0$ to within the accuracy of the calculation.

It is usually of some interest to know how sensitive a Wheatstone bridge is. You can test the sensitivity by making an error in the setting of one of the resistors, then seeing if the galvanometer can detect the resulting current i_6.

Take $R_1 = R_2 = 12\,\Omega$, $R_3 = 18\,\Omega$, $R_4 = 19\,\Omega$, and $R_5 = 3.2\,\Omega$. R_g is still $2.0\,\Omega$ and $\mathcal{E}$ is still 10 V. The balance condition is not met, the current in the galvanometer does not vanish, and the balance equation predicts that $R_3 = 19\,\Omega$ instead of the correct value, $18\,\Omega$. Use the program to solve for i_6. The galvanometer must be able to detect a current of this value or less if the bridge is to measure the unknown resistance with an error of less than $1\,\Omega$. Is this a reasonable current for a galvanometer to detect? [ans: -7.67×10^{-3} A; easily detected]

IV. NOTES

Chapter 34
THE MAGNETIC FIELD

I. BASIC CONCEPTS

Charged particles, when moving, exert magnetic as well as electric forces on each other. Magnetic fields are used to describe magnetic forces and, although the geometry is a little more complicated, the idea is much the same as the idea of using electric fields to describe electric forces. In the next chapter you will learn how a magnetic field is related to the motion of the charges that produce it. Here you will learn about the force exerted by a magnetic field on a moving charge and on a wire carrying an electric current. In each case pay particular attention to what determines the magnitude and direction of the force. As you read the chapter note similarities and differences in electric and magnetic forces.

Magnetic force on a moving charge. _____ charges exert magnetic forces on other _____ charges. The view taken is that a moving charge creates a <u>magnetic field</u> in all of space and this magnetic field exerts a (necessarily magnetic) force on any other moving charge. Charges at rest do not produce magnetic fields nor do magnetic fields exert forces on them. A moving charge does not exert a magnetic force on itself.

The force exerted by the magnetic field **B** on a charge q moving with velocity **v** is given by

$$\mathbf{F} =$$

Notice that the force is written in terms of the vector product $\mathbf{v} \times \mathbf{B}$. You might review Section 3–5 of the text if you do not remember how to determine the magnitude and direction of a vector product. The magnitude of the magnetic force on the moving charge is given by $F =$ _____, where ϕ is the angle between **v** and **B** when they are drawn with their tails at the same point. The direction of $\mathbf{v} \times \mathbf{B}$ is determined by the right-hand rule: _____

Carefully note that the magnetic force is always perpendicular to both the magnetic field and to the velocity of the charge. Because the force is perpendicular to the velocity a static magnetic field cannot change the speed or _____ energy of a charge. It can, however, change the direction of motion.

Also note that the sign of the charge is important for determining the direction of the force. A positive charge and a negative charge moving with the same velocity in the same magnetic field experience magnetic forces in _____ directions.

If a proton is traveling in the positive x direction in a region in which the magnetic field is in the positive z direction, the magnetic force on it is in the _____ direction. If an electron is traveling in the same direction in the same region, the force on it is in the

_____ direction. If either particle is traveling parallel to the magnetic field the force on it is _____.

The magnetic field can be defined in terms of the force on a moving positive test charge. First, the electric force is found by measuring the force on the test charge when it is _____. This force is subtracted vectorially from the force on the test charge when it is moving in order to find the magnetic force. Second, the direction of the magnetic field is found by causing the test charge to move in various directions and seeking the direction for which the magnetic force is _____. Lastly, the test charge is given a velocity perpendicular to the field and the magnetic force on it is found. If its charge is q_0, its speed is v, and the magnitude of the force on it is F, then the magnitude of the magnetic field is given by $B = $ _____.

The SI unit for a magnetic field is _____ and is abbreviated _____. In terms of the units coulomb, kilogram, and second, $1\,\text{T} = 1$ _____. Another unit in common use is the gauss. $1\,\text{T} = $ _____ gauss.

Magnetic field lines. Magnetic field lines are drawn so that at any point the field is tangent to the line through that point and so that the magnitude of the field is proportional to the number of lines through a small area perpendicular to the field. Recall that electric field lines are drawn in the same manner.

The diagram on the right shows the magnetic field lines in a certain region of space. On the diagram label a region of high magnetic field and a region of low magnetic field. Suppose an electron is moving out of the page at the point marked with the symbol • and draw an arrow to indicate the direction of the magnetic force on it.

Magnetic field lines form closed loops. They continue through the interior of a magnet, for example. The end of a magnet from which they emerge is called a _____ pole; the end they enter is called a _____ pole. If the magnet is free to rotate in the earth's magnetic field, the north pole of the magnet will tend to point toward the _____ geographic pole.

The Lorentz force. If both a magnetic field **B** and an electric field **E** act simultaneously on a charge q moving with velocity **v**, the total force on the charge is given by

$$\mathbf{F} = $$

This combined electric and magnetic force is called the Lorentz force.

An electric force can be used to balance a magnetic force on a charge, but its magnitude and direction depend on the velocity of the charge. In particular, if a charge is in a magnetic field **B** and has velocity **v** then the total force is zero if **E** = _____. Notice that this result does not depend on either the sign or magnitude of the charge. Also notice that the balancing electric field is perpendicular to both the magnetic field and the velocity of the charge. An important special case occurs if the velocity of the charge is perpendicular to the magnetic field. Then the magnitude of the electric field is given by $E = $ _____.

Perpendicular electric and magnetic fields form the basis for a <u>velocity selector</u>, used to select charges with a given speed from among a group of charges with a variety of speeds.

Uniform fields **E** and **B**, with known values and perpendicular to each other, are established in a region of space and the charges are incident in a direction that is perpendicular to both fields. Those with a speed given by $v =$ _____ continue straight through the region without being deflected while those with other speeds are deflected. By changing the ratio of the field strengths different sets of charges can be selected.

Crossed electric and magnetic fields were used to measure the charge-to-mass ratio e/m of an electron. Use the space to the right below to draw a diagram and the space to the left to outline the steps in the measurement.

_____
_____
_____
_____
_____
_____
_____
_____
_____

Cyclotron motion. Since a magnetic force is always perpendicular to the velocity of the charge on which it acts it can be used to hold a charge in a circular orbit. All that is required is to fire the charge perpendicularly into a uniform field. Suppose a region of space contains a uniform magnetic field with magnitude B and a particle with charge of magnitude q is given a velocity **v**, perpendicular to the field. Equate the magnitude of the magnetic force (qvB) to the product of the mass and acceleration (mv^2/r) and solve for the radius r of the orbit. In terms of q, m, v, and B the result is

$$r =$$

or, in terms of the magnitude of the momentum p of the charge

$$r =$$

The first of these expressions for r must be modified if the particle is relativistic but the second is correct as it stands.

The time taken by the charge to go once around its orbit does not depend on its speed because the radius of the orbit and hence the distance traveled are directly proportional to the _____. You can see how this comes about by developing an expression for the period. In terms of the radius r and speed v, the period is given by $T =$ _____. Now substitute the expression for r in terms of the magnetic field, mass, charge, and speed. The result, $T = m/qB$, is independent of v. The number of times the charge goes around per unit time, or the frequency of the motion, is given by $\nu = 1/T =$ _____. The angular frequency is given by $\omega = 2\pi\nu =$ _____.

These results are important for the operation of a cyclotron, a type of particle accelerator. The two dees of a cyclotron are diagramed to the right. A uniform magnetic field is everywhere out of the page and an electric field is in the region _____.
A positive charge enters the cyclotron near the center. The _____ field causes the charge to travel in a circular orbit. The _____ field causes the speed of the charge to increase and when it does the charge moves to an orbit of larger radius. Assume the charge goes around four times before leaving the accelerator and draw the path on the diagram. In reality charges circulate many thousands of times.

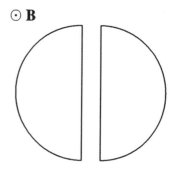

For the electric field to increase the speed of a positive charge each time the charge encounters it, it must be in the direction of motion. For it to increase the speed of a negative charge it must be opposite the direction of motion. In either case its direction must be reversed twice each orbit. This is fairly easy to do since the time between reversals does not change as long as the particle speed is significantly less than the speed of light. At speeds near the speed of light the period does depend on the particle speed and the interval between reversals of the electric field must be _____ as the charge speeds up. Accelerators that do this are called _____.

The Hall effect. A Hall effect experiment is one of the few experiments that can be carried out to determine the sign of the charge carriers in a current-carrying sample. The diagram shows a rectangular sample of width w and thickness t, carrying a current i, in a uniform magnetic field that points out of the page. Assume the current consists of positive charge moving in the direction of the current arrow. The magnetic field pushes them to one side of the sample — mark that side with a series of + signs and the opposite side with a series of − signs.

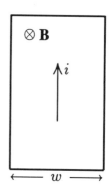

As charge accumulates at the sample sides it creates an electric field in the sample, transverse to the current. For the situation described above the electric field points from the _____ side to the _____ side. Charge continues to accumulate until the electric force on charges in the current is exactly the right strength to balance the magnetic force. In terms of the charge q on each particle, their drift speed v_d, and the magnetic field B, the final electric field strength is given by $E =$ _____. Once this condition is met, usually in times of about 10^{-13} or 10^{-14} s, charge in the current is no longer deflected and no additional charge accumulates at the sample sides.

A potential difference is associated with the transverse electric field. Since the charge carriers are positive, a point on the _____ side of the sample is at a higher potential than the point directly opposite on the other side.

If the current consists of negative charges going in the direction opposite to the current

500   Chapter 34: The Magnetic Field

arrow they are forced to the _____ side by the magnetic field. The final electric field now points from the _____ side toward the _____ side and the _____ side is at the higher potential. Thus the sign of the charge carriers can be found by noting the sign of the deflection of a voltmeter attached across the sample width.

In the usual experiment the transverse potential difference, the current, and the magnetic field are measured and the data is used to calculate the carrier concentration n. No matter what the sign of the carriers the magnitude of the transverse potential difference is given by $V = Ew$, where E is the magnitude of the transverse electric field. If only one type of carrier is present in the current the current is given by $i = qnAv_d$, where $A\,(= wt)$ is the cross-sectional area of the sample (see Eq. 6 of Chapter 32). Substitute $E = V/w$ and $v_d = i/qnwt$ into $E = v_d B$ to find $n = iB/qVt$.

Magnetic force on a current. A current is just a collection of moving charge and so experiences a force when a magnetic field is applied. If the current is in a wire the force is transmitted to the wire itself.

To calculate the force on a wire carrying current i, divide the wire into infinitesimal segments, calculate the force on each segment, then vectorially sum the forces. Let ds be the length of an infinitesimal segment of wire with cross-sectional area A. If n is the concentration of charge carriers and each carrier has charge q then the total charge in the segment is given by $qnA\,\mathrm{d}s$. The magnetic force on the segment is the product of this and the force on a single charge. That is, $\mathrm{d}\mathbf{F} = qnA\mathbf{v}_d \times \mathbf{B}\,\mathrm{d}s$, where $\mathbf{v}_d$ is the drift velocity. Notice that the combination $qnAv_d$ appears in the expression for the magnitude of the force. This is the current i. If we take the vector d$\mathbf{s}$ to be in the direction of $q\mathbf{v}_d$ (i.e. in the direction of the current arrow) then the force can be written in terms of i, d$\mathbf{s}$, and $\mathbf{B}$: $\mathrm{d}\mathbf{F} =$ _____.

For a finite wire the resultant force is given by the integral

$$\mathbf{F} =$$

Check to be sure you have the correct order for the factors in the vector product. For a straight wire of length L in a uniform field $\mathbf{B}$, perpendicular to the wire, the magnitude of the force is $F =$ _____. Explain how the direction of the force is found: _____

For a wire of any shape in a *uniform* magnetic field the magnetic force is given by $\mathbf{F} = i\mathbf{L} \times \mathbf{B}$, where $\mathbf{L}$ is the vector from the end where the current enters to the end where the current exits the wire. Since the field is uniform it can be factored from the integral for the force, with the result $\mathbf{F} = i(\int \mathrm{d}\mathbf{s}) \times \mathbf{B}$. The integral $\int \mathrm{d}\mathbf{s}$ is just $\mathbf{L}$. For a closed loop $\mathbf{L} = 0$ and the total force of a uniform field is zero. The force of a non-uniform field is not zero.

Torque on a current loop. Although a uniform field does not exert a net force on a closed loop carrying current, it may exert a torque. Thus the center of mass of the loop does not accelerate in a uniform field but the loop may have an angular acceleration about the center of mass.

Consider the rectangular loop of wire shown, in the plane of the page and carrying current i. A uniform magnetic field points from left to right. The force on the upper segment is zero since the current there is parallel to the field. The torque exerted on it is also zero. Both the force and torque on the lower segment are zero for the same reason. The force on the left segment has magnitude iaB and is directed into the page. The torque on this segment about the center of the loop has magnitude $iabB/2$ and is directed toward the bottom of the page. The force on the right segment has magnitude _____ and is directed toward _____. The torque on this segment about the center of the loop has magnitude _____ and is directed toward _____. The total force is _____ and the total torque has magnitude _____ and is directed toward _____. The loop tends to turn so its _____ side comes out of the page and its _____ side goes into the page. Use a dotted line to show the axis of rotation on the diagram.

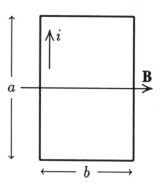

For the orientation shown above the magnitude of the torque is a maximum. It decreases as the loop turns. The diagram on the right shows the view looking down on the loop from the top of the page after it has turned through the angle θ. The magnitude of the torque on it is now given by _____. The torque always tends to orient the loop so it is _____ to the magnetic field and when it has this orientation the torque is _____.

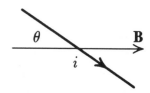

The torque exerted by a uniform field on any planar current loop is easily expressed in terms of the <u>magnetic dipole moment</u> μ of the loop. For a loop with area A, having N turns and carrying current i, the magnitude of the dipole moment is given by $\mu = $ _____. The direction of the dipole moment is determined by the right hand rule: _____
_____

A magnetic dipole moment is not the same as an electric dipole moment. Do not confuse them.

The torque exerted by a field **B** on a loop with magnetic dipole moment μ is given by

$$\tau = $$

Its magnitude is given by _____, where ϕ is the angle between μ and **B** when they are drawn with their tails at the same point.

The magnitude of the dipole moment of the loop diagramed above is $\mu = $ _____. On the diagram above draw a vector representing the dipole moment and label the angle ϕ. The vector product $\mu \times \mathbf{B}$ has magnitude _____ and direction _____. The torque exerted by a uniform field on a current loop always tends to align the dipole moment of the loop with the _____. Your answers should be in agreement with the results of the direct calculation of the torque.

Although magnetic fields are not conservative and a potential energy cannot be associated with a charge in a magnetic field, a potential energy can be associated with a magnetic dipole

μ in a magnetic field **B**. It is given by

$$U = $$

The potential energy is a minimum when the dipole moment is _____ to the magnetic field and is in the _____ direction. It is a maximum when the dipole moment is _____ to the magnetic field and is in the _____ direction. In both these cases the plane of the loop is _____ to the field. If the loop is in the plane of the page and the field is pointing toward you, then the direction of the current is _____ at a minimum of potential energy and is _____ at a maximum.

Suppose a dipole initially makes the angle θ_i with the magnetic field, but then rotates so it makes the angle θ_f. During this rotation the work done on the dipole by the field is given by $W = $ _____.

Many fundamental particles have intrinsic magnetic dipole moments. The dipole moment of an electron is _____ J/T; the magnetic dipole moment of a proton is _____ J/T.

II. PROBLEM SOLVING

You should be able to calculate the magnetic force on a given charge with a given velocity in a given magnetic field. You should also be able to calculate the Lorentz force when both electric and magnetic fields are present.

PROBLEM 1. At one instant an electron is traveling toward the northeast at 7.3×10^5 m/s in a uniform magnetic field of 0.85 T directed to the east. What is the magnetic force on it?

SOLUTION: Use $F = qBv \sin \theta$ with $q = 1.60 \times 10^{-19}$ C, $B = 0.85$ T, $v = 7.3 \times 10^5$ m/s, and $\theta = 45°$. Use the right hand rule to find the direction of the force. Don't forget that the electron is negatively charged.

[ans: 7.03×10^{-14} N, up]

A proton is traveling with the same speed, in the same direction, and in the same field. What is the magnetic force on it?

SOLUTION:

[ans: 7.03×10^{-14} N, down]

Both charges continue moving undeflected toward the northeast because there is also an electric field in the region. What is the electric field?

SOLUTION: Use $\mathbf{E} + \mathbf{v} \times \mathbf{B} = 0$.

[ans: 4.39×10^5 V/m, up]

If the electric field is doubled in magnitude what is the speed of a charge that moves undeflected toward the northeast?

SOLUTION:

[ans: 1.46×10^6 m/s]

PROBLEM 2. A proton (charge = 1.60×10^{-19} C, mass = 1.67×10^{-27} kg) initially moves horizontally at 5.5×10^7 m/s in a region where the magnetic field is upward and has a magnitude of 0.35 T. What is the radius of its orbit? Assume non-relativistic mechanics is valid.

SOLUTION: Use $qvB = mv^2/R$.

[ans: 1.64 m]

Suppose the velocity of the proton is toward the north at one instant of time. In what direction will its velocity be one quarter cycle later?

SOLUTION:

[ans: toward the east]

Find the error made by assuming non-relativistic mechanics: use relativistic mechanics to find the radius.

SOLUTION: Use $qB = p/R$, with $p = mv/\sqrt{1 - v^2/c^2}$.

[ans: 1.67 m]

For the given data the difference is not significant.

What is the period of the motion?

SOLUTION: Use $T = 2\pi R/v$ or $2\pi m/qB$.

[ans: 1.87×10^{-7} s]

Sometimes the kinetic energy is given rather than the speed. If the proton has an energy of 12 MeV what is its speed and the radius of its orbit?

SOLUTION: Assume classical mechanics is valid and use $E = \frac{1}{2}mv^2$ to find the speed, then find the radius as before. You must convert the energy to joules. Recall that $1\,\text{eV} = 1.60 \times 10^{-19}$ J.

[ans: 4.80×10^{-7} m/s; 1.43 m]

PROBLEM 3. Singly ionized deuterons (mass = 3.35×10^{-27} kg, charge = 1.60×10^{-19} C) are accelerated in a cyclotron with an outer radius of 1.50 m and a magnetic field of 0.700 T. What is their speed when they emerge? What is their kinetic energy?

SOLUTION: The final orbit before they emerge is a circle with a radius of 1.50 m. Solve $qvB = mv^2/r$ for v. Use $K = \frac{1}{2}mv^2$ to calculate the kinetic energy.

[ans: 5.03×10^7 m/s; 4.23×10^{-12} J (26.4 MeV)]

What is the period of the motion? What is the frequency of the electric field used to accelerate the deuterons?

SOLUTION:

[ans: 5.26×10^{-8} s; 19.0 MHz]

PROBLEM 4. In a mass spectrometer atoms are ionized at A and the ions are accelerated through a potential difference V. They then enter a region, outlined by dashed lines, where a uniform magnetic field **B** is out of the page and they move on the circular path shown. They emerge from the field a distance d from the entry point. Find an expression for the charge-to-mass ratio q/m of the ions in terms of V, B, and d.

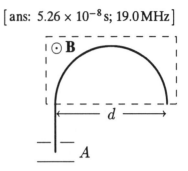

SOLUTION: $qV = \frac{1}{2}mv^2$ can be used to find the speed of the ions and $qB = mv/R$ relates the speed and radius of the path. R, of course is just $d/2$.

[ans: $q/m = 8V/d^2B^2$]

Take $V = 2.5 \times 10^4$ V, $B = 0.75$ T, and $d = 16$ cm. If the atoms are singly ionized what is their mass? Identify the atoms.

SOLUTION:

[ans: 1.15×10^{-26} kg; lithium]

PROBLEM 5. A Hall effect experiment is performed as diagramed in the Basic Concepts section. The sample is 2.3 cm wide and 0.15 mm thick. The current is 18 A and the magnetic field is 0.75 T. A voltmeter attached across the width reads 8.3 μV. Assume the charge carriers are all electrons and calculate their concentration.

SOLUTION: Use $n = iB/qtV$.

[ans: 6.78×10^{28} electrons/m^3]

What is their drift velocity?

SOLUTION: Use $i = env_d A$.

[ans: 4.81×10^{-4} m/s]

What is the magnitude of the transverse electric field?

SOLUTION: Use $E = V/w$.

[ans: 3.61×10^{-4} V/m]

What is the magnitude of the area charge density along a side of the sample and what is the electron concentration (number per unit area) there?

SOLUTION: The sample sides act like capacitor plates and the transverse electric field is related to the area charge density by $E = \sigma/\epsilon_0$.

[ans: 3.2×10^{-15} C/m²; 3.20×10^4 electrons/m²]

Integrate $d\mathbf{F} = i d\mathbf{L} \times \mathbf{B}$ to calculate the force of a magnetic field **B** on a wire carrying current i. Use $\mathbf{F} = i\mathbf{L} \times \mathbf{B}$ to calculate the force on a wire in a uniform field.

PROBLEM 6. A loop of wire is in the form of the triangle shown, with a base of length L_1 and each of the other sides of length L_2. It carries current i in the clockwise direction and is in a uniform magnetic field **B**, into the page. Draw arrows on the diagram to show the directions of the magnetic forces on the three sides. Find expressions for the magnitude and cartesian components of the magnetic force on each side.

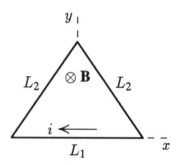

SOLUTION: The force on the left side has magnitude iBL_2. If θ is the angle between the side and the base then the force makes the angle θ with the positive y axis. The x component of the force is $-iBL_2 \sin\theta$ and the y component is $iBL_2 \cos\theta$. Use $\sin\theta = \sqrt{L_2^2 - (L_1/2)^2}/L_2$ and $\cos\theta = L_1/2L_2$. The force on the right side is similar but it has a positive x component.

[ans: left: $F = iBL_2$, $F_x = -iB\sqrt{L_2^2 - (L_1/2)^2}$, $F_y = iBL_1/2$; right: $F = iBL_2$, $F_x = iB\sqrt{L_2^2 - (L_1/2)^2}$, $F_y = iBL_1/2$; bottom: $F = iBL_1$, $F_x = 0$, $F_y = -iBL_1$]

Notice that total field is zero, as it must be for a closed loop in a uniform field.

Suppose the base of the triangle is attached to the sides with an identical spring at each end and the loop is on a horizontal frictionless table top. The springs have their natural lengths when the current is zero. By how much are they extended or compressed when a current of 2.0 A is in the loop and equilibrium is reached? Take $B = 0.85$ T, $L_1 = 45$ cm, and $L_2 = 65$ cm. The spring constant of each spring is 90 N/m.

SOLUTION: The total force acting on the bottom wire is $(2kx - iBL_1)\mathbf{j}$, where x is the elongation of either spring. This vanishes in equilibrium, so $x = iBL_1/2k$.

[ans: 4.25 mm]

PROBLEM 7. A conducting rod has a mass of 85 g and a length of 50 cm. It is suspended from the ceiling by a spring with a spring constant of 50 N/m in such a way that it is horizontal. A horizontal uniform magnetic field of 0.65 T exists in the region. A 2.0-A current is from left to right in the rod and the field points toward you as you look at the rod. The rest of the circuit is outside the magnetic field. When the rod is in equilibrium by how much is the spring extended or compressed from its natural length?

SOLUTION: The gravitational force is mg, down; the spring force is kx, up; and the magnetic force is iBL, down. Here x is the extension of the spring and is negative if the spring is compressed. In equilibrium the total force is zero. Solve for x.

[ans: 2.97 cm]

PROBLEM 8. A conducting rod of mass m is free to slide on horizontal conducting rails, as shown. The generator G maintains a constant current i in the loop and a uniform magnetic field **B** is into the page. If the rod starts with velocity v_0, toward the generator, what is its velocity as a function of time? Neglect friction.

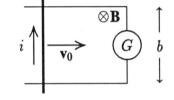

SOLUTION: The magnetic force on the rod is ibB, to the left and the acceleration of the rod is ibB/m. Use constant acceleration kinematics to find an expression for the velocity as a function of time.

[ans: $v_0 - iaBt/m$]

Take $i = 2.0$ A, $b = 75$ cm, $m = 120$ g, $v_0 = 2.8$ m/s, and $B = 0.75$ T. Find the time the rod takes to stop and the distance it travels in stopping.

SOLUTION:

[ans: 0.597 s; 0.836 m]

PROBLEM 9. A rectangular loop of wire is oriented with the lower left corner at the origin and one edge along the x axis, as shown. A magnetic field is into the page and has a magnitude that is given by $B = \alpha y$, where α is a constant. What is the total magnetic force on the loop if it carries current i in the direction shown?

SOLUTION: The force on the lower side is zero since the magnetic field is zero there. The force on the upper side is $iLB = ib\alpha a$, upward. The forces on the left and right sides have equal magnitudes and opposite directions so their sum is zero. In detail, the magnitude of the force on an infinitesimal length is $iB\,dy = i\alpha y\,dy$, so the magnitude of the total force on either of these sides is $i\int_0^a \alpha y\,dy = \frac{1}{2}i\alpha a^2$. The force on the left side is to the left and the force on the right side is to the right.

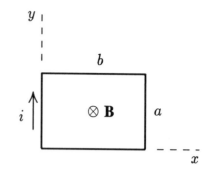

[ans: $iab\alpha$, up]

The net force on a closed loop in a uniform field is zero. Here the field is not uniform and, as you can see, the net force is not zero.

Take $i = 2.0$ A, $a = 15$ cm, $b = 25$ cm, $B = 0.20$ T, and $\alpha = 150$ T/m. What is the magnitude of the total force?

[ans: 2.25 N]

If the loop is in a vertical plane with the y axis upward and has a mass of 175 g, for what value of α will it be in equilibrium when the current is 2.0 A?

SOLUTION: Equate mg and $iab\alpha$, then solve for α.

[ans: 22.9 T/m]

PROBLEM 10. A closed loop of wire is in the form of a semicircle, with its flat side extending from $x = -R$ to $x = +R$ on the x axis, as shown. It carries current i in the direction shown. A magnetic field is everywhere out of the page and its magnitude varies according to $B = \alpha y$, where α is a constant. What is the magnetic force on the wire?

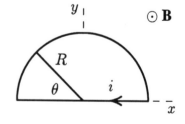

SOLUTION: The magnetic force on the straight portion of the loop is zero since the field is zero there. The force on an infinitesimal length ds of the circular portion is $dF = iB\,ds$. If the radius from the circle center to the infinitesimal length makes the angle θ with the x axis as shown, then $B = \alpha y = \alpha R \sin\theta$ and $ds = R\,d\theta$. So $dF = i\alpha R^2 \sin\theta\,d\theta$. The force is radially inward and its y component is $dB_y = -dB\sin\theta = -i\alpha R^2 \sin^2\theta\,d\theta$. The y component of the total force is the integral of this expression from $\theta = 0$ to $\theta = \pi$ radians. Use $\int \sin^2\theta\,d\theta = -\frac{1}{2}\sin\theta\cos\theta + \theta/2$ to evaluate the integral. Similarly the x component is the integral of $dB_x = dB\cos\theta = i\alpha R^2 \sin\theta\cos\theta\,d\theta$. This integral is zero. You can immediately see that the x component of the total force is zero by comparing the forces on two infinitesimal elements with the same y coordinate, on opposite sides of the semicircle.

[ans: $i\alpha\pi R^2/2$, in the negative y direction]

You should know how to compute the work done by a uniform magnetic field on a magnetic dipole as the dipole turns in the field. Here's an example.

PROBLEM 11. Each edge of a square loop of wire is 15 cm long and the loop carries a current of 1.8 A. What is the magnitude of its magnetic dipole moment?

SOLUTION: Use $\mu = iA = ia^2$, where a is the length of an edge.

[ans: 4.05×10^{-2} A·m^2]

The loop is initially oriented so its plane is perpendicular to a 0.80 T uniform magnetic field, with the field and moment aligned. How much work is done by the magnetic torque if the loop is rotated by 90° around an axis that is perpendicular to the field?

SOLUTION: As the dipole moment turns from the angle θ_1 to the angle θ_2, both measured with respect to the direction of the field, the work W done by the field is given by $W = \int_{\theta_1}^{\theta_2} \tau \, d\theta = \int_{\theta_1}^{\theta_2} \mu B \sin\theta \, d\theta$.

[ans: 3.24×10^{-2} J]

When it is in its new orientation what is the magnitude of the magnetic torque acting on it?

SOLUTION:

[ans: 3.24×10^{-2} N·m]

How much work is done by the magnetic torque if the loop is now rotated by 90° around an axis that is parallel to the field?

SOLUTION:

[ans: 0]

If a problem asks you to calculate the torque of a *uniform* field on a current loop you can use $\tau = \mu \times B$. Remember that the torque always tends to rotate the loop so its dipole moment is aligned with the magnetic field. It is therefore a restoring torque and, in the absence of other torques, causes the loop to oscillate.

PROBLEM 12. A 0.15-kg circular loop is suspended vertically in a 0.60-T uniform horizontal magnetic field so it is free to rotate about a vertical diameter. Initially it is perpendicular to the field but it is rotated slightly and released. If it carries a current of 2.0 A, what is its frequency of oscillation?

SOLUTION: The magnitude of the torque is given by $\tau = \mu B \sin\phi$, where ϕ is the angle between the field and the dipole moment. Since $\tau = I\alpha$, where I is the rotational inertia and α is the angular acceleration, $-\mu B \sin\phi = I\alpha$ and $d^2\phi/dt^2 = -(\mu B/I)\sin\phi$. For small amplitude oscillations $\sin\phi \approx \phi$ in radians, so $d^2\phi/dt^2 = -(\mu B/I)\phi$. Thus the angular frequency is given by $\omega = \sqrt{\mu B/I}$. Use $\mu = iA = \pi R^2 i$, $I = MR^2/2$, and $\nu = \omega/2\pi$. Notice that the loop radius R does not appear in the final expression.

[ans: 1.13 Hz]

III. NOTES

Chapter 35
AMPÈRE'S LAW

I. BASIC CONCEPTS

You will learn to calculate the magnetic field produced by a current using two techniques, one based on the Biot-Savart law and the other based on Ampère's law. When using the first you will sum the fields produced by infinitesimal current elements. Carefully note how the directions of the current and the position vector from a current element to the field point influence the direction of the field. Ampère's law relates the integral of the tangential component of the field around a closed loop to the net current through the loop. Pay attention to the role played by symmetry when you use this law to find the field.

The Biot-Savart law. The diagram on the right shows a portion of a current-carrying wire. Suppose you wish to calculate the magnetic field it produces at point A. First, mark an infinitesimal element ds of the wire in the vicinity of P, in the same direction as _____. Now draw the position vector r for the point A relative to the selected element. Draw an arrow at A to represent the (infinitesimal) field produced there by the element. Be careful about the direction. Finally, write the Biot-Savart law in vector form for the (infinitesimal) magnetic field produced at A by the element:

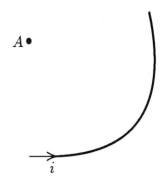

$$d\mathbf{B} =$$

Notice that the field of the element is perpendicular to both ds and r. To find the total field at A sum the contributions from all elements of the wire. The result is the vector integral

$$\mathbf{B} =$$

The Biot-Savart law contains the constant μ_0. It is called the _____ constant and its value is _____ T·m/A. Do not confuse the symbol with that for the magnitude of a magnetic dipole moment (μ without the subscript 0). Your expression for d$\mathbf{B}$ may have contained the unit vector $\mathbf{u}_r$. This vector has magnitude 1 and is directed from _____ to _____. It is unitless.

Chapter 35: Ampère's Law   511

In Section 35–2 of the text the Biot-Savart law is used to find an expression for the magnitude of the magnetic field produced by a long straight wire carrying current i. Each infinitesimal element of the wire produces a field in the same direction so the magnitude of the total field is the sum of the magnitudes of the fields produced by all the elements. Go over the calculation carefully. The diagram shows an infinitesimal element of the wire at x, a distance r from point P. In terms of i and r the magnitude of the field it produces at P is $dB =$ _____. The direction of the field is _____ the page. In terms of x and R, $r =$ _____ and $\sin\theta =$ _____, so $dB =$ _____. Integrate in x from $-\infty$ to $+\infty$. The result is

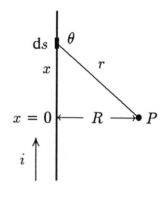

$$B =$$

You should recognize that the field lines around a long straight wire are circles centered on the wire. Three are shown in the diagram. The field has the same magnitude, given by the equation you wrote above, at all points on any given circle. The direction is determined by a right hand rule: if the thumb of the right hand points in the direction of the current arrow then the fingers will curl around the wire in the direction of the magnetic field lines. The current in the wire shown is out of the page. Put arrows on the field lines to show the direction of the field.

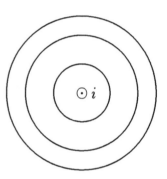

For current carrying wires that are not straight the magnitude of the field if not uniform along a field line. You should be able to make a symmetry argument to show that it is for a straight wire.

The expression for the field of a long straight wire can be combined with the expression for the magnetic force on a wire, developed in the last chapter, to find an expression for the force exerted by one wire on another, parallel to the first. The magnetic field produced by one wire at a point on the other wire is _____ to the second wire. If the wires are separated by a distance d and carry currents i_1 and i_2 then the magnitude of the force per unit length of one on the other is given by

$$F/L =$$

If the currents are in the same direction the wires _____ each other; if the currents are in opposite directions they _____ each other. Notice that the forces of the wires on each other obey Newton's third law: they are equal in magnitude and opposite in direction.

The magnetic field of a circular current loop. The calculation is an excellent illustration of how the Biot-Savart law is used. Pay careful attention to it and notice particularly that the vector integral for the field is evaluated one component at a time. First write the expression for the magnitude dB of the field produced by an infinitesimal element of the current. This involves the distance r from the current element to the field point and the sine of the angle θ between the element d**s** and the vector d**r**. Next determine the direction of the field associated with d**s** and let α be the angle between the field and one of the coordinate axes. d$B \cos\alpha$ is the component of the infinitesimal field along the chosen axis. Integrate d$B \cos\alpha$ over the current to find the component of the total field along the axis. The quantities r, θ, and α may be different for different segments of the current. If they are you must use geometry write them in terms of a single variable of integration. Finally, repeat the process for the other coordinate axes.

Two angles enter the calculation: the angle θ between d**s** and **r** and the angle α between the field and a coordinate axis. The first is used to find an expression for the magnitude of the field produced by an infinitesimal element of current, the second is used to find a component, once the expression for the magnitude is known. Be careful to keep them straight.

Now use the Biot-Savart law to develop an expression for the field on the axis of a circular loop, a distance z from its center. Consider the element d**s** shown. The angle between **r** and d**s** is _____ so the magnitude of the infinitesimal field at P reduces to

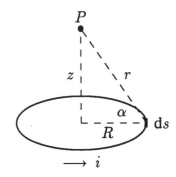

$$dB = $$

Draw a vector at P to indicate the direction of d**B** and verify that it makes the angle α with the axis. The expression for the component of d**B** along the axis is $dB_{\parallel} = $ _____. This must be integrated around the loop.

All quantities in the integral are the same for all segments of the loop, so the integral can be evaluated using $\oint ds = 2\pi R$. The result is $B_{\parallel} = $ _____. In terms of z and R, $r = $ _____ and $\cos\alpha = $ _____. Make these substitutions to obtain

$$B_{\parallel} = $$

A symmetry argument can be made to show all other components vanish.

An important result can be derived from this equation. For points far away from the loop ($z \gg R$) the expression for the field can be written in terms of the magnitude of the magnetic dipole moment μ of the loop: $B = $ _____. This expression is valid for the magnetic field of *any* planer loop, regardless of its shape, for points far away along the axis defined by the direction of the dipole moment.

You should be able to estimate the direction of the field produced in its plane by a current carrying loop of wire. Near any segment of the wire you may treat the segment as a long straight wire and use the right hand rule given above. Since the field decreases with distance take the direction at any point to be roughly in the same direction as the field produced by the nearest segment of the loop, perhaps adjusted for other nearby segments.

For example, the circle on the right represents a current loop, with current i as shown. Points A, B, C, and D are in the plane of the loop. The magnetic field is directed _____ the page at A, _____ the page at B, _____ the page at C, and _____ the page at D. At points far away from the loop the field closely resembles that of a bar magnet (both are dipole fields). The _____ pole is at the upper surface of the loop, where magnetic field lines leave the "magnet" and the _____ pole is at the lower surface of the loop, where field lines enter the "magnet".

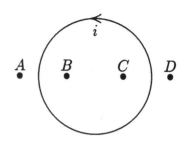

The magnetic dipole moment of a current carrying loop (or a bar magnet) points from the _____ pole toward the _____ pole, in the interior.

Ampère's law. Ampère's law is given by Eq. 19. Copy it here:

The integral on the left side is a path integral around any *closed* path. The current i on the right side is the net current through a surface that is bounded by the path. For example, if the path is formed by the edges of this page then i is the net current through the page. You should recognize that the amperian path need not be the boundary of any physical surface and, in fact, may be purely imaginary. You should recognize that the surface need not be a plane. The sides and bottom of a wastepaper basket form a valid surface with the rim as the boundary.

You should know how to evaluate the right side of the Ampère's law equation when the current is distributed among several wires. First, choose a direction, clockwise or counter-clockwise, to be used in evaluating the integral on the left side of the Ampère's law equation. The choice is immaterial but it must be made since it determines the direction of d**s**. If the tangential component of the field is in the direction of d**s** then the integral is positive, if it is in the opposite direction the integral is negative.

Now curl the fingers of your right hand around the loop in the direction chosen. Your thumb will point in the direction of positive current. Examine each current through the surface and algebraically sum them. If a current arrow is in the direction your thumb pointed it enters the sum with a positive sign; if it is in the opposite direction it enters with a negative sign. If the net current through the path is positive then the average tangential component of **B** is in the direction chosen for d**s**. If the net current is negative then the average tangential component of **B** is in the opposite direction.

A current outside the path does not contribute to the right side of the Ampère's law equation. You should be aware that a current outside the path produces a magnetic field at every point on the path but the integral $\oint \mathbf{B} \cdot d\mathbf{s}$ of the field it produces vanishes.

Only the tangential component of the magnetic field enters. When Ampère's law is used to calculate the magnetic field of a current distribution the path is taken, if possible, to be either parallel or perpendicular to field lines at every point. Then the integral reduces to

$\int B\,ds$, along those parts of the path that are parallel to field lines. If, in addition, the magnitude of the field is constant along the path, then the integral is Bs, where s is the total length of those parts of the path that are parallel to field lines.

Ampère's law can be used to find an expression for the magnetic field produced by a long straight wire carrying current i. The amperian path used is a circle in a plane perpendicular to the wire and centered at the wire. Why? _____

_____

Since the magnetic field is tangent to the circle and has the same magnitude at all points around the circle, the integral $\oint \mathbf{B}\cdot d\mathbf{s}$ is _____, where r is the radius of the circle. In the space below equate this expression to $\mu_0 i$ and solve for B:

Your result should be $B = \mu_0 i/2\pi r$, as before.

Ampère's law can also be used to find the magnetic field *inside* a long straight wire. The diagram shows the cross section of a cylindrical wire of radius R carrying current i, uniformly distributed throughout its cross section. Take the amperian path to be the dotted circle of radius r. The magnetic field is tangent to the circle and has constant magnitude around the circle, so the integral $\oint \mathbf{B}\cdot d\mathbf{s}$ is _____. Not all the current in the wire goes through the dotted circle. In fact, the fraction through the circle is the ratio of the circle area to the wire area: r^2/R^2. Thus the right side of the Ampère's law equation is _____. Equate the two sides and solve for B. The result is $B =$ _____.

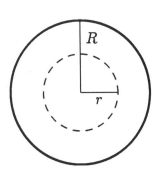

The magnetic field at the center of the wire is _____. The field has it largest value at _____ and this value is given by _____.

Ampère's law can be applied to a solenoid, a cylinder tightly wrapped with a thin wire. For an ideal solenoid (long, with the current approximated by a cylindrical sheet of current rather than a wrapped wire) the magnetic field outside is negligible and the field inside is uniform and is parallel to _____.

The diagram shows the cross section of a solenoid, with the current in each wire coming out of the page at the top and going into the page at the bottom. Use dotted lines to draw the rectangular path you will use to evaluate the integral on the left side of the Ampère's law equation. The integral is $\oint \mathbf{B}\cdot d\mathbf{s} =$ _____, where h is the length of the rectangle. Label it on the diagram.

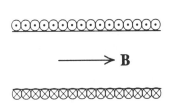

If the solenoid has n turns of wire per unit length then the number of turns that pass through the surface bounded by the amperian path is _____. Each carries current i so the right side of the Ampère's law equation is _____. The equation can be solved for the magnitude of the field, with the result $B =$ _____.

Ampère's law can also be applied to a toroid, with a core shaped like a doughnut and wrapped with a wire, like a solenoid bent so its ends join.

The diagram to the right shows a cross section with the wire omitted for clarity. The magnetic field is confined to the interior of the core and the field lines are concentric circles centered at the center of the hole. Draw the amperian path you will use to find an expression for the field a distance r from the center. The integral $\oint \mathbf{B} \cdot d\mathbf{s}$ is _____ and if there are N turns of wire, each carrying current i, the total current through the surface bounded by the path is _____. Thus the Ampère's law equation is _____ = _____ and $B =$ _____.

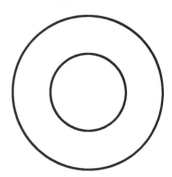

Consider an amperian path in the form of a circle, centered at the center of the hole and inside the hole. The net current through the path is _____ and the field in the "doughnut hole" is _____. Consider a similar path outside the toroid. The net current through this path is _____ and the field outside the toroid is _____.

II. PROBLEM SOLVING

Use $B = \mu_0 i / 2\pi r$ to calculate the magnitude of the field produced by a long straight wire carrying current i. Use the right hand rule to find the direction of the field. If two or more wires are present add the individual fields vectorially.

PROBLEM 1. A power line 30 m above the ground carries a current of 50 A. What is the magnitude of the magnetic field it produces at ground level directly underneath?

SOLUTION:

[ans: 3.33×10^{-7} T]

If the current runs south to north what is the direction of its field directly underneath?

SOLUTION: Point the thumb of your right hand in the direction of the current. Your fingers will curl around the wire in the direction of the magnetic field lines.

[ans: west]

PROBLEM 2. Two long straight wires are 1.5 m apart. The wire on the left carries a current of 1.2 A and the wire on the right carries a current of 1.6 A, both out of the page. What is the magnetic field at the midpoint between them? What is the magnetic field at the point P, 0.75 m to the right of right hand wire? What is the magnitude of the field at the point Q, 1.5 m from the left hand wire?

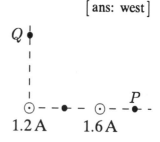

SOLUTION: Use $B = \mu_0 i / 2\pi r$ to calculate the field due to each wire. At the midpoint of the line joining them $r = 0.75$ m for each wire. The left hand wire produces a field that points upward in the diagram and the right hand wire produces a field that points downward, so the total field is the difference in their magnitudes.

At P both wires produce fields that point upward so the total field is the sum of their magnitudes. Now $r = 2.25$ m for the left hand wire and $r = 0.75$ m for the right hand wire.

516   Chapter 35: Ampère's Law

To find the field at Q use $r = 1.5$ m for the left hand wire. Its field points to the left. Use $r = \sqrt{1.5^2 + 1.5^2} = 2.12$ m for the right hand wire. Its field points downward to the left at an angle of $45°$ to the vertical. The horizontal component is $\mu_0 i \cos 45°/2\pi r$, to the left and its vertical component is $\mu_0 i \sin 45°/2\pi r$, downward.

[ans: 1.07×10^{-7} T; 5.33×10^{-7} T; 2.87×10^{-7} T]

When you calculate the force of one wire on another use the magnetic field produced by the first wire at the second, not the total field.

PROBLEM 3. What force per unit length does one wire of the previous problem exert on the other?

SOLUTION:

[ans: 2.56×10^{-7} N/m]

You should be able to derive an expression for the magnetic field produced by a finite straight wire, then use the expression to calculate the field of any combination of straight segments.

PROBLEM 4. A wire segment of length L carries current i. Derive an expression for the magnetic field at point P, a distance R from the segment and a horizontal distance ℓ from one end, as shown.

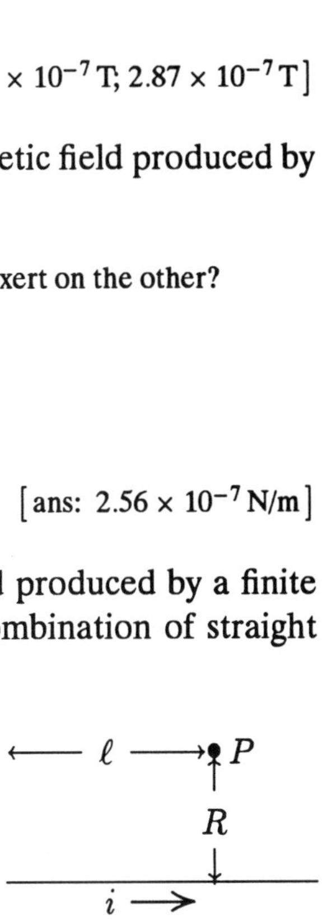

SOLUTION: The derivation is exactly like that leading to Eq. 11 of the text but the limits of integration are from $x = -\ell$ to $x = L - \ell$.

[ans: $\dfrac{\mu_0 i}{4\pi R} \left\{ \dfrac{L-\ell}{[(L-\ell)^2 + R^2]^{1/2}} + \dfrac{\ell}{[\ell^2 + R^2]^{1/2}} \right\}$]

This expression is valid even if ℓ is greater than L or is negative. Then the point P is to the right or left of the current segment in the diagram. No matter what the value of ℓ the field lines are circles centered on the wire and the usual right hand rule can be used to find the field direction.

PROBLEM 5. A loop of wire is in the form of a square, each side having a length of 20 cm. If the current is 2.0 A what is the magnitude of the magnetic field at the center?

SOLUTION: Each side contributes a field that is perpendicular to the plane of the loop and has the same magnitude as the contributions of the other sides. Thus the total field is four times the field due to one side alone. To calculate the contribution of one side take $\ell = L/2$ and $R = L/2$ in the expression you developed in the last problem.

[ans: 1.13×10^{-5} T]

PROBLEM 6. A wire forms a closed loop by following two sides and a diagonal of a square with 20 cm sides, as shown. The current is 2.0 A. What is the magnetic field at P, the fourth corner of the square?

SOLUTION:

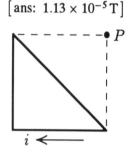

[ans: 2.19×10^{-7} T, out of the page]

You should also be able to find the magnetic field at the center of a circular arc of wire carrying current i.

PROBLEM 7. A segment of wire is in the shape of a circular arc of radius R that subtends the angle θ at its center. If it carries current i what is the magnetic field at the center?

SOLUTION: Consider an infinitesimal part of the arc, with length $ds = R\,d\theta$, where $d\theta$ is measured in radians. The vector **r** is perpendicular to the current arrow so the magnetic field is given by $dB = (\mu_0 i/4\pi R)\,d\theta$. The fields of all segments are in the same direction so the total field is just the algebraic sum of the individual fields.

[ans: $\mu_0 i\theta/4\pi R$]

PROBLEM 8. The current carrying loop shown has a semicircle at one end. If the current is 3.0 A what is the magnetic field at the point P, the center of the semicircle?

SOLUTION:

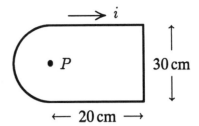

[ans: 1.13×10^{-5} T]

Use $B = \mu_0 i R^2 / 2(R^2 + z^2)^{3/2}$ to find the field produced by a circular loop of radius R carrying current i, at a point on its axis a distance z from its center. Many coils contain several turns of wire, all nearly in the same plane. Since each produces the same field at any point, simply multiply the field of one turn by the number of turns.

PROBLEM 9. Each of two identical circular coils has a radius R and N turns. They are aligned perpendicularly to the line joining their centers, a distance L apart. Each carries current i in the same direction. Find an expression for the magnetic field along the line joining their centers.

SOLUTION: Suppose a point between the coils is a distance z from the center of one coil and a distance $L-z$ from the center of the other. Use $B = \mu_0 N i R^2 / 2(R^2 + z^2)^{3/2}$ for the field of the first and the same equation with z replaced by $L-z$ for the field of the second. The two fields are in the same direction.

[ans: $\dfrac{\mu_0 N i R^2}{2} \left\{ \dfrac{1}{[R^2 + z^2]^{3/2}} + \dfrac{1}{[R^2 + (L-z)^2]^{3/2}} \right\}$]

Show that the field is a minimum at the midpoint of the line joining the centers.

SOLUTION: Evaluate dB/dz for $z = L/2$. You should get 0.

The second derivative $d^2 B/dz^2$ also vanishes at the midpoint provided the distance between the coils is equal to the radius of one of them. Place $L = R$ in the expression for B and evaluate the second derivative for $z = L/2$. You should get 0.

SOLUTION:

Two identical coils, carrying identical currents in the same direction and separated by distance equal to a radius, are known as Helmholtz coils. As your calculation implies, the magnetic field is especially uniform in the region between the coils. This property makes Helmholtz coils particularly useful in the laboratory.

Find an expression in terms of R, i, and N for the magnetic field at the midpoint of the line joining the centers of a pair of Helmholtz coils. Evaluate the expression for $R = 0.60$ m, $i = 2.5$ A, and $N = 100$.

SOLUTION:

[ans: $8\mu_0 Ni/5\sqrt{5}R$; 3.75×10^{-4} T]

You should be able to use $B = \mu_0\mu/2\pi z^3$ to calculate the field of a magnetic dipole μ at a point a distance z from the dipole on a line through the dipole and parallel to the dipole moment. Recall that the magnitude of the dipole moment of a current-carrying loop is given by $\mu = iA$, where A is the area bounded by the loop and i is the current.

PROBLEM 10. A square loop of wire with an edge length of 1.8 cm carries a 2.3 A current. It lies in the xy plane, centered at the origin. What is the magnetic field at $z = 2.0$ m, on the z axis?

SOLUTION: Use $B = \mu_0\mu/2\pi z^3$, with $\mu = ia^2$, where a is the length of an edge. For the dipole approximation to be valid z must be much greater than any dimension of the loop. Since it is about 100 times the square edge, we expect some error in the third significant figure.

[ans: 1.9×10^{-11} T]

Suppose a second loop, with a dipole moment of 8.0×10^{-3} A·m² is on the z axis at $z = 2.0$ m, with its moment in the x direction. What is the magnitude of the torque exerted on it by the first dipole?

SOLUTION: The field of the first dipole is parallel to the z axis, perpendicular to the dipole moment of the second dipole. The magnitude of the torque is given by $\tau = \mu_2 B$, where μ_2 is the magnitude of the second moment.

[ans: 1.5×10^{-13} N·m]

Ampère's law is valid for any closed path. If you know the net current through the area bounded by the path you can easily give the value for the integral $\oint \mathbf{B} \cdot d\mathbf{s}$. If you know the magnetic field at each point on the path you can, in principle, evaluate the integral $\oint \mathbf{B} \cdot d\mathbf{s}$ and find the net current through the area bounded by the path.

PROBLEM 11. Six wires pierce the plane of the page, as shown. Take $i_1 = 0.80$ A, $i_2 = 1.30$ A, $i_3 = 0.60$ A, $i_4 = 1.80$ A, $i_5 = 0.50$ A, and $i_6 = 0.90$ A. Carefully note which are into the page ($\otimes$) and which are out of the page ($\odot$). What is the value of $\oint \mathbf{B} \cdot d\mathbf{s}$ around the dotted circular path?

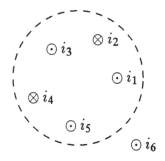

SOLUTION: Use $\oint \mathbf{B} \cdot d\mathbf{s} = \mu_0 i$, where i is the net current through the area bounded by the circle. If current going into the page is taken to be positive then $i = -0.80 + 1.30 - 0.60 + 1.80 - 0.50 = 1.20$ A. Carefully note that i_6 is not included in the sum because it is outside the area.

[ans: 1.51×10^{-6} T·m]

PROBLEM 12. In the region around the origin of a coordinate system the x and y components of the magnetic field are given by $B_x = (3.0y^2) \times 10^{-6}$ and $B_y = (4.0x - 6.0x^2) \times 10^{-6}$, in tesla when x and y are in meters. What is the net current through a square with 25-cm sides, having one corner at the origin and sides extending out the positive x and y axes?

SOLUTION: Let a be the length of a side of the square and integrate around the square in the counterclockwise direction. Along the side from the origin to $x = a$, $y = 0$ the x component of the field is 0, so this side does not contribute to the integral. Along the side from $x = a$, $y = 0$ to $x = a$, $y = a$, $B_y = (4.0a - 6.0a^2) \times 10^{-6}$ and the contribution of this side to the integral is $\int_0^a B_y \, dy = (4.0a^2 - 6.0a^3) \times 10^{-6}$. Along the side from $x = a$, $y = a$ to $x = 0$, $y = a$ the x component of the magnetic field is $B_x = 3.0a^2 \times 10^{-6}$ and the contribution of this side to the integral is $\int_a^0 B_x \, dx = -3.0a^3 \times 10^{-6}$. The y component of the field along the side from $x = 0$, $y = a$ to $x = 0$, $y = 0$ is 0, so this side does not contribute. Thus $\oint \mathbf{B} \cdot d\mathbf{s} = (4.0a^2 - 6.0a^3 - 3.0a^3) \times 10^{-6} = 4.0a^2 - 9.0a^3) \times 10^{-6}$. Equate this to $\mu_0 i$ and solve for i.

[ans: 87.0 mA]

In the text Ampère's law is used to compute the magnetic field inside a long straight cylindrical wire carrying a current that is uniformly distributed throughout its cross section. The same procedure can be used to find the magnetic field if the current distribution is not uniform but still has cylindrical symmetry. That is, if it depends only on distance from the axis of the cylinder.

PROBLEM 13. A long straight cylindrical wire has a radius of 0.15 mm and carries a current of 0.57 A, distributed uniformly across its cross section. What is the magnetic field at a point halfway from the center to the surface? What is the field at the surface?

SOLUTION: The current density is uniform so you use $B = \mu_0 i r / 2\pi R^2$.

[ans: 3.80×10^{-4} T; 7.60×10^{-4} T]

PROBLEM 14. A long straight wire is in the form of a cylindrical shell with inner radius R_a and outer radius R_b. It carries current i out of the paper and uniformly distributed over its cross section. Find an expression for the magnetic field at points in each of the three regions $r < R_a$, $R_a < r < R_b$, and $R_b < r$, where r is the distance from the cylinder axis.

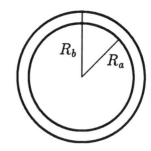

SOLUTION: No matter which region you are considering use a circle of radius r for the amperian path. The field is tangent to the circle and has uniform magnitude around it so $\oint \mathbf{B} \cdot d\mathbf{s} = 2\pi r B$. The current through a circle with radius less than R_a is zero so $B = 0$ in this region. For a point inside the shell not all the current in the shell is through the area bounded by the amperian path. The fraction that is through that area is given by the ratio of the cross-sectional area of the part of the shell within the path to the total shell area. The former is $\pi(r^2 - R_a^2)$ and the latter is $\pi(R_b^2 - R_a^2)$, so the current inside the path is $i(r^2 - R_a^2)/(R_b^2 - R_a^2)$. For a point with $R_a < r$ all the current in the shell is through the area bounded by the amperian path.

[ans: 0; $\mu_0 i (r^2 - R_a^2)/2\pi r(R_b^2 - R_a^2)$; $\mu_0 i/2\pi r$]

Notice that the field is continuous. The expression for the middle region yields $B = 0$ for $r = R_a$ and yields $B = \mu_0 i / 2\pi r$ for $r = R_b$.

PROBLEM 15. A thin wire runs along the cylinder axis of a long straight cylindrical shell. The wire carries current i out of the page and the shell carries the same current but into the page. The current density is uniform in the shell. Find expressions, in terms of the distance r from the cylinder axis, for the magnetic field in the region between the wire and the shell, inside the shell, and outside the shell.

SOLUTION: The problem can be solved by combining the expressions for the fields of a wire and a cylindrical shell. We will, however, apply Ampère's law to the two-conductor system as a whole. For each region use a circle with radius r as the amperian path. The magnetic field is tangent to the circle and has uniform magnitude around it, so $\oint \mathbf{B} \cdot d\mathbf{s} = 2\pi r B$.

For $r < R_a$ the current through the amperian circle is i, so Ampère's law becomes $2\pi r B = \mu_0 i$. For $R_a < r < R_b$ the current through the amperian circle is $i - i[(r^2 - R_a^2)/(R_b^2 - R_a^2)]$, the first term being current in the wire and the second being current in the shell. Since they are in opposite directions they enter with different signs. Once fractions are cleared the expression for the current can be written $i(R_b^2 - r^2)/(R_b^2 - R_a^2)$ and Ampère's law becomes $2\pi r B = \mu_0 i(R_b^2 - r^2)/(R_b^2 - R_a^2)$. The total current through any amperian path totally outside the shell is 0 so the magnetic field is 0 there.

[ans: $\mu_0 i/2\pi r$; $(\mu_0 i/2\pi r)(R_b^2 - r^2)/(R_b^2 - R_a^2)$; 0]

PROBLEM 16. The situation is the same as in the last problem but wire is displaced to the left by d along the dotted line. Find expressions for the magnetic field at any point between the wire and the shell on the dotted line and at any point outside the shell on the dotted line.

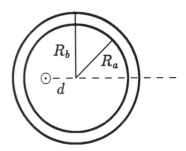

SOLUTION: The geometry here does not have cylindrical symmetry, so Ampère's law cannot usefully be applied to the two-conductor system. But it can be applied to each conductor individually. Use it to find expressions for the fields produced by the wire and by the shell separately, then add them vectorially. If r is the distance from the wire to the field point then $r - d$ is the distance from the center of the shell to the field point.

[ans: $\mu_0 i/2\pi r$; $\mu_0 i d/2\pi r(r-d)$]

PROBLEM 17. A long straight cylindrical wire has a radius R and carries current i, distributed so the current density increases from the wire center to its surface in proportion to the distance from the center. Find an expression for the magnetic field at any point within the wire.

SOLUTION: Take the current density to be $j = Ar$, where A is a constant and r is the distance from the center of the wire. The field lines are circles that are concentric with the wire cross section, so take the amperian path to be a circle of radius r. The magnitude of the field is uniform on this circle so the left side of the Ampère's law equation is $2\pi r B$.

The area of an infinitesimal circular strip that extends from r to $r + dr$ is $2\pi r\, dr$ so the current through the circle is given by the integral $2\pi \int_0^r jr\, dr = 2\pi A \int_0^r r^2\, dr$. Carry out the integration and substitute the result into the Ampère's law equation, then solve for B.

An expression for the constant A in terms of i can be found by requiring that $2\pi \int_0^R jr\, dr = i$, the total current in the wire.

[ans: $\mu_0 i r^2 / 2\pi R^3$]

Notice that this expression yields $B = \mu_0 i / 2\pi R$ for the field at the surface, the same as for a uniform current distribution. The field outside the wire is $\mu_0 i / 2\pi r$, no matter what the current distribution (as long as it is cylindrically symmetric), and this expression yields $\mu_0 i / 2\pi R$ for the field at the surface.

III. COMPUTER PROJECTS

A computer program can be used to integrate the Biot-Savart equation for the magnetic field of a current loop. In general the field is given by the integral around the current loop:

$$\mathbf{B(r)} = \frac{\mu_0}{4\pi} i \int \frac{d\mathbf{r}' \times (\mathbf{r} - \mathbf{r}')}{|\mathbf{r} - \mathbf{r}'|^3},$$

where i is the current, $\mathbf{r}$ is the position vector of the point where the field is $\mathbf{B}$, and $\mathbf{r}'$ is the position vector of a point on the loop.

We consider circular a loop in the xy plane and use an angular variable of integration. Take $\mathbf{r}' = R(\cos\theta\,\mathbf{i} + \sin\theta\,\mathbf{j})$ and $d\mathbf{r}' = R(-\sin\theta\,\mathbf{i} + \cos\theta\,\mathbf{j})\,d\theta$. Since the situation has cylindrical symmetry we can without loss of generality specialize to a point in the yz plane and write $\mathbf{r} = y\mathbf{j} + z\mathbf{k}$. When these substitutions are made the expressions for the components of the field are

$$B_x = 0,$$

$$B_y = \frac{\mu_0}{4\pi}\frac{iz}{R^2}\int_0^{2\pi} \frac{\sin\theta\,d\theta}{\left[1 + (y/R)^2 + (z/R)^2 - 2(y/R)\sin\theta\right]^{3/2}},$$

$$B_z = -\frac{\mu_0}{4\pi}\frac{i}{R}\int_0^{2\pi} \frac{[(y/R)\sin\theta - 1]\,d\theta}{\left[1 + (y/R)^2 + (z/R)^2 - 2(y/R)\sin\theta\right]^{3/2}},$$

Both integrals have the form

$$B_i = \frac{\mu_0 i}{4\pi R} \int_0^{2\pi} f(\theta)\,d\theta.$$

The Simpson's rule program of Chapter 7 can be used to carry out the integrations. Divide the interval in θ from 0 to 2π into N segments, each of length $\Delta\theta = 2\pi/N$. Then B_i is approximated by $(\Delta\theta/3)(2S_e + 4S_o)$, where S_e is the sum of the values of the integrand at the beginning of segments with even labels and S_o is the sum of the values at the beginning of segments with odd labels. The term $f_N - f_0$ does not appear because the integrands have the same value at $\theta = 0$ and $\theta = 2\pi$. N must be an even integer.

To evaluate both integrals (for B_y and B_z) within the same program, let S_{ye} and S_{yo} collect the sums for the y component and S_{ze} and S_{zo} collect the sums for the z component. Here's an outline.

> input number of intervals: N
> replace N with nearest even integer
> calculate interval width: $\Delta\theta = 2\pi/N$
> input coordinates of field point: y, z
> initialize quantities to hold sum of values with even labels: $S_{ye} = 0$, $S_{ze} = 0$
> initialize quantities to hold sum of values with odd labels: $S_{yo} = 0$, $S_{zo} = 0$
> set $\theta = 0$
> **begin loop** over intervals: counter runs from 1 to $N/2$
>   calculate $A = \sin\theta$ for future use
>   calculate $B = \left[1 + (y/R)^2 + (z/R)^2 - 2(y/R)\sin\theta\right]^{3/2}$ for future use

            update sums over even terms
                replace S_{ye} with $S_{ye} + A/B$
                replace S_{ze} with $S_{ze} + [(y/R)A - 1]/B$
            increment θ by $\Delta\theta$
            calculate $A = \sin\theta$ for future use
            calculate $B = [1 + (y/R)^2 + (z/R)^2 - 2(y/R)\sin\theta]^{3/2}$ for future use
            update sums over odd terms
                replace S_{yo} with $S_{yo} + A/B$
                replace S_{zo} with $S_{zo} + [(y/R)A - 1]/B$
            increment θ by $\Delta\theta$
        end loop
        evaluate integrals:
            $B_y = (\mu_0 i z / 4\pi R^2)(\Delta\theta/3)(2S_{ye} + 4S_{yo})$
            $B_z = -(\mu_0 i / 4\pi R)(\Delta\theta/3)(2S_{ze} + 4S_{zo})$
        display result
        go back to input new field coordinates or stop

Running time can be reduced if you add instructions to calculate the new variables $y' = y/R$ and $z' = z/R$ immediately after y and z are read. Then use y' and z' in succeeding instructions. They equations have been written in a convenient form to do this.

The purpose of the first project is to test the program and to obtain a rough idea of the number of intervals that should be used.

PROJECT 1. If the field point is on the z axis the integrals can be evaluated analytically, with the result $B_x = 0$, $B_y = 0$, and

$$B_z = \frac{\mu_0 i R^2}{2(R^2 + z^2)^{3/2}}.$$

Consider a 1.0-m radius loop carrying a current of 1.0 A and use the program to calculate the field at $z = 0$, 0.50, 1.5, and 2.5 m on the z axis. Since the integrand is constant you should obtain the correct answers with $N = 2$. Check your answers by evaluating the analytic expression.

Now find the magnetic field at the point $x = 0$, $y = 0.50$ m, $z = 0$. Start with $N = 2$ and on the next trial double N. Continue to double N until the results of two successive trials agree to 3 significant figures.

Repeat the calculation for the following points along the $x = 0$, $y = 0.50$ m line: $z = 0.10$, 1.0, and 10 m. Note the value of N for which 3 significant figure accuracy is obtained.

A set of Helmholtz coils consists of two identical loops parallel to each other and carrying the same current in the same direction. In the region between the coils the two magnetic fields tend to be in roughly the same direction and when the distance between the coils is equal to the radius of one of the loops, the field in the region between is particularly uniform. For this reason Helmholtz coils are often used to produce magnetic fields in the laboratory. In the following project you use the integration program to investigate the uniformity of the magnetic field between two coils.

PROJECT 2. Two circular loops of wire, each with a radius of 1.0 m, are placed parallel to the xy plane, with their centers on the z axis. The center of one is at $z = 0$ while the center of the other is at $z = 2.0$ m. Each carries a current of 1.0 A in the counterclockwise direction when viewed from the positive z axis. Note that the separation is twice the radius.

Use the program to calculate the magnetic field with 3 significant figure accuracy for field points at $z = 0.60$, 0.80, and 1.0 m, all on the z axis. The last point is at the center of the region between the loops. You will need to run the program twice, once for each loop. For example, the first field point ($z = 0.60$ m) is 0.60 m from the first coil and 1.4 m from the second. Run the program with $z = 0.60$ m, then with $z = -1.4$ m. Vectorially sum the two fields.

Use these calculated fields to test the uniformity of the field along the z axis between the loops. For each of the first two points, subtract the magnitude of the field at the center ($z = 1.0$ m) from the field at the point and divide by the magnitude of the field at the center. If we denote this measure of homogeneity by h, then

$$h(z) = \frac{|B(z) - B_c|}{|B_c|},$$

where B_c is the magnitude of the field at the center of the region.

If h is small the field does not change much with position and is said to be homogeneous. If h is large the field is said to be inhomogeneous. For good laboratory magnets h may be on the order of 10^{-6} or less over several centimeters. The distance between the loops, of course, is usually less than 2 m.

Now consider the same loops, but separated by 1.0 m, the radius of one of them. Use the program to calculate the magnetic field at $z = 0.10, 0.30$, and 0.50 m, all on the z axis. The last point is at the center of the region between the loops. Calculate h for the first two points. Has the field become more or less homogeneous?

PROJECT 3. You can also investigate homogeneity in a transverse direction. Now the field has two non-vanishing components and we define a measure of homogeneity for each:

$$h_z = \frac{|B_z(y) - B_z(y=0)|}{|B_z(y=0)|}$$

$$h_y = \frac{|B_y(y)|}{|B_z(y=0)|}.$$

Both are zero for a perfectly homogeneous field.

Consider the loops of the previous project, with a separation of 2.0 m, and calculate h_y and h_z for $y = 0.20$ and 0.40 m on the line $x = 0$, $z = 1.0$ m.

Now take the separation to be 1.0 m and calculate h_y and h_z for $y = 0.20$ and 0.50 m on the line $x = 0$, $z = 0.50$ m. Has the homogeneity increased or decreased?

If the loops are moved still closer to each other so their separation is less than the radius, the field becomes less uniform. Consider a separation of 0.50 m and calculate h_y and h_z for $y = 0.20$ and 0.40 m along the line $x = 0$, $z = 0.25$ m.

In the following projects you will use an integration program to investigate Ampère's law. The square shown is in the xy plane, is centered at the origin, and has edges of length a. A wire carrying current i pierces the plane of the square at the point on the x axis with coordinate ℓ. It produces a magnetic field at each point on the perimeter of the square (as well as at other points). According to Ampère's law

$$\oint \mathbf{B} \cdot d\mathbf{s} = \mu_0 i,$$

where the integral is around the perimeter and i is the current in the wire. Take the current to be positive if it is out of the page and carry out the integration in the counterclockwise direction around the square.

The magnitude of the magnetic field is given by $B = \mu_0 i/2\pi r'$, where r' is the distance from the wire. The field components are $B_x = -(\mu_0 i/2\pi)y'/(r')^2$ and $B_y = (\mu_0 i/2\pi)x'/(r')^2$. As the diagram shows, $y' = y$, $x' = x - \ell$, and $(r')^2 = (x-\ell)^2 + y^2$ so $B_x = -(\mu_0 i/2\pi)y/[(x-\ell)^2 + y^2]$ and $B_y = (\mu_0 i/2\pi)(x-\ell)/[(x-\ell)^2 + y^2]$.

Consider each of the four sides of the square separately. Across the top $y = a/2$, $ds = dx\,\mathbf{i}$ and

$$\int \mathbf{B} \cdot d\mathbf{s} = -\frac{\mu_0}{2\pi}\frac{a}{2}\int_{a/2}^{-a/2} \frac{dx}{(x-\ell)^2 + (a/2)^2}.$$

The contribution of the bottom is exactly the same. Down the left side $x = a/2$, $ds = dy\,\mathbf{j}$, and

$$\int \mathbf{B} \cdot d\mathbf{s} = \frac{\mu_0 i}{2\pi}\left(\frac{a}{2} + \ell\right)\int_{a/2}^{-a/2} \frac{dy}{(\ell + a/2)^2 + y^2}.$$

Up the right side $x = a/2$, $ds = dy\,\mathbf{j}$, and

$$\int \mathbf{B} \cdot d\mathbf{s} = \frac{\mu_0 i}{2\pi}\left(\frac{a}{2} - \ell\right)\int_{-a/2}^{a/2} \frac{dy}{(-\ell + a/2)^2 + y^2}.$$

Each of the integrals can be evaluated using the Simpson's rule program.

PROJECT 4. Suppose the wire carries a current of 1.0 A, out of the page, and the square has sides of length $a = 2.0$ m. Evaluate $\oint \mathbf{B} \cdot d\mathbf{s}$, one side at a time, for each of the following positions of the wire: $\ell = 0, 0.40, 0.80$, and 1.2 m. The first three points are inside the square and the fourth is outside. Fill in the table below with values of $\int \mathbf{B} \cdot d\mathbf{s}$.

ℓ	0	0.40 m	0.80 m	1.2 m
top				
bottom				
right				
left				
total				

For each of the situations compute $\mu_0 i$, where i is the net current through the square. Compare the result with the totals above.

PROJECT 5. For the square of the previous project evaluate $\oint \mathbf{B} \cdot d\mathbf{s}$ for a current of 2.0 A, out of the page at $\ell = 0$. Evaluate the integral for a current of 1.0 A, out of the page at $\ell = 0.50$ m. Finally, evaluate the integral for a current of 3.0 A, out of the page at $\ell = 0.75$ m. Compare the last answer to the sum of the first two.

Evaluate the integral for the a current of 3.0 A, out of the page at $\ell = 0$. Evaluate the integral for a current of 2.0 A, into the page at $\ell = 0.50$ m. Compare the difference in these results with the value of the integral for a current out of the page at $\ell = 0.40$ m (see the results of Project 4).

IV. NOTES

Chapter 36
FARADAY'S LAW OF INDUCTION

I. BASIC CONCEPTS

As the magnetic flux through any area changes, an electromotive force is generated around the boundary and, if the boundary is conducting, a current is induced. Faraday's law tells us the relationship between the emf and the rate of change of the flux. You will learn that an emf can be generated by changing the magnetic field or by moving a physical object through a magnetic field. You should concentrate on how to calculate the magnetic flux and the emf in each case. In addition, you will learn that a non-conservative electric field is always associated with a changing magnetic field and it is this field that is responsible for the emf when the magnetic field is changing.

Faraday's law. To understand Faraday's law you must first understand <u>magnetic flux</u>. The magnetic flux through any area is proportional to the number of magnetic field lines through that area. Recall that the number of field lines through a small area perpendicular to the field is proportional to the magnitude of the field and that the number of lines through *any* small area is proportional to the component B_n along a normal to the area.

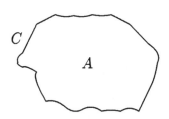

The diagram shows an area A bounded by the curve C. If a magnetic field pierces the plane of the page then there is magnetic flux through the area. It is the integral over the area of the normal component of the magnetic field. Divide the area into a large number of small elements, each with area ΔA, and for each element calculate the quantity $B_n \Delta A$, then sum the results. The magnetic flux is the limit of this sum as the area elements become infinitesimal. That is, the flux through the area is given by the integral $\Phi_B = \int B_n \, dA$.

Usually the infinitesimal area is assigned a direction normal to the surface and written as the vector d**A**. Then $B_n \, dA = \mathbf{B} \cdot d\mathbf{A}$ and the integral for the flux becomes

$$\Phi_B =$$

If the field is uniform over the entire area the flux is given by

$$\Phi_B =$$

where θ is the angle between _____ and _____ . The SI unit of magnetic flux is _____ and is abbreviated _____ .

Faraday's law tells us that whenever the flux through any area is changing with time then an emf is generated around the boundary of that area. Symbolically, the law is

$$\mathcal{E} =$$

where $\mathcal{E}$ is the induced emf. The induced emf is distributed around the boundary and is not localized as is the emf of a battery. Its direction is specified as clockwise or counterclockwise around a loop in the plane of the page. The minus sign that appears in Faraday's law is closely related to the direction of the emf.

Notice that the direction of d**A** is ambiguous since two directions are normal to any surface. For the area shown above the normal directions are into and out of the page. Φ_B has the same magnitude for both choices but is positive for one and negative for the other. You may choose either but you may find thinking about Faraday's law a bit easier if you pick the one for which Φ_B is positive. If you point the thumb of your right hand in the direction chosen for d**A** then your fingers will curl around the boundary in the direction of positive emf. If $\mathcal{E}$, calculated using Faraday's law including the minus sign, turns out to be positive then the emf is in the direction of your fingers. If it turns out to be negative then it is in the opposite direction. If the boundary of the region is conducting and no other emf's are present, the induced current will be in the direction of the induced emf.

Some problems deal with a coil of wire consisting of several identical turns, like a solenoid, with the same magnetic field through each turn. If Φ_B is the flux through each turn and there are N turns, then the total emf induced around the coil is given by

$$\mathcal{E} =$$

The flux through the interior of a loop can be changed by changing the magnitude or direction of the magnetic field, by changing the area of the loop, or by changing the orientation of the loop. No matter how the emf is generated it represents work done on a unit test charge as the test charge is carried around the loop. If the magnetic field is changing in either magnitude or direction, an electric field is generated and the emf is the work per unit charge done by that electric field. If the magnetic field is constant and the loop changes area or orientation, work must be done by an external agent, the agent that is changing the loop. Details are given in Section 7 of the text.

Lenz' law. Lenz' law provides another way to determine the direction of an induced emf. Imagine that the bounding curve is a conducting wire so current flows in it when an emf is induced, the current being in the direction of the emf. The induced current produces a magnetic field everywhere but you should concentrate on the field it produces inside the loop and the flux associated with that field. According to Lenz' law when the flux of the externally applied field changes current flows in the loop in such a way that the flux it produces in the interior of the loop counteracts the change.

Suppose a bar magnet is held above the loop shown above, with its north pole toward the loop. Then the magnetic field points into the page through the loop. If the magnet is moved toward the loop, the magnitude of its field in the plane of the loop increases (the field of a magnet is stronger near the magnet). According to Lenz' law, current will flow in the loop in such a way that its field in the interior of the loop is directed _____ the page. If the magnet is moved away from the loop, the magnitude of the applied field decreases and, as it does, current flows in the loop in such a way that its magnetic field in the interior of the loop is directed _____ the page. Thus the law is used to determine the direction of the field produced by the induced current.

Once you have determined the direction of the magnetic field produced by the induced current you can determine the direction of the current itself and hence that of the emf. Very near any segment of the loop the field is quite similar to the field of a long straight wire; the lines are nearly circles around the segment. You can use the right-hand rule explained in the last chapter: curl your fingers around the segment so they point in the direction of the field in the interior of the loop. That is, they should curl upward through the loop if the field is upward through the loop and they should curl downward if the field is downward. Your thumb will then point along the segment in the direction of the current.

For the situation being considered, the current and the emf are clockwise around the loop if the induced field inside the loop is _____ the page and are counterclockwise if the induced field inside the loop is _____ the page.

Motional emf. An emf is also be generated if all or part of the loop moves in a manner that changes the flux through it. You should realize that if the loop above is moved through a uniform field the flux through it does not change and no emf is generated. If, however, the field is not uniform then the flux changes as the loop moves and an emf is generated. You must find an expression for the flux as a function of the position of the loop, then differentiate it with respect to time to obtain $d\Phi_B/dt$. The emf clearly depends on the velocity of the loop. Either Lenz' law or the sign convention associated with Faraday's law can be used to find the direction of the emf.

If part of the loop moves in such a way that the area changes, an emf is generated even if the field is uniform. The flux through the loop is a function of time because the area is a function of time. In terms of the rate dA/dt with which the area changes in a uniform field B, the emf is given by $\mathcal{E} = $ _____. It is generated only along the moving part, not around the whole loop.

Suppose a wire forms three sides of a rectangle and a rod, free to move on the wire, forms the fourth side, as shown. If W is the fixed dimension of the rectangle, x is the variable dimension, and the magnetic field is uniform and perpendicular to the rectangle, then the flux through it is $\Phi_B = $ _____ and the emf generated around it is _____. Note that the emf is proportional to the speed.

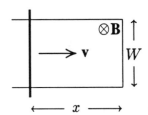

The loop need not consist entirely of physical objects, such as wires. Part of it may be imaginary. An emf is generated along an isolated rod moving in a magnetic field, for example. To calculate it you may complete the loop with imaginary lines. The total emf is along the bar. If the rod is conducting, electrons move in the direction opposite to the emf when the emf first appears. They collect at one end of the rod, leaving the other end positively charged, and as a result an electric field exists along the rod. Very quickly the field prevents further build-up of charge and the current becomes zero. A steady current cannot flow in an isolated rod.

Do not confuse the electric field created by charge at the ends of a moving rod with an electric field associated with a time-varying magnetic field. In the first case the magnetic field is not changing and no electric field is associated with it.

An emf is also generated around a loop that is changing orientation so the angle between its normal and the magnetic field is changing with time. Suppose a circular loop with radius R is rotated about a diameter in a uniform magnetic field, perpendicular to the axis of rotation. The flux through the loop is given by $\Phi_B = BA\cos\theta = \pi R^2 B \cos\theta$, where θ is the angle between the normal to the plane of the loop and the magnetic field. If θ is changing according to $\theta = \omega t$, then $d\Phi_B/dt = $ _____.

You should be able to find the direction of the induced emf (and current if the loop is conducting). When ωt is between 0 and $\pi/2$ the flux is positive and so is the emf. Positive flux tells us that the direction we picked for the normal makes an angle of less than 90° with the magnetic field. If we look from a point where the magnetic field is pointing toward us from the loop, the emf is counterclockwise. When ωt is between 90° and 180° the flux is _____ and the emf is _____. When viewed from the same point the emf is _____. When ωt is between 180° and 270° the flux is _____ and the emf is _____. When viewed from the same point the emf is _____. When ωt is between 270° and 360° the flux is _____ and the emf is _____. When viewed from the same point the emf is _____.

Energy considerations. When a current is induced energy is transferred, via the emf, to the moving charges. Recall that the rate at which a seat of emf $\mathcal{E}$ does work on a current i is given by $P = $ _____. When the emf is associated with a changing magnetic field the energy comes from the agent changing the field; when the emf is motional the energy comes from the agent moving the loop or from the kinetic energy of the loop. In either case the energy is dissipated in the resistance of the loop. You will learn more about the mechanism of energy transfer by a changing magnetic field in the next chapter. Here you learn about energy transfer by a motional emf.

Consider the example above in which one side of rectangular loop moves in a uniform magnetic field and recall that the emf generated is given by $\mathcal{E} = BWv$. If the resistance of the loop is R then the current is given by $i = $ _____, so in terms of B, W, v, and R the emf transfers energy at the rate $P = $ _____. This is precisely the rate at which the external agent does work on the rod to keep it moving at a constant speed.

In terms of i, B, and W the external field **B** exerts a force of magnitude $F = $ _____. The direction of the current is _____ in the diagram and the magnetic field is into the page so the magnetic force on the moving rod is directed to the _____. If the rod is to move with constant velocity an external agent must apply a force of equal magnitude but in the opposite direction. Thus the force of the agent is directed to the _____. Since the rod is moving to the right with speed v the rate at which the agent is doing work is $P = Fv$. In terms of B, v, W, and R, this is $P = $ _____.

The rate at which energy is dissipated in the resistance of the loop is given by $P = i^2 R$ and, in terms of B, v, W, and R, this is $P = $ _____. All of the energy supplied by the agent is dissipated. The magnetic field provides the mechanism of energy transfer but it supplies no energy.

If the agent stops pushing the magnetic field exerts a force that slows the moving rod and eventually stops it. The kinetic energy of the rod is then dissipated in the resistance of the loop. This is the basis of magnetic braking.

Induced electric fields. An electric field is always associated with a changing magnetic field and this electric field is responsible for the induced emf. The relationship between the electric field **E** and the emf around a closed loop is

$$\mathcal{E} = \underline{}$$

Note that the integral is zero for a conservative field, such as the electrostatic field produced by charges at rest. The electric field induced by a changing magnetic field, however, is non-conservative and the integral is not zero.

Suppose a cylindrical region of space contains a uniform magnetic field, directed along the axis of the cylinder, as shown. The field is zero outside the region. If the magnetic field changes with time the lines of the electric field it produces form circles, concentric with the cylinder. You should be able to derive an expression for the magnitude of the electric field at points inside and outside the cylinder.

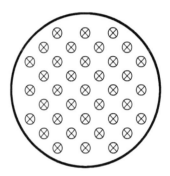

First consider a point inside the cylinder, a distance r from the center. Draw a circle through the point, concentric with the cylinder cross section. The magnetic flux through the circle is given by _____ and in terms of dB/dt its rate of change is given by _____. The magnitude of the emf around the circle is given by _____.

The circle coincides with an electric field line and the electric field has uniform magnitude around the circle. In terms of the magnitude of the electric field and the radius of the circle the emf is given by $\mathcal{E} = $ _____. Equate the two expressions for the emf and solve for the magnitude of the electric field as a function of r. It is $E = $ _____.

Now repeat the calculation for a point outside the cylinder. Draw a circle of radius r. The magnetic flux through this circle is given by _____, where R is the radius of the cylinder. The magnitude of the emf around the circle is given by _____ and the magnitude of the electric field is given as a function of r by $E = $ _____.

If the magnitude of the magnetic field is increasing, the emf (either inside or outside the cylinder) is in the _____ direction and so is the electric field. If the magnitude of the magnetic field is decreasing the emf is in the _____ direction and so is the electric field.

II. PROBLEM SOLVING

Some problems give the rate at which the magnetic field is changing. These are straightforward applications of Faraday's law. Find an expression for the flux, then differentiate it with respect to time. If the field is uniform over the area then the flux is $AB\cos\theta$, where A is the area and θ is the angle between the field and the normal to the area. If the field is not uniform you must evaluate the integral $\int \mathbf{B} \cdot d\mathbf{A}$.

PROBLEM 1. A uniform magnetic field is normal to a circular area with a radius of 15 cm. If the magnitude of the field is increasing at the rate of 75 T/s what emf is generated around the boundary of the area?

SOLUTION: The flux is given by $\Phi_B = \pi R^2 B$ and the magnitude of the emf is given by $\mathcal{E} = d\Phi_B/dt = \pi R^2 dB/dt$. Here R is the radius of the area.

[ans: 5.30 V]

Relative to the externally applied field in what direction is the induced magnetic field? If the area is in the plane of the page and the applied field is into the page in what direction is the induced emf?

SOLUTION:

[ans: opposite; counterclockwise]

If the boundary of the area is a wire with a resistance of 12 Ω, what is the induced current?

SOLUTION:

[ans: 0.442 A]

Suppose the field makes an angle of 35° with the normal to the area. What then is the emf around wire and what is the current induced in the wire?

SOLUTION: The flux is now $\pi R^2 B \cos\theta$, where $\theta = 35°$.

[ans: 4.34 V; 0.363 A]

If the loop consists of several turns of wire you must add the emf's. Suppose the loop has 5 turns, each with a resistance of 12 Ω. What then is the total emf and the induced current when the magnetic field is perpendicular to the area?

SOLUTION:

[ans: 26.5 V; 0.442 A]

PROBLEM 2. Consider a square region of the xy plane, centered at the origin with two sides parallel to the x axis and two sides parallel to the y axis. A magnetic field in the positive z direction has a magnitude given by $B = \alpha x^2 t$, where B is in tesla, x is in meters, and t is in seconds. α is a constant. What is the flux through the square at time t? What is the emf around the boundary of the square?

SOLUTION: Take the normal to the surface to be in the positive z direction. Then an element of area is $a\,dx$ and the flux is given by $\Phi_B = \int_{-a/2}^{a/2} aB\,dx$. The emf, of course, is the derivative of this with respect to time.

[ans: $\alpha a^4 t/12$; $\alpha a^4/12$]

What value should α have to generate a 3.5 mV emf around a square with 12-cm sides?

SOLUTION:

[ans: 2.03×10^2 T/m²·s]

The problem may be written in terms of the current producing the magnetic field rather than in terms of the field itself. Then you must relate the rate of flux change to the rate of current change.

PROBLEM 3. A long straight wire carries a current i that changes with time. It is a distance b from a square loop with edge a, consisting of N turns of wire. Find an expression, in terms of the rate of change di/dt of the current in the straight wire, for the total emf induced around the loop.

SOLUTION: Take the x axis to be to the right, perpendicular to the long wire. Then the field of the current is given by $B = \mu_0 i/2\pi x$ and the flux through each turn is given by $\Phi_B = (\mu_0 i/2\pi) \int_b^{b+a} (a/x)\,dx$. The field is into the page for the current direction shown and the normal to the area was taken to be in that direction.

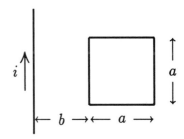

[ans: $N\dfrac{di}{dt}\dfrac{\mu_0 a}{2\pi}\ln\dfrac{b+a}{b}$]

Take $a = 1.8$ cm, $b = 2.0$ mm, $N = 150$ turns, and $di/dt = 175$ A/s. What is the total induced emf? What is the direction of the induced emf?

SOLUTION:

[ans: 0.218 mV; counterclockwise]

PROBLEM 4. Current in a 500-turn/cm solenoid is increasing at the rate of 175 A/s. Inside the solenoid and concentric with it is a 50-turn loop with a radius of 0.60 cm. What emf is induced in the loop?

SOLUTION: Use $B = \mu_0 ni$ for the field of the solenoid and $\Phi_B = NAB = \pi R^2 N B$ for the flux through the loop. Here n is the number of turns per unit length of the solenoid, R is the radius of the loop, and N is the number of turns in the loop.

[ans: 1.24 mV]

PROBLEM 5. A 2000-turn toroid has a square cross section. The length of each side of the square is 8.0 cm and the inner radius of the toroid is 15 cm. A second wire forms a 150-turn winding over a portion of the toroid. If the current in the toroid is changing at the rate of 75 A/s what emf is induced in the second wire?

SOLUTION: The flux through one turn of the second wire is given by $\Phi_B = \mu_0 i N_1 a/2\pi \int_b^{b+a} dx/x$, where a is the length of the square edge, b is the inner radius, and N_1 is the number of turns in the toroid. To find the emf differentiate this with respect to time and multiply by the number of turns N_2 formed by the second wire.

[ans: 0.380 V]

A time-varying flux can be generated through a loop by moving a permanent magnet toward or away from the loop.

PROBLEM 6. A 175-turn circular loop of wire with a radius of 0.10 cm is in the xy plane with its center at the origin. A permanent magnet with a dipole moment of 0.26 A·m² aligned with the z axis, moves along the z axis toward the loop at a speed of 14 m/s. What is the emf induced around the loop when the magnet is 15 cm from it?

SOLUTION: Assume the loop is sufficiently small and far from the magnet that the field produced by the magnet is uniform over the loop. Then $B = \mu_0\mu/2\pi z^3$ is a good approximation for the magnitude of the field at all points inside the loop. Here μ is the dipole moment of the magnet and z is the distance from the dipole to the loop. Since the field is normal to the loop the flux through the loop is given by $\Phi_B = R^2\mu_0\mu/2z^3$ and the rate at which the flux is changing is given by $d\Phi_B/dt = -3R^2\mu_0\mu v/2z^4$, where $v\,(=dz/dt)$ is the speed of the magnet.

[ans: 2.37 μV]

An emf is generated around a loop that moves through a non-uniform field, provided the field has a non-vanishing component normal to the loop. The flux through the loop changes with time because the loop is continually moving into a new region where the magnetic field is different.

PROBLEM 7. A rectangular loop of length L and width W lies with its length along the x axis, as shown. It moves with speed v in the positive x direction in a magnetic field that is into the page and has a magnitude that varies with x according to $B = \alpha x$, where α is a positive constant. What emf is generated around the rectangle?

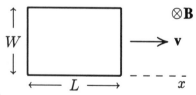

SOLUTION: Take the normal to the area to be into the page and suppose the left end of the loop has coordinate x. Then the flux through the rectangle is given by $W\int_x^{x+L} B\,dx = W\int_x^{x+L}\alpha x\,dx = \frac{1}{2}\alpha LW(2x+L)$. Use Faraday's law to calculate the emf. As the rectangle moves the flux increases so the induced current, if any, must produce a field that is out of the page in the interior of the rectangle.

[ans: αLWv, counterclockwise]

If $W = 15\,\text{cm}$, $L = 25\,\text{cm}$, and $\alpha = 0.18\,\text{T/m}$ how fast must the loop move in order to generate a $9.0\,\text{mV}$ emf?

[ans: 13.3 m/s]

If the loop has 200 turns how fast must it move to generate the same emf?

[ans: 6.67 cm/s]

PROBLEM 8. A 1.8-m wide automobile is traveling at 55 mph in a region where the earth's magnetic field is 0.55 gauss, downward. What emf is induced across the car?

SOLUTION: Apply Faraday's law to a rectangular path consisting of two imaginary lines parallel to the road, a car's width apart, joined by a third imaginary line across the road in front of the car. The car itself forms the fourth side of the rectangle. If the sides have length L then the area of rectangle is LW, where W is the width of the car. The flux through the rectangle is $\Phi_B = BLW$ and the emf is $d\Phi_B/dt = BW\,dL/dt = BWv$, where v is the speed of the car. Convert the car speed to m/s and the field to tesla.

[ans: 2.21 mV]

The emf is induced across the car alone, not around the complete path.

Suppose the magnetic field makes an angle of 30° with the vertical. What then is the induced emf?

SOLUTION:

[ans: 1.92 mV]

PROBLEM 9. In a small electrical generator a 150-turn rectangular loop, with sides of 4.0 cm and 8.0 cm, rotates at 3600 rpm in a 0.95 T magnetic field. The axis of rotation is in the plane of the loop, is through the loop center, and is perpendicular to the field. The total emf induced in the loop can be written $\mathcal{E} = \mathcal{E}_m \sin(\omega t)$. Find values for $\mathcal{E}_m$ and ω.

SOLUTION: If θ is the angle between the magnetic field and the normal to the loop then the flux through one turn of the loop is $\Phi_B = abB\cos\theta$, where a and b are the lengths of the rectangle sides. The emf is given by $\mathcal{E} = N\,d\Phi_B/dt = abBN\sin\theta\,d\theta/dt$, where N is the number of turns. Let $\theta = \omega t$, where ω is the angular speed of the loop, and differentiate the flux with respect to time.

[ans: $\mathcal{E}_m = 172\,\text{V}$; $\omega = 377\,\text{rad/s}$]

You should understand energy relationships when induced current flows. Here are some examples.

PROBLEM 10. Consider the situation described in Problem 7: a rectangular loop of length L and width W, consisting of a single turn of wire, moves with speed v in a magnetic field that is perpendicular to it and has a magnitude given by $B = \alpha x$. If the electrical resistance of the loop is R what is the induced current? What is the net magnetic force on the loop?

SOLUTION: You have already found $\mathcal{E} = \alpha L W v$. Divide by R to find an expression for the current. You should be able to show it is counterclockwise around the loop. If the left end of the loop has coordinate x then the magnetic force on it is $iWB(x) = i\alpha W x$ and the magnetic force on the right end is $-iWB(x + L) = -i\alpha W(x + L)$. The forces on the upper and lower sides are equal in magnitude and opposite in direction so they cancel from the expression for the total force.

$$\left[\text{ans: } \alpha L W v/R;\ \alpha^2 L^2 W^2 v/R\right]$$

Suppose an external force is applied to keep the loop moving with constant velocity. At what rate does this force do work on the loop?

SOLUTION: Use $P = Fv$.

$$\left[\text{ans: } \alpha^2 L^2 W^2 v^2/R\right]$$

At what rate is energy dissipated by the resistance of the loop?

SOLUTION: Use $P = i^2 R$.

$$\left[\text{ans: } \alpha^2 L^2 W^2 v^2/R\right]$$

Take $L = 25$ cm, $W = 15$ cm, $v = 13.3$ m/s, $R = 8.0$ mΩ, and $\alpha = 0.18$ T/m. Calculate the current induced in the loop, the force of an external agent required to keep the loop moving at a constant velocity, and the rate at which the agent does work.

SOLUTION:

$$\left[\text{ans: } 11.2\,\text{A};\ 75.8\,\text{mN};\ 0.101\,\text{W}\right]$$

PROBLEM 11. Suppose the external agent stops pushing on the loop of the last problem. Take the mass of the loop to be m and find an expression for the speed of the loop as a function of time.

SOLUTION: Substitute the expression for the magnetic force into $F = m\,dv/dt$ to obtain $dv/dt = -\beta v$, where $\beta = \alpha^2 W^2 L^2/mR$. Integrate and use v_0 to represent the speed of the loop at $t = 0$, when the agent stops pushing.

$$\left[\text{ans: } v_0 e^{-\beta t}\right]$$

Compare the rate at which the kinetic energy of the loop is changing with the rate of energy dissipation by the loop resistance.

SOLUTION: Since the kinetic energy is $K = \frac{1}{2}mv^2$, its rate of change is $dK/dt = mv\,dv/dt$. The rate of energy dissipation is, of course, $P = i^2R$.

[ans: both are given by $(\alpha^2 W^2 L^2 v_0^2/R)e^{-2\beta t}$]

This is an example of magnetic braking. The motion of the loop in the non-uniform magnetic field induces an emf and a current. The force of the field on the current slows the loop.

Take $L = 25$ cm, $W = 15$ cm, $v = 13.3$ m/s, $R = 8.0\,m\Omega$, and $\alpha = 0.18$ T/m. Calculate the time at which the speed of the loop is half its initial value.

SOLUTION: Substitute $v = v_0/2$ into $v = v_0 e^{-\beta t}$ and solve for t. You should get $t = -(1/\beta)\ln 0.5$.

[ans: 6.09 s]

An electric field is always associated with a changing magnetic field. You should know how to calculate the electric field when the magnetic field has cylindrical symmetry. Then the electric field lines are circles and the integral $\oint \mathbf{E}\cdot d\mathbf{s}$ for the emf can easily be integrated around any circle of radius r to obtain $\mathcal{E} = 2\pi r E$. In the first example the magnetic field is uniform, in the second it is not.

PROBLEM 12. In a cylindrical region of space with a radius of 2.5 m a uniform magnetic field is parallel to the cylinder axis. The field is zero outside the region. If the field is changing at the rate of 35 T/s, what is the electric field at a point 2.0 m from the axis of symmetry? What is the electric field at a point 3.0 m from the axis of symmetry?

SOLUTION: Consider a circle of radius r ($< R$), in a plane that is perpendicular to the magnetic field. Here R is the radius of the cylindrical region. The magnetic flux through the circle is $\pi r^2 B$ and the emf around the circle is $\mathcal{E} = \pi r^2\,dB/dt$. Equate this to $2\pi r E$ and solve for E. Then consider a similar circle, but with $r > R$. The magnetic flux is $\pi R^2 B$ and the emf is $\pi R^2\,dB/dt$. Equate this to $2\pi r E$ and solve for E.

[ans: 35 V/m; 36.5 V/m]

PROBLEM 13. Suppose a magnetic field is in the positive z direction and has a magnitude that is given by $B(r,t) = \alpha r t$, where r is the distance from the z axis, t is the time, and α is a constant. Find an expression for the induced electric field a distance r from the z axis.

SOLUTION: The field has cylindrical symmetry so the electric field lines are circles that are parallel to the xy plane and are centered on the z axis. Consider a circle of radius r. The magnetic flux through it is $\Phi_B = 2\pi \int_0^r B(r,t) r\, dr$ and the electric field is related to the emf by $\mathcal{E} = 2\pi r E$. Carry out the integration for the magnetic flux, differentiate it with respect to time, equate the result to the emf, and solve for E.

[ans: $\alpha r^2/3$]

III. NOTES

Chapter 37
MAGNETIC PROPERTIES OF MATTER

I. BASIC CONCEPTS

The first section of this chapter deals with one of the fundamental laws of electromagnetism: Gauss's law for magnetism. You should understand both the mathematical statement and the important ramifications of the law. Subsequent sections deal with the atomic basis of magnetism and show how the motions of electrons lead to the magnetic properties of materials. Pay particular attention to the intimate relationship between magnetism and angular momentum. Be able to describe the magnetic properties of paramagnetic, diamagnetic, and ferromagnetic materials and explain how these properties arise from electron motion.

Gauss' law for magnetism. Suppose the normal component of the magnetic field is integrated over a *closed* surface (one that completely surrounds a volume). The result is _____. This result is the same no matter what closed surface is considered and no matter what the distribution of current. Write the mathematical statement of the law, Eq. 2 of the text, here:

The direction of the infinitesimal element of area dA is _____. and $\mathbf{B} \cdot \mathbf{dA} = B_n \, dA$, where B_n is the normal component of the magnetic field.

The law does not necessarily mean that the magnetic field is zero at any point on the surface, only that the total flux through any closed surface is zero. This happens because the field is essentially outward over some portions and essentially inward over others. No field lines start or stop in the interior: every line that enters also leaves the volume.

The law can be interpreted to mean that magnetic monopoles do not exist, or at any rate are so rare that their influence has not be detected. Explain what a magnetic monopole is:
_____
_____

If monopoles existed and if a closed surface surrounded one of them the right side of the equation would not be 0. Magnetic field lines would start and stop at monopoles so a net magnetic flux would pass through the surface. On the other hand, magnetic dipoles do exist: a charge circulating around a small loop is an example. They do not violate Gauss' law for magnetism and are, in fact, the fundamental sources of magnetism in matter.

The magnetic field in the exterior of a permanent magnet can be closely approximated by the field that would be produced by a positive monopole (a north pole) at one end and a negative monopole (a south pole) at the other but the field does not actually arise from single monopoles but rather from magnetic dipoles associated with electron motion. As proof that the field is not due to monopoles, the magnet can be cut in half with the result that _____
_____.

This process can be continued to the atomic level, with the same result.

Dipole moments and angular momentum. Review some properties of magnetic dipoles:

1. If current i is in a loop that bounds area A the magnitude of its dipole moment is given by $\mu = $ _____. The direction of the moment is given by a right hand rule. Curl the fingers of your right hand in the direction of _____, then your thumb points in the direction of _____.

2. On the line defined by the moment the magnitude of the magnetic field produced by a magnetic dipole is given by $B = $ _____, where r is the distance from _____ to _____. If a dipole is in the page with its moment normal to the plane of the page, pointing upward, then the direction of the magnetic field at points above the dipole is roughly _____ and the direction of the magnetic field at points below the dipole is roughly _____.

3. An external magnetic field **B** exerts a torque on a magnetic dipole. The torque is given by $\tau = $ _____.

The magnetic dipole moment of a circulating charge is closely related to its angular momentum. Suppose a charge q is traveling with speed v around a circle of radius r. The period of its motion is given by $T = $ _____, the current is given by $i = q/T = $ _____, and the magnitude of its dipole moment is given by $\mu_\ell = iA = $ _____. If its mass is m then the magnitude of its angular momentum is $\ell = $ _____. In terms of ℓ, $\mu_\ell = $ _____. The vector relationship between μ_ℓ and ℓ is:

$$\mu_\ell = $$

If q is positive then μ_ℓ and ℓ are in the same direction; if q is negative they are in opposite directions. Suppose an electron is traveling in a counterclockwise direction around a circular orbit in the plane of the page. The direction of its orbital angular momentum vector is _____ the page and the direction of its magnetic dipole moment is _____ the page.

Every electron has an intrinsic angular momentum, often called its spin angular momentum or spin and denoted by **s**. An intrinsic magnetic dipole moment is associated with spin and is given in terms of **s** by

$$\mu_s = $$

Note that the relationship differs by a factor of 2 from the analogous relationship for orbital angular momentum. Also note that the dipole moment and spin angular momentum are in opposite directions for electrons.

For a collection of electrons, as in an atom, we vectorially sum the individual orbital angular momenta to find the total angular momentum **L** and vectorially sum the individual spin angular momenta to find the total spin angular momentum **S**. The magnetic moment associated with the total orbital angular momentum of an atom is

$$\mu_L = $$

and the moment associated with the total spin angular momentum is

$$\mu_S = $$

The total dipole moment of an atom is the vector sum of these.

Atomic magnetic moments are often measured in units of what is called a <u>Bohr magneton</u> and denoted μ_B. In terms of the Planck constant h and the mass m of an electron, a Bohr magneton is given by $\mu_B =$ _____ and its SI value is _____ J/T.

We are usually interested in the component of the dipole moment along a given axis, in many cases determined by the direction of an applied field. According to quantum mechanics any cartesian component of the orbital angular momentum of either a single electron or a collection of electrons has one of the values $n(h/2\pi)$, where n is an integer and h is the Planck constant. In terms of the Bohr magneton the orbital contribution to a component of the dipole moment has one of the values _____ μ_B. Similarly, any component of the spin angular momentum of a single electron is either $+\frac{1}{2}(h/2\pi)$ or $-\frac{1}{2}(h/2\pi)$. In terms of the Bohr magneton the spin contribution to a component of the dipole moment is either _____ μ_B or _____ μ_B.

Magnetic dipole moments are also associated with the spins of protons and neutrons. The moments, however, are extremely small compared with that of an electron because _____.

In units of a Bohr magneton the magnitude of the intrinsic dipole moment of an electron is _____, the magnitude of the intrinsic dipole moment of a proton is _____, and the magnitude of the intrinsic dipole moment of a neutron is _____. In spite of the smallness of the magnetic moments of protons and neutrons, nuclear magnetism has found important applications in medicine and elsewhere.

Magnetization. In words, the <u>magnetization</u> of a uniformly magnetized object is _____ per unit _____. If the object is not uniformly magnetized you may need the definition of the magnetization at a point. It is the limiting value of the expression you wrote as the volume shrinks to zero around the point.

Some materials, called paramagnetic and diamagnetic, become magnetized only when an external magnetic field is applied; others, such as ferromagnets, can be permanently magnetized. In any event, the dipoles of a magnetized object produce a magnetic field of their own and, to find the total magnetic field at any point, this field must be added vectorially to any applied field that is present. If a magnetic field $\mathbf{B}_0$ is applied to magnetic material the total field is given by $\mathbf{B} = \mathbf{B}_0 + \mathbf{B}_M$, where $\mathbf{B}_M$ is the field produced by the magnetic dipoles of the material. Since the field $\mathbf{B}_M$ depends on the geometry of the sample and the uniformity of the magnetization it is often difficult to calculate. If, however, the sample is in the form of a cylinder and the magnetization $\mathbf{M}$ is uniform and parallel to the axis then $\mathbf{B}_M$ and $\mathbf{M}$ are related by $\mathbf{B}_M =$ _____.

The total field at any point in the interior of the material is often written as the product of a parameter κ_m and the applied field. That is, $\mathbf{B} =$ _____. κ_m is called the _____ constant of the material. For many materials in weak applied fields κ_m is independent of the applied field and the total field is proportional to the applied field. Then the magnetization at any point in the material is also proportional to the total field at that point. For ferromagnets κ_m depends on the applied field and $\mathbf{B}$ is not proportional to $\mathbf{B}_0$. The same statement is true for paramagnets in strong fields.

Chapter 37: Magnetic Properties of Matter **543**

Consider a cylindrical non-ferromagnetic sample, with permeability constant κ_m, in a weak external field $\mathbf{B}_0$ parallel to its axis. In the space below derive expressions for the magnetization and for the field produced in the interior by the dipoles of the material, in terms of κ_m and the applied field:

$$\left[\text{ans: } \mathbf{B}_M = (\kappa_m - 1)\mathbf{B}_0; \mathbf{M} = [(\kappa_m - 1)/\mu_0]\mathbf{B}_0\right]$$

For typical non-ferromagnetic materials the magnitude of $\kappa_m - 1$ is on the order of _____ to _____ .

Paramagnetism. Atoms of paramagnetic materials have permanent dipole moments. When no external field is applied, however, the magnetization is zero because _____ .

When an external field is applied the material becomes magnetized because _____ .

When the applied field is removed the magnetization quickly reduces to zero because _____ .

When an external field is applied the direction of the field produced by dipoles of the material is _____ the direction of the applied field, so the total field in the material is _____ in magnitude than the applied field. Look at Table 2 and note typical values of $\kappa_m - 1$ for paramagnetic materials. The sign of $\kappa_m - 1$ is _____ for all paramagnetic materials. For all of them the field due to magnetization is quite small compared to the applied field.

The permeability constant κ_m for a paramagnetic object depends on the temperature. In fact, $\kappa_m - 1$ is proportional to _____, where T is the absolute temperature. This relationship is usually written in terms of the magnetization for an applied field $\mathbf{B}_0$:

$$\mathbf{M} =$$

where C is called the _____ constant for the material. The relationship, known as _____ law, is valid for small values of B_0/T. Explain qualitatively why the magnetization decreases if the temperature is increased without changing the applied field:

κ_m is independent of the applied field only if the field is small. For large fields the magnetization is no longer proportional to the applied field and, in fact, is less in magnitude than a proportional relationship predicts. For sufficiently high fields the magnetization, rather than κ_m, is independent of the field. The magnetization is then said to be <u>saturated</u>. This occurs when _____ .

544   *Chapter 37: Magnetic Properties of Matter*

On the axes to the right draw a graph of the magnetization as a function of applied field for a typical paramagnetic material. Label the saturation magnetization M_{max} and indicate the linear region where κ_m is independent of **B**.

If the sample contains N dipoles in volume V, each with a dipole moment μ_i, then the saturation value of the magnetization is given by

$$M_{max} = \underline{\hspace{2cm}}$$

When the orientation of a magnetic dipole changes with respect to an external field its energy changes. The potential energy of a dipole μ in a magnetic field B_0 is given by $U = \underline{\hspace{2cm}}$. Suppose a dipole originally makes an angle of 90° with a field. If its orientation changes so it becomes parallel to the field its energy decreases (from 0 to $-\mu B_0$). This means the energy density of a paramagnetic object decreases when an external field is turned on and dipoles align with the field. The energy lost by the material is usually gained by the sources of the external field or flows as heat to the environment of the material.

The change in energy that accompanies a change in magnetization is used in a cooling process known as adiabatic demagnetization. Paramagnetic material is placed in a magnetic field and magnetized. It is then thermally isolated and the external field is turn off. The material demagnetizes and as it does the magnetic dipoles gain energy from $\underline{\hspace{4cm}}$ $\underline{\hspace{10cm}}$.

This results in a decrease in temperature.

Diamagnetism. In the absence of an applied field the atoms of a diamagnetic substance have no magnetic dipole moments, but moments are induced when a field is applied. As an external field is turned on the orbits of the electrons change so the electrons produce an opposing field, in accordance with $\underline{\hspace{2cm}}$ law. The direction of the magnetization is $\underline{\hspace{2cm}}$ that of the applied field and the sign of $\kappa_m - 1$ is $\underline{\hspace{2cm}}$. The total magnetic field in diamagnetic material is $\underline{\hspace{2cm}}$ in magnitude than the applied field. Look at Table 3 for some typical values of $\kappa_m - 1$ for diamagnetic materials.

The effect occurs for all materials, but if the atoms have permanent dipole moments the effect of their alignment with the field dominates.

Ferromagnetism. The atoms of ferromagnetic materials have permanent dipole moments but unlike the atomic moments of paramagnetic materials they *spontaneously* align with each other. List some ferromagnetic materials: $\underline{\hspace{6cm}}$

You should realize that the spontaneously alignment of dipoles in a ferromagnet is *not* due to the torque exerted by one magnetic dipole on another. These torques are not sufficiently strong to overcome thermal agitation that tends to randomize the dipole directions. The torques that align the dipoles have their source in the quantum mechanics of electrons in solids. Above a certain temperature, called its _____ temperature, a ferromagnetic object becomes paramagnetic.

Ferromagnetic materials exhibit <u>hysteresis</u>. On the axes to the right draw a curve that shows the magnetization M for a typical ferromagnet as an applied field B_0 increases from 0 to B_{0f}. Assume the material is unmagnetized to start. Next draw the curve as the applied field decreases from B_{0f} to 0. The magnetization does not retrace the original curve and it does not become 0 when the applied field is 0. The ferromagnet is magnetized even though there is no external field.

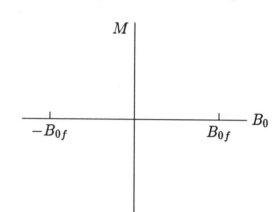

Now suppose the applied field is reversed (has negative values on the graph) and its magnitude is increased to B_{0f}. Draw the curve. Note that a field must be applied opposite to the magnetization in order to reduce the magnetization to 0. Now draw the curve that shows what happens if the applied field is reduced to 0. Finally, draw the curve that shows what happens if the applied field is reversed again and increased to B_{0f}.

Hysteresis is a direct result of the existence of ferromagnetic <u>domains</u>. In a domain the dipole moments are _____ to each other, while dipoles in a neighboring domain are in _____ direction. The total magnetization you plotted above is the vector sum of the magnetizations of all domains. If a ferromagnet is unmagnetized the vector sum of the dipole moments of the domains is _____. When an external field is applied domains with dipoles parallel to the field tend to _____ in size while domains in other directions tend to _____. This amounts to changes in the orientations of dipoles near domain boundaries. Some re-orientation of dipoles within a domain may also take place.

Hysteresis comes about because the growth and shrinkage of domains is not reversible. When an external field is applied to an unmagnetized sample and then turned off the domains do not spontaneously revert to their original sizes.

II. PROBLEM SOLVING

Some problems deal with Gauss' law for magnetism. You should understand that the total magnetic flux through a closed surface is zero and that this can be used to find the flux through part of a surface if the flux through the rest of the surface is known.

PROBLEM 1. A uniform 0.95-T magnetic field is perpendicular to the flat face of a 0.75-m radius hemisphere, as shown. What is the flux through the curved surface?

SOLUTION: Since the field is uniform and perpendicular to the flat face the flux through this portion of the surface is given by $\Phi_B = -\pi R^2 B$. Since the total flux is zero, the flux through the curved surface is given by $\Phi_B = +\pi R^2 B$.

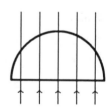

[ans: +1.68 Wb]

PROBLEM 2. Magnetic field lines enter the flat face of a 0.75-m radius hemisphere, then bend to become normal to the curved surface as they leave. If the field has a uniform magnitude of 0.85 T on the curved surface, what is the magnetic flux through the flat surface?

SOLUTION:

[ans: −3.00 Wb]

If the field is also normal to the flat surface and uniform over it, what is its magnitude there?

SOLUTION:

[ans: 1.70 T]

Gauss law for magnetism places restrictions on the magnetic field. For example, if the x component of the field varies with x then the y component must vary with y or the z component must vary with z. If this were not so the law would be violated. Here's an example.

PROBLEM 3. The magnetic field in a region of space around the origin has the form $\mathbf{B} = 0.43x\mathbf{i} + Ay\mathbf{j}$, where A is a constant. $\mathbf{B}$ is in tesla if x and y are in meters. What is the value of A?

SOLUTION: Use Gauss' law for magnetism. Evaluate the integral for the surface of a cube with one corner at the origin and oriented as shown. Take each edge to have length a. The contribution to the integral of the left face is 0 because $B_y = 0$ on this face. The contribution of the right face is $a^2 B_y = a^3 A$. The contribution of the back face is 0 because $B_x = 0$ on this face. The contribution of the front face is $a^2 B_x = 0.43a^3$. The contributions of the top and bottom faces are both 0 because $B_z = 0$. Thus $\oint \mathbf{B} \cdot d\mathbf{A} = 0.43a^3 + a^3 A$. This must vanish. Solve for A.

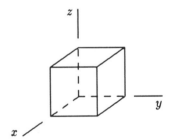

[ans: −0.43 T/m]

You should know how to calculate the orbital contribution to the magnetic dipole moment of a circulating charge, given the radius of its orbit and its speed or, alternatively, given its angular momentum.

PROBLEM 4. An electron is moving around a 4.76-Å radius circle, being held in its orbit by the electric force of a proton. Assume classical mechanics is valid and calculate its speed.

SOLUTION: The force acting on it is $q^2/4\pi\epsilon_0 r^2$ and, according to Newton's second law this must be mv^2/r, where r is the radius of its orbit. So $v^2 = q^2/4\pi\epsilon_0 mr$. Use $q = 1.60 \times 10^{-19}$ C and $m = 9.11 \times 10^{-31}$ kg.

[ans: 7.29×10^5 m/s]

What is the current?

SOLUTION: The period of the motion is given by $T = 2\pi r/v$ and the current is given by $i = q/T$.

[ans: 3.90×10^{-5} A]

What is the magnitude of the orbital contribution to its magnetic dipole moment? Give your answer in SI units and in Bohr magnetons.

SOLUTION: Use $\mu_\ell = iA = \pi r^2 i$.

[ans: 2.78×10^{-23} J/T; $3.00\,\mu_B$]

Its spin angular momentum is parallel to the orbital contribution to its dipole moment. What is the magnitude of its total dipole moment?

SOLUTION: Use $\mu_s = es/m$ and $s = h/4\pi$. The direction of μ_s is opposite to the direction of s and so is opposite to the direction of μ_ℓ. The total dipole moment has magnitude $|\mu_\ell - \mu_s|$.

[ans: 1.85×10^{-23} J/T; $2.00\,\mu_B$]

PROBLEM 5. The orbital and spin angular momenta of an electron are both in the positive z direction. The z component of the orbital angular momentum is $2(h/2\pi)$. What is the magnetic dipole moment of the electron? Give your answer in SI units and in Bohr magnetons.

SOLUTION: The z components of the orbital and spin contributions to the magnetic dipole moment are $\mu_\ell = -(e/2m)\ell_z$ and $\mu_s = -(e/m)s$, where $\ell_z = 2(h/2\pi)$ and $s = \frac{1}{2}(h/2\pi)$.

[ans: -2.78×10^{-23} J/T; $-3.00\,\mu_B$]

Some problems deal with the interaction of magnetic dipoles with each other and with an external field. Here are a few examples.

PROBLEM 6. Two electrons are a distance $d = 4.5$ Å apart, with their intrinsic dipole moments parallel and along the line joining them, as shown. What is the magnetic field produced by electron 1 at the site of electron 2?

SOLUTION: Use $B = (\mu_0/2\pi)(\mu/d^3)$.

[ans: 2.04×10^{-2} T, upward]

What torque does electron 1 exert on electron 2?

SOLUTION:

[ans: 0]

PROBLEM 7. The electrons are now oriented so the dipole moment of electron 2 is perpendicular to the line joining them, as shown. What torque does electron 1 now exert on electron 2? In what direction does the dipole moment of electron 2 tend to turn?

SOLUTION: Use $\tau = \mu \times \mathbf{B}$.

[ans: 1.89×10^{-25} N·m, out of the page; counterclockwise]

What torque does electron 2 exert on electron 1? In what direction does the dipole moment of electron 2 tend to turn?

[ans: 1.89×10^{-25} N·m, into page; clockwise]

Notice that each dipole exerts a torque on the other in such a way that they tend to become aligned.

PROBLEM 8. The spin angular momentum of an electron is aligned with an external field of 1.6 T. Electromagnetic radiation causes the spin to flip so it is directed opposite to the external magnetic field. Was energy absorbed or emitted by the electron? How much?

SOLUTION: Use $U = -\mu \cdot \mathbf{B}$ to calculate the energy of the dipole before and after the flip. Don't forget that μ_s and s are in opposite directions.

[ans: Energy decreases. It is emitted into the radiation; 2.97×10^{-23} J]

You should know how to calculate the total magnetic field and the magnetization when a given external field is applied to a linear paramagnetic or diamagnetic material.

Chapter 37: Magnetic Properties of Matter **549**

PROBLEM 9. Chromium ($\kappa_m - 1 = 3.3 \times 10^{-4}$) fills the interior of a long cylindrical solenoid with 600 turns/cm, carrying a current of 2.3 A. What magnetic field is produced in the interior by the current? What magnetic field is produced in the interior by atomic dipoles in the chromium? What is the magnetization of the chromium?

SOLUTION: Use $B_0 = \mu_0 n i$ to calculate the field of the current and $B_M = (\kappa_m - 1) B_0$ to calculate the field of the dipoles in the chromium. The magnetization of the chromium is given by $M = B_M/\mu_0$. $\mathbf{B}_0$, $\mathbf{B}_M$, and $\mathbf{M}$ are all parallel to the axis of the solenoid.

[ans: 0.173 T; 5.72×10^{-5} T; 45.5 A/m]

If the solenoid has a radius of 0.56 cm, what is the dipole moment per unit length of the chromium?

SOLUTION: The magnetization is uniform so the total dipole moment is the product of the magnetization and the volume.

[ans: 4.49×10^{-3} A·m]

PROBLEM 10. Copper ($\kappa_m - 1 = -9.7 \times 10^{-6}$) fills the interior of a horizontal solenoid carrying a current that by itself produces a magnetic field of 0.67 T, pointing from left to right. What field is produced by dipoles in the copper and what is the magnetization of the copper?

SOLUTION:

[ans: 6.50×10^{-6} T, right to left; 5.17 A/m, right to left]

The atomic mass number of copper is 63.54 and the density of copper is 8.96 g/cm³. What is the average dipole moment of a copper atom in this field?

SOLUTION: The atomic concentration (number of atoms per unit volume) is given by $n = \rho/m$, where ρ is the density and m is the mass of an atom. The mass of an atom in grams is given by the atomic mass number divided by the Avogadro constant. Finally, the average dipole moment of an atom is given by $\mu = M/n$.

[ans: 6.09×10^{-29} A·m²]

PROBLEM 11. The magnetic dipole moment of each atom of a certain paramagnetic material is due entirely to the spin of a single electron, which is either parallel or antiparallel to an applied field. If the atomic concentration is 5.3×10^{28} atoms/m^3 what is the saturation magnetization of this material?

SOLUTION: Saturation magnetization occurs when all dipoles are aligned. Thus $M_{\text{sat}} = n\mu$, where n is the atomic concentration and μ is the magnitude of the dipole moment of a single electron.

[ans: 4.92×10^5 A/m]

When a weak field is applied the magnetization is 3.5×10^3 A/m. How many more dipoles per unit volume are directed parallel to the field than are directed antiparallel to the field? What fraction of the total number of dipoles does this represent?

SOLUTION: If n_+ is the concentration of dipoles directed parallel to the field and n_- is the concentration of dipoles directed antiparallel to the field, then $M = (n_+ - n_-)\mu$. Solve for $n_+ - n_-$.

[ans: 3.78×10^{26} dipoles/m^3; 7.12×10^{-3}]

PROBLEM 12. At a temperature of 30° C the value of $\kappa_m - 1$ for tungsten, a paramagnetic material, is 6.8×10^{-5}. What is its Curie constant?

SOLUTION: Solve $M = (\kappa_m - 1)B_0/\mu_0$ and $M = CB_0/T$ simultaneously for C. T is $273 + 30 = 303$ K.

[ans: 1.64×10^4 m·K/H]

What is the value of $\kappa_m - 1$ at a temperature of 400° C?

SOLUTION:

[ans: 3.06×10^{-5}]

PROBLEM 13. An iron bar 6.0 cm long, with a cross-sectional area of 2.3 m², has a magnetic dipole moment of 25 A·m² when its magnetization is fully saturated. What is its magnetization? What is the dipole moment of an individual iron atom? The density of iron is 7.9 g/cm³ and the atomic mass number of iron is 55.8.

SOLUTION: The dipole moment of the bar is the product of the magnetization and the volume of the bar. To find the dipole moment of a single atom use $M = n\mu$, where n is the concentration of iron atoms. The concentration is given by ρ/m, where m is the mass of an atom.

[ans: 1.81×10^6 A/m; 2.12×10^{-23} J/T]

III. NOTES

Chapter 38
INDUCTANCE

I. BASIC CONCEPTS

Here you learn about inductive circuit elements, which produce emfs via Faraday's law when their currents are changing and which store energy in magnetic fields. Pay attention to the definition of inductance and learn to calculate its value for solenoids and toroids. Learn about the influence of inductance on the current in a circuit and study the calculations of current for series circuits containing an inductor. Also learn what factors determine the energy stored in the magnetic field of an inductor and learn how to calculate it.

Inductance. When the current in a circuit changes the magnetic field it produces and the magnetic flux through the circuit also change. According to Faraday's law this means an emf $\mathcal{E}_L$ is induced around the circuit. The induced emf is proportional to the rate of change of the flux and hence to the rate of change of the current. The inductance L of the circuit is defined by the relationship

$$\mathcal{E}_L = \underline{\hspace{2cm}}$$

L does not depend on the current or its rate of change. In general terms it does depend on

_____

_____

The SI unit of inductance is called the _____ (abbreviated _____). The symbol for an inductor is _____.

A solenoid is an excellent example of an inductor because it concentrates the field in its interior and a current produces a large magnetic flux through its turns. For all practical purposes, the emf induced by a changing current in a circuit containing a solenoid appears in its entirety along the wire of the solenoid.

The diagram shows an inductor carrying current i (the rest of the circuit is not shown). The direction of positive current is from a toward b. If the current is increasing then the emf induced in the inductor is from _____ toward _____; if it is decreasing then the emf is from _____ toward _____.

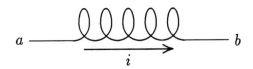

In either case the potential V_b at point b is given by $V_b = V_a - L\,di/dt$, where V_a is the potential at point a and L is the inductance.

If the current rather than its rate of change is known the magnetic field and flux can be computed. If i is the current and Φ_B is the flux through each of N turns then the inductance is given by

$$L = \underline{\hspace{2cm}}$$

For an ideal solenoid of length ℓ and cross-sectional area A, with n turns per unit length and carrying current i, the magnetic field in the interior is given by $B =$ _____, the flux through each turn is given by $\Phi_B =$ _____, and the inductance is given by $L =$ _____. Notice that the inductance depends on the geometry (the length, area, and number of turns per unit length) but not on the current.

If an inductor with inductance L_0 is filled with magnetic material its inductance becomes $L =$ _____, where κ_m is the magnetic permeability of the material.

An LR circuit. The diagram on the right shows a series LR circuit. Take the current to be positive in the direction of the arrow and develop the loop equation in terms of L, R, $\mathcal{E}$, i, and di/dt. Take the electric potential to be 0 at point a. Then the potential at point b is _____ and the potential at point c is given by _____. The potential at point a is this value minus iR and the result must, of course, be zero. Thus the loop equation is

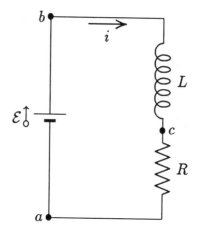

Assume the seat of emf is connected to the circuit at time $t = 0$. The solution to the loop equation is then

$$i(t) =$$

According to this expression, at $t = 0$ the current is $i =$ _____ and long after the seat of emf is connected it is $i =$ _____. The rate at which the current increases to its final value is controlled by the inductive time constant τ_L, which in terms of L and R, is $\tau_L =$ _____. The larger the time constant the slower the rate. A large time constant can be obtained by making _____ large and _____ small. When $t = \tau_L$ the current is about _____ per cent of its final value.

The potential difference across the inductor is given by

$$V_L(t) = L\frac{di}{dt} =$$

and the potential difference across the resistor is given by

$$V_R(t) = iR =$$

On the axes to the left below draw a graph of the current as a function of time. On the time axis mark the approximate position of $t = \tau_L$. On the axes to the right draw graphs of the potential differences across the seat of emf, the resistor, and the inductor, all as functions of time. Here's some information that might help you: $V_R = V_L$ at $t \approx 0.7\tau_L$ and each of these potential differences have the value $\mathcal{E}/2$ at that time. Label the graphs V_R and V_L, as appropriate.

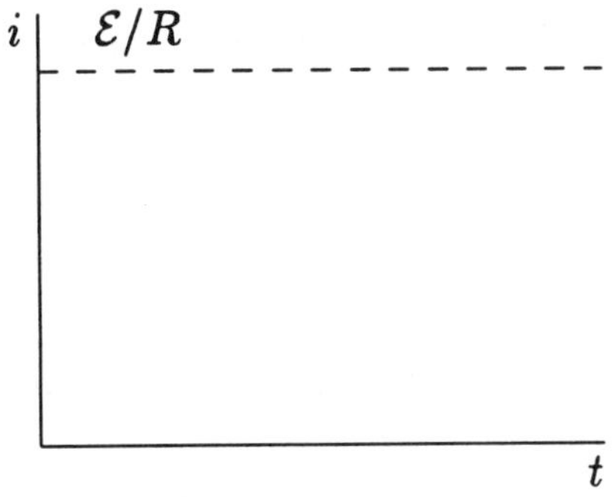

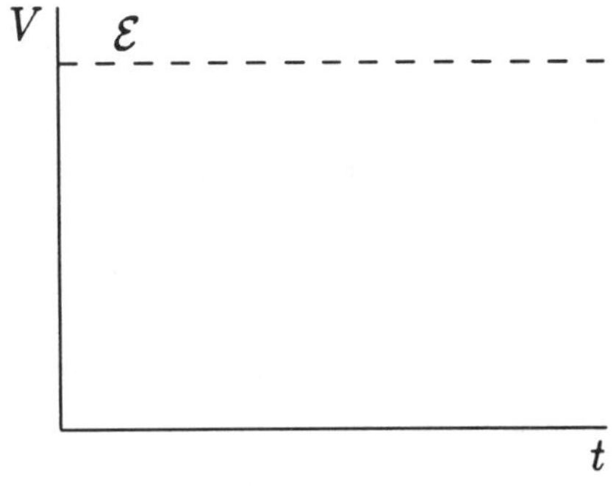

The current is the least but its rate of change is the greatest at $t =$ _____. Then the potential difference across the resistor is _____ and the potential difference across the inductor is _____. A long time after the seat of emf is connected the current is _____ and its rate of change is _____. Then the potential difference across the resistor is _____ and the potential difference across the inductor is _____.

Suppose that the seat of emf is replaced by a wire when the circuit has a steady current i_0. Look at Fig. 4 of the text to see how this might be done. As time goes on we expect the current to decay to 0. The loop equation is

and its solution is

$$i(t) =$$

According to this expression, at $t = 0$ the current is _____ and its rate of change is _____. Then the potential difference across the resistor is _____ and the potential difference across the inductor is _____. A long time after the wire is inserted the current is _____ and its rate of change is _____. Then the potential difference across the resistor is _____ and the potential difference across the inductor is _____. Again the time dependence is controlled by the inductive time constant. When $t = \tau_L$ the current is about _____ per cent of its initial value.

Magnetic energy. Energy must be supplied to build up the current and magnetic field in an inductor, perhaps by a seat of emf. The energy may be considered to be stored in the magnetic field and can be retrieved when the current and field decrease. If current i is in an inductor with inductance L, the energy stored is given by

$$U_B =$$

The loop equation for a series LR circuit, when multiplied by the current, can be written $i\mathcal{E} = Li\,di/dt + i^2 R$. This equation tells us that the rate with which energy is supplied by the seat of emf equals the sum of the rates with which it is dissipated in the resistor and stored in the magnetic field of the inductor. Identify each of the terms:

$i\mathcal{E}$: _____
$Li\,di/dt$: _____
$i^2 R$: _____

Chapter 38: Inductance

The energy density (energy per unit volume) stored in a magnetic field **B** is given by

$u_B =$

The total energy stored in a field is given by the volume integral

$$u_B = \int \frac{1}{2\mu_0} B^2 \, dV.$$

These two equations are universally valid. They hold for every magnetic field, not just the field of a solenoid.

An LC circuit. The circuit you consider consists of an inductor with inductance L and a capacitor with capacitance C, as shown in the diagram to the right. Let q represent the charge on the upper plate of the capacitor and take the current to be positive in the direction of the arrow. The potential difference across the capacitor is given by _____, with the _____ plate at the higher potential if q is positive. The potential difference across the inductor is given by _____, with the _____ end at the higher potential if i is increasing. In terms of q and di/dt the loop equation is _____. Substitute $i = dq/dt$ to obtain the loop equation in terms of q:

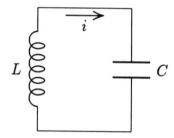

Suppose that at time $t = 0$ the current is zero and charge q_m is on the capacitor, with the upper plate positive. Then the solution to the loop equation is

$q(t) =$

and the current is given by

$i(t) = dq/dt =$

Both q and i oscillate with an angular frequency that is determined by the values of L and C: $\omega =$ _____. Solutions with other phase constants, of the form $q(t) = q_m \cos(\omega t + \phi)$, are valid for other initial conditions. Initial conditions, however, do not influence ω.

No matter what the value of the phase constant, the maximum magnitude i_m of the current is _____ and $i = \pm i_m$ when $q =$ _____. When q is either $+q_m$ or $-q_m$ the current is _____.

Energy is stored in the electric field of the capacitor and in the magnetic field of the inductor. As functions of time the energy in the capacitor is given by

$$U_E(t) = \frac{1}{2}\frac{q^2}{C} =$$

and the energy in the inductor is

$$U_B(t) = \frac{1}{2}Li^2 =$$

Note that the energy is oscillates back and forth between the inductor and the capacitor. The energy is entirely magnetic when the _____ is a maximum and the _____ is zero; it is entirely electric when the _____ is a maximum and the _____ is zero. The sum, $U = U_E + U_B$, is constant and can be written in terms of the current amplitude i_m as $U =$ _____ or in terms of the charge amplitude q_m as $U =$ _____.

556   Chapter 38: Inductance

LRC circuits. If a resistor with resistance R is added in series to the LC circuit shown above, the charge and current amplitudes decay exponentially with time. The loop equation becomes

and its solution is

$$q(t) = $$

where the angular frequency of oscillation is now $\omega' = $ _____. The sum of the electric and magnetic energies is not constant but decreases exponentially because _____
_____.

Each time the charge on the capacitor reaches a maximum it is less than the previous time. Does this violate the principle of charge conservation? _____ Explain: _____
_____.

Suppose an oscillating seat of emf is added in series to the LRC circuit. Take the emf to be $\mathcal{E} = \mathcal{E}_m \cos(\omega'' t)$. After transients die out the current is given as a function of time by

$$i(t) = $$

where i_m is the magnitude of the maximum current. Notice that its angular frequency is the same as that of the source and is not the natural angular frequency. The current amplitude i_m depends on the difference between the imposed angular frequency ω'' and the natural angular frequency ω'. See Fig. 14. As the imposed frequency approaches the natural frequency from either side the current amplitude _____ and reaches its maximum value when $\omega'' = $ _____. This is the condition for _____. Energy is being dissipated in the resistor but it is being supplied by _____ at the same rate.

II. PROBLEM SOLVING

Some problems deal with the inductance of a solenoid. Here are a few.

PROBLEM 1. A 2.5-cm long, 500-turn ideal solenoid has a radius of 8.5 mm. What is the magnetic field in its interior when it carries a current of 2.3 A?

SOLUTION: Use $B = \mu_0 n i$.

[ans: 5.78×10^{-2} T]

Assume the field is uniform and calculate the magnetic flux through each turn.

SOLUTION: Use $\Phi_B = BA = B\pi R^2$, where R is the radius.

[ans: 1.31×10^{-5} Wb]

What is its inductance?

SOLUTION: Use $L = N\Phi_B/i$.

[ans: 2.85 mH]

If the current is changing at 250 A/s what is the emf induced in the solenoid?

SOLUTION: Use $\mathcal{E} = L\,di/dt$.

[ans: 0.713 V]

PROBLEM 2. An ideal solenoid is 2.6 cm long and has a radius of 0.35 cm. How many turns of wire must it have if it is to generate an emf of 1.6 V when the current in it is changing at 350 A/s?

SOLUTION: If ℓ is the length of the solenoid and N is the number of turns, the number of turns per unit length is $n = N/\ell$ and the inductance is given by $L = \mu_0 N^2 A/\ell$. Substitute this into $\mathcal{E} = L\,di/dt$ and solve for N.

[ans: 1570]

PROBLEM 3. A 7.8-mH ideal solenoid has 500 turns. When its current is given by $i(t) = 5.50\,t$, where i is in amperes for t in seconds, it generates an emf of 43 mV. What is its inductance?

SOLUTION: Use $\mathcal{E} = L\,di/dt$.

[ans: 7.82 mH]

Assume the magnetic field is uniform in the interior and calculate the magnetic flux through one turn.

SOLUTION: Use $L = N\Phi_B/i$, where N is the number of turns.

[ans: 1.72×10^{-4} Wb]

You should know how to calculate the inductance of a toroid. The magnetic field at a point in the interior a distance r from the center is tangent to the circle of radius r and its magnitude is given by $B = \mu_0 Ni/2\pi r$, where N is the number of turns and i is the current. Review Section 35–6.

PROBLEM 4. A toroid has N turns, an inner radius R_1, an outer radius R_2, and a square cross section. If it carries current i what is the flux through one turn?

SOLUTION: The thickness of the toroid (measured along a line that is perpendicular to a radius) is $R_2 - R_1$. The magnetic field does not vary along this direction. It does vary along a radius, so you must evaluate an integral to calculate the flux. An infinitesimal element of area perpendicular to the field is $dA = (R_2 - R_1)\,dr$. Thus

$$\Phi_B = \frac{\mu_0 Ni}{2\pi}(R_2 - R_1)\int_{R_1}^{R_2} \frac{dr}{r}.$$

[ans: $(\mu_0 Ni/2\pi)(R_2 - R_1)\ln(R_2/R_1)$]

Take $N = 1500$ turns, $R_1 = 2.50$ cm, and $R_2 = 2.65$ cm. Evaluate this expression for a current of 2.3 A.

SOLUTION:

[ans: 6.03×10^{-8} Wb]

What is the inductance of the toroid?

SOLUTION: Use $L = N\Phi_B/i$.

[ans: 39.3 μH]

Some problems deal with circuits containing inductors. Here are some examples.

PROBLEM 5. At time $t = 0$ a 10-V seat of emf is connected to a series circuit containing a resistor ($R = 2.0\,\Omega$) and an inductor ($L = 15$ mH). What is the time constant for this circuit?

SOLUTION: Use $\tau_L = L/R$.

[ans: 7.50 ms]

Just after the emf is connected what is the current and what is its rate of increase?

SOLUTION: The current is given by $i(t) = (\mathcal{E}/R)[1 - e^{-t/\tau_L}]$ and its rate of change is given by $di/dt = (\mathcal{E}/\tau_L R)e^{-t/\tau_L} = (\mathcal{E}/L)e^{-t/\tau_L}$. Evaluate these expressions for $t = 0$.

[ans: 0; 6.67 × 10² A/s]

At the end of 3.0 ms what is the current? What is the emf across the inductor? What is the potential difference across the resistor?

SOLUTION: Use $i = (\mathcal{E}/R)[1 - e^{-t/\tau_L}]$, $\mathcal{E}_L = L\,di/dt = \mathcal{E}e^{-t/\tau_L}$, and $V_R = iR$.

[ans: 3.49 A; 3.01 V; 6.99 V]

What is the final value of the current?

SOLUTION: A long time after the seat of emf is connected e^{-t/τ_L} is essentially 0 and the current is given by $i = \mathcal{E}/R$.

[ans: 5.00 A]

At what time is the current three-fourths of its final value?

SOLUTION: Solve $0.75 = 1 - e^{-t/\tau_L}$ for t. A little rearranging gives $e^{-t/\tau_L} = 0.25$. Take the natural logarithm of both sides to obtain $-t/\tau_L = \ln 0.25$, or $t = -\tau_L \ln 0.25$.

[ans: 10.4 ms]

At what time is the emf across the inductor three-fourths its initial value?

SOLUTION: The initial value of the emf is $\mathcal{E}$. You need to solve $0.75 = e^{-t/\tau_L}$ for t.

[ans: 2.16 ms]

At what time does the emf across the inductor equal the potential difference across the resistor?

SOLUTION: Now you must solve $\mathcal{E}e^{-t/\tau_L} = \mathcal{E}[1 - e^{-t/\tau_L}]$ for t.

[ans: 5.20 ms]

PROBLEM 6. When the make-before-break switch S is in position a current builds up in the inductor. When it is in position b the current decreases. For one application $L = 7.4\,\text{H}$ and $R = 2.0\,\Omega$. The application requires that the current build up slowly so that it reaches half it final value in 1.5 s and decrease rapidly so it is reduced by half in 5.0 ms. What should the values of R_1 and R_2 be?

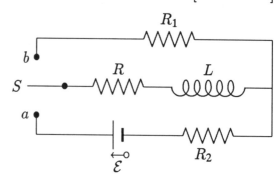

SOLUTION: The time $t_{1/2}$ to reach half the final value is related to the inductive time constant by $t_{1/2} = -\tau_L \ln 0.5 = 0.693\tau_L$. Calculate the time constant for each position of the switch.

When the switch is in position a the time constant is given by $L/(R_2 + R)$; when it is in position B the time constant is given by $L/(R_1 + R)$. Solve for R_1 and R_2.

[ans: 1.02 kΩ; 1.42 Ω]

PROBLEM 7. Consider the circuit shown to the right. Just after switch S is closed what is the current in each resistor and in the inductor? What is the potential difference across each resistor and across the inductor?

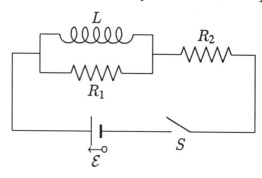

SOLUTION: There is no current in the inductor and it immediately generates an emf from right to left that prevents the buildup of current. All current in the circuit is through R_1 and R_2 and it is $i_1 = i_2 = \mathcal{E}/(R_1 + R_2)$. The potential difference across R_1 is $R_1 i_1$ and the potential difference across R_2 is $i_2 R_2$. The potential difference across the inductor is the same as that across R_1.

[ans: $i_L = 0$; $i_1 = i_2 = \mathcal{E}/(R_1 + R_2)$; $V_L = V_1 = \mathcal{E}R_1/(R_1 + R_2)$; $V_2 = \mathcal{E}R_2/(R_1 + R_2)$]

After the switch has been closed for a long time what is the current in each resistor and in the inductor? What is the potential difference across each resistor and across the inductor?

SOLUTION: After a long time the currents reach their final values and no longer change with time. The potential difference across the inductor is 0. This must be the same as the potential difference across R_1, so the current in R_1 is zero. The circuit acts like a loop containing only the emf $\mathcal{E}$ and the resistor R_2.

$$\left[\text{ans: } i_L = i_2 = \mathcal{E}/R_2; i_1 = 0; V_L = V_1 = 0; V_2 = \mathcal{E}\right]$$

Now find the time constants for the current i_L in the inductor and the current i_1 in R_1.

SOLUTION: First write two loop equations and a junction equation. Take all currents to be positive from left to right in the resistive or inductive elements.

$$\left[\text{ans: } \mathcal{E} - L(di_L/dt) - i_2 R_2 = 0, \mathcal{E} - i_1 R_1 - i_2 R_2 = 0, i_2 = i_1 + i_L\right]$$

Use the second loop equation and the junction equation to eliminate i_1 and i_2 from the first loop equation. Arrange the terms so the resulting equation looks like the differential equation for the current in an LR circuit.

$$\left[\text{ans: } L(di_L/dt) + R_1 R_2 i_L/(R_1 + R_2) = \mathcal{E} R_1/(R_1 + R_2)\right]$$

Identify the time constant.

$$\left[\text{ans: } L(R_1 + R_2)/R_1 R_2\right]$$

Now use the junction equation and the second loop equation to eliminate i_L and i_2 from the first loop equation and find the differential equation obeyed by i_1. What is the time constant for i_1?

$$\left[\text{ans: } L(di_1/dt) + R_1 R_2 i_1/(R_1 + R_2) = 0; L(R_1 + R_2)/R_1 R_2\right]$$

Notice that i_L increases from 0 to $\mathcal{E}/R_2$ while i_1 decreases from $\mathcal{E}/(R_1 + R_2)$ to 0. The current i_2 in R_2 is the sum of these two.

PROBLEM 8. A 15-mH inductor and a 25-μF capacitor form an LC circuit. At time $t = 0$ the charge on one plate of the capacitor is 6.4×10^{-9} C and the current is zero. What are the angular frequency, frequency, and period of the electromagnetic oscillations?

SOLUTION: Use $\omega = 1/\sqrt{LC}$, $\nu = \omega/2\pi$, and $T = 1/\nu$.

$$\left[\text{ans: } 1.63 \times 10^3 \text{ rad/s}; 2.60 \times 10^2 \text{ Hz}; 3.85 \text{ ms}\right]$$

What is the maximum current and when is the magnitude of the current a maximum for the first time?

SOLUTION: Since $q = q_m \cos(\omega t)$ and $i = dq/dt = -\omega q_m \sin(\omega t)$, the magnitude of the maximum current is $i_m = \omega q_m$. The current is a maximum one-fourth of a period after the charge on the capacitor is a maximum.

[ans: $10.5\,\mu\text{A}$; $0.962\,\text{ms}$]

When is the charge on the capacitor one-half its initial value for the first time?

SOLUTION: Solve $q_m/2 = q_m \cos(\omega t)$ for t. First find the angle α in radians for which $\cos\alpha = 1/2$, then divide this angle by ω.

[ans: $0.641\,\text{ms}$]

Notice that this is not one-eighth of a period.

Suppose that at $t = 0$ the current has its maximum value i_m and the charge on the capacitor is 0. Write expressions for the current and charge as functions of time.

SOLUTION: Since $i = i_m$ at $t = 0$ the function that gives the current is $i(t) = i_m \cos(\omega t)$. The charge is given by $q = \int i(t)\,dt$. Choose the constant of integration so $q = 0$ at $t = 0$.

[ans: $i(t) = i_m \cos(\omega t)$; $q(t) = (i_m/\omega)\sin(\omega t)$]

PROBLEM 9. Frequencies in the FM radio band are between 88 MHz and 108 MHz. Tuning is done by means of a fixed inductor ($L = 15\,\text{mH}$) and a variable capacitor. What range of values should the capacitor have?

SOLUTION: Use $\omega = 1/\sqrt{LC}$.

[ans: $1.45 \times 10^{-16}\,\text{F}$ to $2.18 \times 10^{-16}\,\text{F}$]

Some problems deal with energy. The energy stored in the magnetic field of an inductor is given by $U_B = \frac{1}{2}Li^2$, where i is the current in the inductor. Energy is stored at the rate $P_L = dU_B/dt = Li\,di/dt$. Also recall that energy is supplied by a seat of emf at a rate given by $P_\mathcal{E} = i\mathcal{E}$ and is dissipated in a resistor at a rate given by $P_R = i^2 R$.

PROBLEM 10. Starting at time $t = 0$, the current in a 9.7-mH inductor is given by $i(t) = 4.5 + 1.8t$, where i is in amperes for t in seconds. At the end of 3.0 s what energy is stored in the magnetic field of the inductor and what is the rate at which it is increasing?

SOLUTION: Use $U_B = \frac{1}{2}Li^2$ and $P_L = Li\,di/dt$, both evaluated for $t = 3.0\,\text{s}$.

[ans: $0.475\,\text{J}$; $0.173\,\text{W}$]

PROBLEM 11. An LR series circuit has a 8.8-mH inductor, a 5.0 Ω resistor, and a 10-V seat of emf. A switch closes to complete the circuit at time $t = 0$. At $t = 2.0$ ms what is the rate at which the emf is supplying energy, what is the rate at which energy is being dissipated in the resistor, and what is the rate at which energy is being stored in the inductor?

SOLUTION: First find the current using $i(t) = (\mathcal{E}/R)[1 - e^{-t/\tau_L}]$ and the rate of change of the current using $di/dt = (\mathcal{E}/L)e^{-t/\tau_L}$, where $\tau_L = L/R$ is the inductive time constant. Then evaluate $P_\mathcal{E}$, P_R, and P_L.

[ans: $P_\mathcal{E} = 13.6$ W; $P_R = 9.22$ W; $P_L = 4.38$ W]

Note that $P_\mathcal{E} = P_R + P_L$.

During the first 2.0 ms how much energy is stored in the magnetic field of the inductor? How much energy is supplied by the emf? How much energy is dissipated in the resistor?

SOLUTION: Use $U_B = \frac{1}{2}Li^2$ to calculate the energy stored in the inductor. Since the emf is supplying energy at the rate $i\mathcal{E}$ the energy it supplies by the end of time t is given by $\int_0^t i\mathcal{E}\,dt$. Substitute the expression for $i(t)$ and carry out the integration. You should get $(\mathcal{E}^2/R)[t - \tau_L + \tau_L e^{-t/\tau_L}]$. Substitute numerical values. There are now two ways to calculate the energy dissipated in the resistor. You might subtract the energy stored in the inductor from the energy supplied by the emf or you might evaluate the integral $\int_0^t i^2 R\,dt$.

[ans: 8.11×10^{-3} J; 1.61×10^{-2} J; 7.98×10^{-3} J]

PROBLEM 12. A series LR circuit is connected at time $t = 0$ to a constant seat of emf $\mathcal{E}$. Find an expression that gives the fraction of the energy supplied by the emf that is stored in the inductor, as a function of time. Evaluate the expression for $t = \tau_L/2$, τ_L, and $2\tau_L$, where $\tau_L = L/R$ is the inductive time constant.

SOLUTION: The energy stored in the inductor is $\frac{1}{2}Li^2 = \frac{1}{2}(L\mathcal{E}^2/R^2)[1 - e^{-t/\tau_L}]^2$. The energy supplied by the emf is $\int_0^t i\mathcal{E}\,dt = (\mathcal{E}^2/R)[t - \tau_L + \tau_L e^{-t/\tau_L}]$. The ratio is

$$f = \frac{1}{2}\tau_L \frac{[1 - e^{-t/\tau_L}]^2}{t - \tau_L + \tau_L e^{-t/\tau_L}}$$

Evaluate this expression.

[ans: 0.727; 0.543; 0.323]

As time goes on the rate at which energy is stored in the inductor decreases. In fact, when the current reaches its maximum value no more energy is stored. On the other hand, dissipation in the resistor continues.

PROBLEM 13. A 15-mH inductor and a 25-μF capacitor form an LC circuit. At time $t = 0$ the charge on one plate of the capacitor is 6.4×10^{-9} C and the current is zero. See Problem 8 above. What energy is initially stored in the capacitor?

SOLUTION: Use $U_E = \frac{1}{2}q^2/C$.

[ans: 8.19×10^{-13} J]

When the charge is reduced to half its original value what energy is stored in the capacitor? What energy is stored in the inductor?

SOLUTION: According to the solution to Problem 8 the current is $i = -\omega q_m \sin(\omega t)$, where $\omega = 1.63 \times 10^3$ rad/s and $t = 0.641$ ms when $q = q_m/2$. Use $U_B = \frac{1}{2}Li^2$ to calculate the energy stored in the inductor.

[ans: 2.05×10^{-13} J; 6.14×10^{-13} J]

When is the first time the energy in the capacitor is the same as the energy in the inductor?

SOLUTION: When the energies are equal, $\frac{1}{2}Li^2 = \frac{1}{2}q^2/C$. Since $LC = 1/\omega^2$, this means $\omega^2 q_m^2 \sin^2(\omega t) = \omega^2 q_m^2 \cos^2(\omega t)$, or $\tan^2(\omega t) = 1$. Solve for t.

[ans: 0.481 ms]

III. NOTES

Chapter 39
ALTERNATING CURRENT CIRCUITS

I. BASIC CONCEPTS

Here you study in detail a series circuit consisting of a resistor, an inductor, and a capacitor, driven by a sinusoidal emf. The current is an alternating current (AC): it periodically reverses direction. Pay close attention to the relationship between the potential difference across each circuit element and the current in the element. Learn how to compute the current amplitude and phase in terms of the generator emf. You will be able to apply what you learn to many other circuits, including those with elements in parallel.

An AC circuit. The circuit you consider is diagramed on the right. The AC generator is symbolized by _____ and the emf it produces is given by $\mathcal{E} = \mathcal{E}_m \sin(\omega t)$, where ω is its angular frequency. Its frequency is given by $\nu =$ _____. $\mathcal{E}$ is taken to be positive if the upper terminal of the generator is positive and negative if the upper terminal is negative. The arrow on the diagram shows the direction of positive current.

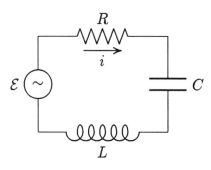

The current is given by $i(t) = i_m \sin(\omega t - \phi)$, where i_m is the current amplitude and ϕ is a phase constant. Note that the current has the same frequency as the generator emf but because the circuit contains a capacitor and an inductor it may not be in phase with the generator emf. If ϕ is between 0 and 90° the current is said to _____ the emf. If it is between 0 and −90° the current _____ the emf.

Quantities with sinusoidal time dependence, such as the generator emf and the current, can be represented by <u>phasors</u>. Tell what a phasor is: _____

The diagram on the right shows the phasor associated with the generator emf at some instant of time. Its length is $\mathcal{E}_m$ and it rotates in the counterclockwise direction. Its projection on the vertical axis is $\mathcal{E}(t) = \mathcal{E}_m \sin(\omega t)$. Label the appropriate angle ωt. Assume ϕ is about 40° and draw the phasor associated with the current. Label its length i_m. Label the angle $\omega t - \phi$ and the angle ϕ between the two phasors.

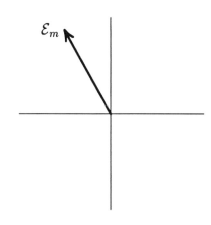

The potential difference V_R across the resistor is given by $V_R = iR = i_m R \sin(\omega t - \phi)$. V_R and i are in phase. The phasor associated with V_R is on the same line as the phasor associated with _____ and its length is $(V_R)_{max} =$ _____.

The potential difference V_L across the inductor is given by $V_L = L\,di/dt = \omega L i_m \cos(\omega t - \phi)$. Use the trigonometric identity $\sin(A + \pi/2) = \cos(A)$ to write this $V_L = \omega L i_m \sin(\omega t - \phi + \pi/2)$. V_L is said to lead the current by $\pi/2$ radians. Its amplitude is given by $(V_L)_{\max} = $ _____. When i is positive and increasing the _____ end of the inductor in the diagram is at a higher potential than the _____ end.

The phasor associated with V_L is $\pi/2$ radians ahead of (more counterclockwise than) the phasor associated with the current. On the axes to the right draw the phasors associated with the current and with the potential difference across the inductor at the instant for which the previous phasor diagram was drawn. Label their lengths i_m and $(V_L)_{\max}$, respectively. Also label the right angle between them.

The potential difference V_C across the capacitor is given by $V_C = q/C$. Now $q = \int i\,dt = -(i_m/\omega)\cos(\omega t - \phi)$, so $V_C = -(i_m/\omega C)\cos(\omega t - \phi)$. Use the trigonometric identity $\sin(A - \pi/2) = -\cos(A)$ to write this $V_C = (i_m/\omega C)\sin(\omega t - \phi - \pi/2)$. V_C is said to lag the current by $\pi/2$ radians. Its amplitude is given by $(V_C)_{\max} = $ _____. When q is positive the _____ plate of the capacitor in the diagram is at a higher potential than the _____ plate.

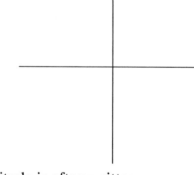

The phasor associated with V_C is $\pi/2$ radians behind (less counterclockwise than) the phasor associated with the current. On the axes to the right draw the phasors associated with the current and with the potential difference across the capacitor at the instant for which the previous phasor diagrams were drawn. Label their lengths i_m and $(V_C)_{\max}$, respectively. Also label the right angle between them.

The relationship between the current and potential amplitude is often written

$$(V_C)_{\max} = i_m X_C$$

for a capacitor and

$$(V_L)_{\max} = i_m X_L$$

for an inductor. Here $X_C = $ _____ and $X_L = $ _____. X_C is called the <u>capacitive reactance</u> and X_L is called the <u>inductive reactance</u>. These relations and $(V_R)_{\max} = i_m R$ for a resistor hold no matter what circuit the elements are in. In addition, the potential difference across an inductor always leads the current in the inductor by $\pi/2$ radians and the potential difference across a capacitor always lags the current into the capacitor by $\pi/2$ radians.

Reactances are not resistances but they play a similar role: they relate the amplitude of the current through a circuit element to the amplitude of the potential difference across the element. Carefully note that the reactances depend on the generator frequency. As the frequency increases X_C _____ and X_L _____. If the frequency is somehow increased without changing i_m the maximum potential difference across the capacitor and the maximum charge on the capacitor both decrease. Explain why this makes sense physically: _____

If the frequency is increased without changing i_m the maximum potential difference across the inductor increases. Explain why this makes sense physically: _____

Now you are ready to put the pieces together and solve for the current amplitude i_m and phase ϕ, given the generator emf. On the axes to the left below draw the phasors for the potential differences across the resistor, capacitor and inductor at the time for which the previous phasor diagrams were drawn.

According to the loop equation for the circuit $\mathcal{E}(t) = V_R(t) + V_L(t) + V_C(t)$. This means the phasors associated with V_R, V_L, and V_C must add like vectors to produce the phasor associated with $\mathcal{E}$. Since the phasors for V_L and V_C are parallel to each other they can be summed easily. Assume $(V_L)_{max} > (V_C)_{max}$ and on the axes to the right below draw phasors associated with V_R and $V_L + V_C$. The first has length $(V_R)_{max}$ and the second has length $[(V_L)_{max} - (V_C)_{max}]$. Now draw the "vector" sum of the two phasors with its tail at the origin. This must be the emf phasor. Label the angle ϕ between it and the phasor associated with V_R. This is also the angle between the emf and current phasors.

The phasors associated with V_R and $V_L + V_C$ form two sides of a right triangle with a hypotenuse of length $\mathcal{E}_m$, so $\mathcal{E}_m^2 = (V_R)_{max}^2 + [(V_L)_{max} - (V_C)_{max}]^2$. Substitute $(V_R)_{max} = i_m R$, $(V_L)_{max} = i_m X_L$, and $(V_C)_{max} = i_m X_C$, then solve for i_m. The result can be written

$$i_m = \mathcal{E}_m / Z,$$

where

$$Z = $$

Z is called the <u>impedance</u> of the circuit.

If the impedance of a circuit is increased without changing the generator emf, the current amplitude _____. The impedance can be increased by increasing the resistance or by

Chapter 39: Alternating Current Circuits   567

increasing the difference between the inductive and capacitive reactances. Suppose $X_L > X_C$. Then as the frequency increases, the impedance _____. If $X_C > X_L$ and the frequency increases, then the impedance initially _____.

According to the phasor diagram the phase constant for the current is given in terms of the potential differences by

$$\tan \phi =$$

and in terms of the resistance and the reactances by

$$\tan \phi =$$

Notice from your phasor diagram that ϕ must be between $-90°$ and $+90°$. Since R is positive the value given for ϕ by your calculator is correct. You will never need to add $180°$.

Give the relation between X_L and X_C for which each of the following occurs:

the current lags the generator emf ($0 < \phi < 90°$) _____

the current leads the generator emf ($-90° < \phi < 0$) _____

the current is in phase with the generator emf ($\phi = 0$) _____

Note that the angular frequency for which the current is in phase with the generator emf is the resonance angular frequency $\omega = \sqrt{1/LC}$. For a given emf amplitude the current amplitude has its greatest value for this angular frequency.

Power. The rate at which energy is supplied by the generator is given by $P_\mathcal{E} = i\mathcal{E}$, the rate at which energy is stored in the capacitor is given by $P_C = iV_C$, the rate at which energy is stored in the inductor is given by $P_L = iV_L$, and the rate at which energy is dissipated in the resistor is given by $P_R = iV_R$. All of these vary with time, being periodic with a period equal to the period of the generator.

Usually the time variations are of no interest. We consider instead averages over a cycle of oscillation. To compute averages you should know that the average of $\sin^2(\omega t - \phi)$ is $1/2$ and the average of $\sin(\omega t - \phi)\cos(\omega t - \phi)$ is 0. See the Mathematical Skills section.

You should be able to show that the averages of P_C and P_L are zero. For example, use $i = i_m \sin(\omega t - \phi)$ and $V_C = -(i_m/\omega C)\cos(\omega t - \phi)$ to obtain $P_C = -(i_m^2/\omega C)\sin(\omega t - \phi)\cos(\omega t - \phi)$. This averages to zero because it contains the product of the sine and cosine of $\omega t - \phi$. In the space below show that the average of P_L is zero:

The power supplied by the generator is $P_\mathcal{E} = i\mathcal{E} = i_m \mathcal{E}_m \sin(\omega t - \phi)\sin(\omega t)$. The first sine function can be expanded as $\sin(\omega t)\cos\phi - \cos(\omega t)\sin\phi$. When the first term is multiplied by $\sin(\omega t)$ and averaged, the result is $\frac{1}{2}\cos\phi$. When the second term is multiplied by $\sin(\omega t)$ and averaged, the result is 0. Thus the average power supplied is

$$\overline{P} = \frac{1}{2}i_m\mathcal{E}_m\cos\phi.$$

This can also be written $\overline{P} = \frac{1}{2}(\mathcal{E}_m^2/Z)\cos\phi$ or $\overline{P} = \frac{1}{2}i_m^2 Z\cos\phi$, where $\mathcal{E}_m = i_m Z$ was used. In the space below use $\tan\phi = (X_L - X_C)/R$ and the trigonometric identity $\cos^2 A = 1/(1 + \tan^2 A)$ to show that $\cos\phi = R/Z$:

Thus the average power supplied by the generator is also given by $\overline{P} = \frac{1}{2}i_m^2 R$. The quantity $\cos\phi$ is called the _____.

The rate at which energy is dissipated in the resistor is $P_R = i^2 R = i_m^2 R \sin^2(\omega t - \phi)$ and its average value is $\overline{P} = \frac{1}{2}i_m^2 R$, the same as the power supplied by the generator. Once transients have died out all energy supplied by the generator is dissipated in the resistor.

For a given emf the greatest power dissipation occurs when the power factor has the value $\cos\phi =$ _____. For this to occur the impedance must be $Z =$ _____, which means the inductive and capacitive reactances must be related by $X_L =$ _____. This is the condition for resonance. If L and C are fixed it occurs if the angular frequency has the value given by $\omega =$ _____. Often a circuit is designed to deliver as much power as possible to a resistive load. L and C are then adjusted so $X_L = X_C$ for the frequency of the generator.

Average power is often expressed in terms of root-mean-square quantities instead of amplitudes. The meaning of the term "root-mean-square" (rms) is _____ _____.

Because the average over a cycle of $\sin^2(\omega t + \phi)$ is 1/2 the rms value of a sinusoidal function of time is the amplitude divided by $\sqrt{2}$. For example, the rms value of $\mathcal{E}(t) = \mathcal{E}_m\sin(\omega t)$ is $\mathcal{E}_{rms} =$ _____. In terms of $\mathcal{E}_{rms}$, $\overline{P} =$ _____ and in terms of i_{rms}, $\overline{P} =$ _____.

II. PROBLEM SOLVING

You should understand quite well the relationship between the potential difference and current for the various circuit elements. Here are some problems that should help.

PROBLEM 1. The current in a 25-mH inductor is given by $i(t) = 35\sin(700t)$, where i is in mA for t in seconds. If the current is positive it enters end a of the inductor and exits end b. What is the potential difference $V_a - V_b$ as a function of time? Write your answer as a sine function.

SOLUTION: Use $V_a - V_b = L\,di/dt = i_m L\omega\cos(\omega t)$. To convert the cosine function to a sine function use $\sin(\omega t + \pi/2) = \cos(\omega t)$.

[ans: $0.613\sin(700t + \pi/2)$ V]

The graph below shows the current $i(t)$ as a function of time. Draw a graph of the potential difference $V_a - V_b$ on the same axes. Label the graphs so you can distinguish them.

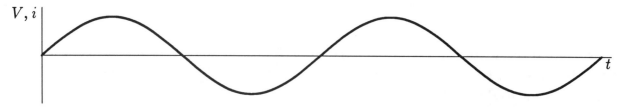

Notice that the potential difference reaches its maximum a quarter cycle before the current reaches its maximum. What is the interval between the time the potential difference reaches a maximum and the time the current next reaches a maximum?

SOLUTION: The potential difference is a maximum when $\omega t + \pi/2 = \pi/2$. The next maximum of the current occurs when $\omega t = \pi/2$ rad. Solve for the difference in these times.

[ans: 2.24 ms]

The phase difference between the current and potential difference is the same no matter what the phase of the current. Suppose the current is given by $i(t) = 35\sin(700t + \pi/2)$, where i is in mA for t in seconds. What then is the potential difference across the inductor?

SOLUTION:

[ans: $0.613\sin(700t + \pi)$ V]

Suppose the potential difference is given by $V_a - V_b = 0.613\sin(700t)$. What then is the current?

SOLUTION:

[ans: $35\sin(700t - \pi/2)$ mA]

The relationship between the current and potential difference amplitudes is usually written $(V_L)_{max} = X_L i_m$, where X_L is the inductive reactance. What is the value of X_L for $\omega = 700$ rad/s?

SOLUTION: Use $X_L = \omega L$.

[ans: $17.5\,\Omega$]

PROBLEM 2. The current entering a 75-μF capacitor is given by $i(t) = 35\sin(700t)$, where i is in mA for t in seconds. What is the potential difference across the capacitor? Take the charge on the capacitor to be zero when the current is a maximum and write your answer as a sine function.

SOLUTION: Use $V_a - V_b = (1/C)\int i(t)\,dt = -(i_m/\omega C)\cos(\omega t)$. The constant of integration is zero if the charge is zero for maximum current. Use $\sin(\omega t - \pi/2) = -\cos(\omega t)$ to write the solution as a sine function.

[ans: $0.667\sin(700t - \pi/2)$ V]

The graph below shows the current $i(t)$ as a function of time. Draw a graph of the potential difference on the same axes.

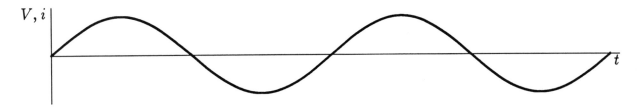

Notice that the potential difference reaches its maximum a quarter cycle after the current reaches its maximum. What is the interval between the time the potential difference reaches a maximum and the time the current next reaches a maximum?

SOLUTION: The current is a maximum for $\omega t = \pi/2$ rad. The potential difference is a maximum for $\omega t - \pi/2 = \pi/2$ rad. Solve for the difference in the times.

[ans: 2.24 ms]

The phase difference between the current and potential difference is the same no matter what the phase of the current. Suppose the current is given by $i(t) = 35\sin(700t + \pi/2)$, where i is in mA for t in seconds. What then is the potential difference across the capacitor?

SOLUTION:

[ans: $0.667\sin(700t)$ V]

Suppose the potential difference is given by $V_a - V_b = 0.667\sin(700t)$. What then is the current?

SOLUTION:

[ans: $35\sin(700t + \pi/2)$ mA]

The relationship between the current and potential difference amplitudes is usually written $(V_C)_{max} = i_m X_C$, where X_C is the capacitive reactance. What is the value of X_C for $\omega = 700$ rad/s?

SOLUTION: Use $X_C = 1/\omega C$.

[ans: 19.0 Ω]

PROBLEM 3. The current in a series LCR circuit is given by $i(t) = 35\sin(\omega t)$, where i is in mA for t in seconds. The resistance is 6.8 Ω. At the angular frequency ω the inductive reactance is 10 Ω and the capacitive reactance is 4.7 Ω. What is the potential difference across each of these circuit elements?

Chapter 39: Alternating Current Circuits 571

SOLUTION: To find the maximum values use $(V_L)_{max} = i_m X_L$, $(V_C)_{max} = i_m X_C$, and $(V_R)_{max} = i_m R$. V_L leads the current by $\pi/2$, V_C lags the current by $\pi/2$, and V_R is in phase with the current.

[ans: $V_L = 0.350 \sin(\omega t + \pi/2)$ V; $V_C = 0.165 \sin(\omega t - \pi/2)$ V; $V_R = 0.238 \sin(\omega t)$ V]

What is the impedance of the circuit and the emf of the generator?

SOLUTION: Use $Z = \sqrt{R^2 + (X_L - X_C)^2}$ to find the impedance and $\mathcal{E}_m = i_m Z$ to find the emf amplitude. If the emf is written $\mathcal{E}(t) = \mathcal{E}_m \sin(\omega t + \phi)$ then $\tan \phi = (X_L - X_C)/R$. Calculate ϕ.

[ans: $8.62\,\Omega$; $0.302 \sin(\omega t + 0.662)$]

PROBLEM 4. A series LCR circuit consists of a 15-Ω resistor, a 2.5-mH inductor, and a 9.8-μF capacitor. The AC generator has a maximum emf of 12 V and a frequency of 750 Hz. What is the inductive reactance, capacitive reactance, and impedance of the circuit?

SOLUTION: Use $X_L = \omega L$, $X_C = 1/\omega C$, and $Z = \sqrt{R^2 + (X_L - X_C)^2}$, where $\omega = 2\pi \nu$.

[ans: $11.8\,\Omega$; $21.7\,\Omega$; $18.0\,\Omega$]

The current in the circuit can be written $i(t) = i_m \sin(\omega t - \phi)$. What are the values of i_m and ϕ?

SOLUTION: Use $i_m = \mathcal{E}_m/Z$ and $\tan \phi = (X_L - X_C)/R$.

[ans: 0.667 A; -0.583 rad]

What is the potential difference across each element?

SOLUTION: Use $V_L = i_m X_L \sin(\omega t - \phi + \pi/2)$, $V_R = i_m R \sin(\omega t)$, and $V_C = i_m X_C \sin(\omega t - \phi - \pi/2)$.

[ans: $V_L = 7.87 \sin(4710 t + 2.15)$ V; $V_R = 10.0 \sin(4710 t)$ V; $V_C = 14.5 \sin(4710 t - 0.988)$ V]

What are the rms vales of the generator emf and the current?

SOLUTION: Use $\mathcal{E}_{rms} = \mathcal{E}_m/\sqrt{2}$ and $i_{rms} = i_m/\sqrt{2}$.

[ans: 8.49 V; 0.472 A]

Some problems deal with energy in AC circuits. Sometimes potential and current amplitudes are given, sometimes rms values are given instead.

PROBLEM 5. A series LCR circuit consists of a 35-mH inductor, a 85-μF capacitor, and a 25-Ω resistor. It is driven by a sinusoidal emf with an amplitude of 12 V and a frequency of 60 Hz. What are the inductive and capacitive reactances? What is the impedance?

SOLUTION: Use $X_L = \omega L$, $X_C = 1/\omega C$, and $Z = \sqrt{R^2 + (X_L - X_C)^2}$, with $\omega = 2\pi \nu$.

[ans: 13.2 Ω; 31.2 Ω; 30.8 Ω]

What are the rms generator emf and rms current?

SOLUTION: Use $\mathcal{E}_{rms} = \mathcal{E}_m/\sqrt{2}$ and $i_{rms} = \mathcal{E}_{rms}/Z$.

[ans: 8.49 V; 0.275 A]

What average power is delivered to the resistor?

SOLUTION: Use $\overline{P} = i_{rms}^2 R$.

[ans: 1.90 W]

What is the power factor?

SOLUTION: Use $\cos\phi = R/Z$.

[ans: 0.811]

Another circuit has exactly the same impedance but a power factor that is half as large. If the generator also has an amplitude of 12 V and a frequency of 60 Hz what power is delivered to the resistor?

SOLUTION: Since $\overline{P} = \frac{1}{2}(\mathcal{E}_m^2/Z)\cos\phi$, the power delivered must be half as large.

[ans: 0.948 W]

What are the resistance and net reactance of the second circuit?

SOLUTION: Since the power factor is halved without changing the impedance, the resistance must be half as much. Use $Z^2 = R^2 + X^2$ to find the reactance $X (= |X_L - X_C|)$.

[ans: 12.5 Ω; 28.2 Ω]

PROBLEM 6. The power delivered depends on the frequency. Suppose the generator has a frequency of 120 Hz. Take $\mathcal{E}_m = 12$ V, $L = 35$ mH, $C = 85\,\mu$F, and $R = 25\,\Omega$, as before. What are the reactances and the impedance?

SOLUTION:

[ans: $X_L = 26.4\,\Omega$; $X_C = 15.6\,\Omega$; $Z = 27.2\,\Omega$]

What is the average power delivered to the resistor?

SOLUTION:

[ans: 2.42 W]

For what frequency is the average power delivered the greatest? What is that average power?

SOLUTION: Z has its minimum value and i_m its maximum value if $X_L = X_C$ or $\omega = 1/\sqrt{LC}$. This is the resonance condition. For this frequency $i_m = \mathcal{E}_m/R$ and $\overline{P} = \frac{1}{2}\mathcal{E}_m^2/R$.

[ans: 92.3 Hz; 2.88 W]

PROBLEM 7. A generator produces a sinusoidal emf with an amplitude of 100 V. What is the rms value of the emf?

SOLUTION: Use $\mathcal{E}_{rms} = \mathcal{E}_m/\sqrt{2}$.

[ans: 70.7 V]

The generator is connected to a circuit with an impedance of 25 Ω. What is the amplitude of the current through the generator and what is its rms value?

SOLUTION: Use $i_m = \mathcal{E}_m/Z$ and $i_{rms} = \mathcal{E}_{rms}/Z$.

[ans: 4.00 A; 2.83 A]

Suppose the circuit consists of a 35-mH inductor, a 95-μF capacitor, and a resistor in series. The frequency is 100 Hz. What is the average energy stored in the inductor?

SOLUTION: The energy stored in the inductor is $\frac{1}{2}Li^2$ and its average value is $\frac{1}{4}Li_m^2 = \frac{1}{2}Li_{rms}^2$.

[ans: 0.140 J]

What is the average energy stored in the capacitor?

SOLUTION: The energy stored in the capacitor is $q^2/2C$ and its average value is $q_m^2/4C = q_{rms}^2/2C = i_{rms}^2/2\omega^2 C$.

[ans: 0.225 J]

Even though no power is delivered to the capacitor or inductor on average after steady state is reached, energy is stored in those elements. It was delivered when the circuit was first turned on, before steady state was reached.

The following problems show how to analyze another type circuit.

PROBLEM 8. An inductor L and a capacitor C are wired in parallel. What is the reactance of the combination?

SOLUTION: The potential difference is the same across the two elements but the currents are different. Let $V(t) = V_m \sin(\omega t)$ be the potential difference, $i_L(t)$ be the current in the inductor, and i_C be the current in the capacitor. The current into the combination is given by $i(t) = i_L(t) + i_C(t)$ and the reactance of the combination is given by $X = V_m/i_m$, where i_m is the current amplitude. You must find i_m in terms of V_m.

The individual currents are given by $i_L = (V_m/X_L)\sin(\omega t - \pi/2)$ and $i_C = (V_m/X_C)\sin(\omega t + \pi/2)$. The first of these can be written $i_L = -(V_m/X_L)\sin(\omega t + \pi/2)$ so $i = V_m(1/X_C - 1/X_L)\sin(\omega t + \pi/2)$.

The phase of the current is $\pi/2$ if $\omega C > 1/\omega L$ and $-\pi/2$ if $\omega C < 1/\omega L$.

[ans: $|\omega C - 1/\omega L|$]

PROBLEM 9. The circuit shown contains an inductor L, a resistor R, and a capacitor C, all in parallel with a sinusoidal seat of emf $\mathcal{E} = \mathcal{E}_m \sin(\omega t)$. $\mathcal{E}$ is taken to be positive when the left terminal of the generator is at a higher potential than the right. Take the currents to be positive in the directions of the arrows and derive expressions for $i_L(t)$, $i_R(t)$, and $i_C(t)$ in terms of $\mathcal{E}_m$ and ω.

SOLUTION: The potential difference is the same for all elements and is $\mathcal{E}(t)$. For the inductor $\mathcal{E}(t) = L\,di_L/dt$ or $i_L(t) = (1/L)\int \mathcal{E}\,dt$. This yields $(i_L)_{max} = \mathcal{E}_m/X_L$. The current, of course, lags the potential difference by $\pi/2$ rad. For the resistor $i_R(t) = \mathcal{E}/R$. For the capacitor $\mathcal{E}(t) = (1/C)\int i_C\,dt$ or $i_C(t) = C\,d\mathcal{E}/dt$. This yields $(i_C)_{max} = \mathcal{E}/X_C$. The current leads the potential difference by $\pi/2$ rad.

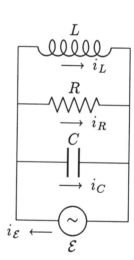

[ans: $i_L = (\mathcal{E}_m/\omega L)\sin(\omega t - \pi/2)$; $i_R = (\mathcal{E}_m/R)\sin(\omega t)$; $i_C = \mathcal{E}_m\omega C \sin(\omega t + \pi/2)$]

Find an expression for the current in the generator.

SOLUTION: You must evaluate the junction equation $i_\mathcal{E} = i_L + i_R + i_C$. Use a phasor diagram to help. On the axes to the left below draw a phasor to represent i_R. Take ωt to be any convenient value, perhaps so the phasor is counterclockwise by about 70° from the positive horizontal axis. Now draw the phasors that represent i_L and i_C. For purposes of drawing the diagram take $(i_L)_{max} > (i_C)_{max}$. On the axes to the right combine the phasors for i_L and i_C into a single phasor and draw the phasors that represent i_R and $(i_L + i_C)$. The first has magnitude $(i_R)_{max}$ and the second has magnitude $[(i_L)_{max} - (i_C)_{max}]$. Also draw the phasor that represents the current i. It is the "vector" sum of the other two phasors.

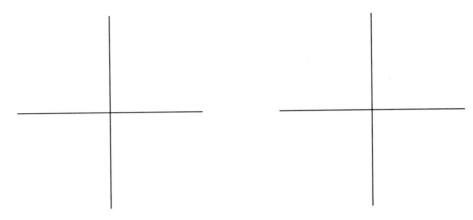

Use the diagram to show that $(i_\mathcal{E})^2_{max} = [(i_L)_{max} - (i_C)_{max}]^2 + (i_R)^2_{max}$. Take ϕ to be the angle between the phasor representing $i_\mathcal{E}$ and the phasor representing i_R. This is also the angle between the phasor representing i and the phasor representing $\mathcal{E}$. Use the diagram to show that $\tan\phi = [(i_L)_{max} - (i_C)_{max}]/i_R$. Make appropriate substitutions to find expressions for $(i_\mathcal{E})_{max}$ and $\tan\phi$ in terms of ω, L, C, and R.

[ans: $i_\mathcal{E} = (i_\mathcal{E})_{max}\sin(\omega t - \phi)$, where $(i_\mathcal{E})_{max} = \mathcal{E}_m\sqrt{(1/\omega L - \omega C)^2 + (1/R)^2}$ and $\tan\phi = (1/\omega L - \omega C)/R$]

What is the impedance of this circuit?

SOLUTION: Use $Z = \mathcal{E}_m/i_{max}$.

[ans: $1/\sqrt{(\omega C - 1/\omega L)^2 + (1/R)^2}$]

III. MATHEMATICAL SKILLS

You should be able to prove that the average over a cycle of $\sin^2(\omega t)$ is $1/2$ and the average over a cycle of $\sin(\omega t)\cos(\omega t)$ is 0. First consider the average of $\sin^2(\omega t)$, given by the integral $I = (1/T)\int_0^T \sin^2(\omega t)\,dt$. Use the trigonometric identity $\cos(2\omega t) = \cos^2(\omega t) - \sin^2(\omega t) = 1 - 2\sin^2(\omega t)$, where $\cos^2(\omega t) + \sin^2(\omega t)$ was used. Thus $\sin^2(\omega t) = \frac{1}{2}[1 - \cos(2\omega t)]$ and $I = (1/2T)\int_0^T [1 - \cos(2\omega t)]\,dt = \frac{1}{2} - (1/4\omega T)\sin(2\omega T)$. Now $2\omega T = 4\pi\nu T = 4\pi$ and $\sin(2\omega T) = \sin(4\pi) = 0$, so $I = 1/2$.

The average of $\sin(\omega t)\cos(\omega t)$ is somewhat easier to find. You must evaluate the integral $I = (1/T)\int_0^T \sin(\omega t)\cos(\omega t)\,dt$ and this is $I = (1/2\omega T)\sin^2(\omega t)|_0^T = 0$. The last result follows from $\sin(\omega T) = \sin(4\pi) = 0$.

IV. NOTES

Chapter 40
MAXWELL'S EQUATIONS

I. BASIC CONCEPTS

The four equations that tell us about the electric and magnetic fields produced by any source are collectively known as Maxwell's equations and are presented together in this chapter so you can see how they complement each other. Most of this material should be a review for you. The new topic deals with displacement current. You should pay close attention to its definition and learn how to calculate it and the magnetic field it produces.

Maxwell's equations. Complete the following statements of Maxwell's equations (see Table 2 of the text):

$$\text{I.} \quad \oint \mathbf{E} \cdot d\mathbf{A} =$$

$$\text{II.} \quad \oint \mathbf{B} \cdot d\mathbf{A} =$$

$$\text{III.} \quad \oint \mathbf{E} \cdot d\mathbf{s} =$$

$$\text{IV.} \quad \oint \mathbf{B} \cdot d\mathbf{s} =$$

Additional terms must be added to some of these equations if magnetic or dielectric materials are present but otherwise they are completely general.

You should understand what each of the Maxwell equations says. Equations I and II contain surface integrals on their left sides and the surface must be specified to evaluate the integral. In each case, dA is a vector whose magnitude is an infinitesimal element of _____ and whose direction is _____ to the surface. In equation I the _____ component of the electric field is integrated over a _____ surface. The symbol q on the right side represents the net charge _____ by the surface. In equation II the right side is zero because there are no _____ .

Earlier in this course you used equation I (Gauss' law for electricity) to find the electric field of a point charge, a disk of charge, a large plate of charge, a line of charge, and other charge configurations. You also used it to show that static charge on a conductor resides on the surface. It tells us that electric field lines can be drawn so they start on positive charge and end on negative charge. Equation II (Gauss' law for magnetism) tells us that magnetic field lines form closed loops.

Equations III and IV contain path integrals on their left sides. In each case, d**s** is a vector whose magnitude is an infinitesimal element of _____ and whose direction is _____ to the path. On the left side of equation III the _____ component of the electric field is integrated around a _____ path. This integral gives the _____ around the path. On the left side of equation IV the _____ component of the magnetic field is integrated around a _____ path. On the right side of equation III, Φ_B is the magnetic flux through the _____ bounded by the path. On the right side of equation IV i is the net _____ through the area and Φ_E is the _____ through the area.

Equation III is Faraday's law. You used it to compute the emf in a loop that is rotating in a magnetic field, the emf generated in a loop by the changing field of a nearby long straight wire, and other emf's generated by changing magnetic fluxes.

Equation IV is Ampère's law. It has been used to find the magnetic field of a long straight wire and the magnetic fields inside solenoids and toroids. For these situations the fields are produced by currents alone and $d\Phi_E/dt = 0$. $\mu_0\epsilon_0 \, d\Phi_E/dt$ is the term introduced in this chapter.

Although these equations have been used in specific examples, chosen in most cases for ease in computation, they are generally valid. The first two are true for *any* closed surface, the second two are true for *any* closed path.

Take special care to distinguish between the various equations. Remember that a changing magnetic field produces an electric field and that *both* a changing electric field and a current produce a magnetic field. Remember that the fields produced appear on the left sides of Maxwell's equations, the sources of the field appear on the right sides. Look at the equations and explain to yourself what each describes.

How can you generate an electric field in a region without having a net charge anywhere?
_____
_____

For this situation which of the Maxwell equations have zero for the value of their right sides? _____ Which do not? _____

How can you generate an electric field in a region without having a current anywhere?
_____
_____

For this situation which of the Maxwell equations have zero for the value of their right sides? _____ Which do not? _____

How can you generate a magnetic field in a region without having a net charge anywhere?
_____
_____

For this situation which of the Maxwell equations have zero for the value of their right sides? _____ Which do not? _____

The displacement current. Equation IV above tells us that there must be a magnetic field around the boundary of any area through which there is a changing electric flux, just as if a current passed through the area. In fact, the quantity $i_d = $ _____ is called the <u>displacement current</u> through the area. You should realize that a displacement current is emphatically not a true current, which consists of moving charge. A displacement current exists in any region of space that contains a changing electric field, whether or not charge is moving in the region. True currents exist only in regions than contain moving charge.

Suppose a cylindrical region with radius R contains a uniform electric field **E** parallel to its axis. A cross section is shown to the right, with the field pointing into the page. First consider the whole cross section. If dA is also into the page the electric flux through the circle of radius R is given by $\Phi_E = $ _____, where E is the magnitude of the field. If the magnitude is changing at the rate dE/dt then the displacement current through the entire cross section is given by $i_d = $ _____.

Now consider the circle of radius r ($< R$). The electric flux through this circle is given by $\Phi_E = $ _____ and the displacement current through it is given by $i_d = $ _____.

Finally consider a circle of radius r ($> R$). The electric flux through this circle is given by $\Phi_E = $ _____ and the displacement current through it is given by $i_d = $ _____.

A magnetic field is associated with a displacement current just as a magnetic field is associated with moving charge. For the cylindrical region considered above, the displacement current is uniformly distributed over every cross section, so the cylinder acts like a long straight wire of radius R with uniform current density. The magnetic field lines are _____ centered at the _____ of the cylinder. Take the left side of Ampère's law to be an integral around a circle in the plane of a cross section and centered at the cylinder axis. Then, in terms of the magnitude B of the magnetic field and the radius r of the circle, $\oint \mathbf{B} \cdot d\mathbf{s} = $ _____. Let i_d be the total displacement current through a cross section of the cylinder. Then, if $r < R$ the displacement current through the amperian circle is given by _____. In terms of i_d the magnitude of the magnetic field is given by $B = $ _____ and in terms of dE/dt it is given by $B = $ _____. If $r > R$ the displacement current through the amperian circle is the same as that through the entire cylinder cross section; that is, it is i_d. For this case the magnitude of the field is given in terms of i_d by $B = $ _____ and in terms of dE/dt by $B = $ _____.

You should be able to find the direction of the magnetic field lines. First chose a direction for dA. It is in one of the two directions that are perpendicular to the surface. This choice determines the signs of Φ_E and $d\Phi_E/dt$. Point the thumb of your right hand in the direction chosen. Your fingers will curl in the direction you *must* use for ds when you evaluate $\oint \mathbf{B} \cdot d\mathbf{s}$. Now evaluate $d\Phi_E/dt$. If it is positive the field lines follow your fingers; if it is negative they go the opposite way.

A charging (or discharging) capacitor provides an excellent example of a displacement current. The charge on the plates produces an electric field in the interior and since the charge is changing so is the field. Consider a parallel plate capacitor with plate area A and let $q(t)$ be the charge on the positive plate at time t. Then the magnitude of the electric field between the plates is given by $E(t) = $ _____. Consider a cross section between the plates, parallel to them, and having the same area as a plate. The electric flux through this cross section is _____ and the total displacement current in the interior of the capacitor is $i_d = $ _____, where i ($= dq/dt$) is the current into the capacitor. You should have shown that $i_d = i$. The total displacement current in the interior is the same as the current in the capacitor wires. The total current (true current plus displacement current) is continuous. Although the true current stops at the plates, the displacement current continues into the interior.

II. PROBLEM SOLVING

Given the electric field as a function of time you should be able to find the value of the integral $\oint \mathbf{B} \cdot d\mathbf{s}$ using Ampère's law. In cases of high symmetry you should be able to solve for the magnetic field.

PROBLEM 1. A uniform electric field exists in a cylindrical region of space with a radius of 5.5 cm. It is parallel to the cylinder axis. If the magnitude of the field is increasing at the rate $dE/dt = 2.5 \times 10^3$ V/m·s what is the induced magnetic field at the circumference of the region? What is the induced magnetic field at a point halfway along a radius from the cylinder axis to the circumference?

SOLUTION: Use Ampère's law in the form $\oint \mathbf{B} \cdot d\mathbf{s} = \mu_0 \epsilon_0 \, d\Phi_E/dt$. Let r be the radius of a circle through the point. If R is the radius of the region then $r = R$ for the first question and $r = R/2$ for the second. The electric flux through the circle is $\Phi_E = \pi r^2 E$ and its rate of change is $d\Phi_E/dt = \pi r^2 \, dE/dt$.

The magnetic field lines are circles centered on the cylinder axis. Integrate the left side of Ampère's law around the circle of radius r. The result is $\oint \mathbf{B} \cdot d\mathbf{s} = 2\pi r B$. Thus $2\pi r B = \mu_0 \epsilon_0 \pi r^2 \, dE/dt$ and $B = \frac{1}{2}\mu_0 \epsilon_0 r \, dE/dt$.

[ans: 7.65×10^{-16} T; 3.82×10^{-16} T]

What is the magnetic field at $r = 11$ cm?

SOLUTION: Consider a circle centered on the axis of the cylinder and through the point. The point is outside the cylinder so the electric flux through the circle is the same as the flux through a cross section of the cylinder, or $\Phi_E = \pi R^2 E$. Ampère's law becomes $2\pi r B = \mu_0 \epsilon_0 \pi R^2 \, dE/dt$. Thus $B = \frac{1}{2}\mu_0 \epsilon_0 (R^2/r) \, dE/dt$.

[ans: 3.82×10^{-16} T]

Suppose the electric field is into the page. What is the direction of the magnetic field?

SOLUTION: Take the area vector to be into the page. Then the electric flux is positive and so is its rate of change. Point the thumb of your right hand in the direction of d**A**. Then the fingers curl in the direction of a positive magnetic field. This should be clockwise. Since $d\Phi_E/dt$ is positive the magnetic field lines curl in the direction of your fingers.

[ans: clockwise]

If the electric field were decreasing in magnitude then $d\Phi_E/dt$ would be negative and the magnetic field would be counterclockwise. What is the direction of the magnetic field if the electric field is out of the page and increasing in magnitude?

[ans: counterclockwise]

Some problems ask you to calculate the displacement current for a given situation. Use $i_d = \epsilon_0 \, d\Phi_E/dt$. In each problem an area is specified and you calculate the displacement current through that area.

PROBLEM 2. For the electric field of the previous problem what is the displacement current through a cross section of the cylindrical region? What is the displacement current through a circle of radius $r = R/2$, in a cross section and centered at the cylinder axis? What is the displacement current through a similar circle but with a radius of $3R$? Here R is radius of the cylindrical region.

SOLUTION: Use $i_d = \epsilon_0 \, d\Phi_E/dt$. For the first two cases $\Phi_E = \pi r^2 E$, where $r = R$ or $r = R/2$. For the third case $\Phi_E = \pi R^2 E$ since the displacement current exists only within the cylindrical region.

[ans: 2.10×10^{-10} A; 5.26×10^{-11} A; 2.10×10^{-10} A]

What is the displacement current density in the cylindrical region?

SOLUTION: The displacement current density $\mathbf{j}_d$ must obey $i_d = \int \mathbf{j}_d \cdot d\mathbf{A}$ for any area. Compare this expression to $i_d = \epsilon_0 \dfrac{d}{dt} \int \mathbf{E} \cdot d\mathbf{A} = \epsilon_0 \int \dfrac{d\mathbf{E}}{dt} \cdot d\mathbf{A}$.

[ans: 2.21×10^{-8} A/m^2]

Many problems deal with the charging or discharging of a capacitor. There is a changing electric flux through any area that is parallel to the plates and at least partially within the capacitor. Thus there is a displacement current through any of these areas. The total displacement current, through the entire cross section, equals the current into the capacitor but the displacement current through a partial cross section is less.

PROBLEM 3. A parallel plate capacitor has circular 2.6-cm radius plates and is being charged by a 4.7-A steady current. What is the rate at which the electric field is changing in the interior? Neglect fringing effects.

SOLUTION: The field is uniform in the interior and its magnitude is given by $E = q/\epsilon_0 A$, where q is the charge on the positive plate and A is the area of a plate. Since $i = dq/dt$, $dE/dt = i/\epsilon_0 A$. Use $A = \pi R^2$ to compute the plate area.

[ans: 2.50×10^{14} V/m·s]

What is the rate of change of the total flux through a cross section of the capacitor?

SOLUTION: Use $d\Phi_E/dt = A\,dE/dt$.

[ans: 5.31×10^{11} V·m/s]

What is the total displacement current through a cross section of the capacitor?

SOLUTION: Use $i_d = \epsilon_0\,d\Phi_E/dt$.

[ans: 4.7 A]

What is the displacement current through a 1.3-cm radius circle parallel to the plates and centered in the interior of the capacitor?

SOLUTION: Use $i_d = \epsilon_0\,d\Phi_E/dt = \epsilon_0 \pi r^2\,dE/dt$, where r is the radius of the circle. Since the area of the circle is one-fourth the area of a plate, the displacement current is one-fourth the total displacement current.

[ans: 1.18 A]

A magnetic field is associated with the displacement current. This field exists both inside and outside the capacitor.

PROBLEM 4. What magnetic field is induced in the interior of the capacitor of the last problem at a point 1.3 cm from the axis of the plates?

SOLUTION: Use $\oint \mathbf{B} \cdot d\mathbf{s} = \mu_0 i_d$, where the integral is around a circle through the point. You should obtain $B = \mu_0 i_d / 2\pi r$. According to the result above $i_d = 1.18$ A.

[ans: 1.81×10^{-5} T]

What is the magnetic field 5.2 cm from the axis of the plates?

SOLUTION: Use $B = \mu_0 i_d / 2\pi r$, where $i_d = 4.7$ A.

[ans: 1.81×10^{-5} A]

The magnetic fields of displacement currents generated in the laboratory are usually quite small and hard to detect. To generate a detectable field an experimenter might place a large amplitude sinusoidal emf across a capacitor. The amplitude of the field is proportional to the frequency of the emf. In the first problem below you will determine the field produced when a capacitor is charged, then in the second problem you will see how a sinusoidal emf helps generate a considerably larger field.

PROBLEM 5. A parallel plate capacitor has capacitance C and circular plates with radius a. It is connected in series with a seat of emf $\mathcal{E}$ and a resistor R. At time $t = 0$ it is uncharged and a switch in the circuit is closed to start the charging process. Find an expression for the magnitude of the magnetic field at a point between the plates a distance r from the axis of symmetry.

SOLUTION: As a function of time the charge on the capacitor is given by $q(t) = q_m \left[1 - e^{-t/\tau}\right]$, where $q_m = C\mathcal{E}$ and $\tau = RC$. See Chapter 33. The current into the capacitor is given by $i(t) = dq/dt = (q_m/\tau)e^{-t/\tau}$ and the total displacement current between the plates is given by the same expression: $i_{d\,\text{total}}(t) = (q_m/\tau)e^{-t/\tau}$. The displacement current density is uniform so the displacement current through a circle of radius r is given by a ratio of areas: $i_d(t) = i_{d\,\text{total}}(r^2/a^2) = (q_m/\tau)(r^2/a^2)e^{-t/\tau}$. According to Ampère's law the magnitude of the magnetic field is given by $B(t) = \mu_0 i_d/2\pi r = (\mu_0 q_m/2\pi\tau)(r/a^2)e^{-t/\tau}$.

[ans: $\dfrac{\mu_0 \mathcal{E}}{2\pi R}\dfrac{r}{a^2}e^{-t/RC}$]

Take $C = 15\,\mu\text{F}$, $R = 250\,\text{k}\Omega$, $\mathcal{E} = 75\,\text{V}$, and $a = 1.3\,\text{cm}$. Find the magnitude of the magnetic field at $r = a/2$, 1.5 s after the switch is closed.

[ans: $1.38 \times 10^{-9}\,\text{T}$]

PROBLEM 6. A parallel plate capacitor has a capacitance of 125 μF and circular plates with a radius of 3.5 cm. A sinusoidal emf with an amplitude of 150 V is connected across it. What should the angular frequency be to produce a magnetic field at the rim with an amplitude equal to the magnitude of the earth's magnetic field (about $5 \times 10^{-5}\,\text{T}$)?

SOLUTION: Take the emf to be $\mathcal{E} = \mathcal{E}_m \sin\omega t$. The charge on the capacitor is given by $q = C\mathcal{E}$ and the current by $i = dq/dt = C\,d\mathcal{E}/dt = C\omega\mathcal{E}_m \cos\omega t$. The magnetic field at the rim is given by $B = \mu_0 i_d/2\pi R = (\mu_0 C\omega\mathcal{E}_m/2\pi R)\cos\omega t$, where R is the radius of a plate. The amplitude of the field is $B_m = \mu_0 C\omega\mathcal{E}_m/2\pi R$.

[ans: 470 rad/s]

Circular capacitor plates are often used in problems because the symmetry allows you to solve for the magnetic field, not just its line integral. Ampère's law, however, holds for every geometry.

PROBLEM 7. In a certain region of space the electric field is uniform and is parallel to the z axis. It changes sinusoidally according to $E_z = E_m \sin\omega t$, where $E_m = 1.7 \times 10^4$ V/m and $\omega = 150$ rad/s. What is the maximum displacement current through a rectangular region of the xy plane with a length of 2.5 cm and a width of 1.7 cm?

SOLUTION: The electric flux is $\Phi_E = EA = E_m LW \sin\omega t$, where L is the length and W is the width of the region. Use $i_d = \epsilon_0 \,d\Phi_E/dt = \omega E_m LW \sin\omega t$. The displacement current is a sinusoidal function of time. Calculate its amplitude.

[ans: $9.60 \times 10^{-9}\,\text{A}$]

What is the value of the integral $\oint \mathbf{B} \cdot d\mathbf{s}$ around the perimeter of the rectangle?

SOLUTION: Use $\oint \mathbf{B} \cdot d\mathbf{s} = \mu_0 i_d$.

[ans: 1.21×10^{-14} T·m]

Note that the magnetic field is not tangent to the rectangle sides so you cannot solve for its magnitude.

III. NOTES

Chapter 41
ELECTROMAGNETIC WAVES

I. BASIC CONCEPTS

Maxwell's equations predict the possibility of electric and magnetic fields that propagate in free space, with the changing magnetic field producing changes in the electric field, via Faraday's law, and the changing electric field producing changes in the magnetic field, via the displacement current term in Ampère's law. Pay close attention to the relationship between the amplitudes of the fields, to the relationship between their phases, and to the relationship between their directions. One important consequence of the equations is that they produce an expression for the wave speed, the speed of light, in terms of ϵ_0 and μ_0.

The electromagnetic spectrum. Electromagnetic waves exist for all wavelengths and frequencies, from very large to very small. Various ranges of wavelengths have been named and you should be familiar the ranges and their names.

1. <u>Gamma</u> radiation extends from the very shortest wavelengths (highest frequencies) to a wavelength of about _____ m (a frequency of about _____ Hz). Some sources of gamma radiation are: _____
_____

2. <u>X-ray</u> radiation extends from a wavelength of about _____ m (a frequency of about _____ Hz) to a wavelength of about _____ m (a frequency of about _____ Hz). Some sources are: _____
_____

3. <u>Ultraviolet</u> radiation extends from a wavelength of about _____ m (a frequency of about _____ Hz) to a wavelength of about _____ m (a frequency of about _____ Hz). Some sources are: _____
_____

4. <u>Visible</u> radiation extends from a wavelength of about _____ m (a frequency of about _____ Hz) to a wavelength of about _____ m (a frequency of about _____ Hz). Some sources are: _____
_____

5. <u>Infrared</u> radiation extends from a wavelength of about _____ m (a frequency of about _____ Hz) to a wavelength of about _____ m (a frequency of about _____ Hz). Some sources are: _____
_____

6. Microwave radiation extends from a wavelength of about _____ m (a frequency of about _____ Hz) to a wavelength of about _____ m (a frequency of about _____ Hz). Some sources are: _____

7. Above this in wavelength lies the TV and radio portion of the spectrum, which extends from a wavelength of about _____ m (a frequency of about _____ Hz) to the longest wavelengths (lowest frequencies). Some sources are: _____

All these electromagnetic radiations are exactly alike except for their frequencies and wavelengths. They are all electromagnetic. That is, they all consist of traveling electric and magnetic fields. They all travel in free space with the same speed, the speed of light c = _____ m/s.

Electromagnetic waves Classically, electromagnetic waves are produced by charges that are _____. Charges at rest or moving with constant velocity do not radiate. In the quantum mechanical picture electromagnetic radiation is produced when a charge (perhaps an electron in an atom or an atomic nucleus) changes its _____ or in reactions of fundamental particles.

Figs. 7, 8, and 9 show the electric and magnetic fields of a radiating antenna consisting of a straight wire in which the current varies sinusoidally. Look at these figures and identify the important characteristics of electromagnetic radiation:

The electric and magnetic fields are perpendicular to each other. At the point on the x axis of Fig. 8 where the most advanced electric field line shown crosses the axis the electric field is in the _____ direction and the magnetic field is in the _____ direction.

The wave travels in a direction that is perpendicular to both the electric and magnetic fields. At the point you just considered the wave is traveling in the _____ direction.

The electric and magnetic fields are in phase. Explain what this means: _____

The wave is linearly polarized. Explain what this means: _____

Very far from the source the wave becomes a plane wave: the electric and magnetic field lines are very close to straight lines at the detector. Suppose the detector is on the x axis so that waves reaching it are traveling in the positive x direction. The diagram shows the fields in the wave near the detector. The electric field is parallel to the y axis and the magnetic field is parallel to the z axis. Mathematically the fields are given by

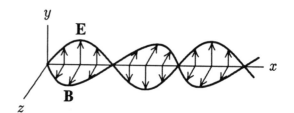

588 *Chapter 41: Electromagnetic Waves*

$$E(x,t) =$$
$$B(x,t) =$$

where $k = 2\pi/\lambda$, λ is the wavelength, and ω is the angular frequency. These quantities are related by $\omega/k = $ _____, where c is the speed of the wave. The phase constants are the same for the two fields.

The electric field is different at some point x and another point $x + dx$, an infinitesimal distance away because a changing magnetic flux penetrates the region between these points. The magnetic field is different at x and $x + dx$ because a changing electric flux penetrates the region between these points. Differences in the fields at different points can be computed using Faraday's and Ampère's laws.

When these laws are applied you find that the amplitudes of the fields are related by

$$B_m =$$

and that the wave speed is determined by the constants ϵ_0 and μ_0:

$$c =$$

You should understand how the laws of electromagnetism lead to these relationships. The diagram on the right shows a small region of space in which a wave is traveling toward the right. The electric fields at two infinitesimally separated points, x and $x + \Delta x$, are shown, as is the magnetic field in the region between. Apply Faraday's law to this situation. To calculate the magnetic flux take dA to be out of the page and to calculate the emf traverse the path shown in the counterclockwise direction. Clearly, $\Phi_B = Bh\Delta x$ and $\oint \mathbf{E} \cdot d\mathbf{s} = h[E(x + \Delta x) - E(x)]$. Why does $E(x + \Delta x)$ enter with a plus sign and $E(x)$ enter with a minus sign? _____

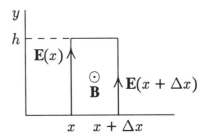

As Δx becomes small $[E(x + \Delta x) - E(x)]/\Delta x$ becomes $\partial E/\partial x$ and Faraday's law yields

$$\frac{\partial E}{\partial x} =$$

The diagram on the right shows the view looking up from underneath the previous diagram. The magnetic field at the two points and the electric field between are shown. Apply Ampère's law to this situation. Take dA to be into the page and traverse the path in the clockwise direction. Clearly, $\Phi_E = Eh\Delta x$ and $\oint \mathbf{B} \cdot d\mathbf{s} = h[B(x) - B(x + \Delta x)]$. Why does $B(x)$ enter with a plus sign and $B(x + \Delta x)$ enter with a minus sign? _____

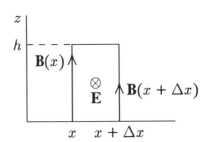

As Δx becomes small $[B(x) - B(x + \Delta x)]/\Delta x$ becomes $-\partial B/\partial x$ and Ampère's law yields

$$\frac{\partial B}{\partial x} = $$

Substitute $E = E_m \sin(kx - \omega t)$ and $B = B_m \sin(kx - \omega t)$ into these two equations and show that $kE_m = \omega B_m$ and that $kB_m = \mu_0 \epsilon_0 E_m$:

Now use $\omega/k = c$ to show that $E_m = cB_m$ and $c^2 = 1/\mu_0\epsilon_0$:

Energy transport. Electromagnetic waves carry energy. The electrical energy per unit volume associated with an electric field **E** is given by $u_E = $ _____ and the magnetic energy per unit volume associated with a magnetic field **B** is given by $u_B = $ _____. You can use $B = E/c$ and $c = 1/\sqrt{\epsilon_0\mu_0}$ to show that for a plane wave $u_E = u_B = EB/2\mu_0 c$ and that the total energy density is $u = EB/\mu_0 c$.

This energy moves with the wave, with speed c. Consider a region of space with infinitesimal width dx and cross-sectional area A, perpendicular to the direction of travel of a plane electromagnetic wave. The volume of the region is $A\,dx$ so, in terms of the energy density u, the electromagnetic energy in the region is $dU = $ _____. All this energy will pass through the area A in time dt, given by $dt = dx/c$. Substitute $dx = c\,dt$ and divide by dt. The energy passing through the area per unit time is $dU/dt = $ _____ and the energy per unit area passing through per unit time is $dU/A\,dt = $ _____. In terms of the field magnitudes E and B, this is $dU/A\,dt = $ _____.

The transport of energy is described in terms of the Poynting vector **S**, defined by the vector product

$$\mathbf{S} = $$

Since **E** and **B** are perpendicular to each other the magnitude of the Poynting vector is given by $S = $ _____. In terms of the magnitude S of the Poynting vector the energy passing through a surface of area A, perpendicular to the direction of travel, per unit time, is given by $dU/dt = $ _____ and the energy passing through per unit area per unit time is given by $dU/A\,dt = $ _____.

For most sinusoidal electromagnetic waves of interest the fields oscillate so rapidly that their instantaneous values cannot be detected or else are not of interest. Energies and energy densities are then characterized by their average over a period of oscillation. The average value of E^2, for example, is $\frac{1}{2}E_m^2$, where E_m is the amplitude.

The underlined _intensity_ of a wave is the magnitude of the Poynting vector averaged over a period of the oscillation. In terms of the electric field amplitude E_m it is $I =$ _____ and in terms of the magnetic field amplitude B_m it is $I =$ _____.

The direction of **S** is the same as the direction in which the wave is traveling. Recall that this direction is intimately connected with the relative signs of the terms kx and ωt in the argument of the trigonometric function that describes the fields. If the wave travels in the negative x direction you write $E_y(x,t) = E_m \sin(kx + \omega t)$ and $B_z(x,t) =$ _____.
Be sure you have selected the appropriate signs so $B_z(x,t)$ represents a wave traveling in the negative x direction and so that $\mathbf{E} \times \mathbf{B}$ is also in the negative x direction. Carefully note that $B_z = (E_m/c)\sin(kx + \omega t)$ is _wrong_ since this would cause $\mathbf{E} \times \mathbf{B}$ to be in the positive x direction.

Electromagnetic waves also carry momentum. If U is the energy in any portion of a wave then

$$p =$$

is the magnitude of the momentum in that portion. The momentum is in the direction of **S**, the direction of propagation. If u is the energy per unit volume then _____ is the momentum per unit volume and if I is the energy transported through an area per unit area per unit time then _____ is the momentum transported through the area per unit area per unit time.

When an electromagnetic wave interacts with a charge the electric field component of the wave does work on the charge and energy is transferred from the wave to the charge. Since the fields exert forces on the charge momentum is also transferred. When electromagnetic radiation is absorbed by a material object its energy and momentum are transferred to the object. Consider a plane wave with time averaged energy density u, incident normally on the plane surface of an object. If the area of the surface is A and the wave is completely absorbed then averaged over a period the energy transferred in time Δt is given by _____ and the momentum transferred is given by _____. The momentum transferred per unit time is the force of the radiation on the object and the force per unit area is the radiation pressure. In terms of u the force is given by _____ and the radiation pressure is given by _____.

If the wave is reflected without loss, as from an ideal mirror, the energy transferred is _____ and the momentum transferred is _____. The radiation pressure is now _____, twice what it would be if the radiation were absorbed. Why? _____

II. PROBLEM SOLVING

Some problems ask you to find the distance traveled by light in a given time or to find the time taken to travel a given distance. The chief purpose is to give you some qualitative feeling for the speed. In each case you use $c = d/t$, where d is the distance and t is the time.

PROBLEM 1. Find the time for light to travel the following distances: (a) across a proton (radius $\approx 10^{-15}$ m); (b) from New York to San Francisco ($\approx 4.5 \times 10^3$ km); (c) from the earth to the moon ($\approx 4 \times 10^8$ m); (d) from the sun to the earth ($\approx 1.5 \times 10^{11}$ m); (e) from the earth to the center of the galaxy ($\approx 10^{20}$ m); (f) from the earth to the Andromeda galaxy ($\approx 2 \times 10^{22}$ m).

SOLUTION:

[ans: (a) 3×10^{-24} s; (b) 1.5×10^{-5} s; (c) 1.3 s; (d) 500 s (8.3 min); (e) 1.1×10^4 yr; (f) 2.1×10^6 yr]

An automobile is traveling at 25 m/s (56 mph). How many times must its speed be doubled before it reaches the speed of light?

SOLUTION: If v is its original speed then $c = 2^n v$. Solve for n by taking the natural logarithm of both sides.

[ans: 23.5]

You should be able to identify the various parameters (k, ω, λ, and ν) that are used to describe a wave and know how they are related ($k = 2\pi/\lambda$, $\omega = 2\pi\nu$, $c = \lambda\nu = \omega/k$). You should also know the relationship between the electric and magnetic field amplitudes ($B_m = E_m/c$).

PROBLEM 2. The electric field component of a plane electromagnetic wave is parallel to the y axis and has a y component that is given by $E_y(x,t) = 2.5 \sin(1.26 \times 10^7 x + \omega t)$, in SI units. What is its wavelength?

SOLUTION: The value of k is 1.26×10^7 m^{-1}. Use $\lambda = 2\pi/k$.

[ans: 4.99×10^{-7} m]

What is the value of ω?

SOLUTION: Use $c = \omega/k$.

[ans: 3.78×10^{15} rad/s]

What is the frequency?

SOLUTION: Use $\omega = 2\pi\nu$.

[ans: 6.01×10^{14} Hz]

In what part of the electromagnetic spectrum does this wave lie?

SOLUTION: See Fig. 1 of the text.

[ans: blue-green region of the visible spectrum]

What is the amplitude of the magnetic field?

SOLUTION: Use $B_m = E_m/c$.

[ans: 8.34×10^{-9} T]

In what direction is the wave traveling?

SOLUTION: Notice that kx and ωt enter the expression for E_y with the same signs.

[ans: negative x direction]

Suppose the magnetic field is parallel to the z axis. When **E** is in the positive y direction what is the direction of **B**?

SOLUTION: Pick the direction for which $\mathbf{E} \times \mathbf{B}$ is in the negative x direction.

[ans: negative z direction]

Write an expression for the magnetic field as a function of x and t.

SOLUTION:

[ans: $B_z(x,t) = -8.34 \times 10^{-9} \sin(1.26 \times 10^7 x + 3.78 \times 10^{15} t)$, in SI units]

PROBLEM 3. A 100-MHz plane electromagnetic wave passes by a square loop of wire, traveling parallel to one edge with its magnetic field perpendicular to the plane of the loop. If the electric field amplitude is 0.75 V/m and the edge of the loop has a length of 1.5 m what is the amplitude of the emf generated around the loop by the magnetic field of the wave?

SOLUTION: Assume a loop edge extends from $x = 0$ to $x = a$, where a is the length of its edge. Take the magnetic field to be $B(x,t) = (E_m/c)\sin(kx - \omega t)$. The magnetic flux through the loop is $\Phi_B = a \int_0^a B\,dx = (aE_m/c)\int_0^a \sin(kx - \omega t)\,dx = -(aE_m/kc)[\cos(ka - \omega t) - \cos(\omega t)]$. The rate of change of the flux is $d\Phi_B/dt = -(a\omega E_m/kc)[\sin(ka - \omega t) + \sin(\omega t)] = aE_m[\sin(ka - \omega t) + \sin(\omega t)]$, where $c = \omega/k$ was used. According to Faraday's law this is the emf.

To find the amplitude use the trigonometric identity $\sin A + \sin B = \sin\frac{1}{2}(A + B)\cos\frac{1}{2}(A - B)$. Thus $\mathcal{E} = 2aE_m \sin(ka/2)\cos(\omega t - ka/2)$. The amplitude is $\mathcal{E}_m = 2aE_m \sin(ka/2)$.

[ans: 2.25 V]

Notice that the emf amplitude depends on the wavelength through the factor $\sin(ka/2)$. The size of the loop was picked so the amplitude is the largest it can be for the incident wave. Circular loops are often used for TV antennas.

Some problems deal with energy and momentum transport. You should know how to calculate the energy density, the energy per unit area crossing a surface per unit time, the intensity, and the momentum per unit area crossing a surface per unit time, all in terms of the fields. Also understand how to use the conservation of energy to relate the power output of a point source to the intensity at a detector.

PROBLEM 4. The intensity of solar radiation received by the earth is about 1340 W/m². If this were all in the form of a single plane wave what would be its electric field amplitude? What would be its magnetic field amplitude?

SOLUTION: Use $I = (1/2\mu_0 c)E_m^2 = (c/2\mu_0)B_m^2$.

[ans: 1.00×10^3 V/m; $3.35\,\mu$T]

What is the total power emitted by the sun? The earth's orbit has a radius of about 1.50×10^{11} m.

SOLUTION: Assume the same intensity in all directions. If I is the intensity at the position of the earth the total power emitted is given by $P = 4\pi R^2 I$, where R is the radius of the earth's orbit: the intensity is multiplied by the surface area of a sphere through the earth and centered at the sun.

[ans: 3.79×10^{26} W]

PROBLEM 5. The electric field amplitude in a plane electromagnetic wave is 1.5 kV/m. What time averaged energy density is stored in the electric field? What time averaged energy density is stored in the magnetic field?

SOLUTION: Use $\bar{u}_E = \epsilon_0 E_m^2/4$ and $\bar{u}_B = B_m^2/4\mu_0 = E_m^2/4\mu_0 c^2 = \epsilon_0 E_m^2/4$. Note that the electric and magnetic energy densities are the same.

[ans: 4.98×10^{-6} J/m³; 4.98×10^{-6} J/m³]

What is the intensity of this wave?

SOLUTION: Use $I = E_m^2/2\mu_0 c$.

[ans: 2.99×10^3 W/m²]

This wave is incident normally on a plane surface with an area of $0.18\,\text{m}^2$ and is completely absorbed. What is the rate of energy absorption by the surface?

SOLUTION: The radiation uniformly illuminates the surface so you simply multiply the intensity by the surface area.

[ans: 538 W]

What radiation pressure is associated with this wave?

SOLUTION: The radiation pressure is the momentum per unit area transferred to the surface per unit time. Since the momentum carried by an electromagnetic wave is U/c, where U is its energy, the radiation pressure is given by I/c, where I is the intensity.

[ans: 9.97×10^{-6} Pa]

What is the time averaged force of the radiation on the surface?

SOLUTION: This is the radiation pressure multiplied by the area of the surface.

[ans: 1.79×10^{-6} N]

If the surface were a perfect reflector what would be the radiation pressure and the time averaged force?

SOLUTION: Now the momentum transfer is twice as great since the wave is reflected and its momentum changes direction. Thus the radiation pressure and time averaged force are each twice as great.

[ans: 1.99×10^{-5} Pa; 3.59×10^{-6} N]

PROBLEM 6. A radio transmitter emits a signal with a time averaged power of 1.0 kW. What are the electric and magnetic field amplitudes 2.0 km away? Assume the signal is equally strong in all directions.

SOLUTION: Since the surface area of a sphere with radius r is $4\pi r^2$ and the transmitted signal is uniform over the sphere, the intensity a distance r from the transmitter is $I = P/4\pi r^2$, where P is the power. Equate this to $E_m^2/2\mu_0 c$ and solve for E_m. The magnetic field amplitude is, of course, E_m/c.

[ans: 0.122 V/m; 4.08×10^{-10} T]

Chapter 41: Electromagnetic Waves

PROBLEM 7. A 2.5-kW laser beam is used to exert pressure on 6.8 mm² of the surface of a plasma. 75 percent of the radiant energy is reflected by the plasma, 15 percent is absorbed, and 10 percent is transmitted through the plasma. What force is exerted on the plasma? What is the radiation pressure?

SOLUTION: Let P be the incident power. The reflected power is $0.75P$ and this exerts a force of $2 \times 0.75P/c$. The absorbed power is $0.15P$ and this exerts a force of $0.15P/c$. The transmitted power does not exert a force. Sum the two forces. The radiation pressure is the total force divided by the area.

[ans: 1.38×10^{-5} N; 2.02 Pa]

III. NOTES

Chapter 42
THE NATURE AND PROPAGATION OF LIGHT

I. BASIC CONCEPTS

In the last chapter you learned that electromagnetic radiation consists of traveling electric and magnetic fields and can be described by Maxwell's equations. In fact, these equations predict the speed of electromagnetic waves. From this general description, based on fundamental principles, you narrow your study to that of visible light, the portion of the electromagnetic spectrum to which our eyes are sensitive. Here you will learn something about sources of visible light, about how the speed of light is measured, and about the Doppler effect for light. This chapter paves the way for the study of optics, which you begin with the next chapter.

The eye as a receptor. Look at Fig. 1 of the text, which shows the sensitivity to light of a typical human eye as a function of wavelength. For the range of wavelengths on the graph the shortest corresponds to the color _____ and the longest corresponds to the color _____. Maximum sensitivity occurs at a wavelength of about _____ nm, corresponding to the color _____. The wavelengths for which the sensitivity is 10 per cent of the maximum sensitivity are about _____ nm and about _____ nm, corresponding to the colors _____ and _____.

Light sources. You should know some of the terms used to describe sources of light. Radiation that is emitted by an object because of its temperature is called _____ radiation. If the radiation is visible the object is said to be _____. On the other hand, emission of light from a cool object is called _____. Normally electrons in these objects must be excited by an external source of energy. The objects then emit light when electrons return to their original states. A <u>fluorescent</u> object is a luminescent object that glows less than about _____ s after the source of the excitation is turned off. A <u>phosphorescent</u> object continues to glow for more than this time. Luminescent objects are often categorized according to the source of the excitation. Give the source for each of the following:
    chemiluminescence: _____
    bioluminescence: _____
    triboluminescence: _____

The speed of light. Until the advent of lasers Fizeau's method provided the most precise measurement of the speed of light. Describe the essence of the method: _____
_____

In more modern experiments lasers were used to determine the speed of light. Tell what was measured and how the speed was computed: _____
_____

Now the speed of light is taken to be exactly $c =$ _____ m/s and the measurements are used to determine the length of a meter. A meter is the distance light travels in _____ s.

When light travels through matter its speed is different from its speed in vacuum. The symbol c is reserved for the speed of light in vacuum, while v is used to denote its speed in a material medium. For linear materials the properties that determine the speed are the _____ constant κ_e, which describes the electric polarization of the material when an electric field is present and the _____ constant, which describes the magnetization of the material when a magnetic field is present. The speed of light in a material medium is given by $v =$ _____ or, when $c = 1/\sqrt{\mu_0 \epsilon_0}$ is used, by

$$v =$$

Most optical materials you will deal with are non-magnetic and for them $\kappa_m = 1$. You should know that κ_e depends on the wavelength of the light, so the speed is different for different wavelengths. This phenomenon is called _____.

The Doppler effect. The Doppler effect occurs for light as well as for sound (see Chapter 20). When the light source is moving toward the observer or the observer is moving toward the source the frequency ν observed is higher than the observed frequency ν_0 when both are at rest. The relationship is

$$\nu =$$

where u is the relative speed of the source and observer. It is positive. An increase in frequency is reflected in a shift toward the blue end of the visible spectrum.

When the source is moving away from the observer or the observer is moving away from the source the observed frequency is less than the frequency when they are at rest. The relationship is

$$\nu =$$

Again u is the relative speed of the observer and source and it is positive. Visible light is shifted toward the red end of the visible spectrum. Remember that these two expressions are valid only if the source or observer are moving along the line joining them.

Notice that the Doppler effect equations for light are quite different from those for sound. The two give nearly the same result, however, when the speed of the source or observer is much less than the speed of the wave (sound or light, as appropriate). The speeds that appear in the Doppler effect equations for sound are measured relative to the medium in which the sound is propagating. Those in the Doppler effect equations for light are measured relative to any inertial frame.

These results can be derived using the theory of special relativity. Special relativity also tells us that a Doppler shift in frequency occurs when the source or observer is moving perpendicularly to the line joining them. If you studied Chapter 21, carefully read Section 42–4 to see how the Lorentz transformation is used to derive the Doppler effect equations. Also write here the expression for the observed frequency when the source or observer are moving with speed u perpendicularly to the line joining them:

$$\nu =$$

Notice that the observed frequency is always less than the frequency that would be observed if both source and observer are at rest.

II. PROBLEM SOLVING

Here are a few examples of problems concerned with measurements of the speed of light and ramifications of its finite value.

PROBLEM 1. The speed of light is measured using the rotating toothed wheel shown here. The beam is aimed into the page at the point marked $\otimes$. It is reflected from a mirror 8500 m behind the wheel and returns along the same path. What angular velocity should the wheel have for the light to return through the adjacent notch?

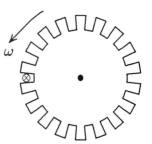

SOLUTION: The time for light to travel from the wheel to the mirror and back again is $\Delta t = 2L/c$, where L is the distance from the wheel to the mirror. During this time the wheel must turn through an angle in radians of $\Delta\theta = 2\pi/n$, where n is the number of teeth on the wheel. The angular velocity of the wheel must be $\omega = \Delta\theta/\Delta t = \pi c/nL$.

[ans: 6.16×10^3 rad/s (980 rev/s)]

In the actual experiment the angular velocity of the wheel is increased from zero until no light from the mirror passes through the wheel. What then is the angular velocity of the wheel?

SOLUTION:

[ans: 3.08×10^3 rad/s (490 rev/s)]

The angular velocity required can be reduced considerably by increasing the number of notches on the wheel. Fizeau used a wheel with 72 notches.

PROBLEM 2. Io, a moon of Jupiter, takes 42.5 h to complete one orbit of that planet. Suppose an orbit is observed from earth at a time when the earth is moving at 29.9 km/s directly away from Jupiter. By how much does the observed period differ from the actual period because the speed of light is finite?

SOLUTION: During one orbit of the moon the earth moves a distance vT away, where T is the period of the moon and v is the speed of the earth. Light coming from the moon at the end of its orbit must travel further by this distance than light coming from the moon at the beginning of its orbit. The time for light to travel this extra distance is vT/c. The observed period is longer than the actual period by this time. We have neglected the distance traveled by the earth while the light is in transit, which is nearly the same for the beginning and end of the orbit. The exact expression for the change in period is $vT/(c-v)$.

[ans: 15.3 s]

If the earth is moving toward Jupiter with the same speed the observed period is shorter than the actual period by the same amount. The actual period can be found by taking the average of the two observed periods. Historically the speed of light was measured by timing the interval for many cycles of the moon and solving for c.

PROBLEM 3. The line from the earth to a certain star is perpendicular to the orbit of the earth. At what angle must a telescope be aimed to see the star? The speed of the earth in its orbit is 29.9 km/s.

SOLUTION: From the point of view of the earth the light has a velocity with a component c along a line that is perpendicular to the orbit and a component v in the plane of the orbit. The telescope must be pointed slightly forward, in the direction of the earth's velocity and the angle θ it must make with the vertical is given by $\tan\theta = v/c$.

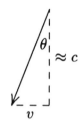

[ans: 20.6 seconds of arc]

This analysis is not relativistically correct. If you studied the relativistic transformation of velocities (Section 21-6) you can work the problem using the correct fundamental principles. Take the speed of light to be c, in the positive y direction and the velocity of the earth to be v, in the positive x direction. Now find the velocity components in a frame that is moving with speed v in the positive x direction. The answer is $c'_x = -v$ and $c'_y = c\sqrt{1 - v^2/c^2}$. The angle the telescope must make with the vertical is given by $\tan\theta = -c'_x/c'_y = v/c\sqrt{1-v^2/c^2}$. The numerical result is essentially the same since v^2/c^2 is so small.

You should know how to compute the speed of light in material media, given the dielectric constant κ_e. Use $v = 1/\sqrt{\mu_0\kappa_e\epsilon_0} = c/\sqrt{\kappa_e}$ (for non-magnetic materials).

PROBLEM 4. The dielectric constant of fused quartz is 2.1514 for light with a wavelength of 436 nm and is 2.1175 for light with a wavelength of 707 nm. Find the speed in fused quartz of light with these wavelengths.

SOLUTION:

[ans: 2.044×10^8 m/s; 2.060×10^8 m/s]

Here are some Doppler effect problems.

PROBLEM 5. A star emits blue light with a wavelength of 480 nm. If it moving away from us with a speed equal to tenth the speed of light what wavelength light do we observe? What is its color?

SOLUTION: The emitted frequency is given by $\nu_0 = c/\lambda_0$, where λ_0 is the emitted wavelength. The observed frequency is given by $\nu = \nu_0\left(\sqrt{1-u/c}\right)/\left(\sqrt{1+u/c}\right)$, where u is the speed of the star. The observed wavelength is given by $\lambda = c/\nu = \lambda_0\left(\sqrt{1+u/c}\right)/\left(\sqrt{1-u/c}\right)$.

[ans: 531 nm; green]

A different star emits light with the same wavelength. Suppose the observed light is red, with a wavelength of 660 nm. What is the speed of the star?

SOLUTION: Solve $\lambda = \lambda_0 \left(\sqrt{1+u/c}\right) / \left(\sqrt{1-u/c}\right)$ for u.

[ans: $0.308c$ (9.24×10^7 m/s)]

PROBLEM 6. Find an expression for the Doppler shift when the relative speed of source and observer is much less than the speed of light. Retain terms that are proportional to u/c but neglect terms that are proportional to $(u/c)^n$, where n is greater than 1.

SOLUTION: Start with $\nu = \nu_0 \left(\sqrt{1+u/c}\right) / \left(\sqrt{1-u/c}\right)$ and use the binomial theorem to expand the quantities $\sqrt{1+u/c}$ and $1/\sqrt{1-u/c}$. You should obtain $\sqrt{1+u/c} \approx 1 + u/2c$ and $1/\sqrt{1-u/c} \approx 1 + u/2c$.

[ans: $\nu = \nu_0(1 + u/c)$]

The fractional change in frequency is given by $\Delta\nu/\nu_0 = u/c$, where $\Delta\nu = \nu - \nu_0$ and u is positive if the source and observer are approaching each other, negative if they are receding from each other.

At low speeds what is the fractional change in frequency for each meter per second of relative speed?

SOLUTION:

[ans: 3.34×10^{-9}]

PROBLEM 7. A stationary source emits light with a frequency of 7.5×10^{14} Hz. It is reflected from a mirror that is perpendicular to the direction of propagation and moving away from the source with speed $3c/4$. The frequency of the reflected light is measured by a stationary detector. What is the result?

SOLUTION: The mirror receives a wave with frequency $\nu = \nu_0 \left(\sqrt{1-u/c}\right) / \left(\sqrt{1+u/c}\right)$, where u is the speed of the mirror. You may think of the mirror as a source of light with this frequency, moving away from the detector. The frequency detected is given by $\nu' = \nu \left(\sqrt{1-u/c}\right) / \left(\sqrt{1+u/c}\right) = \nu_0(1-u/c)/(1+u/c)$.

[ans: 1.07×10^{14} Hz]

The original light (with wavelength 400 nm) is near the boundary between visible and ultraviolet. The reflected light (with wavelength 1430 nm) is in the infrared.

Show that if u is much less than the speed of light then the fractional Doppler shift in frequency is given by $\Delta\nu/\nu_0 = -2u/c$.

SOLUTION:

A stationary radar gun is used by police to measure the speeds of automobiles. Suppose the gun is aimed directly at the back of a car speeding away at 33.5 m/s (75 mph). The reflected radar beam is then detected by the gun. If the gun uses 2450 MHz waves what is the difference in frequency of the emitted and detected signals?

SOLUTION:

[ans: 548 Hz; the detected signal is lower in frequency]

III. NOTES

Chapter 43
REFLECTION AND REFRACTION AT PLANE SURFACES

I. BASIC CONCEPTS

With this chapter you begin the study of optics. In particular, you will learn what happens when light is incident on the boundary between two materials and learn to determine the propagation directions of the reflected and refracted light. First, you should understand what geometrical optics is and when it is valid.

Geometrical and wave optics. In physical optics (or wave optics) the wave nature of light is used to describe and understand optical phenomena. Wave optics is close to the fundamental principles, Maxwell's equations, and is valid for all optical phenomena outside the quantum realm. However, when any obstacles to the light or any openings through which light passes are large compared to the wavelength of the light then details of its wave nature are not important. What is important is the direction of travel of the light. This is the realm of geometrical optics (or ray optics).

Geometrical optics uses rays to describe the path of light. A ray is a line in the direction of _____ of a light wave and at every point is perpendicular to the _____ through that point. The propagation direction changes when light is reflected or enters a region where the wave speed is different.

The laws of reflection and refraction. When light in one medium encounters a boundary with another region, in which the wave speed is different, some light is reflected back into the region of incidence and some is transmitted into the second region. Two laws, the law of reflection and the law of refraction, describe the directions of propagation of the reflected and transmitted light.

The diagram shows a ray in medium 1 incident on the boundary with medium 2. It makes the angle θ_1 with the normal to the boundary. The reflected ray makes the angle θ_1' with the normal and the transmitted ray makes the angle θ_2 with the normal. The law of reflection is

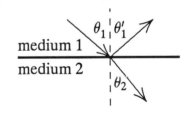

$$\theta_1' =$$

and the law of refraction is

$$n_2 \sin \theta_2 =$$

where n_1 is the index of refraction for medium 1 and n_2 is the index of refraction for medium 2. The angle θ_1 is called the angle of _____, θ_1' is called the angle of _____, and θ_2 is called the angle of _____. The law of refraction is sometimes called Snell's law.

The index of refraction of a medium is the ratio of the speed of light in vacuum to the speed of light in the medium. Specifically, if v is the speed of light in the medium then

$$n =$$

Comparison with the discussion of the last chapter should convince you that the index of refraction is closely related to the dielectric constant κ_e. In fact, for non-magnetic materials $n =$ _____.

Look at Table 1 of the text for some values. Of the materials listed in the table the one with the largest index of refraction is _____, with $n =$ _____. The speed of light is _____ in this material than in any other listed.

Both the law of reflection and the law of refraction can be derived from Maxwell's equations. These equations lead to a geometrical construction, called Huygen's principle, that shows us how to construct the wavefront at some time $t + \Delta t$ given the wavefront at an earlier time t. According to the principle you may think of each point on a wavefront as a point source of spherical waves, called Huygen wavelets. After time Δt the radius of the wavelets will be $v\Delta t$ and the wavefront will be tangent to the wavelets.

The principle is illustrated by the diagram to the right, which shows a plane wavefront at some instant of time. Use the time t to label it. Three spherical Huygen wavelets are shown. Each is centered on the plane wavefront and each has a radius of $c\Delta t$. The position of the plane wavefront at time $t + \Delta t$ is the common tangent to the spherical wavefronts. Draw it on the diagram and label it with the time $t + \Delta t$.

Now use Huygen's principle to prove the law of reflection. Two wavefronts impinging on a reflecting surface are drawn with solid lines; two rays are drawn with dashed lines. Wavefront 1 has reached the surface at B and wavefront 2 has reached the surface at A. Point b on wavefront 2 has not yet reached the reflecting surface but it will arrive at B in time $t = \ell/v_1$, where ℓ is the distance between the wavefronts and v_1 is the wave speed in medium of incidence.

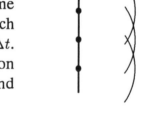

Use Huygen wavelets emanating from points A and B to find the positions of the wavefronts at time $t = \ell/v_1$. Point b of wavefront 2 has reached B so the reflected wavefront goes through that point. Point a of the same wavefront has moved a distance $\ell' = v_1 t$ away from point A on the surface. Draw an arc of the Huygen wavelet through a, centered at A.

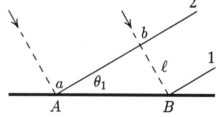

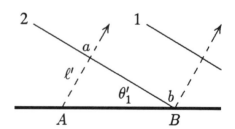

604   Chapter 43: Reflection and Refraction at Plane Surfaces

You should now be able to show that $\theta_1 = \theta_1'$. The two angles are the same because the triangles shown on the two figures are identical: each of two sides of one triangle is equal to a side of the other and an angle of one equals an angle of the other. They have the side AB in common so these sides are equal. What other sides are equal? _____ and _____ Why?

The equal angles are right angles. Mark them on the diagrams.

You must now convince yourself that θ_1 is the angle of incidence and θ_1' is the angle of reflection. On the upper diagram use a dashed line to draw the normal to the surface at point A. The angle of incidence is the angle between this normal and the ray through A. As you can see, that angle plus $90°$ must equal $\theta_1 + 90°$, so the angle of incidence is θ_1. A similar argument shows that θ_1' is the same as the angle between the normal and a ray associated with the reflected wave; that is, it is equal to the angle of reflection.

Now consider the transmitted wave and derive the law of refraction. The diagram shows the two wavefronts in the region on the other side of the boundary, when b has reached B. If the wave speed in this region is v_2, the Huygen wavelet emanating from A has a radius of $v_2 t$ so $\ell' = $ _____. Draw an arc of the wavelet through a. Notice that the line AB is the hypotenuse of a right triangle for which ℓ' is one side. Thus the length of AB is given by $v_2 t / \sin\theta_2$.

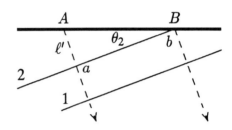

According to the diagram showing the *incident* wavefronts, the length of AB is also given by $v_1 t / \sin\theta_1$. In the space below set the two expressions for AB equal to each other, replace v_1 with c/n_1, replace v_2 with c/n_2, and obtain the law of refraction:

Images in mirrors. When you view light from an object after it has been reflected by a plane mirror the light appears to come from points behind the mirror and, in fact, you see an image that appears to be behind the mirror.

The diagram shows a point source P in front of a mirror and two rays emanating from it. The reflected rays are drawn according to the law of reflection: the angles of incidence and reflection are the same. Draw dotted lines to extend the reflected rays to behind the mirror and use P' to label the point where they intersect. This is the image of the source. When your eyes view the light reflected from the mirror it appears to come from a point source at P'.

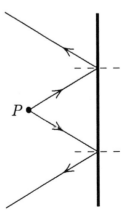

You should realize that any other rays from the source to the mirror could have been drawn. All reflected rays, extended backward to the other side of the mirror, intersect at P'. When you look at mirror only a very narrow bundle of light enters your eyes, but all rays in the bundle diverge from the image.

The image point P' is on the line through the source P that is _____ to the plane of the mirror. It is the same distance behind the mirror as the _____ is in front. If you want to know if you can see the image when your eye is in any given position, draw a line from the image to your eye. If the line intersects the mirror then light reflected from the mirror gets to your eye and you see the image. If the line does not intersect the mirror you do not see the image.

If the source is extended you may think of each point on it as a point source, emitting light in all directions. Only the light that reaches the mirror is reflected, of course. For each point that sends light to the mirror an image point is formed. In many examples an extended source is made up of one or more straight lines. To find the image of a line simply find the images of each end and connect them with a straight line. Clearly the image of an extended source is the same size as the source.

You should be familiar with some of the terms used. The source of light is often called the _____. An image may be real or virtual. In both cases light diverges from the image but if the image is real the light actually _____, while if it is virtual it does not. Images formed by plane mirrors are _____.

The distance from a point source to a mirror is denoted by o (object distance). It is positive. The image position is denoted by i. It is negative (for virtual images) and its magnitude is the distance from the image to the mirror. In terms of these quantities the statement that the image is as far behind the mirror as the object is in front is written

$$o = $$

Refraction and total internal reflection. Refraction is described by Snell's law: $n_1 \sin\theta_1 = n_2 \sin\theta_2$. Recall that $\sin\theta$ increases as θ increases from 0 to 90°. If n_1 is greater than n_2 then θ_1 is less than θ_2. Light incident from a medium with a small index of refraction bends toward the normal as it enters a medium with a higher index of refraction. Light incident from a medium with a large index of refraction bends away from the normal as it enters a medium with a smaller index of refraction. When you shine light into water from air it bends _____ the normal. Light from an underwater source bends _____ the normal as it enters the air.

Total internal reflection can occur when light travels in an optically dense medium (large index of refraction) toward a less optically dense medium (smaller index of refraction). If the angle of incidence is greater than a certain critical value denoted by θ_c no light is transmitted into the less optically dense medium. All the light incident on the boundary is reflected. Suppose $n_1 > n_2$ and light is incident from medium 1 on the boundary with medium 2. Then $\theta_1 = \theta_c$ if $\theta_2 =$ _____ and $\sin\theta_2 =$ _____. In terms of n_1 and n_2, $\theta_c =$ _____.

Total internal reflection does not always occur when the conditions described above are met. If a thin film of low optical density is layered between two materials of higher optical density and the light source is in one of the higher optical density materials, then light does appear in the other high optical density material. Suppose light is passing though a piece of glass and strikes the boundary with air at greater than the critical angle. No light is transmitted. If a second piece of glass is placed very close to the first, near the point where the light strikes, light emerges. This phenomenon is known as _____.

II. PROBLEM SOLVING

You should understand that when light enters a medium with index of refraction n its speed changes to $v = c/n$, its frequency remains the same, and its wavelength changes to $\lambda = v/\nu = c/n\nu = \lambda_0/n$, where λ_0 is its wavelength in vacuum.

PROBLEM 1. The index of refraction of heavy flint glass is 1.675 for light of wavelength 434 nm in vacuum and 1.644 for light of wavelength 656 nm in vacuum. What is the speed, frequency, and wavelength of each of these in the glass?

SOLUTION:

[ans: 434 nm: 1.79×10^8 m/s, 6.91×10^{14} Hz, 259 nm; 656 nm: 1.82×10^8 m/s, 4.57×10^{14} Hz, 399 nm]

PROBLEM 2. A piece of heavy flint glass is placed in water. Light with a wavelength of 500 nm in water enters the glass. What is its frequency and wavelength in the glass? The index of refraction for water is 1.333 and the index of refraction for the glass is 1.650.

SOLUTION: The frequency is the same in the water and glass. To find its value use $\nu = v_w/\lambda_w = c/n_w\lambda_w$, where the subscript refers to water. To calculate the wavelength in the glass use $\lambda_g = v_g/\nu = c/n_g\nu$.

[ans: 4.50×10^{14} Hz; 404 nm]

The law of reflection is straightforward. For a point source find the position of the image by drawing a line from the source to the mirror, perpendicular to the mirror or its extension and locate the image on the line so its distance from the mirror is the same as the distance of the source. You should know that the line you construct may be outside the mirror and not intersect it. Complications may arise from the geometry and you should practice until you are familiar with the techniques.

PROBLEM 3. A point source is centered in front of a 70-cm wide plane mirror. A woman walks along a line parallel to the mirror and close to the source. How far from the source can she walk without losing sight of the image of the source?

SOLUTION: The diagram shows the situation. The source is at S, a distance ℓ from the mirror. Its image is at I, also a distance ℓ from the mirror. W is the width of the mirror. The woman can walk to the point where the line from the image intersects her path (shown dotted). At this point light that is reflected from the edge of the mirror reaches the woman. Beyond this point no reflected light reaches her. The two triangles shown on the diagram are similar (each side of one is parallel to a side of the other), so $(W/2)/\ell = d/2\ell$. Solve for d.

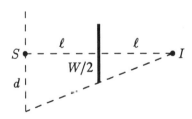

[ans: 70 cm]

Use a colored pen or pencil to draw on the diagram the ray that is reflected from the edge of the mirror. Convince yourself that after reflection it follows the dotted line.

PROBLEM 4. A long rectangular mirror is attached vertically to a wall with its bottom edge on the floor. How much of the bottom portion of the mirror can be covered or removed and still allow you to see your feet when you stand in front of it?

SOLUTION: Suppose your eyes are a distance h above the floor. In the space on the right draw a diagram showing the mirror, the position of your eyes, and the position of your feet. Mark the position of the image of your feet and draw the line from the image to your eyes. The portion of the mirror below the point where this line crosses is not needed to see your feet. Notice that it does not depend on how far from the mirror you stand.

[ans: $h/2$: half the distance from your feet to your eyes]

PROBLEM 5. Suppose the driver's side mirror (M in the diagram) on a car is at eye level and is adjusted so it is perpendicular to the direction of travel. It is 20 cm wide. Take the driver's eye (E) to be 40 cm to the right of the mirror and 50 cm in front of it. A second car (A) is passing so its bumper on the passenger side follows the dotted path shown, 1.2 m to the left of the mirror. For what values of the distance d can the driver see the image of the bumper without moving his eye?

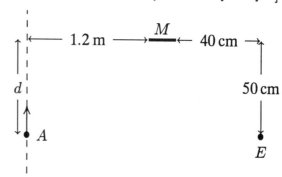

SOLUTION: The image of the bumper is on the dotted line, a distance d on the other side of the mirror. Draw lines from the eye through the edges of the mirror and find their intersections with the path of the bumper. Calculate d for each of them. Similar triangles will help.

[ans: from 1.75 m to 1.0 m]

This is a rather short range. It can be increased substantially by rotating the mirror clockwise through a small angle.

PROBLEM 6. Two mirrors are positioned as shown, making an angle β with each other. A point source S is in front of the bottom of edge of one mirror. Consider a ray from the source that is reflected from both mirrors. Suppose it makes the angle θ_i with the horizontal and find the angle the emerging ray makes with the horizontal.

SOLUTION: Follow the ray and at each reflection point find an expression for the angle of reflection in terms of θ_i and perhaps β. Finally, use geometry to relate the angle between the emerging ray and the horizontal to the angles of reflection and ultimately to θ_i and β.

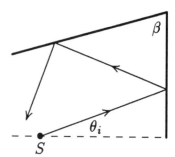

To see the details look at the second diagram, where the normals to the mirrors are drawn as dotted lines and some angles are labeled. In terms of θ_i, $\alpha_1 = $ _____. Since the law of reflection is obeyed at the first mirror $\alpha_2 = $ _____. The normal to the first mirror makes an angle of $90°$ with that mirror so $\alpha_3 = $ _____. Since the interior angles of a triangle sum to $180°$, $\alpha_3 + \alpha_4 + \beta = 180°$ and $\alpha_4 = $ _____ in terms of θ_i and β.

Now consider the reflection at the second mirror. Because the normal makes an angle of $90°$ with the mirror $\alpha_5 = $ _____ and because the law of reflection is obeyed $\alpha_6 = $ _____. Now α_2, α_5, α_6, and α_7 must sum to $180°$, so $\alpha_7 = $ _____. This is the angle between the emerging ray and the horizontal.

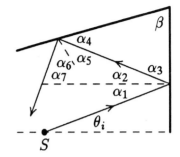

[ans: $180° - 2\beta + \theta_i$]

Suppose the ray enters at $\theta_i = 40°$. What should the angle between the mirrors be if the ray is to emerge straight down?

SOLUTION:

[ans: $65°$]

A corner reflector is a combination of two mirrors arranged so any ray incident on one of them emerges parallel to its original direction after the second reflection. Use the result you obtained above to show that the angle between the mirrors must be $90°$.

Problems involving refraction are also straightforward in principle but are often complicated by the geometry. Don't forget that the angles in the law of refraction are measured with respect to the *normal* to the surface. Here are a few for practice.

PROBLEM 7. A ray of light enters at $45°$ to the normal from air into a block of crown glass, as shown. After passing through the block the transmitted ray enters an adjacent block of flint glass. Each block is 35 cm wide. The index of refraction of crown glass is 1.517 and the index of refraction of flint glass is 1.890. Where (relative to its entry point) and at what angle does the ray emerge?

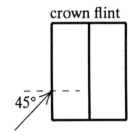

SOLUTION: Sketch the ray. It bends toward the normal as it enters the crown glass, again bends toward the normal as it enters the flint glass, and finally bends away from the normal as it re-enters the air.

Use the law of refraction. First consider refraction at the air-crown glass boundary. Take $n_1 = 1$, $n_2 = 1.517$, and $\theta_1 = 45°$. Solve $n_1 \sin\theta_1 = n_2 \sin\theta_2$ for θ_2. Now consider the boundary between the two glasses. If W is the width of the glass block the ray meets this boundary at a point $W\tan\theta_2$ above its point of entry into the crown glass. The angle of incidence is θ_2. Let $n_3 = 1.890$ and θ_3 be the angle of refraction. Solve $n_2 \sin\theta_2 = n_3 \sin\theta_3$ for θ_3.

Finally consider the flint glass-air boundary. The ray meets this boundary at a point $W\tan\theta_3$ above the point of its entry into the flint glass. The angle of incidence is θ_3. Solve $n_3 \sin\theta_3 = \sin\theta_4$ for θ_4, the angle between the exiting ray and the normal.

[ans: 32.6 cm above the entry point; $45°$]

The ray emerges parallel to the direction of its entry but it is displaced along the block. This is a general result for refraction by blocks of material with parallel plane faces. Consider a single block of material with a width of 70 cm, the combined width of the two glass blocks. What should its index of refraction be for a ray incident at 45° to be displaced by 32.6 cm?

SOLUTION:

[ans: 1.675]

PROBLEM 8. A 3.9-m pole stands vertically in a pool of water 1.6 m deep. Sunlight strikes it at 30° to the horizontal. What is the length of the shadow on the bottom of the pool? The index of refraction of water is 1.33.

SOLUTION: The diagram illustrates the situation. The length of pole above water is $\ell_1 = 2.3$ m, the length below water is $\ell_2 = 1.6$ m. Use $\tan 30° = \ell_1/s_1$ to compute s_1. Use $\sin 60° = n \sin \theta_2$ to compute θ_2, then $\tan \theta_2 = s_2/\ell_2$ to compute s_2. The length of the shadow is $s_1 + s_2$.

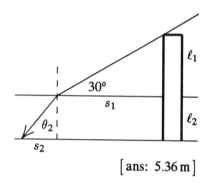

[ans: 5.36 m]

How long is the shadow if the pole is standing on level ground?

SOLUTION:

[ans: 6.75 m]

PROBLEM 9. You look straight down from above the water at a stone on the bottom of a 2.5-m deep pool. At what height does the stone appear to be? Take the index of refraction for water to be 1.33 and assume only an extremely narrow cone of light enters your eye.

SOLUTION: In the diagram the stone is at S but appears to be at S', where the emerging ray, extended backward, intersects the vertical line through the stone. The angles are greatly exaggerated on the diagram. The distances h and ℓ are related by $\tan \theta_1 = \ell/h$. If d is the distance from the surface to S' then $\tan \theta_2 = \ell/d$. Solve these equations for d in terms of h.

Since θ_1 and θ_2 are extremely small you may replace $\tan \theta_1$ by $\sin \theta_1$ and $\tan \theta_2$ by $\sin \theta_2$. Finally, use $n_1 \sin \theta_1 = n_2 \sin \theta_2$ to eliminate the angles from the expression for d. You should get $d = (n_2/n_1)h$.

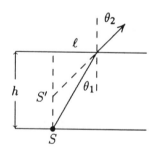

[ans: 1.88 m]

PROBLEM 10. A prism with an apex angle of 60° is made of flint glass, with an index of refraction of 1.650. Light is incident on one side at an angle of 75° to the normal, as shown. Find the angle made by the exiting ray with the normal to the other side of the prism.

SOLUTION: Trace the ray through the prism as indicated by the second diagram. Use $\sin\theta_1 = n\sin\theta_2$ to find θ_2. Since the interior angles of the four-sided figure bounded by the prism sides and their normals must sum to 360° and since two of the angles are right angles, the angle A is $180° - \alpha$. This means $\theta_3 = 180° - \theta_2 - A = \alpha - \theta_2$. Finally, use $n\sin\theta_3 = \sin\theta_4$ to find θ_4.

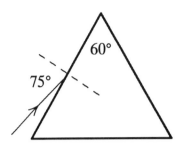

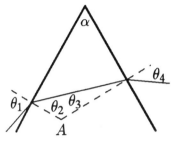

[ans: 42.5°]

One quantity of interest to people who use prisms is the angular deviation of a ray. If the incident ray is extended forward and the emerging ray is extended backward, as shown, this is the angle labelled δ. What is the angle of deviation for the ray and prism of this problem?

SOLUTION: The incident ray makes the angle $90° - \theta_1$ with the prism face where it enters and the angle $180° - (90° - \theta_1 + \alpha)$ with the other face. The emerging ray makes the angle $90° - \theta_4$ with the face it exits. The angle of deflection is the difference of these two angles or $\delta = 90° + \theta_1 - \alpha - 90° + \theta_4 = \theta_1 + \theta_4 - \alpha$.

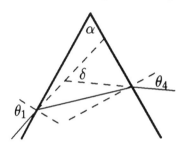

[ans: 57.5°]

Light of different wavelengths is deviated through different angles. This accounts for the array of colors seen when white light is refracted by a prism. The index of refraction given above is for yellow light. What is the angle of deviation for violet light incident at the same angle? The index of refraction for violet light in flint glass is 1.675.

SOLUTION:

[ans: 59.6°]

III. NOTES

Chapter 44
SPHERICAL MIRRORS AND LENSES

I. BASIC CONCEPTS

In this chapter the law of reflection is applied to spherical mirrors and the law of refraction is applied to spherical refracting surfaces. Each type surface forms images of objects placed in front of it and you will learn the relationship between the position of the object, the position of the image, and the radius of curvature of the surface. You will also apply what you learn to lenses with spherical faces and to systems of lenses, such as those used in telescopes, microscopes, and cameras. Pay careful attention to ray tracing techniques. They will help you to visualize image formation and to understand the position and size of the image. When using the equations relating object and image distances be sure to concentrate on the signs of the variables. Knowing how to choose the signs of given quantities and how to interpret the signs of results is vital for obtaining correct answers.

Spherical mirrors. The diagrams below show two spherical mirrors, with their centers of curvature labelled C. A point source of light S is in front of each mirror. The line through C and the center of the mirror is called the _____. Usually angles are measured with respect to this line and distances are measured along it. Light from the source is reflected by the mirror. On each diagram draw a ray from the source to the mirror, about halfway to its edge. Use a dotted line to draw the normal to the mirror at the point where the ray strikes, then draw the reflected ray. The rays should obey the law of reflection at the mirror, but you will probably have some drawing error.

In some cases reflected rays converge to a point on the source side of the mirror. Reflected light then forms a _____ image. In other cases the reflected rays diverge as they leave the mirror but they follow lines that pass through a single point behind the mirror. That is, they form a _____ image. The region in front of the mirror, where the source is located and where real images are formed, is called the R-side of the mirror. The region behind, where virtual images are located, is called the V-side of the mirror. Indicate the R-side and V-side on each of the diagrams above.

Strictly speaking, sharp images are formed only by light with <u>paraxial rays</u>. Define the term: _____

The object distance o is the distance from the source to the mirror, measured along the optic axis. It is positive for a source in front of a single mirror. Indicate o on each of the diagrams above. The image distance i is the distance from the mirror to the image position, measured along the optic axis. A sign is associated with it. It is positive for a _____ image, located _____ the mirror and negative for a _____ image, located _____ the mirror.

The mirror equation relates the object distance o, the image distance i, and the radius of curvature r. It is
$$\frac{1}{o} + \frac{1}{i} =$$
This is often written in terms of the focal length, defined by $f =$ _____. Then
$$\frac{1}{o} + \frac{1}{i} =$$

Signs are associated with the radius of curvature and focal length as well as with object and image distances. Both r and f are positive if the center of curvature is on the _____-side of the mirror and negative if the center of curvature is on the _____-side. A mirror that is concave with respect to the source has a _____ radius of curvature and focal length; a mirror that is convex with respect to the source has a _____ radius of curvature and focal length. On each diagram above indicate whether the mirror is concave or convex with respect to S and give the sign of r and f.

The focal point of a lens is important for tracing rays and graphically finding the position of an image when the object is not on the optic axis. It is on the optic axis, a distance $|f|$ from the mirror. If the mirror is concave it is on the _____-side; if the mirror is convex it is on the _____-side. For each of the mirrors above place a dot at the focal point and label it F.

Incident light with rays that are parallel to the optic axis is reflected so its rays are along lines that pass through _____. If the focal point is on the R-side the rays actually pass through it. If the focal point is on the V-side they do not but their extensions backward into the V-region do. Incident light with rays along lines that pass through the focal point is reflected so its rays are along lines that are _____.

The image is located at the intersection of all paraxial reflected rays. The two special rays mentioned above are easy to locate and draw, so they are often used to locate an image. Another that is used for the same reason is a ray along a line through the center of curvature. Its reflection is along a line that passes through _____.

The diagram shows a spherical mirror with its center of curvature at C. Locate the focal point and mark it F. Locate the image of the arrow at O by drawing two rays both before and after reflection. The first is from the head of the arrow to the mirror and is parallel to the axis. The reflected ray goes through _____. The second is from the head of the arrow through C to the mirror. The reflected ray goes through _____. Draw the image of the arrow with its head at the point where these rays intersect and its tail on the axis.

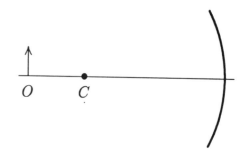

614 Chapter 44: Spherical Mirrors and Lenses

You might also draw a ray from the arrow head through the focal point to the mirror. The reflected ray is parallel to the axis and passes through the intersection of the rays you have already drawn.

Note that the image is real. Light actually passes through points of the image. It is also inverted. If you move the object arrow to the left, away from the mirror, the image moves _____ the mirror. To verify this imagine the object after it is moved to the left and note what happened to the ray through the center of curvature and its intersection with the ray that is parallel to the axis. The latter does not change as the object moves.

Now suppose the object is placed between the focal point and the mirror, as shown. Draw a line from the center of curvature through the head of the arrow to the mirror. The portion between the arrow head and the mirror is a ray. It is reflected through _____. Draw a ray from the arrow head to the mirror, parallel to the axis. It is reflected through _____. Use dotted lines to extend the reflected rays backward into the region behind the mirror, then locate the image of the arrow head at their intersection and draw the image of the arrow.

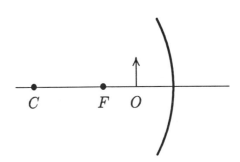

You might also use a line from the focal point through the arrow head to the mirror. The portion between the head and the mirror is a ray and its reflection is parallel to the axis. Its extension passes through the intersection of the dotted lines you have drawn.

Notice that the image is virtual and erect. It is behind the mirror and no light reaches it but reflected light in front of the mirror appears to come from it. If the object arrow is moved closer to the mirror, its image moves _____ the mirror.

The results you found graphically are predicted by the mirror equation. When it is solved for i the result is $i = fo/(o - f)$. If $o > f$ then i is positive: the image is real and is on the R-side. If $o < f$ then i is negative: the image is virtual and is on the V-side. If $o = f$ all paraxial reflected rays are parallel to each other and the image is said to be at infinity.

Now consider a convex mirror. Locate the focal point on the diagram and label it F. Draw a ray from the head of the arrow to the mirror, parallel to the axis. Draw the reflected ray. It is along a line that passes through _____. Draw a ray from the arrow head to the mirror, along the line that passes through C. Draw the reflected ray. It is along a line that passes through _____. Use dotted lines to extend the reflected rays to behind the mirror and draw the image of the arrow at their intersection. It is virtual and erect.

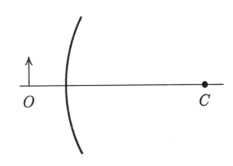

You might also use a line from the head of the arrow through the focal point. The reflected ray is parallel to the axis and its extension behind the mirror should pass through the intersection of the other lines.

Chapter 44: Spherical Mirrors and Lenses   615

The mirror equation also predicts a virtual image. Since f is negative $i = fo/(o-f)$ is negative for any positive value of o.

The lateral magnification m of a mirror is the ratio of the lateral size of the _____ to the lateral size of the _____. The term *lateral size* means the dimension perpendicular to the optic axis. In terms of the object distance o and image distance i the lateral magnification of a spherical mirror is given by

$$m =$$

The negative sign here has special meaning. Values for o and i are substituted with their signs. If m is positive then the image is _____ ; if m is negative then the image is _____. Notice that virtual images are erect and real images are inverted.

Spherical refracting surfaces. The diagrams below show two spherical refracting surfaces, with their centers of curvature labelled C. Each separates a medium with an index of refraction n_1 from a medium with index of refraction n_2. Light from point source S in front of a surface is refracted at the surface and continues into the medium on the other side. Suppose n_2 is greater than n_1, so the rays are bent toward the normal, and for each diagram draw a ray from the source that strikes the surface about halfway up. Use a dotted line to draw the normal to the surface at the point where the ray strikes and show the ray after refraction.

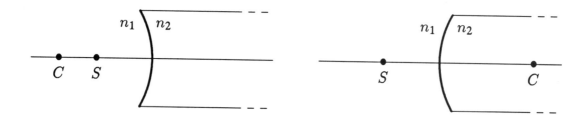

If the light is paraxial it forms a sharp image of the source. If the image is virtual it is formed in the region containing the source and this is called the V-side of the surface. If the image is real it is formed on the side into which light is transmitted. This region is called the R-side of the surface. Label the two sides on each diagram above.

For a single refracting surface the object distance o is positive. Indicate it on each diagram above. The image distance is positive if the image is real and negative if it is virtual. For the surfaces and sources of the diagrams above images with positive values of i are to the _____ of the surface and images with negative values of i are to the _____.

The law of refraction, applied to paraxial light, yields a relationship between the object distance, the image distance, the radius of curvature of the surface, and the indices of refraction for the two sides. It is

$$\frac{n_1}{o} + \frac{n_2}{i} =$$

For this equation to be valid the appropriate sign must be associated with the radius of curvature. If the surface is concave with respect to the source, the center of curvature is on the

_____-side and r is _____. If the surface is convex with respect to the source, the center of curvature is on the _____-side and r is _____. In terms of the labels R and V used to designate the regions the sign convention for r and i is the same as the convention for spherical mirrors. Remember that the R-side is the side the light goes to after striking the surface. For a mirror this is the side of the source; for a refracting surface it is the other side.

If $n_2 > n_1$ then a convex surface forms a real image of an object that is far from it and a virtual image of an object that is near it. Details depend on the values of the indices of refraction and the radius of curvature. A concave surface, on the other hand, always forms a _____ image. If $n_1 > n_2$ then a convex surface always forms a _____ image and a concave surface forms a _____ image for an object that is far from it and a _____ image for an object that is near it.

Thin lenses. A lens consists of two refracting surfaces. The image formed by the first may be considered the object for the second. If the surrounding medium is a vacuum and the thickness of the lens can be neglected, then the object distance, image distance, and focal length are related by

$$\frac{1}{o} + \frac{1}{i} =$$

where the focal length is given in terms of the radii of curvature by

$$\frac{1}{f} =$$

Here r_1 is the radius of curvature of the surface nearer the object, r_2 is the radius of curvature of the surface farther away, and n is the index of refraction of the lens material. The equation is a good approximation if the surrounding medium is air.

Again you should be familiar with the sign convention. Real images are formed on the opposite side of the lens from the object and this side is labelled the _____-side. For these images i is _____. Virtual images are formed on the same side of the lens as the object and this side is labelled the _____-side. For a virtual image i is _____. A surface radius is positive if the center of curvature is on the _____-side and negative if the center of curvature is on the _____-side. You should know that the focal length of a lens does not depend on which side faces the object. Lenses with positive focal lengths are said to be _____, while lenses with negative focal lengths are said to be _____.

For each of the lenses shown below give the sign of each radius, the sign of the focal length, and say if the lens is converging or diverging. Assume the object is to the left of the lens.

sign of r_1: _____ _____ _____ _____
sign of r_2: _____ _____ _____ _____
sign of f: _____ _____ _____ _____
_____ _____ _____ _____

A lens has two focal points, located equal distances $|f|$ on opposite sides of the lens. The first focal point, denoted by F_1, is on the side of the incident light (the V-side) for a converging lens (f positive) and on the side of the refracted light (the R-side) for a diverging lens (f negative). Light rays that are along lines that pass through F_1 are bent by the lens to become parallel to the axis. For a converging lens the rays before refraction actually pass through F_1. For a diverging lens the rays strike the surface, so only their extensions into the R-side pass through F_1. For each of the lenses shown below draw an incident ray along a line that passes through F_1. Also draw the refracted ray, parallel to the axis. For the diverging lens use a dotted line to continue the line of the incident ray to the focal point. Since we have neglected the thickness of the lenses you may assume refraction takes place at the plane through the lens center, perpendicular to the axis. These planes are shown as dotted lines on the diagrams.

The second focal point, denoted by F_2, is on the side of the refracted light (the R-side) for a converging lens (f positive) and on the side of the incident light (the V-side) for a diverging lens (f negative). Light rays that are parallel to the optic axis are bent by the lens to lie along lines that pass through F_2. For a converging lens the rays after refraction actually do pass through F_2. For a diverging lens the backward extension of the rays into the V-side pass through F_2. For each of the lenses shown below draw an incident ray parallel to the optic axis. Also draw the refracted ray, along a line through F_2. For the diverging lens use a dotted line to continue the line of the refracted ray backward to the focal point.

The position of an off-axis image can be found graphically by tracing two or more rays originating at the same point on the object. Two of these might be the rays you drew above: one along a line through F_1 and then parallel to the axis and the other parallel to the axis and then along a line through F_2. The third ray you might use goes through _____ of the lens. It is not refracted.

In terms of the object and image distances the lateral magnification associated with a lens is given by

$$m = $$

an expression that is identical to the expression for a spherical mirror. If m is negative the image is _____; if m is positive the image is _____. A virtual image formed by single thin lens is always _____; a real image is always _____.

618 *Chapter 44: Spherical Mirrors and Lenses*

Angular magnification. Because it takes into account the apparent diminishing of the size of an object with distance, the angular magnification is often a better measure of the usefulness of a lens used for viewing than is the lateral magnification.

If an object has a lateral dimension (perpendicular to the optic axis) of h and is a distance d from the eye, as in Fig. 24 of the text, its angular size is given in radians by $\theta = $ _____. If the object is viewed through a lens and the image is a distance d' from the eye and has a lateral dimension of h' then the angular size of the image is $\theta' = $ _____. The angular magnification of the lens is the ratio of these, or $m_\theta = \theta'/\theta = $ _____, where the last expression gives the angular magnification in terms of h, h', d, and d'.

A single converging lens used as a magnifying glass is usually positioned so the object is just inside the focal point F_1. See Fig. 24 of the text. The image is then virtual and far away. Its lateral size is $h' = mh = -ih/o$, where m is the lateral magnification. Thus $m_\theta = -(i/o)(d/d')$. The eye is close to the lens so $d' \approx |i|$. Furthermore $o \approx f$. Once these substitutions are made the result is $m_\theta = d/f$. Usually d is taken to be the distance to the near point of the eye, about 25 cm, so $m_\theta = 25/f$, where f must be measured in centimeters.

The near point is used since the object is in focus and has its largest angular size when it is this distance from an unaided eye. The angular magnification then tells us how much better the lens is than the best the unaided eye can do.

Lens systems. Telescopes, microscopes, and other optical instruments usually consist of a series of lenses. They can be analyzed graphically by tracing a few rays as they pass through each lens in succession. The lens equation can also be applied to each lens in succession to find the position of the image. Consider a system of two lenses, a distance ℓ apart on the same optic axis, as shown below. Let o_1 be the distance from the object O to the first lens struck. This lens forms an image I_1 at i_1, given by $1/o_1 + 1/i_1 = 1/f_1$, where f_1 is the focal length of the lens. You may think of this image as the object for the second lens. The object distance is $o_2 = \ell - o_1$ and the image distance i_2 is given by $1/o_2 + 1/i_2 = 1/f_2$, where f_2 is the focal length of the second lens. On the diagram identify and label o_1, i_1, o_2, and i_2. Verify that $i_1 + o_2 = \ell$.

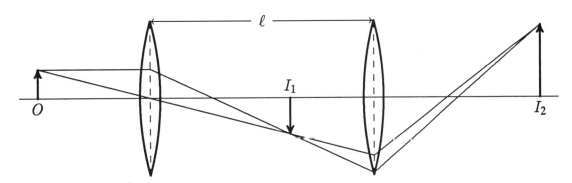

For some systems the object distance for the second lens may be negative. This occurs if the image formed by the first lens is behind the second lens, so the light exiting the first lens is converging as it strikes the second lens. Such objects are called *virtual* objects. The equation relating object and image distances is still valid — just substitute a negative value for o_2.

A simple microscope consists of an objective lens, near the object, and an eyepiece (or ocular). The focal length of the objective lens is small and the lens is positioned so the object lies just outside the _____ of the lens. The image produced by the objective is real, large, inverted, and far from the lens. If the object has a lateral dimension h then the image has a lateral dimension $h' = mh = ih/o$. The magnification is great since i is large and o is small.

The distance s between the second focal point of the objective lens and the first focal point of the eyepiece is called the _____ of the microscope. In terms of this quantity $h' =$ _____, where θ is the angle the ray through the focal point makes with the optic axis. Since $h = f_{ob} \tan\theta$, the magnification of the objective lens is given by $m =$ _____.

The eyepiece is positioned so the image produced by the objective lens falls at its _____ focal point. The eye is placed close to the eyepiece, which then acts as a simple magnifying glass with an angular magnification of $25\,\text{cm}/f_{ey}$, where f_{ey} is the focal length of the eyepiece. The overall angular magnification is given by $M =$ _____. The expression again compares the angular size of the image produced by the microscope with the angular size of the object when it is at the near point of an unaided eye.

Telescopes are used to view objects that are far away. Rays entering the objective lens are essentially parallel to the optic axis and that lens produces an image that is close to the _____ focal point. To compute the angular magnification of a telescope we compare the angular size of the object at its far-away position (not the near point) to the angular size of the image produced by the telescope.

Since the length of the telescope is much less than the object distance the angle subtended by the object at the eye is the same as the angle it subtends at the objective lens. This is the same as the angle subtended at the objective lens by the image produced by that lens, so $\theta_{ob} = h'/f_{ob}$, where h' is the lateral dimension of the image. The angle subtended by the final image at the eye is the same as the angle subtended by the intermediate image at the eyepiece, or $\theta_{ey} = h'/f_{ey}$. The angular magnification of the telescope is given by $m_\theta = \theta_{ob}/\theta_{ey} =$ _____. To obtain a large angular magnification, a telescope should have an objective lens with a _____ focal length and an eyepiece with a _____ focal length.

II. PROBLEM SOLVING

You should be able to find the image positions for objects in front of spherical mirrors, spherical refracting surfaces, and thin lenses. Here are some examples.

PROBLEM 1. What is the position of the image formed by paraxial rays from a point source placed on the optic axis 2.5 cm in front of a 25-cm radius spherical concave mirror?

SOLUTION: Use $f = r/2$ to find the focal length and $1/o + 1/i = 1/f$ to find the image distance i. Since the mirror is concave both r and f are positive.

[ans: 3.13 cm behind the mirror, on the optic axis]

Is the image real or virtual?

SOLUTION: Note that i is negative and the image is behind the mirror.

[ans: virtual]

Suppose the point source is on the optic axis, 15 cm in front of the mirror. Then where is the image? Is it real or virtual?

SOLUTION:

[ans: 75 cm in front of the mirror, on the optic axis; real]

Suppose the point source is 2.5 cm in front of the mirror and 2.5 cm above the optic axis. Then where is the image? Is it real or virtual?

SOLUTION: The object distance is 2.5 cm so the image distance is -3.13 cm, as before. Imagine a line of length ℓ (= 2.5 cm) from the point source to the axis. Its image is a line with a length of $\ell' = m\ell = -i\ell/o$, where m is the lateral magnification. The magnitude of ℓ' gives the distance from the optic axis to the image of the point source. If it is positive the image is above the axis; if it is negative the image is below.

[ans: 3.13 cm behind the mirror, 3.13 cm above the axis; virtual]

Finally, suppose the point source is 15 cm in front of the mirror and 2.5 cm above the axis. Where is its image? Is it real or virtual?

SOLUTION:

[ans: 75 cm in front of the mirror, 12.5 cm below the axis; real]

PROBLEM 2. A 6.0-cm long rod lies along the optic axis of a 36-cm radius concave spherical mirror. One end is 35 cm in front of the mirror and the other is 29 cm in front. What is the length of the image?

SOLUTION: Find the positions of the images of the rod ends. The mirror has a positive radius and focal length.

[ans: 10.4 cm]

Chapter 44: Spherical Mirrors and Lenses

Now the rod is moved so one end is 21 cm in front of the mirror and the other end is 15 cm in front. Describe the image.

SOLUTION: The rod now straddles the focal point of the mirror. The images of points between the focal point and the mirror are virtual while those of points outside the focal point are real. The image of the point at the focal point is at infinity. Again find the positions of the images of the rod ends.

[ans: The image is in two parts: the real part stretches away from the mirror to infinity from a point 126 cm in front; the virtual part stretches away to infinity from the mirror from a point 90 cm behind]

PROBLEM 3. A 5.0-cm long stick is in front of a concave spherical mirror with an 18-cm focal length. One end is on the optic axis 35 cm from the mirror and the other end is positioned so the stick makes an angle of 35° with the axis. What is the length of the image and what angle does the image make with the axis?

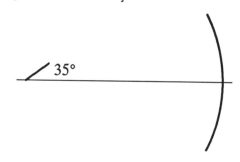

SOLUTION: Find the images of the ends of the stick. The object distance for the end on the axis is $o_1 = 35$ cm. The object distance for the other end is $o_2 = 35 - 5.0\cos 35° = 30.9$ cm. Its distance from the axis is $h = 5.0\sin 35° = 2.87$ cm. Use the expression for the magnification to find the distance its image is from the axis: $h' = ih/o$. The length of the image is given by $\sqrt{(i_2 - i_1)^2 + (h')^2}$ and the angle θ it makes with the axis is given by $\tan\theta = (i_2 - i_1)/h'$.

[ans: 7.25 cm; 33.5°]

PROBLEM 4. A cylinder is made of material with an index of refraction of 1.56. One end is a spherical cap with a radius of 12 cm. Where should a point source of light be placed on the axis in front of the cap so the light refracted by cap travels parallel to the axis?

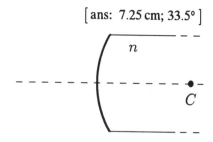

SOLUTION: Put a dot on the axis in front of the surface, label it S, and draw several rays to the surface. Draw them parallel to the axis after refraction. Use $n_1/o + n_2/i = (n_2 - n_1)/r$, with $n_1 = 1$, $n_2 = 1.56$, and $r = +12$ cm. If the rays are parallel to the axis after refraction the image is at infinity, so take $1/i = 0$ and solve for o.

[ans: 21.4 cm in front of the surface]

If light is incident parallel to the axis, where is it focused?

SOLUTION: Now the object distance is infinite. Put $1/o = 0$ and solve for i.

[ans: on the axis, 33.4 cm to the right of the surface]

PROBLEM 5. A cylinder is made of material with an index of refraction of 1.56. One end has a concave spherical indentation with a radius of 12 cm. To what point should incoming rays converge in order to obtain rays that are parallel to the axis after refraction?

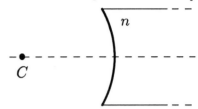

SOLUTION: Use $n_1/o + n_2/i = (n_2 - n_1)/r$, with $n_1 = 1$, $n_2 = 1.56$, and $r = -12$ cm. If the rays are parallel to the axis after refraction the image is at infinity, so take $1/i = 0$ and solve for o.

[ans: to a point on the axis, 21.4 cm behind the surface]

Incident rays that are parallel to the axis are bent by the surface so they lie along lines that diverge from a single point. Where is that point?

SOLUTION:

[ans: on the axis, 33.4 cm in front of the surface]

PROBLEM 6. A point source of light S is located at the center of a 20-cm long cylinder with spherical caps on it ends. The index of refraction is 1.85. What should the radius of each cap be to form an image on the axis 10 cm outside the cap?

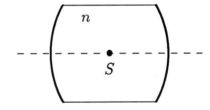

SOLUTION: Consider the right cap. Use $n_1/o + n_2/i = (n_2 - n_1)/r$, with $n_1 = 1.85$, $n_2 = 1$, $o = 10$ cm, and $i = 10$ cm. Solve for r.

[ans: 2.98 cm]

The negative value you obtained for r indicates the surfaces are concave with respect to the point source and are thus as drawn.

PROBLEM 7. The faces of a double concave lens each have a radius of 16 cm. The index of refraction is 1.66. What is the focal length?

SOLUTION: Use $1/f = (n-1)(1/r_1 - 1/r_2)$. The first surface is concave to the incoming light, so $r_1 = -16$ cm. The second surface is convex, so $r_2 = +16$ cm.

[ans: −12.1 cm]

A 1.5-cm long stick is placed perpendicularly to the optic axis, 3.0 cm in front of the lens. Where is the image?

SOLUTION: Use $1/o + 1/i = 1/f$.

[ans: 2.40 cm in front of the lens]

What is the length of the image?

SOLUTION: Use $h' = mh = -ih/o$.

[ans: 1.20 cm]

Is the image real or virtual? Is it erect or inverted?

SOLUTION: To find if it is real or virtual look at the sign of the image distance i. To find if it is erect or inverted look at the sign of the lateral magnification.

[ans: virtual; erect]

PROBLEM 8. The curved surface of a plano-convex lens has a radius of 9.0 cm. Its index of refraction is 1.66. What is its focal length?

SOLUTION: Use $1/f = (n-1)(1/r_1 - 1/r_2)$. The first surface is convex so the incoming light, so $r_1 = 9.0$ cm. The second surface is plane and so has an infinite radius of curvature. Take $1/r_2 = 0$.

[ans: 13.6 cm]

A 1.5-cm long stick is placed perpendicularly to the optic axis, 3.0 cm in front of the lens. Where is the image? What is its length? Is it real or virtual? Is it erect or inverted?

SOLUTION:

[ans: 3.88 cm in front of the lens; 1.92 cm; virtual; erect]

The stick is now placed 25 cm from the lens. Where is the image? What is its length? Is it real or virtual? Is it erect or inverted?

SOLUTION:

[ans: 30.0 cm behind lens; 1.80 cm; real; inverted]

You should know how to find images when more than one refracting or reflecting surface is present. Simply find the image formed by the first surface and use it as the object for the second. Here's an example.

PROBLEM 9. A fish bowl has spherical sides with an inner radius of 16.0 cm and an outer radius of 16.2 cm. A fish swims along the dotted line, which is horizontal and through the center of the bowl. It is viewed from a point on the line, outside the bowl. When it is at the point marked G, 6.0 cm from the inner wall, where is the image formed by paraxial rays? The index of refraction of water is 1.33 and the index of refraction of the glass sides is 1.90.

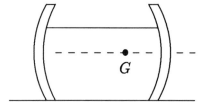

SOLUTION: First consider the surface formed by the water and the inner wall of the bowl. The object distance is $o_1 = 6.0$ cm and the surface radius is -16.0 cm. Set $n_1 = 1.33$ and $n_2 = 1.90$, then solve $n_1/o_1 + n_2/i_1 = (n_2 - n_1)/r$ for i_1. You should get -7.38 cm.

Now consider the surface formed by the outer wall and the air. The object distance is $o_2 = 0.2 - i_1 = 8.91$ cm. Take $n_1 = 1.90$, $n_2 = 1$, and $r = -16.2$ cm, then solve $n_1/o_2 + n_2/i_2 = (n_2 - n_1)/r$ for i_2.

[ans: 5.13 cm from the outer wall, in the water]

What is the lateral magnification of the fish?

SOLUTION: The magnification factor for the first surface is $m_1 = -i_1/o_1$. Refraction at the second surface magnifies the first image by the factor $m_2 = -i_2/o_2$. The overall magnification factor is $m = m_1 m_2 = i_1 i_2/o_1 o_2$.

[ans: 0.832]

Is the image real or virtual? Is it erect or inverted?

SOLUTION: To answer the first question look at the position of the final image. To answer the second look at the sign of the lateral magnification.

[ans: virtual; erect]

PROBLEM 10. Two thin lenses in contact act like a single thin lens. Suppose the focal lengths are f_1 and f_2. What is the focal length of the combination?

SOLUTION: Suppose an object is placed a distance o_1 in front of the combination. Use $1/o_1 + 1/i_1 = 1/f_1$ to find an expression for the image distance associated with the image formed by the first lens. This image becomes the object for the second lens. The object distance is $o_2 = -i_1$ since the lenses are thin. Substitute into $1/o_2 + 1/i_2 = 1/f_2$ and show that the result can be written $1/o_1 + 1/i_2 = 1/f$, where f is independent of o_1 and i_2. f is the focal length of the combination. Solve for f in terms of f_1 and f_2.

[ans: $f_1 f_2/(f_1 + f_2)$]

One of the lenses is a diverging lens with a focal length of -8.0 cm. What focal length lens should be placed in contact with it so the combination is converging with a focal length of $+8.0$ cm?

SOLUTION: Solve $1/f = f_1 f_2/(f_1 + f_2)$ for f_2.

[ans: 4.0 cm]

PROBLEM 11. An object is placed on the optic axis 25 cm from a converging lens with a focal length of 15 cm. Where should a diverging lens with a focal length of -6.0 cm be placed so the final image is real and 45 cm from the first lens?

SOLUTION: Use $1/o_1 + 1/i_1 = 1/f_1$ to find the position of the image formed by the first lens. Here $o_1 = 25$ cm and $f_1 = 15$ cm. If the distance between the two lenses is ℓ then the object distance for the second lens is $o_2 = \ell - i_1$. If d ($= 45$ cm) is the distance from the first lens to the final image then $i_2 = d - \ell$. Substitute these expressions into $1/o_2 + 1/i_2 = 1/f_2$ to obtain $1/(\ell - i_1) + 1/(d - \ell) = 1/f_2$, where f_2 is the focal length of the diverging lens. Solve for ℓ. You will obtain two solutions. One of them, however, corresponds a situation in which the image of the first lens is in front of the second lens and the final image is virtual.

[ans: 33.6 cm behind the first lens]

What is the lateral magnification for this object and lens system? Is the final image erect or inverted?

SOLUTION: The overall magnification is given by the product of the individual magnifications: $m = m_1 m_2 = i_1 i_2 / o_1 o_2$. Examine the sign of the result to determine if the final image is erect or inverted.

[ans: -4.36; inverted]

PROBLEM 12. A converging lens with an 8.0-cm focal length is placed 10 cm in front of a convex mirror with a 12-cm focal length. An object is 10 cm in front of the lens, as shown. What is the position of the final image?

SOLUTION: Let f_1 be the focal length of the lens, f_2 be the focal length of the mirror, and ℓ be the distance between the lens and the mirror. The object distance for the lens is $o_1 = 10$ cm. Use $1/o_1 + 1/i_1 = 1/f_1$ to find the position of the image formed by the lens. The object distance for the mirror is $o_2 = \ell - i_1$. You should find this to be negative, indicating that when light from the lens strikes the mirror it is converging to a point behind the mirror.

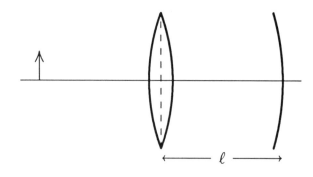

Now use $1/o_2 + 1/i_2 = 1/f_2$ to find the position of the image formed by the mirror. This image is the object for a second passage of the light through the lens. The object distance is $o_3 = \ell - i_2$. Finally use $1/o_3 + 1/i_3 = 1/f_1$ to find the final image position, relative to the lens. The final image is formed by light passing through the lens from right to left, so a positive result means the image is to the left of the lens and a negative result means it is to the right.

[ans: 1.74 cm to the right of the lens]

Is the image real or virtual?

SOLUTION:

[ans: virtual]

Suppose the object arrow is 1.0 cm long. What is the length of the image? Is the image erect or inverted?

SOLUTION: The overall magnification is given by $m = -i_1 i_2 i_3 / o_1 o_2 o_3$.

[ans: 1.39 cm; inverted]

You should understand the concepts of angular size and angular magnification and know how they are applied to optical instruments. These examples may help.

PROBLEM 13. An object with a lateral dimension of 2.0 cm, 91 cm away, is viewed with an unaided eye. What angle does the object subtend at the eye?

SOLUTION: If h is the size of the object and d is the distance between the object and the eye, then the angular size θ of the object is given by $\tan\theta = h/d$ or, because θ is so small, by $\theta = h/d$ in radians.

[ans: 2.20×10^{-2} rad]

A 12-cm focal length converging lens is placed between the object and the eye, 16 cm from the object (and 75 cm from the eye). What is the image distance and what is the lateral size of the image?

SOLUTION: Use $1/o + 1/i = 1/f$ to find the image distance. Use $h' = mh = -ih/o$ to find the image size.

[ans: 48 cm; −6.0 cm]

What angle does the image subtend at the eye?

SOLUTION: Use $\tan\theta' = |h'/d'|$, where d' is the distance from the image to the eye ($75 - 48 = 27$ cm).

[ans: 0.219 rad]

What is the angular magnification?

SOLUTION: Use $m_\theta = \theta'/\theta$.

[ans: 9.95]

PROBLEM 14. A 12.0-cm focal length converging lens is used as a magnifying glass to view a 1.00-cm long object, 11.9 cm away and perpendicular to the optic axis. What is the image distance and lateral size of the image?

SOLUTION:

[ans: -1430 cm; 120 cm]

The eye is placed close to the lens. What angle does the image subtend at the eye?

SOLUTION: Use $\tan\theta' = |h'/i|$.

[ans: 8.38×10^{-2} rad]

What angle does the object subtend at the unaided eye? Assume the object and eye are at the same positions as before.

SOLUTION: Use $\tan\theta = h/o$.

[ans: 8.38×10^{-2} rad]

What angle does the object subtend at the eye if the object is placed at the near point? Assume this is 25 cm away.

SOLUTION: Use $\tan\theta = h/25$ cm.

[ans: 4.00×10^{-2} rad]

What is the angular magnification of the lens, compared to viewing at the near point?

SOLUTION:

[ans: 2.10]

PROBLEM 15. A student builds a small telescope in the laboratory. The objective lens has a focal length of 45 cm and the ocular has a focal length of 3.0 cm. It is used to view an object with a lateral size of 15 cm, 150 m away. What is the angular size of the object to the unaided eye?

SOLUTION: Use $\tan\theta = h/d$, where h is the lateral size of the object and d is the distance from the object to the eye.

[ans: 1.0×10^{-3} rad]

Where is the image formed by the objective lens and what is its lateral size?

SOLUTION: Take the object distance to be 150 m. In principle, the length of the telescope should be subtracted from this distance but the object is so far away that the length of the telescope can be neglected.

[ans: between the lenses, 45.1 cm from the objective; 0.0451 cm]

Where is the final image and what is its lateral size?

SOLUTION: Assume the length of the telescope is equal to the sum of the focal lengths. The object distance for the second lens is then $f_1 + f_2 - i_1$, where i_1 is the image distance for the first lens.

[ans: 63.7 cm in front of the ocular; 1.00 cm]

Notice that essentially no magnification occurs. What angle is subtended by the final image and what is the angular magnification?

SOLUTION: Use $\tan\theta' = |h'/i_2|$, where h' is the lateral size of the final image and i_2 is the final image distance (relative to the ocular).

[ans: 0.0158 rad; 15.8]

This is not quite the ratio of the focal lengths because the image of the objective lens is not quite at its focal point, as it would be if the object were much further away.

III. MATHEMATICAL SKILLS

Geometric optics makes extensive use of the properties of similar triangles. Similar triangles are triangles:

1. such that every angle of one equal an angle of the other, and
2. every side of one is proportional to a side of the other

Two triangles are similar if any of the following are true:

1. any two angles of one are equal to angles of the other
2. an angle of one is equal to an angle of the other and the sides that form that angle in one triangle are proportional to the sides that form the equal angle in the other
3. every side of one is proportional to a side of the other
4. every side of one is either parallel or perpendicular to a side of the other

Use any of these conditions to prove that two triangles are similar. They then are guaranteed to have equal angles and proportional sides.

IV. NOTES

Chapter 45
INTERFERENCE

I. BASIC CONCEPTS

You studied the fundamentals of interference in Chapter 19. Now the results are specialized to light waves and applied to double-slit interference, thin-film interference, and the Michelson interferometer, an important instrument for measuring distances. Pay special attention to the role played by the distances traveled by interfering waves in determining their relative phase.

Fundamental ideas. To find the total disturbance when two or more waves are present you add the waves. Here you consider two sinusoidal waves with the same frequency and traveling in the same direction. The resultant is again a sinusoidal wave. Its amplitude may be as much as the sum of the amplitudes of the individual waves or as small as their difference, depending on their relative phase.

Phasors can be used to sum waves. The diagram on the right shows a phasor used to represent the electric field of a light wave at some point in space. Suppose the field is given by $E = E_0 \sin(\omega t)$, where ω is the angular frequency. The length of the phasor is proportional to the amplitude E_0 and its projection on the vertical axis is proportional to the wave itself. Label the angle ωt and indicate the direction of rotation of the phasor. Its angular speed is _____. Indicate the projection on the vertical axis.

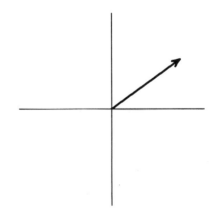

Now draw a second phasor, associated with the field $E = E_0 \sin(\omega t + \phi)$, where ϕ is 75°. Draw it with its tail at the head of the first phasor. Finally draw the phasor that represents the sum of the two waves. Label it E_m.

The phasors are redrawn to the right. Since two sides of the triangle are the same the two angles marked β are equal. Furthermore, since the third angle of the triangle is $180° - \phi$ and the angles of any triangle must sum to 180°, $\beta =$ _____. The dotted line is the perpendicular bisector of the phasor E_m. It is one side of a small right triangle with hypotenuse E_0 and one interior angle β. The other side is $E_m/2$. Thus, $E_m/2 = E_0 \cos \beta$ and, in terms of E_0 and ϕ,

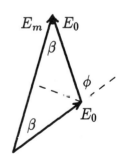

$$E_m = $$

Interference is said to be constructive if $E_m > $ _____ and destructive if $E_m < $ _____. Constructive interference is complete and $E_m = 2E_0$ if $\cos(\phi/2) = $ _____. This means ϕ is a multiple of _____ rad. Both waves have their maximum values at the same time. Destructive interference is complete and $E_m = 0$ if $\cos(\phi/2) = $ _____. This means ϕ is an odd multiple of _____ rad. At all times one wave is the negative of the other.

You should recognize that ϕ in the equations above is the phase of one wave *relative* to the other. Any constant can be added to the phase of both waves without changing the amplitude of the resultant. Said another way, if the waves are given by $E_0 \sin(\omega t + \phi_1)$ and $E_0 \sin(\omega t + \phi_2)$ then $\phi = \phi_1 - \phi_2$. Notice that ϕ can be replaced by $-\phi$ without changing the expression for the amplitude. This is because $\cos(\phi/2) = \cos(-\phi/2)$. Thus you may also take ϕ to be $\phi_2 - \phi_1$.

The electric fields of the two waves are vectors and strictly speaking the waves should be added vectorially. We assume, however, that the waves are plane polarized and the fields are along the same line. Then scalar addition can be used. This is a good approximation for most of the situations considered in this chapter.

Double-slit interference. A monochromatic plane wave is incident normally on a barrier with two slits, as diagramed in Fig. 4 of the text. You can find the intensity at any point on the screen by summing the Huygen wavelets emanating from points within the slits. Note that wavelets arrive at every point on the screen, not just those directly behind the slits. For now assume the slits are so narrow that only one wavelet from each slit is required. Furthermore assume the amplitude of each wavelet is the same at the screen. At the point P in the diagram the wavelet from the upper slit can be written $E_0 \sin(\omega t - kr_1)$ and the wavelet from the lower slit can be written $E_0 \sin(\omega t - kr_2)$, where $k = 2\pi/\lambda$ and λ is the wavelength. The phase difference is $\phi = k(r_2 - r_1) = 2\pi(r_2 - r_1)/\lambda$.

Carefully note that the phase difference arises because the wavelets travel different distances from the slits to the same point on the screen. Also note that because a plane wave is incident normally on the slits the phases of the wavelets are the same at the slits. This is changed if the wave is incident at some other angle or if the slits are covered by transparent materials with different indices of refraction. The expression for ϕ must then be modified.

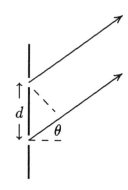

If the screen is far from the slits then rays from the slits to any point on the screen are nearly parallel to each other and the difference in the distances can easily be written in terms of the angle θ between a ray and a line normal to the barrier. The geometry is shown in the diagram to the right. Since the screen is far away the distance from the dotted line between the rays to P is the same along each ray. One of the interior angles of the right triangle formed by the dotted line, the barrier, and the lower ray is also θ. Label it. The lower ray is longer than the upper by $d \sin \theta$, where d is the slit separation. Indicate this distance on the diagram and label it with the expression for its length.

In terms of d, θ, and λ,

$$\phi = $$

The amplitude of the resultant wave at P is denoted by E_θ. In terms of E_0 and ϕ it is given by

$$E_\theta =$$

The intensity I_θ at P is proportional to the square of the amplitude and in terms of ϕ is given by

$$I_\theta =$$

where I_0 is the intensity associated with a single wave. The same result is obtained for a nearby screen if a lens is used to focus light on the screen.

You should recognize that light emanates from each slit in all forward directions but only the portions of wavefronts that follow the rays shown get to point P. Other portions of the wavefronts get to other places on the screen and for them θ has a different value. For different values of θ the phase difference is different and, as a result, so are the resultant amplitude and intensity. Alternating bright and dark regions (fringes) are seen on the screen. Centers of bright fringes (maxima of intensity) occur at points where the phases of the two wavelets differ by _____, where m is an integer. To reach these places one wave travels further than the other by a multiple of _____. That is,

$$d\sin\theta =$$

Centers of dark fringes (minima of intensity) occur at points where the phases of the two wavelets differ by _____, where m is an integer. To reach these places one wave travels further than the other by an odd multiple of _____. That is,

$$d\sin\theta =$$

Notice that complete constructive interference occurs at the point on the screen directly back of the slits, for which $\theta = 0$. The angular separation of the first minima on either side of the central maximum is a measure of the extent to which the intensity pattern is spread on the screen. This is given by $2\theta_0$, where $\sin\theta_0 = \lambda/2d$. As d _____ or λ _____ the angular separation increases and the pattern spreads. If $d = $ _____ then $\theta_0 = 90°$ and no bright fringes appear beyond the central maximum. For any given values of d and λ the number of maxima that are seen can be found by setting θ equal to _____ and solving $d\sin\theta = m\lambda$ for _____.

Coherence. Two sinusoidal waves are said to be coherent if the difference in their phases is constant. Light is emitted from atoms in bursts lasting on the order of _____ s and each burst has a different phase constant associated with it. Thus light from atoms emitting independently of each other is not coherent.

If an extended incoherent source, such as an incandescent lamp, is used to produce light that is incident on a double slit barrier the light from each atom goes through each slit and combines on the other side to form an interference pattern. Light from different atoms, however, form patterns that are shifted with respect to each other, the amount of the shift depending on the separation of the atoms in the source. Describe a technique that can be used to insure that only light from a small region of the source reaches the slits: _____

All the light from another type light source, a _____, is coherent, even though many different atoms are emitting simultaneously. When this light is incident on a double slit it produces an interference pattern without any additional apparatus.

Thin-film interference. If light is incident normally on a thin film, some is reflected from the front surface and some from the back. These two waves interfere and the resultant intensity may be quite large or quite small, depending on the thickness of the film.

To calculate the intensity for a given wavelength light you must be able to find the relative phases of the two waves. For normal incidence the wave reflected from the far surface travels _____ further than the wave reflected from the near surface. Here d is the thickness of the film. In addition, for one or both of the waves the medium beyond the surface of reflection may have a higher index of refraction than the medium of incidence. On reflection the wave then suffers a phase change of _____ radians.

The diagram shows a film of thickness d with a plane wave incident normally. You should be able to show that the phase difference for waves reflected from the two surfaces is given by $\phi = 4\pi d n_2/\lambda$ if $n_1 > n_2 > n_3$ or $n_1 < n_2 < n_3$. In the first case neither wave suffers a phase change on reflection. In the second they both suffer a phase change of π rad.

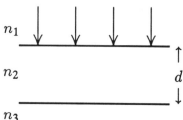

If $n_1 < n_2 > n_3$ or $n_1 > n_2 < n_3$ the phase difference is given by $(4\pi d n_2/\lambda) \pm \pi$. In the first case the wave reflected from the front surface suffers a phase change of π rad but the wave reflected from the back surface does not. In the second case the wave reflected from the back surface suffers a phase change of π but the wave reflected from the front surface does not. Notice that the wavelength for the film must be used to calculate the difference in phases. It is given by λ/n_2, where n_2 is the index of refraction of the film and λ is the wavelength in vacuum. Since you may always add 2π to any phase, it is immaterial whether the sign in front of π is plus or minus.

When white light (a combination of all wavelengths of the visible spectrum) is incident on a thin film the reflected light is colored. It consists chiefly of those wavelengths for which interference of the two reflected waves produces a maximum or nearly a maximum of intensity. Those wavelengths for which interference produces a minimum are missing. This phenomena accounts for the colors of oil films and soap bubbles, for example.

Suppose a film with thickness d has an index of refraction n and is in air (with a smaller index of refraction). Monochromatic light is incident normally on a surface. There is a change in phase on reflection at the _____ surface but not at the _____. The difference in phase of the two reflected waves is $\phi =$ _____. Interference maxima occur for wavelengths given by $\lambda =$ _____ and interference minima occur for wavelengths given by $\lambda =$ _____, where m is an integer. How are these results changed if the film is deposited on glass with a higher index of refraction? _____
_____

When monochromatic light is incident on a thin film with a varying thickness, like a wedge, interference produces bright and dark bands. Bright bands appear in regions for which the film thickness is such that the phase difference between the two reflected waves is close to _____ rad; dark bands appear in regions for which the film thickness is such that the phase difference is close to _____ rad. Here m is an integer.

The Michelson interferometer. The diagram on the right is a schematic drawing of the instrument. Light from a source S is incident on a half-silvered mirror M, where the beam is split. Half is reflected to mirror M_2 where it is reflected back through M to the eye at E. The other half is transmitted to mirror M_1 where it is reflected back to M. There half is reflected to the eye. The two beams reaching the eye interfere. To measure a distance, one of the mirrors (M_2 say) is moved, thereby changing the interference pattern at the eye. As the mirror moves alternately bright and dark fringes are seen. The number of fringes that appear are counted and related to the distance moved by the mirror.

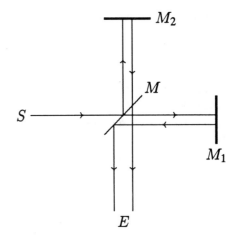

Suppose the distance traveled by the light that strikes mirror M_1 is d_1 and the distance traveled by the light that strikes mirror M_2 is d_2. If the wavelength of the light is λ then the difference in phase of the two waves at the eye is $\phi =$ _____. Suppose d_1 and d_2 happen to have values so a maximum of intensity is produced at the eye. Then mirror M_2 is moved until the intensity is again a maximum. In terms of the wavelength the distance it moved must have been _____. Interferometers have been used to measure distances with an accuracy of about _____ the wavelength of the light used.

The Michelson interferometer has been used extensively in attempts to measure the speed of the earth relative to the ether. Explain what the ether was thought to be: _____

_____

Failure to detect motion led to the abandonment of the ether as a medium for the propagation of light and contributed to the acceptance of the special theory of relativity.

II. PROBLEM SOLVING

You should know how to compute the phase difference ϕ for waves passing through a two-slit barrier and how to use $I_\theta = 4I_0 \cos^2(\phi/2)$ to compute the intensity. You should be able to use the condition $\phi = 2\pi m$ to find the angle θ for which the intensity is a maximum and the condition $\phi = (2m+1)\pi$ to find the angle for which the intensity is a minimum.

PROBLEM 1. Plane waves of red cadmium light (wavelength = 644 nm) are incident normally on a two-slit barrier with a slit separation of 0.015 mm. Consider two wavelets, one from each slit, that travel at an angle of 3.00° to the normal, reaching the same point P on a far-away screen. What is the difference in the distance they travel from the slits to the screen?

SOLUTION: Use $r_2 - r_1 = d\sin\theta$.

[ans: 7.85×10^{-4} mm]

What is their phase difference at the screen?

SOLUTION: Use $\phi = 2\pi(r_2 - r_1)/\lambda$.

[ans: 7.66 rad]

You may subtract multiples of 2π to obtain a result that is less than 2π. Thus 7.66 rad is equivalent to $7.66 - 2\pi = 1.38$ rad.

If no barrier were present the intensity at P would be 5.0×10^{-10} W/m². What is the intensity at P with the barrier in place?

SOLUTION: Use $I_\theta = 4I_0 \cos^2(\phi/2)$, with $I_0 = 5.0 \times 10^{-10}$ W/m².

[ans: 1.19×10^{-9} W/m²]

The intensity with the barrier is slightly more than double the intensity without. Other points on the screen receive much lower intensity.

PROBLEM 2. Plane waves of red cadmium light (wavelength = 644 nm) are incident normally on a barrier with two narrow slits, a distance 0.0150 mm apart. At what angles to the normal are the first 3 intensity maxima on one side of the central maximum?

SOLUTION: Since $\phi = (2\pi d/\lambda)\sin\theta$ for any angle θ and $\phi = 2\pi m$ for a maximum of intensity, the angles at which maxima occur satisfy $\sin\theta = m\lambda/d$. Set $m = 1, 2,$ and 3, in turn, and for each of these values solve for θ.

[ans: 2.46° (4.29×10^{-2} rad); 4.92° (8.60×10^{-2} rad); 7.40° (0.129 rad)]

The interference pattern is viewed on a screen 1.50 m from the slits. How far from the central point on the screen are each of these maxima?

SOLUTION: If D is the slit-to-screen distance and y is the distance from the center point to a maximum then $\tan\theta = y/D$ and $y = D\tan\theta$.

[ans: 0.0644 m; 0.129 m; 0.195 m]

What is the phase difference for the two waves that travel outward at the angle halfway between the normal and the angle for the first minimum?

SOLUTION: The angle for the first minimum is given by $\sin\theta_0 = \lambda/2d$. Find its value and set $\theta = \theta_0/2$. The phase difference is given by $(2\pi d/\lambda)\sin\theta$.

[ans: 1.57 rad]

Compared to the intensity for $\theta = 0$ what is the intensity at the angle halfway between the normal and the angle for the first minimum?

SOLUTION: Use $I_\theta/4I_0 = \cos^2(\phi/2)$, where ϕ is the phase difference you found above.

[ans: 0.500 times as great]

What wavelength light, normally incident on the slits, has it first interference maximum to one side of the central maximum at the same place as red cadmium light has its first interference minimum?

SOLUTION: The first minimum associated with the red light occurs for $\sin\theta = \lambda/2d$, where $\lambda = 644$ nm. The first maximum associated with the other light occurs for $\sin\theta = \lambda'/d$, where λ' is the unknown wavelength. Since these angles are the same, $\lambda/2d = \lambda'/d$ or $\lambda' = \lambda/2$.

[ans: 322 nm]

Does the tenth interference maximum to one side of the central maximum for this light coincide with the tenth interference minimum for the cadmium light? At what angles do these extrema occur?

SOLUTION: For the 322-nm light $\sin\theta' = 10\lambda'/d$. For the cadmium light $\sin\theta = 9.5\lambda/d$. Solve for θ' and θ.

[ans: 12.4°; 24.1°]

If the slit separation is increased the interference pattern narrows. The pattern also depends on the index of refraction of the medium in which the light is propagating. Here are some problems to demonstrate these effects.

PROBLEM 3. Suppose plane waves with a wavelength of 644 nm are incident normally on a barrier with two narrow slits, 0.0300 mm apart. At what angles to the normal are the first 3 intensity maxima?

SOLUTION:

[ans: 1.23° (2.15 × 10⁻² rad); 2.46° (4.29 × 10⁻² rad); 3.69° (6.44 × 10⁻² rad)]

For what wavelength light does the interference pattern for this slit spacing match the pattern produced by 644-nm light incident on slits with a spacing of 0.015 mm?

SOLUTION: The quantity λ/d must be the same to obtain the same pattern. Since d was doubled, λ must also be doubled.

[ans: 1290 nm]

PROBLEM 4. Suppose a two-slit barrier with a slit separation of 0.015 mm and a screen are placed entirely under water with an index of refraction of 1.33. A source of plane light waves with a wavelength of 644 nm (in vacuum) is also placed under water, so the waves are incident normally on the barrier. Find the angles with the normal of rays from the slits for the first 3 maxima on one side of the central maximum.

SOLUTION: For the wavelength in water use $\lambda' = \lambda/n$, where λ is the wavelength in vacuum and n is the index of refraction of water. Maxima occur at values of the angle θ for which $\sin\theta = m\lambda/nd$, where m is an integer.

[ans: 1.85° (3.23 × 10^{-2} rad); 3.70° (6.46 × 10^{-2} rad); 5.56° (9.70 × 10^{-2} rad)]

Because the wavelength is shorter in water than in a vacuum, the interference pattern narrows. Each maximum occurs at a smaller angle than when the experiment is done in a vacuum.

For the problems above plane waves are incident normally on the barrier, so the Huygen wavelets start in phase from the slits. If the angle of incidence is not 0 or if materials with different indices of refraction are placed at the slits then the wavelets do not start in phase and their initial phase difference must be taken into account in calculating the intensity at the screen. Here are some examples.

PROBLEM 5. Plane waves of red cadmium light (wavelength = 644 nm) are incident at 6.00° to the normal on a barrier with two narrow slits, 0.015 mm apart. What is the difference in phase of the two wavelets that leave the slits along normals to the barrier?

SOLUTION: The difference in phase now arises from the difference in distance traveled *before* reaching the slits. This distance is $d\sin\theta_i$, where d is the slit separation and θ_i is the angle of incidence. After passing through the slits the waves travel the same distance to the screen, so the phase difference at the screen is given by $\phi = (2\pi d/\lambda)\sin\theta_i$. You may subtract a multiple of 2π from your answer to obtain a result that is less than 2π.

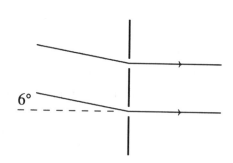

[ans: 15.3 rad (2.73 rad)]

Compared to the intensity in the absence of a barrier what is the intensity on the center line?

SOLUTION: Use $I_\theta/I_0 = 4\cos^2(\phi/2)$.

[ans: 0.167]

At what angle above the normal is the first maximum of intensity?

SOLUTION: If the rays leave the slits at the angle θ then the phase difference at the screen is given by $\phi = (2\pi d/\lambda)(\sin\theta_i + \sin\theta)$. For a maximum of intensity this must be $2\pi m$, where m is an integer. Thus $\sin\theta =$

$m\lambda/d - \sin\theta_i$. You must pick m so $\sin\theta$ has the smallest possible positive value, then solve for θ.

[ans: 1.39°]

PROBLEM 6. A 4.5×10^{-7}-m thick film with an index of refraction of 2.30 is placed in front of the upper slit of a pair, separated by 0.015 mm. Plane waves of cadmium light with a wavelength of 644 nm are incident normally on the slits. What is the phase difference at the screen of wavelets that exit the slits along normals? Assume the amplitudes of the wavelets are the same.

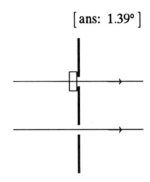

SOLUTION: Light waves through the two slits are in phase at points that are a distance T in front of the slits, where T is the thickness of the film. As it goes through the film light at the upper slit advances in phase by $2\pi T/\lambda' = 2\pi nT/\lambda$, where λ' is the wavelength of the light in the film and n is the index of refraction of the film. Light through the lower slit goes the same distance but in air. It advances in phase by $2\pi T/\lambda$. The difference in phase at the slits is $2\pi(n-1)T/\lambda$. Since we are considering light that travels straight ahead no further changes in the phase difference occur.

[ans: 5.71 rad (0.585 rad)]

Compared to the intensity in the absence of a barrier what is the intensity at the point on the center line directly behind the slits?

SOLUTION: Calculate $I_\theta/I_0 = 4\cos^2(\phi/2)$.

[ans: 2.81 times as great]

At what angle above the normal to the barrier is the first minimum in intensity? At what angle is the first maximum?

SOLUTION: If you calculate the phase difference as the phase of the lower wavelet minus the phase of the upper then $\phi = (2\pi d/\lambda)\sin\theta - 2\pi(n-1)T/\lambda$. For a minimum this must be $\pi(2m+1)$. For a maximum it must be $2\pi m$. In each case chose m to be the integer for which $\sin\theta$ has the smallest possible positive value.

[ans: 1.46°; 0.225°]

Here are some problems dealing with the interference of waves reflected from thin films. Don't forget to examine the medium beyond each reflecting surface to see if there is a change in phase of π rad.

PROBLEM 7. Plane waves of white light are incident normally on a 3.4×10^{-7}-m thick soap bubble. The medium on both sides is air. What visible-light wavelengths are most strongly reflected? What color is the bubble? The index of refraction of the bubble is 1.33.

SOLUTION: Since there is a change of phase on reflection at the front surface but not at the back, the phase difference of the two reflected waves is given by $\phi = (4\pi d n/\lambda) + \pi$, where d is the thickness of the film and n is its index of refraction. For a maximum this must be $2\pi m$, where m is an integer. Thus $\lambda = 4d/(2m-1)$. Try various values of m to find values of λ in the visible region (about 400 nm to about 700 nm).

[ans: 603 nm; orange]

For what visible-light wavelengths does interference produce a minimum in reflected intensity?

SOLUTION: For a minimum $\phi = \pi(2m+1)$, where m is an integer. Thus $\lambda = 2dn/m$.

[ans: 452 nm, blue]

Now suppose the water film is on a glass surface, with index of refraction 1.8. Do any visible-light wavelengths produce complete constructive interference on reflection? If so, which?

SOLUTION: Now both waves undergo a change of phase by π on reflection, so their phase difference is given by $\phi = 4\pi d n/\lambda$. For a maximum this must be $2\pi m$, where m is an integer. Thus $\lambda = 2dn/m$.

[ans: 452 nm]

For what visible-light wavelengths is interference completely destructive?

SOLUTION:

[ans: 603 nm]

PROBLEM 8. A thin film with index of refraction 1.25 is in the form of a wedge 15.0 cm long (ℓ in the diagram) and 6.70×10^{-4} cm thick at its thickest end (T in the diagram). It rests on glass with an index of refraction 1.83. If it is illuminated by plane waves with wavelength 567 nm, at normal incidence, how far from the narrow end do reflection maxima occur? (The diagram greatly exaggerates the angle of the wedge.)

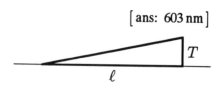

SOLUTION: Changes in phase on reflection occur at both surfaces so the difference in phase of the two reflected waves is given by $\phi = 4\pi d n/\lambda$, where d is the thickness of the film and n is its index of refraction. For an intensity maximum to occur this must be $2\pi m$, where m is an integer. Thus $4\pi d n/\lambda = 2\pi m$ or $d = m\lambda/2n$.

If x is the distance of the maximum from the vertex of the film then $x/\ell = d/T$ or $d = xT/\ell$. Equate the two expressions for d and solve for x. You should get $x = m\ell\lambda/2nT$.

[ans: every 5.08 mm, starting 5.08 mm from the vertex]

Interference phenomena occur in circumstances other than the passage of light through a doubled-slitted barrier. The principle is the same in each situation, however. Find the phase difference of the two waves and use that to calculate the amplitude or intensity or to look for maxima and minima. Often all or part of the phase difference is due to a difference Δd in the distance traveled. Here's an example.

PROBLEM 9. A point source S of light with wavelength λ is a distance y_0 above a mirror M and a distance ℓ to the left of a screen. Two waves arrive at point P on the screen: one directly from the source and one after reflection from the mirror. They travel different distances. Find an expression in terms of y, y_0, ℓ, and λ, for their phase difference.

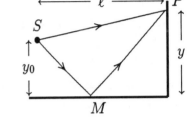

SOLUTION: The direct wave travels a distance d_1 given by $d_1^2 = \ell^2 + (y - y_0)^2$. The reflected wave appears to come from the image of S, a distance y_0 below the mirror. It travels a distance d_2 that is given by $d_2^2 = \ell^2 + (y + y_0)^2$. The reflected wave suffers a change of phase of π on reflection.

[ans: $(2\pi/\lambda)\left[\sqrt{\ell^2 + (y + y_0)^2} - \sqrt{\ell^2 + (y - y_0)^2}\right] - \pi$]

Show that if ℓ is much larger than either y or y_0, then the phase difference is given by $(4\pi y y_0/\ell\lambda) - \pi$.

SOLUTION: Use the binomial expansion $(A + B)^n = A^n + nA^{n-1}B + \ldots$. Take $A = \ell^2$ and $n = 1/2$. For the first radical take $B = (y + y_0)^2$ and for the second take $B = (y - y_0)^2$.

If $y_0 = 0.17$ mm and $\ell = 2.4$ m find the coordinates y of the first 3 maxima produced on the screen by a 500 nm source.

SOLUTION: Set $\phi = 2\pi m$, where m is an integer, then solve for y.

[ans: 1.76 mm, 5.29 mm, 8.82 mm]

III. MATHEMATICAL SKILLS

A few problems deal with the addition of two waves with different amplitudes. Suppose one wave is given by $E_1 \sin(\omega t)$ and the second by $E_2 \sin(\omega t + \phi)$. The phasor diagram is shown to the right. According to the law of cosines, $E_m^2 = E_1^2 + E_2^2 - 2E_1 E_2 \cos \alpha$. Since $\alpha = 180° - \phi$ and $\cos(180° - \phi) = -\cos \phi$, this can be written

$$E_m^2 = E_1^2 + E_2^2 + 2E_1 E_2 \cos \phi$$

Use this equation to calculate the amplitude of the resultant.

If the waves are in phase then $\phi = 0$, $\cos \phi = 1$, and $E_m^2 = E_1^2 + E_2^2 + 2E_1 E_2 = (E_1 + E_2)^2$. If the waves are 180° out of phase then $\cos \phi = -1$ and $E_m^2 = E_1^2 + E_2^2 - 2E_1 E_2 = (E_1 - E_2)^2$. These results should agree with what you expect.

If $E_1 = E_2$ then the general expression reduces to $E_m^2 = 2E_1^2(1 + \cos \phi)$. Now $\cos \phi = \cos(\phi/2 + \phi/2) = \cos^2(\phi/2) - \sin^2(\phi/2) = 2\cos^2(\phi/2) - 1$, where $\cos^2(\phi/2) + \sin^2(\phi/2) = 1$ was used. Thus $E_m^2 = 4E_1^2 \cos^2(\phi/2)$ and $E_m = 2E_1 \cos(\phi/2)$, in agreement with the result you obtained previously.

IV. NOTES

Chapter 46
DIFFRACTION

I. BASIC CONCEPTS

Diffraction is the bending of light as it passes by the edge of an object or through an opening in a barrier. Pay attention to the description of this phenomenon in terms of Huygen wavelets. You should also understand how interference of the wavelets produces bright and dark fringes in the diffraction pattern of an object or opening. When you study diffraction by a double slit pay close attention to the relationship between the single-slit diffraction pattern and the double-slit interference pattern. In particular, know what characteristics of the slits determine the maxima and minima in each pattern.

Diffraction. The phenomenon is described in terms of Huygen wavelets. If there is no barrier the wavelets combine to produce a wave that continues moving in its original direction. If a barrier blocks some of the wavelets then those that are not blocked combine to produce a wave that moves into the geometric shadow. In addition, if the light is coherent, the Huygen wavelets interfere to form a series of bright and dark bands, called the diffraction pattern of the object.

If the viewing screen is very close to an object or aperture, the pattern on the screen is the _____ of the object or aperture. You will consider chiefly situations for which the screen is far away. Then the Huygen wavelets reaching any point on the screen can be considered to be plane waves traveling along rays from near the object or through the aperture to the point on the screen being considered. This is called _____ diffraction. The intermediate case is called _____ diffraction.

Both interference and diffraction arise from the interference of Huygen wavelets and are intimately related. A phenomenon is said to be due to interference if _____

_____

A phenomenon is said to be due to diffraction if _____

_____

Single-slit diffraction. Consider plane waves of monochromatic light incident normally on a barrier with a single slit of width a, as shown in Fig. 5 of the text. To find the intensity at a point P on a screen add the Huygen wavelets emanating from the slit and take the limit as the number of wavelets becomes infinite. We suppose the viewing screen is far away from the slit and consider parallel rays as shown in Fig. 5a.

Each wavelet has an infinitesimal amplitude and a phase that differs from that of a neighboring wavelet by an infinitesimal amount. The phasors form an arc of a circle, as shown to the right. The angle ϕ is the difference in phase of wavelets from the edges of the slit. It is also the angle subtended by the arc at its center. Look at Fig. 11 of the text. If the arc has radius R then its length is given by $E_m = R\phi$, for ϕ in radians. E_m is the sum of the amplitudes of all the wavelets and so is the amplitude at P if they all have the same phase.

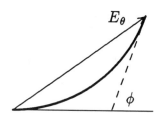

E_θ, the amplitude at P, is the chord of the arc. A little geometry shows that $E_\theta = 2R\sin(\phi/2)$. Eliminating R between these two expressions yields an expression for E_θ in terms of E_m and ϕ:

$$E_\theta =$$

Differences in the phases come about because the wavelets travel different distances to P. If the screen is far away and the slit width is a then the lower wavelet travels $a\sin\theta$ further than the upper wavelet. Label the slit width and mark the distance $a\sin\theta$ on the diagram to the right. If the wavelength of the light is λ then the difference in the phases of these two wavelets is given by

$$\phi =$$

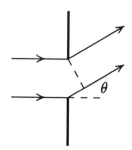

The expression for the amplitude is sometimes written in terms of $\alpha = \phi/2$, rather than in terms of ϕ. Then the amplitude at P is given by

$$E_\theta =$$

where, in terms of a and θ,

$$\alpha =$$

The intensity at P is given by

$$I_\theta =$$

where I_m is the intensity when all wavelets are in phase.

Carefully study Fig. 12 of the text. It shows the intensity as a function of θ for several values of the slit width a. The most prominent feature is a broad central maximum, centered at $\theta = 0$. Note particularly that $\theta = 0$ corresponds to an intensity maximum, not a minimum. In the limit as $\alpha \to 0$, $(\sin\alpha)/\alpha \to 1$. If the slit width is small the central maximum spreads to cover the entire screen and no zeros of intensity occur. For a wide slit the central maximum is narrow and is followed on both sides by secondary maxima. These are narrower and of considerably less intensity than the central maximum. They are roughly midway between zeros of intensity. The number that appear depends on the slit width.

In the space below draw phasor diagrams corresponding to the peak of the central maximum, the first zero, and the second zero to one side of the central maximum. Remember that the total arc length is the same in all cases.

CENTRAL MAXIMUM   FIRST ZERO   SECOND ZERO

The first zero on either side of the central maximum corresponds to $\alpha = \pm\pi$. The angles θ for which these occur are given by

$$\sin\theta = $$

For these angles every wavelet from the upper half of the slit can be paired with a wavelet from the bottom half, emanating from a point $a/2$ away. The phase difference of two wavelets in a pair is _____ and these wavelets sum to _____ .

Notice that as a decreases the angle θ for the first minimum increases. The central maximum is broader for narrow slits than for wide slits and diffraction is more pronounced. If a is greater than λ no zeros of intensity occur ($\sin\theta$ cannot be greater than 1) and the entire screen is within the central maximum.

In the space to the right sketch the phasor diagram corresponding to the first secondary maximum to one side of the central maximum. In terms of the diagram explain why the intensity at a secondary maximum is less than that at the central peak: _____
_____
_____
_____

Double-slit diffraction. Consider two identical slits, each of width a, with a center-to-center separation d. Any point P on a screen is reached by a wave from each slit, the resultant of the Huygen wavelets from that slit. If light is incident normally on the slits and the observation point is far away the wave that reaches it from each slit has amplitude $E_m(\sin\alpha)/\alpha$, where $\alpha = (\pi a/\lambda)\sin\theta$, and the two waves differ in phase by $(2\pi d/\lambda)\sin\theta$. When they are combined the resultant amplitude is

$$E_\theta = $$

where $\beta = (\pi d/\lambda)\sin\theta$.

As you learned in the last chapter interference minima occur for $\beta = (2m+1)\pi/2$ or $\sin\theta = (2m+1)\lambda/2d$. Diffraction minima occur for $\alpha = m\pi$ or $\sin\theta = m\lambda/a$. In each case m is an integer but it may have different values in the two expressions. Since d must be larger than a, the interference minima must be closer together than the diffraction minima. The single-slit diffraction pattern forms an envelope, with the interference pattern inside. Study Fig. 18 of the text.

For any double-slit situation there are 3 parameters you must be aware of: the wavelength λ, the slit width a, and the center-to-center slit separation d. The ratio a/λ controls the width of the central diffraction maximum. It extends from $-\sin^{-1}(\lambda/a)$ to $+\sin^{-1}(\lambda/a)$. The ratio also controls the positions of the secondary diffraction maxima, if any. The ratio d/λ controls the angular positions of the interference maxima and minima.

Finally, the ratio d/a controls how many interference maxima fit within the central diffraction maxima or any of the secondary diffraction maxima. This number is independent of the wavelength. As the wavelength increases the central diffraction maximum widens but the interference pattern also spreads, with the result that just as many interference fringes are within the central diffraction maximum.

Diffraction from a circular aperture. When plane waves pass through a circular aperture and onto a screen a diffraction pattern is formed there. The pattern consists of a bright central disk, followed by a series of alternating dark and bright rings. The first minimum occurs at an angle θ, measured from the normal to the aperture and given by

$$\sin \theta =$$

where d is the diameter of the aperture. This angle can be used as a measure of the angular size of the central disk.

Stars are effectively point sources of light and lenses act like circular apertures. The image of a star formed by a lens is not a point but is broadened by diffraction to a disk and rings. Two stars do not form distinct images if the central disks of their diffraction patterns overlap too much. Describe the Rayleigh criterion for the resolution of two far-away point sources:

_____

_____

II. PROBLEM SOLVING

You should understand how a single-slit diffraction pattern depends on the wavelength of the light and the width of the slit. Here is an example.

PROBLEM 1. A plane wave with a wavelength of 500 nm is incident normally on a slit with a width of 0.025 mm. What is the angular spread of the central maximum, measured from the first minimum on either side?

SOLUTION: The minima occur at $\alpha = \pm\pi$ or $(\pi a/\lambda)\sin\theta = \pm\pi$. Thus $\sin\theta = \pm\lambda/a$. Solve for θ. The angular spread is twice the value you obtain.

[ans: 2.29°]

The same light is incident normally on a slit with half the width of that above. Now what is the angular spread of the central maximum?

SOLUTION:

[ans: 4.58°]

Twice the spread for the narrow slit.

Suppose now that a plane wave with a wavelength of 700 nm is incident normally on each of the two slits described above. For this light what is the spread of the central maximum in each case?

SOLUTION:

[ans: 3.21°; 6.42°]

Notice that in each case the central maximum is wider than before.

Light with wavelength λ is incident normally on a slit with width a. What wavelength light should be incident on a slit with width $2a$ to obtain the same intensity pattern?

SOLUTION: Note that only the ratio λ/a is significant, not the values of λ and a individually.

[ans: 2λ]

PROBLEM 2. A plane wave with a wavelength of 500 nm is incident normally on a slit with a width of 0.025 mm. Compare the angular spread of the central maximum with the angular spread of a secondary maximum adjacent to it.

SOLUTION: The angular spread of the central maximum was calculated above. It is 2.29°. Zeros of intensity occur for $(\pi a/\lambda)\sin\theta = m\pi$. You want the angular separation $\Delta\theta$ of the zeros associated with $m = 1$ and $m = 2$.

[ans: the secondary maximum is half as wide as the central maximum]

Suppose the viewing screen is 2.6 m from the slit. What is the linear width of the central maximum on the screen? What is the linear width of a secondary maximum adjacent to it?

SOLUTION: Rays from the slit making an angle θ with the normal strike the screen a distance $y = D\tan\theta$ from the central peak. Here D is the distance from the slit to the screen. For small angles $\tan\theta$ may be replaced by θ in radians. Then the linear width is given by $\Delta y = D\Delta\theta$.

[ans: central max: 10.4 cm; secondary max: 5.2 cm]

PROBLEM 3. A plane wave is incident normally on a 0.015-mm wide slit. The central maximum exactly covers the forward direction. What is the wavelength? In what part of the electromagnetic spectrum does this wave lie?

SOLUTION: You want the first minimum on each side of the central maximum to occur at $\theta = 90°$. Since $\sin 90° = 1$ this means $\lambda = a$.

[ans: 1.50×10^{-5} m; infrared]

For what wavelength does the central maximum exactly cover the region from $-45°$ to $+45°$?

SOLUTION: Use $\sin\theta = \lambda/a$.

[ans: 1.06×10^{-5} m]

If 500-nm light is incident normally on the slit how many complete secondary bright fringes are produced on each side of the central maximum?

SOLUTION: Complete means that the adjacent zero is at $\theta = 90°$ or less. Count the number of zeros from $\theta = 0$ to $\theta = 90°$. Set $(\pi a/\lambda)\sin\theta = m\pi$, with $\theta = 90°$, and solve for m. If the result is not an integer, round it down to the nearest integer. The number of bright secondary fringes on one side is $m - 1$.

[ans: 29]

PROBLEM 4. Find the smallest two positive angles α for which $(\sin\alpha)/\alpha$ is a local maximum.

SOLUTION: Set the derivative of $(\sin\alpha)/\alpha$ equal to 0. You should find $\tan\alpha = \alpha$. Solve this by systematic trial and error. Don't forget that α must be in radians. One solution is near $3\pi/2$ rad and another is near $5\pi/2$ rad.

[ans: 4.493 rad; 7.725 rad]

Suppose a 500-nm plane wave is incident normally on a 0.020-mm slit. At what angle θ to the normal are each of the first two secondary maxima on one side of the central maximum?

SOLUTION: Solve $\alpha = (\pi a/\lambda)\sin\theta$ for θ, taking α to be each of the results above, in turn.

[ans: 2.05°; 3.52°]

Note that these are not precisely halfway between minima. For each of these maxima find the ratio of the intensity to the intensity at $\theta = 0$.

SOLUTION: Use $I_\theta/I_m = (\sin^2 \alpha)/\alpha^2$.

[ans: 4.72×10^{-2}; 1.65×10^{-2}]

Notice that the intensity at the secondary maxima is much less than at the central maximum.

Just like the two-slit interference patterns of the last chapter, a single-slit diffraction pattern depends on the index of refraction of the material between the slit and screen. Instead of the wavelength λ for a vacuum use the wavelength $\lambda' = \lambda/n$ for the medium. Here n is its index of refraction.

PROBLEM 5. A light source generates plane waves with a wavelength of 500 nm in vacuum. The waves are incident normally on a 0.025-mm wide slit and the pattern is obtained on a screen far away. The medium between the slit and screen is water with an index of refraction of 1.33. What is the angular separation of the minima adjacent to the central maximum?

SOLUTION: The wavelength in water is $\lambda' = \lambda/n = 500/1.33 = 376$ nm. Use $\sin \theta = \lambda'/a$ and double the value you obtain for θ.

[ans: $1.72°$]

Notice that the central maximum has decreased in width compared to the same situation in a vacuum.

Double-slit diffraction problems add the new ingredient of interference. The following problem should help you understand the relationship between the diffraction and interference patterns.

PROBLEM 6. A 430-nm plane wave is incident normally on a barrier with two slits, each 0.015 mm wide, separated by 0.040 mm. How many interference maxima fall within the central diffraction maximum?

SOLUTION: The angular limits of the central diffraction maximum are given by $\sin \theta = \lambda/a$, where a is the slit width. The interference maxima are given by $m\lambda = d \sin \theta$. Use one of these equations to substitute for $\sin \theta$ in the other, then solve for m. You should get $m = d/a$. Round the numerical value for m down to the next smaller integer. This value gives the number of maxima on one side of the central interference maximum. There is an equal number on the other side and the central maximum itself.

[ans: 5]

PROBLEM 7. A double-slit diffraction experiment is designed so that 3 interference maxima occur inside the central diffraction maximum and interference minima coincide with the first diffraction minima on either side of the central maximum. Each slit is 0.020 mm wide. What is the center-to-center slit separation?

SOLUTION: You want the second interference minimum to coincide with the first diffraction minimum. For the interference minimum $(\pi d/\lambda)\sin\theta = (2m+1)\pi/2$, with $m = 1$. So $\sin\theta = (3/2)(\lambda/d)$. For the diffraction minimum $\sin\theta = \lambda/a$. Use one of these equations to eliminate $\sin\theta$ from the other, then solve for d.

[ans: 0.030 mm]

Note that the result does not depend on the wavelength of the light.

Some problems deal with the Rayleigh criterion for the resolution of distant point sources. Here's one.

PROBLEM 8. If the wavelength of light from a star is 450 nm what is the angular diameter of the central bright disk of the image formed by a lens with a 5.0-cm diameter?

SOLUTION: Use $\sin\theta = 1.22\lambda/d$, where d is the diameter of the lens, to calculate the angle θ of the first minimum in the diffraction pattern. The angular diameter of the disk is twice the value you obtain.

[ans: 6.29×10^{-4} degrees]

If the lens has a focal length of 45 cm what is the diameter of the central disk of the image?

SOLUTION: Since the star is far away its image is formed in the focal plane of the lens and the radius R of the first minimum is given by $\tan\theta = R/f$, where f is the focal length of the lens.

[ans: 2.47×10^{-6} m]

III. COMPUTER PROJECTS

The text deals with the diffraction of light by a single slit for the special case when the viewing screen is far away. Then the Huygen wavelets emanating from the slit are essentially plane waves. When the viewing screen is close to the slit or when the slit is wide you must take into account the true spherical nature of the wavelets. The diffraction pattern for such wavelets cannot be described analytically but you can use a computer to describe it numerically. As the slit is widened you will be able to see the diffraction pattern change into the geometrical image of the slit.

The diagram shows a plane wave impinging on a single slit from the left. A spherical wave emanates from each point in the slit. One of them is shown. The viewing screen is to the right, a distance x from the slit. Take the origin to be at the center of the slit and let y' be the coordinate of the point within the slit from which the spherical wave emanates. You will calculate the light intensity at the point on the screen a distance y from the center line.

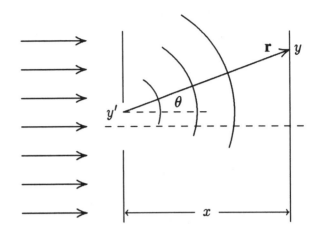

Because we want an expression that is valid no matter what the slit-to-screen distance, we will use an expression for the wavelet emanating from y' that is different from the expression used in the text. In particular, we take the wavelet to have the form

$$E = \frac{B}{r}(1 + \cos\theta)\sin(kr - \omega t),$$

where r is the distance from the wavelet source to the observation point and θ is the angle between the ray from the source to the observation point and the x axis. B is a constant and is chosen so that in the absence of an obstacle the wavelets sum to produce a plane wave with the same amplitude as the original plane wave. Since we are concerned only with the relative intensity we choose its value for computational convenience.

The amplitude of the spherical wavelet is not the same as the amplitude of the incident plane wave. Since there are an infinite number of wavelets with sources in the slit B must be infinitesimal. In addition, the amplitude contains the factor $(1 + \cos\theta)$. The wavelet has a larger amplitude in the forward direction (θ near 0) than in the backward direction (θ near 180°). This factor is important if the sum of the wavelets is to reproduce a wave that is traveling in the same direction as the original wave. If the slit is narrow and the viewing screen is far away then $\cos\theta \approx 1$ for all wavelets that reach the observation point and this factor does not play an important role in determining the intensity pattern. If the slit is wide and the screen is nearby, however, it is important.

The wavelet is spherical and its amplitude decreases as $1/r$. Since the coordinates of the observation point are x and y, $r = [x^2 + (y - y')^2]^{1/2}$ and $\cos\theta = x/r$.

All the wavelets that originate at points in the slit must now be summed to find the total disturbance at the observation point. Since there is a continuous distribution of wavelet sources in the slit, the sum takes the form of an integral. The amplitude of a wavelet must be infinitesimal and we take $B = dy'/a$, where a is the width of the slit. The total disturbance at the observation point is given by

$$E = \frac{1}{a}\int_{-a/2}^{a/2} \frac{(1 + \cos\theta)\sin(kr - \omega t)}{r}\,dy'.$$

Chapter 46: Diffraction   651

The resultant wave has the form $E = E_\theta \sin(-\omega t + \alpha)$, where

$$E_\theta^2 = \frac{1}{a^2}\left[\int_{-a/2}^{a/2} \frac{(1 + \cos\theta)\sin(kr)}{r} dy'\right]^2$$
$$+ \frac{1}{a^2}\left[\int_{-a/2}^{a/2} \frac{(1 + \cos\theta)\cos(kr)}{r} dy'\right]^2$$

The intensity at the observation point is proportional to E_θ^2.

The Simpson's rule integration program of Chapter 7 can be modified to evaluate the integrals. A computer program is given below. In it S_{1e} and S_{1o} collect the sums of even and odd terms, respectively, for the first integral and S_{2e} and S_{2o} collect the even and odd terms, respectively, for the second integral. The program calculates the intensity at a series of observation points starting at $y = 0$ and ending at $y = y_f$, with an interval of Δy. Here's the outline:

> input slit width: a
> input number of segments: N
> replace N with the nearest even integer
> calculate wavelet source interval: $\Delta y' = a/N$
> input wavelength: λ
> calculate $k = 2\pi/\lambda$
> input observation interval and final observation coordinate: $\Delta y, y_f$
> input distance from slit to viewing screen: x
> set $y = 0$
> **begin loop** over observation coordinate
>   set $S_{1e} = 0$, $S_{1o} = 0$, $S_{2e} = 0$, $S_{2o} = 0$, $y' = 0$
>   **begin loop** over wavelet sources; counter runs from 1 to $N/2$
>   calculate r: $r = \left[x^2 + (y - y')^2\right]^{1/2}$
>   calculate $(1 + \cos\theta)/r$: $A = (1 + x/r)/r$
>   update sums of even terms:
>   replace S_{1e} with $S_{1e} + A\sin(kr)$
>   replace S_{2e} with $S_{2e} + A\cos(kr)$
>   increment y' by $\Delta y'$
>   calculate r: $r = \left[x^2 + (y - y')^2\right]^{1/2}$
>   calculate $(1 + \cos\theta)/r$: $A = (1 + x/r)/r$
>   update sums of odd terms:
>   replace S_{1o} with $S_{1o} + A\sin(kr)$
>   replace S_{2o} with $S_{2o} + A\cos(kr)$
>   increment y' by $\Delta y'$
>   **end loop** over wavelet sources
>   calculate integrands at upper and lower limits: $I_{1N}, I_{1 0}, I_{2N}, I_{2 0}$
>   calculate intensity:
> $E_\theta^2 = (\Delta y'/3a)^2[(I_{1N} - I_{1 0} + 2S_{1e} + 4S_{1o})^2$

$$+(I_{2N} - I_{20} + 2S_{2e} + 4S_{2o})^2]$$
   display or plot intensity
   increment y by Δy
   if $y \geq y_f$ exit loop over observation coordinate
  end loop over observation coordinate
  stop

You might write instructions to plot the intensity pattern directly on the monitor screen or you might plot it by hand using data generated by the program.

First use the program to investigate the pattern when the viewing screen is far from the slits. This project also acts as a check on the program. The plot of the intensity as a function of viewing coordinate should agree with the diagram in the text.

PROJECT 1. A 1.0×10^{-4}-m wide slit is illuminated by plane waves with a wavelength of 5.0×10^{-7} m and the intensity pattern is viewed on a screen 1.0 m away. Use the numerical integration program with $N = 16$ to plot E_0^2 every 1.0×10^{-3} m from $y = -15 \times 10^{-3}$ m to $y = +15 \times 10^{-3}$ m. The intensity for negative y is exactly the same as that for positive y, so you need to run the program for positive y only. Indicate on the plot the geometric image of the slit. It extends from $y = -5.0 \times 10^{-5}$ m to $y = +5.0 \times 10^{-5}$ m.

Identify the broad central maximum and any secondary maxima on your graph. Notice that the intensity pattern spreads well beyond the region of the geometric image. In fact, the central bright region alone is roughly 100 times as wide as the geometric image. The appearance of secondary maxima makes the pattern even wider.

The pattern spreads as the slit width is narrowed. Suppose a 7.0×10^{-5}-m wide slit is illuminated by a plane wave with a wavelength of 5.0×10^{-7} m and the intensity pattern is viewed on a screen 1.0 m away. Use the program to plot the intensity as a function of the observation coordinate y. Use the graph to find the coordinates of the minima of intensity closest to the central bright area. Compare the values of these coordinates to those when the slit width is 1.0×10^{-4} m.

An increase in wavelength also produces a broadening of the intensity pattern. Suppose a 1.0×10^{-4}-m wide slit is illuminated by a plane wave with a wavelength of 6.5×10^{-7} m and the intensity pattern is viewed on a screen 1.0 m away. Use the program to plot the intensity as a function of the observation coordinate y. Find the coordinates of the minima of intensity closest to the central bright area. Compare with the coordinates of these minima when the wavelength is 5.0×10^{-7} m and the slit width is the same (1.0×10^{-4} m).

As the slit is widened or the viewing screen is brought closer to the slit, the fringe system narrows. Eventually the central maximum of intensity occupies approximately the region of the geometric image and very little light reaches regions of the geometric shadow. There is still some fringing near the edges of the geometric image, however. Something else happens as the slit is widened: fringes appear in the region of the geometric image. The intensities at minima within the image region are not zero so the fringes are not as noticeable as the fringes of a narrow slit.

The following project is designed to show you how the intensity pattern changes as the slit widens. It requires considerable running time. To shorten the time the number of integration intervals has been selected so the calculation is accurate to only 2 significant figures. If a fast machine is available and higher accuracy is desired, you might double the number of intervals in each case.

PROJECT 2. Plane waves with a wavelength of 5.0×10^{-7} m illuminate a 7.0×10^{-5}-m wide slit and the intensity pattern is viewed on a screen 0.020 m away. Use the program to plot the intensity as a function of the coordinate y of the observation point. Take $N = 16$ and plot points every 2.0×10^{-5} m from $y = -3.6 \times 10^{-4}$ m to $y = +3.6 \times 10^{-4}$ m. Locate the edges of the geometric image on the graph.

Qualitatively the pattern is quite similar to the patterns obtained in the previous project. The bright central region extends a considerable distance beyond the geometric image and this region is followed by a series of secondary fringes. The central maximum is not quite as bright and the dark regions between secondary maxima are slightly broader than for greater slit-to-screen distances.

The slit is now widened to 1.4×10^{-6} m. The wavelength and slit-to-screen distance remain the same. Use the program, with $N = 32$, to plot the intensity every 1.0×10^{-5} m from $y = -2.2 \times 10^{-4}$ m to $y = +2.2 \times 10^{-4}$ m. Locate the edges of the geometric image on the graph.

Notice that the pattern is much more narrow than before. The central bright region is now within the geometric image and the pattern has a shoulder near the edge of the image. This shoulder is a remnant of the first minimum of the pattern for a narrow slit.

The slit is now widened to 2.8×10^{-4} m. Use the program, with $N = 50$, to plot the intensity every 1.0×10^{-5} m from $y = -2.0 \times 10^{-4}$ m to $y = +2.0 \times 10^{-4}$ m. Locate the edges of the geometric image on the graph.

The geometric image is now discernible in the intensity pattern. There are fringes deep within the image but the intensity does not become zero anywhere in that region. There is a gray central area where the intensity is about half that at the maximum but this region merges into a bright region and then, as the edge of geometric image is approached, the intensity falls off rapidly. There is fringing in the neighborhood of the image edge but the illuminated region does not extend very far beyond the edge.

IV. NOTES

Chapter 47
GRATINGS AND SPECTRA

I. BASIC CONCEPTS

This chapter deals with the interference patterns generated by gratings and their use in the study of electromagnetic spectra. The basic ideas are illustrated by a multiple-slit barrier. Your goals should be to learn what the patterns look like and to understand the details of wave interference that leads to these patterns. In addition, learn how to calculate the angles for maximum and minimum intensity. Pay attention to the characteristics of the pattern measured by dispersion and resolving power.

Atoms of crystals act like diffraction gratings for x rays. Learn how the intensity pattern produced can be used to determine the structure of a crystal.

Multiple slits. Consider a barrier in which N parallel slits have been cut, with distance d between adjacent slits. Monochromatic plane waves are incident normally on the barrier and the intensity pattern formed by waves passing through the slits is viewed on a screen far away. The slits are so narrow that single-slit diffraction can be ignored. That is, the interference pattern is well within the central maximum of the single-slit diffraction pattern.

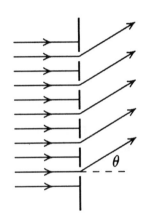

The pattern on the screen consists a series of intense bands, called _____ maxima, with much less intense bands, called _____ maxima, between. It is usually described in terms of the angle θ made with the normal by a ray from the slit system to a point on the screen. On the axes below sketch a graph of the intensity as a function of θ for a small number of slits, 4 or 5, say.

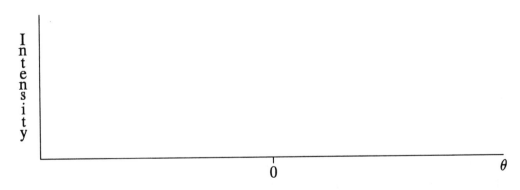

A principal maximum occurs when the phases at the screen of waves from any two adjacent slits are either _____ or differ by a multiple of _____ rad. Since waves from two adjacent slits travel distances that differ by $d\sin\theta$, where d is the slit separation, the condition for a principal maximum is

$$d\sin\theta = $$

where λ is the wavelength. The integer m in this equation is called _____. High order principal maxima occur at _____ angles than low order principal maxima.

Notice that the locations of the principal maxima are determined by the ratio d/λ and are independent of the number of slits. Also notice that the principal maxima occur at different angles for different wavelengths of light. For the same order maximum the angle for red light is _____ than that for violet light. If white light is incident on the barrier, the color of an observed band continuously changes from red at one end to violet at the other.

The width of a principal maximum is indicated by the angular position $\delta\theta$ of an adjacent minimum. For the principal maximum that occurs at angle θ,

$$\delta\theta = $$

Notice that it depends on the number of slits. In fact, as the number of slits increases without change in their separation the width of every maximum _____. Also notice that maxima near the normal (small θ) are _____ than maxima farther away from the normal (larger θ). For most gratings in practical use the principal maxima are so narrow they are called "lines".

Now consider the phasors corresponding to waves reaching a point on the screen, one for each of the N slits. At a principal maximum they are aligned. At an adjacent minimum they form an N-sided regular polygon, with the angle between any two successive phasors given by _____ rad. Thus for either of the minima neighboring the principal maximum of order m the phase difference of waves from any two adjacent slits is _____ rad. For these minima

$$d\sin\theta = $$

If N is sufficiently large other minima lie between any two principal maxima. They occur when the phase difference of waves from adjacent slits is

$$\phi = 2\pi m + (2\pi/N)n,$$

where $n = 1, 2, \ldots, N-1$. Notice that $n = N$ produces the equation for the next principal maximum, of order $m + 1$.

The secondary maxima are between the minima, although *not* exactly halfway between. These maxima are much less intense than the principal maxima and their intensity decreases as the number of slits increases.

Diffraction gratings. A diffraction grating consists of many thousands of closely spaced lines ruled on either a transparent or highly reflecting surface. If the grating is transparent and the transmitted light is viewed, both the resultant of the waves through the rulings and the resultant of the waves through the regions between the rulings form diffraction patterns. If these two resultants had the same phase and amplitude at the grating their sum would nearly be a plane wave and would not correspond to a multiple-slit pattern at the screen. For a transmission grating the amplitudes are very nearly the same but the phases are different and a diffraction pattern is formed. The phase difference arises because _____ _____.

The pattern is just like the multiple-slit pattern you studied above. Light reflected from rulings on a metal surface also produce a multiple-slit diffraction pattern.

Because light with different wavelengths produces principal maxima at different angles, diffraction gratings are often used to analyze the spectra of light sources. Two parameters are used to measure the quality of a diffraction grating. The <u>dispersion</u> of a grating measures the angular separation of principal maxima of the same order for wavelengths differing by $\Delta\lambda$. It is defined by

$$D = $$

and for order m, occurring at angle θ, it is given by

$$D = $$

where d is the slit separation. Notice that dispersion does not depend on the number of rulings. Large dispersion means _____ angular separation.

The second parameter is the <u>resolving power</u>. It measures the difference in wavelength for waves with the angular separation of their principal maxima equal to half the angular width of a principal maximum; that is, for two lines that obey the Raleigh criterion for resolution. Mathematically it is defined by $R = \lambda/\Delta\lambda$, where $\Delta\lambda$ is the difference in wavelength. For a system of N slits the resolving power at the principal maximum of order m is given by

$$R = $$

Be sure you understand the difference between dispersion and resolving power. Describe in words the intensity pattern produced by a grating with large dispersion and small resolving power: _____ _____

Consider a principal maximum near $\theta = 0$ and tell which quantities (of m, d, and N) should be large and which should be small to produce such a pattern: _____

Describe the intensity pattern produced by a grating with small dispersion and large resolving power: _____ _____

Tell which quantities should be large and which should be small to produce such a pattern: _____

X-ray diffraction. Atoms in a crystal form a periodic array in three dimensions and x-ray radiation scattered by their electrons produces a diffraction pattern. Why are x rays with wavelengths on the order of 0.1 nm, rather than visible light with wavelengths on the order of 500 nm, used to form the pattern? _____

_____

Diffraction occurs only when the x rays are incident at certain angles to crystal planes. A crystal plane is a plane that _____

_____.

Suppose monochromatic plane-wave x rays with wavelength λ are incident at an angle θ to a set of crystal planes with separation d. An intense spot is produced if the angle of incidence satisfies

$$2d\sin\theta = $$

where m is an integer. When this condition is met a high intensity beam is radiated only at the angle θ to the planes. The radiation produces a spot on a photographic plate, as in Fig. 14 of the text. Carefully note that θ is *not* the angle between the incident rays and the normal to the planes. Rather, it is the angle between the incident rays and the planes themselves.

The diagram on the right shows the edges of two crystal planes as horizontal dotted lines and two rays reflected from them. If d is the separation of the planes, the lower ray travels a distance $2d\sin\theta$ further than the upper ray. On the diagram draw a normal to the rays through the point of reflection of the upper ray and point out the distance $d\sin\theta$. For constructive interference to occur the difference in the distance traveled must be a multiple of _____, so $2d\sin\theta = m\lambda$. There are many more crystal planes parallel to the ones shown. If x rays from two adjacent planes interfere constructively then x rays from all of them interfere constructively and an intense beam is formed.

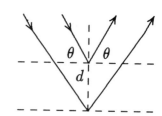

X rays are used to experimentally determine the atomic arrangements in crystals. The idea is to find the orientations and separations of a great many sets of crystal planes and use these to reconstruct the crystal. The symmetry of the diffraction spots is indicative of the symmetry of the crystal and can often be used to great advantage.

Crystals with known atomic arrangements are used as filters to separate x rays of a given wavelength from an incident beam containing a mixture of wavelengths. The crystal is oriented so that only the waves with the desired wavelength form an intense scattered beam.

II. PROBLEM SOLVING

For multiple-slit diffraction patterns you should understand how the angular positions and angular widths of the principal maxima depend on the wavelength and slit separation.

PROBLEM 1. Light with a wavelength of 450 nm is incident normally on a grating with 1.00×10^3 rulings/cm. What are the angular positions of the 3 lowest order principal maxima?

SOLUTION: The ruling separation is 1.00×10^{-3} cm or 1.00×10^{-5} m. Solve $d\sin\theta = m\lambda$, with $m = 0, 1,$ and 2, for θ.

[ans: 0; 2.58°; 5.16°]

Suppose the same light is incident normally on a grating with a slit separation of 2.00×10^{-5} m (twice that of the previous grating). What then are the angular positions of the 3 lowest order principal maxima?

SOLUTION:

[ans: 0; 1.29°; 2.58°]

What wavelength light should be used with the second grating to obtain principal maxima that coincide with those of the first grating when 450 nm light is incident?

SOLUTION: Since the slit separation was doubled the wavelength must be doubled to obtain principal maxima at the same angular positions.

[ans: 900 nm]

Notice that the principal maxima spread in angle when the slit separation is increased and contract when the wavelength is increased.

PROBLEM 2. What slit separation should be used if the first order line is to appear at $\theta = 25°$ for a wavelength of 450 nm?

SOLUTION: Solve $d\sin\theta = m\lambda$ for d.

[ans: 1.06×10^{-6} m]

How many principal maxima appear?

SOLUTION: You want all principal maxima for which $\theta < 90°$. Solve $d\sin\theta = m\lambda$, with $\theta = 90°$, for m. Round down to the nearest integer. The total number of principal maxima is double this number (to account for both sides of the central maximum) plus 1 (to account for the central maximum).

[ans: 5]

This solution ignores the single-slit diffraction pattern. If it is narrow some or all (except the $m = 0$ maximum) may have intensities that are quite low and so cannot be observed.

PROBLEM 3. 500-nm light is incident on a 1.00-cm wide grating with 5.00×10^3 rulings. What is the angular width of the $m = 0$ principal maximum? What is the angular width of the $m = 1$ principal maximum?

SOLUTION: Use $\delta\theta = \lambda/Nd\cos\theta$. You must first solve $d\sin\theta = m\lambda$ for θ. Don't forget that $\delta\theta$ is *half* the angular width in radians.

[ans: 1.00×10^{-4} rad (5.73×10^{-3} deg); 1.03×10^{-4} rad (5.92×10^{-3} deg)]

What are the angular widths of these lines if the grating has 1.50×10^4 rulings?

SOLUTION:

[ans: 1.00×10^{-4} rad (5.73×10^{-3} deg); 1.51×10^{-4} rad (8.66×10^{-3} deg)]

Gratings can be used to measure the wavelength of light and to separate waves with different wavelengths.

PROBLEM 4. Light is incident normally on a grating with a slit separation of 1600 nm. First order lines appear at 17.15°, 20.76°, and 24.62°. What wavelengths does the incident light contain?

SOLUTION: Solve $d \sin \theta = \lambda$ for λ.

[ans: 471.8 nm, 567.1 nm, 666.3 nm]

What is the greatest wavelength light for which this grating forms a line with $m > 0$?

SOLUTION: Take $\theta = 90°$ and $m = 1$.

[ans: 1600 nm, well into the infrared]

If the slit width is taken into account the amplitude of the wave from each slit is proportional to $(1/\beta) \sin \beta$, where $\beta = (\pi a/\lambda) \sin \theta$ and a is the slit width. This single-slit diffraction function forms the envelope for the multiple-slit interference function. Recall that zeros of single-slit diffraction occur when $\beta = \ell \pi$ (or $a \sin \theta = \ell \lambda$), with $\ell = 1, 2, 3, \ldots$. The central maximum extends from $-\sin^{-1}(\lambda/a)$ to $+\sin^{-1}(\lambda/a)$.

PROBLEM 5. A grating has a slit width of 2.3×10^{-6} m and a slit separation of 8.5×10^{-6} m. How many lines are within the central maximum of the single-slit diffraction pattern?

SOLUTION: Use $a \sin \theta = \lambda$ and $d \sin \theta = m\lambda$ to show that $m = d/a$. Round down to the nearest integer, double the result, and add 1.

[ans: 7]

Notice that the answer does not depend on the wavelength.

Suppose another grating has the same slit separation but a different slit width and for it all lines of even order have zero intensity. What is the slit width? These lines are said to be missing.

SOLUTION: Lines occur for $d\sin\theta = m\lambda$ but a line is missing if it coincides with a zero of the single-slit diffraction pattern. That is, a line is missing if $a\sin\theta = \ell\lambda$, where $\ell = 1, 2, 3, \ldots$. Use one of the equations to eliminate $\sin\theta$ from the other and obtain $ma = d\ell$ for a missing line. Since all lines with m even are missing $2a = d$.

[ans: 4.25×10^{-6} m]

The regions between slits have the same width as the slits themselves.

You should know the physical meaning of dispersion and how to calculate it for a grating. Use $D = \Delta\theta/\Delta\lambda$ to relate the dispersion to the spread in angle of two principal maxima of the same order for different wavelengths. Use $D = m/d\cos\theta$ to calculate the dispersion when the order and angle are given. In some cases you may need to use $d\sin\theta = m\lambda$ to calculate the angle. The expression for D assumes the angles for the two wavelengths are so close together you can ignore the difference. If you suspect the difference is significant calculate the angle for each wavelength and use the first expression for D.

PROBLEM 6. A certain grating has 6.5×10^3 rulings and is 1.5 cm wide. What is the angular separation of the $m = 1$ principal maxima for wavelengths of 500 nm and 600 nm?

SOLUTION: Use $d\sin\theta = \lambda$ to calculate θ for each of the two values of the wavelength. The question asks for the difference in the angles.

[ans: 2.56°]

Use this result to calculate the dispersion in the vicinity of the $m = 1$ principal maximum.

SOLUTION: Use $D = \Delta\theta/\Delta\lambda$.

[ans: 2.56×10^7 deg/m (4.46×10^5 rad/m)]

Use $D = m/d\cos\theta$ to calculate the dispersion and compare with the result you just obtained. Take θ to be the angle for the $m = 1$ principal maximum with $\lambda = 550$ nm.

[ans: 4.46×10^5 rad/m]

PROBLEM 7. A certain grating is used to view light containing waves with wavelengths of 550.15 nm and 550.63 nm in first order. If the dispersion is 5.5×10^5 rad/m what is the angular separation of the two lines?

SOLUTION: Solve $D = \Delta\theta/\Delta\lambda$ for $\Delta\theta$.

[ans: 2.64×10^{-4} rad (0.0151°)]

What is the slit separation?

SOLUTION: Solve $d\sin\theta = \lambda$ and $D = 1/d\cos\theta$ simultaneously for d. Use $\sin^2\theta + \cos^2\theta = 1$ to eliminate θ and obtain $d = \sqrt{\lambda^2 + 1/D^2}$.

[ans: 1.90×10^{-6} m]

At what angle do the lines appear?

SOLUTION: Solve $d\sin\theta = \lambda$ for θ.

[ans: 16.8°]

How many rulings must the grating have for these lines to be barely resolvable according to the Rayleigh criterion?

SOLUTION: The resolution is given by $R = \lambda/\Delta\lambda$, where the same order lines for wavelengths λ and $\lambda + \Delta\lambda$ are separated by half the width of a principal maximum. For lines of order m the resolution is given by $R = Nm$, where N is the number of rulings. Solve for N and obtain $N = \lambda/m\Delta\lambda$. Round down to the nearest integer.

[ans: 1146]

How wide must the grating be?

SOLUTION: The width of the grating is given by Nd.

[ans: 2.18 mm]

Suppose the grating is 3.0 cm wide. What wavelength difference can then be resolved in first order at 550 nm?

SOLUTION: Use $\Delta\lambda = \lambda/Nm$, with N given by the width divided by the slit separation.

[ans: 3.48×10^{-2} nm]

The condition for x-ray diffraction by a crystal is $2d\sin\theta = m\lambda$, where θ is the angle between the incoming ray and a crystal plane and d is the separation of the crystal planes being considered. The diffracted beam also makes the angle θ with the planes.

PROBLEM 8. X rays with a wavelength of 0.132 nm are incident on a crystal. The crystal is rotated until first order Bragg reflection is observed from planes with a spacing of 0.353 nm. What angle does the reflected beam make with these planes?

SOLUTION: Solve $2d\sin\theta = \lambda$ for θ.

[ans: 10.8°]

Through what additional angle must the crystal be rotated to observe second order Bragg reflection from the same planes?

SOLUTION: Calculate the angle for second order diffraction and find its difference from the angle for first order diffraction.

[ans: 11.2°]

What is the highest order diffraction that can be observed from these planes?

SOLUTION: Set $\sin\theta = 1$ and solve $2d\sin\theta = m\lambda$ for m. Round down to the nearest integer.

[ans: 5]

PROBLEM 9. Suppose the atoms of a crystal are at the corners of cubes with edge length 6.53×10^{-10} m. The view along a normal to a cube face is shown in the diagram as are the edges of three crystal planes labelled A, B, and C. For each of the planes find the angle of incidence for 0.23-nm x rays in order to obtain first order interference maxima.

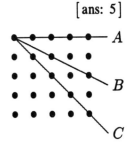

SOLUTION: Let a be the separation of planes parallel to A. For these planes $2d\sin\theta = \lambda$, with $d = a = 6.53 \times 10^{-10}$ m.

[ans: 10.1°]

The diagram to the right shows planes parallel to plane B. The dotted line is perpendicular to the planes and has a length of $\sqrt{5}a$. You may think of it as a vector with an x component of a and a y component of $2a$. It cuts across 5 interplanar gaps so the separation of the planes is $d = (\sqrt{5}/5)a = a/\sqrt{5}$. Now use $2d\sin\theta = m\lambda$.

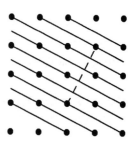

[ans: 23.2°]

The diagram to the right shows planes parallel to plane C. The dotted line is perpendicular to the planes and has a length of $\sqrt{2}a$. You should be able to show that the separation of the planes is $d = a/\sqrt{2}$. Then use $2d\sin\theta = \lambda$.

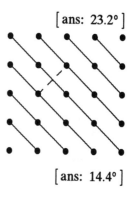

[ans: 14.4°]

III. NOTES

Chapter 48
POLARIZATION

I. BASIC CONCEPTS

You were introduced to the idea of polarization in Chapter 41. Now you will learn about the production, detection, and uses of polarized electromagnetic radiation. Be sure you know how to describe the orientations of the electric fields of linearly polarized light, circularly polarized light, and unpolarized light. Understand what happens to the polarization of each of these types of light waves as they pass through a polarizing sheet or a quarter-wave plate. Pay attention to the polarization and propagation directions of ordinary and extraordinary waves in birefringent materials.

The fundamentals of polarization. The electric field of a <u>linearly polarized</u> electromagnetic wave is always parallel to the same _____. The <u>direction</u> of polarization is the direction of _____ and the <u>plane of polarization</u> is the plane determined by _____ and _____.

You should be able to contrast polarized and unpolarized light. Describe the orientation of the electric field of unpolarized light: _____

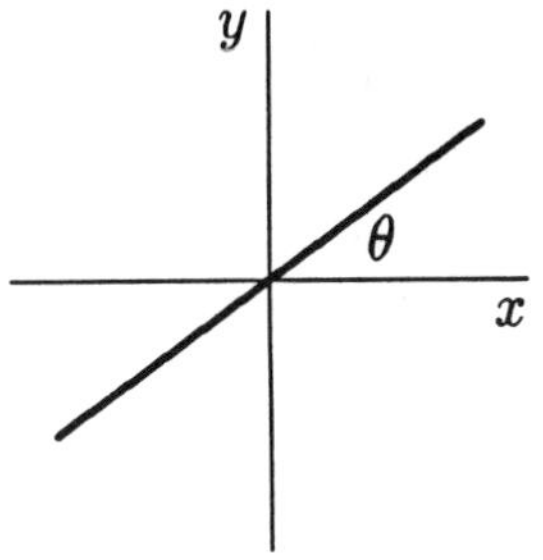

You should recognize that any linearly polarized light wave can be treated as the sum of two waves, polarized in any two mutually orthogonal directions that are perpendicular to the direction of propagation. Suppose a wave with an electric field amplitude E_m is traveling out of the page and is polarized with its electric field along a line that makes an angle θ with the x axis, as shown. You can consider it to be the vector sum of two waves, one with amplitude _____ polarized along the x axis and one with amplitude _____ polarized along the y axis.

For future reference list here the ways mentioned in the chapter for producing polarized light:

_____

Polarization by Polaroid sheets. Polarized light can be produced by shining unpolarized light through a sheet of Polaroid. These sheets contain certain long-chain molecules that are aligned in the manufacturing process. Light with its electric field vector parallel to the molecules is preferentially _____ while light with its electric field perpendicular to the molecules is preferentially _____. A line perpendicular to the molecules is said to be along the polarizing direction of the sheet.

Suppose that linearly polarized light with amplitude E_m and intensity I_m is incident on an ideal polarizing sheet and that its electric field is along a line that makes an angle θ with the polarizing direction. Then the amplitude of the transmitted light is given by $E = $ _____ and its intensity is given by

$$I = $$

This is called the law of Malus. Note that the transmitted amplitude is the component of the incident amplitude along the polarization direction of the sheet. You simply resolved the incident electric field into components parallel and perpendicular to the polarization direction. The parallel component is transmitted while the perpendicular component is absorbed.

If unpolarized light with intensity I_i is incident on an ideal polarizing sheet the intensity of the transmitted light is given by

$$I = $$

Be sure you understand the conditions for which each of these expressions for the transmitted intensity applies. In both cases the transmitted light is linearly polarized with its electric field parallel to _____.

Polarization by reflection. If unpolarized light is incident on a boundary between two different materials, both the reflected and refracted waves are partially polarized for most directions of incidence. There is, however, a special angle of incidence, called the polarizing angle and denoted by θ_p, for which the reflected wave contains only one polarization component. If θ_r is the angle of refraction for incidence at the polarizing angle, then $\theta_p + \theta_r = $ _____. If n_1 is the index of refraction for the medium of incidence and n_2 is the index of refraction for the medium of refraction then Snell's law leads to

$$\tan\theta_p = $$

This is Brewster's law for the polarizing angle. θ_p is sometimes called Brewster's angle.

The direction of polarization for light reflected at the polarizing angle is normal to the plane determined by the _____ and the _____. Carefully note that even at the polarizing angle the refracted light is partially, not completely, polarized. It contains light with all polarization directions but is deficient in light with the polarization of the reflected light.

Double refraction. Unpolarized light incident on a double-refracting (or birefringent) material, such as a calcite crystal, splits into two transmitted beams, named ordinary (o) and extraordinary (e), with mutually perpendicular polarization directions. A single index of refraction n_o is associated with the ordinary beam; its speed is the same for all directions of propagation. The index of refraction for the extraordinary beam varies with the direction of propagation, being equal to n_o for propagation along what is called the _____ axis and equal to n_e for propagation perpendicular to that axis. For other propagation directions the index of refraction has a value between n_o and n_e. n_e is less than n_o for some birefringent materials and greater than n_o for others. For calcite n_e is _____ than n_o.

Each of the diagrams below shows a ray associated with a beam of unpolarized light incident on a calcite crystal. The crystal is cut and oriented differently in each case: note the direction of the optic axis. For each case shown draw and label the ordinary and extraordinary rays associated with the incident ray. Assume they are both in the plane of the page. They may coincide.

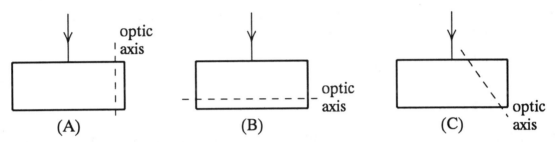

For two of these situations, labelled _____ and _____, both the o-wave and the e-wave follow the same path. These two situations are not exactly the same however. For one of them, namely _____, the o-wave and e-wave have the same speed. For the other, namely _____, they have different speeds. For the situation depicted in _____ the o-wave goes straight through the material but the e-wave is laterally displaced. If an object is viewed through the crystal two images are seen, one formed by o-waves and the other formed by e-waves.

The direction of the o-wave polarization is _____ to the plane of the page while the direction of the e-wave polarization is _____ to the plane of the page. Said another way, if the incident wave is linearly polarized along a line that is normal to the page then only an _____-wave results. If it is linearly polarized along a line that is parallel to the page only an _____-wave results.

Circular polarization. In a circularly polarized wave the electric field vector at any point rotates so its tip follows the circumference of a circle. If the field is in the xy plane its components at a point might be $E_x = E_m \cos(\omega t)$ and $E_y = E_m \sin(\omega t)$, 90° out of phase with each other. These expressions can be written $E_x = E_m \sin(\omega t + \pi/2)$ and $E_y = E_m \sin(\omega t)$.

Circularly polarized light can be produced by shining linearly polarized monochromatic light onto the surface of a birefringent crystal with appropriate thickness and orientation. The light is normally incident and the crystal is oriented so its optic axis is _____ to the direction of propagation. To assure that the e- and o-waves have the same amplitude the polarization direction of the incident light should make an angle of _____ with the optic axis. The minimum thickness of the crystal is such that the difference in the times for the two types of waves to cross it is _____ of a period.

If x is the thickness of the crystal and n_o is the index of refraction for the ordinary wave then the time for this wave to cross the crystal is $t_o = $ _____. If n_e is the index of refraction for the extraordinary wave then the time for this wave to cross is $t_e = $ _____. Set $t_o - t_e = T/4$, where T is the period, given in terms of the wavelength λ in vacuum by $T = \lambda/c$. Thus, in terms of λ, $x =$ _____. Circularly polarized light is also produced if the thickness is increased so the difference in the times is increased by any multiple of T. A crystal used to produce circularly polarized light in this way is called a _____ plate.

If circularly polarized light with an electric field amplitude E_m is incident on a polarizing sheet, the intensity of the transmitted wave is given by $I =$ _____, exactly the same as the transmitted intensity when unpolarized light is incident. A quarter-wave plate cannot be used to distinguished circularly polarized light from unpolarized light. Describe an experiment that can be used to distinguish these two types of light: _____

_____

If circularly polarized light is incident on a quarter-wave plate the transmitted light is polarized _____.

II. PROBLEM SOLVING

Many problems deal with the law of Malus. Note the polarization state of the incident wave and if the wave is linearly polarized note the angle between the line of the electric field and the polarizing direction of the polarizing sheet.

PROBLEM 1. Linearly polarized light with an intensity of 4.7 W/m² is incident on a polarizing sheet. If the electric field of the light is along a line that makes an angle of 55° with the polarizing direction of the sheet, what is the intensity of the transmitted beam?

SOLUTION: Use $I = I_m \cos^2 \theta$ with $\theta = 55°$.

[ans: 1.55 W/m²]

A second polarizing sheet is placed so light transmitted by the first sheet is incident on it. Its polarizing direction makes an angle of 35° with the polarizing direction of the first sheet and an angle of 90° with the electric field incident on the two-sheet system. What is the transmitted intensity?

SOLUTION: The electric field of the wave transmitted by the first sheet is along the polarizing direction of that sheet and makes an angle of 35° with the polarizing direction of the second sheet. Use $I = I_m \cos^2 \theta$ with $\theta = 35°$ and $I_m = 1.55$ W/m².

[ans: 1.04 W/m²]

Notice that light is transmitted even though the polarizing direction of the second sheet is perpendicular to the line of polarization of the incident wave. The first sheet rotates the polarization direction. If the polarizing sheets are interchanged so the order in which the incident light strikes them is reversed what is the transmitted intensity?

SOLUTION:

[ans: 0]

Compare these answers to those for incident light that is unpolarized. Unpolarized light with an intensity of 4.7 W/m² is incident on a polarizing sheet. What is the intensity of the transmitted beam?

SOLUTION: The average of $\cos^2\theta$ over all possible angles is 1/2.

[ans: 2.35 W/m²]

A second polarizing sheet is placed so light transmitted by the first is incident on it. Its polarizing direction makes an angle of 35° with the polarizing direction of the first sheet. What is the transmitted intensity?

SOLUTION: The light leaving the first sheet is polarized along its direction of polarization. Use $I = I_m \cos^2\theta$ with $\theta = 35°$ and $I_m = 2.35$ W/m².

[ans: 1.58 W/m²]

If the sheets are interchanged what is the transmitted intensity?

SOLUTION:

[ans: 1.58 W/m²]

PROBLEM 2. Three polarizing sheets are placed in a stack. The polarizing directions of the first and third are perpendicular to each other. What angle should the polarizing direction of the middle sheet make with the polarizing direction of the first to obtain maximum transmitted intensity when unpolarized light is incident on the first sheet? How then does the transmitted intensity compare with the incident intensity?

SOLUTION: The transmitted intensity is given by $I = \frac{1}{2}I_m \cos^2\theta \cos^2(90° - \theta) = \frac{1}{2}I_m \cos^2\theta \sin^2\theta$, where $\cos(90° - \theta) = \sin\theta$ was used. Now $\sin 2\theta = 2\cos\theta \sin\theta$ so $I = \frac{1}{8}I_m \sin^2 2\theta$. Find the value of θ for which this is a maximum.

[ans: 45°; $I/I_m = 1/8$]

PROBLEM 3. A beam of light is composed of linearly polarized light with amplitude E_m and unpolarized light with amplitude $2E_m$. It is incident on a polarizing sheet. As the sheet is rotated how do the minimum and maximum transmitted intensities compare to the incident intensity?

SOLUTION: The phase of the unpolarized beam changes rapidly so the intensities of the two components are added together, not their amplitudes. The intensity of the incident beam is $(E_m)^2 + (2E_m)^2 = 5E_m^2$. When the polarizing direction of the sheet is aligned with the electric field of the polarized portion of the light the intensity is $(E_m)^2 + \frac{1}{2}(2E_m)^2 = 3E_m^2$. When it is perpendicular to the field the intensity is $\frac{1}{2}(2E_m)^2 = 2E_m^2$.

[ans: 3/5; 2/5]

You should know how to use Brewster's law to compute the angle of incidence for complete polarization of a reflected wave. Here's an example.

PROBLEM 4. Unpolarized monochromatic light is incident from air on the surface of a glass slab with an index of refraction of 1.35. At what angle of incidence is the reflected light completely polarized?

SOLUTION: Use $\tan\theta_p = n_2/n_2$ with $n_1 = 1$ and $n_2 = 1.35$.

[ans: 53.5°]

How would you place a polarizing sheet so no light is transmitted through it?

SOLUTION: The sheet must be placed with its polarizing direction perpendicular to the electric field of the reflected light. Since the field is perpendicular to the plane of incidence the polarizing direction must be in this plane.

[ans: Place the sheet parallel to the surface with its polarizing direction in the plane defined by the direction of incidence and the normal to the surface.]

Suppose now that the light is incident from inside the slab onto the boundary with air. At what angle of incidence is the reflected light completely polarized?

SOLUTION:

[ans: 36.5°]

You should understand the analytical form of the electric field in a circularly polarized wave. Here are some examples.

PROBLEM 5. A monochromatic circularly polarized light wave travels out of the page toward you. Its electric field rotates clockwise. It is right-circularly polarized. Take the z axis to be out of the page and the x and y axes to be in the plane of page, then write expressions for the components of the field as functions of time.

SOLUTION: If the angle between the field and the x axis is $kz - \omega t$, it decreases as time goes on, then becomes negative. This is appropriate for an electric field that is rotating clockwise. Furthermore, the two terms have opposite signs, so the wave travels in the *positive* z direction.

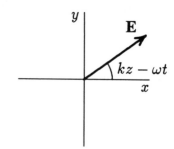

[ans: $E_x = E\cos(kz - \omega t)$, $E_y = E\sin(kz - \omega t)$]

Notice that $\cos(kz - \omega t) = \sin(kz - \omega t + \pi/2)$. The two components are $\pi/2$ radians out of phase.

Suppose the electric field rotates counterclockwise. It is left-circularly polarized. What are the components?

SOLUTION: Let $kz - \omega t$ be the angle between the electric field and the positive y axis. Convince yourself this satisfies the requirements for left-circularly polarized light traveling in the positive z direction.

[ans: $E_x = E\sin(kz - \omega t)$, $E_y = E\cos(kz - \omega t)$]

Suppose the wave is traveling away from you, in the negative z direction, and the electric field rotates clockwise as viewed by you. This is _____-circularly polarized light. What are the components?

SOLUTION:

[ans: $E_x = E\sin(kz + \omega t)$, $E_y = E\cos(kz + \omega t)$]

Finally, suppose the wave is traveling away from you and the electric field rotates counterclockwise as viewed by you. This is _____-circularly polarized light. What are the components?

SOLUTION:

[ans: $E_x = E\cos(kz + \omega t)$, $E_y = E\sin(kz + \omega t)$]

The following example shows how a birefringent crystal can be cut to make a quarter-wave plate.

PROBLEM 6. A wurzite crystal is cut with its optic axis parallel to the surface of incidence. What minimum thickness should it have if it is to be used as a quarter-wave plate for sodium light (589 nm)? The index of refraction for the ordinary wave is 2.356 and the index of refraction for the extraordinary wave is 2.378.

SOLUTION: Use $x = \lambda/4|n_e - n_o|$.

[ans: 6.69×10^{-6} m]

Chapter 48: Polarization

Suppose the phase of the electric field entering the crystal is 0 at time $t = 0$. What is the phase of the electric field of the ordinary wave leaving the crystal at $t = 0$? What is the phase of the electric field of the extraordinary wave leaving the crystal at $t = 0$? Assume the values given for the indices of refraction and wavelength are exact and calculate the phases to 5 significant figures.

SOLUTION: Use $\phi = kx = 2\pi n x/\lambda$, where λ is the wavelength in vacuum, n is the index of refraction, and x is the thickness of the crystal. You will need to re-compute x to obtain high precision.

[ans: 1.6822×10^2 rad; 1.6979×10^2 rad]

Verify that the difference in phase is $\pi/2$ rad, so the combination is circularly polarized if the amplitudes are the same.

III. NOTES

EXAM SUMMARY

Exam number: _____ **Date:** _____ **Chapters:** _____

Definitions:

QUANTITY	DEFINITION
_____	_____
_____	_____
_____	_____
_____	_____
_____	_____
_____	_____

Physical Laws:

_____
_____
_____

Other Important Relationships:

_____
_____
_____
_____
_____

Important Applications:

_____
_____
_____
_____
_____

Notes:

EXAM SUMMARY

Exam number: _____   **Date:** _____   **Chapters:** _____

Definitions:

QUANTITY	DEFINITION
_____	_____
_____	_____
_____	_____
_____	_____
_____	_____
_____	_____

Physical Laws:

_____
_____
_____

Other Important Relationships:

_____
_____
_____
_____
_____
_____

Important Applications:

_____
_____
_____
_____
_____
_____

Notes:

EXAM SUMMARY

Exam number: _____   **Date:** _____   **Chapters:** _____

Definitions:

QUANTITY	DEFINITION
_____	_____
_____	_____
_____	_____
_____	_____
_____	_____
_____	_____

Physical Laws:

_____
_____
_____

Other Important Relationships:

_____
_____
_____
_____
_____

Important Applications:

_____
_____
_____
_____
_____

Notes:

EXAM SUMMARY

Exam number: _____   **Date:** _____   **Chapters:** _____

Definitions:

QUANTITY	DEFINITION
_____	_____
_____	_____
_____	_____
_____	_____
_____	_____

Physical Laws:

_____
_____
_____

Other Important Relationships:

_____
_____
_____
_____
_____

Important Applications:

_____
_____
_____
_____
_____

Notes:

EXAM SUMMARY

Exam number: _____ **Date:** _____ **Chapters:** _____

Definitions:

QUANTITY	DEFINITION
_____	_____
_____	_____
_____	_____
_____	_____
_____	_____
_____	_____

Physical Laws:

_____
_____
_____

Other Important Relationships:

_____
_____
_____
_____
_____

Important Applications:

_____
_____
_____
_____

Notes:

EXAM SUMMARY

Exam number: _____   **Date:** _____   **Chapters:** _____

Definitions:

QUANTITY	DEFINITION

Physical Laws:

_____
_____
_____

Other Important Relationships:

_____
_____
_____
_____
_____

Important Applications:

_____
_____
_____
_____
_____

Notes: